AF389503

Physiological and Molecular Characterization of Crop Resistance to Abiotic Stresses

Physiological and Molecular Characterization of Crop Resistance to Abiotic Stresses

Editors

Monica Boscaiu
Ana Fita

MDPI • Basel • Beijing • Wuhan • Barcelona • Belgrade • Manchester • Tokyo • Cluj • Tianjin

Editors
Monica Boscaiu
Universitat Politècnica de València
Spain

Ana Fita
Universitat Politècnica de València
Spain

Editorial Office
MDPI
St. Alban-Anlage 66
4052 Basel, Switzerland

This is a reprint of articles from the Special Issue published online in the open access journal *Agronomy* (ISSN 2073-4395) (available at: https://www.mdpi.com/journal/agronomy/special_issues/Physiological_Molecular_Crop_Abiotic).

For citation purposes, cite each article independently as indicated on the article page online and as indicated below:

LastName, A.A.; LastName, B.B.; LastName, C.C. Article Title. *Journal Name* **Year**, *Article Number, Page Range.*

ISBN 978-3-03943-458-9 (Hbk)
ISBN 978-3-03943-459-6 (PDF)

Contents

About the Editors . . . . **ix**

Preface to "Physiological and Molecular Characterization of Crop Resistance to Abiotic
Stresses" . . . . **xi**

Monica Boscaiu and Ana Fita
Physiological and Molecular Characterization of Crop Resistance to Abiotic Stresses
Reprinted from: *Agronomy* **2020**, *10*, 1308, doi:10.3390/agronomy10091308 . . . . **1**

Yingying Li, Qiuqiu Zhang, Lina Ou, Dezhong Ji, Tao Liu, Rongmeng Lan, Xiangyang Li and
Linhong Jin
Response to the Cold Stress Signaling of the Tea Plant (*Camellia sinensis*) Elicited by
Chitosan Oligosaccharide
Reprinted from: *Agronomy* **2020**, *10*, 915, doi:10.3390/agronomy10060915 . . . . **9**

Sugenith Arteaga, Lourdes Yabor, María José Díez, Jaime Prohens, Monica Boscaiu
and Oscar Vicente
The Use of Proline in Screening for Tolerance to Drought and Salinity in Common Bean
(*Phaseolus vulgaris* L.) Genotypes
Reprinted from: *Agronomy* **2020**, *10*, 817, doi:10.3390/agronomy10060817 . . . . **21**

Marta Muñoz, Natalia Torres-Pagán, Rosa Peiró, Rubén Guijarro,
Adela M. Sánchez-Moreiras and Mercedes Verdeguer
Phytotoxic Effects of Three Natural Compounds: Pelargonic Acid, Carvacrol, and Cinnamic
Aldehyde, against Problematic Weeds in Mediterranean Crops
Reprinted from: *Agronomy* **2020**, *10*, 791, doi:10.3390/agronomy10060791 . . . . **37**

María Isabel Martínez-Nieto, Elena Estrelles, Josefa Prieto-Mossi, Josep Roselló
and Pilar Soriano
Resilience Capacity Assessment of the Traditional Lima Bean (*Phaseolus lunatus* L.) Landraces
Facing Climate Change
Reprinted from: *Agronomy* **2020**, *10*, 758, doi:10.3390/agronomy10060758 . . . . **57**

Marco Brenes, Andrea Solana, Monica Boscaiu, Ana Fita, Oscar Vicente, Ángeles Calatayud,
Jaime Prohens and Mariola Plazas
Physiological and Biochemical Responses to Salt Stress in Cultivated Eggplant (*Solanum
melongena* L.) and in *S. insanum* L., a Close Wild Relative
Reprinted from: *Agronomy* **2020**, *10*, 651, doi:10.3390/agronomy10050651 . . . . **73**

Leandro Pereira-Dias, Daniel Gil-Villar, Vincente Castell-Zeising, Ana Quiñones,
Ángeles Calatayud, Adrián Rodríguez-Burruezo and Ana Fita
Main Root Adaptations in Pepper Germplasm (*Capsicum* spp.)   to Phosphorus
Low-Input Conditions
Reprinted from: *Agronomy* **2020**, *10*, 637, doi:10.3390/agronomy10050637 . . . . **93**

Renata M. Sumalan, Sorin I. Ciulca, Mariana A. Poiana, Diana Moigradean, Isidora Radulov,
Monica Negrea, Manuela E. Crisan, Lucian Copolovici and Radu L. Sumalan
The Antioxidant Profile Evaluation of Some Tomato Landraces with Soil Salinity Tolerance
Correlated with High Nutraceuticaland Functional Value
Reprinted from: *Agronomy* **2020**, *10*, 500, doi:10.3390/agronomy10040500 . . . . **113**

Ali Anwar, Jun Wang, Xianchang Yu, Chaoxing He and Yansu Li
Substrate Application of 5-Aminolevulinic Acid Enhanced Low-temperature and Weak-light
Stress Tolerance in Cucumber (*Cucumis sativus* L.)
Reprinted from: *Agronomy* 2020, *10*, 472, doi:10.3390/agronomy10040472 . . . . **129**

**Hamideh Fatemi, Chokri Zaghdoud, Pedro A. Nortes, Micaela Carvajal
and Maria del Carmen Martínez-Ballesta**
Differential Aquaporin Response to Distinct Effects of Two Zn Concentrations after Foliar
Application in Pak Choi (*Brassica rapa* L.) Plants
Reprinted from: *Agronomy* 2020, *10*, 450, doi:10.3390/agronomy10030450 . . . . **141**

**Ricardo Gil-Ortiz, Miguel Ángel Naranjo, Antonio Ruiz-Navarro, Marcos Caballero-Molada,
Sergio Atares, Carlos García and Oscar Vicente**
New Eco-Friendly Polymeric-Coated Urea Fertilizers Enhanced Crop Yield in Wheat
Reprinted from: *Agronomy* 2020, *10*, 438, doi:10.3390/agronomy10030438 . . . . **159**

**Charles Nelimor, Baffour Badu-Apraku, Antonia Yarney Tetteh, Ana Luísa Garcia-Oliveira
and Assanvo Simon-Pierre N'guetta**
Assessing the Potential of Extra-Early Maturing Landraces for Improving Tolerance to Drought,
Heat, and Both Combined Stresses in Maize
Reprinted from: *Agronomy* 2020, *10*, 318, doi:10.3390/agronomy10030318 . . . . **175**

**Shuangjie Jia, Hongwei Li, Yanping Jiang, Yulou Tang, Guoqiang Zhao, Yinglei Zhang,
Shenjiao Yang, Husen Qiu, Yongchao Wang, Jiameng Guo, Qinghua Yang and Ruixin Shao**
Transcriptomic Analysis of Female Panicles Reveals Gene Expression Responses to Drought
Stress in Maize (*Zea mays* L.)
Reprinted from: *Agronomy* 2020, *10*, 313, doi:10.3390/agronomy10020313 . . . . **199**

Muhammad Zeeshan, Meiqin Lu, Shafaque Sehar, Paul Holford and Feibo Wu
Comparison of Biochemical, Anatomical, Morphological, and Physiological Responses to
Salinity Stress in Wheat and Barley Genotypes Deferring in Salinity Tolerance
Reprinted from: *Agronomy* 2020, *10*, 127, doi:10.3390/agronomy10010127 . . . . **215**

Cuiyu Liu, Yujie Zhao, Xueqing Zhao, Jinping Wang, Mengmeng Gu and Zhaohe Yuan
Transcriptomic Profiling of Pomegranate Provides Insights into Salt Tolerance
Reprinted from: *Agronomy* 2020, *10*, 44, doi:10.3390/agronomy10010044 . . . . **231**

**Khaled A. Abdelaal, Lamiaa M. EL-Maghraby, Hosam Elansary, Yaser M. Hafez,
Eid I. Ibrahim, Mostafa El-Banna, Mohamed El-Esawi and Amr Elkelish**
Treatment of Sweet Pepper with Stress Tolerance-Inducing Compounds Alleviates
Salinity Stress Oxidative Damage by Mediating the Physio-Biochemical Activities and
Antioxidant Systems
Reprinted from: *Agronomy* 2020, *10*, 26, doi:10.3390/agronomy10010026 . . . . **249**

Aleck Kondwakwenda, Julia Sibiya, Rebecca Zengeni, Cousin Musvosvi and Samson Tesfay
Screening of Provitamin-A Maize Inbred Lines for Drought Tolerance: Beta-Carotene Content
and Secondary Traits
Reprinted from: *Agronomy* 2019, *9*, 692, doi:10.3390/agronomy9110692 . . . . **265**

**Emuejevoke D. Vwioko, Mohamed A. El-Esawi, Marcus E. Imoni, Abdullah A. Al-Ghamdi,
Hayssam M. Ali, Mostafa M. El-Sheekh, Emad A. Abdeldaym and Monerah A. Al-Dosary**
Sodium Azide Priming Enhances Waterlogging Stress Tolerance in Okra
(*Abelmoschus esculentus* L.)
Reprinted from: *Agronomy* 2019, *9*, 679, doi:10.3390/agronomy9110679 . . . . **283**

Misganaw Wassie, Weihong Zhang, Qiang Zhang, Kang Ji and Liang Chen
Effect of Heat Stress on Growth and Physiological Traits of Alfalfa (*Medicago sativa* L.) and a
Comprehensive Evaluation for Heat Tolerance
Reprinted from: *Agronomy* **2019**, *9*, 597, doi:10.3390/agronomy9100597 . . . . **299**

**Zehao Hou, Junliang Yin, Yifei Lu, Jinghan Song, Shuping Wang, Shudong Wei,
Zhixiong Liu, Yingxin Zhang and Zhengwu Fang**
Transcriptomic Analysis Reveals the Temporal and Spatial Changes in Physiological Process
and Gene Expression in Common Buckwheat (*Fagopyrum esculentum* Moench) Grown under
Drought Stress
Reprinted from: *Agronomy* **2019**, *9*, 569, doi:10.3390/agronomy9100569 . . . . **319**

**Nuengsap Thangthong, Sanun Jogloy, Tasanai Punjansing, Craig K. Kvien, Thawan Kesmala
and Nimitr Vorasoot**
Changes in Root Anatomy of Peanut (*Arachis hypogaea* L.) under Different Durations of Early
Season Drought
Reprinted from: *Agronomy* **2019**, *9*, 215, doi:10.3390/agronomy9050215 . . . . **337**

**Bui Manh Minh, Nguyen Thuy Linh, Ha Hong Hanh, Le Thi Thu Hien, Nguyen Xuan Thang,
Nong Van Hai and Huynh Thi Thu Hue**
A *LEA* Gene from a Vietnamese Maize Landrace Can Enhance the Drought Tolerance of
Transgenic Maize and Tobacco
Reprinted from: *Agronomy* **2019**, *9*, 62, doi:10.3390/agronomy9020062 . . . . **355**

**Narges Moradtalab, Roghieh Hajiboland, Nasser Aliasgharzad, Tobias E. Hartmann and
Günter Neumann**
Silicon and the Association with an Arbuscular-Mycorrhizal Fungus (*Rhizophagus clarus*)
Mitigate the Adverse Effects of Drought Stress on Strawberry
Reprinted from: *Agronomy* **2019**, *9*, 41, doi:10.3390/agronomy9010041 . . . . **367**

Olga Mayoral, Jordi Solbes, José Cantó and Tatiana Pina
What Has Been Thought and Taught on the Lunar Influence on Plants in Agriculture?
Perspective from Physics and Biology
Reprinted from: *Agronomy* **2020**, *10*, 955, doi:10.3390/agronomy10070955 . . . . **387**

**Toi Ketehouli, Kue Foka Idrice Carther, Muhammad Noman, Fa-Wei Wang, Xiao-Wei Li and
Hai-Yan Li**
Adaptation of Plants to Salt Stress: Characterization of Na^+ and K^+ Transporters and Role of
CBL Gene Family in Regulating Salt Stress Response
Reprinted from: *Agronomy* **2019**, *9*, 687, doi:10.3390/agronomy9110687 . . . . **409**

Eszter Nemeskéri and Lajos Helyes
Physiological Responses of Selected Vegetable Crop Species to Water Stress
Reprinted from: *Agronomy* **2019**, *9*, 447, doi:10.3390/agronomy9080447 . . . . **441**

**Kazi Khayrul Bashar, Md. Zablul Tareq, Md. Ruhul Amin, Ummay Honi,
Md. Tahjib-Ul-Arif, Md. Abu Sadat and Quazi Md. Mosaddeque Hossen**
Phytohormone-Mediated Stomatal Response, Escape and Quiescence Strategies in Plants under
Flooding Stress
Reprinted from: *Agronomy* **2019**, *9*, 43, doi:10.3390/agronomy9020043 . . . . **461**

About the Editors

Monica Boscaiu is a Professor of Botany at the Department of Agroforest Ecosystems at the Polytechnic University of Valencia. She graduated from the Faculty of Biology at the University Babes-Bolyai in Cluj-Napoca, Romania, and obtained her Ph.D. degree in Botany at the University of Vienna, Austria. Since 1996, she lived in Spain, where she was first employed at the Botanical Garden. In 2002, she joined the Polytechnic University of Valencia (UPV), where she teaches different subjects related to botany and plant ecology. In 2019, she obtained a Full Professor position at the Agronomic School of the same university. She is the leader of the Phytochemical Resources group at the Mediterranean Agroforestry Institute (UPV). In the last 20 years, she published over 120 scientific articles related to the responses of wild plants and crops to abiotic stress factors and supervised numerous graduate and master's theses, as well as several Spanish and international Ph.D. students.

Ana Fita was born in 1979 in Valencia (Spain), where she graduated with a degree in Agriculture Engineering before obtaining her Ph.D. in Plant Breeding at the Polytechnic University of Valencia. At present, she is an Associate Professor at the same university. For 15 years, she has been teaching a wide variety of courses related to genetics, plant breeding, and plant genetic resources, such as 'Genetics and Plant Breeding', 'Plant Genetic Resources', 'In Vitro Culture and Plant Transformation', and 'Plant Variety Laws and Commercial Production of Plant Material', for graduate and master's students at UPV. She has supervised several graduate and master's theses and has participated in seminars in several European Universities. In addition, Ana Fita leads the Root Breeding group at the Institute for the Preservation and Improvement of Valencian Agrodiversity, COMAV, at UPV. Her research includes breeding vegetables for improved root systems and improving our understanding of root architecture and its relationship with resistance to biotic and abiotic stresses. At the moment, she is studying the genotype/soil/root interaction in different vegetables and the effects of abiotic stress on fruit quality. She has received awards from the American Society for Horticultural Science (ASHS) and the Valencian Government (Excellence Teaching Award at University Level 2010).

Preface to "Physiological and Molecular Characterization of Crop Resistance to Abiotic Stresses"

Abiotic stress represents the main constraint for agriculture, affecting plant growth and productivity worldwide. Yield losses in agriculture will be potentiated in the future by global warming, increasing contamination, and reduced availability of fertile land. The challenge for agriculture of the present and future is that of increasing the food supply for a continuously growing human population under environmental conditions that are deteriorating in many areas of the world. Minimizing the effects of diverse types of abiotic stresses represents a matter of general concern. Abiotic stress in plants is a vast subject that can be addressed from different points of view and includes many different components, mainly environmental factors (e.g., soil, water, climate, irradiation, and even the influence of the moon). Plants have evolved a series of physiological and molecular mechanisms of response that may (or may not) allow them to adapt to and survive this broad range of stressful conditions. Understanding those mechanisms will help us to improve our interventions towards more sustainable and efficient agriculture. The papers included in this Special Issue cover a broad range of topics related to the effects of different abiotic stress types on crop plants, at the morphological, physiological, biochemical, and molecular levels, and the mechanisms of defense of the plants against these stresses. The methods employed were also diverse, from the analysis of agronomic traits based on morphological characteristics to omics approaches and the use of transgenics. Special attention was given to the screening for stress tolerance in local landraces, stress alleviation using different strategies, and the proposal of practical solutions for the agriculture of the (near) future, threatened by global warming and environmental pollution. The editors wish to thank the contributors, reviewers, and the editorial staff of MDPI for their professionalism.

Monica Boscaiu, Ana Fita

Editors

Editorial

Physiological and Molecular Characterization of Crop Resistance to Abiotic Stresses

Monica Boscaiu [1],* and Ana Fita [2]

[1] Mediterranean Agroforestry Institute (IAM), Universitat Politècnica de València, Camino de Vera 14, 46022 Valencia, Spain

[2] Institute for the Conservation and Improvement of Valencian Agrodiversity (COMAV), Universitat Politècnica de València, Camino de Vera 14, 46022 Valencia, Spain; anfifer@btc.upv.es

* Correspondence: mobosnea@eaf.upv.es; Tel.: +34-963-879-253

Received: 3 August 2020; Accepted: 26 August 2020; Published: 2 September 2020

Abstract: Abiotic stress represents a main constraint for agriculture, affecting plant growth and productivity. Drought and soil salinity, especially, are major causes of reduction of crop yields and food production worldwide. It is not unexpected, therefore, that the study of plant responses to abiotic stress and stress tolerance mechanisms is one of the most active research fields in plant biology. This Special Issue compiles 22 research papers and 4 reviews covering different aspects of these responses and mechanisms, addressing environmental stress factors such as drought, salinity, flooding, heat and cold stress, deficiency or toxicity of compounds in the soil (e.g., macro and micronutrients), and combination of different stresses. The approaches used are also diverse, including, among others, the analysis of agronomic traits based on morphological characteristics, physiological and biochemical studies, and transcriptomics or transgenics. Despite its complexity, we believe that this Special Issue provides a useful overview of the topic, including basic information on the mechanisms of abiotic stress tolerance as well as practical aspects such as the alleviation of the deleterious effects of stress by different means, or the use of local landraces as a source of genetic material adapted to combined stresses. This knowledge should help to develop the agriculture of the (near) future, sustainable and better adapted to the conditions ahead, in a scenario of global warming and environmental pollution.

Keywords: salinity; drought; heat stress; flooding; nutrient stress; ROS; cold stress

---

1. Introduction

Abiotic stress represents the main constraint for agriculture, affecting plant growth and productivity worldwide. Yield losses in agriculture will be potentiated in the future by global warming, increasing contamination, and reduced availability of fertile land [1]. The challenge of the present and future agriculture is to increase the food supply for a continuously growing human population under environmental conditions that are deteriorating in many areas of the world. Minimizing the effects of diverse types of abiotic stresses represents a matter of general concern [2].

The study of abiotic stress tolerance mechanisms is one of the most active lines of research in plant biology, given its undoubted academic interest and practical implications in agriculture. The different types of abiotic stresses imposed by the environment usually are interconnected and often have an osmotic component, affecting plant cell homeostasis [3].

To counteract abiotic stress, plants activate a series of stress responses, which are shared by both sensitive and tolerant plants as they use the same basic effectors [4]. The knowledge of the limits of tolerance to abiotic stress of different crops, and the understanding of their mechanisms of response to increasing environmental constraints are gaining importance in agronomic research [5]. Research on crop abiotic stress responses is diverse, as plants undergo specific changes in their gene expression, metabolism, and physiology in response to different environmental stress conditions [6].

In this Special Issue, 22 research papers and 4 reviews are presented covering different aspects of the responses of plants to abiotic stresses and their mechanisms of tolerance. However, what is considered abiotic stress? We can define it as any physical or chemical constraint to the potential development and growth of a plant not involving interactions with other living organisms. Abiotic stress in plants is a vast subject, which can be addressed from different points of view and includes many different components, mainly environmental factors, for instance: soil, water, climate, irradiation—even the moon influence! Plants have evolved a series of physiological and molecular mechanisms of response that may (or may not) allow them to adapt to and survive this broad range of stressful conditions. Understanding those mechanisms will help us to improve our interventions towards a more sustainable and efficient agriculture.

2. Drought and Salinity

Drought and salinity are major abiotic stresses that affect agricultural yields worldwide. The more frequent, longer, and more intense dry periods in many regions of the world, due to global warming, are associated with increasing salinization of land cultivated under irrigation. About 20% of irrigated land in the world, producing one-third of the global food, is affected by secondary salinization of the soil [7]. Drought and salinity have a common osmotic component and early responses to these two types of stress are practically identical [8]. Besides, salt stress causes ionic stress and Na^+ toxicity [3]. Like other types of stress, drought and salinity or their combination may trigger growth inhibition, including, for example, disturbances in mineral nutrition, alteration of membrane permeability and cellular osmotic balance, generation of oxidative stress by increasing reactive oxygen species (ROS) levels, or inhibition of different enzyme activities [9–11].

In the Special Issue is included a review on physiological changes under drought conditions that influence yields in several vegetable crops summarizing changes in the stomatal conductance and chlorophyll content of leaves for individual plants, but also the utility of water stress indices and spectral vegetation indices for predicting yields [12]. An overview by Ketehouli et al. [13] on the effects of salinity on plants and their tolerance mechanisms with particular emphasis on K^+ and Na^+ homeostasis and transport and their regulation is also here included.

Plants defense against abiotic stress starts within their roots [3], and a well-developed root system is essential to provide water uptake [12]. The ability of plants to change their root anatomy was found to improve water uptake and transport in peanut and, therefore, may be considered as a relevant drought tolerance mechanism in this species [14].

This Special Issue includes several papers on morphological, physiological, and biochemical responses to these two types of stress or their combination, and their use in screening for stress-tolerant cultivars. Increased activities of ROS-scavenging enzymes and a more balanced Na^+: K^+ ratio was reported as the main mechanism of tolerance in wheat and barley [15]. Accumulation of proline and monovalent cations was related to salt tolerance mechanism in cultivated eggplant and its wild relative *Solanum insanum* [16]. Of special interest is the screening of neglected varieties and local landraces, as they can be a valuable source of allelic richness. Landraces evolved due to selection of traits specifically adapted to local conditions, often suboptimal or even highly stressful [17]. Therefore, such genotypes may enhance agronomic production under the foreseeable restrictive conditions imposed by climate change [2]. Proline was the marker used for screening of beans tolerant to water and salt stress [18], or antioxidant for salt-tolerant tomatoes with high nutraceutical value [19]. Proline and chlorophyll contents, in combination with several morphological and physiological traits, are optimal markers for screening drought tolerance in provitamin A maize, used in sub-Saharan Africa to combat vitamin A deficiency [20].

The irruption of transcriptomics, metabolomics, high-throughput DNA sequencing and high-density microarrays in the analysis of plants' responses to stress have brought new insights and allowed a better understanding on plants reactions to stressful conditions [21]. The stress-responding genes and their regulation pattern under drought were analyzed in common buckwheat cotyledons and roots [22] and female panicles in maize [23], and under salinity in roots and leaves of pomegranate [24].

Others papers published here deal with mitigation of the effects of drought in different crops, such as the synergistic effect of silicon and inoculation with an arbuscular mycorrhizal fungus on strawberries [25], transfer of a LEA gene of a Vietnamese maize landrace to transgenic maize and tobacco [26], and that of salinity by salicylic acid, yeast extract, and proline in sweet pepper [27].

3. Other Significant But Less Studied Stresses

Global warming alters the rainfall regime in many areas of the world [28], leading to increased floods and poorly drained, waterlogged soils; these conditions have a negative effect on crops by reducing oxygen availability for roots and soil microorganisms [29]. Escape and resilience strategies under flooding stress are presented in an extensive review, concluding that plants maintain their internal homeostasis by balancing hormonal cross-talk under excess water stress [30]. Besides, some treatments can help plants to cope with the stressful effects of waterlogging, for example, seed priming by sodium azide (NaN_3) was found to enhance the performance of okra plants under waterlogged conditions [31].

Extreme temperatures pose another challenge for crops. Irregular weather patterns have increased their occurrence in the present climatic conditions; for example, more frequent heat waves are now reported worldwide [28]. One paper deals with the effect of heat stress in alfalfa and extensively discusses the effects of heat on plants [31]). In addition, cold is also a common stress which triggers sophisticated events that alter the biochemical composition of cells in order to protect them from damage [32,33]. Again, some treatments can reduce the negative effects of low temperatures. This is the case of studies on the physiological performance of plants, in which cold stress was alleviated by chitosan via enhancing the photosynthesis and carbon process in tea plant [34], or by 5-Aminolevulinic in cucumber [35].

4. Combination of Different Stresses

Usually, abiotic stresses come together. The association of drought and salinity is well known, but also that of drought with high temperatures. When different stresses combine, plants need to adjust their physiology to those specific conditions. Landraces, through their long process of farmers' selection in a pre-intensive agriculture period, offer a great opportunity to find appropriate combinations of genes and phenotypes tolerant to complex situations. The most stressful period in the Mediterranean region is summer, when drought is associated with increased temperatures, including heat waves, which are increasingly more frequent in recent years [36]. A comparative study on the responses of local landraces and a commercial cultivar of *Phaseolus lunatus* L. to different temperature and water stress regimes is presented here. The results indicated a better response and a marked competitiveness of one local cultivar [37]. Effects on agronomic traits of the same stresses and their combination was analyzed in African landraces of maize compared with drought and/or heat-tolerant lines [38], and some local landraces proved to be good candidates for improving stress tolerance in this crop.

5. Soil Constrains

Besides soil salinity, discussed above, there are several other soil constraints with an important impact on agriculture [39]. Of special interest are those related to nutrient conditions in the soil, such as soil P immobilization. Phosphorus is an essential element for plants, but is lacking in 40% of arable land. This nutrient is normally applied as P-enriched fertilizers, which contribute to increased eutrophication of water bodies [40]. Therefore, screening for cultivars with a good performance under low P-input conditions is of interest, as shown by an analysis of morphological traits in relation to P accumulation in pepper cultivars [41]. Zinc is a microelement necessary for plants, animals, and humans; when it is not present in the soil in sufficient amounts, it is necessary either to use varieties with a better uptake of this micronutrient, or its external application in the form of fertilizers and foliar sprays [42]. However, when in excess it has a toxic effect for plants [43]. Morphological and physiological traits, in combination with the transcriptional regulation of aquaporin isoforms expression, were analyzed in pak choi subjected to two Zn concentrations [44].

Nitrogen is necessary for plant development; it is required in large quantities and, therefore, supplied to crops in fertilizers [45]. Nevertheless, an excessive N application was reported to decrease ROS scavenging ability, and to cause significant metabolic changes in wheat [46]. In the same species, the use of new ecofriendly polymeric-coated urea fertilizers insured a balanced proportion of N with beneficial effects [47].

Another paper deals with abiotic stress in crops imposed by treatments with herbicides and explores the possibility to control weeds with three natural compounds, analyzing the phytotoxic effects that they produce in weeds. The tree products demonstrated great possibilities as sustainable tools for integrated weed management [48].

Finally, this special issue also includes a review on some questions and beliefs that still impregnate a large part of agricultural traditions and agronomic practices, according to which the different lunar phases are beneficial or stressful to plant growth and development [49]. To address the possible link between the phases of the moon and agriculture from a scientific perspective, the authors analyzed physics and biology research papers and handbooks, focusing on those abiotic factors that have a proved influence on plant growth, searching specifically for any that could explain the influence of the moon on plant growth. They did not find any reliable, science-based evidence for such a relationship.

6. Conclusions

The papers included in this special issue cover a broad range of topics related to the effects on crop plants of different types of abiotic stress, at the morphological, physiological, biochemical, and molecular levels, and the mechanisms of defense of the plants against these stresses. The methods employed were also diverse, from the analysis of agronomic traits based on morphological characteristics to omics approaches and the use of transgenics. Special attention was given to the screening for stress tolerance in local landraces, stress alleviation using different strategies, and the proposal of practical solutions for the agriculture of the (near) future, threatened by global warming and environmental pollution.

Author Contributions: M.B. and A.F. equally contributed to organizing the special issue, editorial work, and writing this editorial. All authors have read and agreed to the published version of the manuscript.

Funding: No external funding was obtained.

Acknowledgments: Many thanks to the authors, reviewers, and to the editorial staff of MDPI for their professionalism.

Conflicts of Interest: The authors declare no conflict of interest.

References

1. Fedoroff, N.V.; Battisti, D.S.; Beachy, R.N.; Cooper, P.J.; Fischhoff, D.A.; Hodges, C.N.; Knauf, V.C.; Lobell, D.; Mazur, B.J.; Molden, D.; et al. Radically rethinking agriculture for the 21st century. *Science* **2010**, *327*, 833–834. [CrossRef]
2. Fita, A.; Rodriguez-Burruezo, A.; Boscaiu, M.; Prohens, J.; Vicente, O. Breeding and domesticating crops adapted to drought and salinity: A new paradigm for increasing food production. *Front. Plant Sci.* **2015**, *6*, 978. [CrossRef] [PubMed]
3. Gull, A.; Lone, A.A.; Islam Wani, N.U. Biotic and abiotic stresses in plants. In *Abiotic and Biotic Stress in Plants*; de Oliveira, A.B., Ed.; IntechOpen: London, UK, 2019.
4. Zhu, J.K. Plant salt tolerance. *Trends Plant Sci.* **2001**, *6*, 66–71. [CrossRef]
5. Shah, T.M.; Imran, M.; Atta, B.M.; Ashraf, M.I.; Hameed, A.; Waqar, I.; Shafiq, M.; Hussain, K.; Naveed, M.; Aslam, M.; et al. Selection and screening of drought tolerant high yielding chickpea genotypes based on physio-biochemical indices and multi-environmental yield trials. *BMC Plant Biol.* **2020**, *20*, 171. [CrossRef]
6. Zhu, J.K. Abiotic stress signaling and responses in plants. *Cell* **2016**, *167*, 313–324. [CrossRef]
7. Machado, R.M.A.; Serralheiro, R.P. Soil salinity: Effect on vegetable crop growth. Management practices to prevent and mitigate soil salinization. *Horticulture* **2017**, *3*, 30. [CrossRef]

8. Munns, R. Comparative physiology of salt and water stress. *Plant Cell Environ.* **2002**, *25*, 239–250. [CrossRef] [PubMed]

9. Munns, R.; Tester, M. Mechanisms of salinity tolerance. *Annu. Rev. Plant Biol.* **2008**, *59*, 651–681. [CrossRef]

10. Khan, A.; Pan, X.; Najeeb, U.; Yuen Tan, D.K.; Fahad, S.; Zahoor, R.; Luo, H. Coping with drought: Stress and adaptive mechanisms, and management through cultural and molecular alternatives in cotton as vital constituents for plant stress resilience and fitness. *Biol. Res.* **2018**, *51*, 47. [CrossRef]

11. Hernández, J.A. Salinity tolerance in plants: Trends and perspectives. *Int. J. Mol. Sci.* **2019**, *20*, 2408. [CrossRef]

12. Nemeskéri, E.; Helyes, L. Physiological responses of selected vegetable crop species to water stress. *Agronomy* **2019**, *9*, 447. [CrossRef]

13. Ketehouli, T.; Idrice Carther, K.F.; Noman, M.; Wang, F.-W.; Li, X.-W.; Li, H.-Y. Adaptation of plants to salt stress: Characterization of Na$^+$ and K$^+$ transporters and role of CBL gene family in regulating salt stress response. *Agronomy* **2019**, *9*, 687. [CrossRef]

14. Thangthong, N.; Jogloy, S.; Punjansing, T.; Kvien, C.K.; Kesmala, T.; Vorasoot, N. Changes in root anatomy of peanut (*Arachis hypogaea* L.) under different durations of early season drought. *Agronomy* **2019**, *9*, 215. [CrossRef]

15. Zeeshan, M.; Lu, M.; Sehar, S.; Holford, P.; Wu, F. Comparison of biochemical, anatomical, morphological, and physiological responses to salinity stress in wheat and barley genotypes deferring in salinity tolerance. *Agronomy* **2020**, *10*, 127. [CrossRef]

16. Brenes, M.; Solana, A.; Boscaiu, M.; Fita, A.; Vicente, O.; Calatayud, Á.; Prohens, J.; Plazas, M. Physiological and biochemical responses to salt stress in cultivated eggplant (*Solanum melongena* L.) and in *S. insanum* L., a close wild relative. *Agronomy* **2020**, *10*, 651. [CrossRef]

17. Fess, T.L.; Kotcon, J.B.; Benedito, V.A. Crop breeding for low input agriculture: A sustainable response to feed a growing world population. *Sustainability* **2011**, *3*, 1742–1772. [CrossRef]

18. Arteaga, S.; Yabor, L.; Díez, M.J.; Prohens, J.; Boscaiu, M.; Vicente, O. The use of proline in screening for tolerance to drought and salinity in common bean (*Phaseolus vulgaris* L.) genotypes. *Agronomy* **2020**, *10*, 817. [CrossRef]

19. Sumalan, R.M.; Ciulca, S.I.; Poiana, M.A.; Moigradean, D.; Radulov, I.; Negrea, M.; Crisan, M.E.; Copolovici, L.; Sumalan, R.L. The antioxidant profile evaluation of some tomato landraces with soil salinity tolerance correlated with high nutraceutical and functional value. *Agronomy* **2020**, *10*, 500. [CrossRef]

20. Kondwakwenda, A.; Sibiya, J.; Zengeni, R.; Musvosvi, C.; Tesfay, S. Screening of provitamin-A maize inbred lines for drought tolerance: Beta-carotene content and secondary traits. *Agronomy* **2019**, *9*, 692. [CrossRef]

21. Urano, K.; Kurihara, Y.; Seki, M.; Shinozaki, K. 'Omics' analyses of regulatory networks in plant abiotic stress responses. *Curr. Opin. Plant Biol.* **2010**, *13*, 132–138. [CrossRef]

22. Hou, Z.; Yin, J.; Lu, Y.; Song, J.; Wang, S.; Wei, S.; Liu, Z.; Zhang, Y.; Fang, Z. Transcriptomic analysis reveals the temporal and spatial changes in physiological process and gene expression in common buckwheat (*Fagopyrum esculentum* Moench) grown under drought Stress. *Agronomy* **2019**, *9*, 569. [CrossRef]

23. Jia, S.; Li, H.; Jiang, Y.; Tang, Y.; Zhao, G.; Zhang, Y.; Yang, S.; Qiu, H.; Wang, Y.; Guo, J.; et al. Transcriptomic analysis of female panicles reveals gene expression responses to drought stress in maize (*Zea mays* L.). *Agronomy* **2020**, *10*, 313. [CrossRef]

24. Liu, C.; Zhao, Y.; Zhao, X.; Wang, J.; Gu, M.; Yuan, Z. Transcriptomic profiling of pomegranate provides insights into salt tolerance. *Agronomy* **2020**, *10*, 44. [CrossRef]

25. Moradtalab, N.; Hajiboland, R.; Aliasgharzad, N.; Hartmann, T.E.; Neumann, G. Silicon and the association with an arbuscular-mycorrhizal fungus (*Rhizophagus clarus*) mitigate the adverse effects of drought stress on strawberry. *Agronomy* **2019**, *9*, 41. [CrossRef]

26. Minh, B.M.; Linh, N.T.; Hanh, H.H.; Hien, L.T.T.; Thang, N.X.; Hai, N.V.; Hue, H.T.T. A LEA gene from a Vietnamese maize landrace can enhance the drought tolerance of transgenic maize and tobacco. *Agronomy* **2019**, *9*, 62. [CrossRef]

27. Abdelaal, K.A.; EL-Maghraby, L.M.; Elansary, H.; Hafez, Y.M.; Ibrahim, E.I.; El-Banna, M.; El-Esawi, M.; Elkelish, A. Treatment of sweet pepper with stress tolerance-inducing compounds alleviates salinity stress oxidative damage by mediating the physio-biochemical activities and antioxidant systems. *Agronomy* **2020**, *10*, 26. [CrossRef]

28. IPCC. *Climate Change. 2014: Impacts, Adaptation, and Vulnerability. Part B: Regional Aspects. Contribution of Working Group II to the Fifth Assessment Report of the Intergovernmental Panel on Climate Change*; Cambridge University Press: Cambridge, UK; New York, NY, USA, 2014.

29. Loreti, E.; van Veen, H.; Perata, P. Plant responses to flooding stress. *Curr. Opin. Plant Biol.* **2016**, *33*, 64–71. [CrossRef]

30. Bashar, K.K.; Tareq, M.Z.; Amin, M.R.; Honi, U.; Tahjib-Ul-Arif, M.; Sadat, M.A.; Hossen, Q.M.M. Phytohormone-mediated stomatal response, escape and quiescence strategies in plants under flooding stress. *Agronomy* **2019**, *9*, 43. [CrossRef]

31. Vwioko, E.D.; El-Esawi, M.A.; Imoni, M.E.; Al-Ghamdi, A.A.; Ali, H.M.; El-Sheekh, M.M.; Abdeldaym, E.A.; Al-Dosary, M.A. Sodium azide priming enhances waterlogging stress tolerance in okra (*Abelmoschus esculentus* L.). *Agronomy* **2019**, *9*, 679. [CrossRef]

32. Wassie, M.; Zhang, W.; Zhang, Q.; Ji, K.; Chen, L. Effect of heat stress on growth and physiological traits of alfalfa (*Medicago sativa* L.) and a comprehensive evaluation for heat tolerance. *Agronomy* **2019**, *9*, 597. [CrossRef]

33. Eremina, M.; Rozhon, W.; Poppenberger, B. Hormonal control of cold stress responses in plants. *Cell. Mol. Life Sci.* **2016**, *73*, 797–810. [CrossRef] [PubMed]

34. Li, Y.; Zhang, Q.; Ou, L.; Ji, D.; Liu, T.; Lan, R.; Li, X.; Jin, L. Response to the cold stress signaling of the tea plant (*Camellia sinensis*) elicited by chitosan oligosaccharide. *Agronomy* **2020**, *10*, 915. [CrossRef]

35. Anwar, A.; Wang, J.; Yu, X.; He, C.; Li, Y. Substrate application of 5-aminolevulinic acid enhanced low-temperature and weak-light stress tolerance in cucumber (*Cucumis sativus* L.). *Agronomy* **2020**, *10*, 472. [CrossRef]

36. Diffenbaugh, N.S.; Pal, J.S.; Giorgi, F.; Gao, X. Heat stress intensification in the Mediterranean climate change hotspot. *Geophys. Res. Lett.* **2007**, *34*, 11. [CrossRef]

37. Martínez-Nieto, M.I.; Estrelles, E.; Prieto-Mossi, J.; Roselló, J.; Soriano, P. Resilience capacity assessment of the traditional Lima Bean (*Phaseolus lunatus* L.) landraces facing climate change. *Agronomy* **2020**, *10*, 758. [CrossRef]

38. Nelimor, C.; Badu-Apraku, B.; Tetteh, A.Y.; Garcia-Oliveira, A.L.; N'guetta, A.P. Assessing the potential of extra-early maturing landraces for improving tolerance to drought, heat, and both combined stresses in maize. *Agronomy* **2020**, *10*, 318. [CrossRef]

39. Probert, M.E.; Keating, B.A. What soil constraints should be included in crop and forest models? *Agric. Ecosyst. Environ.* **2000**, *82*, 273–281. [CrossRef]

40. Kauranne, L.-M.; Kemppainen, M. Urgent need for action in the Baltic sea area. In *Phosphorus in Agriculture: 100% Zero*; Springer: Dordrecht, The Netherlands, 2016; pp. 1–6.

41. Pereira-Dias, L.; Gil-Villar, D.; Castell-Zeising, V.; Quiñones, A.; Calatayud, A.; Rodríguez-Burruezo, A.; Fita, A. Main root adaptations in pepper germplasm (*Capsicum* spp.) to phosphorus low-input conditions. *Agronomy* **2020**, *10*, 637. [CrossRef]

42. Hefferon, K. Biotechnological approaches for generating zinc-enriched crops to combat malnutrition. *Nutrients* **2019**, *11*, 253. [CrossRef]

43. Szopiski, M.; Sitko, K.; Gierón, Z.; Rusinowski, S.; Corso, M.; Hermans, C.; Verbruggen, N.; Małkowski, E. Toxic effects of Cd and Zn on the photosynthetic apparatus of the *Arabidopsis halleri* and *Arabidopsis arenosa* Pseudo-Metallophytes. *Front. Plant Sci.* **2019**, *10*, 748. [CrossRef]

44. Fatemi, H.; Zaghdoud, C.; Norteempes, P.A.; Carvajal, M.; Martínez-Ballesta, M.C. Differential aquaporin response to distinct effects of two Zn concentrations after foliar application in pak choi (*Brassica rapa* L.) plants. *Agronomy* **2020**, *10*, 450. [CrossRef]

45. Leghari, S.J.; Wahocho, N.A.; Laghari, G.M.; Talpur, K.H.; Wahocho, S.A.; Lashari, A.A. Role of nitrogen for plant growth and development: A review. *Adv. Environ. Biol.* **2016**, *10*, 209–2018.

46. Kong, L.; Xie, Y.; Hu, L.; Si, J.; Wang, Z. Excessive nitrogen application dampens antioxidant capacity and grain filling in wheat as revealed by metabolic and physiological analyses. *Sci. Rep.* **2017**, *7*, 43363. [CrossRef] [PubMed]

47. Gil-Ortiz, R.; Naranjo, M.Á.; Ruiz-Navarro, A.; Caballero-Molada, M.; Atares, S.; García, C.; Vicente, O. New eco-friendly polymeric-coated urea fertilizers enhanced crop yield in wheat. *Agronomy* **2020**, *10*, 438. [CrossRef]
48. Muñoz, M.; Torres-Pagán, N.; Peiró, R.; Guijarro, R.; Sánchez-Moreiras, A.M.; Verdeguer, M. Phytotoxic effects of three natural compounds: Pelargonic acid, carvacrol, and cinnamic aldehyde, against problematic weeds in Mediterranean crops. *Agronomy* **2020**, *10*, 791. [CrossRef]
49. Mayoral, O.; Solbes, J.; Cantó, J.; Pina, T. What has been thought and taught on the lunar influence on plants in agriculture? Perspective from physics and biology. *Agronomy* **2020**, *10*, 955. [CrossRef]

 *agronomy*

Article

Response to the Cold Stress Signaling of the Tea Plant (*Camellia sinensis*) Elicited by Chitosan Oligosaccharide

Yingying Li, Qiuqiu Zhang, Lina Ou, Dezhong Ji, Tao Liu, Rongmeng Lan, Xiangyang Li and Linhong Jin *

State Key Laboratory Breeding Base of Green Pesticide and Agricultural Bioengineering, Key Laboratory of Green Pesticide and Agricultural Bioengineering, Ministry of Education, Guizhou University, Huaxi District, Guiyang 550025, China; gs.yingyingli17@gzu.edu.cn (Y.L.); gs.zhangqq18@gzu.edu.cn (Q.Z.); gs.lnou17@gzu.edu.cn (L.O.); gs.dzji19@gzu.edu.cn (D.J.); gs.taoliu18@gzu.edu.cn (T.L.); gs.rmlan19@gzu.edu.cn (R.L.); xyli1@gzu.edu.cn (X.L.)
* Correspondence: lhjin@gzu.edu.cn; Tel.: +86-851-362-0521; Fax: +86-851-362-2211

Received: 26 April 2020; Accepted: 22 June 2020; Published: 26 June 2020

Abstract: Cold stress caused by a low temperature is a significant threat to tea production. The application of chitosan oligosaccharide (COS) can alleviate the effect of low temperature stress on tea plants. However, how COS affects the cold stress signaling in tea plants is still unclear. In this study, we investigated the level of physiological indicators in tea leaves treated with COS, and then the molecular response to the cold stress of tea leaves treated with COS was analyzed by transcriptomics with RNA-Sequencing (RNA-Seq). The results show that the activity of superoxide dismutase (SOD) activity, peroxidase (POD) activity, content of chlorophyll and soluble sugar in tea leaves in COS-treated tea plant were significantly increased and that photosynthesis and carbon metabolism were enriched. Besides, our results suggest that COS may impact to the cold stress signaling via enhancing the photosynthesis and carbon process. Our research provides valuable information for the mechanisms of COS application in tea plants under cold stress.

Keywords: tea plant; cold stress; chitosan oligosaccharide; physiological response; transcriptome

---

1. Introduction

The tea plant (*Camellia sinensis* (L.) O. Kuntze) is one of the most important commercial beverage crops in the world and an important revenue source in tea-producing countries [1]. The tea production in over 50 countries has reached over 5.95 million tons on 4.1 million hectares around the world [2]. Among them, the cultivar 'Anji Baicha' is a special green-revertible albino mutant widely cultivated in China, especially in Zhejiang, Hubei and Guizhou provinces, which exhibits periodic albinism during the development of young shoots [3,4]. It is rare and represent precious tea germplasm because of it special flavor, and also has high levels of total amino acids and low levels of polyphenols, which differs from conventional tea [3–8]. In addition, it has a higher commercial value than green tea [4].

The tea plant can grow in different agroclimates and adapted to optimal temperature of 18 to 30 °C and pH ranging from 4.5 to 5.5, but the thermophilic nature of tea plants confines their growth to temperate area [9–11]. Furthermore, tea plants that are exposed to a low temperature, such as a sudden frost in fall or early spring, may be at risk of cold stress [12]. Cold environment can adversely affect tea plants on their growth, development, and spatial distribution with decreasing yield and quality, which is one of the factors restricting the healthy development of the tea industry [13–15]. So, it is significant to explore the ways to improve the cold resistance of tea plants. Some studies have reported that the cold resistance of tea plant can be effectively improved by cultivating cold-resistant tea plant varieties (e.g., Fudingdabai, Shuchazao), cold acclimation of tea plant and the application of exogenous substances [16–19].

Chitosan oligosaccharide (COS) prepared from chitosan, is an environmentally friendly plant growth regulator and stress tolerance inducer [20–24]. Chitosan is a linear polysaccharide composed of β-1,4-glucosamines. The hydrolysis of the glycosidic chitosan chains yields oligosaccharides, including the water-soluble oligochitosan [21,22]. Chitosan and COS have a rich history of being researched for applications in agriculture, primarily for plant defense and yield increase [23,24]. As a natural biocontroller and elicitor of defense responses, COS can boost the innate ability of plants to defend themselves by stimulating secondary metabolite synthesis, and increasing the chlorophyll content and photosynthetic ability [20,21], enrich the soluble sugar in plant [25], and enhancing the activities of antioxidant enzymes [25–27]. COS stimulated the signaling pathways involved in disease resistance in rice [28], and its role in tobacco mosaic virus (TMV) resistance in *Arabidopsis* has been investigated [29]. And studies have shown that COS enhances carbon metabolism, nitrogen metabolism, photosynthesis, and defense against abiotic stress in plants [30]. As reported, COS was able to mitigate the effects of abiotic stresses in plant, including salt, cold and drought [25–27,31,32]. The mechanism of COS in increasing abiotic stress tolerances was summarized as: enhancing the activities of antioxidant enzymes [25], photosynthesis, and stimulate secondary metabolite synthesis [31]. For example, COS has been applied to wheat seedlings for improved chilling tolerance by enhancing antioxidant activities of superoxide dismutase (SOD) and peroxidase (POD) and increasing content of chlorophyll.

These physiological responses of plants elicited by COS are closely related to the regulation of plant gene expression. Transcriptome sequencing has been widely applied to tea plant, which is has the advantage of highly accurate, highly efficient and sensitive profiling in recent years [33]. RNA sequencing (RNA-Seq) technology for measuring transcriptomes of organisms can analyze genes related to abiotic and biotic stress responses, growth, development and metabolites [34–37], to improve our understanding of the molecular mechanism of the tea plant [13–16,38], and RNA-Seq will also be a valuable tool to reveal the role of exogenous substances in tea plant cold resistance.

Though many investigators provided valuable information to cold stress in tea plant, the action mode of COS eliciting responses to cold stress of tea plant is unclear. Therefore, in this report, we studied the effect of exogenous COS on the molecular mechanism of tea plant under low temperature stress. Herein, the physiological parameters of tea plants with and without COS-treatment were compared. The molecular response to cold resistance within tea plant was analyzed by RNA-Seq technology. This research improves the understanding of the cold resistance mechanism of COS-treated tea plant and provides important guidance for COS application under low temperature stress.

2. Materials and Methods

2.1. Plant Materials and Cold Treatments

Two-year-old albino tea cultivar (*Camellia sinensis* (L.) O. Kuntze cv. 'Anji Baicha') were used in the experiment from AnShun County, Guizhou Province, China. Additionally, the tea plants were transplanted into the plastic pot. Plants were grown in a growth chamber at the experimental of Guizhou University, Guizhou Province, China (16 h day/8 h night at 25 °C/20 °C and relative humidity of 70%). After a month, tea plants were treated with 10 mL of following elicitors by surface spraying with sterile distilled water (control, CK), or with 1.25 mL/L COS solution (COS comes from Hainan Zhengye Zhongnong High-tech Co., Ltd., Haikou, Hainan Province, China). After 24 h, the two groups of tea plants were separately maintained in a chamber at −4 and −8 °C at cold treatment for 24 h, with one group maintained under normal room temperature conditions. Three independent biological repeats were collected for each treatment. Fresh leaves from the stable stage (re-greening stage) of chlorophyll development of Anji Baicha were harvested at 24 h and frozen immediately in liquid nitrogen and stored at −80 °C for further study.

2.2. Physiological Response Assay

Physiological indexes of tea leaves (containing 1st, 2nd, 3rd leaf and old leaves), involving the activities of SOD and POD, and content of chlorophyll and soluble sugar, were determined. Additionally, the assay kits used included the SOD assay kit, the POD assay kit, the chlorophyll assay kit, the soluble sugar assay kit (Solarbio, Cat. No. BC0175, BC0095, BC0995, BC0035, respectively, Beijing, China). All assays were performed according to the manufacturer's instructions.

2.3. cDNA Library Construction and Sequencing

We selected tea leaves from control and COS treatment on −4 °C for RNA-Seq analysis. Total RNA was extracted from tea leaves using TRIzol reagent (Invitrogen, Carlsbad, CA, USA) following the manufacturer's instruction. Poly (A) + mRNA was purified with oligo (dT) beads. The mRNA was randomly cut into short fragments using Fragmentation Buffer, which were used as a template for the short fragment mRNA, first-strand cDNA was synthesized with 6 bp random primers, and then the Buffer, dNTPs and DNA polymerase I were added to synthesize the second-strand cDNA. RNA Integrity was confirmed using 1.5% agarose gel. RNA quality was checked by a NanoDrop^{TM} OneC spectrophotometer (Thermo Fisher Scientific, New York, NY, USA). RNA qualified was measured by Qubit^{TM} RNA BR Assay Kit in Qubit® 2.0 (Life Technologies, Carlsbad, CA, USA). The cDNA library construction and Illumina sequencing of the samples were performed using a 150 bp paired-end Illumina Nova-seq 6000 (Illumina, San Diego, CA, USA) by Seqhealth Technology Co., Ltd. (Wuhan, China).

2.4. RNA-Seq Data Analysis

The raw reads were first filtered to obtain the clean reads by removing the adaptor sequences, unknown sequences "N" and low-quality reads using Trimmomatic (version 0.36). After filtering, the clean reads were mapped to the reference genome of *Camellia sinensis* using STATR software (version 2.5.3a).

2.5. Identification of Differentially Expressed Genes

The expression levels of each gene were calculated and normalized by the corresponding Reads Per Kilobase of transcript per Million mapped reads (RPKM). The RPKM method can eliminate the influence of gene length and sequencing amount differences on gene expression. FeatureCounts (version 1.5.1) was used to count the read numbers mapped to each gene [39]. Additionally, differentially expressed genes (DEGs) were identified with the edge R package (version 3.12.1) [40]. The resulting p-values were adjusted using Benjamini and Hochberg's method for controlling the false discovery rate (FDR). Genes with p-value < 0.05 and a logarithm two-fold change |$\log_2$FC| > 1 were defined as DEGs.

2.6. Gene Ontology and KEGG Pathway Analysis

Gene ontology (GO) analysis and Kyoto encyclopedia of genes and genomes (KEGG) enrichment analysis of DEGs were both implemented by KEGG orthology based annotation system (KOBAS) software (version 2.1.1) with p-value < 0.05 to judge statistically significant enrichment [41].

2.7. Quantitative RT-PCR (qRT-PCR) Analysis

To verify the RNA-Seq analysis, we randomly selected five unigenes and used qRT-PCR to confirm their participation in the high-temperature reaction. RT-qPCR was conducted on ABI ViiA^{TM} 7 Real-Time PCR System (Applied Biosystems, Foster, CA, USA) using GoTaq® qPCR Master Mix (Promega, Madison, WI, USA). The PCR amplifications were consisted of 95 °C for 3 min, followed by 40 cycles of 95 °C for 15 s, 60 °C for 30 s, and then 72 °C for 30 s. Gene expression was normalized using the glyceraldehyde-3-phosphate dehydrogenase (GADPH) as an internal reference gene, and the relative changes of gene expression were calculated using the $2^{-\Delta\Delta Ct}$ method. The list of primers is presented in Table S1.

2.8. Statistical Analysis

Data were expressed as the mean ± standard error, and the data were subjected to one-way analysis of variance (ANOVA) ($p < 0.05$) followed by a significant difference test (LSD) using SPSS statistics v17.0 (SPSS Inc., Chicago, IL, USA).

3. Results

3.1. Physiological Parameter Response to a Low Temperature

To analyze the effects of COS on tea plant growth, we measured the change in activity of SOD, and POD enzymes and content of chlorophyll, soluble sugar in COS-treated tea plant and their respond to low temperature stress, with sterile distilled water served as control. As shown in Figure 1, under a low temperature, the tea plant responds to cold stress with all the physiological parameters changed and COS-enhanced freeze protection. As in the control group, a low temperature caused increases in those physiological parameters. As shown in Figure 1A, the enzyme activity of SOD was significantly increased by 24.04% at −4 °C and 32.68% at −8 °C. Similarly, the enzyme activity of POD was significantly increased by 38.05% at −4 °C and 8.81% at −8 °C. Cold stress significantly reduced the chlorophyll content by 20.18% and 21.96% at −4 and −8 °C, respectively (Figure 1C). Moreover, soluble sugar content was significantly increased by 29.87% at −4 °C and 28.16% at −8 °C, respectively (Figure 1D). The results show that cold stress consistently increased SOD and POD activity, and soluble sugar content, when the temperature was switched from 25 °C to −4 °C or −8 °C, but POD activity was highest at −4 °C.

When exogenous COS was used, it consistently enhanced SOD and POD activities, and the soluble sugar content and chlorophyll content in the tea plant. For example, COS improved SOD activity by 11.75% at 25 °C, 25.93% at −4 °C and 9.21% at −8 °C, respectively, as compared with the control. Similarly, POD activity was enhanced by 19.91%, 19.23% and 30.09% on 25 °C, −4 °C and −8 °C, respectively.

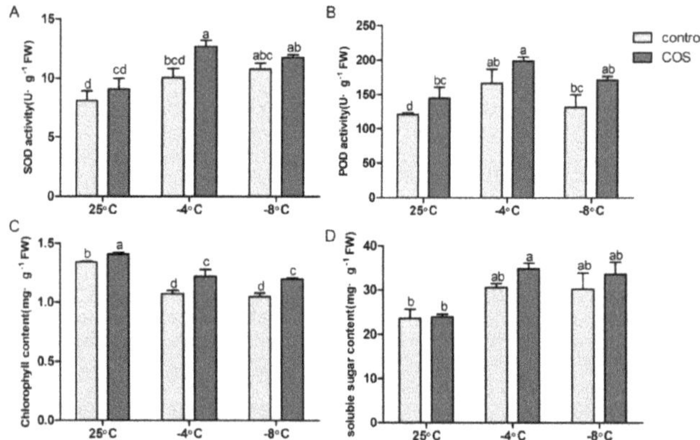

Figure 1. Effect of chitosan oligosaccharide (COS) on physiological parameters of tea leaves. (**A**) Superoxide dismutase (SOD) activity; (**B**) peroxidase (POD); (**C**) chlorophyll content; (**D**) soluble sugar content. The data represent the means ± SD of three replicates samples. Different letters indicate significant differences at $p < 0.05$.

For all the tested parameters, the effects of COS were more pronounced under cold stress. When tea plants were treated with COS combined with cold stress, SOD enhanced by 56.21% and 44.91% at −4 and −8 °C, respectively. Similarly, POD increased 37.26% and 18.04%. The content of soluble sugar also increased by 45.22% and 40.25% at −4 and −8 °C, respectively. Chlorophyll content was decreased by 13.47% and 14.99%, respectively. The results show that COS treatment consistently increased chlorophyll content, but three parameters of SOD, POD and soluble sugar were highest at −4 °C of cold stress combined with COS.

3.2. Transcriptome Sequencing and Assembly

To understand the response of the tea plant to cold stress and the effect of COS on the molecular level, we compared the transcriptomes between COS treatment and the control group at −4 °C by RNA-Seq. Replicate samples of the control group (ConT3_1/2/3) and COS-treatment group (TreT3_1/2/3) were included in this study. We obtained 5.59–6.60 million raw reads in control and 5.79–6.77 million raw reads in the COS-treatment group. After filtering and removing low-quality reads, the clean reads were limited 5.26–6.21 million and 5.45–6.34 million, respectively. Of these clean reads, the GC content was 46.46–47.21% and the Q30 values were over 98.45%. The ratio of total mapped reads between the control and COS-treatment groups was 94.69–94.90% and 94.85–95.20% for *Camellia sinensis* according to the Genome Database. Unique mapped reads were 91.48–92.10% in the control group and 88.02–90.66% in the COS-treatment group (Table 1).

Table 1. Statistical analyses and mapping results of RNA sequencing reads.

Sample	ConT3_1	ConT3_2	ConT3_3	TreT3_1	TreT3_2	TreT3_3
Raw reads	55,965,032	56,476,808	66,044,722	57,864,054	67,743,104	65,453,870
Clean reads	52,619,470	53,061,678	62,155,236	54,555,936	63,422,124	61,416,118
Q30 (%)	98.45	98.45	98.70	98.65	98.55	98.45
GC content (%)	46.60	46.46	46.63	46.82	46.83	47.21
Total reads	44,163,580	43,980,650	52,344,630	45,455,332	52,920,720	51,188,834
Total mapped	41,828,592 (94.71%)	41,644,005 (94.69%)	49,676,907 (94.90%)	43,274,546 (95.20%)	50,292,573 (95.03%)	48,551,418 (94.85%)
Unique mapped	38,522,223 (92.10%)	38,095,551 (91.48%)	45,663,412 (91.92%)	39,232,023 (90.66%)	45,576,183 (90.62%)	42,734,470 (88.02%)

3.3. Differentially Expressed Genes Analysis

In order to verify the correlation of gene expression level between samples, we demonstrated that the biological repeatability between samples was great through spearman correlation analysis based on the RPKM of different samples. Genes with p-value < 0.05 and $|\log_2(\text{Foldchange})| > 1$ were defined as differentially expressed genes between control and COS. There were identified 4503 differentially expressed genes (DEGs) between the control and COS, including 1605 up-regulated and 2898 down-regulated genes in the leaves of tea plant (Figure 2 and Table S2).

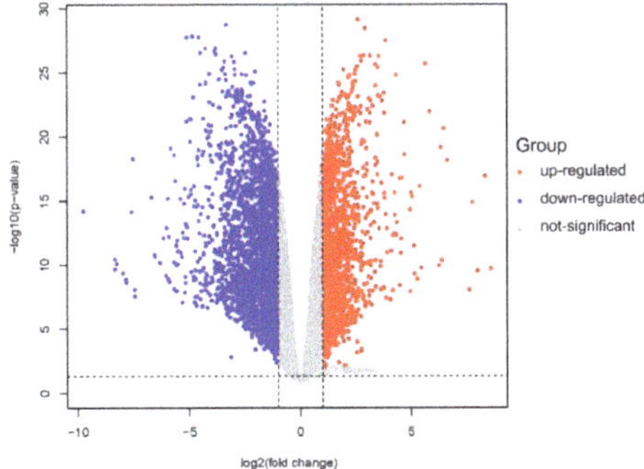

Figure 2. Volcano plot of differentially expressed genes (DEGs) showed up-regulated and down-regulated between control and COS under −4 °C treatment. The red dots represent up-regulated genes, the blue dots represent down-regulated genes, and the gray dots represent no significant difference. The horizontal coordinates indicate the change in multiple expression, the longitudinal coordinates indicate the magnitude of differences.

3.4. Gene Ontology (GO) Annotation

The differentially expressed mRNAs were analyzed by GO enrichment, as shown in Figure 3 and Table S3. The differentially expressed genes were mostly enriched in biological process (Figure 3). In the biological process categorization, functional enrichment mainly focuses on metabolic processes and nutrient synthesis processes, such as "single-organism biosynthetic process" (GO: 0044711), "metabolic process" (GO: 0008152), "carbohydrate metabolic process" (GO: 0005975) and "carbohydrate derivative biosynthetic process" (GO: 1901137). The molecular function category includes the expression of transmembrane transporters and catalytic enzyme-related genes, such as "catalytic activity" (GO: 0003824), "transporter activity" (GO: 0005215), "transmembrane transporter activity" (GO: 0022857), and "ion transmembrane transporter activity" (GO: 0015075). Besides, "serine-type endopeptidase activity" (GO: 0004252) was mostly enriched in the molecular function category.

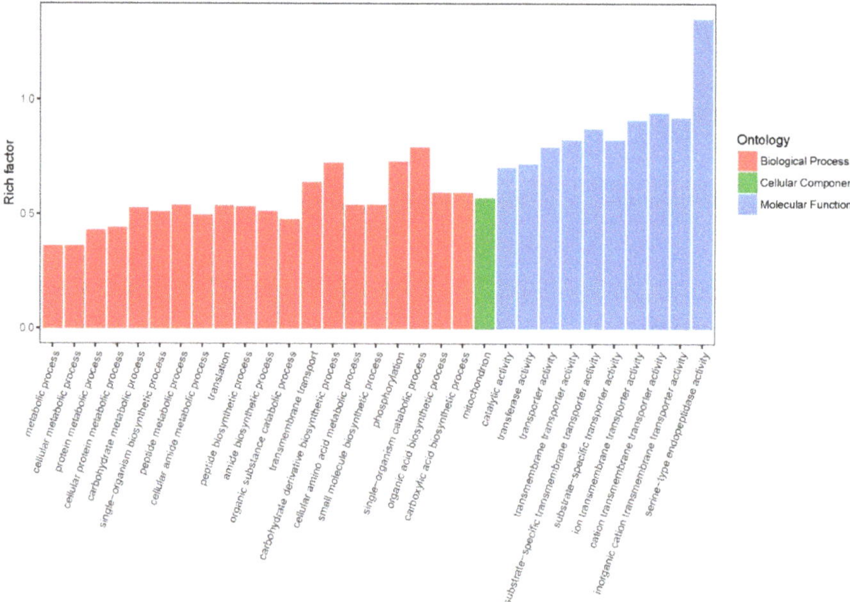

Figure 3. Gene ontology (GO) classification analysis based on DEGs induced by COS under −4 °C treatment. The horizontal coordinates indicate GO terms, the longitudinal coordinates indicate rich factor, rich factor represents the ratio between the number of different genes enriched in the term and the background genes in GO term.

3.5. Kyoto Encyclopedia of Genes and Genomes (KEGG) Pathway Annotation

The KEGG enrichment scatter plot is a graphical representation of the statistical analyses that visualizes the pathway enrichment (Figure 4). The degree of KEGG enrichment was measured in terms of richness factor, *p*-value, and the number of genes in the pathway. The important enriched pathways with high generation, low *p*-value and large numbers of genes are shown in the Figure 4 and Table S4. As shown in Figure 4, these enriched pathways, including "photosynthesis" (ko00195), "carbon fixation in photosynthetic organisms" (ko00710), "photosynthesis–antenna proteins" (ko00196), "ribosome" (ko03010), "carbon metabolism" (ko01200).

Compared with the control group, 71 genes were significantly induced to up-regulated by COS treatment, including PSII, PSI, cytochrome b6/f complex, photosynthethic electron transport and F-type ATPase (Table S5). In the carbon metabolism pathway, a total of 77 genes were differentially expressed, including 52 up-regulated and 25 down-regulated (Table S6). A total of 43 genes were assigned to the plant hormone signal transduction pathway, including 16 genes that were up-regulated in auxin, abscisic acid, ethylene, salicylic acid (Table S7). These results suggest that the addition of COS at a low temperature have a complex effect on biological process and metabolism of the tea plant.

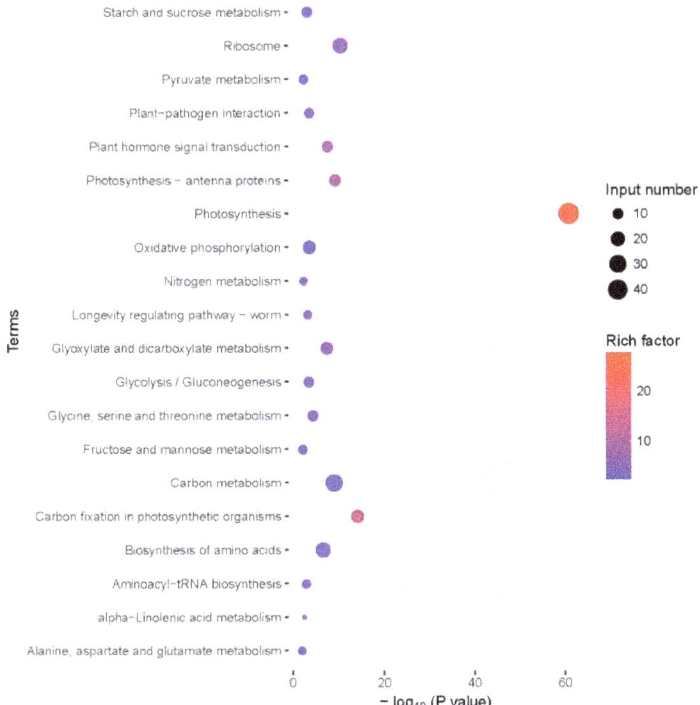

Figure 4. Kyoto Encyclopedia of Genes and Genomes (KEGG) enrichment analysis based on DEGs induced by COS under −4 °C treatment. The significance of enrichment is shown on the horizontal coordinates (represented by −$\log_{10}$ (*p*-value), the greater the value, the more significant the enrichment), and the KEGG pathway is shown on the longitudinal coordinates. The size of the dots indicates the number of different genes contained in the KEGG pathway, and the color of the dots indicates the degree of rich factor enrichment.

3.6. qRT-PCR Validation of Differentially Expressed Transcripts from RNA-Seq

Five transcripts were randomly selected for qRT-PCR analysis, which used to confirm validity and accuracy the RNA-Seq data. The results show that the trend of qRT-PCR is consistent with the results of RNA-Seq in Figure S1.

4. Discussion

Cold stress affects photosynthetic activities and metabolic functions in plants, which further affected growth, development, and metabolism. It has a negative effect on the yield and quality of tea. Anji Baicha is a temperature-sensitive albino tea cultivar. When the environment temperature is below 20 °C in early spring, the white shoots phenomenon will appear. After about two weeks, the plant gradually turns as green, as does those of common tea cultivars [4–6]. The change of leaf color was mainly due to chloroplast development in the albescent stage, the etioplast–chloroplast transition was blocked, and the accumulation of chlorophyll was inhibited under low temperature [4–8,37]. In this study, we chose Anji Baicha in the stable stage of chlorophyll development as a research object, the results revealed that COS could enhance antioxidant activity, increase accumulation of sugar content and chlorophyll content in tea plant. It is confirmed that COS could play an important role in improving stress tolerance of Anji Baicha.

Cold stress can cause excessive production of reactive oxygen species (ROS), disrupt the normal physiological and metabolic balance of plants, lead to the increase of membrane lipid peroxidation and damage to vital biomolecules [42,43]. Plants have evolved complex mechanisms to combat against the damage induced by ROS, including improve the antioxidant enzymes [44,45]. In this study, under cold stress, the tea plant natively reacted to protect themselves by increasing the activity of SOD and POD enzyme, and the application of COS provided external assistance plant. Chlorophyll content in COS-treated tea plant was higher than in control, which indicated that COS application mitigated the cold-induced decline in chlorophyll content. Soluble sugar can maintain the osmotic balance, and the soluble sugar in COS treated tea plant was higher than that without COS treatment, suggesting that COS can stabilize cell membrane and enhance cold resistance of plant. Those results indicated that the utilization of COS can positively affect these physiological parameters in tea plants, and beneficially regulate the natural defense system and improve growth and developmental processes of tea plants under cold stress. Moreover, this was also demonstrated in wheat seedlings where the application of COS could enhance the activities of antioxidant enzymes and the content of chlorophyll and alleviate the damage of abiotic stress in wheat [25–27,46]. In wheat, COS could enhance the activities of antioxidant enzymes and the content of chlorophyll, alleviate plant the damage of abiotic stress [25–27,46]. These differentially expressed genes indicate that the application of COS has complex effects on metabolism and signaling pathways of tea plants at low temperature. From RNA sequencing, we found that COS significantly altered the level of gene expression involved in photosynthesis and carbon metabolism under cold stress.

The up-regulated differentially expressed genes could be important for the pathology and biological processes of response to cold stress. Chlorophyll content is an important parameter frequently used to indicate chloroplast development, and which is sensitive to abiotic stresses [47]. COS can increase chlorophyll content under cold stress, which is consistent with the observations from RNA-Seq. Compared with the control group, COS treatment may increase the photosynthesis of plants by significantly up-regulating photosystem I (PSI), photosystem II (PSII)-related genes (Table S5). In the PSII core complex, PsbR is an important link, which can stabilize the assembly of the oxygen-evolving complex protein PsbP [48]. In the present study, PsbR was up-regulated, which was consistent with the action of chitosan heptamer response in wheat seedling [49]. Besides, Chlorophyll a/b-binding protein can participate in light uptake, transfer energy to the reaction centers of the photosystem I and photosystem II, and regulate the excitation energy distribution to maintain the structure of the thylakoid membrane [50], and all of 23 chlorophyll a/b-binding protein genes were also up-regulated, which can imply the recovery of photosynthesis activities by COS treatment under cold stress [51]. These results indicate that COS may enhance photosynthesis via the upregulation of related proteins to improve the cold resistance of tea plant.

In the carbon metabolism pathway, genes encoding ribulose bisphosphate carboxylase small subunit (rbcS), phosphoglycerate kinase, glyceraldehyde-3-phosphate dehydrogenase, triosephosphate isomerase were up-regulated significantly (Table S6). RbcS is one of the subunits of Ribulose-1,5-bisphosphate carboxylase/oxygenase (RuBisCo), and the activity of rbcS decreased to inhibit photosynthesis under cold stress [52]. This result was consistent with previous research demonstrating the application of COS to regulate the photosynthetic mechanism and carbon metabolism and thereby the plant growth [53].

During plant development, the response of plants to endogenous and environmental signals is mediated by several hormones, which are involved in almost every aspect of plant growth. For example, plants respond very quickly to auxin, including cell growth and the activation of multiple auxin-responsive genes [53]. Indole-3-acetic acid (GH3) and the ethylene receptor (ETR) were up-regulated genes in the plant hormone signal transduction pathway (Table S7). GH3 is an important response gene of auxin-responsive protein (IAA), which can encode a class of IAA-amido synthetases responsible for balancing endogenous free IAA content and plays an important role in IAA-regulated plant growth and development [54,55]. The ETR responds to ethylene and abscisic acid (ABA) signaling. ETR is the most important ethylene receptor protein in plants, and the lack of ETR will hinder the transduction of ethylene signal cascade reaction, resulting in the insensitivity to ethylene in plant [56–58].

The application COS can improve antioxidant enzyme activities, and the content of chlorophyll and soluble sugar. Besides, compared with the control group, the addition of COS significantly changed the photosynthesis pathway and carbon metabolism of tea plants under low temperature stress, which may contribute to COS' ability to improve the cold tolerance of tea plants. These results may represent that COS participates in the specific regulatory mechanism related to cold adaptation in the cold resistance of Anji Baicha. As for the comparison of cold resistance between Anji Baicha and other tea plants (e.g., Xiaoxueya, Fudingdabai), we are further carrying out relevant experimental verification.

5. Conclusions

In summary, low temperature will impact the key physiological and developmental processes that determine the yield of tea. This study indicates that the utilization of COS can positively affect these physiological parameters in tea plants by improving antioxidant enzyme activities, and the content of chlorophyll and soluble sugar. Hence, COS can beneficially regulate the natural defense system and improve the growth and developmental processes of tea plants under cold stress. With transcriptome sequencing and differentially expressed genes analysis, we identified 1605 up-regulated and 2898 down-regulated genes in COS compared to the control, and photosynthesis and the carbon metabolism pathway of enrichment may play a role in the COS-improved cold resistance of a tea plant. The results may provide the foundation for further research on the regulation mechanism of COS on plant cold tolerance.

Supplementary Materials: The following are available online at http://www.mdpi.com/2073-4395/10/6/915/s1, Table S1: Primer sequences used for qRT-PCR. Table S2: The list of different expression genes. Table S3: GO enrichment list of different expression genes. Table S4: KEGG pathway enrichment list of different expression genes. Table S5: Differentially expressed genes in photosynthesis related pathway. Table S6: Differentially expressed genes in carbon metabolism pathway. Table S7: Differentially expressed genes in plant hormone signal transduction pathway. Figure S1. Verification of relative expression levels of DEGs in transcriptome date by qRT-PCR between control and COS.

Author Contributions: Y.L. conducted the experiments; Y.L., L.O. and D.J. designed and performed the experiments; Y.L., Q.Z., T.L. and R.L. analyzed the data; X.L. and L.J. conceived and supervised the project. All authors have read and agreed to the published version of the manuscript.

Funding: This research was funded by science and technology project of Guizhou province ([2015]5020) and scientific research projects of major agricultural industries of Guizhou province ([2019]006).

Acknowledgments: We are grateful to Xia Zhou, Guizhou University, for the fruitful discussions and helpful comments on earlier draft.

Conflicts of Interest: The authors declare no conflict of interest.

References

1. Singh, H.R.; Hazarika, P. Biotechnological Approaches for Tea Improvement. In *Biotechnological Progress and Beverage Consumption*; Academic Press: Cambridge, MA, USA, 2020; pp. 111–148.

2. FAOSTAT-Food and Agriculture Organization of the United Nations Statistics Division. Available online: http://faostat3.fao.org/home/E (accessed on 31 December 2018).

3. Ma, C.L.; Chen, L.; Wang, X.; Jin, J.Q.; Ma, J.Q.; Yao, M.Z. Differential expression analysis of different albescent stages of 'Anji Baicha' (*Camellia sinensis* (L.) O. Kuntze) using cDNA microarray. *Sci. Hortic.* **2012**, *148*, 246–254. [CrossRef]

4. Du, Y.Y.; Liang, Y.R.; Wang, H.; Wang, K.R.; Lu, J.L.; Zhang, G.H.; Lin, W.P.; Li, M.; Fang, Q.Y. A study on the chemical composition of albino tea cultivars. *J. Hort. Sci. Biotechnol.* **2006**, *81*, 809–812. [CrossRef]

5. Cheng, H.; Li, S.F.; Chen, M.; Yu, F.L.; Yan, J.; Liu, Y.M.; Chen, L.A. Physiological and biochemical essence of the extraordinary characters of Anji Baicha. *J. Tea Sci.* **1999**, *19*, 87–92.

6. Du, Y.Y.; Chen, H.; Zhong, W.L.; Wu, L.Y.; Ye, J.H.; Lin, C.; Zheng, X.Q.; Lu, J.L.; Liang, Y.R. Effect of temperature on accumulation of chlorophylls and leaf ultrastructure of low temperature induced albino tea plant. *Afr. J. Biotechnol.* **2008**, *7*, 1881–1885. [CrossRef]

7. Feng, L.; Gao, M.J.; Hou, R.Y.; Hu, X.Y.; Zhang, L.; Wan, X.C.; Wei, S. Determination of quality constituents in the young leaves of albino tea cultivars. *Food Chem.* **2014**, *155*, 98–104. [CrossRef]

8. Wei, K.; Wang, L.Y.; Zhou, J.; He, W.; Zeng, J.M.; Jiang, Y.W.; Cheng, H. Comparison of catechins and purine alkaloids in albino and normal green tea cultivars (*Camellia sinensis* L.) by HPLC. *Food Chem.* **2012**, *130*, 720–724. [CrossRef]

9. Shen, J.; Wang, Y.; Chen, C.; Ding, Z.; Hu, J.; Zheng, C.; Li, Y. Metabolite profiling of tea (*Camellia sinensis* L.) leaves in winter. *Sci. Hortic.* **2015**, *192*, 1–9. [CrossRef]

10. Wang, L.; Cao, H.; Qian, W.; Yao, L.; Hao, X.; Li, N.; Yang, Y.; Wang, X. Identification of a novel bZIP transcription factor in *Camellia sinensis* as a negative regulator of freezing tolerance in transgenic *Arabidopsis*. *Ann. Bot.* **2017**, *119*, 1195–1209. [CrossRef] [PubMed]

11. Zhang, Q.W.; Li, T.Y.; Wang, Q.S.; LeCompte, J.; Harkess, R.L.; Bi, G.H. Screening tea cultivars for novel climates: Plant growth and leaf quality of *Camellia sinensis* cultivars grown in Mississippi, United States. *Front. Plant Sci.* **2020**, *11*, 280. [CrossRef]

12. Li, X.; Ahammed, G.; Li, Z.; Zhang, L.; Wei, J.; Yan, P.; Zhang, L.; Han, W. Freezing stress deteriorates tea quality of new flush by inducing photosynthetic inhibition and oxidative stress in mature leaves. *Sci. Hortic.* **2018**, *230*, 155–160. [CrossRef]

13. Wang, X.C.; Zhao, Q.Y.; Ma, C.L.; Zhang, Z.H.; Cao, H.L.; Kong, Y.M.; Yue, C.; Hao, X.Y.; Chen, L.; Ma, J.Q.; et al. Global transcriptome profiles of *Camellia sinensis* during cold acclimation. *BMC Genom.* **2013**, *14*, 415. [CrossRef] [PubMed]

14. Yin, Y.; Ma, Q.; Zhu, Z.; Cui, Q.; Chen, C.; Chen, X.; Fang, W.; Li, X. Functional analysis of CsCBF3 transcription factor in tea plant (*Camellia sinensis*) under cold stress. *Plant Growth Regul.* **2016**, *80*, 335–343. [CrossRef]

15. Zhang, Y.; Zhu, X.; Chen, X.; Song, C.; Zou, Z.; Wang, Z.; Wang, M.; Fang, W.; Li, X. Identification and characterization of cold-responsive microRNAs in tea plant (*Camellia sinensis*) and their targets using high-throughput sequencing and degradome analysis. *BMC Plant Biol.* **2014**, *14*, 271. [CrossRef] [PubMed]

16. Yang, Y.J.; Zheng, L.Y.; Wang, X.C. Effect of cold acclimation and ABA on cold hardiness contents of proline in tea plants. *J. Tea Sci.* **2004**, *24*, 177–182.

17. Li, Y.Y.; Wang, X.W.; Ban, Q.Y.; Zhu, X.X.; Jiang, C.J.; Wei, C.L.; Bennetzen, J.L. Comparative transcriptomic analysis reveals gene expression associated with cold adaptation in the tea plant *Camellia sinensis*. *BMC Genom.* **2019**, *20*, 624. [CrossRef] [PubMed]

18. Ban, Q.Y.; Wang, X.W.; Pan, C.; Wang, Y.W.; Kong, L.; Jiang, H.G.; Xu, Y.Q.; Wang, W.Z.; Pan, Y.T.; Li, Y.Y.; et al. Comparative analysis of the response and gene regulation in cold resistant and susceptible tea plants. *PLoS ONE* **2017**, *12*, e0188514. [CrossRef] [PubMed]

19. Li, J.H.; Yang, Y.Q.; Sun, K.; Chen, Y.; Chen, X.; Li, X.H. Exogenous Melatonin Enhances Cold, Salt and Drought Stress Tolerance by Improving Antioxidant Defense in Tea Plant (Camellia sinensis (L.) O. Kuntze). *Molecules* **2019**, *24*, 1826. [CrossRef]

20. Cabrera, J.; Wégria, G.; Onderwater, R.; González, G.; Nápoles, M.; Falcón-Rodríguez, A.; Costales, D.; Rogers, H.; Diosdado, E.; González, S.; et al. Practical use of oligosaccharins in agriculture. *Acta. Hortic.* **2013**, *1009*, 195–212. [CrossRef]

21. Kim, S.; Rajapakse, N. Enzymatic production and biological activities of chitosan oligosaccharides (COS): A review. *Carbohydr. Polym.* **2005**, *62*, 357–368. [CrossRef]

22. Yin, H.; Du, Y.G.; Zhang, J.Z. Low molecular weight and oligomeric chitosans and their bioactivities. *Curr. Top. Med. Chem.* **2009**, *9*, 1546–1559. [CrossRef]

23. Wang, M.Y.; Chen, Y.C.; Zhang, R.; Wang, W.X.; Zhao, X.M.; Du, Y.G.; Yin, H. Effects of chitosan oligosaccharides on the yield components and production quality of different wheat cultivars (*Triticum aestivum* L.) in Northwest China. *Field Crop. Res.* **2015**, *172*, 11–20. [CrossRef]

24. Chatelain, P.G.; Pintado, M.E.; Vasconcelos, M.W. Evaluation of chitooligosaccharide application on mineral accumulation and plant growth in *Phaseolus vulgaris*. *Plant Sci.* **2014**, *215*, 134–140. [CrossRef] [PubMed]

25. Zou, P.; Tian, X.Y.; Dong, B.; Zhang, C.S. Size effects of chitooligomers with certain degrees of polymerization on the chilling tolerance of wheat seedlings. *Carbohydr. Polym.* **2017**, *160*, 194–202. [CrossRef] [PubMed]

26. Ma, L.J.; Li, Y.Y.; Yu, C.M.; Wang, Y.; Li, X.M.; Li, N.; Chen, Q.; Bu, N. Alleviation of exogenous oligochitosan on wheat seedlings growth under salt stress. *Protoplasma* **2012**, *249*, 393–399. [CrossRef] [PubMed]

27. Zou, P.; Li, K.C.; Liu, S.; Xing, R.; Qin, Y.K.; Yu, H.K.; Zhao, M.M.; Li, P.C. Effect of chitooligosaccharides with different degrees of acetylation on wheat seedlings under salt stress. *Carbohydr. Polym.* **2015**, *126*, 62–69. [CrossRef] [PubMed]

28. Yang, A.M.; Yu, L.; Chen, Z.; Zhang, S.X.; Shi, J.; Zhao, X.Z.; Yang, Y.Y.; Hu, D.Y.; Song, B.A. Label-free quantitative proteomic analysis of chitosan oligosaccharide-treated rice infected with southern rice black-streaked dwarf virus. *Viruses* **2017**, *9*, 115. [CrossRef]

29. Jia, X.C.; Meng, Q.S.; Zeng, H.H.; Wang, W.X.; Yin, H. Chitosan oligosaccharide induces resistance to tobacco mosaic virus in *Arabidopsis* via the salicylic acid-mediated signalling pathway. *Sci. Rep.* **2016**, *6*, 26144–26155. [CrossRef]

30. Ahmed, K.; Khan, M.; Siddiqui, H.; Jahan, A. Chitosan and its oligosaccharides, a promising option for sustainable crop production-a review. *Carbohydr. Polym.* **2020**, *227*, 115331. [CrossRef]

31. Cheplick, S.; Sarkar, D.; Bhowmik, P.C.; Shetty, K. Improved resilience and metabolic response of transplanted blackberry plugs using chitosan oligosaccharide elicitor treatment. *Can. J. Plant Sci.* **2017**, *98*, 717–731. [CrossRef]

32. Zeng, D.F.; Luo, X.R. Physiological effects of chitosan coating on wheat growth and activities of protective enzyme with drought tolerance. *Open J. Soil Sci.* **2012**, *2*, 282–288. [CrossRef]

33. Hu, Z.H.; Tang, B.; Wu, Q.; Zheng, J.; Leng, P.S.; Zhang, K.Z. Transcriptome sequencing analysis reveals a difference in monoterpene biosynthesis between scented *Lilium* 'Siberia' and unscented *Lilium* 'Novano'. *Front. Plant Sci.* **2017**, *8*, 1351. [CrossRef] [PubMed]

34. Wang, W.D.; Xin, H.H.; Wang, M.L.; Ma, Q.P.; Wang, L.; Kaleri, N.A.; Wang, Y.H.; Li, X.H. Transcriptomic analysis reveals the molecular mechanisms of drought-stress-induced decreases in *Camellia sinensis* leaf quality. *Front. Plant Sci.* **2016**, *7*, 385. [CrossRef] [PubMed]

35. Hao, X.Y.; Tang, H.; Wang, B.; Yue, C.; Wang, L.; Zeng, J.M.; Yang, Y.J.; Wang, X.C. Integrative transcriptional and metabolic analyses provide insights into cold spell response mechanisms in young shoots of the tea plant. *Tree Physiol.* **2018**, *38*, 1655–1671. [CrossRef] [PubMed]

36. Paul, A.; Jha, A.; Bhardwaj, S.; Singh, S.; Shankar, R.; Kumar, S. RNA-seq-mediated transcriptome analysis of actively growing and winter dormant shoots identifies non-deciduous habit of evergreen tree tea during winters. *Sci. Rep.* **2014**, *4*, 5932. [CrossRef] [PubMed]

37. Li, C.F.; Xu, Y.X.; Ma, J.Q.; Jin, J.Q.; Huang, D.J.; Yao, M.Z.; Ma, C.L.; Chen, L. Biochemical and transcriptomic analyses reveal different metabolite biosynthesis profiles among three color and developmental stages in 'Anji Baicha' (*Camellia sinensis*). *BMC Plant Biol.* **2016**, *16*, 195. [CrossRef] [PubMed]

38. Wei, C.L.; Yang, H.; Wang, S.B.; Zhao, J.; Liu, C.; Gao, L.P.; Xia, E.H.; Lu, Y.; Tai, Y.L.; She, G.B.; et al. Draft genome sequence of *Camellia sinensis* var. *sinensis* provides insights into the evolution of the tea genome and tea quality. *Proc. Natl. Acad. Sci. USA* **2018**, *115*, 201719622. [CrossRef] [PubMed]

39. Liao, Y.; Smyth, G.; Shi, W. featureCounts: An efficient general-purpose program for assigning sequence reads to genomic features. *Bioinformatics* **2014**, *30*, 923–930. [CrossRef]

40. Robinson, M.; McCarthy, D.; Smyth, G. edgeR: A bioconductor package for differential expression analysis of digital gene expression data. *Bioinformatics* **2010**, *26*, 139–140. [CrossRef]

41. Xie, C.; Mao, X.; Huang, J.; Ding, Y.; Wu, J.; Dong, S.; Kong, L.; Gao, G.; Li, C.; Wei, L. KOBAS 2.0: A web server for annotation and identification of enriched pathways and diseases. *Nucleic Acids Res.* **2011**, *39*, 316–322. [CrossRef]

42. Apel, K.; Hirt, H. Reactive oxygen species: Metabolism, oxidative stress, and signal transduction. *Annu. Rev. Plant Biol.* **2004**, *55*, 373–399. [CrossRef]

43. Gill, S.S.; Tuteja, N. Reactive oxygen species and antioxidant machinery in abiotic stress tolerance in crop plants. *Plant Physiol. Biochem.* **2010**, *48*, 909–930. [CrossRef] [PubMed]

44. Chen, J.N.; Huang, M.; Cao, F.B.; Pardha-Saradhi, P.; Zou, Y.B. Urea application promotes amino acid metabolism and membrane lipid peroxidation in *Azolla*. *PLoS ONE* **2017**, *12*, e0185230. [CrossRef] [PubMed]

45. Zhou, C.Z.; Zhu, C.; Fu, H.F.; Li, X.Z.; Chen, L.; Lin, Y.L.; Lai, Z.X.; Guo, Y.Q. Genome-wide investigation of superoxide dismutase (SOD) gene family and their regulatory miRNAs reveal the involvement in abiotic stress and hormone response in tea plant (*Camellia sinensis*). *PLoS ONE* **2019**, *14*, e0223609. [CrossRef] [PubMed]

46. Zou, P.; Li, K.C.; Liu, S.; He, X.F.; Xing, R.; Zhang, X.Q.; Li, P.C. Effect of sulfated chitooligosaccharides on wheat seedlings (Triticum aestivum L.) under saltstress. *J. Agric. Food Chem.* **2016**, *64*, 2815–2821. [CrossRef] [PubMed]

47. Anwar, A.; Yan, Y.; Liu, Y.; Li, Y.; Yu, X. 5-aminolevulinic acid improves nutrient uptake and endogenous hormone accumulation, enhancing low-temperature stress tolerance in cucumbers. *Int. J. Mol. Sci.* **2018**, *19*, 3379. [CrossRef] [PubMed]

48. Suorsa, M.; Sirpio, S.; Allahverdiyeva, Y.; Paakkarinen, V.; Mamedov, F.; Styring, S.; Aro, E.M. PsbR, a missing link in the assembly of the oxygen-evolving complex of plant photosystem II. *J. Biol. Chem.* **2006**, *281*, 145–150. [CrossRef]

49. Zhang, X.Q.; Li, K.C.; Xing, R.; Liu, S.; Chen, X.L.; Yang, H.Y.; Li, P.C. MiRNA and mRNA expression profiles reveal insight into the chitosan-mediated regulation of plant growth. *J. Agric. Food Chem.* **2018**, *66*, 3810–3822. [CrossRef] [PubMed]

50. Li, X.W.; Zhu, Y.X.; Chen, C.Y.; Geng, Z.J.; Li, X.Y.; Ye, T.T.; Mao, X.N.; Du, F. Cloning and characterization of two chlorophyll A/B binding protein genes and analysis of their gene family in *Camellia sinensis*. *Sci. Rep.* **2020**, *10*, 4602. [CrossRef]

51. Jiang, X.F.; Zhao, H.; Guo, F.; Shi, X.P.; Ye, C.; Y, P.X.; Liu, B.Y.; Ni, D.J. Transcriptomic analysis reveals mechanism of light-sensitive albinism in tea plant *Camellia sinensis* 'Huangjinju'. *BMC Plant Biol.* **2020**, *20*, 216. [CrossRef]

52. Sharma, A.; Kumar, V.; Shahzad, B.; Ramakrishnan, M.; Sidhu, G.P.S.; Bali, A.S.; Handa, N.; Kapoor, D.; Yadav, P.; Khanna, K.; et al. Photosynthetic response of plants under different abiotic stresses: A review. *J. Plant Growth Regul.* **2020**, *39*, 509–531. [CrossRef]

53. Chamnanmanoontham, N.; Pongprayoon, W.; Pichayangkura, R.; Roytrakul, S.; Chadchawan, S. Chitosan enhances rice seedling growth via gene expression network between nucleus and chloroplast. *Plant Growth Regul.* **2015**, *75*, 101–114. [CrossRef]

54. Abel, S.; Nguyen, M.D.; Theologis, A. The PS-IAA4/5-like family of early inducible mRNAs in *Arabidopsis thaliana*. *J. Mol. Biol.* **1995**, *251*, 533–549. [CrossRef] [PubMed]

55. Feng, S.; Yue, R.; Tao, S.; Yang, Y.; Zhang, L.; Xu, M.; Wang, H.; Shen, C. Genome-wide identification, expression analysis of auxin-responsive GH3 family genes in maize (*Zea mays* L.) under abiotic stresses. *J. Integr. Plant Biol.* **2015**, *57*, 783–795. [CrossRef] [PubMed]

56. Solano, R.; Ecker, J.R. Ethylene gas: Perception, signaling and response. *Curr. Opin. Plant Biol.* **1998**, *1*, 393–398. [CrossRef]

57. Chomczynski, P.; Sacchi, N. Single-step method of RNA isolation by acid guanidinium thiocyanate-phenol-chloroform extraction. *Anal. Biochem.* **1987**, *162*, 156–159. [CrossRef]

58. La Camera, S.; Gouzerh, G.; Dhondt, S.; Hoffmann, L.; Fritig, B.; Legrand, M.; Heitz, T. Metabolic reprogramming in plant innate immunity: The contributions of phenylpropanoid and oxylipin pathways. *Immunol. Rev.* **2004**, *198*, 267–284. [CrossRef]

Article

The Use of Proline in Screening for Tolerance to Drought and Salinity in Common Bean (*Phaseolus vulgaris* L.) Genotypes

Sugenith Arteaga [1], Lourdes Yabor [1,2], María José Díez [3], Jaime Prohens [3], Monica Boscaiu [4,*] and Oscar Vicente [3]

[1] Institute for Plant Molecular and Cell Biology (IBMCP, UPV-CSIC), Universitat Politècnica de València, Camino de Vera s/n, 46022 Valencia, Spain; suarcas@alumni.upv.es (S.A.); lyabor@bioplantas.cu (L.Y.)

[2] Permanent Address: Laboratory for Plant Breeding and Conservation of Genetic Resources, Bioplant Center, University of Ciego de Avila, Ciego de Ávila 69450, Cuba

[3] Institute for the Conservation and Improvement of Valencian Agrodiversity (COMAV, UPV), Universitat Politècnica de València, Camino de Vera s/n, 46022 Valencia, Spain; mdiezni@btc.upv.es (M.J.D.); jprohens@btc.upv.es (J.P.); ovicente@upvnet.upv.es (O.V.)

[4] Mediterranean Agroforestry Institute (IAM, UPV), Universitat Politècnica de València, Camino de Vera s/n, 46022 Valencia, Spain

[*] Correspondence: mobosnea@eaf.upv.es; Tel.: +34-963-879-253

Received: 30 April 2020; Accepted: 5 June 2020; Published: 9 June 2020

Abstract: The selection of stress-resistant cultivars, to be used in breeding programmes aimed at enhancing the drought and salt tolerance of our major crops, is an urgent need for agriculture in a climate change scenario. In the present study, the responses to water deficit and salt stress treatments, regarding growth inhibition and leaf proline (Pro) contents, were analysed in 47 *Phaseolus vulgaris* genotypes of different origins. A two-way analysis of variance (ANOVA), Pearson moment correlations and principal component analyses (PCAs) were performed on all measured traits, to assess the general responses to stress of the investigated genotypes. For most analysed growth variables and Pro, the effects of cultivar, treatment and their interactions were highly significant ($p < 0.001$); the root morphological traits, stem diameter and the number of leaves were mostly due to uncontrolled variation, whereas the variation of fresh weight and water content of stems and leaves was clearly induced by stress. Under our experimental conditions, the average effects of salt stress on plant growth were relatively weaker than those of water deficit. In both cases, however, growth inhibition was mostly reflected in the stress-induced reduction of fresh weight and water contents of stems and leaves. Pro, on the other hand, was the only variable showing a negative correlation with all growth parameters, but particularly with those of stems and leaves mentioned above, as indicated by the Pearson correlation coefficients and the loading plots of the PCAs. Therefore, in common beans, higher stress-induced accumulation of Pro is unequivocally associated with a stronger inhibition of growth; that is, with a higher sensitivity to stress of the corresponding cultivar. We propose the use of Pro as a suitable biochemical marker for simple, rapid, large-scale screenings of bean genotypes, to exclude the most sensitive, those accumulating higher Pro concentrations in response to water or salt stress treatments.

Keywords: abiotic stress biomarkers; bean landraces; osmolytes; plant breeding; salt stress; salt stress tolerance; water deficit; water stress tolerance

---

1. Introduction

Drought and soil salinity are amongst the most restrictive environmental factors affecting agriculture worldwide. Even moderate degrees of water deficit or salt stress can lead to a reduction of

50–70% in average yields in most crops when compared with registered record yields [1–3]. Drought, brought about by the scarcity of rain, affects more than half of the agricultural land of our planet and is often linked to secondary salinisation of farmland due to intensive irrigation [4,5]. Cropland salinisation is becoming one of the major constrains for agriculture in many parts of the world, especially in arid and semi-arid regions. At the beginning of this century, it was estimated that around 20% of the irrigated lands were salinised [6], but this figure is increasing yearly, mainly due to anthropogenic alterations, such as irrigation with brackish water or the abusive and indiscriminate use of chemical fertilisers [4]. On the other hand, the scarcity of good-quality water for irrigation, mainly as a consequence of the effects of global warming, will mean more-significant crop losses in the near future, which will especially affect subsistence agriculture in developing countries [7]. Legumes are some of the most important crops, representing a significant component of the human diet. Globally, legumes complement cereal crops as the main sources of plant minerals and proteins [8]. Among the leguminous crops, *Phaseolus* L. is a large and diverse genus comprising about 70 American species [9], five of which have been domesticated (*Phaseolus vulgaris* L., *Phaseolus dumosus* Macfady, *Phaseolus coccineus* L., *Phaseolus acutifolius* A. Gray and *Phaseolus lunatus* L); moreover, a few additional species show signs of incipient domestication [10].

The common bean (*P. vulgaris*) is the most-consumed legume in human nutrition; it is an essential component of the diet, especially in developing countries, as a source of proteins, vitamins, minerals and fibre [8,11]. The species has a natural distribution area from northern Mexico to northwestern Argentina. It was domesticated independently in Central America and the Andes [12,13], but now it is cultivated practically all over the world. Beans from both origins were introduced to Spain in the 16th century [14–16], where they had to adapt to the new environmental conditions, which were very different from those in their native areas. The cropping system in small farms, spread in proximal areas, allowed the genetic flow between genotypes of Mesoamerican and Andean origin [17]. Due to centuries of bean cultivation, the Iberian Peninsula has become a secondary centre of diversification of this species [18].

Phaseolus vulgaris is not considered as very tolerant to water stress [19]; nevertheless, it is cultivated under diverse environmental conditions, including relatively dry areas [20,21]. In fact, globally, only a small percentage, around 7%, of the cropland planted with common bean receives adequate rainfall [11], and in some areas, drought causes yield losses of up to 80% [22] Like practically all cultivated plants, the bean is a glycophyte, sensitive to soil salinity even at electric conductivity values below 2 dS·m^{-1} [23]. However, just as there are cultivars that are more resistant to water stress, some respond better to high soil salinity [24,25].

As for other common crops, many bean genotypes no longer grown in the fields or cultivated only locally at a small scale (landraces, local varieties, heirlooms or minor commercial cultivars) are available from small farmers or germplasm banks and represent a rich source of genetic variability. Landraces appeared over time due to selection of traits specifically adapted to local conditions, often suboptimal or even highly stressful. Therefore, such genotypes are probably more competitive in low-input agriculture and represent a source of allelic richness that may enhance agronomic production under the foreseeable restrictive conditions imposed by climate change [26]. There is an increasing interest for the recovery of local landraces by consumers and markets, not only concerning global warming but also because of the commercial demand for local products, considered as tastier and healthier [27]. Unfortunately, many autochthonous varieties have been lost, and many others are at risk of extinction, due to genetic erosion. Screening this type of varieties for tolerance to stresses represents an interesting strategic path for the agriculture of the future.

The screening of a large number of genotypes would be greatly facilitated by identifying a suitable stress biomarker, easily quantified by simple, rapid and non-destructive assays, and unequivocally associated to the relative resistance of the cultivars to water deficit or salt stress. Proline (Pro), one of the commonest plant osmolytes [28,29], could be an appropriate candidate because a significant increase in Pro contents in response to water deficit, high salinity or other stressful conditions has been detected in beans—as in many other species. However, it is not yet clear whether Pro accumulation in *P. vulgaris* is associated with enhanced or reduced tolerance to stress since contradictory results are available in the literature. Some reports correlated higher Pro contents with a relatively higher stress tolerance when comparing different bean cultivars [30–37], whereas in other cases higher Pro concentrations were measured in the relatively more stress-sensitive cultivars [38–40]. All these studies were based on the comparison of a few genotypes. Only a wider analysis, based on a considerably higher number of cultivars, grown under the same experimental conditions and subjected to the same stress treatments, could establish whether responses to stress based on Pro accumulation are relevant, or not, for stress tolerance in *P. vulgaris*, and how Pro could be used as a reliable abiotic stress biomarker in this species.

Based on the ideas mentioned above, we have applied specific water deficit and salt stress treatments, under controlled greenhouse conditions, to a relatively large number of common bean cultivars, obtained from germplasm banks. The aims of this study were (i) to determine the overall response of the analysed genotypes to controlled water and salt stress treatments, (ii) to establish the role of Pro in bean stress responses, either as a mere stress biomarker or as an osmolyte directly involved on stress tolerance mechanisms and (iii) based on the results obtained, to propose Pro as a suitable biochemical marker for the rapid selection of bean cultivars with a (relatively) higher tolerance (or sensitivity) to drought or salinity.

2. Materials and Methods

2.1. Plant Material

The study included 47 accessions of common bean (*P. vulgaris*), from Spain (23), Colombia (19) and Cuba (5), provided by the Germplasm Bank of Universitat Politècnica de València (UPV), the International Center for Tropical Agriculture (CIAT) and the Bioplants Center, University of Ciego de Ávila, respectively.

Spanish genotypes are represented by local landraces, with geographic origins indicated in Table 1. Materials from Cuba are commercial varieties or experimental lines from INIFAT (Alexander Humboldt Institute for Basic Research in Tropical Agriculture) or IIHDL (Liliana Dimitrova Horticultural Research Institute, La Habana, Cuba), and those from Colombia are lines reported to be relatively resistant to drought and high temperatures.

Table 1. Origin of the analysed *Phaseolus vulgaris* accessions and duration of the applied stress treatments.

Abbreviation	Treatment (Weeks)	Genebank Code	Country	Origin	Cultivar Name
Sp 1	2	BGV000143	Spain	Lecina, Huesca	Judía amarilla de enrame
Sp 2	2	BGV001191	Spain	Velez Rubio, Almería	Judía
Sp 3	2	BGV001581	Spain	Mercado el Olivar, Palma de Mallorca	Judia de careta
Sp 4	2	BGV003176	Spain	Barlovento, Santa Cruz de Tenerife	Judia blanca mantecosa
Sp 5	2	BGV003616	Spain	La Bañeza, León	
Sp 6	2	BGV003941	Spain	AldeaNueva de Barbarroya, Toledo	Judía larguilla
Sp 7	2	BGV004159	Spain	Plascencia, Cáceris	
Sp 8	2	BGV011254	Spain	Las Presillas, Puente Viesgo, Cantabria	Garrafal oro
Sp 9	2	BGV013605	Spain	Campo, Huesca	Negra
Co 10	2	INB-39	Colombia	-	
Co 11	2	INB-40	Colombia	-	
Co 12	2	INB-42	Colombia	-	
Co 13	2	INB-43	Colombia	-	
Co 14	2	INB-48	Colombia	-	
Co 15	2	INB-48I	Colombia	-	
Cu 16	2	V-71	Cuba	INIFAT	Bolita 11 [a]
Cu 17	2	E-125	Cuba	IIHLD	E-125 [b]
Cu 18	2	Milagro VIII	Cuba	INIFAT	Milagro Villareño [a]
Sp 19	3	BGV001167	Spain	Chirivel, Almeria	Judia
Sp 20	3	BGV001169	Spain	Laujar de Andarax, Almeria	Judia mocha
Sp 21	3	BGV001182	Spain	Juviles, Granada	Alubias
Sp 22	3	BGV003610	Spain	Ponferrada, León	

Table 1. *Cont.*

Abbreviation	Treatment (Weeks)	Genebank Code	Country	Origin	Cultivar Name
Sp 23	3	BGV003614	Spain	La Bañeza, León	
Sp 24	3	BGV003618	Spain	La Bañeza, León	
Sp 25	3	BGV004161	Spain	Plasencia, Cáceres	
Sp 26	3	BGV004466	Spain	Bilbao, Vizcaya	Alubias pintas
Sp 27	3	BGV011235	Spain	Beranga, Hazas de Cesto, Cantabria	Carica
Sp 28	3	BGV013603	Spain	Beceite, Teruel	Judia de Franco
Sp 29	3	BGV013609	Spain	Centenero, Huesca	Judia Fartapobres
Sp 30	3	BGV014980	Spain	Alcorisa, Teruel	De tabilla ancha
Sp 31	3	BGV015856	Spain	Alicante	Habichuela del barco
Sp 32	3	BGV015859	Spain	Albarracín	Judia
Co 33	3	ALB-74	Colombia	-	
Co 34	3	INB-35	Colombia	-	
Co 35	3	INB-38	Colombia	-	
Co 36	3	INB-41	Colombia	-	
Co 37	3	INB-44	Colombia	-	
Co 38	3	INB-45	Colombia	-	
Co 39	3	INB-46	Colombia	-	
Co 40	3	INB-47	Colombia	-	
Co 41	3	SEF-9	Colombia	-	
Co 42	3	SEF-52	Colombia	-	
Co 43	3	SEF-53	Colombia	-	
Co 44	3	SEF-55	Colombia	-	
Co 45	3	SEF-56	Colombia	-	
Cu 46	3	V-13	Cuba	INIFAT	P 2240 [b]
Cu 47	3	V-51	Cuba	INIFAT	P 186 [b]

a: commercial varieties; b: experimental lines.

2.2. Plant Growth and Stress Treatments

The plants were obtained by seed germination. Several seeds of each genotype were germinated in trays with peat, perlite and vermiculite (2:1:1). When the first trifoliate true leaves were formed, the seedlings were transplanted to individual 1.6 L-pots with the same substrate in the greenhouse; Hoagland's nutrient solution [41] was used for irrigation. When the plants reached a height of at least 20 cm and had two to five true leaves, plants were selected for the treatments and placed in 55 × 40 cm plastic trays (10 pots per tray). Irrigation was performed twice a week by adding to each tray 1.5 L deionised water or a 150 mM NaCl solution, for the control and salt stress treatments, respectively. The water stress treatment was applied by completely withholding irrigation of the plants. Five individual plants (biological replicas) of each genotype were used per treatment. Treatments were stopped after two weeks for 18 genotypes when plants showed clear wilting and general decline symptoms in the water deficit treatment, but before plant mortality was observed; salt treatments of these cultivars were stopped at the same time. The remaining, relatively more resistant 29 genotypes were treated for an additional week. The two groups of plants were analysed independently. All treatments were carried out under controlled conditions in the greenhouse: long-day photoperiod (16 h of light), temperature set at 23 °C during the day and 17 °C at night. Once the treatments were finished, whole plants were harvested, collecting separately their roots, stems and leaves. Several growth parameters were measured in all plants: the diameter of the stem (SD), the length of the roots (RL) and stems (SL), the number of trifoliate leaves (Lno) and the fresh weight of roots (RFW), stems (SFW) and leaves (LFW).

Part of the fresh material of roots, stems and leaves was weighed (FW), placed at 65 °C in an oven for three days, and weighed again to determine the dry weight (DW). The water content percentage (WC%) of the three organs was calculated according to the formula:

$$WC\ (\%) = [(FW - DW)/Fw] \times 100 \tag{1}$$

2.3. Quantification of Proline Contents

Leaf Pro concentrations were quantified using dry plant material, according to the ninhydrin-acetic acid method [42]. Pro was extracted in a 3% (*w/v*) aqueous sulfosalicylic acid solution; the sample was mixed with the acid ninhydrin solution, incubated for 1 h at 95 °C, cooled on ice and extracted with toluene. Samples with known Pro amounts were assayed in parallel to obtain a standard curve.

The absorbance of the supernatants was read at 520 nm using toluene as a blank. Pro concentration was finally expressed as μmol g^{-1} DW.

2.4. Statistical Analysis

Plants from the two- and three-week treatments were analysed separately. A two-way analysis of variance (ANOVA) was performed for all determined traits, to check the effects of the 'cultivar' and 'treatment' factors, and the interaction between treatment and genotype. Pearson moment correlations were also performed for all measured parameters, and a principal component analysis (PCA) was used to check the similarity between the responses to the different types of stress within each cultivar, and the similarity between accessions. Data were analysed using Statgraphics Centurion v.16 software (Statpoint Technologies, Warrenton, VA, USA).

3. Results

3.1. Analysis of Variance of Registered Traits

Some cultivars (18) were apparently more sensitive to both salt and water stress and therefore treatments were stopped after two weeks. For the remaining genotypes (29), treatments were extended to three weeks (Table 1). All growth parameters and the leaf Pro concentration of control and stressed plants, for each cultivar, are summarised in Supplementary Table S1. Notwithstanding quantitative differences between genotypes, the overall picture is that plants of most cultivars were affected by both types of stress, water deficit and salinity, which inhibited growth as indicated by the general relative reduction observed in the measured morphological variables. Under the specific stress conditions applied in the experiments, in most cases, growth inhibition was more accentuated in the water-stressed plants than in the salt-stressed ones. Again for most cultivars, leaf Pro contents increased significantly in response to both types of stress. To assess the general responses to stress of the selected cultivars, a two-way ANOVA was performed considering the effect on each parameter of cultivar and treatment, and their interaction (Table 2).

Table 2. Two-way analysis of variance (ANOVA) of cultivar, treatment and their interactions for the parameters considered. Numbers represent percentages of the sum of squares at the 5% confidence level. Abbreviations: RL, root length; RFW, root fresh weight; RWC, root water content; SD, stem diameter; SL, stem length; SFW, stem fresh weight; SWC, stem water content; Lno, leaf number; LFW, leaf fresh weight; LWC, leaf water content; Pro, proline content. Asterisks indicate the degree of significance: ** $p < 0.01$; *** $p < 0.001$, ns = not significant.

Trait	Two Weeks				Three Weeks			
	Cultivar	Treatment	Interaction	Residual	Cultivar	Treatment	Interaction	Residual
RL	22.14 ***	18.42 ***	9.24 ns	50.19	32.90 ***	10.43 ***	21.52 ***	35.16
RFW	33.01 ***	17.20 ***	18.45 ***	31.34	24.88 ***	4.54 ***	15.05 **	55.53
RWC	4.34 ***	70.52 ***	14.75 ***	10.39	22.23 ***	52.05 ***	13.58 ***	12.15
SD	24.77 ***	27.08 ***	13.15 ***	35.00	14.17 ***	1.06 **	23.61 ***	61.16
SL	52.63 ***	11.77 ***	9.85 ***	25.76	42.71 ***	18.30 ***	18.17 ***	20.81
SFW	19.51 ***	47.29 ***	13.84 ***	19.36	26.50 ***	38.21 ***	15.10 ***	20.19
SWC	30.07 ***	33.12 ***	16.07 ***	20.74	31.64 ***	21.15 ***	32.13 ***	15.08
Lno	24.67 ***	32.12 ***	10.16 ***	33.05	14.69 ***	9.06 ***	21.14 ***	55.12
LFW	15.70 ***	53.55 ***	16.85 ***	13.89	32.54 ***	32.42 ***	23.8 ***	11.24
LWC	25.53 ***	37.59 ***	14.52 ***	22.36	32.50 ***	24.36 ***	20.21 ***	22.93
Pro	30.46 ***	28.92 ***	18.89 ***	21.72	40.52 ***	20.08 ***	15.61 ***	23.79

For most analysed variables, the effects of cultivar, treatment and their interactions were highly significant ($p < 0.001$). The only non-significant value was found in the two-week treatment and the trait 'root length', for the interaction cultivar × treatment. In plants subjected to the two-week treatment, relatively stronger contributions to the sum of squares were those of 'cultivar' for the variables root fresh weight (RFW) and stem length (SL), and 'treatment' for root water content (RWC), stem fresh weight

(SFW), leaf fresh weight (LFW) and leaf water content (LWC). For stem water content (SWC) and Pro, both factors, cultivar and treatment, contributed similarly to the sum of squares (SS). On the other hand, most of the variation observed for root length (RL), and stem diameter (SD) was due to uncontrolled variation, as shown by the higher SS percentage of the residual (Table 2).

The ANOVA of data obtained from the three-weeks-treated plants showed somewhat different results. The effect of 'cultivar' was the most substantial contributor to SS for the variables SL, LWC and Pro, and that of 'treatment' for RWC and SFW. The relative contributions of cultivar and treatment were similar for LFW, and those of cultivar and the interaction of both factors, for SWC. The most-significant contribution to variation of RL, RFW, SD and the number of leaves (Lno) is accounted for by the residual source of variation.

Disregarding the individual responses to water and salt stress of the selected bean genotypes, which vary quantitatively (Table S1), a general analysis was performed, including all cultivars and using the mean values calculated for all measured growth variables and Pro contents (Table 3). After the water stress treatments, either for two or three weeks, all morphological parameters determined in the stressed plants showed a significant decrease with respect to the corresponding values of the well-watered controls. The strongest reductions, down to less than 30% of the controls, were observed for root and leaf fresh weight. The effect of water deficit was relatively weaker regarding the reduction of root length and stem parameters (SD, SL and SWC), especially in the three-week treatments. Leaf Pro concentration, on the contrary, significantly increased in response to water stress, about 2.7-fold and 2.1-fold, as average, for the plants treated for two and three weeks, respectively (Table 3).

Table 3. Mean values and percentages with respect to the control (%) of traits measured in *Phaseolus vulgaris* cultivars after two and three weeks of control (C), water stress (WS) (withholding of irrigation) and salt stress (SS) (150 mM NaCl) treatments. Abbreviations: RL, root length; RFW, root fresh weight; RWC, root water content; SD, stem diameter; SL, stem length; SFW, stem fresh weight; SWC, stem water content; Lno, leaf number; LFW, leaf fresh weight; LWC, leaf water content; Pro, proline content. Different letters (lowercase for two-week and capital for three-week treatments) indicate significant differences between treatments for each trait, according to the Tukey test, at the 95% confidence level.

Trait	Two Weeks			Three Weeks		
	C	WS	SS	C	WS	SS
RL (cm)	36.04c	23.24a	30.39b	29.50C	21.14A	25.73B
%		64.48	84.32		71.66	87.22
RFW (g)	3.22b	0.44a	3.36b	2.78C	0.69A	1.85B
%		13.66	104.35		24.82	66.55
RWC (%)	85.70b	31.94a	82.08b	84.07B	42.11A	85.28B
%		37.27	95.78		50.09	101.44
SD (mm)	3.87c	2.91a	3.55b	3.89C	3.18A	3.45B
%		75.19	91.73		81.75	88.69
SL(cm)	148.63b	109.95a	115.75a	139.90B	95.43A	90.85A
%		73.98	77.88		68.21	64.94
SFW (g)	10.40c	2.82a	5.37b	9.00C	3.18A	4.63B
%		27.12	51.63		35.33	51.44
SWC (%)	82.14b	56.84a	78.31b	82.82B	64.06A	79.47B
%		69.20	95.34		77.35	95.96
Lno	12.43b	6.47a	7.42a	13.28B	7.35A	7.98A
%		52.05	59.69		55.35	60.09
LFW (g)	22.73c	2.57a	5.94b	18.21B	4.08A	5.40A
%		11.31	26.13		22.41	29.65
LWC	84.04c	38.39a	58.58b	81.62C	49.31A	55.48B
%		45.68	69.70		60.41	67.97
Pro (μmol g^{-1} DW)	31.67a	86.61b	82.74b	25.89A	53.57B	68.29C
%		273.48	261.26		206.91	263.77

Under the specific conditions of our experiments, salt stress had a smaller effect than water deficit on the average growth inhibition of the bean cultivars, reflected mostly in a sharp reduction

(>70%) of the leaf fresh weight with respect to the control, followed by that of stem fresh weight (about 50%). Other parameters, such as root and stem water content, stem diameter or root fresh weight (in the two-week treatment) did not change significantly or decreased only slightly in response to increased salinity. The mean values calculated for most growth variables were similar for both treatment times. Pro contents also rose significantly, about 2.6-fold over control values, in the salt-treated plants (Table 3).

3.2. Correlation Analysis

Pearson moment correlation between the analysed traits for salt and water stress are presented separately for plants from the two-week (Figure 1a) and three-week (Figure 1b) treatments. Correlations between all morphological variables were in most cases positive, for both stresses and the two treatment times, although the correlation coefficients varied widely, from $r < 0.1$ to $r > 0.9$. Considering specifically the two-week treatment, the strongest correlations ($r \geq 0.8$) for the salt stress treatment were found between root water content (RWC) and stem fresh weight (SFW) or water content (SWC); or between SFW, leaf fresh weight (LFW) and the number of leaves (Lno) (Figure 1a). Under conditions of water stress, the strongest positive correlations were also found between SFW, LFW and Lno; between water contents of roots, stems and leaves (RWC/SWC/LWC) or between SFW and stem diameter (SD) (Figure 1a). On the other hand, Pro contents showed negative correlations with all growth parameters (except for RWC in the salt stress treatment), most significantly with leaf water content, but also with LFW and stem growth parameters (SFW and SWC). Correlations followed a similar pattern for both types of stress but were weaker (lower '*r*' values) in the case of salt stress (Figure 1a).

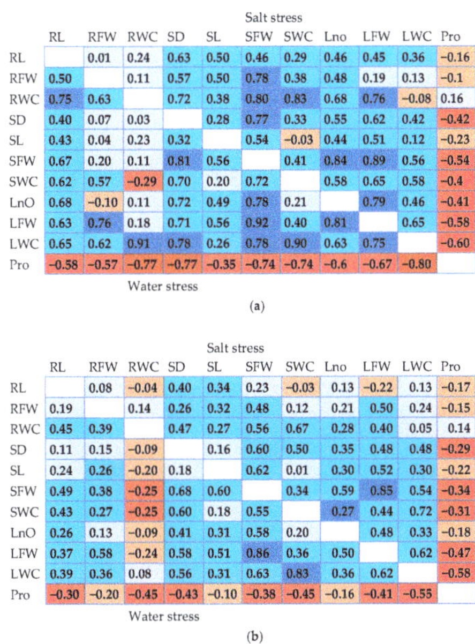

Salt stress

	RL	RFW	RWC	SD	SL	SFW	SWC	Lno	LFW	LWC	Pro
RL		0.01	0.24	0.63	0.50	0.46	0.29	0.46	0.45	0.36	−0.16
RFW	0.50		0.11	0.57	0.50	0.78	0.38	0.48	0.19	0.13	−0.1
RWC	0.75	0.63		0.72	0.38	0.80	0.83	0.68	0.76	−0.08	0.16
SD	0.40	0.07	0.03		0.28	0.77	0.33	0.55	0.62	0.42	−0.42
SL	0.43	0.04	0.23	0.32		0.54	−0.03	0.44	0.51	0.12	−0.23
SFW	0.67	0.20	0.11	0.81	0.56		0.41	0.84	0.89	0.56	−0.54
SWC	0.62	0.57	−0.29	0.70	0.20	0.72		0.58	0.65	0.58	−0.4
LnO	0.68	−0.10	0.11	0.72	0.49	0.78	0.21		0.79	0.46	−0.41
LFW	0.63	0.76	0.18	0.71	0.56	0.92	0.40	0.81		0.65	−0.58
LWC	0.65	0.62	0.91	0.78	0.26	0.78	0.90	0.63	0.75		−0.60
Pro	−0.58	−0.57	−0.77	−0.77	−0.35	−0.74	−0.74	−0.6	−0.67	−0.80	

Water stress

(a)

Salt stress

	RL	RFW	RWC	SD	SL	SFW	SWC	Lno	LFW	LWC	Pro
RL		0.08	−0.04	0.40	0.34	0.23	−0.03	0.13	−0.22	0.13	−0.17
RFW	0.19		0.14	0.26	0.32	0.48	0.12	0.21	0.50	0.24	−0.15
RWC	0.45	0.39		0.47	0.27	0.56	0.67	0.28	0.40	0.05	0.14
SD	0.11	0.15	−0.09		0.16	0.60	0.50	0.35	0.48	0.48	−0.29
SL	0.24	0.26	−0.20	0.18		0.62	0.01	0.30	0.52	0.30	−0.22
SFW	0.49	0.38	−0.25	0.68	0.60		0.34	0.59	0.85	0.54	−0.34
SWC	0.43	0.27	−0.25	0.60	0.18	0.55		0.27	0.44	0.72	−0.31
LnO	0.26	0.13	−0.09	0.41	0.31	0.58	0.20		0.48	0.33	−0.18
LFW	0.37	0.58	−0.24	0.58	0.51	0.86	0.36	0.50		0.62	−0.47
LWC	0.39	0.36	0.08	0.56	0.31	0.63	0.83	0.36	0.62		−0.58
Pro	−0.30	−0.20	−0.45	−0.43	−0.10	−0.38	−0.45	−0.16	−0.41	−0.55	

Water stress

(b)

Figure 1. Heatmap of Pearson moment correlation coefficients (r) between the analysed traits in *Phaseolus vulgaris* cultivars submitted to two weeks (**a**) and three weeks (**b**) of water and salt stresses. Dark blue denotes high correlation ($r \rightarrow 1$), dark red high negative correlation ($r \rightarrow -1$). Abbreviations: RL, root length; RFW, root fresh weight; RWC, root water content; SD, stem diameter; SL, stem length; SFW, stem fresh weight; SWC, stem water content; Lno, leaf number; LFW, leaf fresh weight; LWC, leaf water content; Pro, proline content.

Correlations between the different measured variables, generally positive for growth parameters and negative between Pro contents and the rest of variables, were maintained, qualitatively, when comparing the two- and three-week treatments, and for both stresses, but with lower relative significance for the longer treatment time (Figure 1b).

3.3. Principal Component Analysis (PCA)

A PCA was performed, separately for the cultivars subjected to the two-week and three-week treatments, and including the mean values of all measured parameters and the three applied conditions (control, water stress and salt stress) (Table 4, Figure 2).

Table 4. Component weights in the PCA performed on cultivars subjected to two and three weeks of treatment. Abbreviations: RL, root length; RFW, root fresh weight; RWC, root water content; SD, stem diameter; SL, stem length; SFW, stem fresh weight; SWC, stem water content; Lno, leaf number; LFW, leaf fresh weight; LWC, leaf water content; Pro, proline.

Trait	Two Weeks		Three Weeks	
	Component 1	Component 2	Component 1	Component 2
RL	0.320	−0.054	0.243	0.109
RFW	0.222	0.329	0.282	−0.017
RWC	0.318	0.267	0.250	0.529
SL	0.201	−0.578	0.262	−0.470
SFW	0.388	−0.178	0.421	−0.217
SWC	0.325	0.436	0.319	0.491
Lno	0.322	−0.374	0.286	−0.282
LFW	0.370	−0.227	0.406	−0.264
LWC	0.354	0.260	0.384	0.218
Pro	−0.291	−0.053	−0.240	−0.030

The PCA corresponding to the two-week treatments detected two components with Eigenvalues higher than 1, which explained 70.1% of the total variability of data (56.5% and 13.6% for the first and second components, respectively). All growth parameters—most significantly the fresh weights of stems (SFW) and leaves (LFW), followed by the water contents of both organs (LWC and SWC)—were positively correlated with the first component, whereas the only one negatively correlated was Pro concentration in leaves. Regarding the second component, some morphological variables (especially SWC and RFW) were positively correlated, whereas for others (e.g., SL or Lno) the correlation was negative (Table 4, Figure 2a).

Two components with an Eigenvalue higher than one were also detected in the PCA corresponding to the three-week treatments, the first explaining 44.3% and the second 14.0% of the total variability; that is, together explaining 58.3% of the total variation. Correlations of the different variables followed similar patterns to those observed for the cultivars treated for two weeks, for example regarding the negative correlation of Pro with the first component, and the positive correlations of all growth variables, with SFW and LFW showing the highest significance (Table 4, Figure 2b).

The 18 cultivars from the shorter treatment period (Figure 3a) were dispersed onto the two axes of the scatterplot, indicating high variability in the selected genotypes. There was, however, good separation between the different treatments, not only when looking individually at each cultivar, but also considering the overall behaviour of all genotypes. Plants from the control (green symbols) and water stress (pink) treatments were clearly separated, with almost no overlapping between the two conditions. Those symbols (blue) corresponding to the salt stress treatments appear located in the scatterplot in-between the control and water stress samples, which was in agreement with the weaker effect (on average) of the salt treatments as compared to water deficit, under the specific conditions used in our experiments. The scatterplot corresponding to the 29 cultivars that were subjected to the more prolonged (three-week) treatment (Figure 3b) showed the same general picture, maybe with more overlapping of the water- and salt-stressed plants. More-significant dispersion of the scores was found, for both treatment times, in the controls indicating a high variability of morphological traits of the different cultivars. Under salt stress, the separation between scores was not so pronounced as under water stress, suggesting a more homogeneous general response of the bean genotypes to salinity than to drought, at least under the conditions of our experiments (Figure 3).

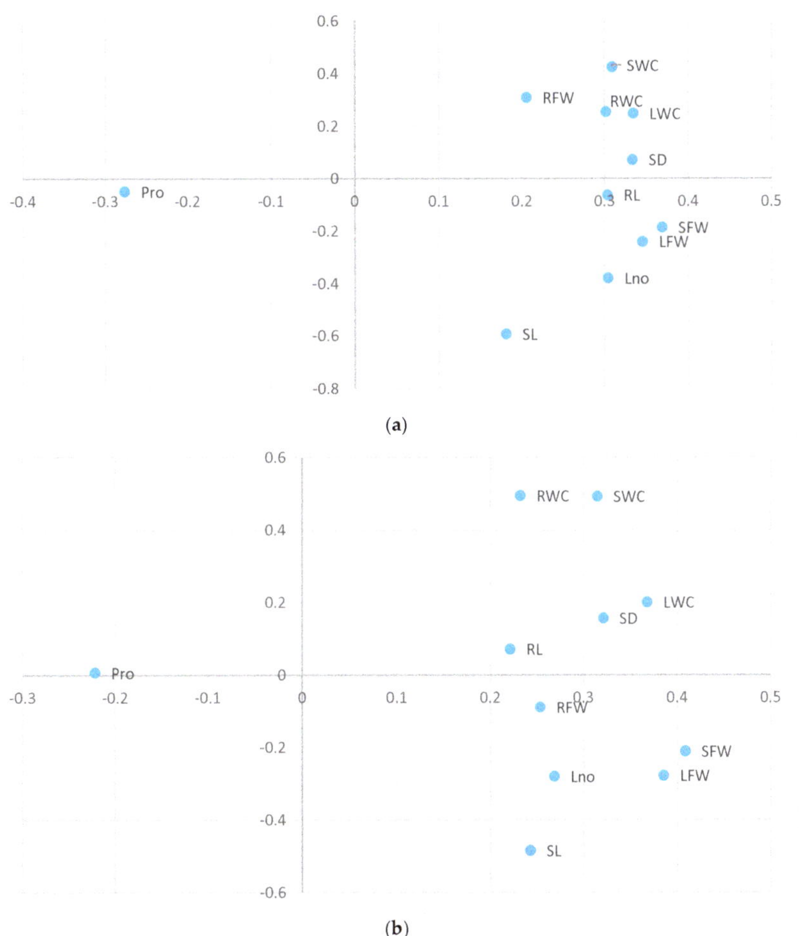

Figure 2. Loading plot of the principal component analysis (PCA) conducted with the analysed traits, in *P. vulgaris* cultivars subjected to control, water deficit and salt stress treatments. Two-week treatments (**a**); 56.5% and 13.6% of the total variability are explained by the first (*x*-axis) and the second (*y*-axis) components, respectively. Three-week treatments (**b**); 44.3% and 14.0% of the total variability are explained by the first (*x*-axis) and the second (*y*-axis) components, respectively. Abbreviations: RL, root length; RFW, root fresh weight; RWC, root water content; SD, stem diameter; SL, stem length; SFW, stem fresh weight; SWC, stem water content; Lno, leaf number; LFW, leaf fresh weight; LWC, leaf water content; Pro, proline.

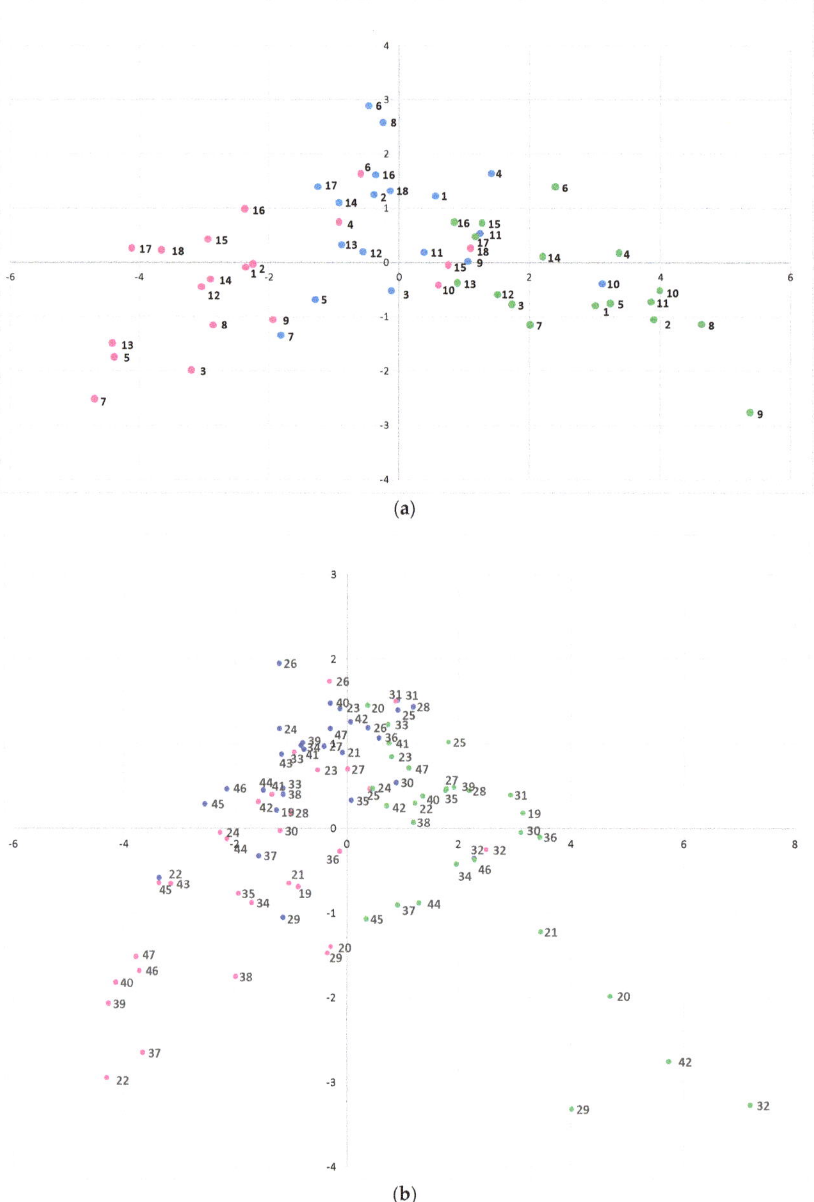

(a)

(b)

Figure 3. Scatter plot of the PCA scores. Plants treated for two (**a**) or three weeks (**b**); control (green), water deficit (pink) and salt stress (blue) treatments. (**a**) 1–9, cultivars from Spain; 10–15, from Colombia and 16–18, from Cuba and (**b**) 19–32, cultivars from Spain; 33–45, from Colombia and 46 and 47 from Cuba.

Based on the PCA scatter plot in the two-week trial (Figure 3a), we identified four accessions (7, 5, 13 and 17) with highly negative values for the first component (i.e., with high concentrations of Pro and low values for growth and water content parameters), both for the water deficit and salinity treatments; these cultivars can be considered as highly susceptible to both stresses. On the other hand, three accessions (25, 31 and 32) were detected in the three-week scatter plot (Figure 3b), showing positive values for the first component (low Pro contents and limited growth inhibition), both for the drought and salt stress treatments, indicating that these accessions can be considered as the most tolerant to both stresses. Similarly, the relative position of other accessions along the *x*-axis should allow a ranking of their tolerance to water deficit and to salinity, within each group of cultivars (treated for two or three weeks).

4. Discussion

In the present study, responses to drought and salinity have been analysed in 47 *Phaseolus vulgaris* genotypes of different origins. Large variability was observed in the size and morphology of the plants of the different bean cultivars—as seen when comparing their growth parameters (Supplementary Table S1) individually and also by their dispersion in the PCA scatterplots (Figure 3)—making it difficult to determine, at first sight, the variables that are more relevant for assessing the relative degree of stress-induced growth inhibition and, therefore, for ranking the different cultivars according to their relative sensitivity or resistance to water deficit and salt stress. However, the statistical analyses performed with all experimental data provided a clear overall picture of the responses to stress of the *P. vulgaris* cultivars. Both 'cultivar' and 'treatment', as well as their interaction, had a highly-significant effect on (practically) all growth traits analysed, and on Pro contents, for the two- and three-week treatments of both water deficit and salt stress. In all cases, growth inhibition was mostly reflected in the stress-induced reduction of fresh weight and water contents of stems (SFW and SWC) and leaves (LFW and LWC), as reported in the same species [39,40] or other species of this genus [43]. These parameters are the growth variables most significantly correlated, positively, with the first principal component in the PCA. Pro, on the other hand, was the only variable showing a negative correlation with all growth variables, but particularly with those of stems and leaves mentioned above—as indicated by the Pearson correlation coefficients and the loading plots of the PCAs.

When comparing the stress tolerance of related taxa, for example, different cultivars of a particular crop, measurements of growth parameters are often complemented with the determination of several biochemical stress markers, associated with increased (or lower) tolerance; they include compatible solutes or osmolytes [44–47]. Proline (Pro) is a common osmolyte in plants, which accumulates in response to different types of abiotic stress, including drought and salinity, in a variety of plant species [28,48–50]. Besides its role in cellular osmotic adjustment, Pro has additional functions as 'osmoprotectant'; it directly stabilises sub-cellular structures, such as membranes and proteins, scavenges free radicals buffering redox potential, alleviates cellular acidosis and acts as a signalling molecule in the responses to stress [51,52]. Proline also plays essential roles in the absence of stress, being involved in many developmental processes; for example, Pro concentration increases during pollen and seed maturation. However, Pro can be toxic for certain tissues if it is partially catabolised to pyrroline-5-carboxylate (P5C), leading to apoptosis [53]. Considering the multiple functions of Pro, it is logical to assume that Pro accumulation would be associated with higher stress tolerance, and this has indeed been demonstrated for many plants, both wild species [54,55] and crops [43,56]. However, other comparative studies on related taxa, such as species of the same genus or cultivars or varieties of the same species, revealed higher Pro accumulation under stress in the less-tolerant genotypes [57,58]. There is some confusion, often found in the literature, between the concepts of 'stress responses' and 'stress tolerance'. Even though stress tolerance mechanisms are based on specific stress responses, not all responses are relevant for tolerance. On this line, Pro accumulation can be considered as a general 'response' to abiotic stress in many plant species, but Pro may or may not be involved in stress tolerance mechanisms, depending on the species.

Common bean is clearly a Pro accumulator species, as numerous reports have shown significant increases in Pro contents in *Phaseolus* plants in response to either salt stress [38,39,59] or water stress [40,60,61] treatments. Also, Pro appears to be a good bioindicator in other types of stress in beans, such as that induced by excess nitrogen dosage [62], herbicides [63] or heavy metals [64].

Moreover, exogenous application of Pro was shown to alleviate the salt stress deleterious effects in beans [65]. However, there are some contradictory data in the literature regarding the function of Pro in the mechanisms of stress tolerance in *Phaseolus*. Some published reports indicated higher Pro contents in more drought-tolerant [30,34–37] or salt-tolerant [31–33] cultivars than in less tolerant ones; that is, Pro accumulation correlates positively with the degree of stress resistance, suggesting a direct contribution to stress tolerance mechanisms. Other reports, on the contrary, showed that, under stress conditions, the less tolerant genotypes had a higher concentration of this osmolyte than the more resistant cultivars [38–40,66]; therefore, in this case, Pro is simply a marker of the level of stress affecting the plants, accumulating at higher concentrations in the more stressed—the more sensitive—cultivars, but is not directly involved in the mechanisms of tolerance. This was also the conclusion of previous work from our laboratory, comparing three commercial cultivars (two of *P. vulgaris* and one of *P. coccineus*) and one Spanish common bean landrace [39,40]. All these latter studies, based on the comparison of a few bean genotypes, generally some commercial cultivars, have been confirmed in the present work, using a much larger number of cultivars of different origins and an extensive statistical analysis of the experimental data.

Our results showed a strong negative correlation of Pro levels and growth variables, especially the fresh weight and water content of the aboveground organs of the plants; these are the most relevant parameters to evaluate the inhibition of growth induced under water deficit and high salinity conditions. Therefore, there is an unequivocal association of higher Pro contents with stronger growth inhibition; that is, with a higher sensitivity to stress of the bean cultivars.

5. Conclusions

Phaseolus vulgaris cannot be considered as drought- or salt-tolerant. It is even more sensitive to stress than many other crops such as barley or cowpea [67,68]. However, amongst the extremely high number of available genotypes of *P. vulgaris*, some will show a relatively higher resistance and could be used as parental lines in bean breeding programmes aimed at enhancing stress tolerance in this major crop. The identification of common bean accessions in the extremes of variation for susceptibility and tolerance to water deficit and salinity is of great interest for further studies on the physiological mechanisms of tolerance to both stresses. Also, the development of segregating generations after hybridisation between both types of materials can lead to the identification of genomic regions involved in tolerance to these stresses.

Proline concentrations in stressed plants can be determined by a simple and rapid spectrophotometric assay, requiring only small amounts of leaf material. From a practical point of view, our results support the use of Pro as a biochemical marker for the initial, large-scale screening of bean cultivars, to exclude the most sensitive, those accumulating higher Pro concentrations in response to water or salt stress.

Supplementary Materials: The following are available online at http://www.mdpi.com/2073-4395/10/6/817/s1, Table S1: Variation of morphological parameters and proline concentrations in 47 accessions of common bean (*Phaseolus vulgaris*) under salt stress and water stress.

Author Contributions: Conceptualization, M.B. and O.V.; methodology, S.A. and L.Y.; software, J.P.; validation, M.J.D. and J.P.; formal analysis, L.Y.; investigation, S.A. and L.Y.; resources, O.V.; data curation, S.A.; writing—original draft preparation, S.A. and M.B.; writing—review and editing, M.J.D., J.P. and O.V.; visualization, S.A.; supervision, M.B. and O.V.; project administration, M.B. and O.V.; funding acquisition, O.V. All authors have read and agreed to the published version of the manuscript.

Funding: This research received no external funding.

Acknowledgments: We are indebted to Steve Beebe, from Alliance Biodiversity CIAT, for providing the Colombian bean cultivars and for his helpful comments on the manuscript.

Conflicts of Interest: The authors declare no conflict of interest.

References

1. Mantri, N.; Patade, V.; Penna, S.; Ford, R.; Pang, E. Abiotic stress responses in plants: Present and future. In *Abiotic Stress Responses in Plants: Metabolism, Productivity and Sustainability*; Ahmad, P., Prasad, M.N.V., Eds.; Springer: New York, NY, USA, 2012; pp. 1–19. [CrossRef]
2. Zörb, C.; Geilfus, C.-M.; Dietz, K.J. Salinity and crop yield. *Plant Biol.* **2019**, *21*, 31–38. [CrossRef]
3. Osakabe, Y.; Osakabe, K.; Shinozaki, K.; Tran, L. Response of plants to water stress. *Front. Plant Sci.* **2014**, *5*, 86. [CrossRef] [PubMed]
4. Fita, A.; Rodríguez-Burruezo, A.; Boscaiu, M.; Prohens, J.; Vicente, O. Breeding and domesticating crops adapted to drought and salinity: A new paradigm for increasing food production. *Front. Plant Sci.* **2015**, *6*, 978. [CrossRef] [PubMed]
5. Shahid, S.A.; Zaman, M.; Heng, L. Soil salinity: Historical perspectives and a world overview of the problem. In *Guideline for Salinity Assessment, Mitigation and Adaptation Using Nuclear and Related Techniques*; Zaman, M., Shahid, S.A., Eds.; Springer: Cham, Switzerland, 2018; pp. 43–53. [CrossRef]
6. Flowers, T.; Flowers, S. Why does salinity pose such a difficult problem for plant breeders? *Agric. Water Manag.* **2005**, *78*, 15–24. [CrossRef]
7. Morton, J.F. The impact of climate change on smallholder and subsistence agriculture. *Proc. Natl. Acad. Sci. USA* **2007**, *104*, 19680–19685. [CrossRef]
8. Bellucci, E.; Bitocchi, E.; Rau, D.; Rodriguez, M.; Biagetti, E.; Giardini, A.; Attene, G.; Nanni, L.; Papa, R. Genomics of origin, domestication and evolution of *Phaseolus vulgaris*. In *Genomics of Plant Genetics Resources*; Tuberosa, R., Graner, A., Frison, E., Eds.; Springer: Dordrecht, The Netherlands, 2014; pp. 483–507. [CrossRef]
9. Freytag, G.F.; Debouck, D.G. *Taxonomy, Distribution, and Ecology of the genus Phaseolus (Leguminosae-Papilionoideae) in North America, Mexico and Central America*; Botanical Research Institute of Texas (BRIT): Forth Worth, TX, USA, 2002; pp. 1–298.
10. Delgado-Salinas, A.; Bibler, R.; Lavin, M. Phylogeny of the genus *Phaseolus* (Leguminosae): A recent diversification in an ancient landscape. *Syst. Bot.* **2006**, *31*, 779–791. [CrossRef]
11. Broughton, W.J.; Hernández, G.; Blair, M.; Beebe, S.; Gepts, P.; Vanderleyden, J. Beans (*Phaseolus* spp.)—Model food legumes. *Plant Soil* **2003**, *252*, 55–128. [CrossRef]
12. Gepts, P.; Debouck, D.G. Origin, domestication, and evolution of the common bean, *Phaseolus vulgaris*. In *Common Beans: Research for Crop Improvement*; van Schoonhoven, A., Voysest, O., Eds.; Cab Intern: Wallingford, UK, 1991; pp. 7–53.
13. Rendón-Anaya, M.; Montero-Vargas, J.M.; Saburido-Álvarez, S.; Vlasova, A.; Capella-Gutierrez, S.; Ordaz-Ortiz, J.J.; Aguilar, O.M.; Vianello-Brondani, R.P.; Santalla, M.; Delaye, L.; et al. Genomic history of the origin and domestication of common bean unveils its closest sister species. *Genome Biol.* **2017**, *18*, 60. [CrossRef]
14. Ortwin-Sauer, C. *The Early Spanish Men*; University of California Press: Berkeley, CA, USA, 1966.
15. Brucher, O.B.; Brucher, H. The South American wild bean (*Phaseolus aborigeneus* Burk.) as ancestor of the common bean. *Econ. Bot.* **1976**, *30*, 257–272. [CrossRef]
16. Debouck, D.G.; Smartt, J. Bean. In *Evolution of Crop Plants*, 2nd ed.; Smartt, J., Simmonds, N.W., Eds.; Longman Scientific and Technical: Harlow, UK, 1995; pp. 287–296.
17. Arteaga, S.; Yabor, L.; Torres, J.; Solbes, E.; Muñoz, E.; Díez, M.J.; Vicente, O.; Boscaiu, M. Morphological and agronomic characterization of Spanish landraces of *Phaseolus vulgaris* L. *Agriculture* **2019**, *9*, 149. [CrossRef]
18. Pinheiro, C.; Baeta, J.P.; Pereira, A.M.; Dominguez, H.; Ricardo, C. Mineral elements correlations in a Portuguese germplasm collection of *Phaseolus vulgaris*. Integrating Legume Biology for sustainable Agriculture. In Proceedings of the 6th European Conference on Grain Legumes, Lisbon, Portugal, 12–16 November 2007; pp. 125–126.
19. Molina, J.; Moda-Cirino, V.; Da Silva Fonseca, N.J.; Faria, R.; Destro, D. Response of common bean cultivars and lines to water stress. *Crop Breed. Appl. Biotechnol.* **2001**, *1*. [CrossRef]
20. Graham, P.; Ranalli, P. Common bean (*Phaseolus vulgaris* L.). *Field Crop Res.* **1997**, *53*, 131–146. [CrossRef]
21. Singh, S.P. Drought resistance in the race Durango dry bean landraces and cultivars. *Agron. J.* **2007**, *99*, 1219–1225. [CrossRef]

22. Cuellar-Ortiz, S.; Arrieta-Montiel, M.; Acosta-Gallegos, J.; Covar-Rubias, A. Relationship between carbohydrate partitioning and drought resistance in common bean. *Plant Cell Environ.* **2008**, *31*, 1399–1409. [CrossRef]

23. Maas, E.; Hoffman, G. Crop salt tolerance-current assessment. *J. Irrig. Drain. Eng.* **1977**, *103*, 115–134.

24. Gama, P.; Inanaga, S.; Tanaka, K.; Nakazawa, R. Physiological response of common bean (*Phaseolus vulgaris* L.) seedlings to salinity stress. *Afr. J. Biotechnol.* **2007**, *6*, 79–88.

25. Zhumabayeva, B.A.; Biotechnology, K.A.; Aytasheva, Z.G.; Dzhangalina, E.D.; Esen, A.; Lebedeva, L.P. Screening of domestic common bean cultivar for salt tolerance during in vitro cell cultivation. *Int. J. Biol.* **2019**, *12*, 94–102. [CrossRef]

26. Fess, T.L.; Kotcon, J.B.; Benedito, V.A. Crop breeding for low input agriculture: A sustainable response to feed a growing world population. *Sustainability* **2011**, *3*, 1742–1772. [CrossRef]

27. Hurtado, M.; Vilanova, S.; Plazas, M.; Gramazio, P.; Andújar, I.; Herraiz, F.J.; Prohens, J. Enhancing conservation and use of local vegetable landraces: The Almagro eggplant (*Solanum melongena* L.) case study. *Genet. Resour. Crop Evol.* **2014**, *61*, 787–795. [CrossRef]

28. Szabados, L.; Savouré, A. Proline: A multifunctional amino acid. *Trends Plant Sci.* **2010**, *15*, 89–97. [CrossRef]

29. Verslues, P.E.; Sharma, S. Proline metabolism and its implications for plant-environment interaction. *Arabidopsis Book* **2010**, *8*, e0140. [CrossRef] [PubMed]

30. Kapuya, J.A.; Barendse, G.W.M.; Linskens, H.F. Water stress tolerance and proline accumulation in *Phaseolus vulgaris* L. *Acta Bot. Neerl.* **1985**, *34*, 293–300. [CrossRef]

31. Misra, N.; Gupta, A.K. Effect of salt stress on proline metabolism in two high yielding genotypes of green gram. *Plant Sci.* **2005**, *169*, 331–339. [CrossRef]

32. Cárdenas-Avila, M.; Verde-Star, J.; Maiti, R.; Foroughbakhch, R.; Gámez-González, H.; Martínez-Lozano, S.; Núñez-González, M.; García Díaz, G.; Hernández-Piñero, J.; Morales-Vallarta, M. Variability in accumulation of free proline on in vitro calli of four bean (*Phaseolus vulgaris* L.) cultivars exposed to salinity and induced moisture stress. *Phyton* **2006**, *75*, 103–108.

33. Kaymakonova, M.; Stoeva, N. Physiological responses of bean plants (*Phaseolus vulg.* L.) to salt stress. *Gen. Appl. Plant Physiol.* **2008**, *34*, 177–188.

34. Herrera Flores, T.S.; Ortíz Cereceres, J.; Delgado Alvarado, A.; Acosta Galleros, J.A. Growth and, proline and carbohydrate content of bean seedlings subjected to drought stress. *Rev. Mexicana Cienc. Agric.* **2012**, *3*, 713–725.

35. Ghanbari, A.A.; Mousavi, S.H.; Mousapou Gorji, A.; Rao, I. Effects of water stress on leaves and seeds of bean (*Phaseolus vulgaris* L.). *Turk. J. Field Crops* **2013**, *18*, 73–77.

36. Kusvuran, S.; Dasgan, H.Y. Effects of drought stress on physiological and biochemical changes in *Phaseolus vulgaris* L. *Legume Res.* **2017**, *40*, 55–62.

37. Wang, Q.; Ang, Q.; Lin, F.; Wei, S.H.; Meng, X.X.; Yin, Z.G.; Guo, Y.F.; Yang, G.D. Effects of drought stress on endogenous hormones and osmotic regulatory substances of common bean (*Phaseolus vulgaris* L.) at seedling stage. *Appl. Ecol. Environ. Res.* **2019**, *17*, 4447–4457. [CrossRef]

38. Jiménez-Bremont, J.F.; Becerra-Flora, A.; Hernández-Lucero, E.; Rodríguez-Kessler, M.; Acosta-Gallegos, J.; Ramírez Pimentel, J. Proline accumulation in two bean cultivars under salt stress and the effect of polyamines and ornithine. *Biol. Plant.* **2006**, *50*, 763–766. [CrossRef]

39. Al Hassan, M.; Morosan, M.; López-Gresa, M.P.; Prohens, J.; Vicente, O.; Boscaiu, M. Salinity-induced variation in biochemical markers provides insight into the mechanisms of salt tolerance in common (*Phaseolus vulgaris*) and runner (*P. coccineus*) beans. *Int. J. Mol. Sci.* **2016**, *17*, 1582. [CrossRef] [PubMed]

40. Morosan, M.; Al Hassan, M.; Naranjo, M.; Lopez-Gresa, M.P.; Vicente, O. Comparative analysis of drought responses in *Phaseolus vulgaris* (common bean) and *P. coccineus* (runner bean) cultivar. *EuroBiotech J.* **2017**, *1*, 247–252. [CrossRef]

41. Hoagland, D.; Arnon, D. The water-culture method for growing plants without soil. *Circ. Califor. Agric. Exp. Stat.* **1950**, *347*, 32–63.

42. Bates, L.S.; Waldren, R.P.; Teare, I.D. Rapid determination of free proline for water stress studies. *Plant Soil* **1973**, *39*, 205–207. [CrossRef]

43. Arteaga, S.; Al Hassan, M.; Wijesinghe, C.; Yabor, L.; Llinares, J.; Boscaiu, M.; Vicente, O. Screening for Salt Tolerance in Four Local Varieties of *Phaseolus lunatus* from Spain. *Agriculture* **2018**, *8*, 201. [CrossRef]

44. Andrade, E.; Ribeiro, V.; Azvedo, C.; Chiorato, A.; Williams, T.; Carbonell, S. Biochemical indicators of drought tolerance in the common bean (*Phaseolus vulgaris* L.). *Euphytica* **2016**, *210*, 277–289. [CrossRef]

45. Bacha, H.; Tekaya, M.; Drine, S.; Guasami, F.; Touil, L.; Enneb, H.; Triki, T.; Cheour, F.; Ferchichi, A. Impact of salt stress on morpho-physiological and biochemical parameters of *Solanum lycopersicum* cv. Microtom leaves. *S. Afr. J. Bot.* **2017**, *108*, 364–369. [CrossRef]

46. Sen, A.; Ozturk, I.; Yaycili, O.; Alikamanoglu, S. Drought tolerance in irradiated wheat studied by genetic and biochemical markers. *J. Plant Growth Regul.* **2017**, *36*, 669–676. [CrossRef]

47. Koźmińska, A.; Wiszniewska, A.; Hanus-Fajerska, E.; Boscaiu, M.; Al Hassan, M.; Halecki, W.; Vicente, O. Identification of salt and drought biochemical stress markers in several *Silene vulgaris* populations. *Sustainability* **2019**, *11*, 800. [CrossRef]

48. Verbruggen, N.; Hermans, C. Proline accumulation in plants: A review. *Amino Acids* **2008**, *35*, 753–759. [CrossRef]

49. Grigore, M.N.; Boscaiu, M.; Vicente, O. Assessment of the relevance of osmolyte biosynthesis for salt tolerance of halophytes under natural conditions. *Eur. J. Plant Sci. Biotechnol.* **2011**, *5*, 12–19.

50. Parvaiz, A.S.; Satyawati, S. Salt stress and phyto-biochemical responses of plants—A review. *Plant Soil Environ.* **2008**, *54*, 89–99. [CrossRef]

51. Hayat, S.; Hayat, Q.; Alyemeni, M.N.; Wani, A.S.; Pichtel, J.; Ahmad, A. Role of proline under changing environments: A review. *Plant Signal. Behav.* **2012**, *7*, 1456–1466. [CrossRef]

52. Rana, V.; Ram, S.; Nehra, K. Proline biosynthesis and its role in abiotic stress. *Int. J. Agric. Res. Innov. Technol.* **2017**, *6*.

53. Kavi Kishor, P.; Sreenivasulu, N. Is proline accumulation per se correlated with stress tolerance or is proline homeostasis a more critical issue? *Plant Cell Environ.* **2014**, *37*, 300–311. [CrossRef] [PubMed]

54. Al Hassan, M.; López-Gresa, M.P.; Boscaiu, M.; Vicente, O. Stress tolerance mechanisms in *Juncus*: Responses to salinity and drought in three *Juncus* species adapted to different natural environments. *Funct. Plant Biol.* **2016**, *43*, 949–960. [CrossRef]

55. Al Hassan, M.; Pacurar, A.; López-Gresa, M.P.; Donat-Torres, M.; Llinares, J.; Boscaiu, M.; Vicente, O. Effects of salt stress on three ecologically distinct *Plantago* species. *PLoS ONE* **2016**, *11*, e0160236. [CrossRef]

56. Plazas, M.; Nguyen, H.; González-Orenga, S.; Fita, A.; Vicente, O.; Prohens, J.; Boscaiu, M. Comparative analysis of the responses to water stress in eggplant (*Solanum melongena*) cultivars. *Plant Physiol. Biochem.* **2019**, *143*, 72–82. [CrossRef]

57. Chen, Z.; Cuin, T.; Zhou, M.; Twomei, A.; Naidu, B.; Shabala, S. Compatible solute accumulation and stress-mitigating effects in barley genotypes contrasting in their salt tolerance. *J. Exp. Bot.* **2007**, *58*, 4245–4255. [CrossRef]

58. Koźmińska, A.; Al Hassan, M.; Hanus-Fajerska, E.; Naranjo, M.A.; Vicente, O.; Boscaiu, M. Comparative analysis of water deficit and salt tolerance mechanisms in Silene. *S. Afr. J. Bot.* **2018**, *117*, 193–206. [CrossRef]

59. Nagesh, B.; Devaraj, V. High temperature and salt stress response in French bean (*Phaseolus vulgaris*). *Austr. J. Crop Sci.* **2008**, *2*, 40–42.

60. Ashraf, M.; Iram, A. Drought stress induced changes in some organic substances in nodules and other plant parts of two potential legumes differing in salt tolerance. *Flora* **2005**, *200*, 535–546. [CrossRef]

61. Rosales, M.A.; Ocampo, O.; Rodríguez-Valentín, R.; Olvera-Carrillo, Y.; Acosta-Gallegos, J.; Covarrubias, A.A. Physiological analysis of common bean (*Phaseolus vulgaris* L.) cultivars uncovers characteristics related to terminal drought resistance. *Plant Physiol. Biochem.* **2012**, *56*, 24–34. [CrossRef] [PubMed]

62. Sánchez, E.; Ruiz, J.M.; López-Lefebre, L.R.; Rivero, R.M.; García, P.C.; Romero, L. Proline metabolism in response to highest nitrogen dosages in green bean plants (*Phaseolus vulgaris* L. cv Strike). *J. Plant Physiol.* **2001**, *158*, 593–598.

63. Mackay, C.E.; Hall, C.; Hofstra, G.; Fletcher, R.A. Uniconazole-induced changes in abscisic acid, total amino acids, and proline in *Phaseolus vulgaris*. *Pestic. Biochem. Phys.* **1990**, *37*, 74–82. [CrossRef]

64. Zengin, F.K.; Munzuroglu, O. Effects of some heavy metals on content of chlorophyll, proline and some antioxidant chemicals in bean (*Phaseolus vulgaris* L.) seedlings. *Acta Biol. Cracov. Bot.* **2005**, *47*, 157–164.

65. Abdelhamid, M.T.; Rady, M.M.; Osman, A.S.; Abdalla, M.A. Exogenous application of proline alleviates salt-induced oxidative stress in *Phaseolus vulgaris* L. plants. *J. Hortic. Sci. Biotech.* **2013**, *88*, 439–446. [CrossRef]

66. Domínguez, A.; Yunel Pérez, Y.; Alemán, S.; Sosa, M.; Fuentes, L.; Darias, R.; Demey, J.; Rea, R.; Sosa, D. Respuesta de cultivares de *Phaseolus vulgaris* L. al estrés por sequía. *Biot. Veg.* **2014**, *14*, 29–36.

67. Gürel, F.; Öztürk, Z.N.; Uçarlı, C.; Rosellini, D. Barley genes as tools to confer abiotic stress tolerance in crops. *Front Plant Sci.* **2016**, *7*, 1137. [CrossRef]

68. Yoshida, J.; Tomooka, N.; Khaing, T.Y.; Sunil Shantha, P.G.; Naito, H.; Matsuda, Y.; Ehara, H. Unique responses of three highly salt-tolerant wild *Vigna* species against salt stress. *Plant Prod. Sci.* **2020**, *23*, 114–128. [CrossRef]

Article

Phytotoxic Effects of Three Natural Compounds: Pelargonic Acid, Carvacrol, and Cinnamic Aldehyde, against Problematic Weeds in Mediterranean Crops

Marta Muñoz [1,2], Natalia Torres-Pagán [1], Rosa Peiró [3], Rubén Guijarro [2], Adela M. Sánchez-Moreiras [4,5] and Mercedes Verdeguer [1,*]

[1] Instituto Agroforestal Mediterráneo (IAM), Universitat Politècnica de València, Camino de Vera s/n, 46022 Valencia, Spain; mmunoz@seipasa.com (M.M.); natorpa@etsiamn.upv.es (N.T.-P.)
[2] SEIPASA S.A. C/Ciudad Darío, Polígono Industrial La Creu naves 1-3-5, L'Alcudia, 46250 Valencia, Spain; rguijarro@seipasa.com
[3] Centro de Conservación y Mejora de la Agrodiversidad Valenciana (COMAV), Universitat Politècnica de València, Camino de Vera s/n, 46022 Valencia, Spain; ropeibar@btc.upv.es
[4] Department of Plant Biology and Soil Science, Faculty of Biology, University of Vigo, Campus Lagoas-Marcosende s/n, 36310 Vigo, Spain; adela@uvigo.es
[5] CITACA. Agri-Food Research and Transfer Cluster, Campus da Auga. University of Vigo, 32004 Ourense, Spain
* Correspondence: merversa@doctor.upv.es; Tel.: +34-9-6387-7000

Received: 28 April 2020; Accepted: 27 May 2020; Published: 2 June 2020

Abstract: Weeds and herbicides are important stress factors for crops. Weeds are responsible for great losses in crop yields, more than 50% in some crops if left uncontrolled. Herbicides have been used as the main method for weed control since their development after the Second World War. It is necessary to find alternatives to synthetic herbicides that can be incorporated in an Integrated Weed Management Program, to produce crops subjected to less stress in a more sustainable way. In this work, three natural products: pelargonic acid (PA), carvacrol (CV), and cinnamic aldehyde (CA) were evaluated, under greenhouse conditions in postemergence assays, against problematic weeds in Mediterranean crops *Amaranthus retroflexus*, *Avena fatua*, *Portulaca oleracea*, and *Erigeron bonariensis*, to determine their phytotoxic potential. The three products showed a potent herbicidal activity, reaching high efficacy (plant death) and damage level in all species, being PA the most effective at all doses applied, followed by CA and CV. These products could be good candidates for bioherbicides formulations.

Keywords: weeds; abiotic stress; natural herbicides; secondary metabolites; postemergence; phytotoxicity

1. Introduction

One of the main challenges for the agriculture in this 21st century is to be capable to feed the increasing world population in a sustainable way, because natural resources are becoming even more scarce [1]. Crop protection measures can prevent yield losses due to pests [2]. Herbicides have been the most used method to control weeds since their development, at the end of the Second World War because they are effective and economical [3,4].

Herbicides cause stress in crops and can make them more susceptible to other pests [5]. Other problems derived from the overuse of herbicides are environmental pollution, toxicity for nontarget organisms, and the development of herbicide-resistant weed biotypes [6]. In the latest 10 years, integrated weed management (IWM) strategies have been promoted worldwide [7,8] to control weeds. They consist of a combination of methods: cultural, mechanical, physical, biological, biotechnological, and chemical. In Europe, IWM has been promoted through the European Union Directive 2009/128/EC [8].

The society is demanding new solutions for weed control and "greener" weed management products. The use of natural products as bioherbicides could be one alternative to reduce the stress that synthetic herbicides promote in crops and all their negative impacts aforementioned. Bioherbicides could be incorporated in IPM programs as an innovative weed control method. They are less persistent than synthetic herbicides and are potentially more environmentally friendly and safe [9] and also, they have different modes of action, which can prevent the development of herbicide-resistant weed biotypes [10].

Bailey [11] defined bioherbicides as products of natural origin for weed control. The EPA (USA Environmental Protection Agency), considers three categories of biopesticides: (1) biochemical pesticides, which include naturally occurring substances that control pests; (2) microbial pesticides or biocontrol agents, which are microorganisms that control pests; and (3) plant-incorporated protectants, or PIPs, which are pesticide substances produced by plants that contain added genetic material) [10]. In recent years, the search for natural substances that can act as bioherbicides has been very extensive.

The weeds selected for this study were *Amaranthus retroflexus* L., *Avena fatua* L., *Portulaca oleracea* L., and *Erigeron bonariensis* L. because of their importance in many crops worldwide and their difficult management. *A. fatua* is a very important weed mainly in cereals and also in other crops around the world [12], and this weed is on the fourth position in resistance to herbicides worldwide, having developed resistance to nine different modes of action [13]. *A. retroflexus* is a serious and aggressive weed in summer crops, with cosmopolite distribution [14]. It has developed resistance to five modes of action and is on the eight position worldwide in resistance to herbicides [13]. *E. bonariensis*, which can be found both in summer or winter crops, especially with no-tillage practices [15], is on the ninth position in resistance to herbicides worldwide, with resistance to four modes of action. *P. oleracea*, which is a summer weed difficult to control in Mediterranean crops [16], has developed resistance only to two modes of action [13]. *A. fatua* and *E. bonariensis* have developed resistance to glyphosate, which is the herbicide most commonly used around the world [13,17].

There are several examples of natural products that have been tested as potential bioherbicides to control *A. fatua*, *A. retroflexus*, *E. bonariensis*, and *P. oleracea*, mainly essential oils (EOs) [14,18–26], or extracts from plants with different solvents [27–29], or their isolated compounds [30,31]. Most studies have been carried out only in in vitro conditions. Of the weeds considered, *A. retroflexus* has been the most tested. In vitro studies with EOs from *Artemisia vulgaris*, *Mentha spicata*, *Ocimum basilicum*, *Salvia officinalis*, and *Thymbra spicata* from Turkey demonstrated high phytotoxic effects on seed germination and seedling growth of *A. retroflexus*, with stronger effects with higher doses [18]. EOs from *Tanacetum* species growing in Turkey, rich in oxygenated monoterpenes, inhibited completely *A. retroflexus* germination in in vitro assays [19]. In addition, EOs from *Nepeta meyeri*, with high content in oxygenated monoterpenes controlled completely *A. retroflexus* germination [20]. The phytotoxic potential of 12 EOs was studied in vitro against *A. retroflexus* and *A. fatua*, and the most phytotoxic EOs were those constituted mainly by oxygenated monoterpenes [21]. Other EOs which showed strong herbicidal potential against *A. retroflexus* seed germination and seedling growth were *Rosmarinus officinalis*, *Satureja hortensis*, and *Laurus nobilis* [14], and a nanoemulsion of *S. hortensis* EO was tested against *A. retroflexus* in greenhouse conditions killing the weed at 4000 μL/mL dose [22]. *P. oleracea* germination was completely inhibited by *Eucalyptus camaldulensis* EO in in vitro conditions [23]. The application of leaf extracts (obtained using water, methanol, and ethanol as solvents) of cultivated *Cynara cardunculus* in in vitro bioassays inhibited seed germination and germination time in *A. retroflexus* and *P. oleracea* [27].

Different natural compounds have demonstrated herbicidal potential against the germination and seedling growth of *A. fatua*, such as EOs from *Artemisia herba-alba* [24] and *Eucalyptus citriodora* EOs [25] and extracts from *Sapindus mukorossi*, which inhibited *A. fatua* and *A. retroflexus* growth in vitro and in pots [28] or from *Iris sibirica* rhizomes [29].

EOs from *Thymbra capitata*, *Mentha piperita*, *Eucalyptus camaldulensis*, and *Santolina chamaecyparissus* were tested in vivo against *E. bonariensis*. *T. capitata* EO, with high content in carvacrol, was the most effective to control *E. bonariensis*, showing an excellent potential to develop bioherbicide formulations [26].

Some studies carried out in recent years relate the herbicidal activity of plant extracts or EOs to their composition in monoterpenes, and these substances are postulated as the future of natural herbicide components [32–35]. For example, eugenol, a monoterpene that can be found in many EOs as the major compound, like in *Syzygium aromaticum* EO, has shown strong phytotoxic potential against *A. retroflexus* [30] and *A. fatua* [31]. In *A. fatua*, eugenol inhibited its seedling growth, affecting more the roots than the coleoptiles. In addition, sesquiterpenes, secondary metabolites in plants, present in some EOs, have demonstrated strong herbicidal activity [36,37].

The natural products studied on this work for their potential as bioherbicides were pelargonic acid, trans-cinnamaldehyde and carvacrol. Pelargonic acid (PA) ($CH_3(CH_2)_7CO_2H$, n-nonanoic acid), which is present as esters in the EO of *Pelargonium* spp., is a saturated fatty acid with nine carbons in its structure [28–40]. PA and its salts are used like active ingredients in bioherbicide formulations for garden and professional uses worldwide. They are applied as burndown herbicides, which in a short time, attack cell membranes, causing cell leakage, followed by breakdown of membrane acyl lipids [41], and finally causing visible effects of desiccation of green areas of the weeds [38]. All the symptoms caused by PA on weeds involve extreme phytotoxicity for the plants and their cells, which rapidly begin to oxidize, causing necrotic lesions on aerial parts of plants [42,43].

Herbicidal fatty acids have been used for a long time in weed management, and some of them are used as natural herbicides. Still, the high dosage and the high cost are some of the drawbacks of its practical application in the current agriculture. In 2015, the bioherbicide Beloukha® was authorized as plant protection product to be marketed in Europe [44]. It is derived from oleic acid from different origin. Actually, it is authorized also for markets in USA and Canada. This work aims to find an optimal formulation of PA capable to be effective at reduced doses compared to the existing products in the market.

Trans-cinnamaldehyde (CA) (C_9H_8O) is one of the major components of two different cinnamon species (*Cinnamomum zeylanicum* and *Cinnamomum cassia*) and their EOs [45–48]. This compound has shown strong antioxidant properties and is responsible for various observed biological activities of cinnamon like bactericidal, fungicidal, or acaricidal [49–52]. The antimicrobial activity of CA is well known, however, its potential as bioherbicide has been less studied. Despite that, recent research demonstrated the herbicidal activity of CA against *Echinochloa crus-galli* by reducing the fresh weight and growth of this important weed [53]. To our knowledge, the mode of action of CA on weeds has not been elucidated.

The third natural compound evaluated was carvacrol (CV), a phenolic monoterpene frequently present on EOs obtained from many species belonging to Lamiaceae family like *Thymus* spp., *Thymbra* spp., and *Origanum* spp. [34]. CV presents antimicrobial properties that make it helpful for controlling diseases in crop protection [54–58]. In relation to its mode of action, CV exhibited membrane-disrupting activity that was dependent on long exposure at high concentration [33]. Postemergence exposure of plants to high concentrations of CV causes severe phytotoxicity. One of the effects associated with the mode of action of CV is the reduction of weed growth [22,41,54].

This work is a collaboration between the Universitat Politècnica de València (UPV) and the company Seipasa S.A., which develops and commercializes biopesticides, with the purpose to manage agricultural ecosystems in a more sustainable way. The objective of the present study was to evaluate the herbicidal potential of the natural compounds pelargonic acid, trans-cinnamaldehyde, and carvacrol against important cosmopolite weeds (*Amaranthus retroflexus* L., *Portulaca oleracea* L., *Erigeron bonariensis* L., and *Avena fatua* L.) as an alternative to synthetic herbicides to reduce the abiotic stress that they cause on crops. Effective compounds were formulated as emulsifiable concentrates (ECs) by Seipasa S.A., and evaluated for their postemergence herbicidal activity in greenhouse conditions in the UPV (Spain).

2. Materials and Methods

2.1. Postemergence Herbicidal Assays against Targeted Weed Species

2.1.1. Weeds

Seeds of *Amaranthus retroflexus* L., *Portulaca oleracea* L., and *Avena fatua* L. purchased from Herbiseed (Reading, UK) (year of collection 2017), which have been previously tested in a plant growth chamber EGCHS series from Equitec (Madrid, Spain) (30 ± 0.1 °C, 16 h light and 20 ± 0.1 °C, 8 h dark for *A. retroflexus* and *P. oleracea*; 23.0 ± 0.1 °C, 8 h light and 18.0 ± 0.1 °C 16 h dark for *A. fatua*) to assure their germination viability, were sown in pots (8 × 8 × 7 cm) filled with 2 cm of perlite and 5 cm of soil collected from a citrus orchard nontreated with herbicides. In Figure 1, the location (39°37′24.8″ N, 0°17′25.6″ W Puzol, Valencia province, Spain) and a view of the citrus orchard (0.4 ha) from which the soil was collected is reported. Table 1 shows the main physical characteristics of the soil used for the experiments.

Figure 1. Location (**A**) and view (**B**) of the citrus orchard where the soil for the herbicidal tests was collected.

Table 1. Physical properties of the soil used for the experiments [59].

Soil Properties
Clay 21.85%
Silt 47.55%
Sand 30.60%

Erigeron bonariensis L. seeds were collected from an ecological weed management persimmon orchard located in Carlet (Valencia province, Spain) in July 2018. They were previously tested in the plant growth chamber described before (30 ± 0.1 °C, 16 h light and 20 ± 0.1 °C 8 h dark) to assure their germination capability and after that, sown in plastic pots filled with a mix of three-fourth peat and one-fourth perlite instead of soil because it was very difficult to germinate the seeds on the soil, as *E. bonariensis* germinates better in lighter soils [60] and, therefore, the properties of the soil collected from the citrus orchard (Table 1) did not fit the needs for their germination.

All weeds were irrigated by capillarity from trays (43 cm × 28 cm × 65 cm) placed under the pots and filled with water, until the plants were ready for the herbicidal experiments.

2.1.2. Treatments

Ten pots were prepared for each treatment, described in Table 2. The treatments were applied when plants reached the phenological stage of 2-3-true leaves, corresponding to stage 12-13 BBCH (Biologische Bundesanstalt, Bundessortenamt und Chemische Industrie) scale for the monocotyledonous *A. fatua*, and 3-4-true leaves, corresponding to stage 13-14 BBCH scale for the dicotyledonous *A. retroflexus* and *P. oleracea* and in rosette stage for *E. bonariensis*, stage 14-15 BBCH scale (Figure 2). Pelargonic acid, cinnamic aldehyde and carvacrol were provided formulated as emulsifiable concentrates (ECs) by the company Seipasa S.A. (L'Alcudia, Valencia province, Spain). Beloukha® was purchased from Ferlasa (Museros, Valencia province, Spain) and Roundup® Ultra Plus was purchased from Cooperativa Agrícola Nuestra Señora del Oreto (CANSO, L'Alcudia, Valencia province, Spain).

Table 2. Treatments tested.

	Treatments	Abbreviations
T1	Control treated with water	CW
T2	Pelargonic acid 5%	PA5
T3	Pelargonic acid 8%	PA8
T4	Pelargonic acid 10%	PA10
T5	Cinnamic aldehyde 6%	AC6
T6	Cinnamic aldehyde 12%	AC12
T7	Cinnamic aldehyde 24%	AC24
T8	Carvacrol 8%	CV8
T9	Carvacrol 16%	CV16
T10	Carvacrol 32%	CV32
T11	Bioherbicide reference: pelargonic acid (Beloukha® 8%)	BE
T12	Chemical reference: glyphosate (Roundup® Ultra Plus 10%)	GL

Figure 2. Pots ready for the postemergence treatments. (**A**) *A. fatua*, (**B**) *A. retroflexus*, (**C**) *P. oleracea*, and (**D**) *E. bonariensis*.

In Table 3, the dates of the herbicidal tests and the greenhouse conditions during the experimental periods are reported. Data were registered using a HOBO U23 Pro v2 data logger (Onset Computer Corporation, Bourne, MA, USA).

Table 3. Greenhouse conditions during the herbicidal tests.

Species	Starting-End Date	Temperature (°C)			Relative Humidity (%)		
		Mean	Max.	Min.	Mean	Max.	Min.
P. oleracea	August 9, 2018–September 9, 2018	28.03	38.39	22.87	68.04	87.03	37.18
A. retroflexus	September 2, 2018–October 2, 2018	26.38	35.42	19.82	70.91	85.88	31.14
A. fatua	December 3, 2018–January 3, 2019	18.57	25.72	12.75	57.87	75.56	29.84
E. bonariensis	February 15, 2019–March 15, 2019	22.62	27.16	17.99	45.88	50.26	40.40

2.2. Evaluation of the Herbicidal Activity of Each Natural Product

During the experiments, images from the plants were taken 24 h and 3, 7, 15, and 30 days after the treatments application to be processed with Digimizer v.4.6.1 software (MedCalc Software, Ostend, Belgium, 2005–2016).

To evaluate the herbicidal activity, two variables were measured for each plant: the efficacy, which was scored 0 if the plant was alive and 100 if the plant was dead, and the damage level, which was assessed between 0 and 4 as reported in Table 4 and Figure 3. The efficacy and damage level for each treatment were calculated as the mean of the 10 treated plants.

Table 4. Damage level assessment.

Level of Damage	
0	Undamaged plant
1	Plant with slight damage
2	Plant with severe damage
3	Dead plant
4	Regrown plant

Figure 3. Damage scale for each species: (**A**) *A. fatua*, (**B**) *P. oleracea*, (**C**) *A. retroflexus*, and (**D**) *E. bonariensis*.

2.3. Statistical Analyses

Data were processed using Statgraphics® Centurion XVII (StatPoint Technologies Inc., Warrenton, VA, USA) software. A multifactor analysis of variance (ANOVA) was performed on efficacy and damage level including species, treatments, time after treatments application, and their double significant interactions as effects, followed by Fisher's multiple comparison test (LSD intervals, least significant difference, at $p \leq 0.05$) for the separation of the means.

3. Results and Discussion

3.1. Efficacy of Pelargonic Acid, Cinnamic Aldehyde, and Carvacrol against Target Weeds

A. retroflexus was the weed species most susceptible to the treatments tested, with 73.50 efficacy (Table 5). No significant differences were observed between the other species, which showed around 55 efficacies. The fact that all species tested were susceptible to all treatments with natural products assayed confirm that they could be a more sustainable alternative to synthetic herbicides, and they also offer new modes of action to control weeds that have developed resistant biotypes to many herbicides.

Table 5. Efficacy according to the species, time, and treatment.

Species	Efficacy
Portulaca oleracea	56.17 ± 1.11 b
Amaranthus retroflexus	73.50 ± 1.11 a
Avena fatua	54.83 ± 1.11 b
Erigeron bonariensis	55.67 ± 1.11 b
Time (Days after application)	**Efficacy**
1	41.67 ± 1.24 c
3	81.88 ± 1.24 b
7	87.08 ± 1.24 a
15	89.58 ± 1.24 a
Treatment	**Efficacy**
Control treated with water	4.00 ± 1.92 g
Pelargonic acid 5%	70.50 ± 1.92 b
Pelargonic acid 8%	73.50 ± 1.92 ab
Pelargonic acid 10%	74.50 ± 1.92 ab
Cinnamic aldehyde 6%	53.50 ± 1.92 e
Cinnamic aldehyde 12%	70.00 ± 1.92 bc
Cinnamic aldehyde 24%	70.00 ± 1.92 bc
Carvacrol 8%	60.50 ± 1.92 d
Carvacrol 16%	64.50 ± 1.92 d
Carvacrol 32%	65.00 ± 1.92 cd
Bioherbicide reference: pelargonic acid (Beloukha® 8%)	78.50 ± 1.92 a
Chemical reference: glyphosate (Roundup® Ultra Plus 10%)	36.00 ± 1.92 f

Values are efficacy ± standard error. Means followed by different letters in the same column differ significantly ($p \leq 0.05$).

Efficacy increased with time after treatments application, with values close to 90 between 7 and 15 days (Table 5). This happened because PA, at all doses applied, and the higher doses of CA and CV acted very quickly in the treated species, causing the death of all plants between 24 h and 3 days after application of treatment (Figures 4–7, Tables S1–S4). The same happened for the bioherbicide reference BE (as PA was also the active compound on it), while GL acted more slowly, depending on the species against which it was applied; it killed *A. retroflexus* plants after 3 days, *A. fatua* and *P. oleracea* after 15 days, and *E. bonariensis* after 30 days (Figures 4–7, Tables S1-S4). It has been reported that weed damage caused by PA can be observed visually few hours after application [61]. Thymol, *trans*-cinnamaldehyde, eugenol, farnesol, and nerolidol were tested in postemergence in *E. crus-galli* applied at two-leaf stage, and significantly reduced the shoot growth and the fresh and dry weight 2 days after the foliar treatments with 0.5%, 1.0%, and 2.0% concentrations. All treatments except thymol controlled the weed completely when applied at 1.0% and 2.0% [52]. The concentrations of CA used in this work were higher, and this could explain the quicker toxic effect observed on weeds. It is also remarkable that weed species displayed different sensitivity to low doses of CA; *E. bonariensis* and *P. oleracea* showed more resistance to this compound than the other weeds tested (Figures 4–7, Tables S1-S4), as the lowest concentration (6%) used took more time (15 days) to kill all the plants in *E. bonariensis* than in *A. retroflexus* (24 h) or *A. fatua* (3 days), whereas in *P. oleracea*, this dose reached 50 efficacy, i.e., only 50% of plants were dead at the end of the experiment (30 days). Previous studies also confirmed the rapid activity of carvacrol in plants; in a greenhouse experiment, a nanoemulsion (NE) of *Satureja hortensis* L. EO, rich in carvacrol (55.6%), was applied against *A. retroflexus* and *C. album*, and after 30 min, the weeds were exhibiting injury symptoms, reaching the maximum lethality within 24 h of treatment application. The lethality percentage was dependent on the doses applied and the species against which NE was applied [21]. As observed with CA, also weed species showed different sensibility to CV application, especially at the lower dose, which took more time to control the weeds (Figures 4–7, Tables S1–S4): *A. retroflexus* was the more sensitive species, being controlled by all doses 24 h after application of treatment (Figure 4, Table S1), whereas in *A. fatua* and *E. bonariensis*, the lowest dose took 7 and 15 days, respectively, to reach 100 efficacy (Figures 5 and 6, Tables S2 and S3), being again *P. oleracea* the most resistant weed species, 7 days after treatment application, all plants were killed in all CV treatments, although then some regrew 15 and 30 days after treatments application (Figure 7, Table S4).

All the treatments managed to control the weed species tested, and the results of the treatments were statistically significant compared to CW (Table 5). The most effective treatment was the PA formulation at 10%, achieving 74.50 efficacy. This treatment did not show significant differences compared to the results obtained by the commercial product used as biological reference, also containing PA as active ingredient, which obtained an efficacy of 78.50. Moreover, there were no significant statistical differences in the efficacy between the three doses of the PA-based formulations (5%, 8%, and 10%). The next most effective treatment was the CA-based formulation, which exhibited the same efficacy values for the two higher doses applied (12% and 24%), while the lowest dose (6%) had significant less efficacy. This can be explained by the different sensitivity of the weed species to low doses of CA, as commented above. Finally, the treatments with carvacrol did not show significant differences in efficacy between doses, but with the control, and were also very effective, reaching an efficacy between 60.50 and 65.00 (Table 5).

All treatments tested with natural products showed higher efficacy for the control of weeds than GL, which showed efficacy values of 36. This was because of its slower activity. Mechanism of action of GL is by affecting the enzyme 5-enolpyruvlyshikimate-3-phosphate synthase (EPSPS), and it is the only herbicide with this mode of action. The inhibition of EPSPS reduces levels of amino acids needed for the synthesis of proteins, cell walls, and secondary plant products. In addition, the inhibition of EPSPS causes deregulation of the shikimic acid pathway, promoting the disruption of plant carbon metabolism [62]. GL is translocated in plants and differential responses of weed species may be caused by differences in herbicide translocation, i.e., weeds capable to translocate GL more efficiently are more severely damaged [63]. In field experiments conducted for 2 years, it was verified that GL controlled more effectively *A. retroflexus* than other species [64], which supports our results. Decreased herbicide translocation to the meristem causes reduced glyphosate efficacy [65]. The necessity of being translocated explains the slow effect of GL compared with the natural compounds, as ^{14}C translocation throughout the plant demonstrated that glyphosate took 3 days to reach and accumulate in the meristematic tips of the roots and shoots [66]."

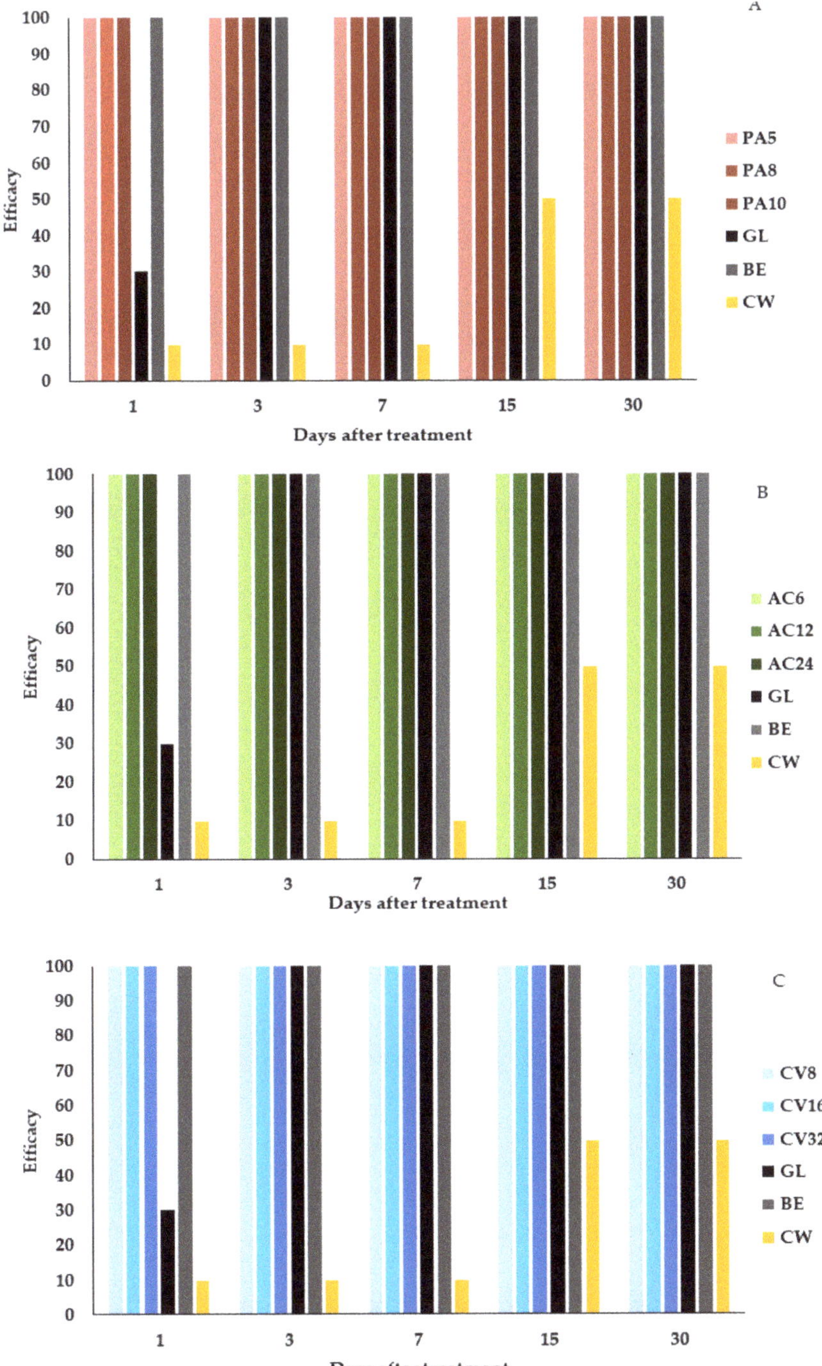

Figure 4. Evolution of efficacy of the tested treatments (**A**) pelargonic acid, (**B**) cinnamic aldehyde and (**C**) carvacrol in *A. retroflexus* during 30 days after application.

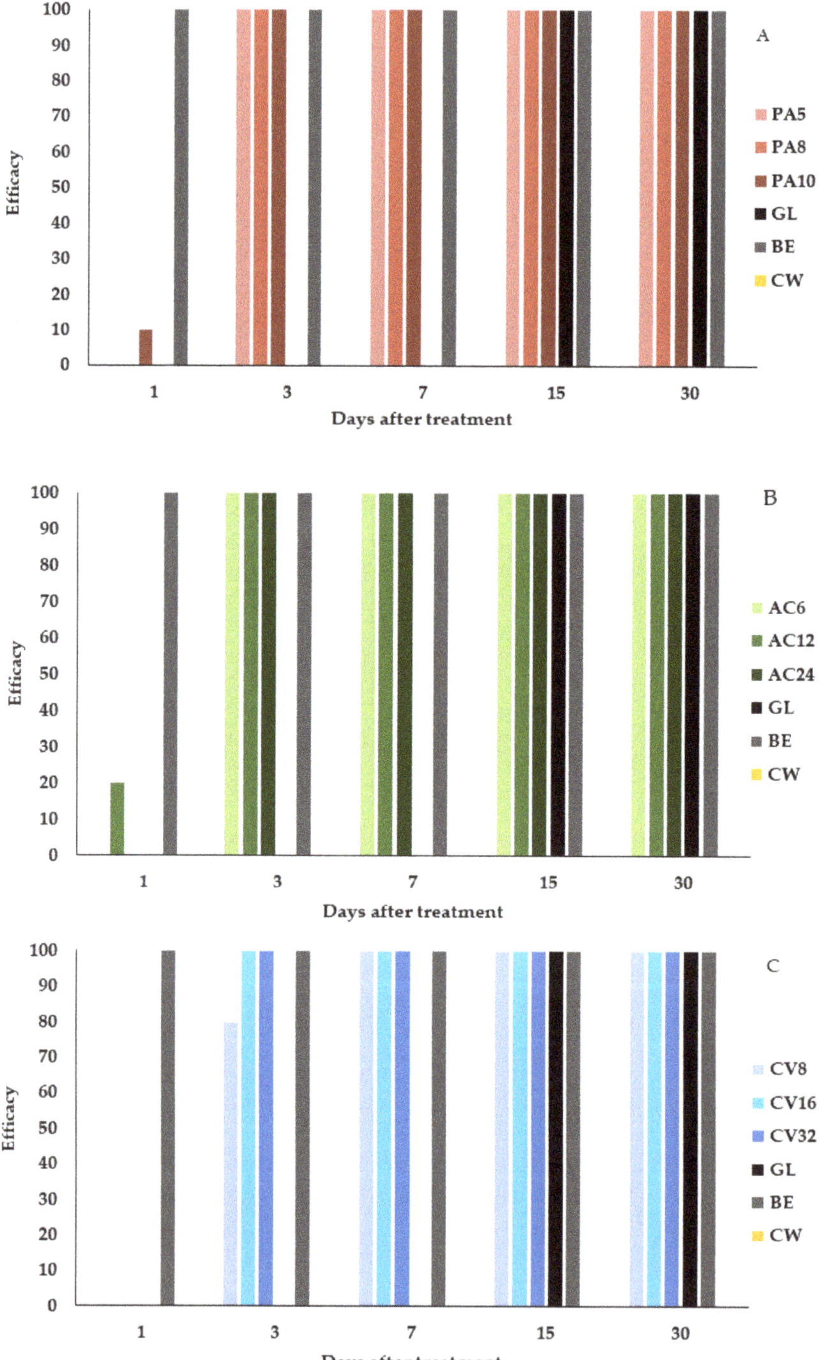

Figure 5. Evolution of efficacy of the tested treatments (**A**) pelargonic acid, (**B**) cinnamic aldehyde, and (**C**) carvacrol in *A. fatua* during 30 days after their application.

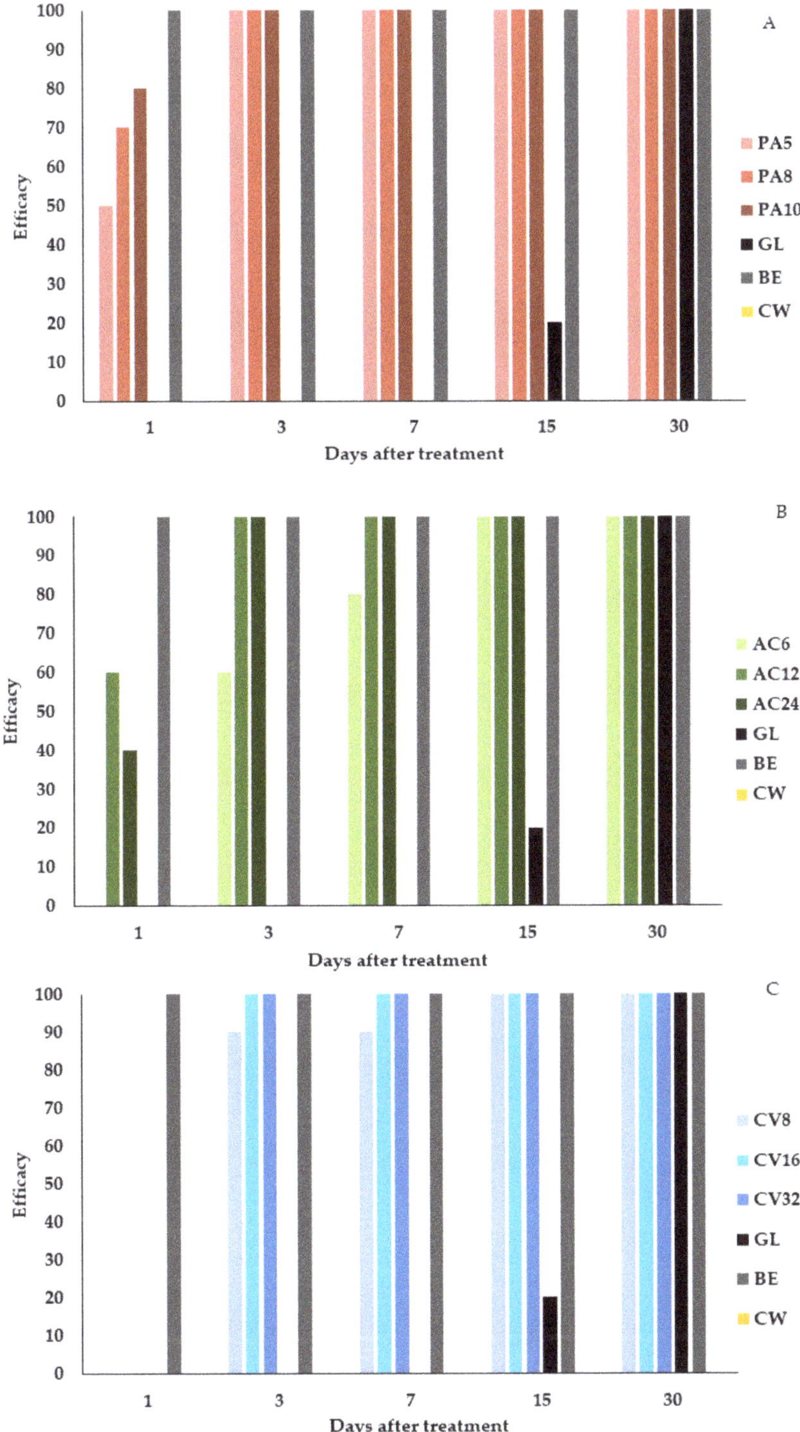

Figure 6. Evolution of efficacy of the tested treatments (**A**) pelargonic acid, (**B**) cinnamic aldehyde, and (**C**) carvacrol in *E. bonariensis* during 30 days after their application.

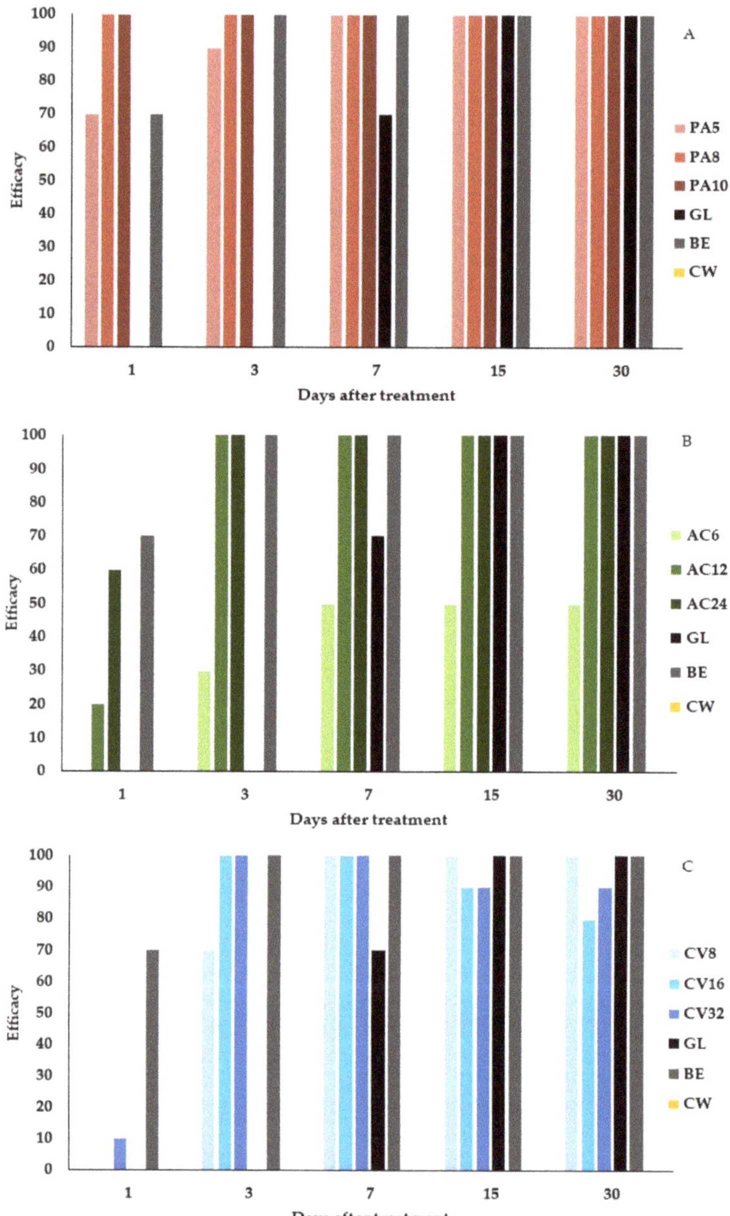

Figure 7. Evolution of efficacy of the tested treatments (**A**) pelargonic acid, (**B**) cinnamic aldehyde, and (**C**) carvacrol in *P. oleracea* during 30 days after their application.

3.1.1. Efficacy of Pelargonic Acid, Cinnamic Aldehyde, and Carvacrol on *A. retroflexus*

In the species *A. retroflexus* (Figure 4, Table S1) all the treatments tested obtained 100 efficacy (all treated plants were dead) one day after the application of the treatment, except for the chemical reference. The treatment with GL managed to control the species on the third day after its application. In this trial, there was a relevant percentage of mortality in the CW, especially at the end of the trial.

3.1.2. Efficacy of Pelargonic Acid, Cinnamic Aldehyde, and Carvacrol on *A. fatua*

All the tested treatments managed to control completely the species *A. fatua* from the third day after application (Figure 5, Table S2), except CV6, which achieved 100 efficacy after 7 days, and GL, which reached 100 efficacy 15 days after application. The treatments that showed phytotoxic effects more quickly were, starting from the first day after application, the bioherbicide reference (BE), AC12, and PA10.

3.1.3. Efficacy of Pelargonic Acid, Cinnamic Aldehyde, and Carvacrol on *E. bonariensis*

All treatments were able to control *E. bonariensis* (Figure 6, Table S3). The higher doses of the treatments performed with CA- and CV-based formulations achieved a total control of this species faster than their lower doses. It should be noted that despite this, all of them managed to control it completely 15 days after the application. The bioherbicide reference (BE) reached 100 efficacy 24 h after its application, instead GL took 30 days to reach 100 efficacy (death of all treated plants).

3.1.4. Efficacy of Pelargonic Acid, Cinnamic Aldehyde, and Carvacrol on *P. oleracea*

The most effective treatments to control *P. oleracea* were the three treatments carried out with the PA-based formulation (PA5, PA8, and PA10) (Figure 7, Table S4). A dose effect was observed in this species for the tested natural products, being higher doses more effective and showing phytotoxic effects faster than lower ones. The treatment AC6 reached 50 efficacy at the end of the experiment (30 days after application), while the higher doses of this compound (AC12 and AC24) killed all plants after 3 days of application. The treatments CV8, CV16, and CV32 decreased their efficacy from day 7, when some of the evaluated plants regrew. It should be noted that the treatment with the chemical reference, GL, exhibited a slower action than the rest of the treatments with natural products, showing phytotoxic effects on this species between 7 and 15 days after application.

When analyzing the effect of the interaction between species and time after treatments with respect to efficacy, the species that showed the highest sensitivity most rapidly was *A. retroflexus*. On the other hand, the species that took longer to show phytotoxic effects was *A. fatua*. However, at the end of the trials, all species showed high mortality rates, which were slightly higher in *A. retroflexus* and *A. fatua* than in *P. oleracea* and *E. bonariensis* (Figure 8).

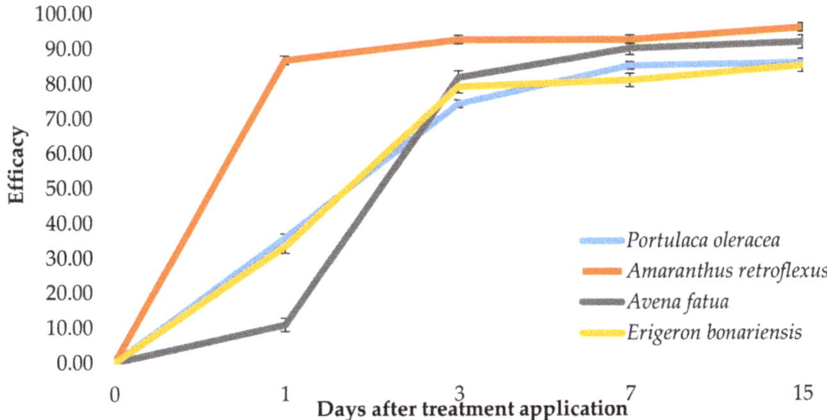

Figure 8. Effect of the interaction between treatment and days after treatment application in the efficacy per species.

3.2. Damage Level of Pelargonic Acid, Cinnamic Aldehyde, and Carvacrol against Target Weeds

A. retroflexus was the species which presented higher damage level, followed by *P. oleracea* and *A. fatua* (without significant differences between them), and finally *E. bonariensis* (Table 6). All species exhibited damage level near 2 or higher, which means severe damage (Table 4). It is important to consider the damage level caused by the treatments on the weed species in addition to their efficacy because it represents the state of the plants that were not killed. If the plants remaining alive were more damaged, it would mean that in field conditions, they would be less competitive with crops, causing less stress to them.

Table 6. Damage level depending on the species, time after application, and treatment.

Species	Level of Damage
Portulaca oleracea	1.98 ± 0.02 b
Amaranthus retroflexus	2.24 ± 0.02 a
Avena fatua	1.96 ± 0.02 bc
Erigeron bonariensis	1.92 ± 0.02 c
Time (Days after Application)	**Level of Damage**
0	0.00 ± 0.02 e
1	2.08 ± 0.02 d
3	2.59 ± 0.02 c
7	2.68 ± 0.02 b
15	2.78 ± 0.02 a
Treatment	**Damage level**
Control treated with water	0.16 ± 0.04 g
Pelargonic acid 5%	2.31 ± 0.04 abc
Pelargonic acid 8%	2.34 ± 0.04 ab
Pelargonic acid 10%	2.35 ± 0.04 ab
Cinnamic aldehyde 6%	2.13 ± 0.04 e
Cinnamic aldehyde 12%	2.30 ± 0.04 abc
Cinnamic aldehyde 24%	2.30 ± 0.04 abc
Carvacrol 8%	2.18 ± 0.04 de
Carvacrol 16%	2.23 ± 0.04 cd
Carvacrol 32%	2.25 ± 0.04 bcd
Bioherbicide reference: pelargonic acid (Beloukha 8%)	2.39 ± 0.04 a
Chemical reference: glyphosate (Roundup Ultra Plus 10%)	1.40 ± 0.03 f

Values are mean of damage level ± standard error (ten replicates). Different letters in the same column indicate significant differences ($p \leq 0.05$).

Throughout time, more severe levels of damage were reached as more days after treatment applications passed, with significant differences in the damage level assessment between different days after the applications (Table 6). All the treatments tested successfully controlled the weed species inducing a high level of damage compared with CW. The treatments that showed the strongest phytotoxicity on weeds were PA10 and BE, with no significant differences between them. PA10 showed no significant differences with the other two doses of PA-based formulations tested (PA5 and PA8), neither with the two highest doses of CA based formulations tested (CA12 and CA24) nor with the highest doses of CV tested (CV32) (Table 6).

The damage level increased in all species with time after treatments (Figure 9). *A. retroflexus* was confirmed as the most susceptible species to the treatments, as it showed a higher level of damage than the other species 24 h after the treatments were administrated. No differences between species were observed 15 days after treatment, as all showed similar levels of damage.

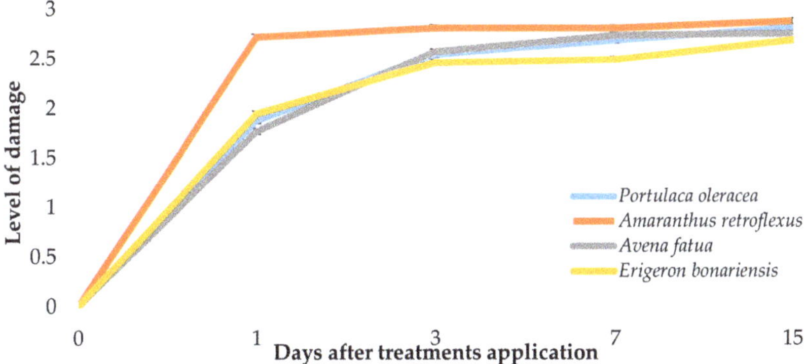

Figure 9. Effect of treatment and time after treatment interaction on damage level.

The effects induced by the different treatments on *E. bonariensis* 24 h after their administration are presented in Figure 10. This species is shown because of its intermediate response to all treatments as compared with *A. retroflexus* that was more sensitive or *P. oleracea*, which was more resistant and because phytotoxic effects can be better visualized in it than in *A. fatua*. The intermediate concentration tested for PA, CV, and CA is shown to be representative of the effects of the other concentrations tested. All the natural compounds tested caused more severe plant damage than the synthetic herbicide GL 1 day after treatment. The effects of 8% PA were very similar to those induced by the positive bioherbicide control Beloukha (also containing PA as active compound). Probably due to the effect of PA, the cuticles exhibited alteration on membrane permeability and peroxidation of thylakoid membranes [67] and leaves appeared desiccated, with reduced photosynthetic pigments but without punctual damages on the leaves, which resulted in a stoppage of growth and development of the whole plant. In contrast, CV-treated leaves showed signs of dehydration, resulting in curling and punctual damages on the leaves with increased necrotic spots related to application spots, which could be due to the disruption of cell membranes [68]. Finally, CA treatment resulted in growth reduction and loss of photosynthetic pigments, which could be related to oxidative damage induced by this compound. This oxidative damage has to be further investigated as no mode of action of CA has been reported in the literature up to now.

Bioherbicides are new products on the international markets and consequently, the processes for obtaining natural raw materials are not yet very efficient or the final cost of its extraction is elevated compared to synthetics. This fact affects the final cost of these formulated products, making them more expensive in some cases than conventional herbicides for farmers. Nevertheless, it is important to evaluate the cost–benefit factor of bioherbicides, including sustainability, reduction of soil and water contamination, or the absence of residues on crops. In line with legal framework, policies, and global sustainability objectives, the higher price of bioherbicides justifies the benefits that can be achieved with their implementation [69]. On the other hand, the rapid action, broad spectrum, and eco-friendly profile make bioherbicides molecules more attractive to the pesticide market, which is increasingly concerned with the sustainability of treatments applied in agriculture. Herbicide market is expected to reach a value of \$37.99 billion by 2025 [38]. Improving the efficiency of raw material extraction, decreasing the applied doses per hectare using improved formulations, as well as combining active substances in search of synergies may be the future of new sustainable herbicides.

The natural products tested, PA, CV, and CA, performed strong herbicidal activity in all the treated weeds, causing high lethality and damage levels; hence, they demonstrated that they could be good candidates for bioherbicides formulations. Further investigations should focus on determining the dose–response of different weed species to these compounds in order to find the optimal doses, which is very important in the context of integrated weed management and sustainable agriculture.

Another key point is to find out the optimum phenological stage in which the products should be applied to weeds and crops, to achieve the maximum phytotoxic effect on weeds minimizing their phytotoxic effects and consequent stress on crops. A better understanding of their mode of action could lead to a more efficient administration. Finally, different combinations between these natural products could be a powerful tool for weed management. Their synergies and antagonisms must be also considered and studied.

Figure 10. Images of *Erigeron bonariensis* plants 24 h after treatment applications.

4. Conclusions

The natural products PA, CV, and CA showed great herbicidal activity against the weeds *A. retroflexus*, *A. fatua*, *E. bonariensis*, and *P. oleracea* and could be good candidates for bioherbicides formulations. *A. retroflexus* was the most sensitive weed to all the applied treatments. For CV and CA, the higher doses applied exhibited greater and quicker phytotoxic effects than the lowest, with different responses in the weed species, while there were no significant differences in the herbicidal activity between the tested doses of PA. This study demonstrates that natural products could be sustainable as well as effective alternatives to synthetic herbicides, and they contribute to integrated weed management.

Supplementary Materials: The following are available online at http://www.mdpi.com/2073-4395/10/6/791/s1, Table S1. Efficacy of the tested treatments on *A. retroflexus* after 1, 3, 7, 15 and 30 days of application. Table S2. Efficacy of the tested treatments on *A. fatua* after 1, 3, 7, 15 and 30 days of application. Table S3. Efficacy of the tested treatments on *E. bonariensis* after 1, 3, 7, 15 and 30 days of application. Table S4. Efficacy of the tested treatments on *P. oleracea* after 1, 3, 7, 15 and 30 days of application.

Author Contributions: Conceptualization, M.V., M.M., and A.M.S.-M.; methodology M.V., M.M., N.T.-P., and R.G.; formal analysis, M.V., M.M., and N.T.-P.; investigation, M.V., M.M., A.M.S.-M., R.P., N.T.-P., and R.G.; resources, M.V., M.M., and R.G.; data curation, N.T., M.M., R.G. and R.P.; writing—original draft preparation, M.V., M.M., and N.T.-P.; writing—review and editing, M.V., A.M.S.-M., and R.P.; visualization, M.V., M.M., N.T.-P., A.M.S.-M., R.P., and R.G.; supervision, M.V., M.M., and A.M.S.-M.; project administration, M.V.; and funding acquisition, M.V. All authors have read and agreed to the published version of the manuscript.

Funding: This research was funded by SEIPASA.

Acknowledgments: Thanks to Vicente Estornell Campos and the Library staff from Polytechnic University of Valencia that assisted us to get some helpful references.

Conflicts of Interest: The authors declare no conflict of interest.

References

1. Vos, R.; Bellù, L.G. Global trends and challenges to food and agriculture into the 21st century. In *Sustainable Food and Agriculture: An Integrated Approach*; Campanhola, C., Pandey, S., Eds.; Academic Press: London, UK, 2019; pp. 11–30. [CrossRef]
2. Oerke, E.C. Crop losses to pests. *J. Agric. Sci.* **2006**, *144*, 31–43. [CrossRef]
3. Vats, S. Herbicides: History, classification and genetic manipulation of plants for herbicide resistance. In *Sustainable Agriculture Reviews*; Lichtfouse, E., Ed.; Springer: Cham, Switzerland, 2015; Volume 15. [CrossRef]
4. Heap, I.M. Global perspective of herbicide-resistant weeds. *Pest Manag. Sci.* **2014**, *70*, 1306–1315. [CrossRef] [PubMed]
5. Bagavathiannan, M.; Singh, V.; Govindasamy, P.; Abugho, S.B.; Liu, R. Impact of concurrent weed or herbicide stress with other biotic and abiotic stressors on crop production. In *Plant Tolerance to Individual and Concurrent Stresses*; Senthil-Kumar, M., Ed.; Springer: New Delhi, India, 2017; pp. 33–45. [CrossRef]
6. Abbas, T.; Zahir, Z.A.; Naveed, M.; Kremer, R.J. Limitations of existing weed control practices necessitate development of alternative techniques based on biological approaches. In *Advances in Agronomy*; Sparks, D.L., Ed.; Academic Press: Cambridge, MA, USA, 2018; Volume 147, pp. 239–280. [CrossRef]
7. World Health Organization; Food and Agriculture Organization of the United Nations. The International Code of Conduct on Pesticide Management. Rome. 2014. Available online: http://www.fao.org/agriculture/crops/thematic-sitemap/theme/pests/code/en/ (accessed on 22 April 2020).
8. Villa, F.; Cappitelli, F.; Cortesi, P.; Kunova, A. Fungal biofilms: Targets for the development of novel strategies in plant disease management. *Front. Microbiol.* **2017**, *8*, 654–664. [CrossRef] [PubMed]
9. De Mastro, G.; Fracchiolla, M.; Verdini, L.; Montemurro, P. Oregano and its potential use as bioherbicide. *Acta Hortic.* **2006**, *723*, 335–346. [CrossRef]
10. Seiber, J.N.; Coats, J.; Duke, S.O.; Gross, A.D. Biopesticides: State of the art and future opportunities. *J. Agric. Food Chem.* **2014**, *62*, 11613–11619. [CrossRef] [PubMed]
11. Bailey, K.L. The bioherbicide approach to weed control using plant pathogens. In *Integrated Pest Management*; Abrol, D.P., Ed.; Academic Press, Elsevier: San Diego, CA, USA, 2014; pp. 245–266.
12. Dahiya, A.; Sharma, R.; Sindhu, S.; Sindhu, S.S. Resource partitioning in the rhizosphere by inoculated *Bacilluss* pp. towards growth stimulation of wheat and suppression of wild oat (*Avena fatua* L.) weed. *Physiol. Mol. Biol. Plants* **2019**, *25*, 1483–1495. [CrossRef]
13. Heap, I. The International Herbicide-Resistant Weed Database. Available online: www.weedscience.org (accessed on 25 April 2020).
14. Hazrati, H.; Saharkhiz, M.J.; Moein, M.; Khoshghalb, H. Phytotoxic effects of several essential oils on two weed species and tomato. *Biocatal. Agric. Biotechnol.* **2018**, *13*, 204–212. [CrossRef]
15. Bajwa, A.A.; Sadia, S.; Ali, H.H.; Jabran, K.; Peerzada, A.M.; Chauhan, B.B. Biology and management of two important Conyza weeds: A global review. *Environ. Sci. Pollut. Res.* **2016**, *23*, 24694–24710. [CrossRef]
16. Graziani, F.; Onofri, A.; Pannacci, E.; Tei, F.; Guiducci, M. Size and composition of weed seedbank in long-term organic and conventional low-input cropping systems. *Eur. J. Agron.* **2012**, *39*, 52–56. [CrossRef]
17. Benbrook, C.M. Trends in glyphosate herbicide use in the United States and globally. *Environ. Sci. Eur.* **2016**, *28*, 3–18. [CrossRef]
18. Önen, H.; Özer, Z.; Telci, I. Bioherbicidal effects of some plant essential oils on different weed species. *J. Plant Dis. Prot.* **2002**, *18*, 597–605.
19. Salamci, E.; Kordali, S.; Kotan, R.; Cakir, A.; Kaya, Y. Chemical compositions, antimicrobial and herbicidal effects of essential oils isolated from Turkish *Tanacetum aucheranum* and *Tanacetum chiliophyllum* var. *chiliophyllum*. *Biochem. Syst. Ecol.* **2007**, *35*, 569–581. [CrossRef]
20. Mutlu, S.; Atici, Ö.; Esim, N. Bioherbicidal effects of essential oils of *nepeta meyeri* benth. On weed spp. *Allelopath. J.* **2010**, *26*, 291–300.
21. Synowiec, A.; Kalemba, D.; Drozdek, E.; Bocianowski, J. Phytotoxic potential of essential oils from temperate climate plants against the germination of selected weeds and crops. *J. Pest Sci.* **2017**, *90*, 407–419. [CrossRef]
22. Hazrati, H.; Saharkhiz, M.J.; Niakousari, M.; Moein, M. Natural herbicide activity of *Satureja hortensis* L. essential oil nanoemulsion on the seed germination and morphophysiological features of two important weed species. *Ecotoxicol. Environ. Saf.* **2017**. [CrossRef]

23. Verdeguer, M.; Blazquez, M.A.; Boira, H. Phytotoxic effects of *Lantana camara*, Eucalyptus camaldulensis and *Eriocephalus africanus* essential oils in weeds of Mediterranean summer crops. *Biochem. Syst. Ecol.* **2009**, *37*, 362–369. [CrossRef]

24. Benarab, H.; Fenni, M.; Louadj, Y.; Boukhabti, H.; Ramdani, M. Allelopathic activity of essential oil extracts from *Artemisia herba-alba* Asso. on seed and seedling germination of weed and wheat crops. *Acta Sci. Nat.* **2020**, *7*, 86–97. [CrossRef]

25. Benchaa, S.; Hazzit, M.; Abdelkrim, H. Allelopathic Effect of *Eucalyptus citriodora* essential oil and its potential use as bioherbicide. *Chem. Biodivers.* **2018**, *15*, e1800202. [CrossRef]

26. Verdeguer, M.; Castañeda, L.G.; Torres-Pagan, N.; Llorens-Molina, J.A.; Carrubba, A. Control of Erigeron bonariensis with *Thymbra capitata*, *Mentha piperita*, *Eucalyptus camaldulensis*, and *Santolina chamaecyparissus* essential oils. *Molecules* **2020**, *25*, 562. [CrossRef]

27. Scavo, A.; Pandino, G.; Restuccia, A.; Mauromicale, G. Leaf extracts of cultivated cardoon as potential bioherbicide. *Sci. Hortic.* **2020**, 109024. [CrossRef]

28. Ma, S.; Fu, L.; He, S.; Lu, X.; Wu, Y.; Ma, Z.; Zhang, X. Potent herbicidal activity of *Sapindus mukorossi* Gaertn. against *Avena fatua* L. and *Amaranthus retroflexus* L. *Ind. Crops Prod.* **2018**, *122*, 1–6. [CrossRef]

29. Pacanoski, Z.; Mehmeti, A. Allelopathic effect of Siberian iris (*Iris sibirica*) on the early growth of wild oat (*Avena fatua*) and Canada thistle (*Cirsium arvense*). *J. Cent. Eur. Agric.* **2019**, *20*, 1179–1187. [CrossRef]

30. Bainard, L.D.; Isman, M.B.; Upadhyaya, M.K. Phytotoxicity of clove oil and its primary constituent eugenol and the role of leaf epicuticular wax in the susceptibility to these essential oils. *Weed Sci.* **2006**, *54*, 833–837. [CrossRef]

31. Ahuja, N.; Singh, H.P.; Batish, D.R.; Kohli, R.K. Eugenol-inhibited root growth in *Avena fatua* involves ROS-mediated oxidative damage. *Pestic. Biochem. Phys.* **2015**, *118*, 64–70. [CrossRef]

32. Vaughn, S.F.; Spencer, G.F. Volatile Monoterpenes as Potential Parent Structures for New Herbicides[1]. *Weed Sci.* **1993**, *41*, 114–119. [CrossRef]

33. Chaimovitsh, D.; Shachter, A.; Abu-Abied, M.; Rubin, B.; Sadot, E.; Dudai, N. Herbicidal Activity of Monoterpenes is associated with disruption of microtubule functionality and membrane integrity. *Weed Sci.* **2017**, *65*, 19–30. [CrossRef]

34. Amri, I.; Lamia, H.; Mohsen, H.; Bassem, J. Review on the phytotoxic effects of essential oils and their individual components: News approach for weed management. *Int. J. Appl. Biol. Pharm. Technol.* **2013**, *4*, 96–114.

35. Verdeguer, M.; García-Rellán, D.; Boira, H.; Pérez, E.; Gandolfo, S.; Blázquez, M.A. Herbicidal activity of Peumus boldus and Drimys winterii essential oils from Chile. *Molecules* **2011**, *16*, 403–411. [CrossRef]

36. Saad, M.M.G.; Abdelgaleil, S.A.M.; Suganuma, T. Herbicidal potential of pseudoguaninolide sesquiterpenes on wild oat, *Avena fatua* L. *Biochem. Syst. Ecol.* **2012**, *44*, 333–337. [CrossRef]

37. Araniti, F.; Sánchez-Moreiras, A.M.; Graña, E.; Reigosa, M.J.; Abenavoli, M.R. Terpenoid trans-caryophyllene inhibits weed germination and induces plant water status alteration and oxidative damage in adult Arabidopsis. *Plant Biol. (Stuttg)* **2017**, *19*, 79–89. [CrossRef]

38. Ciriminna, R.; Fidalgo, A.; Ilharco, L.M.; Pagliaro, M. Herbicides based on pelargonic acid: Herbicides of the bioeconomy. *Biofuels Bioprod. Biorefining* **2019**, *13*, 1476–1482. [CrossRef]

39. Coleman, R.; Penner, D. Organic acid enhancement of pelargonic acid. *Weed Technol.* **2008**, *22*, 38–41. [CrossRef]

40. Crmaric, I.; Keller, M.; Krauss, J.; Delabays, N. Efficacy of natural fatty acid based herbicides on mixed weed stands. *Jul. Kühn Arch.* **2018**, *458*, 327–332. [CrossRef]

41. Dayan, F.E.; Duke, S.O. Natural compounds as next-generation herbicides. *Plant Physiol.* **2014**, *166*, 1090–1105. [CrossRef]

42. Croteau, R.; Kutchan, T.M.; Lewis, N.G. Natural products (secondary metabolites). In *Biochemistry & Molecular Biology of Plants*, 2nd ed.; Buchanan, B.B., Gruissem, W., Jones, R.L., Eds.; Wiley: Rockville, MD, USA, 2000; pp. 1250–1318.

43. Lebecque, S.; Lins, L.; Dayan, F.E.; Fauconnier, M.L.; Deleu, M. Interactions between natural herbicides and lipid bilayers mimicking the plant plasma membrane. *Front. Plant Sci.* **2019**, *10*, 329–340. [CrossRef]

44. Cordeau, S.; Triolet, M.; Wayman, S.; Steinberg, C.; Guillemin, J.P. Bioherbicides: Dead in the water? A review of the existing products for integrated weed management. *Crop Prot.* **2016**, *87*, 44–49. [CrossRef]

45. Gruenwald, J.; Freder, J.; Armbruester, N. Cinnamon and health. *Crit. Rev. Food Sci. Nutr.* **2010**, *50*, 822–834. [CrossRef]

46. Ranasinghe, P.; Pigera, S.; Premakumara, G.S.; Galappaththy, P.; Constantine, G.R.; Katulanda, P. Medicinal properties of "true" cinnamon (*Cinnamomum zeylanicum*): A systematic review. *BMC Complement. Altern. Med.* **2013**, *13*, 275. [CrossRef]

47. Doyle, A.A.; Krämer, T.; Kavanagh, K.; Stephens, J.C. Cinnamaldehydes: Synthesis, antibacterial evaluation, and the effect of molecular structure on antibacterial activity. *Results Chem.* **2019**, *1*, 100013–100018. [CrossRef]

48. Chericoni, S.; Prieto, J.M.; Iacopini, P.; Cioni, P.; Morelli, I. In vitro activity of the essential oil of *Cinnamomum zeylanicum* and eugenol in peroxynitrite-induced oxidative processes. *J. Agric. Food Chem.* **2005**, *53*, 4762–4765. [CrossRef]

49. Viazis, S.; Akhtar, M.; Feirtag, J.; Diez-Gonzalez, F. Reduction of Escherichia coli O157:H7 viability on leafy green vegetables by treatment with a bacteriophage mixture and trans-cinnamaldehyde. *Food Microbiol.* **2011**, *28*, 149–157. [CrossRef]

50. Kwon, J.A.; Yu, C.B.; Park, H.D. Bacteriocidal effects and inhibition of cell separation of cinnamic aldehyde on *Bacillus cereus*. *Lett. Appl. Microbiol.* **2003**, *37*, 61–65. [CrossRef]

51. Friedman, M. Chemistry, antimicrobial mechanisms, and antibiotic activities of cinnamaldehyde against pathogenic bacteria in animal feeds and human foods. *J. Agric. Food. Chem.* **2017**, *65*, 10406–10423. [CrossRef]

52. Kim, H.K.; Kim, J.R.; Ahn, Y.J. Acaricidal activity of cinnamaldehyde and its congeners against *Tyrophagus putrescentiae* (Acari: Acaridae). *J. Stored Prod. Res.* **2004**, *40*, 55–63. [CrossRef]

53. Saad, M.M.G.; Gouda, N.A.A.; Abdelgaleil, S.A.M. Bioherbicidal activity of terpenes and phenylpropenes against *Echinochloa crus-galli*. *J. Environ. Sci. Health B* **2019**, *54*, 954–963. [CrossRef]

54. Roselló, J.; Sempere, F.; Sanz-Berzosa, I.; Chiralt, A.; Santamarina, M.P. Antifungal activity and potential use of essential oils against *Fusarium culmorum* and *Fusarium verticillioides*. *J. Essent. Oil Bear Plants* **2015**, *18*, 359–367. [CrossRef]

55. Santamarina, M.; Ibáñez, M.; Marqués, M.; Roselló, J.; Giménez, S.; Blázquez, M. Bioactivity of essential oils in phytopathogenic and post-harvest fungi control. *Nat. Prod. Res.* **2017**, *31*, 2675–2679. [CrossRef]

56. Krepker, M.; Shemesh, R.; Danin Poleg, Y.; Kashi, Y.; Vaxman, A.; Segal, E. Active food packaging films with synergistic antimicrobial activity. *Food Control* **2017**, *76*, 117–126. [CrossRef]

57. Ye, H.; Shen, S.; Xu, J.; Lin, S.; Yuan, Y.; Jones, G.S. Synergistic interactions of cinnamaldehyde in combination with carvacrol against food-borne bacteria. *Food Control* **2013**, *34*, 619–623. [CrossRef]

58. De Sousa, J.P.; de Azerêdo, G.A.; de Araújo Torres, R.; da Silva Vasconcelos, M.A.; da Conceição, M.L.; de Souza, E.L. Synergies of carvacrol and 1,8-cineole to inhibit bacteria associated with minimally processed vegetables. *Int. J. Food Microbiol.* **2012**, *154*, 145–151. [CrossRef]

59. Oddo, M. Effects of Different weed Control Practices on Soil Quality in Mediterranean Crops. Ph.D. Thesis, Universita' degli Studi di Palermo, Palermo, Italy, Polytechnic University of Valencia, Valencia, Spain, 2 October 2017.

60. Wu, H.; Walker, S.; Rollin, M.J.; Tan, D.K.Y.; Robinson, G.; Werth, J. Germination, persistence, and emergence of flaxleaf fleabane (*Conyza bonariensis* [L.] Cronquist). *Weed Biol. Manag.* **2007**, *7*, 192–199. [CrossRef]

61. Webber, C.L.; Shrefler, J.W. Pelargonic acid weed control parameters. *HortScience* **2006**, *41*, 1034. [CrossRef]

62. Velini, E.D.; Duke, S.O.; Trindade, M.B.; Meschede, D.K.; Carbonari, C.A. Modo de acao do Glyphosate (Mode of Action of Glyphosate in Portuguese). In *Glyphosate*; Velini, E.D., Meschede, D.K., Carbonari, C.A., Trindade, M.L.B., Eds.; Fundaçã de Estudos e Pesquisas Agricoloas e Florestais: Botucato-SP, Brazil, 2009; pp. 113–133.

63. Hoss, N.; Al-Khatib, K.; Peterson, D.; Loughin, T. Efficacy of glyphosate, glufosinate, and imazethapyr on selected weed species. *Weed Sci.* **2003**, *51*, 110–117. [CrossRef]

64. Jordan, D.; York, A.; Griffin, J.; Clay, P.; Vidrine, P.; Reynolds, D. Influence of Application Variables on Efficacy of Glyphosate. *Weed Technol.* **1997**, *11*, 354–362. [CrossRef]

65. Mithila, J.; Swanton, C.J.; Blackshaw, R.E.; Cathcart, R.J.; Hall, J.C. Physiological basis for reduced glyphosate efficacy on weeds grown under low soil nitrogen. *Weed Sci.* **2008**, *56*, 12–17. [CrossRef]

66. Sandberg, C.L.; Meggitt, W.F.; Penner, D. Absorption, translocation and metabolism of ^{14}C-glyphosate in several weed species. *Weed Res.* **1980**, *20*, 195–200. [CrossRef]

67. Lederer, B.; Fujimori, T.; Tsujino, Y.; Wakabayashi, K.; Bögera, P. Phytotoxic activity of middle-chain fatty acids II: Peroxidation and membrane effects. *Pestic. Biochem. Physiol.* **2004**, *80*, 151–156. [CrossRef]

68. Albuquerque, C.C.; Camara, T.R.; Sant'ana, A.E.G.; Ulisses, C.; Willadino, L.; Marcelino Júnior, C. Effects of the essential oil of *Lippia gracilis* Schauer on caulinary shoots of heliconia cultivated in vitro. *Rev. Bras. Plantas Med.* **2012**, *14*, 26–33. [CrossRef]
69. Hasanuzzaman, M.; Mohsin, S.M.; Borhannuddin Bhuyan, M.H.M.; Farha Bhuiyan, T.; Anee, T.I.; Awal, A.; Masud, C.; Nahar, K. Phytotoxicity, environmental and health hazards of herbicides: Challenges and ways forward. In *Agrochemicals Detection, Treatment and Remediation. Pesticides and Chemical Fertilizers*; Vara Prasad, M.N., Ed.; Butterworth Heinemann: Hyderabad, India, 2020; pp. 55–99. [CrossRef]

Article

Resilience Capacity Assessment of the Traditional Lima Bean (*Phaseolus lunatus* L.) Landraces Facing Climate Change

María Isabel Martínez-Nieto [1], Elena Estrelles [1], Josefa Prieto-Mossi [1], Josep Roselló [2] and Pilar Soriano [1,*]

[1] Jardí Botànic-ICBiBE, Universitat de València, Quart 80, 46008 Valencia, Spain; maria.isabel.martinez@uv.es (M.I.M.-N.); elena.estrelles@uv.es (E.E.); josefa.prieto-mossi@uv.es (J.P.-M.)
[2] Estación Experimental Agraria de Carcaixent, Carretera CV-5950 (Camino Del Barranquet), Carcaixent, 46740 Valencia, Spain; joseprosello@gmail.com
*   Correspondence: pilar.soriano@uv.es

Received: 30 March 2020; Accepted: 18 May 2020; Published: 26 May 2020

Abstract: Agriculture is highly exposed to climate warming, and promoting traditional cultivars constitutes an adaptive farming mechanism from climate change impacts. This study compared seed traits and adaptability in the germinative process, through temperature and drought response, between a commercial cultivar and Mediterranean *Phaseolus lunatus* L. landraces. Genetic and phylogenetic analyses were conducted to characterize local cultivars. Optimal germination temperature, and water stress tolerance, with increasing polyethylene glycol (PEG) concentrations, were initially evaluated. Base temperature, thermal time, base potential and hydrotime were calculated to compare the thermal and hydric responses and competitiveness among cultivars. Eight molecular markers were analyzed to calculate polymorphism and divergence parameters, of which three, together with South American species accessions, were used to construct a Bayesian phylogeny. No major differences were found in seed traits, rather different bicolored patterns. A preference for high temperatures and fast germination were observed. The 'Pintat' landrace showed marked competitiveness compared to the commercial cultivar when faced with temperature and drought tolerance. No genetic differences were found among the Valencian landraces and the phylogeny confirmed their Andean origin. Promoting landraces for their greater resilience is a tool to help overcome the worldwide challenge deriving from climate change and loss of agrobiodiversity.

Keywords: *Phaseolus*; landrace; seed; germination; drought tolerance; genetic approach; sustainable agriculture; climate change

---

1. Introduction

Agriculture is highly exposed to environmental changes, such as climate warming and aridification, as farming activities depend directly on climate conditions. Indeed, the role of agriculture is fundamentally improving natural resources management, rural development, food production and preserving environmental heritage by the conservation of seminatural habitats, landscape and biodiversity [1,2].

Accordingly, the cultivation and conservation of traditional landraces and crop diversification can be effective adaptation strategies to respond to these changing conditions [3], mainly given the increase in aridity and rainfall unpredictability that derive from these changes in environmental conditions.

Loss of crop diversity is a worldwide challenge. Modern cultivars have replaced local landraces, which are now threatened in food production systems, including cultural heritage, local knowledge and traditional farmer skills. This decline, supported by worldwide globalization, leads to reduced agrobiodiversity on a massive scale, and mainly in developed countries where the industrial food system moves towards genetic uniformity. With the disappearance of traditional species and cultivars, wide ranges of unharvested species also disappear. Promoting local cultivars, which are theoretically more competitive, is one of the major adaptive mechanisms of agriculture to climate change impacts [4–6].

Legumes, specifically *Phaseolus lunatus*, are considered one of the most valuable sources of nutrients in developing countries [7,8]. *Phaseolus lunatus* (Fabaceae), commonly known as "lima bean", and locally termed "garrofó", is the second genus *Phaseolus* species to follow *P. vulgaris* in terms of its economic interest. Attention is paid mainly to its food use worldwide [9], even though other relevant aspects are under study, such as the role on plant protection of the cyanogenic glycosides present in the seeds of this species [10,11].

Cultivated varieties have a South American origin and initially concentrated in northern Peru, where an in-depth selection was developed by the Inca civilization for a long time [12]. According to current germplasm and herbarium records, the conspecific wild ancestor of lima bean is widely distributed from Mexico to Argentina [13]. These landraces are classified into two major groups, Mesoamerican and Andean, according to their geographic origin and seed characteristics [14].

Although it originally comes from Mesoamerica and the Andes, it is currently cultivated throughout Latin America, the southern United States, Canada, and many other world regions including Mediterranean countries, where it is associated with local gastronomy.

On the coasts of the Mediterranean Basin, it is cultivated in warm sunny places in deep well-drained soil. Its strong roots allow plants to thrive on lands where other legumes cannot. It is a highly demanding crop with special requirements. These plants have a type IV climbing growth habit [15] with considerable vegetative development, which means they need a structure that supports, ventilates and illuminates their branches. Nowadays, this cultivation is maintained only for the value of its tender pods and dried grains, and for its special link with traditional cuisine. Currently, the traditional cultivars of this species are being replaced with commercial varieties and represent a testimonial crop in small areas on the European continent.

Hence the present research intends to compare and assess the resistance and adaptability of local cultivars and a commercial variety to face the environmental alterations deriving from climate change. The commercial cultivar is imported from Peru and can be purchased in most retail stores. Primitive landraces, known as 'Pintat', 'Ull de Perdiu' and 'Cella Negra', are traditionally used in the western Mediterranean Basin and are especially cultivated in east Spain (Valencian Community). The use of this species in the eastern Iberian Peninsula in that traditional cuisine is very ancient. Today, we only have references to using these four cultivars in the last 100 years in this region. Our main aim in this work was to recover forgotten crops for the future. In fact, some of the studied cultivars, in particular 'Cella Negra', have practically disappeared today and it has been very difficult to find seeds of this plant.

Furthermore, barcoding is a method to identify taxonomic units using short DNA sequences that allow the determination of the genetic polymorphism and divergences between them. The aim is to identify a region or a combination of regions capable of discriminating taxonomic units, such as species, subspecies, cultivars, or even gene lineages within species [16] and references therein]. Although chloroplast DNA barcoding is utilized mainly to identify plant species, its application can be extended to the food industry, evolution studies and forensics [17]. Various regions of the plastid genome have been proposed to serve as DNA barcodes in plants, such as those put forward by Shaw et al. [18] or Taberlet et al. [19], internal transcribed spacers (ITS) [20] or other specific genes like FRO1 and Phs7 used in legumes phylogenetics by Diniz et al. [21]. This method has been useful in Leguminosae phylogenetics and wild gene pool identifications in *Phaseolus lunatus* [16,21–23]. Thus, it might be a useful tool for typifying local landraces.

This study focuses on seed characterization and providing new information about seed response to temperature and water stress tolerance, estimated during the germinative process, in the

Phaseolus lunatus traditional cultivars from Mediterranean Europe in line with the future global warming and water deficit scenario.

It also aims to characterize molecularly cultivars—by determining the genetic polymorphism and divergences among local, traditional and commercial, as well as American accessions—of *P. lunatus* in an attempt to genetically delimitate landraces, and to find the potential correlation of these genetic characteristics and germination responses. Moreover, the phylogenetic origin of the Valencian cultivars is studied as part of its molecular characterization.

2. Materials and Methods

Four lima bean (*Phaseolus lunatus*) cultivars were tested, three of which were from local Valencian traditional crops ('Pintat', 'Ull de Perdiu' and 'Cella Negra'), mainly provided by the Estación Experimental Agraria de Carcaixent (EEA-Carcaixent) (province of Valencia, Spain). A fourth commercial cultivar imported from Peru to Spain (hereinafter referred to as 'Peru') was bought for the study. The seeds provided by the EEA-Carcaixent were collected during the previous season, nearly one year before starting the germination tests. We did not collect data on the seeds of commercial origin.

2.1. Seed Features

Seed dimensions were measured on a digital image using the ImageJ software [24]. Seed weight was determined by an Orion Cahn C-33 microbalance. All the data were obtained from $n = 50$ seeds from each cultivar.

In order to detect differences in variance levels and to identify homogeneous groups, a one-way ANOVA and Tukey's test ($p < 0.05$) were applied, respectively, for each parameter among the different cultivars.

2.2. Germination Assays

Seed germination assays were performed with the 'Pintat' and the commercial cultivar, 'Peru', for the low seed availability of the rarest landraces, 'Ull de Perdiu' and 'Cella Negra'. Sporadic tests were conducted with them to provide the initial data for future studies. Data were included as Supplementary data.

Tests were carried out using four replications of 10–15 seeds (depending on seed availability) per treatment for each cultivar. Tests were conducted on 14-cm diameter Petri dishes with paper filters kept in climate-controlled cabinets. Illumination was provided by daylight fluorescent tubes with a 12-h photoperiod and a mean irradiance of 100 μmol·m^{-2}·s^{-1}. The germination process was evaluated for 15 days. Germinated seeds were counted daily.

Firstly, the optimum germination conditions for successive experiments were set. Temperature screening, using six constant temperatures (15 °C, 20 °C, 25 °C, 30 °C, 35 °C, 40 °C), was applied to determine the optimal germination temperature.

The water stress effect was evaluated by the controlled osmotic potential levels generated using polyethylene glycol (PEG 6000) solution at 30 °C according to Villela et al. [25] to obtain 0 (control), −1, −2, −3, −4 and −5 bar. In order to minimize the evaporation and concentration of the effect of solutions and to maintain the known osmotic potential stable, seeds were moistened every 24 h with fresh PEG solutions and plates were kept in double plastic zip lock bags. After 15 days, non-germinated seeds were transferred to distilled water. Thereby, germination capacity recovery was tested to check the potential influence of PEG exposure on seed germination behavior of *Phaseolus lunatus* cultivars.

Germination Percentage and Mean Germination Time (MGT) were considered to compare seed responses. The base temperature (Tb), by back extrapolation [26], and the thermal time requirement [27] were also calculated to compare thermal responses. Then, the base potential (Ψb) and hydrotime (θ) for each cultivar were calculated [28,29].

Variance levels and homogeneous groups were determined by the one-way ANOVA and Tukey's test ($p < 0.05$), respectively, for each parameter among cultivars.

2.3. Genetic Assays

2.3.1. Plant Material and DNA Extraction

The plant material used in the molecular analysis was obtained from the seeds germinated in Germplasm Bank (UV) or Estación Experimental Agraria (EEA-Carcaixent). Eight individuals of each cultivar were analyzed, except for 'Cella Negra', where only five individuals were available at the time of when the genetic assays were done. For the 'Ull de Perdiu' and 'Pintat', we studied two different samples; one seed accession was obtained from the EEA-Carcaixent, while the other was bought from a traditional market. All the 'Cella Negra' seeds came from EEA-Carcaixent and all the 'Peru' ones were obtained from a market as local farmers do not traditionally cultivate them. All the accessions were identified according to seeds' distinctive morphological features. DNA was extracted from young leaves using Doyle and Doyle [30] protocol, modified by Soltis laboratory (2002; https://www.floridamuseum.ufl.edu/wp-content/uploads/sites/95/2014/02/CTAB-DNA-Extraction.pdf). In order to phylogenetically locate the Valencian cultivars, all the accessions provided by Serrano-Serrano et al. [22] in NCBI were used.

2.3.2. Molecular Analyses

A pool of five chloroplast and three nuclear markers (see Table 1) was analyzed to characterize the Valencian landraces. These markers were variable in other studies related specifically to *Phaseolus lunatus* or *Phaseolus* spp. [21–23]. A standard PCR protocol following GoTaq® Polymerase (Promega, Madison, WI, USA) instructions was used for all the markers, except for *Phs7* and *FRO3*, which were amplified following Diniz et al. [21]. The PCR products were purified using the Real Clean PCR Kit (Durviz, Valencia, Spain) and sequenced in an ABI 3100 Genetic Analyzer with the ABI BigDye Terminator Cycle Sequencing Ready Reaction Kit (Applied Biosystems, Foster City, CA, USA).

Table 1. The markers analyzed for the *P. lunatus* Valencian cultivars, primer names, Tm (primer melting temperature) and original references in which they were described.

Marker	Primer Names	Tm	Reference
atpB-rbcL	atpB-f rbcL-r	55	[31]
trnL-trnF	trnL (UAA) 3′ exon f trnF (GAA) r	58	[19]
trnL intron + *trnL-trnF*	trnL (UAA) 5′ exon f trnF (GAA) r	58	[19]
rpoB-trnC	rpoB-f trnC-r	55	[18]
psbA-trnH	psba-f trnH-r	56	[32]
ITS	ITS1-f ITS4-r	58	[20]
Phs7	Phs7-f Phs7-r	62	[21]
FRO3	FRO3-f FRO3-r	62	[21]

2.3.3. Phylogenetic Analyses

As Serrano-Serrano et al. [22] used ITS, *Atpb-rbcL* and *trnL-trnF* fragments to construct a wide *P. lunatus* phylogeny, these fragments were employed to locate the origin of the Valencian cultivars. Two individuals of each Valencian landrace and all the accessions provided by Serrano-Serrano et al. [22] in NCBI were analyzed. MAFFT v. 7.402 [33,34] was utilized to generate a multiple sequence alignment. The preconfigured MAFFT strategy, which favors accuracy with the FFT-NS-I algorithm (an iterative refinement method that performs 1000 iterations), and default parameters were selected. The ambiguously aligned regions were automatically dealt with using GBlocks v. 091b [35] by implementing the least stringent parameters, but allowing for gaps in 50% of sequences. (NCBI accession numbers: MT072230–MT072258, ITS; MT080626–MT080654, *atpB-rbcL*; MT090972–MT091000, *trnL-trnF*; MT110491–MT110519, *rpoB-trnC*; MT124955–MT124983, *psbA-trnH*; MT154089– MT154117, phs7; MT154118–MT154146, FRO3).

A Bayesian phylogenetic MCMC analysis was run with MrBayes v. 3.2.2 [36]. Indels were coded with SeqState v. 1.4.1 [37] according to modified complex coding. The coded indels were considered to be a partition of standard data (states = 0, 1, 2, 3, ?), with the gamma rate and hyperprior fixed at 1.0 to allow different stationary state frequency proportions to be explored by the MCMC procedure. The optimal substitution models for the nucleotide section were inferred with PartitionFinder2 [38] by considering a model with linked branch lengths for the codificant and non-codificant regions of nrITS and chloroplast fragments, respectively, and using the Bayesian information criterion (BIC). Finally, three partitions were considered: two within the ITS (ITS1 + ITS2 and 5.8S), as well as *Atpb-rbcL + trnL-trnF*. This analysis favored the HKY + G model for the ITS1 + ITS2 partition, K80 + I for 5.8S, and also GTR + I for the chloroplast region. Then, a MrBayes analysis was conducted with two parallel and simultaneous four-chain runs, executed over 5×10^6 generations, starting with a random tree, and sampling after every 500th step. The first 25% of the data was discarded as burn-in. The 50% majority-rule consensus tree and the corresponding posterior probabilities were calculated from the remaining trees. Chain convergence was assessed by ensuring that the average standard deviation or split frequencies (ASDSF) values were below 0.01, and the potential scale reduction factor (PSRF) values approached 1.00. iTOL v. 4.4.2 [39,40] was used to construct the 50% majority rule consensus tree. The programs MAFFT, MrBayes and PartitionFinder2 were hosted at the CIPRES Science Gateway [41].

2.3.4. DNA Polymorphism and Divergence

The MAFFT original alignment without outgroups was employed to evaluate DNA polymorphism and divergence by taking in account the studied Valencian cultivars and all the accessions, including those used by Serrano-Serrano et al. [22], respectively. All the analyzed markers were utilized to study the Valencian landraces, as well as ITS, *Atpb-rbcL* and *trnL-trnF*, for the whole analysis. Five parameters were calculated by DnaSP v. 6 [42]: segregating sites (s), nucleotide diversity (π), number of haplotypes (h), haplotype diversity (H_d), and the nucleotide genetic differentiation estimate K_{st}.

3. Results

3.1. Seed Features

The seed dimensions of these four *Phaseolus lunatus* cultivars were similar (Table 2). It is noteworthy that the 'Pintat' seeds obtained higher values for the length and width parameters, and had a more rounded contour. The thickness analysis indicated significant differences among cultivars, with the lowest values for the traditional landraces. The 'Peru' and 'Pintat' seeds were the heaviest, while the 'Cella Negra' seeds were lightweight.

Seed coat color is an important consumer trait. In this group, it is a relevant distinctive character for these traditional cultivars (Figure 1; Table 2). The studied commercial cultivar, identified herein as 'Peru', has a completely white seed coat showing no type of pigmentation. The traditional 'Pintat' depicts an irregular spotted pigmentation over the whole external cover, from dark maroon to brown, depending on the maturation stage. The 'Ull de Perdiu' cultivar has a characteristic black eye surrounding the hilum seed zone. Finally, the cultivar known as locally 'Cella Negra' is identified by having a dark brown to black seed tip close to the embryo radicle lobe.

Table 2. Seed morphological features for the different studied cultivars. Length (L), width (W) and their relation (L/W), thickness, as well as weight and color trait of seed coat, are indicated. The same letters indicate homogeneous groups among temperatures ($p < 0.05$) for each cultivar.

	'Peru'	'Pintat'	'Ull de Perdiu'	'Cella Negra'
L (mm)	25.3 ± 0.21 b	26.4 ± 0.13 a	24.8 ± 0.17 b	25.0 ± 0.18 b
W (mm)	15.5 ± 0.14 b	17.3 ± 0.09 a	15.7 ± 0.12 b	15.9 ± 0.14 b
L/W	1.63 ± 0.16 a	1.53 ± 0.10 b	1.58 ± 0.14 ab	1.58 ± 0.21 ab
Thickness (mm)	6.72 ± 0.45 a	5.60 ± 0.42 c	5.96 ± 0.81 bc	6.37 ± 0.72 ab
Weight (g)	1.82 ± 0.07 a	1.81 ± 0.10 a	1.75 ± 0.39 ab	1.54 ± 0.15 b
Pigmentation	No	Yes	Yes	Yes

Figure 1. Seed morphological traits and pigmentation for the studied *Phaseolus lunatus* cultivars; 'Peru'; 'Pintat'; 'Ull de Perdiu'; 'Cella Negra'.

3.2. Germination Assays

3.2.1. Germination Response to Temperature

High germination percentages were achieved at almost all the tested temperatures. The lowest values were for 35 °C in the two studied cultivars, while no germination was observed in any of them above this temperature.

After taking into account the values obtained for the germination percentage and mean germination time, the optimal germination temperature for the studied group of *Phaseolus lunatus* cultivars was set at 30 °C (Figure 2; Table 3). Good results for germination percentages were also obtained at 15 °C and 25 °C, mainly for the 'Pintat' cultivar, but germination was slower in both cases. The values with the same letters did not significantly differ at the 5% level. No significant differences were found when comparing germination velocities among the cultivars at each specific temperature.

The regression lines, indicating the response of germination velocity to increasing temperature (Figure 3), showed a steeper slope for the local 'Pintat' cultivar than for the commercial one, labelled as 'Peru', given the shorter mean germination time; i.e., faster germination. This effect became evident at the temperatures exceeding 19 °C. When the thermal time, S and Tb parameters were calculated from the regression line equations, the 'Pintat' seeds gave values of 131.6 °C·day^{-1} and 5.2 °C respectively, with 185 °C·day^{-1} and −15.0 °C for 'Peru'.

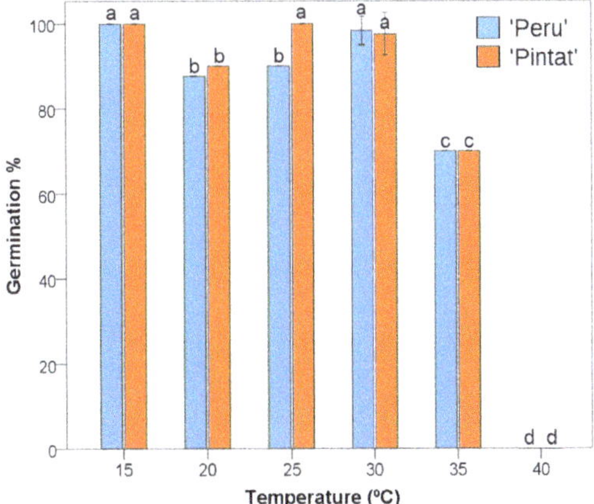

Figure 2. The germination percentage values obtained at different temperatures for the studied *Phaseolus lunatus* cultivars. The same letters indicate homogeneous groups ($p < 0.05$).

Table 3. Mean germination time (days) at different temperatures (°C) for 'Pintat' and 'Peru' cultivars. The same letters indicate homogeneous groups among temperatures ($p < 0.05$) for each cultivar.

	15 °C	20 °C	25 °C	30 °C	35 °C	40 °C
'Peru'	5.6 ± 0.2 b	5.9 ± 0.5 b	5.0 ± 0.3 b	3.9 ± 0.6 a	5.1 ± 0.6 b	-
'Pintat'	6.1 ± 0.2 cd	5.6 ± 0.2 c	4.6 ± 0.2 b	3.6 ± 0.4 a	6.3 ± 0.5 d	-

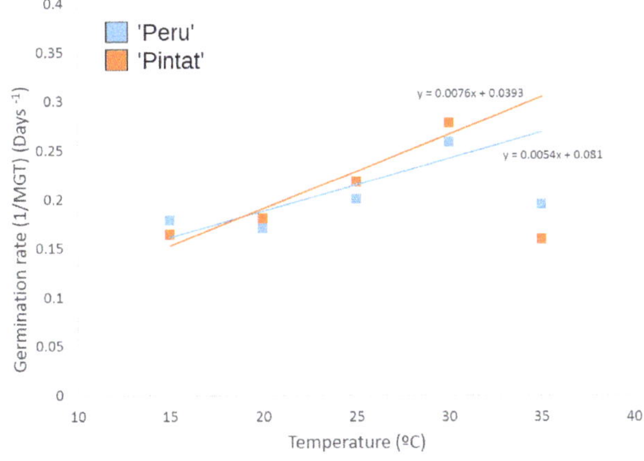

Figure 3. The linear regression of the germination rates (MGT) related to the tested temperatures for two cultivars: 'Pintat' and 'Peru'.

3.2.2. Germination Response to Drought Stress

Characteristically, germination was affected by rising PEG concentrations. In both cases, a drastic reduction in germination was recorded from −4 bars, and no germination took place at −5 bar. However, Figure 4 and Table 4 show better tolerance to induced water stress for the 'Pintat' cultivar, which obtained higher germination percentages and velocity under all the tested conditions. At −2 bar, no significant differences appeared in relation to the control for 'Pintat' landrace, while germination lowered by 28.8% for the cultivar 'Peru'.

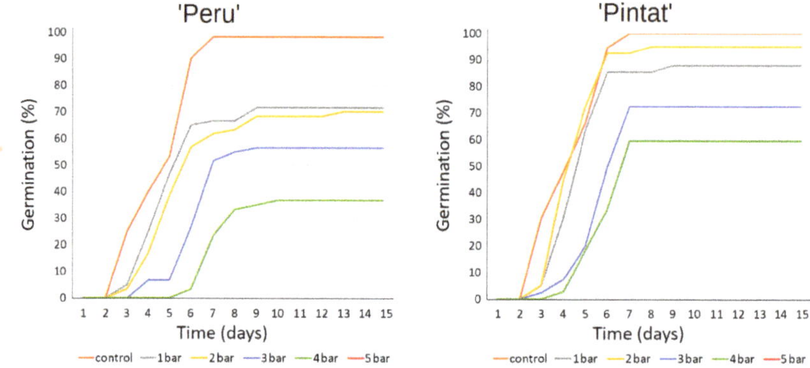

Figure 4. The accumulative germination percentages of the studied cultivars at increasing osmotic pressures obtained with PEG from 0, the control, to a maximum of −5 bar.

Table 4. Mean germination time (MGT), expressed as days, for the *Phaseolus lunatus* cultivars analyzed at increasing PEG 6000 concentrations. The same letters indicate homogeneous groups among cultivars and the tested concentrations ($p < 0.05$).

	Osmotic Potential (Bar)				
	0	**−1**	**−2**	**−3**	**−4**
'Peru'	3.9 ± 0.6 a	4.2 ± 0.4 ab	4.8 ± 1.2 ab	5.5 ± 0.6 abc	6.5 ± 0.6 bc
'Pintat'	3.6 ± 0.4 a	3.9 ± 0.8 a	3.8 ± 0.2 a	5.1 ± 0.2 ab	5.2 ± 0.6 ab

The germination test conducted at increasing water stress pointed out differences in the seeds of the studied cultivars for their physiological potential to face water deficit. A drastic drop in germination was recorded at −4 and −5 bars. The cultivar 'Pintat' demonstrated better tolerance to water stress, which obtained values above 50% for the germination percentage for all the tested osmotic potentials up to −4 bar.

The 'Pintat' cultivar displayed a faster response to germination velocity under all the conditions, and only showed a clear decrease from −2 bar (Table 4; Figure 5).

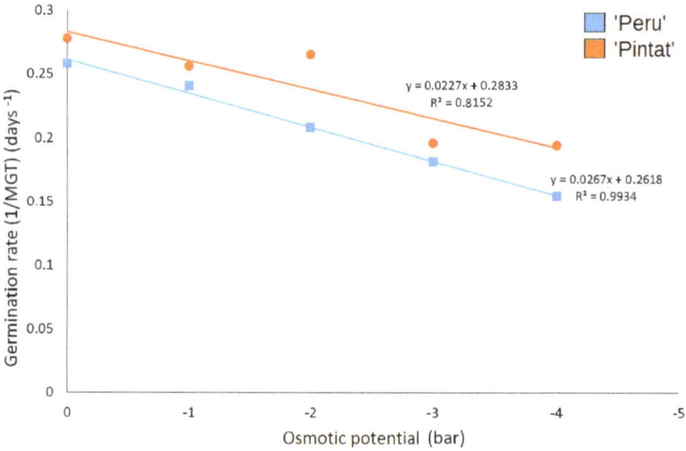

Figure 5. The relation between osmotic potential (bar) and germination rate (1/MGT) for the studied cultivars at 30 °C.

The hydrotime calculated from the linear regression slope was 37.5 and 44.1 bar·day for the cultivars 'Peru' and 'Pintat', respectively. The theoretical values calculated for the minimum osmotic potential (Ψb) at which radicle emergence was prevented were respectively −9.8 and −12.5 bar for these same cultivars. When PEG exposure ended, non-germinated seeds were transferred to the non-stressed medium. After 15 days of incubation in distilled water, no recovery was observed at any tested concentration.

3.3. Genetic Assays

The dataset herein considered comprised new 29 sequences, including three nuclear and five chloroplastic concatenated fragments that belong to the four more common *P. lunatus* cultivars in Spain. The phylogenetic analyses included the ITS, *Atpb-rbcL* and *trnL-trnF* fragments, two individuals of each Valencian landrace and all the accessions provided by Serrano-Serrano et al. [22] in NCBI. Seventy-eight individuals were analyzed. The MAFFT algorithm produced an alignment of 1828 bp with outgroups and 1410 without them. After the automatic removal of ambiguously aligned positions in GBlocks v. 0.91b, 97% (1781 nucleotides) of the original length, 13 selected blocks were kept after taking the outgroups into account. This final alignment included 110 variable positions, of which 73 were parsimony informative and 37 were singletons. The MrBayes analysis reached an average standard deviation of split frequencies of 0.01 after 156 generations. The resulting topology is presented in Figure 6, where the Valencian cultivars were clustered in the AI gene pool, together with the Andean Cordillera accessions from Ecuador and Peru with high clade support (BI $\geq$ 0.9). These landraces also formed a high supported clade inside the AI gene pool (BI = 0.98). The main groups also displayed good clade support (BI $\geq$ 0.9), except for the MII gene pool (BI = 0.61), which was clustered in a wider and well-supported Mesoamerican group, split inside.

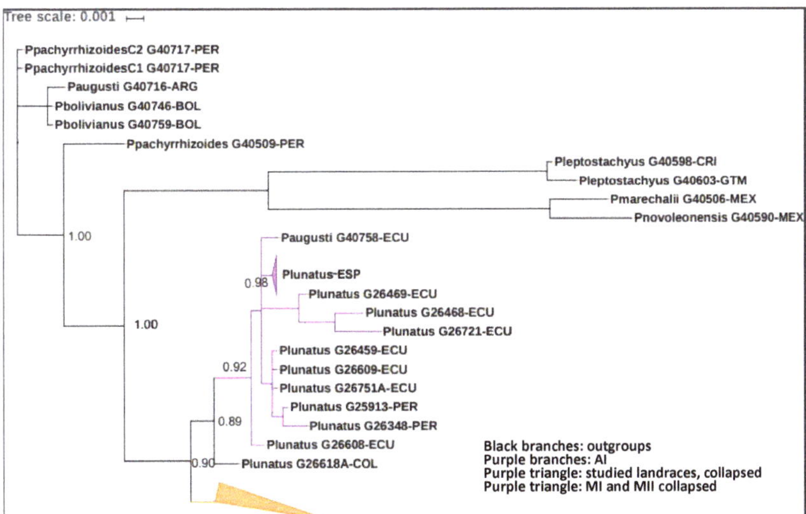

Figure 6. Phylogram depicting the phylogenetic relations among the *P. lunatus* accessions from Spain and South America obtained with MrBayes and based on nrITS and cpDNA data. Support values are given for the main nodes (BI). Colors correspond to the gene pools for wild *P. lunatus*: black branches belong to outgroups, purple branches to AI (Andean I) and the purple triangle inside represents collapsed clades of Valencian landraces (local and 'Peru') as they were almost genetically identical, the orange triangle represents the MI (Mesoamerican I) and MII (Mesoamerican II) collapsed clades as they did not provide any relevant information for our purposes. The whole tree is shown in the Supplementary Material (Figure S2). COL = Colombia, ECU = Ecuador, PER = Peru, ESP = Spain.

Polymorphism and divergence analyses were conducted throughout two groups: only the Valencian cultivars and Valencian and South American cultivars from Serrano-Serrano et al. [22], excluding outgroups. All the analyzed fragments were used in the 29 sequences of the Valencian group with very low genetic diversity estimates. There were no gaps and 4080 sites, of which only four were variable and none showed any pattern of change. This group presented nine haplotypes, 3.8×10^{-4} of nucleotide diversity and a non-significant genetic differentiation estimate K_{st} of 0.022. The group including the South American varieties comprised 67 sequences of concatenated ITS, *Atpb-rbcL* and *trnL-trnF* fragments, 1800 sites and 1413 sites excluding gaps, 44 of which were variable. The group showed 37 haplotypes, a nucleotide diversity of 3.89×10^{-3} and a significant genetic differentiation estimate K_{st} of 0.528 (Table 5).

Table 5. The polymorphism and divergence data of the two *P. lunatus* cultivars groups. The Valencian landraces included the most frequently used cultivars in Spain ('Peru', 'Pintat', 'Ull de Perdiu', 'Cella Negra'). The Valencian + South American group included the Valencian and South American accessions from Serrano-Serrano et al. [22]. N: number of individuals, n: number of sites, n': number of sites excluding sites with gaps/missing data, S: number of variable sites, h: number of haplotypes, Π (s.d.): nucleotide diversity and standard deviation in brackets, H_d (s.d.): haplotype diversity and standard deviation in brackets, K_{st}: genetic differentiation estimate and its *p*-value (n.s.: non-significant, ***: $p < 0.001$).

	N	n	n′	S	H	Π (s.d.)	H_d (s.d.)	K_{st}
Valencian cultivars	29	4080	4080	4	9	$3.8 \cdot \times 10^{-4}$ (4×10^{-5})	0.862 (0.035)	0.022 n.s.
Valencian + S. American Accessions	67	1800	1413	44	37	$3.9 \cdot \times 10^{-3}$ (2.5×10^{-4})	0.954 $(2 \cdot 10^{-4})$	0.528 ***

4. Discussion

A landrace differs from a variety that has been selectively modified to improve particular characteristics. These traditional landraces, cultivated continuously for years, are severely threatened by genetic extinction because they are replaced with modern varieties, selected mainly for their higher productivity instead of their resistance to climate change consequences [43].

Currently, the commercial white-seed bean ('Peru') is the cheapest and the most widely sold among lima beans in the Valencian Community, and probably the only one known to most people. Seeds of 'Pintat', and rarely of 'Ull de Perdiu' are sold only in a few local markets, while the cultivar 'Cella Negra' has practically disappeared. The EEA-Carcaixent conserves and multiplies a few accessions of the cultivar 'Cella Negra' for its preservation, from the few seeds that it has been able to find from some farmers who still cultivate it for their own use. We focused our research according to the assumption that the commercial predominance of the different cultivars is not a question based on consumer preferences, but on local farmers' low profitability.

Local landraces are associated with one specific geographical location and, therefore, present climatic adaptability. They are generally better adapted to abiotic stress than modern cultivars [44] and supporting the recovery of their cultivation can mean advantages to face the climate change threat, especially if consumer demand increases. Hence, this climatic adaptability reveals the need to conserve the landrace germplasm as a means to provide information about adaptations to drought and heat stress, and because it constitutes a tool to identify stress-tolerant alleles to improve productivity when faced with climate change [45,46].

Baudoin [47], after thoroughly reviewing the diversity of *Phaseolus lunatus*, already indicated this species as an underexploited crop with a very high cultivation potential given its ability to withstand several types of stress, including severe drought. Moreover, the necessity of carrying out preservation programs for germplasm banks of wild forms and landraces was highlighted.

Regarding seed morphology, clear variability that depends on cultivars is described in the literature. Additionally, variations in dimensions, test patterns and color in cultivars from different countries are known as Potato with small rounded seeds, along with Sieva, with medium-sized reniform seeds, while Andean ones are known as Big Lima and have large, but flat, seeds [14].

In the four studied landraces, no major differences in seed dimensions appeared. Seeds have the morphological characteristics of most of the individuals cultivated for commercial purposes, mainly with big, attractive and nutritional seeds, which indicate their Andean origin.

Regarding seed color and according to bibliographic references, the most frequent seed coat color of the cultivated plants differs depending on the considered geographic area. White seeds are one of the most frequently found among Cuban cultivars [48], while a predominant bicolored pattern is observed in the cultivars grown in Peru [12]. The studied landraces exhibit different bicolored patterns, which also agrees with their geographical provenance.

The seeds of the studied landraces underwent fast germination with no primary dormancy trait, even though dormancy was detected in some colored lima beans [49]. The marked preference for high temperatures in this species stood out, as clearly evidenced by our results with an optimal germination response at 30 °C. Indeed, Polock and Toole [50] and Polock [51] indicated that temperatures below 25 °C in the imbibition phase can be harmful. Other authors have indicated a good response of lima beans at high temperatures (25 °C and 30 °C), even when they were exposed to different salt concentration levels [52].

When we compared germination behavior of the commercial cultivar and the landrace 'Pintat' for the different tested temperature regimes, a stronger competitiveness of the local cultivar was observed from 19 °C to the optimum temperature, close to 30 °C.

Drought is one of the most important problems in agriculture as it leads to reduced yields and loss of crops. Water availability is essential for plants, as they need a good water supply throughout their life cycle. Therefore, water deficit in plants affects all phases of their development, physiological processes, growth and production which, under extreme conditions, can lead to plants dying [43,53,54]. Like the exposed response to temperatures, our results also support the hypothesis of the higher tolerance of those landraces cultivated for years that better adapt to changes in environmental conditions deriving from the Mediterranean climate. In fact, the 'Pintat' landrace was the most tolerant to water stress, simulated by lowering the osmotic potential of PEG solutions. In fact, the thermal time and hydrotime parameters have proven to be good discrimination tools to identify drought- and high temperature-tolerant common bean cultivars [55].

Conversely, DNA barcoding has not found any differences between the local and commercial cultivars used in the Valencian Community, not even when using different sorts: chloroplastic, nuclear, codificant and non-codificant markers. Nevertheless, their origin can be clearly situated. Recent phylogenetic studies have used genome-wide SNP markers polymorphisms [56] to indicate that the wild lima bean is structured into three gene pools, as previously proposed by Serrano-Serrano et al. [22]: the Mesoamerican one (MI); the Mesoamerican two (MII); the Andean one (AI). Their geographic ranges do not generally overlap. In addition, Chacón-Sánchez and Martínez-Castillo [56] also suggest the existence of another Andean gene pool (AII) in central Colombia. Our phylogenetic analyses, based on the data of Serrano-Serrano et al. [22], placed the Valencian cultivars in AI in relation to the 'Big Lima' morphology. These cultivars were phylogenetically grouped with the Andean Cordillera accessions from Ecuador, this being the domestication area of the Andean gene pool located between Ecuador and northern Peru [22,57].

For the Mesoamerican landraces, recent evidence indicates a scenario of a single domestication event in the gene pool MI for all the Mesoamerican landraces, perhaps in central-western Mexico, and the subsequent admixture among landraces and wild populations within the distribution range of gene pool MII, which gave rise to the MII landraces [56]. Therefore, and according to our results, these previous studies have shown that domestication was accompanied by strong founder effects that decreased the genetic diversity of the landraces in the Andes and MI of Mesoamerica. Thus, low polymorphism and divergence statistics have been found in the cultivars used in the Valencian Community (Spain), even between traditional ('Pintat', 'Ull de Perdiu', 'Cella Negra') and commercial ones ('Peru').

However, they all came from the same original gene pool in which a split occurred, as the earliest, when Europeans arrived in America 500 years ago, which is a negligible time in evolutionary terms.

Although other genome-wide barcoding techniques can be used [56,58], the different responses of these genetically close landraces can be explained by epigenetic mechanisms or by a few genes that play a relevant role in crop stress responses [59]. Indeed, rather than DNA barcoding, the search for relevant genes and local landrace alleles related to water stress tolerance could lead to new research works to help preserve these cultivars, by identifying the particular genetic features and their purity. When considering crop tolerance to overcome climate change-related stresses, natural variance among different cultivars can act as genetic reservoir for adaptation capability [60]. This idea, combined with an interest in providing added value to local landraces to defend their use recovery and agro-biodiversity conservation, could supply key future tools that promote local activities to face climate change effects on crops in order to contribute to the auto-sustainability of agronomy activities.

Supplementary Materials: The following are available online at http://www.mdpi.com/2073-4395/10/6/758/s1.

Author Contributions: Conceptualization, E.E., J.R. and P.S.; Investigation, M.I.M.-N., E.E. and P.S.; Methodology, M.I.M.-N., E.E., J.P.-M. and P.S.; Project administration, E.E. and P.S.; Visualization, M.I.M.-N., E.E. and P.S.; Writing—Original draft, M.I.M.-N., E.E. and P.S.; Writing—Review & Editing, M.I.M.-N., E.E., J.P.-M., J.R. and P.S. All authors have read and agreed to the published version of the manuscript.

Funding: This research received no external funding.

Acknowledgments: The authors wish to thank the Estación Experimental Agraria de Carcaixent (Valencia, Spain) for its support, particularly Fernando Amorós Ortega for providing us with some of the plant materials (seeds and leaves) used in the experiments. The authors sincerely acknowledge the valuable comments, corrections and suggestions made by anonymous reviewers that have significantly improve the manuscript.

Conflicts of Interest: The authors declare no conflict of interest.

References

1. Crop Production Statistics at Regional Level, EUROSTAT. 2014. Available online: http://ec.europa.eu/eurostat/statistics-explained/index.php/Agricultural_production_-_crops (accessed on 20 December 2019).
2. Knox, J.; Daccache, A.; Hess, T.; Haro, D. Meta-analysis of climate impacts and uncertainty on crop yields in Europe. *Environ. Res. Lett.* **2016**, *11*, 113004. [CrossRef]
3. Climate Change and Agriculture. Houses of Parliament. POSTnote 600. Available online: http://researchbriefings.files.parliament.uk/documents/POST-PN-0600/POST-PN-0600.pdf (accessed on 18 December 2019).
4. Arteaga, S.; Al Hassan, M.; Chaminda Bandara, W.; Yabor, L.; Llinares, J.; Boscaiu, M.; Vicente, O. Screening for salt tolerance in four local varieties of *Phaseolus lunatus* from Spain. *Agriculture* **2018**, *8*, 201. [CrossRef]
5. van de Wouw, M.; Kik, C.; van Hintum, T.; van Treuren, R.; Visser, B. Genetic erosion in crops: Concept, research results and challenges. *Plant Genet. Resour.* **2010**, *8*, 1–15. [CrossRef]
6. What is agrobiodiversity? *Food and Agricultural Organization of the UN*; FAO: Rome, Italy, 2004. Available online: http://www.fao.org/3/a-y5609e.pdf (accessed on 18 December 2019).
7. Seidu, K.T.; Osundahunsi, O.F.; Osamudiamen, P.M. Nutrients assessment of some lima bean varieties grown in southwest Nigeria. *Int. Food Res. J.* **2018**, *25*, 848–853.
8. Seidu, K.T.; Osundahunsi, O.F.; Olaleye, M.T.; Oluwalana, I.B. Amino acid composition, mineral contents and protein solubility of some lima bean (*Phaseolus lunatus* L. Walp) seeds coat. *Food Res. Int.* **2015**, *73*, 130–134. [CrossRef]
9. López-Alcocer, J.J.; Lépiz-Ildefonso, R.; González-Eguiarte, D.R.; Rodríguez-Macías, R.; López-Alcocer, E. Morphological variability of wild *Phaseolus lunatus* from the western region of Mexico. *Rev. Fitotec. Mex.* **2016**, *39*, 49–58.
10. Cuny, M.A.; La Forgia, D.; Desurmont, G.A.; Glauser, G.; Benrey, B. Role of cyanogenic glycosides in the seeds of wild lima bean, *Phaseolus lunatus*: Defense, plant nutrition or both? *Planta* **2019**, *250*, 1281–1292. [CrossRef]
11. Gleadow, R.M.; Woodrow, I.E. Constraints on effectiveness of cyanogenic glycosides in herbivore defense. *J. Chem. Ecol.* **2002**, *28*, 1301–1313. [CrossRef]
12. Pesantes-Vera, M.F.; León-Alcántara, E.; De La Cruz-Araujo, E.; Rodríguez-Soto, J.C. Variabilidad morfo-agronómica en poblaciones de pallar, *Phaseolus lunatus*, cultivado en condiciones de Costa de la Provincia de Trujillo (Perú). *REBIOL* **2015**, *35*, 29–38.

13. Debouck, D.G. *Notes sur Les Différents Taxons de Phaseolus à Partir des Herbiers. Cahiers de Phaséologie—Section Paniculati*; International Center for Tropical Agriculture (CIAT): Cali, Colombia, 2008; p. 233. Available online: http://www.ciat.cgiar.org/urg (accessed on 11 February 2019).
14. Baudet, J.C. The taxonomic status of the cultivated types of Lima bean (*Phaseolus lunatus* L.). *Trop. Grain Legume Bull.* **1977**, *7*, 29–30.
15. Checa, O.; Ceballos, H.; Blair, M.W. Generation means analysis of climbing ability in common bean (*Phaseolus vulgaris* L.). *J. Hered.* **2006**, *97*, 456–465. [CrossRef] [PubMed]
16. Madesis, P.; Ganopoulos, I.; Ralli, P.; Tsaftaris, A. Barcoding the major Mediterranean leguminous crops by combining universal chloroplast and nuclear DNA sequence targets. *Genet. Mol. Res* **2012**, *11*, 2548–2558. [CrossRef] [PubMed]
17. Ferri, G.; Alù, M.; Corradini, B.; Beduschi, G. Forensic botany: Species identification of botanical trace evidence using a multigene barcoding approach. *Int. J. Legal Med.* **2009**, *123*, 395–401. [CrossRef] [PubMed]
18. Shaw, J.; Lickey, E.B.; Beck, J.T.; Farmer, S.B.; Liu, W.; Miller, J.; Siripun, K.C.; Winder, C.T.; Schilling, E.E.; Small, R.L. The tortoise and the hare II: Relative utility of 21 noncoding chloroplast DNA sequences for phylogenetic analysis. *Am. J. Bot.* **2005**, *92*, 142–166. [CrossRef]
19. Taberlet, P.; Gielly, L.; Pautou, G.; Bouvet, J. Universal primers for amplification of three non-coding regions of chloroplast DNA. *Plant Mol. Biol.* **1991**, *17*, 1105–1109. [CrossRef]
20. White, T.J.; Bruns, T.D.; Lee, S.; Taylor, J. Amplification and direct sequencing of fungal ribosomal RNA genes for phylogenetics. In *PCR Protocols*; Innis, M.A., Gelfand, D.H., Sninsky, J.J., White, T.J., Eds.; Academic Press: San Diego, CA, USA, 1990; pp. 315–322.
21. Diniz, A.L.; Zucchi, M.I.; Santini, L.; Benchimol-Reis, L.L.; Fungaro, M.H.P.; Vieira, M.L.C. Nucleotide diversity based on phaseolin and iron reductase genes in common bean accessions of different geographical origins. *Genome* **2014**, *57*, 69–77. [CrossRef]
22. Serrano-Serrano, M.L.; Hernández-Torres, J.; Castillo-Villamizar, G.; Debouck, D.G.; Sánchez, M.I.C. Gene pools in wild Lima bean (*Phaseolus lunatus* L.) from the Americas: Evidences for an Andean origin and past migrations. *Mol. Phylogenet. Evol.* **2010**, *54*, 76–87. [CrossRef]
23. Nicolè, S.; Erickson, D.L.; Ambrosi, D.; Bellucci, E.; Lucchin, M.; Papa, R.; Kress, W.J.; Barcaccia, G. Biodiversity studies in *Phaseolus* species by DNA barcoding. *Genome* **2011**, *54*, 529–545. [CrossRef] [PubMed]
24. Rasband, W.S. ImageJ, US National Institutes of Health, Bethesda, Maryland, USA, 1997–2018. Available online: https://imagej.nih.gov/ij/ (accessed on 16 September 2019).
25. Villela, F.A.; Doni Filho, L.; Sequeira, E.L. Tabela de potencial osmótico em função da concentração de polietileno glicol 6000 e da temperatura. *Pesqui. Agropecu. Bras.* **1991**, *26*, 1957–1968.
26. García-Huidobro, J.; Monteith, J.L.; Squire, G.R. Time, temperature and germination of pearl millet. *J. Exp. Bot.* **1982**, *33*, 288–296. [CrossRef]
27. Trudgill, D.L. Why do tropical poikilothermic organisms tend to have higher threshold temperatures for development than temperate ones? *Funct. Ecol.* **1995**, *9*, 136–137.
28. Kebreab, E.; Murdoch, A.J. Modelling the effects of water stress and temperature on germination rate of *Orobanche aegyptiaca* seeds. *J. Exp. Bot.* **1999**, *50*, 655–664. [CrossRef]
29. Bradford, K.J. A water relations analysis of seed germination rates. *Plant Physiol.* **1990**, *94*, 840–849. [CrossRef] [PubMed]
30. Doyle, J.J.; Doyle, J.L. A rapid procedure for DNA purification from small quantities of fresh leaf tissue. *Phytochem. Bull.* **1987**, *19*, 11–15.
31. Chiang, T.Y.; Schaal, B.A.; Peng, C.I. Universal primers for amplification and sequencing a noncoding spacer between the *atpB* and *rbcL* genes of chloroplast DNA. *Bot. Bull. Acad. Sin.* **1998**, *39*, 245–250.
32. Sang, T.; Crawford, D.J.; Stuessy, T.F. Chloroplast DNA phylogeny, reticulate evolution, and biogeography of *Paeonia* (Paeoniaceae). *Am. J. Bot.* **1997**, *84*, 1120–1136. [CrossRef]
33. Katoh, K.; Standley, D.M. MAFFT multiple sequence alignment software version 7: Improvements in performance and usability. *Mol. Biol. Evol.* **2013**, *30*, 772–780. [CrossRef]
34. Katoh, K.; Misawa, K.; Kuma, K.; Miyata, T. MAFFT: A novel method for rapid multiple sequence alignment based on fast Fourier transform. *Nucleic Acids Res.* **2002**, *30*, 3059–3066. [CrossRef]
35. Castresana, J. Selection of conserved blocks from multiple alignments for their use in phylogenetic analysis. *Mol. Biol. Evol.* **2000**, *17*, 540–552. [CrossRef]

36. Ronquist, F.; Teslenko, M.; Van der Mark, P.; Ayres, D.L.; Darling, A.; Höhna, S.; Larget, B.; Liu, L.; Suchard, M.A.; Huelsenbeck, J.P. MrBayes 3.2: Efficient Bayesian phylogenetic inference and model choice across a large model space. *Syst. Biol.* **2012**, *61*, 539–542. [CrossRef]

37. Müller, K. Seqstate: Primer design and sequences statistics for phylogenetic DNA data sets. *Appl. Bioinf.* **2005**, *4*, 65–69. [CrossRef] [PubMed]

38. Lanfear, R.; Frandsen, P.B.; Wright, A.M.; Senfeld, T.; Calcott, B. PartitionFinder 2: New methods for selecting partitioned models of evolution for molecular and morphological phylogenetic analyses. *Mol. Biol. Evol.* **2017**, *34*, 772–773. [CrossRef] [PubMed]

39. Letunic, I.; Bork, P. Interactive Tree Of Life (iTOL) v4: Recent updates and new developments. *Nucleic Acids Res.* **2019**, *47*, 256–259. [CrossRef] [PubMed]

40. Letunic, I.; Bork, P. Interactive Tree Of Life (iTOL): An online tool for phylogenetic tree display and annotation. *Bioinformatics* **2006**, *23*, 127–128. [CrossRef]

41. Miller, M.A.; Pfeiffer, W.; Schwartz, T. Creating the CIPRES science gateway for inference of large phylogenetic trees. In Proceedings of the Gateway Computing Environments Workshop (GCE), New Orleans, LA, USA, 14 November 2010; IEEE: New York, NY, USA, 2010; pp. 45–52.

42. Rozas, J.; Ferrer-Mata, A.; Sánchez-DelBarrio, J.C.; Guirao-Rico, S.; Librado, P.; Ramos-Onsins, S.E.; Sánchez-Gracia, A. DnaSP 6: DNA sequence polymorphism analysis of large data sets. *Mol. Biol. Evol.* **2017**, *34*, 3299–3302. [CrossRef]

43. Fita, A.; Rodríguez-Burruezo, A.; Boscaiu, M.; Prohens, J.; Vicente, O. Breeding and domesticating crops adapted to drought and salinity: A new paradigm for increasing food production. *Front. Plant Sci.* **2015**, *6*, 978. [CrossRef]

44. Mohammadi, R.; Haghparast, R.; Sadeghzadeh, B.; Ahmadi, H.; Solimani, K.; Amri, A. Adaptation patterns and yield stability of durum wheat landraces to highland cold rainfed areas of Iran. *Crop Sci.* **2014**, *54*, 944–954. [CrossRef]

45. Dwivedi, S.L.; Ceccarelli, S.; Blair, M.W.; Upadhyaya, H.D.; Are, A.K.; Ortiz, R. Landrace germplasm for improving yield and abiotic stress adaptation. *Trends Plant Sci.* **2016**, *21*, 31–42. [CrossRef]

46. Lopes, M.S.; El-Basyoni, I.; Baenziger, P.S.; Singh, S.; Royo, C.; Ozbek, K.; Aktas, H.; Ozer, E.; Ozdemir, F.; Manickavelu, A.; et al. Exploiting genetic diversity from landraces in wheat breeding for adaptation to climate change. *J. Exp. Bot.* **2015**, *66*, 3477–3486. [CrossRef]

47. Baudoin, J.P. Genetic Resources, Domestication and Evolution of Lima Bean, *Phaseolus lunatus*. In *Genetic Resources of Phaseolus Beans. Current Plant Science and Biotechnology in Agriculture*; Gepts, P., Ed.; Springer: Dordrecht, The Netherlands, 1988; Volume 6, pp. 393–407.

48. Castiñeiras, L.; Guzmán, F.A.; Duque, M.C.; Shagarodsky, T.; Cristóbal, R.; De Vicente, M.C. AFLPs and morphological diversity of *Phaseolus lunatus* L. in Cuban home gardens: Approaches to recovering the lost *ex situ* collection. *Biodivers. Conserv.* **2007**, *16*, 2847–2865. [CrossRef]

49. Wahua, T.A.T.; Tariah, N.M. Seed treatment to enhance germination of coloured lima bean (*Phaseolus lunatus* L.). *Field Crops Res.* **1984**, *8*, 361–369. [CrossRef]

50. Pollock, B.M.; Toole, V.K. Imbibition period as the critical temperature sensitive stage in germination of lima bean seeds. *Plant Physiol.* **1966**, *41*, 221–229. [CrossRef] [PubMed]

51. Pollock, B.M. Imbibition temperature sensitivity of lima bean seeds controlled by initial seed moisture. *Plant Physiol.* **1969**, *44*, 907–911. [CrossRef] [PubMed]

52. Rodrigues Do Nascimento, M.G.; Ursulino Alves, E.; Mauricio da Silva, M.L.; Marques Rodrigues, C. Lima bean (*Phaseolus lunatus* L.) seeds exposed to different salt concentrations and temperatures. *Rev. Caatinga* **2017**, *30*, 738–747. [CrossRef]

53. Sedlar, A.; Kidrič, M.; Šuštar-Vozlič, J.; Pipan, B.; Zadražnik, T.; Meglič, V. Drought Stress Response in Agricultural Plants: A Case Study of Common Bean (*Phaseolus vulgaris* L.). In *Drought-Detection and Solutions*; Ondrasek, G., Ed.; IntechOpen: London, UK, 2019. Available online: https://www.intechopen.com/books/drought-detection-and-solutions/drought-stress-response-in-agricultural-plants-a-case-study-of-common-bean-em-phaseolus-vulgaris-em- (accessed on 21 November 2019).

54. da Silva, E.C.; de Albuquerque, M.B.; de Azevedo Neto, A.D.; da Silva Junior, C.D. Drought and its consequences to plants—from individuals to ecosystem. In *Responses of Organisms to Water Stress*; Akinci, S., Ed.; Intechopen: London, UK, 2013. Available online: https://www.intechopen.com/books/responsesof-organisms-to-water-stress/drought-and-its-consequences-toplants-from-individual-to-ecosystem (accessed on 10 September 2019).

55. Cardoso, V.J.M.; Bianconi, A. Hydrotime model can describe the response of common bean (*Phaseolus vulgaris* L.) seeds to temperature and reduced water potential. *Acta Sci. Biol. Sci.* **2013**, *35*, 255–261. [CrossRef]

56. Chacón-Sánchez, M.I.; Martínez-Castillo, J. Testing domestication scenarios of lima bean (*Phaseolus lunatus* L.) in Mesoamerica: Insights from genome-wide genetic markers. *Front. Plant Sci.* **2017**, *8*, 1551. [CrossRef]

57. Motta-Aldana, J.R.; Serrano-Serrano, M.L.; Hernández-Torres, J.; Castillo-Villamizar, G.; Debouck, D.G. Multiple origins of Lima bean landraces in the Americas: Evidence from chloroplast and nuclear DNA polymorphisms. *Crop Sci.* **2010**, *50*, 1773–1787. [CrossRef]

58. Abu-Zaitoun, S.Y.; Chandrasekhar, K.; Assili, S.; Shtaya, M.J.; Jamous, R.M.; Mallah, O.B.; Nashef, K.; Sela, H.; Distelfeld, A.; Alhajaj, N.; et al. Unlocking the genetic diversity within a Middle-East panel of durum wheat landraces for adaptation to semi-arid climate. *Agronomy* **2018**, *8*, 233. [CrossRef]

59. Fortes, A.M.; Gallusci, P. Plant stress responses and phenotypic plasticity in the epigenomics era: Perspectives on the grapevine scenario, a model for perennial crop plants. *Front. Plant Sci.* **2017**, *8*, 82. [CrossRef]

60. Pereira, A. Plant abiotic stress challenges from the changing environment. *Front. Plant Sci.* **2016**, *7*, 1123. [CrossRef]

Article

Physiological and Biochemical Responses to Salt Stress in Cultivated Eggplant (*Solanum melongena* L.) and in *S. insanum* L., a Close Wild Relative

Marco Brenes [1,2], Andrea Solana [1], Monica Boscaiu [3], Ana Fita [1], Oscar Vicente [1], Ángeles Calatayud [4], Jaime Prohens [1] and Mariola Plazas [1,*]

[1] Institute for the Conservation and Improvement of Valencian Agrodiversity (COMAV), Universitat Politècnica de València, Camino de Vera 14, 46022 Valencia, Spain; marcob2103@gmail.com (M.B.); ansogar4@posgrado.upv.es (A.S.); anfifer@btc.upv.es (A.F.); ovicente@upvnet.upv.es (O.V.); jprohens@btc.upv.es (J.P.)

[2] Faculty of Biology, Instituto Tecnológico de Costa Rica, Avenida 14, calle 5, Cartago 30101, Costa Rica

[3] Mediterranean Agroforestry Institute (IAM), Universitat Politècnica de València, Camino de Vera 14, 46022 Valencia, Spain; mobosnea@eaf.upv.es

[4] Horticulture Department, Valencian Institute for Agriculture Research (IVIA), CV-315, Km 10.7, 46113 Moncada, Valencia, Spain; calatayud_ang@gva.es

* Correspondence: maplaav@btc.upv.es; Tel.: +34-96-387-9424

Received: 31 March 2020; Accepted: 17 April 2020; Published: 4 May 2020

Abstract: Eggplant (*Solanum melongena*) has been described as moderately sensitive to salinity. We characterised the responses to salt stress of eggplant and *S. insanum*, its putative wild ancestor. Young plants of two accessions of both species were watered for 25 days with an irrigation solution containing NaCl at concentrations of 0 (control), 50, 100, 200, and 300 mM. Plant growth, photosynthetic activity, concentrations of photosynthetic pigments, K^+, Na^+, and Cl^- ions, proline, total soluble sugars, malondialdehyde, total phenolics, and total flavonoids, as well as superoxide dismutase, catalase, and glutathione reductase specific activities, were quantified. Salt stress-induced reduction of growth was greater in *S. melongena* than in *S. insanum*. The photosynthetic activity decreased in both species, except for substomatal CO_2 concentration (Ci) in *S. insanum*, although the photosynthetic pigments were not degraded in the presence of NaCl. The levels of Na^+ and Cl^- increased in roots and leaves with increasing NaCl doses, but leaf K^+ concentrations were maintained, indicating a relative stress tolerance in the two accessions, which also did not seem to suffer a remarkable degree of salt-induced oxidative stress. Our results suggest that the higher salt tolerance of *S. insanum* mostly lies in its ability to accumulate higher concentrations of proline and, to a lesser extent, Na^+ and Cl^-. The results obtained indicate that *S. insanum* is a good candidate for improving salt tolerance in eggplant through breeding and introgression programmes.

Keywords: eggplant; wild relative; vegetative growth; photosynthesis; ion homeostasis; osmolytes; oxidative stress

---

1. Introduction

Soil salinity affects over 1000 million ha of land throughout the world [1,2], and it continuously increases worldwide, affecting large areas of arable land [3]. The effects of soil salinity on plants vary depending on weather conditions, light intensity, soil characteristics, and species or taxonomic groups [4], but most crops are glycophytes and, therefore, are not able to grow on saline soils. Generally, growth of glycophytes is completely inhibited at salt concentrations in soil equivalent to 100–200 mM NaCl, eventually resulting in the death of the plant [5].

Eggplant (*Solanum melongena* L.) is one of the most popular vegetable crops throughout the world and, especially in Southeast Asia [6], and is moderately sensitive to salinity [7]. Eggplant fruits have a low calories content and contain high concentrations of phenolic acids, beneficial for human health [8,9]. Eggplant is cultivated on more than 1.86 million hectares and its annual production is over 54 million tonnes [6]. *Solanum melongena* can be crossed with a wide range of wild relatives from the primary, secondary, and tertiary genepools [10], and backcrossing to *S. melongena* of the interspecific hybrids for introgression breeding can result in the incorporation of traits from wild species into the eggplant genepool and in the broadening of the genetic basis of the crop [11–13]. Therefore, identifying sources of variation for tolerance to salinity among eggplant wild relatives, some of which grow in harsh environments, including areas prone to salinity [14], can contribute to breeding eggplant for higher tolerance to salinity. One of the most promising species for introgression breeding in eggplant is *S. insanum* L., which is the wild ancestor of eggplant and grows in a wide range of soil conditions [15]. Interspecific hybrids between *S. melongena* and *S. insanum* as well as backcrosses of the hybrids to *S. melongena*, are easily obtained and are highly fertile [10,11,16], which facilitates the transfer of traits from *S. insanum* to *S. melongena*.

To our knowledge, the responses of *S. insanum* under conditions of salt stress have not yet been studied. Data on physiological and biochemical traits under stressful conditions could be used as selection criteria for possible breeding programmes [17]. This study aims to determine the level of tolerance to salinity of *S. insanum*, comparing it to *S. melongena* by analysing the variation of growth traits, photosynthesis, and biochemical responses associated with tolerance to salinity, such as levels of ions accumulated in different tissues, osmolytes, and antioxidants. The results will provide relevant information on *S. insanum* as a possible source of variation of tolerance to salinity, for eggplant breeding.

2. Materials and Methods

2.1. Plant Material and Experimental Layout

The plant material used was provided by the Institute for the Conservation and Improvement of Valencian Agrodiversity (COMAV-UPV). *Solanum melongena* accession MEL1 originates from Ivory Coast, and *S. insanum* INS2 from Sri Lanka. *Solanum melongena* MEL1 was chosen as this accession is of particular interest for breeding as it has an excellent fruit set and shows a high degree of success in interspecific hybridisation [10,11]. Seeds were germinated following a shortened version of a protocol developed by Ranil et al. [18]. Briefly, seeds were soaked first for 24 h in water and for an additional 24 h in a 500 ppm solution of gibberellic acid (GA_3), and then placed in Petri dishes on filter paper moistened with a solution of 1000 ppm KNO_3 and subjected to a heat shock treatment at 37 °C for 24 h. The Petri dishes were transferred to a growth chamber under conditions of 16 h light/8 h darkness at 25 °C until germination was completed. Once germinated, seedlings were placed in seedbeds and kept under the same conditions of light and temperature for two weeks. Seedlings homogenous in size were selected and transplanted to small pots and, subsequently, to 1.3 L pots with 500 g of Huminsubstrat N3 (Klasmann-Deilmann, Geeste, Germany) commercial substrate. The plants were transferred to a greenhouse with benches and controlled temperature (maximum of 30 °C and minimum of 15 °C) for acclimatisation for 20 days, and when plants developed 6–8 fully expanded leaves, the stress treatments were started. Five plants of each species, each one corresponding to a biological replica, were irrigated every four days with 1.25 L of NaCl solutions (final concentrations: 50, 100, 200, and 300 mM NaCl dissolved in deionised water) or deionised water for the control plants, for 25 days, and several non-destructive growth parameters were measured in all plants (stem length, stem diameter, and number of leaves). Runoff water after irrigation was allowed to freely drain. Measurements for physiological, biochemical, and ion content parameters were based on one technical replicate.

2.2. Electrical Conductivity of the Substrate

Electrical conductivity of the substrate was measured in a 1:5 suspension ($EC_{1:5}$). At the end of the treatments, after removing the plants from the pots, the remaining substrate was dried in an oven at 65 °C for four days and a soil/water (1:5) suspension was prepared in deionised water and stirred for 1 h at 600 rpm and 21 °C. EC was measured with a Crison Conductivity-meter 522 (Crison Instruments SA, Barcelona, Spain) and expressed in dS m^{-1}.

2.3. Gaseous Exchange

At the end of the stress period (25 days), the CO_2 assimilation rate (A_N, µmol CO_2 m^{-2} s^{-1}), stomatal conductance to water vapor (gs, mol H_2O m^{-2} s^{-1}), substomatal CO_2 concentration (Ci, µmol CO_2 mol^{-1} air), and transpiration rate (E, mmol H_2O m^{-2} s^{-1}) were measured in one of the fully developed leaves of each plant using a portable LI-COR 6400 infrared gas analyser (Li-Cor Inc., Lincoln, NE, USA).

2.4. Evaluation of Growth Parameters

To assess the effect of salt stress on the two species, several growth parameters were analysed at the end of the treatments: fresh weight of roots (RFW), stems (SFW), and leaves (LFW); length of roots (RL) and stems (SL); stem diameter (SD); and area of the largest leaf (LA). Stem elongation (SE), stem thickening (ST), and increase in the number of leaves (Lno) were calculated as the difference between the final and initial values of stem length, stem diameter, and number of leaves, respectively, in the same plant. The water content of roots (RWC), stems (SWC), and leaves (LWC) was determined by weighing a part of fresh material, drying it for four days at 60 °C, and weighing it again; the humidity percentage was calculated with the following formula: [(Fresh weight − Dry weight)/Fresh weight] * 100.

2.5. Ion Quantification

Contents of potassium (K^+), sodium (Na^+), and chloride (Cl^-) were determined in roots and leaves. Samples of 50 mg of ground dry plant material in 15 mL of deionised water were heated at 95 °C for one hour, followed by cooling on ice and filtration through a 0.45 µm nylon filter [19]. The Na^+ and K^+ content was quantified with a PFP7 flame photometer (Jenway Inc., Burlington, VT, USA), and the Cl^- content was determined using a chlorimeter (Sherwood, model 926, Cambridge, UK).

2.6. Quantification of Photosynthetic Pigments

The content of chlorophyll a (Chl a), chlorophyll b (Chl b), and carotenoids (Caro) was determined using the methodology described by Lichtenthaler and Wellburn [20]. Pigments were extracted from 50 mg fresh plant material using 10 mL of ice-cold 80% acetone (v/v), and the extracts were diluted 10 times using the same solvent. The absorbance was measured at 470, 645, and 663 nm (A_{470}, A_{645}, and A_{663}, respectively), and the following formulas were used to calculate the different pigments:

$$\text{Chl a } (\mu g \ mL^{-1}) = 12.21 \times A_{663} - 2.81 \times A_{646} \tag{1}$$

$$\text{Chl b } (\mu g \ mL^{\pm 1}) = 20.13 \times A_{646} - 5.03 \times A_{663} \tag{2}$$

$$\text{Caro } (\mu g \ mL^{-1}) = (1000 \times A_{470} - 3.27 \times [\text{Chl a}] - 104 \times [\text{Chl b}])/227 \tag{3}$$

2.7. Quantification of Osmolytes

The quantification of free proline (Pro) was carried out following the acetic acid-ninhydrin method [21]. An aqueous solution (2 mL) of 3% (w/v) sulfosalicylic acid was added to 50 mg freshly ground plant material (from each biological replica). One volume of extract was mixed with one volume of ninhydrin acid and one volume of glacial acetic acid, and then the mix was placed in a water bath at 95 °C for one hour, and subsequently cooled for 10 min on ice and extracted with toluene. The absorbance of the organic phase was determined at 520 nm using toluene as the blank.

Total soluble sugars (TSSs) were measured according to the methodology described in [22]. Fresh leaf material (50 mg) was ground and mixed with 3 mL of 80% (v/v) methanol on a rocker shaker for 24 h, and the extract was recovered by centrifugation; concentrated sulfuric acid and 5% phenol were added to the supernatant and the absorbance was measured at 490 nm. TSS contents were expressed as 'mg equivalent of glucose' per g dry weight (DW).

2.8. Measurement of Malondialdehyde (MDA) and Antioxidant Compounds

MDA, total phenolic compounds (TPCs), and total flavonoids (TFs) were measured in plant extracts prepared from 50 mg ground fresh leaf material using 80% (v/v) methanol. For MDA quantification, extracts were mixed with 0.5% thiobarbituric acid (TBA) prepared in 20% trichloroacetic acid (TCA), or with 20% TCA without TBA for the controls, and then incubated at 95°C for 20 min, cooled on ice, and centrifuged at $12,000 \times g$ for 10 min at 4 °C [23]. The absorbance of the supernatants was measured at 532 nm. The non-specific absorbance at 600 and 440 nm was subtracted, and MDA concentration was determined using the equations included in [23], based on the extinction coefficient of the MDA-TBA adduct at 532 nm. The concentration of MDA was expressed as nmol g^{-1} DW.

TPCs were measured using the Folin–Ciocalteu reagent [24]. Methanol extracts were mixed with Na_2CO_3 and the reagent and, after 90 min of incubation in the dark, the absorbance was measured at 765 nm. A standard reaction was performed in parallel using known amounts of gallic acid (GA), and TPC contents were reported as equivalents of GA (mg eq. GA g^{-1} DW).

Total flavonoids (TFs) were quantified according to the method described by Zhisen et al. [25], based on the nitration of aromatic rings containing a catechol group. Methanol extracts of each sample were reacted with $NaNO_2$ and $AlCl_3$ under alkaline conditions, and the absorbance at 510 nm was measured. The concentration of TFs was expressed as equivalents of catechin, used as the standard (mg eq. C g^{-1} DW).

2.9. Antioxidant Enzyme Activities

The activities of superoxide dismutase (SOD), catalase (CAT), and glutathione reductase (GR) were measured in crude protein extracts prepared from frozen (−70 °C) leaf material, as previously described [26]. Enzyme activities in the extracts were expressed as 'specific activities', in units per mg of protein.

SOD activity in the protein extracts was determined as described by Beyer and Fridovich [27], following the inhibition of nitroblue tetrazolium (NBT) photoreduction by measuring the absorbance of the sample at 560 nm. The reaction mixtures contained riboflavin as the source of superoxide radicals. One SOD unit was defined as the amount of enzyme causing 50% inhibition of NBT photoreduction under the assay conditions.

CAT activity was measured by the decrease in absorbance at 240 nm, which accompanies the consumption of H_2O_2 added to protein extracts [28]. One CAT unit was defined as the amount of enzyme that will decompose one mmol of H_2O_2 per minute at 25 °C.

The protocol of Conell and Mullet [29] was used for the GR assays, following the oxidation of NADPH (the cofactor in the GR-catalysed reduction of oxidised glutathione (GSSG)) by the decrease in absorbance at 340 nm. One GR unit was defined as the amount of enzyme that will oxidise one mmol of NADPH per minute at 25 °C.

2.10. Statistical Analysis

Data were analysed using the software Statgraphics Centurion v. XVI (Statpoint Technologies Inc., Warrenton, VA, USA). The significance of the differences between treatments (for each species), between species (for each treatment) and their interaction were evaluated through a two-factorial analysis of variance (ANOVA) for traits related to plant growth, photosynthetic pigments, photosynthesis parameters, osmolytes, MDA, and antioxidants. For ion accumulation, an additional factor (organ) was included and a three-way factor analysis of variance (ANOVA) (treatment, species, and organ) was performed. Post-hoc comparisons were made using the Tukey Honestly Significant Difference (HSD) test at $p < 0.05$ for the effects of treatment within species (and combinations of species and organ in the case of ions). All the parameters measured in plants of the control and salt stress treatments were subjected to multivariate analysis through a principal component analysis (PCA).

3. Results

3.1. Substrate Electrical Conductivity

Electrical conductivity of the substrate increased in parallel to the concentration of NaCl, applied in a similar manner in both species, as indicated by the analysis of variance, which detected significant differences only between treatments, but not between the two species. EC reached the highest levels at the end of the treatments (19.19 dS m^{-1} for *S. melongena* and 23.66 dS m^{-1} for *S. insanum*) in the pots watered with 300 mM NaCl (Figure 1).

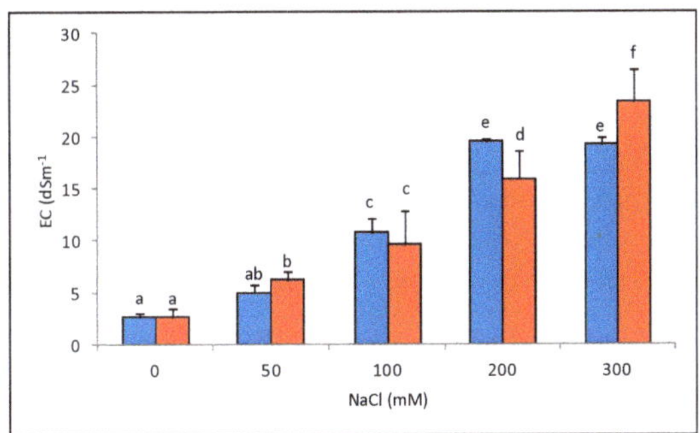

Figure 1. Electrical conductivity (EC$_{1:5}$) of the pot substrates after 25 days of treatment with the indicated NaCl concentrations, in *Solanum melongena* (blue) and *S. insanum* (red). Same letters indicate homogeneous groups between combinations of treatments for EC according to the Tukey test ($p < 0.05$, $n = 5$).

3.2. Analysis of Morphological and Photosynthetic Parameters

Salt stress inhibited the growth of the two species, in a concentration-dependent manner. Several growth parameters were determined in control and salt-stressed plants, at the end of the treatments, and a two-way ANOVA was performed, considering the effect of treatment, species, and their interaction (Table 1). The effect of 'species' was significant for most of the parameters, except root length (RL), some stem traits [stem elongation (SE), thickening (ST), fresh weight (SFW), water content (SWC)], total fresh weight (TFW), and chlorophyll a (Chl a). The effect of 'treatment' was significant for all traits analysed, except water content of roots (RWC), stems (SWC), and leaves (LWC), as well chlorophylls a and b (Chl a and Chl b). The interaction of the two factors was significant only for stem elongation (SE), the increase in leaf number (Lno), leaf fresh weight (LFW), and the area of the largest leaf (Table 1).

Table 1. Two-way analysis of variance (ANOVA) of species, treatment, and their interactions, for the indicated parameters. Numbers shown represent percentages of the sum of squares (SS).

	Abbr.	Treatment [a]	Species [a]	Interaction [a]	Residual
Root length	RL	11.42 *	1.32	20.77	66.48
Root fresh weight	RFW	6.36	24.43 ***	2.60	66.60
Root water content	RWC	55.17 ***	14.41 ***	5.12	25.29
Stem elongation	SE	68.59 ***	2.26	6.44 *	22.71

Table 1. *Cont.*

	Abbr.	Treatment [a]	Species [a]	Interaction [a]	Residual
Stem thickening	ST	63.70 ***	0.93	2.32	32.96
Stem fresh weight	SFW	56.05 ***	0.26	3.30	40.54
Stem water content	SWC	5.04	4.83	4.66	85.47
Increase in no. of leaves	Lno	34.91 ***	29.19 ***	10.20 **	39.82
Leaf area	LA	34.69 ***	29.19 ***	10.20 **	25.69
Leaf fresh weight	LFW	43.15 ***	10.54 **	11.11 *	35.20
Leaf water content	LWC	7.45	7.10 *	17.34	68.11
Total fresh weight	TFW	44.77 ***	0.15	5.98	49.09
Chlorophyll a	Chl a	14.09	3.33	4.58	78.00
Chlorophyll b	Chl b	12.86	8.37 *	9.39	69.43
Carotenoids	Caro	18.92 *	13.35 ***	4.19	63.40
Photosynthestic rate	A_N	23.29 ***	37.82 ***	4.75	34.58
Stomatal conductance	gs	16.22 *	29.92 ***	3.63	50.23
Int. CO_2 concentration	Ci	29.10 **	17.22 ***	1.11	52.57
Transpiration rate	E	17.92 **	39.17 **	2.35	40.56

[a] ***, **, and * indicate significant at $p < 0.001$, $p < 0.01$, and $p < 0.05$, respectively.

At the root level, the effect of salt was more pronounced in *S. melongena*, as root length (RL) and root fresh weight (RFW) did not vary significantly in *S. insanum* (Table 2). In both species, the water content of the roots increased with salinity. Growth of the stems was affected by salinity, but the water content was maintained stable in both species. Regarding the analysed leaf parameters, all showed a significant decrease in salt-treated plants of *S. melongena*, whereas in S. *insanum*, their variation was not significant, except for the increase in the number of leaves (Lno). When considering the total fresh weight (TFW), the reduction was significant only in the cultivated eggplant, but not in the wild species, in which, at the lowest concentration applied, TFW even increased, although the variation was not statistically significant in relation to the control. In both species, the water content of the leaves (LWC) did not vary significantly with the treatments (Table 2). Moreover, the variation between treatments of Chl a and Chl b was non-significant, whereas carotenoids decreased only in *S. insanum*. Stomatal conductance (gs), internal concentration of CO_2 (Ci), and transpiration (E) decreased in *S. melongena*, but not in *S. insanum*; photosynthesis rate (A_N), on the other hand, showed a significant reduction in both species (Table 2).

Table 2. Growth responses and photosynthetic parameters in *Solanum melongena* (MEL) and *S. insanum* (INS) after 25 days of treatment with the indicated NaCl concentrations.

Trait	Taxa	Treatment (mM NaCl) 0	50	100	200	300
RL	MEL	26.2 ± 1.3 c	26.7 ± 2.2 c	24.6 ± 1.7 bc	19.8 ± 1.7 ab	17.8 ± 0.6 a
	INS	21.0 ± 3.5 A	20.2 ± 0.5 A	24.0 ± 1.1 A	20.1 ± 2.5 A	24.2 ± 2.9 A
RFW	MEL	9.0 ± 0.8 b	9.8 ± 0.5 b	9.1 ± 0.7 b	9.1 ± 0.5 b	6.9 ± 0.1 a
	INS	10.8 ± 2.1 A	11.9 ± 0.8 A	11.2 ± 0.3 A	11.4 ± 0.9 A	10.9 ± 0.8 A
RWC	MEL	71.2 ± 1.2 a	76.8 ± 0.7 b	78.8 ± 0.7 bc	80.4 ± 0.4 c	79.5 ± 0.5 bc
	INS	65.5 ± 3.6 A	68.4 ± 1.2 AB	76.0 ± 1.0 BC	78.0 ± 0.5 C	77.7 ± 0.6 C
SE	MEL	6.3 ± 1.9 c	5.7 ± 1.1 bc	5.8 ± 0.5 c	3.7 ± 0.5 ab	2.5 ± 0.2 a
	INS	8.3 ± 0.6 C	6.7 ± 0.2 C	4.9 ± 0.8 BC	3.7 ± 0.2 AB	3.3 ± 0.4 A
ST	MEL	3.5 ± 0.2 c	2.9 ± 0.2 bc	2.3 ± 0.1 bc	1.6 ± 0.3 ab	1.3 ± 0.3 a
	INS	3.2 ± 0.5 B	2.2 ± 0.4 AB	2.2 ± 0.3 AB	1.8 ± 0.2 AB	1.1 ± 0.1 A
SFW	MEL	4.4 ± 0.9 b	4.7 ± 0.2 b	3.9 ± 0.5 ab	2.6 ± 0.2 ab	1.9 ± 0.3 a
	INS	5.5 ± 1.0 C	4.7 ± 0.2 BC	3.5 ± 0.1 B	2.5 ± 0.1 AB	2.1 ± 0.1 A
SWC	MEL	68.4 ± 6.2 a	78.1 ± 0.7 a	76.8 ± 3.9 a	73.6 ± 3.5 a	75.1 ± 5.6 a
	INS	70.8 ± 2.3 A	70.6 ± 1.1 A	72.8 ± 1.7 A	71.4 ± 0.7 A	67.6 ± 1.4 A

Table 2. *Cont.*

| Trait | Taxa | Treatment (mM NaCl) | | | | |
		0	50	100	200	300
Lno	MEL	2.4 ± 0.2 ab	2.6 ± 0.2 ab	3.0 ± 0.0 b	1.8 ± 0.2 a	1.8 ± 0.2 a
	INS	2.2 ± 0.4 C	1.8 ± 0.4 BC	1.4 ± 0.2 B	0.8 ± 0.4 AB	−0.4 ± 0.2 A
LA	MEL	205.7 ± 12.0 c	161.6 ± 8.4 b	149.2 ± 4.1 ab	132.0 ± 3.9 ab	119.4 ± 8.6 a
	INS	143.4 ± 17.0 A	148.6 ± 3.7 A	139.8 ± 8.2 A	136.6 ± 8.2 A	105.5 ± 9.1 A
LFW	MEL	22.4 ± 1.9 d	19.3 ± 1.0 cd	16.6 ± 0.8 bc	12.9 ± 1.1 b	9.1 ± 0.7 a
	INS	13.8 ± 3.2 A	14.8 ± 1.1 A	14.4 ± 0.7 A	12.0 ± 0.3 A	9.6 ± 1.1 A
LWC	MEL	81.3 ± 2.3 a	84.4 ± 4.8 a	85.3 ± 1.7 a	81.9 ± 2.7 a	72.2 ± 2.8 a
	INS	75.5 ± 4.5 A	78.3 ± 0.8 A	79.5 ± 0.7 A	80.8 ± 0.3 A	82.2 ± 0.7 A
TFW	MEL	35.8 ± 2.8 d	33.8 ± 1.5 cd	29.7 ± 3.2 c	24.5 ±0.8 b	18.0 ± 1.3 a
	INS	30.1 ± 6.0 a	31.5 ± 1.9 a	29.1 ± 2.1 a	25.9 ± 0.9 a	22.7 ± 1.8 a
Chl a	MEL	9.4 ± 2.2 a	8.4 ± 1.1 a	11.0 ± 1.1 a	5.7 ± 1.1 a	8.1 ± 0.7 a
	INS	7.6 ± 1.9 A	6.8 ± 1.9 A	8.7 ± 1.1 A	7.2 ± 0.5 A	6.5 ± 0.5 A
Chl b	MEL	4.3 ± 1.0 a	3.8 ± 0.6 a	5.7 ± 1.4 a	2.2 ± 0.3 a	3.6 ± 0.4 a
	INS	3.3 ± 0.8 A	2.5 ± 0.7 A	2.9 ± 0.4 A	2.7 ± 0.8 A	3.1 ± 0.3 A
Caro	MEL	1.6 ± 0.4 a	1.3 ± 0.5 a	1.3 ± 0.4 a	0.9 ± 0.4 a	1.3 ± 0.3 a
	INS	2.1 ± 0.3 B	1.9 ± 0.2 AB	1.7 ± 0.2 AB	1.4 ± 0.1 AB	1.2 ± 0.1 A
A_N	MEL	8.5 ± 1.1 b	9.9 ± 1.3 b	9.0 ± 0.8 b	5.4 ± 0.9 ab	3.4 ± 0.3 a
	INS	18.9 ± 3.3 B	15.4 ± 1.5 AB	18.2 ± 2.0 AB	15.8 ± 1.9 AB	7.7 ± 2.9 A
gs	MEL	0.11 ± 0.00 b	0.11 ± 0.00 b	0.09 ± 0.01 ab	0.05 ± 0.00 ab	0.03 ± 0.00 a
	INS	0.24 ± 0.10 A	0.26 ± 0.10 A	0.30 ± 0.15 A	0.21 ± 0.10 A	0.09 ± 0.03 A
C	MEL	234.6 ± 17.0 b	230.0 ± 8.5 b	213.8 ± 5.6 b	195.0 ± 5.7 ab	182.2 ± 9.7 a
	INS	274.6 ± 24.0 A	248.2 ± 13.0 A	243.2 ± 13.0 A	222.0 ± 14.0 A	219.2 ± 5.3 A
E	MEL	2.6 ± 0.5 b	2.7 ± 0.4 b	2.5 ± 0.1 b	1.5 ± 0.2 ab	0.9 ± 0.1a
	INS	4.7 ± 1.0 A	5.3 ± 0.8 A	5.8 ± 0.8 A	4.8 ± 0.8 A	2.6 ± 0.4 A

Mean ± SE values are shown ($n = 5$). Same letters within each row (lowercase for *S. melongena* and capital letters for *S. insanum*) indicate homogeneous groups between treatments for each species, according to the Tukey HSD test ($p < 0.05$). Abbreviations: root length (RL; cm), root fresh weight (RFW; g), root water content (RWC; %), stem elongation (SE; cm), stem thickening (ST; mm), stem fresh weight (SFW; g), stem water content (SWC, %), increase in the number of leaves (Lno), area of the largest leaf (LA; cm^2), leaf fresh weight (LFW; g), leaf water content (LWC; %), total fresh weight (TFW; g), chlorophyll a (Chl a; mg g^{-1} dry weight (DW)), chlorophyll b (Chl b; mg g^{-1} DW), carotenoids (Caro; mg g^{-1} DW), photosynthetic rate (A_N; μmol CO$_2$ m^{-2} s^{-1}), stomatal conductance (gs; mol H$_2$O m^{-2} s^{-1}), internal concentration of CO$_2$ (Ci; μmol CO$_2$ mol^{-1} air), and transpiration rate (E; mmol H$_2$O m^{-2} s^{-1}).

For an easier estimation of the pattern of variation of growth parameters in the two species, the variation of fresh weight and water content in the roots, stems, and leaves of the plants subjected to the salt treatments is shown in Figure 2, as percentages of the values measured in the corresponding non-stressed controls. In general, both fresh weight (FW) and water content (WC) showed a relatively smaller reduction in *S. insanum* than in *S. melongena*, at least in roots and leaves, and more pronounced at the highest salt concentration tested (Figure 2).

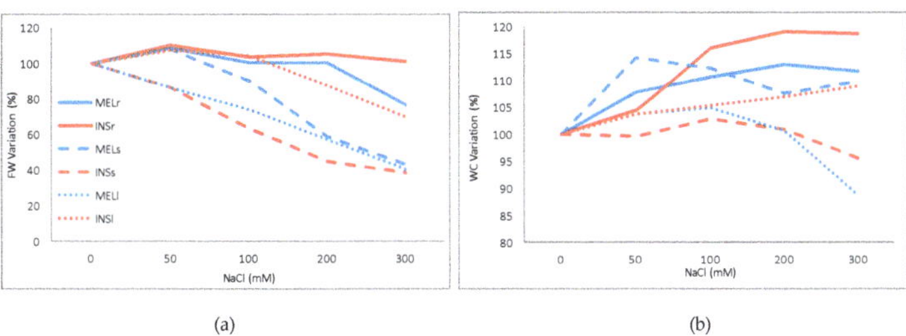

(a) (b)

Figure 2. Reduction of fresh weight (FW) (**a**) and water content (WC) (**b**) in roots (MELr and INSr), stems (MELs and INSs), and leaves (MELl and INSl) of *Solanum melongena* (MEL; blue lines) and *S. insanum* (INS; red lines) plants after 25 days of salt treatments at the indicated NaCl concentrations. Values are shown as percentages of the corresponding controls (0 mM NaCl).

3.3. Ion Accumulation

To analyse the changes in ion contents in the plants, in response to the salt treatments, a multifactorial ANOVA was performed, considering the effect of the treatment, species, organs of the plants (roots vs. leaves), and their interactions (Table 3). In the case of Na^+ and Cl^- contents and the K^+/Na^+ ratio, the main effect was that of the treatment, which was highly significant for all traits, whereas the 'species' factor was significant only for Cl^- and K^+. The effect of the 'organ' variable was significant for Cl^-, K^+, and the K^+/Na^+ ratio, but it was by far the greatest contributor to the sums of squares for K^+, as leaves of both species contain considerably higher concentrations of K^+ than the roots. Some significant double and triple interactions were detected, for example, between 'treatment' and 'species' or between 'treatment' and 'organ' for Na^+ and Cl^-, but their contribution to the sums of squares was generally low (below 3.5%), except for the interaction between 'treatment' and 'organ' for the K^+/Na^+ ratio (Table 3).

Table 3. Factorial analysis of variance (ANOVA) considering the effect of treatment (A), species (B), organ (C), and their interactions (A × B; A × C; B × C; A × B × C) on ions (Na^+, Cl^-, K^+) contents and the K^+/Na^+ ratio, in *Solanum melongena* and *S. insanum*. Numbers represent percentages of sum of squares (SS).

Ion Contents and K^+/Na^+ Ratio	A [a]	B [a]	C [a]	AB [a]	AC [a]	BC [a]	ABC [a]	Residuals
Na^+	84.70 ***	0.24	0.04	2.00 **	1.16 *	1.2 **	1.45 **	8.61
Cl^-	79.30 ***	1.85 ***	1.81 ***	2.52 **	1.29 *	1.01 *	2.13 *	10.07
K^+	2.30 ***	6.10 ***	70.20 ***	0.58	2.14 *	0.61	3.39 **	14.69
K^+/Na^+	71.27 ***	0.04	15.54 ***	0.25	10.42 ***	0.01	0.16	2.26

[a] ***, **, and * indicate significant at $p < 0.001$, $p < 0.01$, and $p < 0.05$, respectively.

In both species, Na^+ and Cl^- concentrations increased in parallel to the increase in external salinity, in the roots and the leaves of the plants (Figure 3a,b). The pattern of variation was similar in the two species, as were, in general, the contents of both ions in roots and leaves for each NaCl concentration tested, except that *S. insanum* accumulated higher levels of Na^+ and Cl^- in leaves than in roots at high salinity (200–300 mM NaCl). On the contrary, K^+ levels remained generally steady in response to the salt treatments, in roots and leaves of the two species, and in all cases, significantly higher in the leaves (Figure 3c). The salt-induced increase in Na^+ concentrations, accompanied by no significant changes of K^+ contents, led to a significant decrease of the K^+/Na^+ ratio in both species, especially in the leaves, where the initial values in the controls were higher than in roots (Figure 3d).

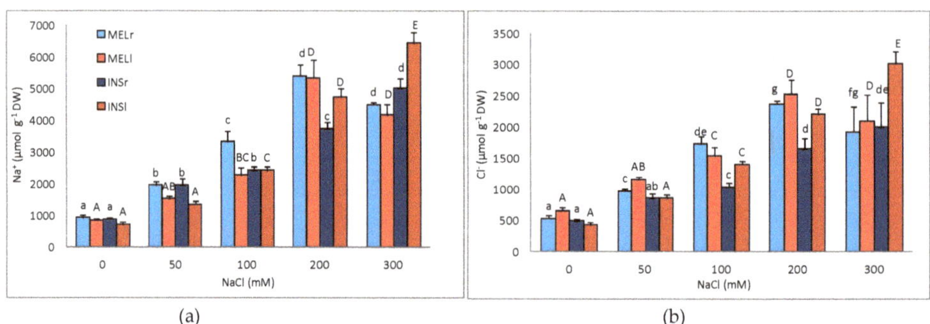

(a)

(b)

Figure 3. *Cont.*

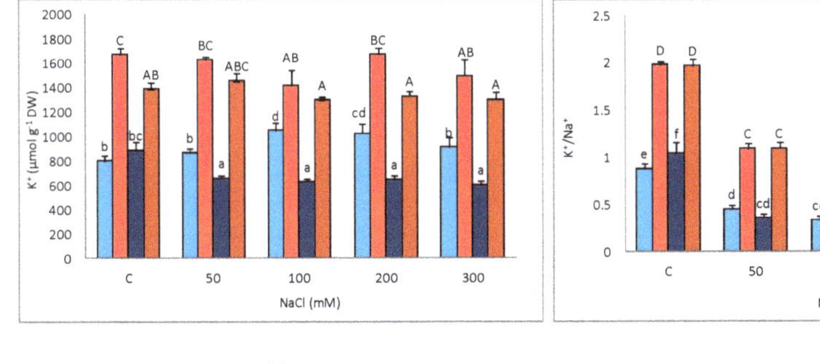

(c)   (d)

Figure 3. Na$^+$ (**a**), Cl$^-$ (**b**), and K$^+$ (**c**) contents and K$^+$/Na$^+$ ratio (**d**) in roots (MELr and INSr) and leaves (MELl and INSl) in *Solanum melongena* (blue) and *S. insanum* (red), after 25 days of treatments with the indicated NaCl concentrations. Mean ± SE values are shown (*n* = 5). Same letters (lowercase for roots, or uppercase for leaves) indicate homogeneous groups between combinations of treatments, according to the Tukey HSD test (*p* < 0.05).

3.4. Osmolytes, MDA, and Antioxidants

A two-way ANOVA was performed to analyse the effects of the variables 'treatment' and 'species', as well as their interaction, on different biochemical parameters related to the general responses of plants to salt stress (Table 4). This analysis revealed a strong effect of 'treatment', but also a significant effect of 'species' and their interaction for proline. In the case of TSS, however, only the 'species' factor and its interaction with 'treatment' were significant. MDA showed a significant variation according to the treatment and the species; for total phenolic compounds (TPCs), the two factors and their interaction were significant, although the strongest contribution to the sums of squares was that of 'species'. For total flavonoids (TFs), the only significant effect was owing to the treatment. Regarding the antioxidant enzymatic activities, the two factors, treatment and species, as well as their interaction, were significant for SOD, whereas only the species effect was significant for CA, and no significant factor was detected for GR. It is remarkable that, for all biochemical compounds analysed, except proline, and for the three enzymatic activities, the percentage of the sum of square of residuals was the most important contributor to the sums of squares, indicating a high influence of uncontrolled residual variation (Table 4).

Table 4. Two-way analysis of variance (ANOVA) of treatment, species, and their interactions for the parameters considered. Numbers represent percentages of sum of squares (SS) at the 5% confidence level. Abbreviations: proline (Pro), total soluble sugars (TSSs), malondialdehyde (MDA), total phenolic compounds (TPCs), total flavonoids (TFs), superoxide dismutase (SOD), catalase (CAT), and glutathione reductase (GR).

Trait	Treatment [a]	Species [a]	Interaction [a]	Residual
Pro	63.60 ***	18.20 ***	14.06 ***	4.29
TSS	5.31	22.87 ***	18.76 *	53.05
MDA	25.29 ***	29.23 ***	6.48	38.98
TPC	8.19 *	28.90 ***	15.32 *	47.57
TF	27.01 **	1.83	4.30	66.85
SOD	13.76 *	21.11 ***	16.92 *	48.07
CAT	2.82	13.21 *	10.26	73.24
GR	16.50	1.59	11.30	70.60

[a] ***, **, and * indicate significant at $p < 0.001$, $p < 0.01$, and $p < 0.05$, respectively.

Leaf proline (Pro) levels increased significantly in the two species in response to the salt stress treatments. In *S. melongena*, Pro contents were lower than in *S. insanum* at all tested salinities, reaching a peak in the presence of 200 mM NaCl, and decreasing at 300 mM NaCl. In *S. insanum*, Pro increased gradually in parallel to the external NaCl concentration, reaching levels about 10-fold higher than in the control at 300 mM NaCl (Figure 4a). Contrary to Pro, total soluble sugars (TSSs) in leaves showed a slight increase in salt-treated *S. melongena* plants, but the difference with the control was significant only in the presence of 200 mM NaCl. Average TSS contents were substantially higher in *S. insanum* than in *S. melongena* plants, in the control and at low salinity, to decrease at higher NaCl concentrations; however, the differences with the non-stressed controls were non-significant (Figure 4b).

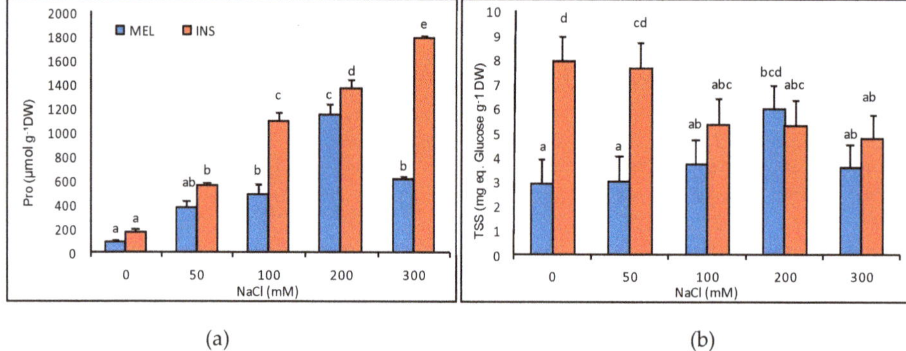

Figure 4. Proline (Pro) (**a**) and total soluble sugars (TSSs) (**b**) contents in *Solanum melongena* (blue) and *S. insanum* (red) after 25 days of treatments with the indicated NaCl concentrations. Mean ± SE values are shown ($n = 5$). Same letters (lowercase for *S. melongena* and capital for *S. insanum*) indicate homogeneous groups between combinations of treatments, according to the Tukey HSD test ($p < 0.05$).

Malondialdehyde (MDA) is regarded as a reliable marker of oxidative stress, as it is a product of peroxidation of unsaturated fatty acids, indicating damage to cell membranes by 'reactive oxygen species' (ROS) in plants and animals [30]. However, its levels did not increase in salt-treated plants as compared with the controls, neither in *S. melongena* nor in *S. insanum*; on the contrary, leaf MDA contents slightly decreased in response to increasing salinity in plants of the two species (Table 5). A similar decreasing trend was observed for the mean values of the analysed antioxidant compounds, TPC and TF, although the differences with the non-stressed controls were not statistically significant in *S. melongena* (Table 5). Moreover, no significant salt-induced differences in specific activity could be detected in the assays of the antioxidant enzymes, SOD, CAT, and GR. When comparing the two species, higher MDA, TPC, and TF contents and higher specific enzyme activities were generally observed in *S. insanum*, at each external salt concentration tested (Table 5).

Table 5. Malondialdehyde (MDA), total phenolic compounds (TPCs), total flavonoids (TFs), and activity of the antioxidant enzymes: superoxide dismutase (SOD), catalase (CAT), and glutathione reductase (GR) in *S. melongena* (MEL) and *S. insanum* (INS) after 25 days of treatment with the indicated NaCl concentrations.

Trait	Taxa	Treatment (mM NaCl)				
		0	**50**	**100**	**200**	**300**
MDA	MEL	145.7 ± 12.2 [b]	134.4 ± 9.1 [ab]	114.7 ± 4.1 [ab]	107.3 ± 7.5 [a]	107.9 ± 7.8 [a]
	INS	207.1 ±2 4.6 [B]	143.9 ± 11.7 [A]	145.3 ± 7.8 [A]	145.2 ± 4.2 [A]	162.1 ± 8.1 [AB]
TPC	MEL	12.3 ± 0.7 [a]	11.1 ± 2.1 [a]	6.3 ± 0.6 [a]	5.7 ± 0.4 [a]	7.8 ± 1.3 [a]
	INS	15.5 ± 0.8 [B]	10.7 ± 1.8 [AB]	7.3 ± 0.5 [A]	6.3 ± 0.4 [A]	10.6 ± 0.4 [AB]

Table 5. *Cont.*

Trait	Taxa	Treatment (mM NaCl)				
		0	50	100	200	300
TF	MEL	9.5 ± 2.3 [a]	7.8 ± 3.4 [a]	9.1 ± 2.9 [a]	5.0 ± 0.7 [a]	6.0 ± 0.6 [a]
	INS	15.5 ± 1.8 [B]	10.5 ± 1.9 [AB]	6.5 ± 0.9 [A]	5.5 ± 0.6 [A]	8.7 ± 0.7 [A]
SOD	MEL	377.5 ± 46.9 [a]	272.4 ± 24.2 [a]	180.5 ± 30.8 [a]	315.0 ± 20.3 [a]	412.8 ± 75.0 [a]
	INS	1181.0 ± 263.0 [A]	464.9 ± 134.0 [A]	915.0 ± 221.0 [A]	586.3 ± 163.0 [A]	315.6 ± 71.0 [A]
CAT	MEL	280.2 ± 92.8 [a]	453.4 ± 84.5 [a]	304.3 ± 69.7 [a]	514.1 ± 96.2 [a]	413.2 ± 131.0 [a]
	INS	1135.0 ± 416.0 [A]	523.9 ± 80.4 [A]	7214.0 ± 140.0 [A]	560.8 ± 85.4 [A]	692.1 ± 147 [A]
GR	MEL	2419.0 ± 454.0 [a]	1937.0 ± 384.0 [a]	1468.0 ± 433.0 [a]	1426.0 ± 268.0 [a]	1484.0 ± 263.0 [a]
	INS	2881 ± 684 [A]	1523 ± 197 [A]	3571 ± 764 [A]	1390 ± 144 [A]	1114 ± 77 [A]

Units: MDA (nmol g^{-1} DW), TPC (mg eq. GA g^{-1} DW), TF (mg eq. C g^{-1} DW), and enzymatic activity (U g^{-1} protein). Mean $\pm$ SE values are shown ($n = 5$). Same letters within each row (lowercase for *S. melongena* and capital letters for *S. insanum*) indicate homogeneous groups between treatments for each species according to the Tukey HSD test ($p < 0.05$).

3.5. Principal Component Analysis

A principal component analysis (PCA) was performed, including all analysed traits in all individuals (Figure 5). Eight components with an Eigenvalue greater than one were identified, which overall explained 82.6% of the total variability; the first and second principal components accounted for 33.0% and 15.4% of the total variation, respectively. The first principal component displays positive correlations with growth parameters of stem (SE, ST, and SFW) and leaves (LA, LFW, and Lno), as well as with total fresh weight (TFW); carotenoids (Caro); photosynthetic parameters (A_N, Ci, E, gs); K in leaves (Kl); the ratio K/Na in roots (K/Nar) and leaves (K/Nal); as well as MDA, TP, and TF contents. On the other hand, this first PC is negatively correlated with the levels of Na^+ and Cl^- in roots and leaves (Nar, Nal, Clr, Cll), and with Pro and root water content (RWC). The second component displays strong positive correlations with Pro, some photosynthesis parameters (A_N, E, gs), TSS, and CAT and SOD activities, whereas it is negatively correlated with root water content (RWC), leaf traits (LA, LFW, Lno), chlorophylls a and b (Chl a and Chl b), and K^+ contents in roots (Kr) and leaves (Kl) (Figure 5a).

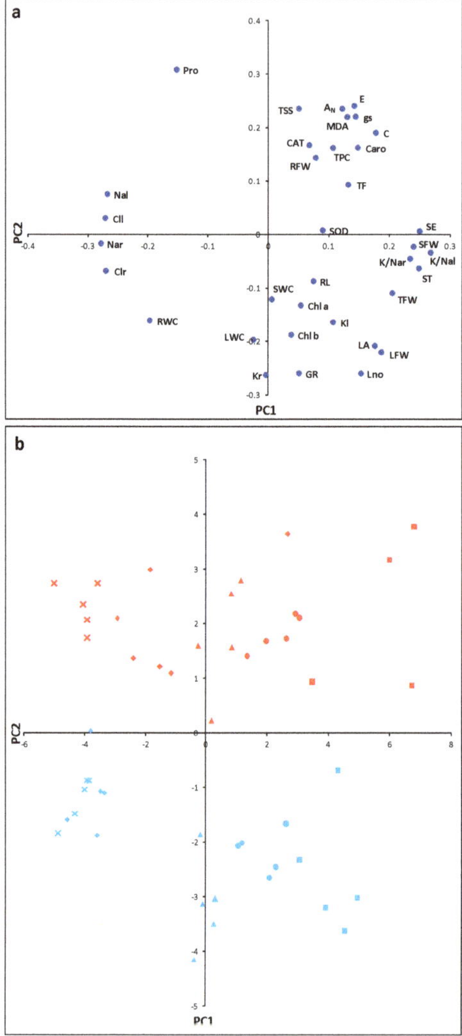

Figure 5. Loading plot (**a**) and scatterplot (**b**) of the principal component analysis (PCA) including all the analysed traits in *Solanum melongena* and *S. incanum* plants subjected for 25 days to salt treatments. The first (PC1; X-axis) and second (PC2; Y-axis) principal components accounted for 33.0% and 15.4% of the total variation, respectively. Abbreviations in the loading plot (**a**) are as follows: root length (RL), root fresh weight (RFW), root water content (RWC), stem elongation (SE), stem thickening (ST), stem fresh weight (SFW), stem water content (SWC), leaf number increment (Lno), maximal leaf area (LA), leaf fresh weight (LFW), leaf water content (LWC), total fresh weight (TFW), chlorophyll a (Chla), chlorophyll b (Chlb), carotenoids (Caro), photosynthetic rate (A_N), internal concentration of CO_2 (Ci), transpiration (E), stomatal conductance (gs), sodium in roots (Nar), sodium in leaves (Nal), potassium in roots (Kr), potassium in leaves (Kl), chloride in roots (Clr), chloride in leaves (Cll), ratio potassium/sodium in roots (K/Nar), ratio potassium/sodium in leaves (K/Nal), proline (Pro), total soluble sugars (TSS), malondialdehyde (MDA), total phenolic compounds (TPC), total flavonoids (TF), superoxide dismutase (SOD), catalase (CAT), and glutathione reductase (GR). Plants of *S. melongena* and of *S. insanum* are represented in blue and red, respectively, in the scatter plot (**b**). Salt treatments are represented by different symbols: 0 (■), 50 (●), 100 (▲), 200 (◆), and 300 (×) mM NaCl.

The 50 individuals analysed were dispersed onto the two axis of the PCA scatterplot (Figure 5b), indicating a clear separation of the applied treatments along the first principal component (X-axis), and of the two species along the second principal component (Y-axis). Plants subjected to the different salt treatments are distributed along the X-axis, from higher positive values (non-stressed controls), to higher negative values (300 mM NaCl), with almost no overlapping of the different treatments, except for the 200 and 300 mM NaCl in *S. melongena*. Samples from moderate salinity treatments (50–100 mM NaCl for *S. melongena* and 100–200 mM NaCl for *S. insanum*) are located in the scatterplot in intermediate positions, closer to '0' in the X-axis. This pattern of distribution validates the homogeneity of responses within each treatment in the two species. Regarding the second principal component, except for one sample per species, *S. insanum* samples are located in the positive part of the Y-axis, whereas *S. melongena* samples have negative values for this component.

4. Discussion and Conclusions

Eggplant is a glycophyte and, as such, responds to increased salinity by a reduction in growth parameters and yield, being generally considered as moderately sensitive (or moderately resistant) to salt stress [7,17,31,32], as other cultivated species of the same genus [33]. However, this crop is characterised by a large variation of phenotypical, biochemical, and physiological traits, which is related to differences between cultivars in their responses to biotic [34] or abiotic stresses, including drought and salinity [35–37]. Therefore, the use of more stress-tolerant cultivars of eggplants on marginal lands or on salinised soils is a realistic challenge for the future, considering that global warming is generating an increased rate of secondary salinisation [38]. Soils are considered as saline when their EC (in a soil saturated paste) is above 4 dS m^{-1}; this electric conductivity corresponds to approximately 40 mM NaCl, generating an osmotic pressure of 0.2 MPa, which significantly reduces the yield of most crops [39]. These values cannot be directly compared with our results as we measured the substrate EC in soil/water (1:5) suspensions (EC$_{1:5}$), not in saturated soil pastes. Nevertheless, in our experiments, all concentrations of NaCl applied were higher than 40 mM, ranging from 50 to 300 mM NaCl. After 25 days of treatments, the salinity of the substrate in pots exposed to the higher concentrations of salt was clearly beyond that normally occurring on salinised soils. All plants survived the salt treatments, but, as expected, growth of stressed plants was reduced in comparison with those from the control treatments in the two investigated species, *S. melongena*, the cultivated eggplant, and its wild relative *S. insanum*.

The analysis of several growth parameters indicated that, in general, the degree of salt-induced growth inhibition was relatively lower in *S. insanum* than in *S. melongena*. One of the most reliable growth variables, when ranking stress tolerance in different cultivars or related species, is the variation of fresh weight (FW) of the plants [36,40,41]. The analysis of this parameter clearly indicated a better tolerance to high salinity in *S. insanum* as the FW of all vegetative organs (roots, stems, and leaves) showed a lesser reduction than in *S. melongena* in the presence of 200 mM and, especially, 300 mM NaCl. Under the 50 mM and 100 mM NaCl treatments, RFW and LFW even slightly increased in the wild species, indicating that these low concentrations have an inhibitory effect only on stem growth. A smaller increase, also non-significant, was registered under 50 mM NaCl for the leaf area (LA) and total fresh weight (TFW) in this species. The highest concentration of 300 mM was not lethal, as all individuals survived until the end of the experiment, but its effect was considerably stronger on *S. melongena*, as shown by a 60% reduction of the total fresh weight (TFW) as compared with only a 30% reduction in *S. insanum*. Special attention is required for the analysis of the root growth parameters because, apparently, all salt treatments stimulated root growth in *S. insanum*. On the contrary, although lower salt concentrations had a positive effect of root growth in *S. melongena*, under the 300 mM NaCl treatment, root length (RL) and root fresh weight (RFW) were significantly reduced. Therefore, the development of more vigorous roots under salt stress represents an important adaptative trait in *S. insanum*. The water content (WC) of vegetative organs, particularly leaves, is another useful indicator of the relative salt tolerance of related taxa. The more tolerant species or cultivars are usually resistant to salt-induced leaf dehydration, or at least the degree of water loss is lower than in the more sensitive ones [42,43]. Indeed, this has also been observed comparing different eggplant cultivars, with those more stress-tolerant showing higher leaf water contents under salt stress conditions [37]. It is worth mentioning that the specific eggplant cultivar used in the present work, MEL1, although more sensitive to salt stress than *S. insanum* INS2, is nevertheless quite tolerant to salinity, at least much more than other common crops such as *Phaseolus* cultivars [42]; all plants survived the salt treatments, even at 300 mM NaCl, and a significant growth inhibition was only observed at the highest salinities tested.

Salt stress reduces photosynthesis, which is one of the major reasons for growth inhibition [44,45]. One of the first effects of abiotic stress is the closure of stomata, which helps in reducing the water loss, but also limits the intake of CO_2. Therefore, in C3 plants (like the two species studied here), C assimilation decreases in such conditions [46]. The photosynthetic rate may also decrease owing to the degradation of chlorophylls or the inhibition of photosynthetic enzymes caused by toxic ions. The photosynthesis rate (A_N) decreased in the two species, but only in plants treated with the highest NaCl concentrations, not at lower salinities, as has been reported in different eggplant cultivars [47]. The internal concentration of CO_2 (Ci) and the transpiration (E) were reduced in *S. melongena* plants in response to the salt treatments, which is associated with a decrease in stomatal conductance (gs); this has also been observed in other cultivars of eggplant [48,49]. In *S. insanum*, however, salt stress did not induce any significant change in the above-mentioned photosynthetic parameters. On the other hand, in both species, chlorophylls a and b levels remained constant, for the control and all salt treatments, contrary to previous reports in eggplant [48,50]. The maintenance of a high assimilation rate in *S. insanum* may rely on its better developed root system, which allowed a higher water uptake under stressful conditions and a lower need for a restriction in transpiration (E), reflected in a higher stomatal conductance (gs) and internal concentration of CO_2 (C). Taken together, these results point to a slightly higher salt tolerance of *S. insanum* INS2, as compared with *S. melongena* MEL1.

Regarding ion accumulation, a significant increase in Na^+ and Cl^- contents was registered in parallel to increasing external salinity, at 100 mM and higher NaCl concentrations, both in roots and leaves and in plants of the two species; similar results have been previously reported in different eggplant cultivars [7,48,49]. Glycophytes typically respond to salt stress trying to limit the accumulation of toxic ions in the leaves, either reducing their absorption by the roots or blocking their transport to the aerial parts of the plant [50]; these mechanisms are effective only at low or moderate salinities, and once a certain threshold—dependent on the tolerance of each specific genotype—is exceeded, Na^+ and Cl^- concentrations increase in the leaves. In our experiments, no inhibition of Na^+ or Cl^- transport from roots to leaves was observed because, generally, their concentration in roots was not higher than in leaves. In *S. melongena*, the concentration of the two ions was practically identical in roots and leaves, at each salinity level (except for Na^+ at 100 mM NaCl). Interestingly, in *S. insanum* plants treated with 200 or 300 mM NaCl, Na^+ concentrations in leaves were substantially higher than in roots, and the same pattern was observed for Cl^- at 100 mM and higher NaCl concentrations. This suggests that, in this species, high salinity activates the transport of these ions from roots to leaves, where they could contribute to cellular osmotic balance as inorganic osmolytes. This is not a common behaviour of glycophytes like eggplant, but represents one of the most relevant mechanisms of salt tolerance in dicotyledonous halophytes [51,52], which could also be operative in *S. insanum*, contributing to its relative higher tolerance, enhanced also by a more developed root system that allows a higher ion uptake.

Potassium homeostasis is also critical for salt tolerance, which includes as a key mechanism the intracellular retention of K^+ in the presence of high external salinities [53,54], as this cation is essential in plant metabolism. An increase in Na^+ concentration is generally accompanied by a decrease of K^+, as both cations compete for the same membrane transport proteins [55]. Furthermore, high Na^+ levels produce a depolarisation of the plasma membrane, which induces K^+-efflux from cells by activating voltage-dependent outward rectifying channels [56,57]. Many reports indicated a reduction of K^+ in conditions of salt stress in eggplant, as expected [32,48,49]. In our experiments, however, no significant changes in root or leaf K$^+$ concentrations were observed in response to the salt treatments. Maintenance of constant K$^+$ levels, despite the increase in Na$^+$ concentrations, probably also contributes to salt tolerance, in this case, in both tested genotypes, *S. melongena* MEL1 and *S. insanum* INS2. Further studies will be required to elucidate the specific ion transporters involved in these regulatory mechanisms.

Another general response to salt stress is the synthesis of Pro, one the commonest osmolytes in plants, which, besides osmotic adjustment, plays an important role in ROS detoxification and maintenance of membrane integrity under stress [58,59]. Pro accumulation may be simply a biomarker of the level of stress affecting a plant, reaching higher concentrations in the more stressed individuals, as has been shown in some comparative studies on related genotypes [42]. On the contrary, Pro can be directly involved in the mechanisms of tolerance to stress, so that higher contents correlate with higher tolerance [40]. Comparative analyses of different eggplant cultivars have provided mixed results; in some cases, the more stress-tolerant genotypes accumulated higher Pro concentrations [35,36,60], but in other studies, higher levels were found in the more sensitive ones [32]. Our results clearly showed higher Pro levels in *S. insanum* than in *S. melongena* in all experimental conditions, but especially in the presence of the highest salinity tested, 300 mM NaCl, thus correlating with the relative salt tolerance of the two investigated species.

Although soluble sugars play a role in osmoregulation under stress conditions in many plant species [61], their levels did not vary significantly in response to the salt treatments in *S. insanum*, and were similar in the two species at high salinities. Therefore, TSS contents do not correlate with the degree of salt tolerance, and probably do not play any relevant role in the responses to salt stress of the two species studied here.

Mechanisms of salt tolerance based mostly on the accumulation of Pro, for osmotic adjustment and as 'osmoprotector'—with the possible contribution of Na$^+$ and Cl$^-$ as inorganic osmolytes in the case of *S. insanum*—appear to be efficient enough to avoid the generation of oxidative stress under the specific conditions used in our experiments. A common effect of high salinity, as well as other abiotic stresses, is the increase in the concentration of ROS, leading to secondary oxidative stress [62]. That did not occur in the present work, as shown by the determination of MDA contents, which did not increase in response to the salt treatments. Consequently, the activation of antioxidant systems, enzymatic and non-enzymatic, was also not detected, as the plants did not need to counteract any salt-induced oxidative stress. Generally, this behaviour is not observed in glycophytes, but has been reported for many halophytes [26,63].

In conclusion, our results from plant growth, photosynthetic parameters, and biochemical stress markers measurements indicate that *S. insanum* displays greater tolerance to moderate salt stress than *S. melongena*, mostly because of its ability to accumulate higher concentrations of Pro and, to a lesser extent, Na$^+$ and Cl$^-$ in the leaves, especially at high external salinities. Given that *S. insanum* and *S. melongena* are fully cross-compatible [10,16], and introgression breeding from *S. insanum* into *S. melongena* is relatively easy [11], we suggest that *S. insanum* can contribute to the development of *S. melongena* cultivars with increased salt tolerance. It remains to be evaluated if *S. insanum* could also be useful as a rootstock for eggplant under conditions of salinity. Therefore, the use of *S. insanum* in eggplant breeding and rootstock development may make an effective contribution to extending cultivation of eggplant in cultivated lands that are affected, or will be in the future, by soil salinity.

Author Contributions: Conceptualization, O.V. and J.P.; Data curation, M.B. (Marco Brenes) and M.B. (Monica Boscaiu); Formal analysis, J.P.; Funding acquisition, O.V. and J.P.; Investigation, M.P.; Methodology, M.B. (Marco Brenes) and A.S.; Project administration, O.V.; Resources, O.V. and J.P.; Software, M.B. (Monica Boscaiu); Supervision, M.P.; Validation, A.F. and A.C.; Visualization, M.B. (Monica Boscaiu) and M.P.; Writing—original draft, M.B. (Monica Boscaiu) and J.P.; Writing—review & editing, A.F., O.V., A.C., and M.P. All authors have read and agreed to the published version of the manuscript

Funding: This work was undertaken as part of the initiative "Adapting Agriculture to Climate Change: Collecting, Protecting and Preparing Crop Wild Relatives", which is supported by the Government of Norway and managed by the Global Crop Diversity Trust. For further information, see the project website: http://cwrdiversity.org/. Funding was also received from Ministerio de Ciencia, Innovación y Universidades, Agencia Estatal de Investigación and Fondo Europeo de Desarrollo Regional (grant RTI-2018-094592-B-100 from MCIU/AEI/FEDER, UE), European

Union's Horizon 2020 Research and Innovation Programme under grant agreement No. 677379 (Linking genetic resources, genomes, and phenotypes of Solanaceous crops; G2P-SOL) and Vicerrectorado de Investigación, Innovación y Transferencia de la Universitat Politècnica de València (Ayuda a Primeros Proyectos de Investigación; PAID-06-18). Mariola Plazas is grateful to Generalitat Valenciana and Fondo Social Europeo for a post-doctoral grant (APOSTD/2018/014). Marco Brenes is indebted to the Faculty of Biology of the Costa Rica Institute of Technology for partially supporting his stay in Valencia ("Fondo Solidario y Desarrollo Estudiantil").

Conflicts of Interest: The authors declare no conflict of interest. The funders had no role in the design of the study; in the collection, analyses, or interpretation of data; in the writing of the manuscript; or in the decision to publish the results.

References

1. Wicke, B.; Smeets, E.; Dornburg, V.; Vashev, B.; Gaiser, T.; Turkenburg, W.; Faaij, A. The global technical and economic potential of bioenergy from salt-affected soils. *Energy Environ. Sci.* **2011**, *8*, 2669–2681. [CrossRef]
2. Qureshi, A.S.; Mohammed, M.; Daba, A.W.; Hailu, B.; Belay, G.; Tesfaye, A.; Ertebo, A-M. Improving agricultural productivity on salt-affected soils in Ethiopia: Farmers' perceptions and proposals. *Afr. J. Agric. Res.* **2019**, *14*, 897–906. [CrossRef]
3. Daliakopoulos, I.; Tsanis, I.K.; Koutroulis, A.; Kourgialas, N.; Varouchakis, E.; Karatzas, G.; Ritsema, C. The threat of soil salinity: A European scale review. *Sci. Total Environ.* **2016**, *573*, 727–739. [CrossRef] [PubMed]
4. Larcher, W. *Physiological Plant Ecology*, 4th ed.; Springer: Berlin/Heidelberg, Germany; New York, NY, USA, 2003.
5. Tang, X.; Mu, X.; Shao, H.; Wang, H.; Brestic, M. Global plant-responding mechanisms to salt stress: Physiological and molecular levels and implications in biotechnology. *Crit. Rev. Biotechnol.* **2015**, *35*, 425–437. [CrossRef]
6. Food and Agriculture Organization of the United Nations (FAO). *FAOSTAT: Food and Agriculture Data*; Food and Agriculture Organization of the United Nations: Rome, Italy, 2020. Available online: http: //www.fao.org/faostat/en/ (accessed on 14 February 2020).
7. Ünlükara, A.; Kurunç, A.; Kesmez, G.D.; Yurtseven, E.; Suarez, D.L. Effects of salinity on eggplant (*Solanum melongena* L.) growth and evapotranspiration. *Irrig. Drain.* **2010**, *59*, 203–214. [CrossRef]
8. Mennella, G.; Lo Scalzo, R.; Fibiani, M.; D'Alessandro, A.; Francese, G.; Toppino, L.; Acciarri, N.; de Almeida, A.E.; Rotino, G.L. Chemical and bioactive quality traits during fruit ripening in eggplant (*S. melongena* L.) and allied species. *J. Agric. Food Chem.* **2012**, *60*, 11821–11831. [CrossRef]
9. Plazas, M.; López Gresa, M.P.; Vilanova, S.; Torres, C.; Hurtado, M.; Gramazio, P.; Andújar, I.; Herráiz, F.J.; Bellés, J.M.; Prohens, J. Diversity and relationships in key traits for functional and apparent quality in a collection of eggplant: Fruit phenolics content, antioxidant activity, polyphenol oxidase activity, and browning. *J. Agric. Food Chem.* **2013**, *61*, 8871–8879. [CrossRef]
10. Plazas, M.; Vilanova, S.; Gramazio, P.; Rodríguez-Burruezo, A.; Fita, A.; Herráiz, F.J.; Prohens, J. Interspecific hybridization between eggplant and wild relatives from different genepools. *J. Am. Soc. Hort. Sci.* **2016**, *141*, 34–44. [CrossRef]
11. Kouassi, B.; Prohens, J.; Gramazio, P.; Kouassi, A.B.; Vilanova, S.; Galán-Ávila, A.; Herráiz, F.J.; Kouassi, A.; Seguí-Simarro, J.M.; Plazas, M. Development of backcross generations and new interspecific hybrid combinations for introgression breeding in eggplant (*Solanum melongena*). *Sci. Hortic.* **2016**, *213*, 199–207. [CrossRef]
12. Gramazio, P.; Prohens, J.; Plazas, M.; Mangino, G.; Herraiz, F.J.; Vilanova, S. Development and genetic characterization of advanced backcross materials and an introgression line population of *Solanum incanum* in a *S. melongena* background. *Front. Plant Sci.* **2017**, *8*, 1477. [CrossRef]
13. García-Fortea, E.; Gramazio, P.; Vilanova, S.; Fita, A.; Mangino, G.; Villanueva, G.; Arrones, A.; Knapp, S.; Prohens, J.; Plazas, M. First successful backcrossing towards eggplant (*Solanum melongena*) of a New World species, the silverleaf nightshade (*S. elaeagnifolium*), and characterization of interspecific hybrids and backcrosses. *Sci. Hortic.* **2019**, *246*, 563–573. [CrossRef]
14. Knapp, S.; Vorontsova, M.S. A revision of the "African non-spiny" clade of *Solanum* L. (*Solanum* sections *Afrosolanum* Bitter, *Benderianum* Bitter, *Lemurisolanum* Bitter, *Lyciosolanum* Bitter, *Macronesiotes* Bitter, and *Quadrangulare* Bitter: *Solanaceae*). *PhytoKeys* **2016**, *66*, 1–142. [CrossRef] [PubMed]

15. Ranil, R.H.; Prohens, J.; Aubriot, X.; Niran, H.M.; Plazas, M.; Fonseka, R.M.; Vilanova, S.; Gramazio, P.; Knapp, S. *Solanum insanum* L. (subgenus *Leptostemonum* Bitter, Solanaceae), the neglected wild progenitor of eggplant (*S. melongena* L.): A review of taxonomy, characteristics and uses aimed at its enhancement for improved eggplant breeding. *Genet. Resour. Crop Evol.* **2017**, *64*, 1707–1722. [CrossRef]

16. Davidar, P.; Snow, A.A.; Rajkumar, M.; Pasquet, R.; Daunay, M.C.; Mutegi, E. The potential for crop to wild hybridization in eggplant (*Solanum melongena*; Solanaceae) in southern India. *Am. J. Bot.* **2015**, *102*, 129–139. [CrossRef] [PubMed]

17. Akinci, I.E.; Akinci, S.; Yilmaz, K.; Dikici, H. Response of eggplant varieties (*Solanum melongena*) to salinity in germination and seedling stages. *N. Zeal. J. Crop Hort. Sci.* **2004**, *32*, 193–200. [CrossRef]

18. Ranil, R.H.; Niran, H.M.L.; Plazas, M.; Fonseka, R.M.; Fonsekad, H.H.; Vilanova, S.; Andújar, I.; Gramazio, P.; Fita, A.; Prohens, J. Improving seed germination of the eggplant rootstock *Solanum torvum* by testing multiple factors using an orthogonal array design. *Sci. Hortic.* **2015**, *193*, 174–181. [CrossRef]

19. Weimberg, R. Solute adjustments in leaves of two species of wheat at two different stages of growth in response to salinity. *Physiol. Plant.* **1987**, *70*, 381–388. [CrossRef]

20. Lichenthaler, H.K.; Wellburn, A.R. Determinations of total carotenoids and chlorophylls a and b of leaf extracts in different solvents. *Biochem. Soc. Trans.* **1983**, *11*, 591–592. [CrossRef]

21. Bates, L.S.; Waldren, R.P.; Teare, I.D. Rapid determination of free proline for water stress studies. *Plant Soil* **1973**, *39*, 205–207. [CrossRef]

22. Dubois, M.; Gilles, K.A.; Hamilton, J.K.; Rebers, P.A.; Smith, F. Colorimetric method for determination of sugars and related substances. *Anal. Chem.* **1956**, *28*, 350–356. [CrossRef]

23. Hodges, D.M.; Delong, J.M.; Forney, C.F.; Prange, R.K. Improving the thiobarbituric acid-reactive-substances assay for estimating lipid peroxidation in plant tissues containing anthocyanin and other interfering compounds. *Planta* **1999**, *207*, 604–611. [CrossRef]

24. Blainski, A.; Lopes, G.C.; Palazzodemello, J.C. Application and analysis of the Folin Ciocalteu method for the determination of the total phenolic content from *Limonium brasiliense* L. *Molecules* **2013**, *18*, 6852–6865. [CrossRef] [PubMed]

25. Zhishen, J.; Mengcheng, T.; Jianming, W. The determination of flavonoid contents in mulberry and their scavenging effects on superoxide radicals. *Food Chem.* **1999**, *64*, 555–559. [CrossRef]

26. Gil, R.; Bautista, I.; Boscaiu, M.; Lidón, A.; Wankhade, S.; Sánchez, H.; Llinares, J.; Vicente, O. Responses of five Mediterranean halophytes to seasonal changes in environmental conditions. *AoB Plants* **2014**, *6*, plu049. [CrossRef]

27. Aebi, H. Catalase in vitro. *Methods Enzymol.* **1984**, *105*, 121–126. [CrossRef]

28. Beyer, W.F., Jr.; Fridovich, I. Assaying for superoxide dismutase activity: Some large consequences of minor changes in conditions. *Anal. Biochem.* **1987**, *161*, 559–566. [CrossRef]

29. Conell, J.P.; Mullet, J.E. Pea chloroplast glutathione reductase: Purification and characterization. *Plant Physiol.* **1986**, *82*, 351–356. [CrossRef]

30. Del Rio, D.; Stewart, A.J.; Pellegrini, N. A review of recent studies on malondialdehyde as toxic molecule and biological marker of oxidative stress. *Nutr. Metab. Cardiovas.* **2005**, *15*, 316–328. [CrossRef]

31. Zayova, E.; Philipov, P.; Nedev, T.; Stoeva, D. Response of in vitro cultivated eggplant (*Solanum melongena* L.) to salt and drought stress. *AgroLife Sci. J.* **2017**, *6*, 276–282.

32. Hannachi, S.; Van Labeke, M.C. Salt stress affects germination, seedling growth and physiological responses differentially in eggplant cultivars (*Solanum melongena* L.). *Sci. Hortic.* **2018**, *228*, 56–65. [CrossRef]

33. Foolad, M.R. Recent advances in genetics of salt tolerance in tomato. *Plant Cell Tissue Organ Cult.* **2004**, *76*, 101–119. [CrossRef]

34. Rotino, G.L.; Sala, T.; Toppino, L. Eggplant. In *Alien Gene Transfer in Crop Plants, Volume 2: Achievements and Impacts*; Pratap, A., Kumar, J., Eds.; Springer Science+Business Media: New York, NY, USA, 2014.

35. Plazas, M.; Nguyen, T.; González-Orenga, S.; Fita, A.; Vicente, O.; Prohens, J.; Boscaiu, M. Comparative analysis of the responses to water stress in eggplant (*Solanum melongena*) cultivars. *Plant Physiol. Biochem.* **2019**, *143*, 72–82. [CrossRef] [PubMed]

36. Mustafa, Z.; Ayyub, C.M.; Amjad, M.; Ahmad, R. Assessment of biochemical and ionic attributes against salt stress in eggplant (*Solanum melongena* L.) genotypes. *J. Anim. Plant Sci.* **2017**, *27*, 503–509.

37. Hanachi, S.; Van Labeke, M.C.; Mehouachi, T. Application of chlorophyll fluorescence to screen eggplant (*Solanum melongena* L.) cultivars for salt tolerance. *Photosynthetica* **2014**, *52*, 57–62. [CrossRef]

38. Intergovernmental panel on climate change (IPCC). *Climate Change 2014: Impacts, Adaptation, and Vulnerability*; 5th Assessment Report, WGII; Cambridge University Press: Cambridge, UK, 2015. Available online: http://www.ipcc.ch/report/ar5/wg2/ (accessed on 18 March 2020).

39. Richards, L. Diagnosis and improvement of saline and alkali soils. *Soil Sci.* **1954**, *78*, 154. [CrossRef]

40. Al Hassan, M.; López-Gresa, M.P.; Boscaiu, M.; Vicente, O. Stress tolerance mechanisms in *Juncus*: Responses to salinity and drought in three *Juncus* species adapted to different natural environments. *Funct. Plant Biol.* **2016**, *43*, 949–960. [CrossRef]

41. González-Orenga, S.; Ferrer-Gallego, P.P.; Laguna, E.; López-Gresa, M.P.; Donat-Torres, M.P.; Verdeguer, M.; Vicente, O.; Boscaiu, M. Insights on salt tolerance of two endemic *Limonium* species from Spain. *Metabolites* **2019**, *9*, 294. [CrossRef]

42. Al Hassan, M.; Morosan, M.; López-Gresa, M.D.P.; Prohens, J.; Vicente, O.; Boscaiu, M. Salinity-induced variation in biochemical markers provides insight into the mechanisms of salt tolerance in common (*Phaseolus vulgaris*) and runner (*P. coccineus*) beans. *Int. J. Mol. Sci.* **2016**, *17*, 1582. [CrossRef]

43. Al Hassan, M.; Pacurar, A.; López-Gresa, M.P.; Donat-Torres, M.P.; Llinares, J.V.; Boscaiu, M.; Vicente, O. Effects of salt stress on three ecologically distinct *Plantago* species. *PLoS ONE* **2016**, *11*, e0160236. [CrossRef]

44. Jamil, M.; Rehman, S.U.; Lee, K.J.; Kim, J.M.; Kim, H.; Rha, E.S. Salinity reduced growth PS2 photochemistry and chlorophyll content in radish. *Sci. Agric.* **2007**, *64*, 111–118. [CrossRef]

45. Shrivastava, P.; Kumar, R. Soil salinity: A serious environmental issue and plant growth promoting bacteria as one of the tools for its alleviation. *Saudi J. Biol. Sci.* **2015**, *22*, 123–131. [CrossRef] [PubMed]

46. Acosta-Motos, J.; Ortuño, M.; Bernal-Vicente, A.; Diaz-Vivancos, P.; Sanchez-Blanco, M.; Hernandez, J. Plant responses to salt stress: Adaptive mechanisms. *Agronomy* **2017**, *7*, 18. [CrossRef]

47. Wu, X.; Zhu, Z.; Li, X.; Zha, D. Effects of cytokinin on photosynthetic gas exchange, chlorophyll fluorescence parameters and antioxidative system in seedlings of eggplant (*Solanum melongena* L.) under salinity stress. *Acta Physiol. Plant.* **2012**, *34*, 2105–2114. [CrossRef]

48. Shaheen, S.; Naseer, S.; Ashraf, M.; Akram, N.A. Salt stress affects water relations, photosynthesis, and oxidative defense mechanisms in *Solanum melongena* L. *J. Plant Interact.* **2013**, *8*, 85–96. [CrossRef]

49. Shahbaz, M.; Mushtaq, Z.; Andaz, F.; Masood, A. Does proline application ameliorate adverse effects of salt stress on growth, ions and photosynthetic ability of eggplant (*Solanum melongena* L.)? *Sci. Hortic.* **2013**, *164*, 507–511. [CrossRef]

50. Volkov, V. Salinity tolerance in plants. Quantitative approach to ion transport starting from halophytes and stepping to genetic and protein engineering for manipulating ion fluxes. *Front. Plant Sci.* **2015**, *6*. [CrossRef]

51. Flowers, T.J.; Colmer, T.D. Salinity tolerance in halophytes. *New Phytol.* **2008**, *179*, 945–963. [CrossRef]

52. Munns, R.; Tester, M. Mechanisms of salinity tolerance. *Annu. Rev. Plant. Biol.* **2008**, *59*, 651–681. [CrossRef]

53. Wu, H.; Zhang, X.; Giraldo, J.P.; Shabala, S. It is not all about sodium: Revealing tissue specificity and signalling roles of potassium in plant responses to salt stress. *Plant Soil* **2018**, *431*, 1–17. [CrossRef]

54. Assaha, D.V.M.; Ueda, A.; Saneoka, H.; Al-Yahyai, R.; Yaish, M.W. The role of Na$^+$ and K$^+$ transporters in salt stress adaptation in glycophytes. *Front. Physiol.* **2017**, *8*, 509. [CrossRef]

55. Almeida, D.M.; Oliveira, M.M.; Saibo, N.J.M. Regulation of Na$^+$ and K+ homeostasis in plants: Towards improved salt stress tolerance in crop plants. *Genet. Mol. Biol.* **2017**, *40*, 326–345. [CrossRef] [PubMed]

56. Flowers, T.; Troke, P.F.; Yeo, A.R. The mechanism of salt tolerance in halophytes. *Annu. Rev. Plant Physiol.* **1977**, *28*, 89–121. [CrossRef]

57. Greenway, H.; Munns, R. Mechanisms of salt tolerance in non-halophytes. *Annu. Rev. Plant Biol.* **1980**, *31*, 149–190. [CrossRef]

58. Verbruggen, N.; Hermans, C. Proline accumulation in plants: A review. *Amino Acids* **2008**, *35*, 753–759. [CrossRef]

59. Gupta, B.; Huang, B. Mechanism of salinity tolerance in plants: Physiological, biochemical, and molecular characterization. *Int. J. Genom.* **2014**, *2014*, 701596. [CrossRef]

60. Sarker, B.C.; Hara, M.; Uemura, M. Proline synthesis, physiological responses and biomass yield of eggplants during and after repetitive soil moisture stress. *Sci. Hortic.* **2005**, *103*, 387–402. [CrossRef]

61. Gil, R.; Boscaiu, M.T.; Lull, C.; Bautista, I.; Lidón, A.; Vicente, O. Are soluble carbohydrates ecologically relevant for salt tolerance in halophytes? *Funct. Plant Biol.* **2013**, *40*, 805–818. [CrossRef]

62. Apel, K.; Hirt, H. Reactive oxygen species: Metabolism, oxidative stress, and signal transduction. *Annu. Rev. Plant Biol.* **2004**, *55*, 373–399. [CrossRef]

63. Bose, J.; Rodrigo-Moreno, A.; Shabala, S. ROS homeostasis in halophytes in the context of salinity stress tolerance. *J. Exp. Bot.* **2014**, *65*, 1241–1257. [CrossRef]

Article

Main Root Adaptations in Pepper Germplasm (*Capsicum* spp.) to Phosphorus Low-Input Conditions

Leandro Pereira-Dias [1], Daniel Gil-Villar [1], Vincente Castell-Zeising [2], Ana Quiñones [3], Ángeles Calatayud [3], Adrián Rodríguez-Burruezo [1] and Ana Fita [1,*]

[1] Instituto de Conservación y Mejora de la Agrodiversidad Valenciana (COMAV), Universitat Politècnica de València, 46022 Valencia, Spain; leapedia@etsiamn.upv.es (L.P.-D.); dagivil@upv.es (D.G.-V.); adrodbur@upv.es (A.R.-B.)

[2] Departamento de Producción Vegetal, Universitat Politècnica de València, 46022 Valencia, Spain; vcastell@prv.upv.es

[3] Instituto Valenciano de Investigaciones Agrarias (IVIA), 46113 Moncada, Valencia, Spain; quinones_ana@gva.es (A.Q.); calatayud_ang@gva.es (Á.C.)

* Correspondence: anfifer@btc.upv.es; Tel.: +34-963-879-418

Received: 27 March 2020; Accepted: 19 April 2020; Published: 1 May 2020

Abstract: Agriculture will face many challenges regarding food security and sustainability. Improving phosphorus use efficiency is of paramount importance to face the needs of a growing population while decreasing the toll on the environment. Pepper (*Capsicum* spp.) is widely cultivated around the world; hence, any breakthrough in this field would have a major impact in agricultural systems. Herein, the response to phosphorus low-input conditions is reported for 25 pepper accessions regarding phosphorus use efficiency, biomass and root traits. Results suggest a differential response from different plant organs to phosphorus starvation. Roots presented the lowest phosphorus levels, possibly due to mobilizations towards above-ground organs. Accessions showed a wide range of variability regarding efficiency parameters, offering the possibility of selecting materials for different inputs. Accessions bol_144 and fra_DLL showed an interesting phosphorus efficiency ratio under low-input conditions, whereas mex_scm and sp_piq showed high phosphorus uptake efficiency and mex_pas and sp_bola the highest values for phosphorus use efficiency. Phosphorus low-input conditions favored root instead of aerial growth, enabling increases of root total length, proportion of root length dedicated to fine roots and root specific length while decreasing roots' average diameter. Positive correlation was found between fine roots and phosphorus efficiency parameters, reinforcing the importance of this adaptation to biomass yield under low-input conditions. This work provides relevant first insights into pepper's response to phosphorus low-input conditions.

Keywords: *Capsicum annuum*; root structure; root hairs; phosphorus use efficiency; P-starvation; abiotic stress; macrominerals; nutrient; breeding

---

1. Introduction

Agriculture will face many challenges in the next generations, especially those related to food security and agricultural sustainability [1,2]. On one hand, intensive agriculture has a significant impact on the environment, contributing to soil erosion, soil salinization, eutrophication and contamination of water bodies, and biodiversity reduction [3,4]. On the other hand, agricultural systems need to be improved in order to cope with requirements of an increasing population as well as the impact of climate change consequences [1,5].

In both cases, one of the most critical resources involved is phosphorus (P), an inorganic mineral with a major role within the physiochemical processes of plants [6,7]. Since almost 40% of the world's arable land lacks of P or the soil properties to make it available for crops, P absence is a major constraint to food production all around the world [8–10]. Until now, application of P-enriched fertilizers has been the main strategy to face its deficiency in soils despite the severe contaminants emissions associated to its production [3,9,11]. In addition, only 15 to 40% of the added P is taken up by crops [3,9,12], while the remaining ends up being washed down through the soil, contributing to eutrophication of water bodies [13,14]. Furthermore, as costs of extraction increase and rock-phosphate reserves decline, P is becoming an extremely expensive resource that is already unaffordable in many regions of the globe [10]. As demand for P-enriched fertilizers is going to increase in the next decades, the control for P supply will be a source of conflicts [7,9]. Therefore, there is a need for P low-input adapted varieties.

The response to P-starvation conditions has been studied for a few model organisms and some economically important crops, such as soybean, maize, sunflower, brassica or melon over the last decades [15–19]. As a result, researchers have linked several root traits to a greater performance under low P conditions [20]. Thus, morphological changes, such as the increment of number of root hairs and higher root branching [15,18,21], as well as physiological changes, such as cellular structure alteration, enhanced phosphatases enzyme activity and organic acids production and root P transporters enhanced expression [12,16,22,23], are adaptations expressed under P-starvation conditions. The exploitation of these plant adaptations could have a remarkable impact on the reduction of chemical fertilizers inputs in agricultural systems [12,24].

Peppers (*Capsicum* spp.) are one of the most relevant vegetables, grown in almost all temperate and tropical regions of the world [25]. Food and Agriculture Organization of the United Nations (FAO) last available data estimates around 40×10^6 t of pepper produced each year [26]. Therefore, improving pepper for its uptake and use of P would significantly reduce the need for P-fertilizer applications [3,12]. Notwithstanding, the development of improved *Capsicum* varieties for P low-input conditions is a challenging goal and is conditioned by both the availability of genetic variability within *Capsicum* and the understanding of the mechanisms underlying the response. Regarding the first point, *Capsicum* spp., particularly *Capsicum annuum* L., is remarkably diverse, as well as adapted to a wide range of environments and, therefore, tolerant to several abiotic stresses [27–30]. However, pepper fundamentals regarding this subject have never been studied. Hence, we believe that an exhaustive characterization of pepper germplasm for its responses under P low-input conditions is of paramount importance in order to recognize the variability within the genus, to enhance our understanding regarding the responses activated under such conditions and, finally, to link those responses to the genomic regions controlling them. Herein, the characterization of the main root adaptations of pepper accessions to low P conditions was established as a main goal, as a first step towards the identification of elite individuals for future pepper breeding programs.

2. Materials and Methods

2.1. Germplasm

A collection of 25 pepper accessions, encompassing 22 *Capsicum annuum*, two *Capsicum chinense* and one *Capsicum frutescens* accessions, comprising a wide range of variability for fruit shape, fruit pungency, fruit color, biotic resistances and adaptation to the environments, was studied herein [31] (Table 1). The considered collection belongs to the Instituto Universitario de Conservación y Mejora de la Agrodiversidad Valenciana (COMAV) Germplasm Bank (Universitat Politècnica de València, Spain) and to the COMAV *Capsicum* breeding group, and was selected based on previous experiments, where an interesting performance and diversity for several relevant root and P uptake traits was observed [32].

Table 1. List of the 25 accessions and corresponding abbreviation, species, varietal status, origin, fruit shape, fruit taste, fruit color and trial year.

Abbreviation	Species	Accession (UPV Genebank Code)	Origin	Fruit Shape	Fruit Taste	Fruit Color	Trial
	Traditional varieties						
fra_DLL	*Capsicum annuum*	Doux Long des Landes	France (INRA-GEVES, F. Jourdan)	Cayenne, long-sized	Sweet	Red	Trial 2
mex_096D	*Capsicum annuum*	Chile Ancho Poblano	Mexico, Aguascalientes	Triangular, Pochard's C4 type	Hot	Red	Trial 2
mex_103B	*Capsicum annuum*	Chile Ancho Poblano	Mexico, Aguascalientes	Triangular, Pochard's C4 type	Hot	Red	Trial 2
mex_pas	*Capsicum annuum*	Pasilla Bajío	Mexico, Reymer Seeds	Cayenne, long-sized	Hot	Brown	Trial 1 and Trial 2
mex_ng	*Capsicum annuum*	Numex Garnet	Mexico, Aguascalientes	Elongated, Pochard's C2 type	Sweet	Red	Trial 2
mu_esp	*Capsicum annuum*	Jalapeno Espinalteco	Mexico/USA (P. W. Bosland)	Jalapeno	Hot	Red	Trial 1 and Trial 2
sp_060	*Capsicum annuum*	Pimiento morrón de bola (BGV00060)	Spain, Zamora	Round, Pochard's F type	Sweet	Red	Trial 2
sp_11814	*Capsicum annuum*	Dulce Italiano (BGV11814)	Spain, León	Elongated, Pochard's C2 type	Sweet	Red	Trial 2
sp_bola	*Capsicum annuum*	Pimiento de bola, ñora	Spain, Murcia (P.D.O. Pimentón Murcia)	Round, Pochard's N type	Sweet	Red	Trial 1 and Trial 2
sp_lam	*Capsicum annuum*	Lamuyo	Spain, Valencia	Blocky, Pochard's B1 or B2 type	Sweet	Red	Trial 2
sp_piq	*Capsicum annuum*	Pimiento Piquillo de Lodosa	Spain, Navarra (P.D.O. Piquillo Lodosa)	Triangular, Pochard's C4 type	Sweet	Red	Trial 1 and Trial 2
usa_chi	*Capsicum annuum*	Chimayó	USA, New Mexico (P. W. Bosland)	Blocky small-sized, Pochard's B4 type	Hot	Red	Trial 1
usa_conq	*Capsicum annuum*	Numex Conquistador	USA, New Mexico	Elongated, Pochard's C2 type	Sweet	Red	Trial 2
usa_jap	*Capsicum annuum*	Chile Japonés	USA, New Mexico	Cayenne, very short-sized	Hot	Red	Trial 2
usa_numex	*Capsicum annuum*	Numex X	USA, New Mexico	Elongated, Pochard's C2 type	Hot	Red	Trial 2
usa_sandia	*Capsicum annuum*	Numex Sandia (BGV13293)	USA, New Mexico	Elongated, Pochard's C2 type	Hot	Red	Trial 2
	Experimental lines						
mex_scm	*Capsicum annuum*	Serrano Criollo de Morellos (SCM334)	Mexico	Serrano	Hot	Red	Trial 1 and Trial 2
sp_cwr	*Capsicum annuum*	California Wonder	Spain, Valencia (COMAV)	Blocky, Pochard's A type	Sweet	Red	Trial 1
	Commercial hybrids (F₁)						
sp_anc	*Capsicum annuum*	Ancares	Spain (Ramiro Arnedo)	Blocky, Pochard's B1 or B2 type	Sweet	Red	Trial 2
sp_cat	*Capsicum annuum*	Catedral	Spain (Zeraim Ibérica)	Blocky, Pochard's A type	Sweet	Red	Trial 1
sp_lobo	*Capsicum annuum*	El Lobo	Spain (Zeraim Ibérica)	Blocky, Pochard's A type	Sweet	Red	Trial 2
sp_mel	*Capsicum annuum*	Melchor	Spain (Ramiro Arnedo)	Blocky, Pochard's A type	Sweet	Red	Trial 1
	Other Capsicums						
bol_037	*Capsicum chinense*	Bol-37R (BGV007644)	Chuquisaca, Bolivia	Triangular, small-sized, thin flesh	Hot	Red	Trial 1
bol_144	*Capsicum baccatum*	Bol-144 (BGV007751)	Bolivia, Santa Cruz	Cayenne, very short-sized	Hot	Red	Trial 1
eq_973	*Capsicum chinense*	ECU-973	Ecuador, Napo	Triangular, small-sized, thin flesh	Hot	Red	Trial 1

2.2. Germination and Cultivation Conditions

Seeds were surface sterilized with a 30% NaClO solution (v:v) for five minutes, followed by rinsing with steril deionized water, and transfered to individual Petri dishes containing a wet layer of cotton under a filter paper disk. Two drops of 2% Tetramethylthiuram disulphide solution were added to each Petri dish to prevent fungal proliferation. Petri dishes were kept under germination chamber conditions until two-cotyledon stage. Seedlings were then transferred to seedling trays filled with Neuhaus N3 substrate (Klasmann-Dellmann GmbH, Geeste, Germany), kept under heated nursery conditions until the five leaves stage and, finally, transplanted to the greenhouse.

The experiment was carried out in two years. In the first year (from now on Trial 1), 12 accessions were trialed and the five most interesting genotypes were re-trialed in the second year (from now on Trial 2), against 13 new accessions (Table 1). In both trial years, plants were grown for 60 days under a mesh greenhouse, during the spring-summer cycle, on COMAV experimental fields (Universitat Politècnica de València Vera Campus GPS coordinates: 39°28′56.33″ N; 0°20′10.88″ W). Transplant was carried out in June and the experiment was finished in August. Nine (Trial 1) and six (Trial 2) plants, per accession and treatment, were grown in 15 L plastic pots filled with substrate made by mixing a part of soil with a part of sand (1:1) and arranged into a completely randomized design with six rows. Pots were spaced 1.2 m between rows and 0.40 m inside rows, while a drip irrigation system provided water and nutrient solutions to cover the plants' water and nutritional requirements. Individual plants were trained with vertical strings, according to standard local practices for pepper. Plants were not pruned during the experiment in order to avoid interference with biomass yield. Likewise, phytosanitary treatments against whiteflies, spider mites, aphids and caterpillars were applied accordingly to population levels.

Plants were subjected to two treatments. On one hand, control treatment was applied using a standard solution providing all elements (Table S1). On the other hand, stress treatment (from now on NoP) was applied using similar solution to the control treatment except for P carrying ions, which were removed from the formulation of the solution (Table S1).

2.3. Sample Preparation

After the 60 days period plants were harvested for processing. Shoot and fruits were processed separately in order to assess effects of P deprivation on both tissues. Each tissue was put into individual paper bags and dried at 70 °C, until constant weight was achieved, in a Raypa ID-150 oven (R. Espinar S.L., Barcelona, Spain). At this point, shoot (SW, g) and fruit (FW, g) dry weights were determined, and those tissues were ground into a thin powder, using a domestic Taurus coffee grinder (Taurus Group, Oliana, Spain), for later mineral content analysis. Furthermore, all plants' roots were separated from substrate by gently washing them with running tap water and processed separately from other tissues [33]. This was done by hand, one root at a time (Figure 1).

For Trial 1 ($n = 9$), root hairs ($\varnothing < 0.5$ mm) were separated from lateral roots ($\varnothing > 0.5$ mm) and dried at 70 °C in order to obtain root hairs dry weight (RHW, g) (Figure 1). It is important to note that what is referred here as root hairs does not correctly translate to the root anatomical definition of root hairs; instead, it includes root hairs and some fine tertiary and lower order roots. However, herein it is useful to differentiate between the evaluated root parts. In the same way, lateral roots are mainly secondary roots; however, as can be seen in the picture Figure 1C, they can also include a portion of tertiary roots, as it was impossible to separate all in such a large amount of samples. Lateral roots were scanned, using an Epson Expression 1640XL G650C scanner (Seiko Epson Corp., Suwa, Japan), and resulting images were analyzed by WinRIZHO^{TM} Pro 2.3 software (Regent Instruments Inc., Québec, QC, Canada). Lateral root total length (LRL, m), lateral root average diameter (LRAD, mm) and total length of lateral roots with diameter under (LRL$_{<1mm}$, m) and above (LRL$_{>1mm}$, m) 1 mm were determined based on said images for each plant included in the experiment. Finally, scanned lateral roots were dried in order to obtain lateral roots dry weight (LRW, g) and ground for mineral content determination (Figure 1). From those measurements, several parameters were calculated in order to better characterize plants' performance. Hence, for trial 1, total root dry weight (RW, g) was determined as the sum of RHW and LRW and, therefore, total biomass dry weight (BW, g) was calculated as the sum of RW, SW and FW. In addition, root to shoot weight ratio (R/S) was calculated by dividing RW by SW; the percentage of root dry weight devoted to root hairs (RHW%) was calculated by the division of RHW by RW. Furthermore, the proportion of root length devoted to fine lateral roots (PLFR, %) was defined as the ratio between $LRL_{<1mm}$ and LRL. Finally, lateral root specific length (LRSL, m/g) was calculated by dividing LRL by LRW.

For Trial 2 (n = 6), roots were entirely scanned (Figure 1). In order to fully capture a root's morphometrics, individual roots were properly spread over several transparent acetate sheets (Figure 1) and analyzed by WinRIZHO^{TM} Pro 2.3 software (Regent Instruments Inc., Canada). Root total length (RTL, m), total root average diameter (TAD, mm) and total length of roots with diameters under ($RL_{<1mm}$, m) and above ($RL_{>1mm}$, m) 1 mm, were determined for each plant. Finally, the scanned roots were dried until constant weight was achieved and ground to a powder as in Trial 1. Root hairs dry weight (RHW, g), lateral roots dry weight (LRW, g), total root dry weight (RW, g), total biomass dry weight (BW, g), root to shoot ratio (R/S), percentage of root dry weight devoted to root hairs (RHW%) and root specific length (RSL, m/g) were determined as in Trial 1. Finally, the proportion of root length devoted to fine lateral roots was determined, that is, including root hairs and roots below 1 mm (PFR, %), as the ratio between $RL_{<1mm}$ and RTL.

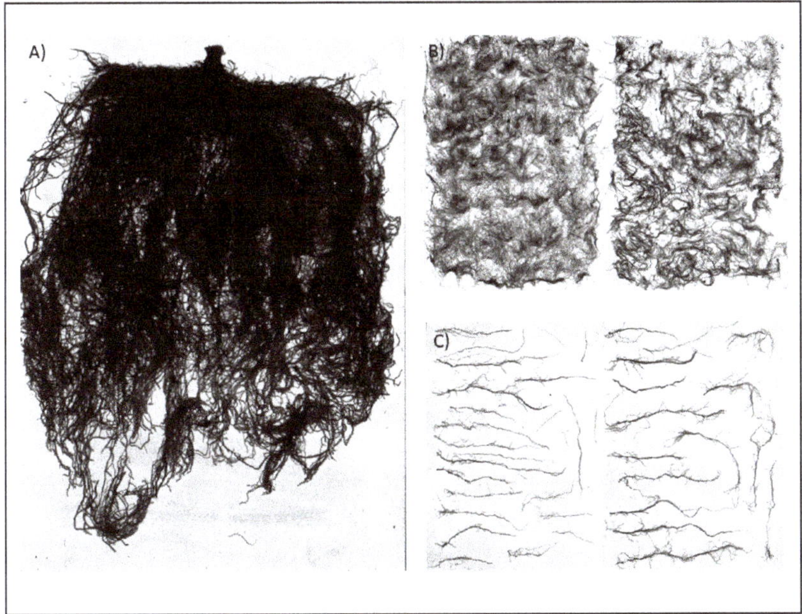

Figure 1. Illustration of the roots along the scanning process (from a representative sample). Individual root systems were separated from the soil with running tap water and taken to the laboratory to be scanned and dried. In Trial 1, whole roots (**A**) were separated into (**B**) root hairs (Ø < 0.5 mm) and (**C**) lateral roots (Ø > 0.5 mm). Root hairs (**B**) were only weighed while lateral roots (**C**) were scanned and weighed. In Trial 2, whole roots (**A**) were also separated into root hairs (**B**) and lateral roots (**C**) and both were scanned and weighed.

2.4. Tissue Mineral Concentration Assessment

Before mineral content determination, samples were mineralized [34]. Thus, 2 g of powdered plant tissue were calcined for 2 h in a muffle at 450 °C. Ashes resulting from mineralization were let to cool down, weighted and then hydrated with 2 mL of distilled water followed by addition of 2 mL of concentrated HCl (Scharlau, Valencia, Spain). At this point, the solution was heated on a hot plate, until first fumes appeared, and then filtered with Whatman filter paper (Sigma-Aldrich, St. Louis, MI, USA). Finally, distilled water was added in order to make up to 100 mL volume [34].

In Trial 1 ($n = 4$), phosphorus (P), potassium (K), calcium (Ca), magnesium (Mg), sodium (Na) and sulfur (S) concentration (g 100 g^{-1} DW) in different plant tissues (root, shoot and fruits, $[Mineral]_{Tissue}$) was determined by Inductively Coupled Plasma-Atomic Emission Spectrometry (ICP-AES; iCAp-AES 6000, Thermo Scientific, Cambridge UK). Samples were digested for 24 h by adding 10 mL 65% HNO_3 solution (Panreac Quimica S.A.U., Barcelona, Spain) to 0.5 g dried material, in a 25 mL open vessel. Digested samples were then boiled at 120 °C for 10 min followed by another 25 min at 170 °C. Finally, samples were cooled, 2 mL of 70% $HClO_4$ was added (Panreac Quimica S.A.U., Barcelona, Spain) and were then heated at 200 °C for 40 min. At this point, samples were transferred to a flask and volume was brought up to 25 mL with distilled water.

For Trial 2 ($n = 6$), leaves' P-concentration ($[P]_{Shoot}$) was determined by colorimetric reaction (MAPA, 1994). This method is based on absorbance measurement at 430 nm of each sample in acid solution and on the presence of vanadium (V^{5+}) and molybdenum (Mo^{6+}) ions. Under these conditions, phosphoric acid forms a phosphomolybvanadate complex that gives yellow coloration. Hence, 5 mL of mineralized solution were pipetted into a new 25 mL volumetric flask, followed by the addition of 5 mL of nitro-vanado-molybdic reagent. Volume was then brought up to 25 mL with distilled water. Prior to mineral concentration determination, a standard curve was constructed with standards 0, 2, 4, 6, 8, 10 and 12 µg of P mL^{-1} prepared from an initial solution of 20 µg of P mL^{-1}. Sample P concentration was determined using a 6305 model UV/V spectrophotometer (Jenway, Gransmore Green, England, UK) at 430 nm against a standard curve.

2.5. Phosphorus Uptake and Use Efficiency Parameters

In order to better characterize treatment effect on accessions performance, several widely-used P uptake and P use efficiency parameters (PUE) were calculated based on previous works [18,21] (Table 2).

Table 2. P uptake and P use efficiency (PUE) parameters used in this experiment and corresponding abbreviation, formula and expressed units. Dry weight (DW), total biomass dry weight biomass weight (BW).

Parameter	Abbreviation	Formula [1]	Units
Tissue total P content	RootP, ShootP, FruitP	$[P]_{Tissue} \times DW_{Tissue}$	G
Plant total P content	PTP [2]	$[P]_{Root} \times DW_{Root} + [P]_{Shoot} \times DW_{Shoot} + [P]_{Fruit} \times DW_{Fruit}$	mg P
P uptake efficiency	PUpE [3]	$([P]_{Control} \times BW_{Control}) - ([P]_{NoP} \times BW_{NoP})$	mg P
P utilization efficiency	PUtE[3]	$(BW_{Control} - BW_{NoP}) / (([P]_{Control} \times BW_{Control}) - ([P]_{NoP} \times BW_{NoP}))$	g DW g^{-1} P
Physiological P use efficiency	PPUE	$BW_{Control}/[P]_{Control}$ and $BW_{NoP}/[P]_{NoP}$	g^2 DW g^{-1} P
P efficiency ratio	PER	$BW_{Control}/ ([P]_{Control} \times BW_{Control})$ and $BW_{NoP}/ ([P]_{NoP} \times BW_{NoP})$	g DW g^{-1} P

[1] P concentration ($[P]$), Dry weight (DW), total biomass dry weight (BW) [2] Note that for Trial 2 only $[P]_{shoot}$ was measured, therefore PTP was obtained as $[P]_{shoot} \times BW$; [3] Note that $[P]$ in Trial 1 is the weighted average $[P]$ among different tissues, whereas in Trial 2 $[P] = [P]_{shoot}$.

2.6. Statistical Analysis

Two-way factorial analysis of variance (ANOVA) was performed using individual plant values in order to assess accession and treatment effects and interaction significance [35]. In addition,

Student-Newman-Keuls post-hoc multiple range test ($p < 0.05$) was used to detect significant differences among accession means for all evaluated traits. Finally, trait differences between treatments ($\mu_{NoP}-\mu_{Control}$) were used to perform multivariate Principal Component Analysis (PCA) using Euclidean pairwise distances. In addition, traits variation (%) between control and NoP conditions was calculated as $\left(\frac{\mu_{NoP}-\mu_{Control}}{\mu_{Control}}\right) \times 100\%$. All statistical analysis were performed using Statgraphics Centurion XVII (StatPoint Technologies, Warrenton, VA, USA) and plotted using R package ggplot2 [36,37].

3. Results

3.1. General Treatment Effect on P and Other Minerals Concentrations for Trial 1

P concentration ([P]) in plant tissues is an important indicator of both treatment effectiveness and accession's capability to make the most with the available resources. In Trial 1 ($n = 4$), plants cultivated under NoP conditions showed significantly lower [P] compared to control plants. This behavior was statistically significant for all three sampled tissues (Table S2). For $[P]_{Roots}$, there was a reduction from 0.56 g P 100 g^{-1} DW, when cultivated under control conditions, to 0.10 g P 100 g^{-1} DW (-81.76%) when cultivated under NoP conditions (Table S2). For $[P]_{Shoot}$, values decreased from 0.18 g 100 g^{-1} DW to 0.12 g P 100 g^{-1} DW (-29.31%) for control and NoP conditions, respectively; this the organ is less affected by the treatment (Table S2). Finally, fruit P levels dropped from 0.26 g P 100 g^{-1} DW, when irrigated with control solution, to 0.17 g P 100 g^{-1} DW (-35.18%) when NoP solution treatment was applied (Table S2).

Concentration of other macrominerals was determined in order to assess possible deficiencies induced by the applied treatments. Regarding that, significant differences between treatments were observed, particularly for K and Mg, probably due to the differences in the nutrient solutions and as a result of plant ionic adjustments. Despite that, mineral concentrations were within the normal range for pepper (Table S2) [6,38].

3.2. Treatment Effect on P Accumulation and Efficiency Parameters for Trial 1 by Accessions

Accessions were significantly affected by the NoP treatment, but not all to the same extent, as shown by the two-way ANOVA (Table S3) and the accession mean values for the evaluated traits (Table S4). For example, $[P]_{Root}$ dropped as much as 91.62% and 90.53% for bol_037 and sp_cwr accessions, respectively, while for bol_144 the reduction was considerably lower (74.12%) (Table 3). Regarding $[P]_{Shoot}$, the most affected accessions were sp_piq (-45.26%) and sp_cwr (-45.03%), while some accessions experienced no statistically significant reduction of their shoot P concentration, e.g., bol_037, bol_144, eq_973, mex_pas, sp_bola and sp_cat (Table 3 and Table S4). Finally, for $[P]_{Fruit}$, only sp_bola showed no statistical difference between both treatments, whereas the remaining accessions represented significant reductions of around 35% (Table 3).

In terms of P tissue accumulation, significant differences were found among accessions. Thus, despite P-deficient plants had on average 86.16% less accumulated P in the root (RootP) than control plants, 56.12% less P in the shoots (ShootP), and 34.32% less in fruits (FruitP), some genotypes, such as bol_144 or eq_973, increased the amount of fruit accumulated P (although this was not statistically significant). Overall, plant total P (PTP) was reduced by 63.30%; however, several genotypes experienced no statistically significant reduction of this trait, e.g., bol_037, bol_144_eq_973 and sp_bola, while others, like sp_mel, were highly affected (Table 3 and Table S4).

Furthermore, in order to evaluate how efficient genotypes were under these conditions parameters of physiological P use efficiency (PPUE), P efficiency ratio (PER), P uptake efficiency (PUpE) and P utilization efficiency (PUtE), were calculated [18,21]. Overall, PPUE was on average 36.26% higher under NoP conditions (Table 3). However, accessions' behavior for this parameter was extremely variable and significant differences between treatments were only found for two accessions, mu_esp (41.31%) and sp_cwr (97.53%) (Table 3 and Table S4). On the other hand, PER's behavior was more consistent and NoP treatment produced a generalized increase, averaging at 87.64% (Table 3). In this case, only two accessions showed no significant differences, bol_144 and sp_bola (Table 3 and Table S4). Interestingly, the best performers in terms of increasing PER from control to NoP conditions were California type accessions (sp_cat, sp_mel and sp_cwr) and *Capsicum chinense* accession eq_973.

In addition, an interesting amount of variability was observed for P uptake efficiency (PUpE) and P utilization efficiency (PUtE) parameters (Figure 2). PUpE refers to the increase of internal P when it is available in the environment, whereas PUtE measures the ability of a genotype to increase its biomass per unit of internal P. Both measures always compare two conditions differing in P levels. PUpE averaged 96.56 mg P for the whole collection, where accessions mex_scm (128.77 mg P), mu_esp (144.43 mg P), sp_cat (130.06 mg P), sp_piq (124.53 mg P) and usa_chi (108.70 mg P) showed the highest values of the experiment (Figure 2). PUtE values ranged from 29.77 g DW g^{-1} P (bol_144) to 404.95 g DW g^{-1} P (mex_pas) and averaged 186.55 g DW g^{-1} P. Accessions mex_pas (404.95 g DW g^{-1} P), usa_chi (316.77 g DW g^{-1} P) and sp_bola (273.16 g DW g^{-1} P) presented the most interesting results (Figure 2).

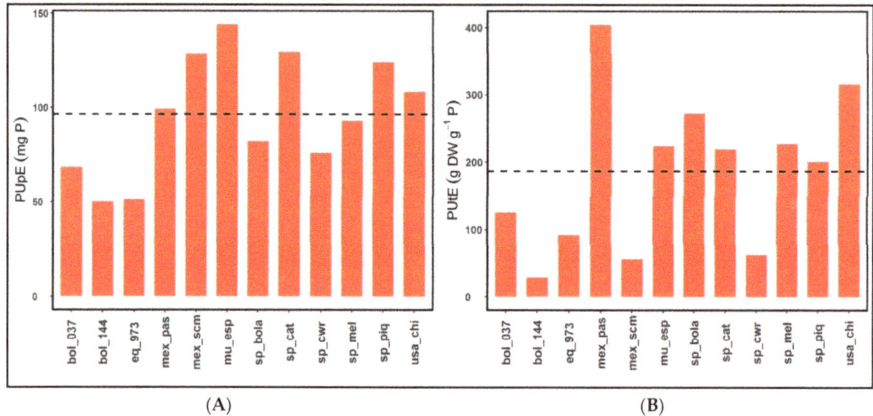

Figure 2. (**A**) Average PUpE (P Uptake Efficiency) and PUtE (P Utilisation Efficiency) values for the 12 accessions (*n* = 4) studied in Trial 1. (**B**) The black dashed line represents the average value for the whole collection for both PUpE (96.56 mg P) and PUtE (186.55 g DW g^{-1} P) parameters.

3.3. Treatment Effect on Root and Shoot Biomass and Morphometrics for Trial 1 by Accessions

P is a major factor controlling root structure and architecture [20]. Hence, in order to understand the possible effects on plant morphology, root structure and architecture resulting from lack of P, we compared several biomass and root traits.

Thus, roots dry weight (RW) showed a significant generalized decrease (−24.52%) when genotypes were cultivated under NoP conditions compared to control plants (Table 3, Tables S3 and S4). This weight difference was more evident in lateral roots (LRW), which, on average, weighed 26.07% less, while root hairs (RHW) weight was 18.34% lower than under control conditions. Notwithstanding, the greatest weight difference was observed for shoot dry weight (SW), −36.04% under NoP conditions (Table 3, Table S3 and Table S4). Despite that, taking a closer look at the treatment effect on biomass by accession, it is observed that only three accessions reduced it significantly in all organs: mex_pas, mu_esp and usa_chi. The rest of the accessions also reduced their biomass but not so systematically (Table 3 and Table S4). Interestingly, accession mex_scm, presented similar RW and SW under both treatments, while presenting the heaviest root system and shoot within the collection under NoP conditions (Table S4). Finally, root to shoot ratio (R/S) was positively affected under NoP treatment. This parameter increased by 20.94%, on average, although only usa_chi showed statistically significant differences between treatments (+22.73%) (Table 3), apparently achieved by reducing a shoot's weight instead of increasing a root's weight (Table S4).

Table 3. Accession behavior given by differences (%) between control and NoP conditions for Trial 1. Twenty different P accumulation and efficiency ($n = 4$), biomass and root traits ($n = 9$) and parameters were considered.

	P Accumulation and Efficiency Traits									Biomass Traits						Root Traits				
Accession	[P]$_{Root}$	[P]$_{Shoot}$	[P]$_{Fruit}$	RootP	ShootP	FruitP	PTP	PPUE	PER	RW	LRW	RHW	SW	BW	R/S	LRL	LRAD	RHW%	PLFR	LRSL
bol_037	−91.62 *	−32.10	−32.56 *	−93.37 *	−37.32	−25.00	−54.72	95.23	106.37 *	−12.91	−8.50	−18.68	−22.16	−21.67	25.41	33.30	−8.59	−6.42	5.21	38.14
bol_144	−74.12 *	−34.29	−27.41 *	−86.11 *	−55.45	98.52	−51.66	90.59	78.15	−26.94	−36.58	−1.37	−8.13	−4.08	−16.22	15.14	−21.81 *	51.06	9.56	112.94
eq_973	−87.67 *	−34.32	−35.19 *	−87.60 *	−41.15	55.56	−53.92	112.72	115.10 *	−21.00	−30.62	−3.38	−16.21	−14.17	6.67	49.55	−16.80	39.34	6.02	78.20
mex_pas	−78.01 *	1.51	−19.10 *	−86.13 *	−52.54 *	−59.45	−63.36 *	−27.82	36.65 *	−36.24 *	−42.38 *	−26.69	−59.65 *	−54.20 *	27.89	−3.07	−9.83	12.06	6.63	70.84
mex_scm	−84.98 *	−44.35 *	−37.94 *	−85.62 *	−62.45 *	−24.63	−57.57 *	50.86	96.13 *	0.65	9.15	−8.95	−18.18	−12.62	21.37	8.86	7.98	−11.28	1.63	16.26
mu_esp	−77.38 *	−32.15 *	−33.93 *	−88.61 *	−77.91 *	−72.38	−76.17 *	−41.31 *	51.20 *	−46.63 *	−29.39	−52.87 *	−56.71 *	−55.84 *	9.88	26.88	−8.57	−12.76	7.36	115.25
sp_bola	−77.51 *	−4.11	−17.47	−75.94 *	−41.77	−53.31	−55.46	−0.75	38.47	−19.85	−30.22 *	−5.64	−43.10	−40.06	27.73	2.12	−10.84	2.89	8.55	57.72
sp_cat	−83.53 *	−24.27	−45.63 *	−83.02 *	−56.11	−78.90 *	−73.96 *	27.80	108.33 *	−2.79	−23.62	31.54	−33.26	−41.92	50.00	3.73	−13.32	25.62	6.58	20.20
sp_cwr	−90.53 *	−45.03 *	−47.21 *	−92.86 *	−55.30 *	−46.11	−61.87 *	97.53 *	132.69 *	−19.90 *	−14.68	−26.20	−16.17	−11.06	−6.58	65.62 *	−7.13	−6.42	4.17	91.05
sp_mel	−83.67 *	−35.14 *	−49.43 *	−87.27 *	−66.67	−74.04	−77.17 *	24.96	137.24 *	−30.17	−35.41	−21.27	−58.18 *	−48.90	63.81	−10.47	−3.90	11.05	3.12	54.84
sp_piq	−75.52 *	−45.26 *	−44.11 *	−78.51 *	−59.93 *	−70.57	−67.78 *	24.31	98.40 *	−25.14	−17.60	−32.96	−39.38 *	−38.84 *	18.59	−19.76	4.76	−14.57	0.98	6.84
usa_chi	−76.56 *	−22.25	−32.20 *	−88.89 *	−66.88 *	−61.56 *	−65.95 *	−18.94	52.95 *	−53.35 *	−53.03 *	−53.63 *	−61.36 *	−59.42 *	22.73 *	27.85	12.59	−1.07	−1.29	142.63
Global mean	−81.76 *	−29.31 *	−35.18 *	−86.16 *	−56.12 *	−34.32 *	−63.30 *	36.26 *	87.64 *	−24.52 *	−26.07 *	−18.34 *	−36.04 *	−33.56 *	20.94 *	16.65 *	−6.29 *	7.46	4.88	67.08

Root P concentration ([P]$_{Root}$), shoot P concentration ([P]$_{Shoot}$), fruit P concentration ([P]$_{Fruit}$), root total P content (Root P), shoot total P content (ShootP), fruit total P content (FruitP), plant total P content (PTP), physiological P use efficiency (PPUE), P efficiency ratio (PER), total root dry weight (RW), lateral root dry weight (LRW), root hairs dry weight (RHW), shoot dry weight (SW), total biomass dry weight (BW), root to shoot ratio (R/S), lateral root total length (LRL), lateral root average diameter (LRAD), percentage of root dry weight devoted to root hairs (RHW%), proportion of root length devoted to fine lateral roots (PLFR) and lateral root specific length (LRSL). * Indicates significant differences between treatments for that accession and trait.

Regarding root morphology traits, treatment and accession effects showed significant influence over most traits, except for the percentage of root dry weight devoted to root hairs (RHW%), for which significant differences between treatments were not detected (Table S3), despite there being differences among accessions. In addition, for lateral root specific length (LRSL), there was a significant accession per treatment interaction (Table S3). The significant effects of the NoP conditions on pepper's roots where to increase: the lateral root length (LRL), by 16.65%, the proportion of root length devoted to fine lateral roots (PLFR), by 4.88%, and the lateral root specific length (LRSL), by 67.08%, and to decrease the root average diameter by 6.29%.

Regardless of the general treatment effect, there were significantly different responses among genotypes (Table 3 and Table S4). It is worth to mention the significant increase of percentage of root length devoted to fine lateral roots (PLFR) and lateral root specific length (LRSL) observed in mu_esp and sp_bola (Table 3), with mu_esp having the higher absolute values for these traits of the whole collection under NoP conditions. Another interesting response was presented by accession bol_144, which outperformed the other genotypes for reducing its roots average diameter (21.81%) and increase PLFR and LRSL under the NoP treatment.

3.4. Principal Components Analysis for Trial 1

Principal components analysis (PCA) was pursued in order to determine possible correlations between the response of the different measured traits to different inputs of P (% of increase or decrease, as in Table 3), trying to demonstrate how accessions differed in terms of response to different P levels. The first two principal components (PC) explained in combination 59.79% of the total variability (Figure 3A). Response in terms of total biomass dry weight (BW), physiological P use efficiency (PPUE), fruit total P content (FruitP), total shoot dry weight (SW), plant total P content (PTP), P efficiency ratio (PER) and root hairs dry weight (RHW), and values for P uptake efficiency (PUpE) and P utilization efficiency (PUtE) were the traits that contributed the most to the positive component of PC1, which explained 36.96% of the total variation. Response of lateral root average diameter (LRAD) and root total P content (RootP) were negatively correlated to PC1 (Figure 3A). Therefore, accessions plotted in the extreme right of the graph (Figure 3B), such as usa_chi, mu_esp, mex_pas and sp_piq, have in common that they have a great reduction in biomass when passing from control to NoP, and have good PUpE and PUtE. In other words, those are accessions that react very positively to any P addition to the soil but probably will not be appropriate to cultivate on poor soils (Figure 3B). At the same time, accession plotted at the upper most left part of the graph (Figure 3B), such as bol_144 and eq_973, are grouped for having high reductions in the total amount of P in the roots with a high reduction in the diameter of the roots (LRAD) as adaptation to low P, while having little difference in biomass under the two assayed conditions. In addition, PC2 explained 22.83% of variability with the response of lateral root dry weight (LRW), lateral root average diameter (LRAD), root dry weight (RW) and P utilization efficiency (PUtE) being the traits that contributed the most to it. Conversely, shoot P concentration ([P]$_\text{Shoot}$), fruit P concentration ([P]$_\text{Fruit}$) and shoot total P content (ShootP) were negatively correlated with PC2 (Figure 3A). Hence, accession located in the top part of the graph, such as mex_pas and sp_bola (Figure 3B), change the allocation of root resources, reducing the lateral root weight and diameter in situation of P restriction, maintaining the P level status of the shoots. On the contrary, the accessions located on the lower part of the graph such as mex_scm stand out by changing the level of P of the shoots and fruits ([P]$_\text{Shoot}$, [P]$_\text{Fruit}$, and ShootP) as a strategy to adapt to the low P conditions without modifying the lateral root morphology or size (Figure 3A). Interestingly, there was a cluster of parameters, such as the response in terms of [P]$_\text{Root}$, R/S and LRSL, indicating correlations among them.

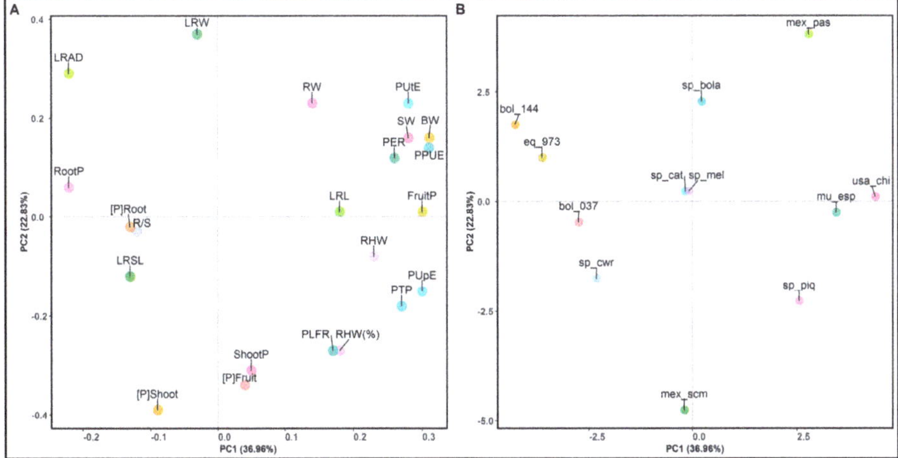

Figure 3. Principal Component Analysis (PCA) for the first two components based on trait differences between treatments for Trial 1. (**A**) Correlation between traits and the first two principal components. (**B**) Distribution of accessions based on studied traits. P tissue concentration traits [P]$_{Tissue}$, P tissue total content traits RootP, ShootP, FruitP, plant total P content (PTP) trait, efficiency parameters PPUE (physiological P use efficiency), PER (P efficiency ratio), PUpE (P uptake efficiency) and PUtE (P utilization efficiency) and morphometric traits RW (root dry weight), LRW (lateral root dry weight), RHW (root hairs dry weight), SW (shoot dry weight), BW (total biomass dry weight), R/S (root to shoot ratio), LRL (lateral root length), LRAD (lateral root average diameter), RHW% (root hairs dry weight %), PLFR (proportion of length dedicated to fine roots) and LRSL (lateral root specific length) were considered.

Bearing these results, the second trial was designed. In it, five accessions from Trial 1 (mu_esp, mex_pas, sp_bola, sp_piq and mex_scm) were re-trialed and used as a comparison standard against 13 new *C. annuum* accessions. These genotypes were selected based on their above average P uptake efficiency (PUpE) and P utilization efficiency (PUtE) scores and differential behavior against the lack of P. Note that an insufficient number of seeds to re-trial sp_cat and sp_mel, and the poor germination of usa_chi dictated their exclusion of Trial 2. The second trial was focused on checking the diversity within the germplasm belonging to *Capsicum annuum* species; for that reason, bol_144 and eq_973 were not selected, despite their interesting features. In this case, only P from the shoots was analyzed by a colorimetric protocol as a faster general measure of the P status of the plant, instead of a multi-elemental analysis by tissue. In addition, root hairs weight clustered together with P efficiency parameters was analyzed, and it was demonstrated that the lateral roots increase their length and reduce their diameter; thus, it was decided to analyze the root hairs' behavior as well. Both lateral and hair roots were scanned and analyzed (Figure 1).

3.5. Treatment Effect on P Accumulation and Efficiency Parameters for Trial 2 by Accessions

As in Trial 1, ANOVA showed that accession and treatment effects significantly affected P-related traits (Table S5). Interestingly, for physiological P use efficiency (PPUE) the accession effect was more important than treatment (Table S5). Remarkably, accession per treatment interaction was significant for a plant's total P content (PTP), PPUE and P efficiency ratio (PER) (Table S5). Accessions' individual variation between treatments are shown in Table 4 as $\left(\frac{\mu_{NoP} - \mu_{Control}}{\mu_{Control}} \right) \times 100\%$ negative then indicating lower values under NoPtraits. To consult the accessions' mean values per treatment, please refer to Table S6.

In Trial 2 ($n = 6$), all accessions but two showed significant differences between treatments for shoot P concentration ([P]$_{Shoot}$), plant total P content (PTP) and P efficiency ratio (PER) showing an average reduction of −31.5% and −66.17%, and an increase of 49.26%, respectively (Table 4). Accessions mex_096D and sp_piq stood out for their substantial [P]$_{Shoot}$ reduction and high PER value. In addition, accession sp_piq showed a significant reduction of its PTP level (−86.93%), which, along with mex_scm (−84.16%) and usa_sandia (−87.88%), represented the highest reductions of the whole collection (Table 4). Contrarily, sp_lam and sp_lobo showed no differences between treatments regarding PTP (Table 4 and Table S6). In the case of PPUE, NoP treatment presented an average reduction of 24.98%; however, significant differences were only detected for six accessions and with extremely erratic behavior within the collection; some accessions showed a reduction up to 72.75% (mex_scm), while others showed increases up to 45.04% (usa_jap).

Regarding P uptake efficiency (PUpE), average value was 298 mg P, ranging from 63 mg P (sp_lobo) to 796 mg P (usa_sandia). Accessions presenting above the average mean values were mex_096D, mex_103B, mex_ng, usa_conq and the re-trialed mex_scm and sp_piq (Figure 4). Contrarily to what happened in Trial 1, mu_esp was above the average for PUpE in this trial. Finally, average P utilization efficiency (PUtE) was 110 g DW g^{-1} P, while the minimum observed value was 43 g DW g^{-1} P (sp_lam) and the maximum was 183 g DW g^{-1} P (mex_pas). Like in Trial 1, accessions mex_pas (183 g DW g^{-1} P) and sp_bola (147 g DW g^{-1} P) presented the best performance for this parameter (Figure 4).

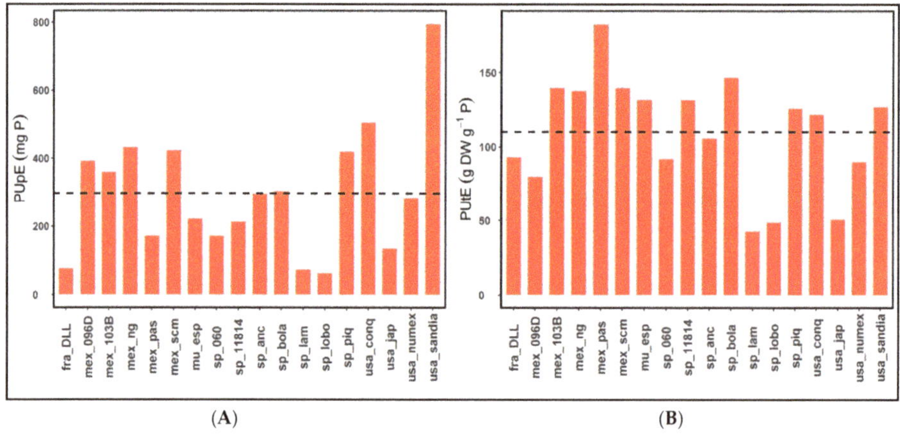

(A)   (B)

Figure 4. (**A**) Average P uptake efficiency (PUpE) and P utilization efficiency (PUtE) values for the 12 accessions studied in Trial 2 ($n = 6$). (**B**) Black dashed line represents average value for the whole collection for both PUpE (298 mg P) and PUtE (110 g DW g^{-1} P) parameters.

3.6. Treatment Effect on Root and Shoot Biomass and Morphometrics for Trial 2 by Accessions

In trial 2 ($n = 6$), multi-factorial ANOVA detected significant accession and treatment effects as well as the accession per treatment interaction for all biomass traits except root to shoot ratio (R/S) (Table S5). As expected, the effect of the NoP treatment led to lower dry weight accumulation of all sampled organs. This time, the most affected organs were the roots (RW, −52.96%) and root hairs (RHW, −59.10%) (Table 4). The genotypes usa_sandia, mex_scm and sp_piq showed the highest biomass reduction when passing from control to NoP. On the contrary, fra_DLL, sp_lam, sp_lobo and usa_jap showed no statistical differences between treatments, although it is important to note that fra_DLL, sp_lobo, and sp_lam displayed the smallest plants within the collection for both treatments, which could explain their results. Accession usa_jap, on the other hand, showed medium-sized plants (Table S6).

Table 4. Accession behavior given by differences (%) between the control and NoP conditions for Trial 2. Twenty different P accumulation and efficiency, biomass and root traits ($n = 6$) and parameters were considered.

| Accession | P Accumulation and Efficiency Traits | | | | | Biomass Traits | | | | | Root Traits | | | | |
	[P] Shoot	PTP	PPUE	PER	RW	LRW	RHW	SW	BW	R/S	RTL	TAD	RHW%	PFR	RSL
fra_DLL	−31.92 *	−42.13 *	25.76	48.75 *	−31.04	−20.87	−34.20	−14.49	−15.61	−20.51	6.27	4.23	−6.84	−0.31	58.00
mex_096D	−44.10 *	−74.54 *	8.80	87.88 *	−42.62 *	−29.17 *	−50.76 *	−49.18 *	−48.86 *	27.27	−0.69	2.96	−19.79 *	−2.46	71.79 *
mex_103B	−25.45 *	−68.35 *	−42.41 *	33.91 *	−63.97 *	−57.22 *	−68.38 *	−56.97 *	−57.48 *	−4.26	−10.44	−7.53 *	−10.29	5.91 *	131.10 *
mex_ng	−38.75 *	−80.58 *	−53.53	66.18 *	−70.25 *	−68.60 *	−71.38 *	−69.72 *	−69.75 *	7.95	−0.06	−13.30 *	−7.35	9.79 *	229.58 *
mex_pas	−11.63	−54.71 *	−39.68 *	14.69	−45.43 *	−37.30 *	−52.56 *	−48.13 *	−47.97 *	13.16	6.22	−0.80	−11.40	1.83	81.25 *
mex_scm	−24.46 *	−84.16 *	−72.75 *	30.99 *	−71.09 *	−53.17 *	−79.62 *	−79.40 *	−78.97 *	60.71	−15.45	−0.72	−21.79	2.51	96.42
mu_esp	−28.78 *	−76.55 *	−45.93	41.15 *	−58.48 *	−36.78 *	−66.55 *	−64.64 *	−64.02 *	33.93	83.39	−8.37 *	−16.18	6.53 *	293.57 *
sp_060	−23.21 *	−57.00 *	−19.84	30.93 *	−58.36 *	−32.94 *	−64.94 *	−41.59 *	−41.30 *	−24.24 *	17.90	−1.52	−18.90	5.90	181.31 *
sp_11814	−27.72 *	−66.80 *	−40.61	34.34 *	−54.26 *	−9.84 *	−65.36 *	−55.36 *	−54.65 *	−6.46	110.07	−0.85	−21.27 *	3.37	282.31 *
sp_anc	−35.55 *	−66.88 *	−20.37	53.60 *	−47.93 *	−37.11 *	−53.30 *	−48.66 *	−48.08 *	2.94	−9.82	−5.51	−7.61	6.16	44.62
sp_bola	−29.23 *	−78.82 *	−59.09 *	41.52 *	−63.73 *	−6.21 *	−73.15 *	−70.90 *	−70.35 *	23.08	52.67	−3.10	−26.14 *	4.62	261.22 *
sp_lam	−30.84 *	−36.71	29.12	42.54 *	−35.96	−24.57	−42.23	−7.24	−10.66	−30.56 *	−7.73	2.45	−8.40	0.22	38.68
sp_lobo	−24.83	−35.31	16.75	31.29	−13.58	−9.56	−15.06	−14.14	−10.95	6.90	20.75	−7.69	−0.93	8.97 *	27.40
sp_piq	−42.48 *	−86.93 *	−50.75 *	91.29 *	−72.33 *	−51.80 *	−79.12 *	−75.78 *	−75.48 *	8.33	4.58	3.29	−21.96 *	2.35	220.45 *
usa_conq	−38.95 *	−79.74 *	−47.09	61.46 *	−59.27 *	−41.08 *	−65.12 *	−67.50 *	−66.97 *	13.04	3.73	−8.02	−14.25	8.23	122.83 *
usa_jap	−39.15 *	−49.02 *	45.04	68.30 *	−48.83	−23.47	−60.76	−13.55	−15.53	−45.95 *	9.01	−0.44	−21.29	0.00	72.51 *
usa_numex	−33.47 *	−64.88 *	−17.21	50.09 *	−49.78 *	−38.25 *	−53.04 *	−45.86 *	−46.10 *	−9.52	−39.33 *	8.76	−7.89	−3.70	21.77
usa_sandia	−36.52 *	−87.88 *	−65.94 *	57.77 *	−66.39 *	−59.05 *	−68.20 *	−80.15 *	−79.31 *	66.67	10.55	1.80	−10.35	0.10	262.06
Global mean	−31.50 *	−66.17 *	−24.98 *	49.26 *	−52.96 *	−35.39 *	−59.10 *	−50.18 *	−50.11 *	6.80	13.42	−1.91	−14.03 *	3.33 *	138.71 *

Shoot P concentration ([P]$_{Shoot}$), plant total P content (PTP), physiological P use efficiency (PPUE), P efficiency ratio (PER), total root dry weight (RW), lateral root dry weight (LRW), root hairs dry weight (RHW), shoot dry weight (SW), total biomass dry weight (BW), root to shoot ratio (R/S), root total length (RTL), total root average diameter (TRAD), percentage of root dry weight devoted to root hairs (RHW%), proportion of root length devoted to fine roots (PFR) and root specific length (RSL). * Indicates significant differences between treatments for that accession and trait.

All root parameters were significantly affected by the accession effect, while only a percentage of root hairs (RHW%), proportion of root length devoted to fine roots (PFR) and root specific length (RSL) were significantly affected by the treatment. For root total length (RTL), accession usa_numex (−39.33%) was the only genotype that significantly reduced its root length, while the general population tendency was to increase it (Table 4). Accessions mex_103B (−7.53%), mex_ng (−13.30%) and mu_esp (−8.37%) significantly decreased their total average diameter under NoP conditions (Table 4) in accordance with the population general trend (−1.91%). Likewise, the percentage of root dry weight devoted to root hairs (RHW%) was 14.03% lower under P-stress conditions, with accessions mex_096D (−19.79%), sp_11814 (−21.27%), sp_bola (−26.14%) and sp_piq (−21.96%) being the significantly affected ones (Table 4). Contrarily, the proportion of root length devoted to fine roots (PFR) showed a slight increase under NoP (3.33%), compared to control conditions, although only four accessions were significantly affected. Thus, accessions mex_103B (5.91%), mex_ng (9.79%), mu_esp (6.53%) and sp_lobo (8.97%) significantly increased this parameter under NoP conditions (Table 4). Ultimately, root specific length (RSL) was, on average, 138.71% higher under NoP conditions. Most accessions were significantly affected by the treatment; mu_esp (293.57%) and sp_11814 (282.31%) were the accessions with a higher increase for root specific length (Table 4).

3.7. Principal Components Analysis for Trial 2

The first two PCs explained 63.84% of total variation for Trial 2 (Figure 5). PC1 explained 47.48% of the total variation and was defined by the response of total biomass dry weight (BW), shoot dry weight (SW), physiological P use efficiency (PPUE), plant total P content (PTP), root dry weight (RW), root hair dry weight (RHW) and absolute values for P uptake efficiency (PUpE), while the traits that most contributed negatively were the response of the root to shoot ratio (R/S) and root specific length (RSL) (Figure 5A). PC2 on the other hand explained 16.36% of total variability and was positively correlated with response of shoot P concentration ([P]$_{Shoot}$), proportion of root length devoted to fine lateral roots (PFR) and root total length (RTL), while being negatively correlated with the total root average diameter (TAD) and P efficiency ratio (PER) (Figure 5A).

Based on those results, accessions located at the right side of the graph (usa_sandia, mex_ng, sp_piq, usa_conq, mex_scm and mex_103B), presented an important biomass reduction under NoP conditions and, at the same time, interesting PUpE and PUtE values and an increase at the R/S and RSL level when in NoP, indicating that these are good candidates for high input conditions due to their excellent response to the addition of P through fertilization (Table 4 and Figure 5B). On that matter, usa_sandia stood out for its impressive PUpE values and high increase of R/S and high increase of RSL (Figure 4B). On the opposite side, sp_lam, fra_DLL and usa_jap accessions were located, with negative values of R/S and relative low increase of RSL (Figure 5B) and poor values for PUpE and PUtE. From this group, usa_jap and fra_DLL had good values of biomass under NoP (Table S6). Altogether, this indicates that they perform well under NoP conditions but do not improve with additional units of P. Furthermore, in the upper part of the graph, accessions usa_numex and mex_096D were characterized by decreasing the P concentration in the shoot ([P]$_{Shoot}$), and thus, increasing PER, and having higher root's total length (RTL), proportion of length dedicated to fine roots (PFR) and root diameter (TAD) in the control than in NoP conditions (Table 4 and Figure 5B). Finally, on the bottom part of the plot, accessions mu_esp, mex_pas, sp_11814 and sp_bola were positively correlated with changes in TAD and PER and high values of PUtE, indicating a tendency to reduce their roots' average diameter while maintaining the shoot [P] concentration (Table 4 and Figure 5B).

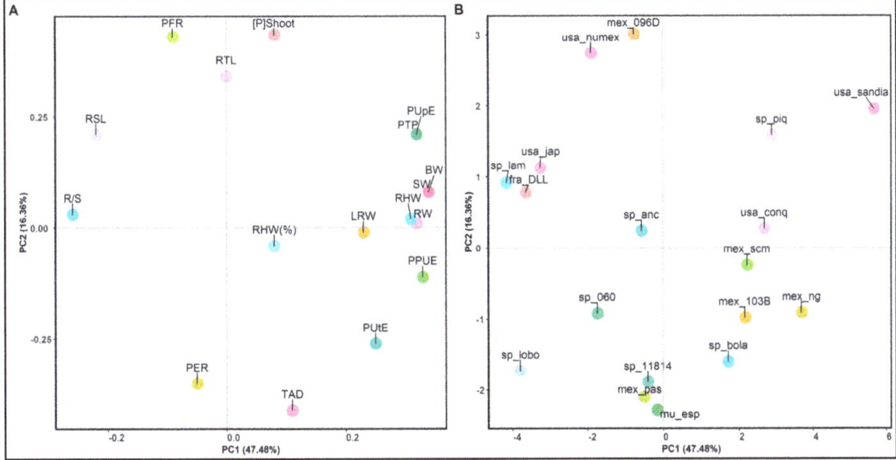

Figure 5. PCA based on trait increments between treatments for the Trial 2 experiment. (**A**) Correlation between traits and the first two principal components. (**B**) Distribution of accessions based on the studied traits. P tissue concentration traits [P]$_{Shoot}$, P total plant content (PTP) trait, efficiency parameters PPUE (physiological P use efficiency), PER (P efficiency ratio), PUpE (P uptake efficiency) and PUtE (P utilization efficiency) and morphometric traits RW (root dry weight), LRW (lateral root dry weight), RHW (root hairs dry weight), SW (shoot dry weight), BW (total biomass dry weight), R/S (root to shoot ratio), RTL (root total length), TAD (root total average diameter), RHW% (root hairs dry weight %), PFR (proportion of length dedicated to fine roots) and RSL (root specific length) were considered.

4. Discussion

4.1. Peppers Change Their Mineral Homeostasis and re-Allocate Their P Reserves to Adjust to Low-P Conditions

A comparison of P (root, shoot and fruit in Trial 1 and just shoot in Trial 2) concentrations provided relevant information on the impact of the different levels of P on pepper. There is evidence to suggest that pepper plant organs require P in different amounts, and the minimum levels are drastically different between tissues. Regarding that, roots presented the highest drop of P concentrations between treatments, indicating that they are able to mobilize P in order to benefit above-ground biomass. This response has been described in other crops, in which physiological and morphological changes, such as changes in root porosity and aerenchyma proportion, have been reported as mechanisms for reducing both the metabolic expenses and P requirements of the root system, while maintaining the foraging ability [15,22,39]. Interestingly, there were also differences among genotypes on P-tissue allocation, which opens the door to breeding materials with minimal P levels in the fruits and less need for fertilization without hampering production. For instance, some authors believe that we consume more P than required for a healthy diet, and often in the form of phytate, which is not fully absorbed by the human digestive system [40,41].

Homeostatic processes by which plants take up, transport and store nutrients are not independent, and therefore, the absence or excess of some elements can affect how the rest are processed, as was observed herein [6,38,39]. However, despite some significant differences between treatments for other macro minerals and tissue combinations, the values observed for this experiment are within the normal range, and therefore, no deficiency or excess was detected apart from P [6,42].

4.2. P Efficiency Parameters Measure Different Aspects of the Plant Response

The use of parameters to describe a plant's mineral uptake and use efficiency is a widely adopted practice in this scientific field [18,21]. Thus, physiological P use efficiency (PPUE) provides information on how productive a genotype may be, based on its tissues P concentration under a specific treatment; hence, high values indicate higher efficiency transforming absorbed P into biomass. Under these conditions, accessions mex_pas (control) and bol_144 (NoP) presented the highest PPUE for Trial 1, whereas in Trial 2, usa_sandia and fra_DLL presented the highest values for control and NoP, respectively. These results indicate a differential response, making these accessions interesting candidates for different P-fertilizers input conditions (e.g., high and low). Interestingly, the general response of increasing PPUE from control to NoP was not observed for trial 2. It is important to point out that although it is the same parameter, it was calculated in a different way depending on the trial. For Trial 1, concentration of P was a mean of all plant tissues whereas for Trial 2 it was extrapolated from shoot only, which may have caused a behavior distortion. P efficiency ratio (PER), on the other hand, relates the amount of yielded biomass with the amount of accumulated P in the plant; thus, high values indicate a higher ability to generate biomass with less P. Thus, bol_144 (Trial 1) and fra_DLL (Trial 2) are extremely efficient genotypes, especially under low-input conditions. These results indicate an interesting ability to use every unit of absorbed P and convert it into biomass and suggest that aptitude should be used in low-input systems.

Regarding P uptake efficiency (PUpE), accessions mex_scm and sp_piq showed an above average performance in both trails, although in the Trial 2 both usa_conq and usa_sandia showed higher values. This indicates that these accessions responded well to fertilization and were able to take up high amounts of P when it is present. In terms of P utilization efficiency (PUtE), accessions mex_pas and sp_bola showed the highest values in both trials, indicating that they are able to use the absorbed P into biomass generation more efficiently than the rest of the accessions. Furthermore, genetic variation regarding P acquisition and use efficiency has been widely reported in soybean, maize, sunflower, brassica and melon [15,18,21,43,44]. However, to our knowledge, this is the first work that provides such information for pepper germplasm. Herein, a wide range of variability is reported regarding P efficiency parameters, as well as several combinations among them, offering numerous possibilities for breeding for improved P uptake and P use efficiency parameters (PUE). Several authors have reported independence between uptake and use efficiency, which enables the improvement of both as well as selecting materials with different purposes (e.g., high- and low-input environments) [12,18,21,44]. These results seem to point towards that idea, since both parameters were located separately in both trials' PCA.

4.3. Modifications at Root Level

Many species promote root instead of aerial growth in order to enhance foraging capability [15,17,21]. In this experiment, a loss of root mass was observed under NoP conditions; however, this reduction was lower than that of the aerial part. This resulted in an increased root to shoot ratio under NoP conditions compared to control plants. The results indicate that, apart from lower biomass accumulation and redistribution of it, there are important modifications, particularly at root level, that help the plant to cope with P-stress. This was also observed in previous works with other crops for P-starvation conditions [20].

Morphological adaptations to low P concentrations in the soil aim at enhancing P acquisition by enabling exploitation of a greater soil volume, as well as enhancing P uptake without significantly increasing metabolic costs [17,18,45,46]. This is achieved mainly by the stimulation of root hairs [15,18,45], by halting secondary growth of the root and promoting primary growth and elongation [46] or increasing porosity and aerenchyma in roots [22]. Herein, lateral root length (LRL), but not total root length (RTL), increased under NoP conditions. It seems that lateral root elongation was a key response of the plant to reach possible P patches in the soil. This response has been described as an adaptive response to low P in *Phaseolus vulgaris* [46]. Other parameters, such as the lateral root specific length (LRSL), root specific length (RSL), percentage of fine roots (PFR) and percentage of fine lateral roots (PFLR), were higher under NoP, whereas the LRAD was lower. Therefore, pepper genotypes react to low P by producing thinner and lighter roots (with less carbon cost), which is in concordance with the literature [15,18,45,47]. On that regard, bol_144 (Trial 1) and mex_ng (Trial 2) stood out for their significant reduction of root average diameter while increasing the proportion of length dedicated to thinner roots under NoP conditions. In addition, accessions such as mu_esp and sp_bola showed a significant stimulation of their root specific length and proportion of length dedicated to thinner roots under the NoP conditions, despite presenting a lower root weight than under control conditions.

Although the percentage of root hair weight (RHW%) decreased in Trial 2, and was not significantly different in Trial 1, it must be pointed out that this measure includes fine roots and not specifically root hairs; therefore, it must be investigated if root hairs are modified in pepper under contrasting P conditions. Analyzing roots is a difficult task, and specific protocols must be set up to increase accuracy of root traits' study. The differences regarding root scanning and the analysis procedure between trials indicate that the first methodology (scanning just lateral roots) was more effective in finding root responses, since scanning all the fine roots has technical limitations.

4.4. A Wide Range of Responses to Breed Efficient Genotypes

Despite the general responses of pepper to low P described in the previous section, there was a wide range of responses depending on the accession studied. PCA's projection showed a widely differentiated behavior among accessions, creating several accession clusters depending on their overall response to NoP. For example, sp_piq, mex_scm, usa_sandia, usa_conquistador and mex_ng showed high PUpE values associated with increases in root to shoot ratio and root specific length. Sp_bola, mex_pas and mu_esp were associated with high PUtE values, no changes in their concentration of P in the shoots, reduction of the root diameter and an increase of percentage of fine roots and root total length. On the other hand, there were accessions that were poorly responsive to the changes in P levels, such as bol_37 and eq_973, or sp_lam, fra_DLL or usa_jap. Results indicated that some accessions were more suited to grow under low input conditions (bol_144, eq_973 and usa_jap) and others were highly responsive to increasing amounts of P available in the soil (sp_piq, sp_pas and mu_esp). It was also observed that P uptake efficiency and P utilization efficiency seem to be controlled independently, and here, this is demonstrated by the positioning of both parameters in opposite quadrants of the PCAs' second component, and accessions with contrasting levels.

On that matter, the availability of diversity is of paramount importance for crop breeding, enabling the combination of several favorable traits or behaviors in a single genotype, which in return can be a more effective solution than to have those traits in separate genotypes. For example, Miguel et al. [48] demonstrated, in common bean, that combining shallow basal roots and long root hairs yielded a larger effect regarding P acquisition than their additive effects separately. Breeding for efficient genotypes needs an accurate definition of the target to be improved; this is not the same as improving the ability to grow under low inputs than reacting favorably to P addition. Defining the best ideotype to each condition and the combination of different adaptation opens the possibility to breed towards different goals [12,18,21,44].

5. Conclusions

Herein, a diverse collection of 25 pepper accessions has been characterized for their behavior under P low-input conditions. A considerable amount of diversity has been reported for the response to phosphorus low-input conditions for several phosphorus uptake and use efficiency parameters, and root and biomass traits. Evidence suggests that P low-input conditions play an important role in the plant's tissues allocation for this mineral and that different organs show different critical levels of phosphorus. In addition, the responses of this collection indicate the existence of genetic diversity, which may be used in breeding programs to generate materials with different applications.

Accessions bol_144 and fra_DLL showed promising results for low-input conditions, whereas mex_scm, sp_piq, usa_conq and usa_sandia were on the opposite spectrum and are probably best used under high-input conditions due to their uptake efficiency. In addition, mex_pas and sp_bola showed the best results regarding P use efficiency. Finally, P low-input conditions proved to be an important factor controlling root morphology. Under these conditions, roots presented longer and thinner roots. These traits correlated to a higher efficiency and biomass accumulation under P-starving conditions. This work provides relevant first insights into pepper's response to phosphorus low-input conditions. More works are needed in order to dissect the mechanisms controlling the response, and consequently, to be introgressed into new materials.

Supplementary Materials: The following are available online at http://www.mdpi.com/2073-4395/10/5/637/s1, Table S1: Ion concentrations for irrigation water, Control and NoP solutions used in both Trial 1 and Trial 2. Table S2: Effect of Control and NoP treatments on P, K, Ca, Mg, Na and S (g/100g DW) concentrations for root, shoot and fruit tissues for Trial 1. Overall mean values, standard deviation and *p*-value for each plant tissue and treatment are provided. Table S3: Trial 1 multi-factor ANOVA's mean square values of accession and treatment effects, their interaction, and error for P concentration traits $[P]_{Root}$, $[P]_{Shoot}$, $[P]_{Fruit}$, P content traits RootP (g P), ShootP (g P), FruitP (g P), PTP (mg P), P efficiency parameters PPUE (g^2 DW g^{-1} P) and PER (g DW g^{-1} P) and for morphometric traits RW (g), LRW (g), RHW(g), SW (g), BW (g), R/S, LRL (m), LRAD (mm), RHW% (%), PLFR (%) and LRSL (m/g). Table S4: Trial 1 mean values and standard deviation for P accumulation and efficiency traits and parameters, and biomass and root traits and parameters. Table S5: Trial 2 multi-factor ANOVA's mean square values of accession (A) and treatment (T) effects, their interaction and error (E) for P concentration trait $[P]$Shoot, P content trait PTP (mg P), for efficiency parameters PPUE (g^2 DW g^{-1} P) and PER (g DW g^{-1} P) and for morphometric traits RW (g), LRW (g), RHW (g), SW (g), BW (g), R/S, RTL (m), TAD (mm), RHW%, PFR (%) and RSL (m/g). Table S6: Trial 2 mean values and standard deviation for P accumulation and efficiency traits and parameters, and biomass and root traits and parameters.

Author Contributions: Conceptualization, methodology and validation: A.F., Á.C. and A.R.-B.; Data curation: L.P.-D. and D.G.-V.; Formal analysis and investigation: L.P.-D., D.G.-V., V.C.-Z. and A.Q.; Resources, funding acquisition and project administration: A.F. and Á.C.; Writing–original draft: L.P.-D. and A.F.; Writing—review and editing: L.P.-D., D.G.-V., V.C.-Z., A.Q., Á.C., A.R.-B. and A.F. All authors have read and agreed to the published version of the manuscript.

Funding: This research was funded by FEDER-Funds and INIA, grant number RTA2013-00022-C02-02. The APC was self-funded.

Acknowledgments: Authors thank seed providers included in Table 1, such as P.W. Bosland, François Jourdan and the different Regulatory Boards of the PDOs and GPIs included in this study. Additionally, we want to thank Jose J. Luna for his advice on Mexican peppers.

Conflicts of Interest: The authors declare no conflict of interest. The funders had no role in the design of the study; in the collection, analyses or interpretation of data; in the writing of the manuscript, or in the decision to publish the results.

References

1. Jaggard, K.W.; Qi, A.; Ober, S. Possible changes to arable crop yields by 2050. *Philos. Trans. R. Soc. B Biol. Sci.* **2010**, *365*, 2835–2851. [CrossRef] [PubMed]

2. Grafton, R.Q.; Daugbjerg, C.; Qureshi, M.E. Towards food security by 2050. *Food Secur.* **2015**, *7*, 179–183. [CrossRef]

3. Tilman, D.; Cassman, K.G.; Matson, P.A.; Naylor, R.; Polasky, S. Agricultural sustainability and intensive production practices. *Nature* **2002**, *418*, 671–677. [CrossRef] [PubMed]

4. Tsiafouli, M.A.; Thébault, E.; Sgardelis, S.P.; de Ruiter, P.C.; van der Putten, W.H.; Birkhofer, K.; Hemerik, L.; de Vries, F.T.; Bardgett, R.D.; Brady, M.V.; et al. Intensive agriculture reduces soil biodiversity across Europe. *Glob. Chang. Biol.* **2015**, *21*, 973–985. [CrossRef] [PubMed]

5. Raza, A.; Razzaq, A.; Mehmood, S.S.; Zou, X.; Zhang, X.; Lv, Y.; Xu, J. Impact of climate change on crops adaptation and strategies to tackle its outcome: A review. *Plants* **2019**, *8*, 34. [CrossRef] [PubMed]

6. Jones, J.B.J. *Plant Nutrition and Soil Fertility Manual*, 2nd ed.; Press, C., Ed.; Taylor & Francis Group: Boca Raton, FL, USA, 2012; ISBN 9781439816103.

7. Schnug, E.; Haneklaus, S.H. Assessing the plant phosphorus status. In *Phosphorus in Agriculture: 100% Zero*; Springer: Dordrecht, The Netherlands, 2016; pp. 95–125.

8. Vance, C.P.; Uhde-Stone, C.; Allan, D.L. Phosphorus acquisition and use: Critical adaptations by plants for securing a nonrenewable resource. *New Phytol.* **2003**, *157*, 423–447. [CrossRef]

9. Cordell, D.; Drangert, J.O.; White, S. The story of phosphorus: Global food security and food for thought. *Glob. Environ. Chang.* **2009**, *19*, 292–305. [CrossRef]

10. Mogollón, J.M.; Beusen, A.H.W.; van Grinsven, H.J.M.; Westhoek, H.; Bouwman, A.F. Future agricultural phosphorus demand according to the shared socioeconomic pathways. *Glob. Environ. Chang.* **2018**, *50*, 149–163. [CrossRef]

11. Schnug, E.; Haneklaus, S.H. The enigma of fertilizer phosphorus utilization. In *Phosphorus in Agriculture: 100% Zero*; Springer: Dordrecht, The Netherlands, 2016; pp. 7–26.

12. Lynch, J.P. Roots of the second green revolution. *Aust. J. Bot.* **2007**, *55*, 493–512. [CrossRef]

13. Fernández, J.M.; Selma, M.A.E. Estimación de la contaminación agrícola en el Mar Menor mediante un modelo de simulación dinámica. In *Proceedings of the El Agua y Sus Usos Agrarios*; Universidad de Zaragoza: Zaragoza, Spain, 1998; Volume 9, pp. 1–9.

14. Kauranne, L.-M.; Kemppainen, M. Urgent need for action in the Baltic sea area. In *Phosphorus in Agriculture: 100% Zero*; Springer: Dordrecht, The Netherlands, 2016; pp. 1–6.

15. Fernández, M.C.; Rubio, G. Root morphological traits related to phosphorus-uptake efficiency of soybean, sunflower, and maize. *J. Plant Nutr. Soil Sci.* **2015**, *178*, 807–815. [CrossRef]

16. Fita, A.; Bowen, H.C.; Hayden, R.M.; Nuez, F.; Picó, B.; Hammond, J.P. Diversity in expression of Phosphorus (P) responsive genes in *Cucumis melo* L. *PLoS ONE* **2012**, *7*, e35387. [CrossRef]

17. Li, J.; Xie, Y.; Dai, A.; Liu, L.; Li, Z. Root and shoot traits responses to phosphorus deficiency and QTL analysis at seedling stage using introgression lines of rice. *J. Genet. Genom.* **2009**, *36*, 173–183. [CrossRef]

18. Hammond, J.P.; Broadley, M.R.; White, P.J.; King, G.J.; Bowen, H.C.; Hayden, R.M.; Meacham, M.C.; Mead, A.; Overs, T.; Spracklen, W.P.; et al. Shoot yield drives phosphorus use efficiency in *Brassica oleracea* and correlates with root architecture traits. *J. Exp. Bot.* **2009**, *60*, 1953–1968. [CrossRef]

19. Lynch, J.P.; Brown, K.M. Topsoil foraging—An architectural adaptation of plants to low phosphorus availability. *Plant Soil* **2001**, *237*, 225–237. [CrossRef]

20. Niu, Y.F.; Chai, R.S.; Jin, G.L.; Wang, H.; Tang, C.X.; Zhang, Y.S. Responses of root architecture development to low phosphorus availability: A review. *Ann. Bot.* **2013**, *112*, 391–408. [CrossRef]

21. Fita, A.; Nuez, F.; Picó, B. Diversity in root architecture and response to P deficiency in seedlings of *Cucumis melo* L. *Euphytica* **2011**, *181*, 323–339. [CrossRef]

22. Fan, M.; Zhu, J.; Richards, C.; Brown, K.M.; Lynch, J.P. Physiological roles for aerenchyma in phosphorus-stressed roots. *Funct. Plant Biol.* **2003**, *30*, 493–506. [CrossRef]

23. Richardson, A.E.; Lynch, J.P.; Ryan, P.R.; Delhaize, E.; Smith, F.A.; Smith, S.E.; Harvey, P.R.; Ryan, M.H.; Veneklaas, E.J.; Lambers, H.; et al. Plant and microbial strategies to improve the phosphorus efficiency of agriculture. *Plant Soil* **2011**, *349*, 121–156. [CrossRef]

24. van de Wiel, C.C.M.; van der Linden, C.G.; Scholten, O.E. Improving phosphorus use efficiency in agriculture: Opportunities for breeding. *Euphytica* **2016**, *207*, 1–22. [CrossRef]

25. Bosland, P.W.; Votava, E.J. *Peppers: Vegetable and Spice Capsicums*; CABI Publishing: Wallingford, Oxon, UK, 2012; ISBN 178064020X.

26. Food and Agriculture Organization of the United Nations. *FAOSTAT Statistics Database*; FAO: Rome, Italy, 2019.

27. DeWitt, D.; Bosland, P.W. *Peppers of the World: An Identification Guide*; Ten Speed Press: Berkeley, CA, USA, 1996; ISBN 0898158400.

28. Sahitya, U.L.; Krishna, M.S.R.; Suneetha, P. Integrated approaches to study the drought tolerance mechanism in hot pepper (*Capsicum annuum* L.). *Physiol. Mol. Biol. Plants* **2019**, *25*, 637–647. [CrossRef]

29. Hwang, E.-W.; Kim, K.-A.; Park, S.-C.; Jeong, M.-J.; Byun, M.-O.; Kwon, H.-B. Expression profiles of hot pepper (*Capsicum annuum*) genes under cold stress conditions. *J. Biosci.* **2005**, *30*, 657–667. [CrossRef] [PubMed]

30. Jing, H.; Li, C.; Ma, F.; Ma, J.-H.; Khan, A.; Wang, X.; Zhao, L.-Y.; Gong, Z.-H.; Chen, R.-G. Genome-wide identification, expression diversication of dehydrin gene family and characterization of *CaDHN3* in pepper (*Capsicum annuum* L.). *PLoS ONE* **2016**, *11*, e0161073. [CrossRef] [PubMed]

31. Pereira-Dias, L.; Vilanova, S.; Fita, A.; Prohens, J.; Rodríguez-Burruezo, A. Genetic diversity, population structure, and relationships in a collection of pepper (*Capsicum* spp.) landraces from the Spanish centre of diversity revealed by genotyping-by-sequencing (GBS). *Hortic. Res.* **2019**, *6*, 54. [CrossRef]

32. Fita, A.; Alonso, J.; Martínez, I.; Avilés, J.; Mateu, M.; Rodríguez-Burruezo, A. Evaluating Capsicum spp. root architecture under field conditions. In Proceedings of the Breakthroughs in the Genetics and Breeding of Capsicum and Eggplant; Lanteri, S., Rotino, G.L., Eds.; Università degli Studi di Torino: Torino, Italy, 2013; pp. 373–376.

33. Fita, A.; Picó, B.; Roig, C.; Nuez, F. Performance of *Cucumis melo* ssp. agrestis as a rootstock for melon. *J. Hortic. Sci. Biotechnol.* **2007**, *82*, 184–190. [CrossRef]

34. Ministerio de Agricultura, Pesca y Alimentación. *Métodos Oficiales de Análisis*; MAPA: Madrid, Spain, 1994.

35. Hills, T.M.L.; Jackson, F. *Agricultural Experimentation: Design and Analysis*; Wiley: New York, NY, USA, 1978; ISBN 978-0-471-02352-4.

36. R Development Core Team. *R: A Language and Environment for Statistical Computing*; R Foundation for Statistical Computing: Vienna, Austria, 2009.

37. Wickham, H. *ggplot2: Elegant Graphics for Data Analysis*; Springer International Publishing: Basel, Switzerland, 2016; ISBN 978-3-319-24277-4.

38. Bouain, N.; Shahzad, Z.; Rouached, A.; Khan, G.A.; Berthomieu, P.; Abdelly, C.; Poirier, Y.; Rouached, H. Phosphate and zinc transport and signalling in plants: Toward a better understanding of their homeostasis interaction. *J. Exp. Bot.* **2014**, *65*, 5725–5741. [CrossRef] [PubMed]

39. Ham, B.-K.; Chen, J.; Yan, Y.; Lucas, W.J. Insights into plant phosphate sensing and signaling. *Curr. Opin. Biotechnol.* **2018**, *49*, 1–9. [CrossRef] [PubMed]

40. Rose, T.J.; Pariasca-Tanaka, J.; Rose, M.T.; Fukuta, Y.; Wissuwa, M. Genotypic variation in grain phosphorus concentration, and opportunities to improve P-use efficiency in rice. *Field Crop. Res.* **2010**, *119*, 154–160. [CrossRef]

41. Bryant, R.J.; Dorsch, J.A.; Peterson, K.L.; Rutger, J.N.; Raboy, V. Phosphorus and mineral concentrations in whole grain and milled low phytic acid (*lpa*) 1-1 rice. *Cereal Chem.* **2005**, *82*, 517–522. [CrossRef]

42. Russo, V.M. *Peppers: Botany, Production and Uses*; Russo, V.M., Ed.; CABI: Wallingford, UK, 2012; ISBN 9781845937676.

43. Akhtar, M.S.; Oki, Y.; Adachi, T. Genetic variability in phosphorus acquisition and utilization efficiency from sparingly soluble P-sources by *Brassica* cultivars under P-stress environment. *J. Agron. Crop Sci.* **2008**, *194*, 380–392. [CrossRef]

44. Hu, Y.; Ye, X.; Shi, L.; Duan, H.; Xu, F. Genotypic differences in root morphology and phosphorus uptake kinetics in *Brassica napus* under low phosphorus supply. *J. Plant Nutr.* **2010**, *33*, 889–901. [CrossRef]

45. Bates, T.R.; Lynch, J.P. Root hairs confer a competitive advantage under low phosphorus availability. *Plant Soil* **2001**, *236*, 243–250. [CrossRef]

46. Strock, C.F.; Morrow de la Riva, L.; Lynch, J.P. Reduction in root secondary growth as a strategy for phosphorus acquisition. *Plant Physiol.* **2018**, *176*, 691–703. [CrossRef] [PubMed]

47. López-Bucio, J.; Cruz-Ramírez, A.; Herrera-Estrella, L. The role of nutrient availability in regulating root architecture. *Curr. Opin. Plant Biol.* **2003**, *6*, 280–287. [CrossRef]

48. Miguel, M.A.; Postma, J.A.; Lynch, J.P. Phene synergism between root hair length and basal root growth angle for phosphorus acquisition. *Plant Physiol.* **2015**, *167*, 1430–1439. [CrossRef]

Article

The Antioxidant Profile Evaluation of Some Tomato Landraces with Soil Salinity Tolerance Correlated with High Nutraceuticaland Functional Value

Renata M. Sumalan [1], Sorin I. Ciulca [1], Mariana A. Poiana [2], Diana Moigradean [2], Isidora Radulov [3], Monica Negrea [2], Manuela E. Crisan [4], Lucian Copolovici [5] and Radu L. Sumalan [1,*]

[1] Faculty of Horticulture and Forestry, Banat's University of Agricultural Sciences and Veterinary Medicine "King Michael I of Romania" from Timisoara, CaleaAradului 119, 300645 Timisoara, Romania; srenata_maria@yahoo.com (R.M.S); c_i_sorin@yahoo.com (S.I.C.)

[2] Faculty of Food Processing, Banat's University of Agricultural Sciences and Veterinary Medicine "King Michael I of Romania"from Timisoara, CaleaAradului 119, 300645 Timisoara, Romania; atenapoiana@yahoo.com (M.A.P.); dimodean@yahoo.com (D.M.); negrea_monica2000@yahoo.com (M.N.)

[3] Faculty of Agriculture, Banat's University of Agricultural Sciences and Veterinary Medicine "King Michael I of Romania" from Timisoara, CaleaAradului 119, 300645 Timisoara, Romania; isidoraradulov@yahoo.com

[4] "CoriolanDragulescu" Institute of Chemistry, 24 Mihai Viteazul Blvd., 300223 Timisoara, Romania; mdorosencu@yahoo.com

[5] Faculty of Food Engineering, Tourism and Environmental Protection, Research Center in Technical and Natural Sciences, "AurelVlaicu" University, Elena Dragoi 2, 310330 Arad, Romania; lucian.copolovici@uav.ro

* Correspondence: sumalanagro@yahoo.com; Tel.: +40-723547363

Received: 23 March 2020; Accepted: 31 March 2020; Published: 2 April 2020

Abstract: Romania has a wide variety of local landraces and heirloom genotypes. Our study aims to assess the performance of twenty halotolerant tomato landraces, collected from areas with medium and high levels of soil salinity, in terms ofthe accumulation of antioxidant compounds in fruits and to cluster them according to their nutraceutical components. The tomatoes used in the study were harvested once they had attained full ripeness and then analyzed for lycopene (Lyc), ascorbic acid content (AsA), total phenolic content (TPC), and total antioxidant capacity (TAC). The results revealed major differences between genotypes in terms of nutraceutical values. According to principal component analysis, the tomato landraces were grouped into five clusters, characterized by different proportions of compounds with antioxidant activity. The high/moderate nutritional values of Lyc, TAC, TPC, and AsA were obtained from varieties taken from local lands with high soil salinity, over 6.5 dS m^{-1}. These findings support the idea that metabolites and secondary antioxidants are involved in the process of stress adaptation, thereby increasing salinity tolerance in tomatoes. Our results show that there are tomato landraces with a tolerance of adaptation to conditions of high soil salinity and provide information on their ability to synthesize molecules with antioxidant functions that protect plants against oxidative damage.

Keywords: tomato cultivars; salinity tolerance; antioxidant activity; lycopene; ascorbic acid; total polyphenols content

---

1. Introduction

The commercial production of tomato (*Solanum lycopersicum* L.) in the developed regions of the world mostly concerns modern varieties of the fruit, and, more often than not, genetically uniform F1 hybrids that have a high yield, greater tolerance to diseases, and a long shelf life are chosen [1]. Landraces represent an important alternative, as they constitute a reservoir of genetic diversity, with their important abiotic stress tolerance and high fruit quality. Tomato landraces contain valuable alleles uncommon in highlight germplasms [2]; therefore, these local populations represent a valuable resource of genetic traits that can be used in breeding programs for the improvement of the crop [3].

Intensive tomato cultivation technologies require genotypes with good productivity, handling, transport, and storage properties, while the nutraceutical properties are passed on to the secondary level. Nevertheless, consumption of the traditional plant foods that have antioxidant content naturally occurring may be a better strategy to improve the human health status than the consumption of artificial antioxidant products [4]. From this perspective, an appropriate selection of tomato cultivar would help to achieve a higher antioxidant intake with the potential to produce significant health benefits. Considering the growing demand for tasty tomatoes and being rich in phytochemicals, a detailed characterization of the biologically active compounds completed with total antioxidant capacityevaluation should be performed. That is why the tomato landraces and their relatives are of great importance in breeding programs [5–7]. Besides, landraces have low requirements for inputs, which contribute to the development of environmentally-friendly technologies [8–10].

The genetic resources of cultivated plants that come from the soil salinity-affected areas have a major importance due to drought tolerance. The salinity is associated with the physiological drought that may induce the growth of the bioactive compound content with antioxidant properties [11,12]. The information about genetic variability specific to local populations is assumed to be limited [13–15] because traditional locally grown cultivars should not be considered rigorously homogenous.

The tomato taste is correlated with some bioactive components. Some authors consider that AsA (ascorbic acid), TAC (total antioxidant capacity), and TPC (total phenols content) have a direct impact on tomato taste [5], while others assign the taste to the ratio between sugar content and acidity [16–19]. The high content in valuable phytonutrients depends equally on technological, genetic, and storage factors [16,20–25]. Besides dietary fibers and carbohydrates, compounds such as lycopene, β-carotene, ascorbic acid, and polyphenols provide high levels of antioxidants. For this reason, the consumption of raw or processed tomatoes contributes to good activity into an organism by maintaining oxidative stressat a low level [26,27].

Lycopene is the most important and recognized phyto bioactive-compound of tomatoes. It is a carotenoid pigment that is less bio-available compared to β-carotene and lutein [28,29]. In addition to lycopene, the tomato fruits contain vitamins A and C, other carotenoids whose action interacted with those of polyphenols, resulting in an overall benefit on human health [30,31]. The ripening stage of tomato fruits represents a decisive factor regarding the establishment of nutritional values of thereof. From this point of view, the tomatoes harvested at technological ripeness revealed low quantities of lycopene while the content of ascorbic acid was variable, depending on genotype [32].

The aim of this paperis to characterize some tomato local landraces, originating from areas with saline soils located in western Romania (Banat region), concerning the total antioxidantcapacity and the potential of biosynthesis and accumulation of some antioxidant biocomponents, such as phenolic compounds, lycopene and ascorbic acid.

In this work, we test the hypothesis that local tomato landraces originating from areas with soil salinity show a higher biosynthesis capacity of some compounds with antioxidant activity, and this property is maintained even under cultivation on non-saline soil conditions.

2. Materials and Methods

2.1. Plant and Soil Analyses

Twenty halotolerant tomato landraces, collected from local farmers ofthe country-side situated in areas with soils affected by different levels of sodicity from westernRomania (Banat region) were analyzed. Saline-sodic soils are high in exchangeable sodium and low in total soluble salts. The level of ESP (exchangeable sodium percentage) in these soils is 15 or more, which tends to destroy their structure by dispersing the particle, and electrical conductivity (EC) is over 4 dS m^{-1}. Most saline soils in the collecting area have a clayed loamy texture, medium glomerular structure and moderately to high salinization level determined by using an EC-meter (model consort C933, producer De Bruine Instruments bvba, Belgium) determination using the EC 1:2*w/v* method [33] (Table 1).

Seeds and soil samples were collected in the period 2012–2015, and the specimen is available at Plant Physiology Department, Faculty of Horticulture and Forestry, Timisoara. Previously, each local landraces was characterized in morpho-physiologically and genetically manner [34], for each of them was drawn up an identification sheet in which source, specific cultivation technology, productivity, shape, and color of tomato fruits were noticed. Based on productivity traits, high tolerance to salinity and minimum growth requirements proven in the summer of 2015, twenty tomato landraces were selected for assessment of nutraceutical traits (Table 1).

The genotypes were open field cultivated in a plain site located on the northern side of Timisoara (45°78′N; 21°21′E), on cambicchernozem soil [35]. The soil had the following physico-chemical characteristics: clay 402 g kg^{-1}; sand 330 g kg^{-1}; loam 268 g kg^{-1}; organic matter 26.8 g kg^{-1}; pH 6.26; total N_2 g kg^{-1}; available P_2O_5 20.52mg kg^{-1}; exchangeable K_2O 117 mg kg^{-1}; sulfates (mobile in water) 105.6 mg kg^{-1}; sodium (mobile in water) 366.7mg kg^{-1}; calcium (water-soluble) 270.5 mg kg^{-1}; magnesium (soluble in water) 60.8 mg kg^{-1}.

A randomized complete block experimental design with three replicates was used in the field during the spring–summer season. Plants were transplanted at the four-leaf stage on 18 April 2017 in plots of 24 m^2 (6 × 4 m) at 1.66 plants m^{-2}. The average monthly temperatures ranged between 11.6 °C in April to 27.4 °C in July. Plants were spaced 1.2 m between the rows and 0.5 m within the row and watered with a drip irrigation system. Plants were trained with canes and cultivated using a traditional horticultural practice in the area for local tomato varieties. All genotypes have an indeterminate plant growth type.

Table 1. Geographical origin, EC of origin soils, and the main morphological and productive characteristics of tomato landraces.

Genotype Code	Site	GPS Coordinates (lat/long)	Soil EC(dSm^{-1})	Fruit Shape [1]	Tomatoes Weight Average (g) [2]	Full Ripeness Color [3]
CN26	Crai Nou	45°29′17″N/21°0′1″E	6.86	flattened	352.47	light red
PN	Peciu Nou	45°36′54″N/21°01′54″E	5.63	flattened	184.66	light red
Gi	Giera	45°25′21″N/20°57′25″E	5.25	circular	124.18	red
L-189a	Lovrin	45°57′03″N/20°46′32″E	5.02	obovate	73.75	light red
C-102	Cruceni	45°28′23″N/20°52′44″E	7.04	circular	133.55	light red
Pe	Periam	46°01′41″N/20°53′35″E	6.86	flattened	295.76	red
Gr	Gradinari	45°06′16″N/21°34′59″E	4.38	circular	273.00	light red
DV	Dudestii Vechi	46°04′55″N20°26′55″E	5.80	obovate	164.92	red
Ch	Cheglevici	46°6′40″N/20°26′56″E	6.04	flattened	264.54	red
C-60pr12	Cherestur	46°7′60″N/20°22′60″E	5.65	ovate	150.64	light red
Ch-165	Cheglevici	46°6′40″N/20°26′56″E	6.29	circular	105.44	light red
Li	Livezile	45°23′09″N/21°02′43″E	6.44	cilindric	133.62	yellow
L-189b	Lovrin	21°02′43″E/20°46′32″E	6.58	flattened	136.11	light red
Ru	Rudna	45°29′54″N/21°0′31″E	4.50	obovate	81.73	red
SS180	Sanmartinu Sarbesc	45°36′23″N/20°57′38″E	4.47	cordate	303.09	red
T673	Tarnova	45°20′06″N/22°00′08″E	4.11	flattened	309.57	red
T370	Tarnova	45°20′06″N/22°00′08″E	4.23	flattened	348.42	red
IM/pusta	Iecea Mare/Pusta	45°50′51″N/20°54′08″E	4.18	flattened	392.60	red
SS	Sanmartinu Sarbesc	45°36′23″N/20°57′38″E	4.30	flattened	185.59	light red
CN-254	Crai Nou	45°29′17″N/21°0′1″E	7.21	obovate	150.33	red

[1] According to UPOV (International Union for the Protection of New Varieties of Plants) classification in tomatoes fruit shape [36]; [2] data were collected from 15 fruits; [3] USDA (United States Department of Agriculture) tomato ripeness color chart.

2.2. Tomatoes Samples Preparation

From each genotype, samplings were taken at different harvesting times only when tomatoes were at a fully physiological ripening stage. Fifteen fruits were randomly taken from each replication in order to compose the average tomato sample. Evaluations of shape, weight, and colors, as well as chemical analysis, were done. The fruits were stored in polyethylene bags and kept in freezing conditions at −18 °C until performing the chemical analysis. All analyses werecarried out in triplicate.

2.3. Chemical Analysis

The chemical analysis consisted of assessing of TAC, TPC, AsA, and Lyc content from each tomatoes landrace. Prior analysis, the frozen samples were kept in refrigeration condition (4–6 °C) for 6 h and then homogenized in a Bosch Blender (MMB42G0B, 700 W, Germany) for 1 min. Three replicates were prepared from each average sample.

2.3.1. Extract Preparation

Briefly, 10 g of blended tomato sample was mixed with 20 mL ethyl alcohol 70% (*v/v*) for 2 h at 25 °C, then, the mixture was filtered and the clear extract was used for the analysis of TAC and TPC.

2.3.2. Reagents and Equipment

All chemicals and reagents were analytical grade or purest quality purchased from Merckand Fluka. Deionized water was used.

TAC Evaluation

TAC of tomatoes was evaluated by ferric reducing antioxidant power (FRAP) assay, according to the method described by Benzie and Strain [37]. This method supposed the reduction of Fe^{3+}-TPTZ (2,4,6-tris(2-pyridyl)-s-triazine) complex to ferrous form at low pH. The ferrous tripyridyltriazine complex has an intense blue color monitored at a wavelength of 593 nm. An aqueous solution of Fe^{2+} with a concentration in the range of 0.1 to 0.8 mM/L was used for the calibration curve preparation. The absorption was measured at $\lambda = 593$ nm after 15 min of incubation at 25 °C using the UV–vis spectrophotometer SPECORD 205 (Analytic Jena, Germany). TAC was expressed as μM Fe^{2+} equivalents·100 g^{-1} FW (fresh weight).

TPC Determination

TPC in tomato samples was evaluated following the Folin–Ciocalteu colorimetric method described by Singleton and Rossi [38]. For analysis, it was used the tomato ethanol extracts diluted 1/10 with bidistilled water. For calibration curve preparation, 0.5 mL aliquot of aqueous gallic acid solution with a concentration in the range 0.2–1.2 μM/mL were mixed with 2.5 mL Folin–Ciocalteu reagent (diluted ten-fold with bidistilled water) and 2.0 mL sodium carbonate (7.5%). The absorption was read at $\lambda = 750$ nm after 2 h of incubation at 20 °C. TPC in mg gallic acid equivalents (GAE)·100 g^{-1} FW was calculated.

AsA Content

The AsA content of tomato samples was measured on the base of the AOAC method [39] by titration with 2,6-dichlorophenolindophenol sodium salt. For this purpose, 10 g of blended tomato sample was mixed with 10 mL bidistilled water for 2 h at 25 °C, then, the mixture was filtered, and the clear extract was used for analysis. Further, 5 mL of extract was diluted with 10 mL bidistilled water, then 1 mL HCl 1N was added and the mixture was titrated with 2,6-dichlorphenolindophenol sodium solution 1 mM in an acid medium (pH = 4). The results are expressed as mg ascorbic acid·100 g^{-1} FW.

Determination of Lyc Content

Lyc was extracted from tomato samples with a hexane–ethanol–acetone (2:1:1) mixture in agreement with the method describes by Sharma and Le Maguer [40]. Briefly, 1 g of tomato blended sample was mixed with 25 mL of the previously mentioned mixture and then placed on a rotary mixer for 30 min. Further, 10 mL of bidistilled water was added, and the mixture was stirred for another 2 min. The obtained mixture solution was separated into two distinct polar and non-polar layers. The absorbance was measured at 502 nm, using hexane as a blank. The Lyc content of tomato samples was calculated on the base of its specific extinction coefficient (E 1%, 1 cm) of 3150 at 502 nm [41]. The Lyc concentration was expressed as mg 100 g^{-1} FW.

2.4. Statistical Analysis

The experimental data were statistically processed using ANOVA, and the means were compared using the multiple range test [42]. The significance of differences was expressed based on letters, being considered as significant the differences between genotypes marked with different letters.

The clustering of genotypes was carried out using the UPGMA (unweighted pair group method with arithmetic mean), with the NEIGHBOR program of PHYLIP package, version 3.5c. [43]. Average intra and inter-cluster distances (D2) were estimated [44], and the percent contribution of each character to the total divergence was calculated by ranking each character on the basis of transformed uncorrelated values. To display the performance of each landrace for each of four nutritional traits in a single graph, the basic principle of the biplot technique was used [45,46].

3. Results

3.1. Fruit Morphological Traits

All the measured fruit morphological traits (shape, weight, and color), showed a large range of phenotypic variation among the 20 tomato landraces.

Fruit shape is one of the most important qualities which can be determined with the naked eye, being used to identify local tomato populations. Globally, there is a huge variety of fruit shapes on tomatoes, specific to different landraces, a fact reported by many studies conducted in different geographic areas and time periods [2,15,36,47–52].

As for the traits related to color and flesh color of fruits, ten genotypes had *red* fruit, nine genotypes had a *light red* color, and one had *yellow* fruits.

Regarding the tomato fruit weight, a number of eight genotypes had large fruits (≥200 g), of which five were very large (≥300 g), ten landraces had average fruits weighing between 100 and 200 g, and only two formed small fruits (below 100 g). The comparative analysis of the form of fruit and their weight shows that six of the local populations with high weight fruit have a flattened shape.

These results are also in line with other studies [53] confirming that, in the process of improving tomatoes, people prefer to increase the size and mass of the fruit, causing and modifying the round shape (wild species) with a flattened one in the most forms cultivated for fresh consumption or elongated for industrialization. On the other hand, some studies [15] found that, in the case of some local Italian and South American tomatoes populations, the flattened form of the fruit has been associated with a small and average weight (50–150 g). Obviously, the fruit mass is a genetically controlled process, but it depends to a large extent on the specific pedo-climatic conditions and applied technology.

Therefore, the study of genetic variability inlocal tomato landraces will be able to provide additional information on the genetic, physiological, and biochemical mechanisms that are based on the correlation of the shape, size, and weight of the fruit, thus contributing to the identification of the alleles and new ecotypes with superior properties in terms of productivity, adaptation, quality, and nutritional value.

3.2. Antioxidants and Nutraceutical Component Analysis

Antioxidants are redox buffers that interact with ROS (reactive oxidative species) and can manifest as a metabolic interface that regulates adaptation responses or programmed cell death [54]. The low values of F: 120 and 126, for significant differences at $p < 0.01$, show that there is a lower variability between landraces for TPC and AsA.

3.2.1. Assessment of TAC

Statistical data analysis concerning the performance of TAC of tomato fruit samples reveals significant differences between landraces (Table 2). The quantity and proportion between bioactive compounds with antioxidant capacity depend on the plant's genotype and post-harvest storage conditions [55,56]. The reducing ability recorded for the 20 landraces varies from 561.61 to 240.75 μM Fe^{2+} 100 g^{-1} FW with an amplitude of variation of 320.86 between landraces (Table 2).

Table 2. Total antioxidant capacity (TAC) of tomato fruit samples.

Genotype Code	FRAP μM $Fe^{2\pm}$ 100 g^{-1} FW
CN-26	413.27 ± 3.63 d
PN	305.44 ± 2.99 g
Gi	349.03 ± 3.76 f
L-189a	386.49 ± 3.91 e
C-102	420.97 ± 2.13 d
Pe	451.76 ± 1.36 c
Gr	384.03 ± 3.19 e
DV	411.36 ± 2.02
Ch	274.00 ± 1.53 i
C-60pr12	240.75 ± 1.51 j
Ch-165	304.67 ± 1.65 g
Li	415.14 ± 1.99 d
L-189b	506.51 ± 2.65 b
Ru	300.05 ± 2.42 gh
SS-180	289.21 ± 1.58 h
T-673	287.27 ± 1.60 hi
T-370	407.56 ± 2.31 d
IM/pusta	307.69 ± 1.62 g
SS	387.05 ± 1.92 e
CN-254	561.61 ± 4.37 a
Mean	*370.19 ± 10.48*
Cochran's C Test	*0.145; p = 1.00*
Bartlett's Test	*1.300; p = 0.974*
LSD5%	*13.78*

Data are mean ± SE, n = 3. Values within columns with different letters are significantly different (p = 0.05).

The high value of TAC was determined for CN-254, 561.61 μM Fe^{2+} 100 g^{-1} FW, followed by L-189b with 506.51 μM Fe^{2+} 100 g^{-1} FW. Both landraces have recorded statistically higher differences related to mean. In the second category fall the local populations Pe (451.76 μM Fe^{2+} 100 g^{-1} FW), C-102 (420.97 μM Fe^{2+} 100 g^{-1} FW), and CN-26 (413.27 μM Fe^{2+} 100 g^{-1} FW). The lowest value of FRAP (240.75 μM Fe^{2+} 100 g^{-1} FW) was determined for genotype C-60pr12. It should be noted that all five cultivars of the first two categories were collected from the areas with high concentrations of soil salinity between 7.21 dS m^{-1} for CN-254 and 6.58 dS m^{-1} at L-189b.

The recorded TAC values are in agreement with those reported by other authors who have used the same method. Thus, mean values of TAC of 506 μM Fe^{2+} 100 g^{-1} FW [57] were reported, while other authors reported limits between 387 and 493 μM Fe^{2+} 100 g^{-1} FW [31] for tomatoes fruits.

Phenolic compounds play an important role in the antioxidant capacity of tomato fruit. In addition, other bioactive compounds such as ascorbic acid and lycopene, the main tomato carotenoid, play a major role in the antioxidant capacity of tomato fruit samples. The ascorbic acid is considered the most important water-soluble antioxidant, with a significant contribution to the antioxidant cellular defense against oxidative stress. FRAP assay also measures the antioxidant capacity of ascorbic acid besides that of phenolic compounds. The phenolic compounds act synergistically with ascorbic acid in order to preserve and regenerate the antioxidant species. Lycopene is one of the most powerful antioxidants among the dietary carotenoids. The total antioxidant capacity of tomato fruit is the result of a combination of different compounds having synergistic and antagonistic effects.

3.2.2. Assessment of TPC, Lyc, and AsAContent

The quantity and quality of phenolic compounds determined in tomato fruit varies in relation to the genotype but also depends on environmental and technology factors [58]. Data from Table 3 reveal that the TPC values range from 51.49 to 123.32 mg GAE·100 g^{-1} FW. The highest value was registered for C-102 (123.32 30 mg GAE·100 g^{-1} FW), followed by L-189b with 117.77 mg GAE·100 g^{-1} FW and CN-254 (114.89 mg GAE·100 g^{-1} FW), but the differences between them have no statistical significance.

The TPC values for the local tomato populations collected from the saline areas are higher compared to other studies [31,56,59,60]. TPC valuesrecorded for industrial processing cultivars(green Ronaldo and red cherry Pera) were between 18.69 to 55.86 mg GAE 100 g^{-1} FW [56] and in the range 19.7–21.1 mg GAE 100 g^{-1} FW [61], while for fresh tomato consumption ranging from 50.86 to 53.88 mg GAE 100 g^{-1} FWwere reported [31].Also, Martinez-Valverde reported TPC values between 25.9 and 49.8 mg GAE 100 g^{-1} FW for determinations of some commercial varieties of tomatoes from supermarkets [59]. The lowest values of TPC in our cultivars were determined for IM/Pusta with 51.49 mg GAE 100 g^{-1} FW, followed by genotype Ch with 53.06 mg GAE 100 g^{-1} FW.

The value of TPC for our landraces (especially C-102, L-189b, and CN-254) are higher than those obtained by other researchers for different hybrids. The highest value of TPC for CN-254 justifies the large TAC value determined for this local population, poly-phenols being well known for their antioxidant activity [30,62]. By correlating the results to the fact that these local populations have adapted over time to high levels of soil salinity, we can assume that they have developed genetic, biochemical, and physiological mechanisms to synthesize some of the antioxidant-related co-agents, to allow survival under specific stress conditions. It seems that the high pressure of the soil salinity influenced the plants which were obliged to adapt by increasing the synthesis capacities of polyphenols and indirectly intensifying of antioxidant processes. These traits appear to be manifested in the conditions of their cultivation outside specific saline areas. Indeed in a recent work, it has been shown that the traditional variety of tomato improved cations homeostasis and increased sucrose content in the fruits as a part of the salt stress tolerance mechanism [63].

Raw tomatoes contain usually Lyc between 3 to 10 mg 100 g^{-1}, but field-grown tomatoes appear to contain higher levels of this compound, ranging from 5.2 to 23.6 mg 100 g^{-1} FW than greenhouse-grown tomatoes (0.1 and 10.8 mg 100 g^{-1} FW) [31].

In our research, the Lyc content recorded a high variability of 13.13 mg 100 g^{-1} FW amplitude of variation (Table 3). The highest value of Lyc was determined for SS-180 (18.43 mg 100 g^{-1} FW) followed by Gr (17.37 mg 100 g^{-1} FW) and T-673 (16.61 mg 100 g^{-1} FW). At the same time, significantly lower mean values of the Lyc content were noticed for landraces Li (5.30 mg 100 g^{-1} FW), Ch-165 (6.79 mg 100 g^{-1} FW), and C-60pr12 (6.81 mg 100 g^{-1} FW) that have recorded statistically assured differences related to mean of experience.

Comparing our results concerning the Lyc content of tomato fruit with other studies [64–67], we found that the local populations collected by the salt-crops cultivated under non-saline conditions recorded higher values. Most previous studies reported values of Lyccontent in fresh tomato fruits ranging from 3 to 10 mg 100 g^{-1} FW. Higher amounts of Lyc (around 25 mg 100 g^{-1} FW) were reported in the cherry tomatoes IIHR-249-1 line [68].

Although most research confirms the growth of Lyccontent in salinity conditions [69–71], there are a few studies that infirm this hypothesis [23,72]. Therefore, the genetic requirements and the specific conditions in which the genotype was cultivated are decisive factors on the capacities of biosynthesis of Lyc, alongside other evident factors such as ripening stage and cultivation technologies (e.g., field/greenhouse, organic/non-organic) [32].

The results obtained through this research do not show a direct connection between the soil salinity level in the harvesting area and the amount of Lyc of the various local populations. Cultivars with high levels of Lyc come from areas with moderate levels of soil salinity (E C $\leq$ 4.5 dS m^{-1}).

The introduction of Lyc as a phyto-compound in the human diet, due to multiple antioxidant properties, led to obtaining products enriched in this biomolecule [73]. From a nutritional point of view and based on recommendations on the consumption of 35 mg Lyc/day [74], this can be achieved by eating 200 g of tomatoes from SS 180 or GR local population for the daily requirement of an adult.

AsA is an antioxidant with an electron donor role in many important reactions [75,76]. It plays a key role in photosynthesis protection during salinity stress. It has been shown that on salinity stress conditions, mutations with deficiency of AsA synthesis accumulate very large amounts of H_2O_2, which coincides with an important decrease of the reduced AsA. The accumulation of H_2O_2 in the foliar apparatus of tomato plants determines the inhibition of photosynthesis by reducing the amount of chlorophyll, CO_2 assimilation, and decrease in PS II activities [54].

Table 3. Lycopene (Lyc),ascorbic acid (AsA),and total phenolic content (TPC)of tomato fruit samples.

Genotype Code	Lycopene mg 100 g^{-1} FW	Ascorbic Acid mg 100 g^{-1} FW	Total Phenols mg GAE 100 g^{-1} FW
CN26	10.72 ± 0.19 hi	16.93 ± 0.20 de	69.15 ± 1.28 f
PN	7.93 ± 0.12 k	15.21 ± 0.14 ij	92.88 ± 1.83 c
Gi	11.43 ± 0.17 g	15.94 ± 0.17 h	93.22 ± 1.71 c
L-189a	11.20 ± 0.13 gh	16.44 ± 0.19 fg	93.68 ± 2.18 c
C-102	12.57 ± 0.12 f	17.37 ± 0.16 bc	123.32 ± 2.10 a
Pe	12.51 ± 0.10 f	17.70 ± 0.18 b	91.76 ± 2.19 cd
Gr	17.37 ± 0.14b	16.56 ± 0.14 ef	86.58 ± 1.67 cd
DV	12.29 ± 0.11 f	16.99 ± 0.16 cde	91.37 ± 1.70 cd
Ch	9.30 ± 0.08 j	13.98 ± 0.16 l	53.06 ± 0.73 g
C-60pr12	6.81 ± 0.06 l	13.17 ± 0.12 m	112.96 ± 1.74 b
Ch-165	6.79 ± 0.07 l	15.07 ± 0.13 ij	75.19 ± 1.53 ef
Li	5.30 ± 0.07 m	17.06 ± 0.11 cd	88.26 ± 2.12 cd
L-189b	10.73 ± 0.12 hi	19.23 ± 0.16 a	117.77 ± 2.53 ab
Ru	12.87 ± 0.17 ef	14.81 ± 0.13 jk	89.29 ± 2.09 cd
SS-180	18.43 ± 0.15 a	14.54 ± 0.11 kl	73.19 ± 1.17 f
T-673	16.61 ± 0.13c	14.31 ± 0.13 l	71.79 ± 1.01 f
T-370	12.83 ± 0.11 ef	16.10 ± 0.18 gh	82.69 ± 1.83 de
IM/pusta	13.23 ± 0.10 e	15.30 ± 0.13 i	51.49 ± 1.19 g
SS	14.00 ± 0.13 d	16.72 ± 0.15 def	93.48 ± 1.46 c
CN-254	10.18 ± 0.08 i	20.15 ± 0.17 a	114.89 ± 2.16 ab
means	*11.65 ± 0.44*	*16.18 ± 0.22*	*88.30 ± 2.49*
Cochran's C Test	*0.128; p = 1.00*	*0.086; p = 1.00*	*0.102; p = 1.00*
Bartlett's Test	*1.191; p = 0.998*	*1.057; p = 1.00*	*1.162; p = 0.999*
LSD5%	*0.65*	*1.56*	*9.49*

Data are mean ±SE, n = 3. Values within columns with different letters are significantly different (p = 0.05).

Consequently, our results show that AsA content of tomato fruit has not a large amplitude, the determined quantities ranging from 13.17 to 20.15 mg 100 g^{-1} FW. Particularly notable are CN-254 and L-189b landraces, having the highest values of AsA content of 20.15 and 19.23 mg 100 g^{-1} FW, respectively, followed by Pe (17.70 mg 100 g^{-1} FW) and C-102 (17.37 100 g^{-1} FW).

The analysis of the results confirms that the above-presented genotypes with high synthesis capacity of the AsA have manifested very good qualities of TAC and TPC (Table 2).The results prove that the AsA content is conditioned by genotype to a minimal extent. High AsAcontent was reported in F1 tomato hybrid for which heterosis effect was manifested, the maximum content determined of 36.3 mg 100 g^{-1} FW being over mid parental values [20,77].

It was demonstrated that AsA increase is linked to the adopted cultivation system. Studies on tomato growing technologies have highlighted the importance of nutrient availability for tomato plant requirements for AsA accumulation [78].Therefore, there are numerous reports of increasing the content of AsA in tomato fruit subject to various types of stress [79–84], as well as some results reported in which stress resulted in the reduction of AsA [85,86]. The contradictory results can be attributed to the genetic differences regarding the sensitivity of the different genotypes to the oxidative stress manifested by salinity.

It is known that there are universal mechanisms as responses to the action of stress factors, but their relative impacts can vary from one species to another and within the same species from one genotype to another, depending, actually, on specific metabolic background.

3.3. Comparison of Tomato Landraces for Nutraceutical Traits

Using the UPGMA method for 20 variables of tomato landraces, a dendrogram was designed that has identified five groups (clusters) on the basis of coefficients similarity for bioactive compounds (Figure 1). The first cluster brings together nine landraces: CN26, Gi, L-189a, DV, SS, T370, Pe, Gr, and C-102. These are characterized by higher TAC and AsA values, while TPC and Lyc have recorded values above landrace average. Three landraces are grouped in the second cluster, namely: PN, Ch-165, and Li; these recorded low values of Lyc but medium values of TAC, TPC, and AsA content.

In Cluster III, there are grouped five landraces characterized by a high Lyc, while TAC and AsA have lower values than average, and TPC has the lowest values. In Cluster IV, one single landraceis noticed; C-60pr12 has a high value of TPC but a low one of Lyc, while TAC and AsA contents are lower than mean values of landraces.

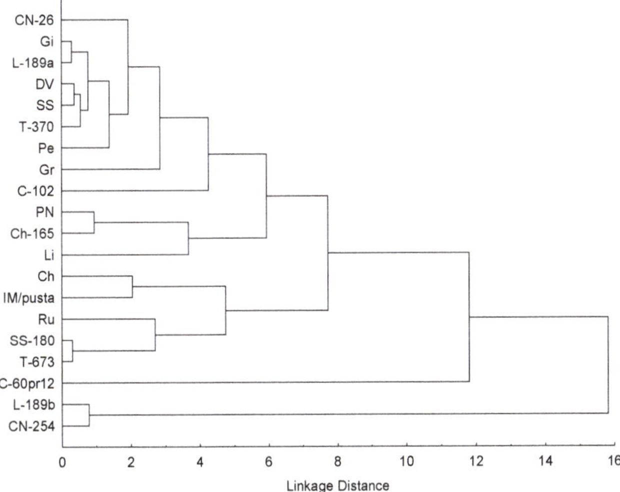

Figure 1. Dendrogram of tomato landraces for quality traits.

The landraces L-189b and CN-254 are grouped in Cluster V, which has recorded higher TAC, TPC, and AsA values, but medium Lyc content. Regarding the contribution of different traits to diversity between clusters, it was found that TPC is the most important by 45%, contributing to divergence; meanwhile, AsA has the lowest contribution, respectively, 13.5% (Table 4). Therefore, studied tomato landraces can be distinguished between each other to a much lesser extent in terms of AsA content.

Table 4. Cluster's mean for five traits of tomato landraces and the contribution of each trait to the total divergence.

Traits	Clusters					Times Ranked First	Contribution to Divergence (%)
	I	II	III	IV	V		
TAC	401.28	341.75	291.77	240.75	534.06	35	17.5
TPC	91.79	85.44	67.76	112.96	116.33	90	45.0
Lyc	12.77	6.67	14.09	6.81	10.46	48	24.0
AsA	16.75	15.78	14.59	13.17	19.69	27	13.5

It seems that this bioactive compound might not be important in landraces recognition and might oscillate depending on the tomato growing technology. Our observation is also supported by other researches regarding the accumulation of AsA in tomatoes; major differences have been found among the individual samples but not between tomato varieties. It seems that light exposure of tomato fruits directly affects the accumulation of phyto-compound [87]. Analyzing the contribution of different traits to inside cluster diversity, it was found that the highest diversity exists between the landraces Ch, IM/pusta, Ru, SS180, and T-673 grouped in Cluster III (D^2 = 3.62), while between landraces L-189b and CN 254 from Cluster V, a high similarity (D^2 = 0.78) was recorded to all nutritional components (Table 5).

According to the inter-cluster distances, it was observed that the landraces L-189b and CN254 from Cluster V differ significantly to the landraces from other clusters except for the nine landraces of Cluster I. Also, the landrace Ch-60pr12, characterized by low values of these quality traits, differs significantly to the landraces from Clusters I and III.

Table 5. Average intra- (bold diagonal) and inter-cluster (off diagonal) D2 values.

Cluster	Landraces	I	II	III	IV	V
I	CN-26; Gi; L-189a; DV; SS; T-370; Pe; Gr; C102	2.30	5.94	7.42	13.22 *	8.71
II	PN; Ch-165; Li		2.76	8.63	6.70	15.42 **
III	Ch; IM/pusta; Ru; SS180; T-673			3.62	12.31 *	26.36 ***
IV	Ch-60pr12				0.00	28.31 ***
V	L-189b; CN254					0.78

$\chi2 = 9.49$ ($p = 0.05$); $\chi2 = 13.28$ ($p = 0.01$); $\chi2 = 18.47$ ($p = 0.001$). The data show correlation index values; * $p < 0.05$, ** $p < 0.01$, *** $p < 0.001$.

The four principal components account for the whole variability among the studied tomato landraces for the analyzed quality traits (Table 6). The first principal component (PC1) has a major contribution of 59.84% to the total variation. Only Lyc (0.144) contributed positively to PC1, while the other traits contributed negatively to this principal component.

Table 6. Eigen vectors and eigen values of the first four principal components for quality traits of tomato landraces.

Traits	PC1	PC2	PC3	PC4
TAC	−0.956	0.171	0.225	−0.081
TPC	−0.737	−0.253	−0.627	0.000
Lyc	0.144	0.959	−0.244	0.000
AsA	−0.957	0.169	0.221	0.082
Eigen value	2.394	1.041	0.552	0.013
Cumulative eigen value	2.394	3.435	3.987	4.000
Proportion variance	59.84	26.02	13.80	0.33
Cumulative variance	59.84	85.86	99.67	100.00

The second principal component (PC2) accounted for 26.02% of the total variation, with positive support of Lyc (0.959), TAC (0.171), and AsA (0.169), while TPC (−0.253) has a negative involvement. The third principal component (PC3) showed 13.80% of the overall variation and was positively associated with TAC (0.225), and AsA, as well as TPC (0.627) and Lyc (−0.244), were negatively associated with PC3. The fourth principal component (PC4) depicted a low proportion of the whole variability (0.33%), indicating the strongest discriminatory power of these two principal components (Figure 2). The biplot reveals a broad dispersion of the landraces and explained 85.86% of the variability. Negative values at PC1 indicate landraces with high TAC, ascorbic acid, and TPC.In this regard were highlighted the landraces from Cluster V, L-189a, C-102, and CN-245 have the highest values for TAC, AsA, and TPC, but a medium Lyc content. Positive values for PC2 belong to landraces having a high Lyc amount, SS-180, T-673, IM/pusta, and Ru, from Cluster III. The negative values of PC2 associated with positive values of PC1 are characteristic of the landraces with low levels of these traits like PN, Ch-165, and Ch-60pr12, grouped in Clusters II and IV, respectively. The landrace Li shows a higher TAC compared to PN and Ch-165 from Cluster II, thus being the main contributor to intra-cluster diversity. According to the dendrogram (Figure 1), it was noticed that there are different landraces groups inside Cluster I. Thus, landrace Gr with a high Lyc content exceeds the average values for TAC and ascorbic acid content, while SS-180 and T-673 with a high content of Lyc are associated with low values for the other qualitative traits.

The nutritional value of tomatoes proven by the content in bioactive compounds with high value is influenced by several factors. Many studies on tomato cultivars highlighted that variation of both abiotic [23,29,60] and technological [10,25,78,88] factors have a decisive effect on increasing the nutritional value of tomatoes.

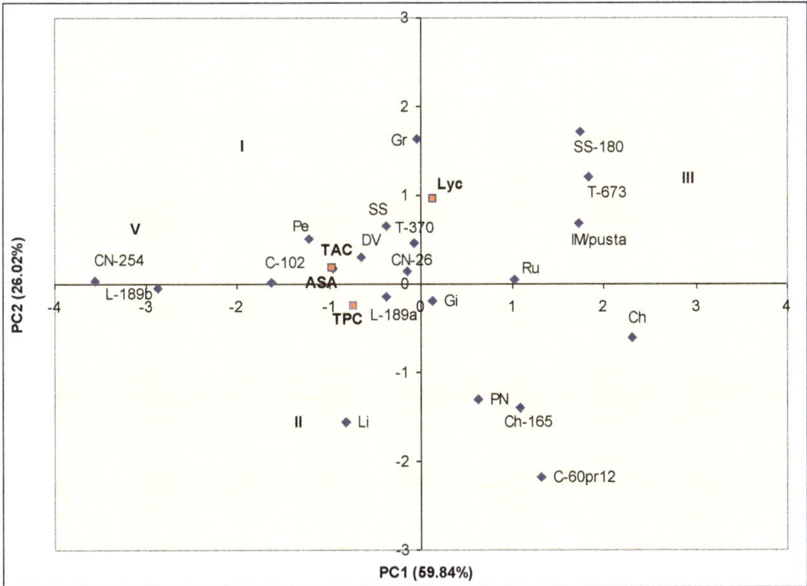

Figure 2. Biplot of the first two principal components for 20 tomato landraces and four quality traits.

4. Conclusions

The results reveal that the 20 tomato landraces with tolerance to salinity have high potential in phyto-compound accumulation with high antioxidant levels. The ratio between these is different. Even if it is widely accepted nowadays that the idea ofphyto-chemicals with high nutraceutical value depend on plants' genetic information, environmental factors may alter the expression of these genes.

The research confirms the hypothesis that tomato landraces with tolerance to soil salinity have a higher ability to accumulate in ripe fruits large amounts of antioxidants such as phenolic compounds and carotenoids. The largest amounts of antioxidants were recorded in that local populations originating from the areas with a high level of soil salinity, whose electrical conductance was over 6.5 dS m^{-1}.

Correlating the results to the fact that these tomato landraces have adapted over time to high levels of soil salinity, we can assume that they have developed genetic, biochemical, and physiological mechanisms to synthesize some of the antioxidant-related co-agents, to allow survival under specific stress conditions. Under the higher soil-salinity conditions, the plants were more obliged to adapt by increasing the synthesis capacities of polyphenols and indirectly intensifying of antioxidant processes. These traits seem to be manifested in the conditions of their cultivation outside specific saline areas. The compositional evaluation highlighted that tomato halo-tolerant landraces are an inexhaustible resource of variability with nutraceutical properties that have been proven. These resources can be exploited in breeding programs or could be cultivated in traditional farms that adopt ecological technology.

The results of the TAC, TPC, Lyc, and AsA determinations can be considered the cumulative response of the genetic fund interaction with all the interactive effects occurring during the maturation phases of the fruit. This approach can provide meaningful information on the modeling of the nutritional quality of tomato fruit and also provides interesting insights into the metabolic capacities of the old local populations that have adapted to the conditions of high soil salinity. However, additional functional research is still needed to link the direct determinations with genetic and metabolic analyses.

Author Contributions: Conceptualization, R.M.S. and R.L.S.; methodology, M.A.P., R.L.S., R.M.S., and L.C.; software, S.I.C.; validation, M.E.C., L.C., and I.R.; formal analysis, L.C.; investigation, R.L.S., S.I.C., R.M.S., D.M., and M.N., resources, R.L.S.; data curation, D.M., M.N., and S.I.C.; writing—original draft preparation, R.M.S. and R.L.S.; writing—review and editing, M.E.C. and L.C.; project administration, R.L.S. and I.R.; funding acquisition, I.R. All authors have read and agreed to the published version of the manuscript.

Funding: This research was supported by Romanian National Authority for Scientific Research, CNDI-UEFISCDI project number PN-II-PT-PCCA-2011-3.1-0965.

Acknowledgments: The authors thanks for the financial support for publishing throught the project "Ensuring excellence in RDI activities within USAMVBT", code 35PFE/2018 financed by the Ministry of Research and Innovation (MCI) through Program 1–Development of the national research and development system, Subprogram 1.2–Institutional performance, Institutional development projects–Projects to fund excellence in RDI; Special thanks to all persons from the villages who were very kind to let us visit their gardens and supplied us with the necessary plant material.

Conflicts of Interest: The authors declare no conflict of interest.

References

1. Díez, M.J.; Nuez, F. *Vegetables II Fabaceae, Liliaceae, Solanaceae, and Umbelliferae*; Prohens, J., Nuez, F., Eds.; Springer: New York, NY, USA, 2008; pp. 249–323. [CrossRef]

2. Sardaro, M.L.S.; Marmiroli, M.; Maestri, E.; Marmiroli, N. Genetic characterization of Italian tomato varieties and their traceability in tomatofood products. *Food Sci. Nutr.* **2013**, *1*, 54–62. [CrossRef] [PubMed]

3. Shaye, N.A.; Migdadi, H.; Charbaji, A.; Alsayegh, S.; Daoud, S.; Al-Anazi, W.; Alghamdi, S. Genetic variation among Saudi tomato (*Solanum lycopersicum* L.) landraces studied using SDS-PAGE and SRAP markers. *Saudi J. Biol. Sci.* **2018**, *25*, 1007–1015. [CrossRef] [PubMed]

4. Abete, I.; Perez-Cornago, A.; Navas-Carretero, S.; Bondia-Pons, I.; Zulet, M.A.; Martinez, J.A. A regular lycopene enriched tomato sauce consumption influences antioxidant status of healthy young-subjects: A crossover study. *J. Funct. Foods* **2013**, *5*, 28–35. [CrossRef]

5. Figàs, M.R.; Prohens, J.; Raigón, M.D.; Fita, A.; García-Martínez, M.D.; Casanova, C.; Borràs, D.; Plazas, M.; Andújar, I.; Soler, S. Characterization of composition traits related to organoleptic and functional quality for the differentiation, selection and enhancement of local varieties of tomato from different cultivar groups. *Food Chem.* **2015**, *187*, 517–524. [CrossRef]

6. Scalzo, J.; Politi, A.; Pellegrini, N.; Mezzetti, B.; Battino, M. Plant genotype affects total antioxidant capacity and phenolic contents in fruit. *Nutrition* **2005**, *21*, 207–213. [CrossRef]

7. Figàs, M.R.; Prohens, J.; Raigón, M.D.; Fernández-de-Córdova, P.; Fita, A.; Soler, S. Characterization of a collection of local varieties of tomato (*Solanum lycopersicum* L.) using conventional descriptors and the high-throughput phenomics tool Tomato Analyzer. *Genet. Resour. Crop Evol.* **2015**, *62*, 189–204. [CrossRef]

8. Schreiner, M. Vegetable crop management strategies to increase the quantity of phytochemicals. *Eur. J. Nutr.* **2005**, *44*, 85–94. [CrossRef]

9. Ceccarelli, S. Landraces: Importance and use in breeding and environmentally friendly agronomic systems. In *Agrobiodiversity Conservation: Securing the Diversity of Crop Wild Relatives and Landraces*; Maxted, N., EhsanDulloo, M., Ford-Lloyd, B.V., Frese, L., Iriondo, J.M., Pinheiro de Carvalho, M.A.A., Eds.; CAB International: Wallingford, UK, 2012; Chapter 15; pp. 103–117. [CrossRef]

10. Zoran, I.S.; Nikolaos, K.; Ljubomir, Š. Tomato Fruit Quality from Organic and Conventional Production. In *Organic Agriculture towards Sustainability*; Pilipavicius, V., Ed.; In Tech: Dubrovnik, Croatia, 2014; Chapter 7; pp. 17–169. [CrossRef]

11. Jacobo-Velázquez, D.A.; Cisneros-Zevallos, L. An alternative use of horticultural crops: Stressed plants as biofactories of bioactive phenolic compounds. *Agriculture* **2012**, *2*, 259–271. [CrossRef]

12. Rozema, J.; Schat, H. Salt tolerance of halophytes, research questions reviewed in the perspective of saline agriculture. *Environ. Exp. Bot.* **2013**, *92*, 83–95. [CrossRef]

13. Mazzucato, A.; Ficcadenti, N.; Caioni, M.; Mosconi, P.; Piccinini, E.; Rami, V.; Sanampudi, R.; Sestili, S.; Ferrari, V. Genetic diversity and distinctiveness in tomato (*Solanum lycopersicum* L.) landraces: The Italian case study of 'A peraAbruzzese'. *Sci. Hortic.* **2010**, *125*, 55–62. [CrossRef]

14. Terzopoulos, P.J.; Bebeli, P.J. Phenotypic diversity in Greek tomato (*Solanum lycopersicum* L.) landraces. *Sci. Hortic.* **2010**, *126*, 138–144. [CrossRef]

15. Sacco, A.; Ruggieri, V.; Parisi, M.; Festa, G.; Rigano, M.M.; Picarella, M.E.; Mazzucato, A.; Barone, A. Exploring a Tomato Landraces Collection for Fruit-Related Traits by the Aid of a High-Throughput Genomic Platform. *PLoS ONE* **2015**, *10*, e0137139. [CrossRef] [PubMed]

16. Beckles, D.M. Factors affecting the postharvest soluble solids and sugar content of tomato (*Solanum lycopersicum* L.) fruit. *Postharvest Biol. Technol.* **2012**, *63*, 129–140. [CrossRef]

17. Oliveira, S.M.; Brandao, T.R.S.; Silva, C.L.M. Influence of drying processes and pretreatments on nutritional and bioactive characteristics of dried vegetables: A review. *Food Eng. Rev.* **2016**, *8*, 134–163. [CrossRef]

18. Causse, M.; Friguet, C.; Coiret, C.; Lépicier, M.; Navez, B.; Lee, M.; Holthuysen, L.; Sinesio, F.; Moneta, E.; Grandillo, S. Consumer preferences for fresh tomato at the European scale: A common segmentation on taste and firmness. *J. Food Sci.* **2010**, *75*, S531–S541. [CrossRef]

19. Siddiqui, M.W.; Ayala-Zavala, J.F.; Dhua, R.S. Genotypic variation intomatoes affecting processing and antioxidant properties. *Crit. Rev. Food Sci.Nutr.* **2015**, *55*, 1819–1835. [CrossRef]

20. Garg, N.; Cheema, D.S. Assessment of fruit quality attributes of tomato hybrids involving ripening mutants under high temperature conditions. *Sci. Hortic.* **2011**, *131*, 29–38. [CrossRef]

21. Fernandez-Ruiz, V.; Olives, A.I.; Camara, M.; Sanchez-Mata, M.C.; Torija, M.E. Mineral and trace elements content in 30 accessions of tomato fruits (*Solanum lycopersicum* L.) and wild relatives (*Solanum pimpinellifolium* L., *Solanum cheesmaniae* L. Riley, and *Solanum habrochaites* S. Knapp & D.M. Spooner). *Biol. Trace Elem. Res.* **2011**, *141*, 329–339. [CrossRef]

22. Antunes, M.D.C.; Rodrigues, D.; Pantazis, V.; Cavaco, A.M.; Siomos, A.; Miguel, G. Nutritional quality changes of fresh-cut tomato during shelf life. *Food Sci. Biotechnol.* **2013**, *22*, 1–8. [CrossRef]

23. Hala, E.M.A.; Ghada, S.M.I. Tomato fruit quality as influenced by salinity and nitric oxide. *Turk. J. Bot.* **2014**, *38*, 122–129. [CrossRef]

24. Barros, L.; Duenas, M.; Pinela, J.; Carvalho, A.M.; Buelga, C.S.; Ferreira, I.C. Characterization and quantification of phenolic compounds in four tomato (*Lycopersicon esculentum* L.) farmers' varieties in northeastern Portugal homegardens. *Plant Foods Hum. Nutr.* **2012**, *67*, 229–234. [CrossRef] [PubMed]

25. Doncean, A.; Sumalan, R.M.; Beinsan, C.; Gergen, I.; Sumalan, R.L. Influence of different types and mixtures of composts on quality of tomatoes fruits (*Solanum lycopersicum* L.). *J. Hyg. Eng. Des.* **2015**, *8*, 40–47.

26. Gitenay, D.; Lyan, B.; Rambeau, M.; Mazur, A.; Rock, E. Comparison of lycopene and tomato effects on biomarkers of oxidative stress in vitamin E deficient rats. *Eur. J. Nutr.* **2007**, *46*, 468–475. [CrossRef] [PubMed]

27. Pinela, J.; Barros, L.; Carvalho, A.M.; Ferreira, I.C. Nutritional composition and antioxidant activity of four tomato (*Lycopersicon esculentum* L.) farmer' varieties in Northeastern Portugal homegardens. *Food Chem. Toxicol.* **2012**, *50*, 829–834. [CrossRef]

28. Khoo, H.E.; Prasad, K.N.; Kong, K.W.; Jiang, Y.; Ismail, A. Carotenoids and their isomers: Color pigments in fruits and vegetables. *Molecules* **2011**, *16*, 1710–1738. [CrossRef]

29. Srivastava, S.; Srivastava, A.K. Lycopene; chemistry, biosynthesis, metabolism and degradation under various abiotic parameters. *J. Food Sci. Technol.* **2015**, *52*, 41–53. [CrossRef]

30. Chiva-Blanch, G.; Visioli, F. Polyphenols and health: Moving beyond antioxidants. *J. Berry Res.* **2012**, *2*, 63–71. [CrossRef]

31. Raiola, A.; Del Giudice, R.; Monti, D.M.; Tenore, G.C.; Barone, A.; Rigano, M.M. Bioactive compound content and cytotoxic effect on human cancer cells of fresh and processed yellow tomatoes. *Molecules* **2015**, *21*, 33–47. [CrossRef]

32. Opara, U.L.; Al-Ani, M.R.; Al-Rahbi, M. Effect of fruit ripening stage on physico-chemical properties, nutritional composition and antioxidant components of tomato (*Lycopersicum esculentum*) cultivars. *Food Bioprocess Technol.* **2012**, *5*, 3236–3243. [CrossRef]

33. Corwin, D.L.; Lesch, S.M. Application of soil electrical conductivity to precision agriculture: Theory principles and guideline. *Agron. J.* **2003**, *95*, 455–471. [CrossRef]

34. Sumalan, R.L.; Popescu, I.; Schmidt, B.; Sumalan, R.M.; Popescu, C.; Gaspar, S. Salt tolerant tomatoes local landraces from Romania–Preserving the genetic resources for future sustainable agriculture. *J. Biotehnol.* **2015**, *208*, S18. [CrossRef]

35. IUSS Working Group WRB. World Reference Base for Soil Resources 2014, update 2015. In *International Soil Classification System for Naming Soils and Creating Legends for Soil Maps*; World Soil Resources Reports No. 106; FAO: Rome, Italy, 2015.

36. UPOV—International Union for the Protection of New Varieties of Plants–Geneva. TOMATO UPOV Code: SOLAN_LYC *Solanum lycopersicum* L. Guidelines for the Conduct of Tests for Distinctness, Uniformity and Stability. 2018. Available online: https://www.upov.int/edocs/tgdocs/en/tg044.pdf (accessed on 27 March 2020).

37. Benzie, I.F.; Strain, J.J. Ferric reducing ability of plasma (FRAP) as a measure of "antioxidant power": The FRAP assay. *Anal. Biochem.* **1996**, *239*, 70–76. [CrossRef] [PubMed]

38. Singleton, V.L.; Orthofer, R.; Lamuela-Raventos, R.M. Analysis of total phenols and other oxidation substrates and antioxidants by means of Folin-Ciocalteu reagent. *Methods Enzymol.* **1999**, *299*, 152–178. [CrossRef]

39. AOAC. Vitamin C (ascorbic acid) in vitamin preparations and juices. In *AOAC: Official Methods of Analysis*, 15th ed.; Helrich, K., Ed.; Association of Official Analytical Chemists, Inc.: Arlington County, VA, USA, 1990; Volume 1, pp. 1058–1059.

40. Sharma, S.K.; Le Maguer, M. Lycopene in tomatoes and tomato pulp fractions. *Ital. J. Food Sci.* **1996**, *8*, 107–113.

41. Toor, R.K. Influence of different types of fertilizers on the major antioxidant components of tomatoes. *J. Food Compos. Anal.* **2006**, *19*, 20–27. [CrossRef]

42. Ciulca, S. *Metodologii de Experimentare în Agricultura și Biologie*; Editura Agroprint: Timisoara, Romania, 2006; pp. 33–53.

43. Felsenstein, J. *PHYLIP (Phylogeny Inference Package) Version 3.5c*; Distributed by the author; Department of Genetics, University of Washington: Seattle, WA, USA, 1993.

44. Singh, R.K.; Chaudhary, B.D. *Biometrical Methods in Quantitative Genetic Analysis*; Kalyani Publishers: Ludhiana, India, 1979; pp. 191–200.

45. Yan, W.; Hunt, L.A.; Sheng, Q.; Szlavnics, Z. Cultivar evaluation and mega-environment investigation based on the GGE biplot. *Crop Sci.* **2000**, *40*, 597–605. [CrossRef]

46. Yan, W.; Kang, M.S. *GGE Biplot Analysis: A Graphical Tool for Breeders, Geneticists, and Agronomists*, 1st ed.; CRC Press: Boca Raton, FL, USA, 2002; p. 144.

47. Corrado, G.; Piffanelli, P.; Caramante, M.; Coppola, M.; Rao, R. SNP genotyping reveals genetic diversity between cultivated landraces and contemporary varieties of tomato. *BMC Genom.* **2013**, *14*, 835–848. [CrossRef]

48. Cebola-Cornejo, J.; Rosello, S.; Nuez, F. Phenotypic and genetic diversity of Spanish tomato landraces. *Sci. Hortic.* **2013**, *162*, 150–164. [CrossRef]

49. Knapp, S.; Peralta, I.E. The tomato (*Solanum lycopersicum* L. Solanaceae) and its botanical relatives. In *The Tomato Genome (Compendium of Plant Genomes)*; Causse, M., Giovannoni, J., Bouzayen, M., Zouine, M., Eds.; Springer: Berlin/Heidelberg, Germany, 2016; pp. 7–21. [CrossRef]

50. Corrado, G.; Caramante, M.; Piffanelli, P.; Rao, R. Genetic diversity in Italian tomato landraces: Implications for the development of a core collection. *Sci. Hortic.* **2014**, *168*, 138–144. [CrossRef]

51. Tomescu, D.; Sumalan, R.L.; Copolovici, L.; Copolovici, D. The influence of soil salinity on volatile organic compounds emission and photosynthetic parameters of *Solanum lycopersicum* L. varieties. *Open Life Sci.* **2017**, *12*, 135–142. [CrossRef]

52. Salim, M.M.R.; Rashid, M.H.; Hossain, M.M.; Zakaria, M. Morphological characterization of tomato (*Solanum lycopersicum* L.) genotypes. *J. Saudi Soc. Agric. Sci.* **2018**, in press. [CrossRef]

53. Tanksley, S.D. The genetic, developmental, and molecular bases of fruit size and shape variation in tomato. *Plant Cell* **2004**, *16*, S181–S189. [CrossRef]

54. Miller, G.; Suzuki, N.; Ciftci-Yilmaz, S.; Mittler, R. Reactive oxygen species homeostasis and signaling during drought and salinity stresses. *Plant Cell Environ.* **2010**, *33*, 453–467. [CrossRef]

55. Passam, H.C.; Karapanos, I.C.; Bebeli, P.J.; Savvas, D. A review of recent research on tomato nutrition, breeding and post-harvest technology with reference to fruit quality. *Eur. J. Plant Sci. Biotechnol.* **2007**, *1*, 1–21.

56. García-Valverde, V.; Navarro-González, I.; García-Alonso, J.; Periago, M.J. Antioxidant bioactive compounds in selected industrial processing and fresh consumption tomato cultivars. *Food Bioprocess Technol.* **2013**, *6*, 391–402. [CrossRef]

57. Deng, G.-F.; Lin, X.; Xu, X.-R.; Gao, L.-L.; Xie, J.-F.; Li, H.-B. Antioxidant capacities and total phenolic contents of 56 vegetables. *J. Funct. Foods* **2013**, *5*, 260–266. [CrossRef]

58. Ilahy, R.; Hdider, C.; Lenucci, M.S.; Tlili, I.; Dalessandro, G. Antioxidant activity and bioactive compound changes during fruit ripening of high-lycopene tomato cultivars. *J. Food Compos. Anal.* **2011**, *24*, 588–595. [CrossRef]

59. Martinez-Valverde, I.; Periago, M.J.; Provan, G.; Chesson, A. Phenolic compounds, lycopene and antioxidant activity in commercial varieties of tomato (*Lycopersicum esculentum*). *J. Sci. Food Agric.* **2002**, *82*, 323–330. [CrossRef]

60. Drakou, M.; Birmpa, A.; Koutelidakis, A.E.; Komaitis, M.; Panagou, E.Z.; Kapsokefalou, M. Total antioxidant capacity, total phenolic content and iron and zinc dialyzability in selected Greek varieties of table olives, tomatoes and legumes from conventional and organic farming. *Int. J. Food Sci. Nutr.* **2015**, *66*, 197–202. [CrossRef] [PubMed]

61. Hernández, V.; Hellín, P.; Fenoll, J.; Flores, P.J. Increased temperature produces changes in the bioactive composition of tomato, depending on its developmental stage. *J. Agric. Food Chem.* **2015**, *63*, 2378–2382. [CrossRef]

62. Lairon, D. Nutritional quality and safety of organic food. A review. *Agron. Sustain. Dev.* **2010**, *30*, 33–41. [CrossRef]

63. Massaretto, I.L.; Albaladejo, I.; Purgatto, E.; Flores, F.B.; Plasencia, F.; Egea-Fernanadez, J.M.; Bolarin, M.C.; Egea, I. Recovering tomato landraces to simultaneously fruit yield and nutritional quality against salt stress. *Front. Plant Sci.* **2018**, *9*, 1–17. [CrossRef] [PubMed]

64. Alba, R.; Cordonnier-Pratt, M.M.; Pratt, L.H. Fruit-localized phytochromes regulate lycopene accumulation independently of ethylene production in tomato. *Plant Physiol.* **2000**, *123*, 363–370. [CrossRef] [PubMed]

65. Rosati, C.; Aquilani, R.; Dharmapuri, S.; Pallara, P.; Marusic, C.; Tavazza, R.; Bouvier, F.; Camara, B.; Giuliano, G. Metabolic engineering of beta-carotene and lycopene content in tomato fruit. *Plant J.* **2000**, *24*, 413–419. [CrossRef] [PubMed]

66. Giovannetti, M.; Avio, L.; Barale, L.; Ceccarelli, N.; Cristofani, R.; Iezzi, A.; Mignolli, F.; Picciarelli, P.; Pinto, B.; Reali, D.; et al. Nutraceutical value and safety of tomato fruits produced by mycorrhizal plants. *Br. J. Nutr.* **2012**, *107*, 242–251. [CrossRef]

67. Perveen, R.; Rasul-Suleria, H.A.; Anjum, F.M.; Butt, M.S.; Pasha, I.; Sarfraz, A. Tomato (*Solanumlycopersicum*) carotenoids and lycopenes chemistry; metabolism, absorption, nutrition, and allied health claims—A comprehensive review. *Crit. Rev. Food Sci. Nutr.* **2015**, *55*, 919–929. [CrossRef]

68. Kavitha, P.; Shivashankara, K.S.; Rao, V.K.; Sadashiva, A.T.; Ravishankar, K.V.; Sathish, G.J. Genotypic variability for antioxidant and quality parameters among tomato cultivars, hybrids, cherry tomatoes and wild species. *J. Sci. Food Agric.* **2014**, *94*, 993–999. [CrossRef]

69. Borghesi, E.; González-Miret, M.L.; Escudero-Gilete, M.L.; Malorgio, F.; Heredia, F.J.; Meléndez-Martínez, A.J. Effects of Salinity Stress on Carotenoids, Anthocyanins, and Color of Diverse Tomato Genotypes. *J. Agric. Food Chem.* **2011**, *59*, 11676–11682. [CrossRef]

70. Giannakoula, A.E.; Ilias, A.F. The effect of water stress and salinity on growth and physiology of tomato (*Lycopersicon Esculentum* Mill.). *Arch. Biol. Sci.* **2013**, *65*, 611–620. [CrossRef]

71. Ehret, D.L.; Usher, K.; Helmer, T.; Block, G.; Steinke, D.; Frey, B.; Kuang, T.; Diarra, M. Tomato Fruit Antioxidants in Relation to Salinity and Greenhouse Climate. *J. Agric. Food Chem.* **2013**, *61*, 1138–1145. [CrossRef]

72. Van Meulebroek, L.; Hanssens, J.; Steppe, K.; Vanhaecke, L. Metabolic Fingerprinting to Assess the Impact of Salinity on Carotenoid Content in Developing Tomato Fruits. *Int. J. Mol. Sci.* **2016**, *17*, 821. [CrossRef] [PubMed]

73. Kaur, S.; Das, M. Functional foods: An overview. *Food Sci. Biotechnol.* **2011**, *20*, 861–875. [CrossRef]

74. Rao, A.V.; Agarwal, S. Role of antioxidant lycopene in cancer and heart disease. *J. Am. Coll. Nutr.* **2000**, *19*, 563–569. [CrossRef] [PubMed]

75. Ivanov, B.; Asada, K.; Kramer, D.M.; Edwards, G. Characterization of photosynthetic electron transport in bundle sheath cells of maize. I. Ascorbate effectively stimulates cyclic electron flow around PSI. *Planta* **2005**, *220*, 572–581. [CrossRef]

76. Shao, H.B.; Chu, L.Y.; Shao, M.A.; Jaleel, C.A.; Mi, H.M. Higher plant antioxidants and redox signaling under environmental stresses. *Comptes Rendus Biol.* **2008**, *331*, 433–441. [CrossRef]

77. Bodnarescu, F.; Sumalan, R.M.; Ciulca, S.; Copolovici, L.; Sumalan, R.L. The influence of parental lines on lycopene and ß-carotene content in tomato F1 hybrids (*Solanum lycopersicum* L.). *Res. J. Agric. Sci.* **2018**, *50*, 90–97.

78. Rossi, F.; Godani, F.; Bertuzzi, T.; Trevisan, M.; Ferrari, F.; Gatti, S. Health-promoting substances and heavy metal content in tomatoes grown with different farming techniques. *Eur. J. Nutr.* **2008**, *47*, 266–272. [CrossRef]

79. Carnovale, E. La qualitanutrizionaledeiprodottidell'agricolturabiologica. *Italus Hortus* **1999**, *6*, 41–44.

80. Lucarini, M.; Carbonaro, M.; Nicoli, S.; Aguzzi, A.; Cappelloni, M.; Ruggeri, S.; DiLullo, G.; Gambelli, L.; Carnovale, E. Endogenous markers for organic versus conventional plant products. In *Agrifood Quality II: Quality Managements of Fruits and Vegetables, Session VI: Quality Assessment*; Hagg, M., Ahvenainen, R., Evers, A.M., Tiilikkala, K., Eds.; Royal Society of Chemistry: Cambridge, UK, 1999; pp. 306–310.

81. Asami, D.K.; Hong, Y.J.; Barrett, D.M.; Mitchell, A.E. Comparison of the total phenolic and ascorbic acid content of freeze-dried and air-dried marionberry, strawberry and corn grown using conventional, organic and sustainable agricultural practices. *J. Agric. Food Chem.* **2003**, *51*, 1237–1241. [CrossRef]

82. Caris-Veyrat, C.; Amiot, M.J.; Tyssandier, V.; Grasselly, D.; Buret, M.; Mikolajczak, M.; Guilland, J.C.; Bouteloup-Demange, C.; Borel, P. Influence of organic versus conventional agricultural practice on the antioxidant microconstituent content of tomatoes and derived pureed; consequences on antioxidant plasma status in humans. *J. Agric. Food Chem.* **2004**, *52*, 6503–6509. [CrossRef]

83. Auclair, L.; Zee, J.A.; Karam, A.; Rochat, E. Valeur nutritive qualitéorganoleptique et productivite des tomatoes de serre en function de leur mode de production: Biologique-conventionel-hydroponique. *Sci. Aliment.* **1995**, *15*, 511–528.

84. Dumas, Y.; Dadomo, A.; Di Lucca, G.; Grolier, P. Effects of environmental factors and agricultural techniques on antioxidant content of tomatoes. *J. Sci. Food Agric.* **2003**, *83*, 369–382. [CrossRef]

85. Dorais, M.; Ehret, D.L.; Papadopoulos, A.P. Tomato (*Solanum lycopersicum*) health components: From the seed to the consumer. *Phytochem. Rev.* **2008**, *7*, 231–250. [CrossRef]

86. Fanasca, S.; Martino, A.; Heuvelink, E.; Stanghellini, C. Effect of electrical conductivity, fruit pruning, and truss position on quality in greenhouse tomato fruit. *J. Hort. Sci. Biotechnol.* **2007**, *82*, 488–494. [CrossRef]

87. Helyes, L.; Pék, Z.; Lugasi, A. Function of the variety technological traits and growing conditions on fruit components of tomato (*Lycopersicon Lycopersicum* L Karsten). *Acta Aliment.* **2008**, *37*, 427–436. [CrossRef]

88. Hallmann, E.; Lipowski, J.; Marszałek, K.; Rembiałkowska, E. The seasonal variation in bioactive compounds content in juice from organic and non-organic tomatoes. *Plant Foods Hum. Nutr.* **2013**, *68*, 171–176. [CrossRef]

Article

Substrate Application of 5-Aminolevulinic Acid Enhanced Low-temperature and Weak-light Stress Tolerance in Cucumber (*Cucumis sativus* L.)

Ali Anwar [1,2,†], Jun Wang [1,†], Xianchang Yu [1], Chaoxing He [1] and Yansu Li [1,*]

[1] Institute of Vegetables and Flowers, Chinese Academy of Agricultural Sciences, Beijing 100081, China; anwarsnu@aol.com (A.A.); wangjun01@caas.cn (J.W.); xcyu1962@163.com (X.Y.); hechaoxing@caas.cn (C.H.)
[2] Graduate School of International Agricultural Technology and Crop Biotechnology Institute/Green Bio Science & Technology, Seoul National University, Pyeongchang 25354, Korea
*   Correspondence: liyansu@caas.cn
†   These authors contributed equally to this work.

Received: 9 March 2020; Accepted: 27 March 2020; Published: 29 March 2020

Abstract: 5-Aminolevulinic acid (ALA) is a type of nonprotein amino acid that promotes plant stress tolerance. However, the underlying physiological and biochemical mechanisms are not fully understood. We investigated the role of ALA in low-temperature and weak-light stress tolerance in cucumber seedlings. Seedlings grown in different ALA treatments (0, 10, 20, or 30 mg ALA·kg^{-1} added to substrate) were exposed to low temperature (16/8 °C light/dark) and weak light (180 μmol·m^{-2}·s^{-1} photosynthetically active radiation) for two weeks. Treatment with ALA significantly alleviated the inhibition of plant growth, and enhanced leaf area, and fresh and dry weight of the seedlings under low-temperature and weak-light stress. Moreover, ALA increased chlorophyll (Chl) *a*, Chl *b*, and Chl *a+b* contents. Net photosynthesis rate, stomatal conductance, transpiration rate, photochemical quenching, non-photochemical quenching, actual photochemical efficiency of photosystem II, and electron transport rate were significantly increased in ALA-treated seedlings. In addition, ALA increased root activity and antioxidant enzyme (superoxide dismutase, peroxidase, and catalase) activities, and reduced reactive oxygen species (hydrogen peroxide and superoxide radical) and malondialdehyde accumulation in the root and leaf of cucumber seedlings. These findings suggested that ALA incorporation in the substrate alleviated the adverse effects of low-temperature and weak-light stress, and improved Chl contents, photosynthetic capacity, and antioxidant enzyme activities, and thus enhanced cucumber seedling growth.

Keywords: ALA; abiotic stress; chlorophyll; photosynthesis; antioxidant enzyme

1. Introduction

Cucumber (*Cucumis sativus* L.), a member of the *Cucurbitaceae* family, is an important vegetable widely cultivated and consumed around the world [1]. Plants are challenged by numerous environmental stresses (e.g., high and low temperatures, salinity, light, drought, and heavy metal stress) that affect plant growth and productivity [2,3]. Low temperature and low-light stress are the most critical environmental factors that influence cucumber production in a solar greenhouse [2,4]. Plant exposed to low temperature and light stress exhibit a number of physiological and biochemical abnormalities, including reduction in chlorophyll biosynthesis, photosynthetic capacity, carbohydrate and nitrogen metabolism, nutrient uptake and accumulation, and overproduction of reactive oxygen species (ROS) [5]. Accumulation of ROS negatively affects enzyme activities, biosynthesis of carbohydrates, DNA, and proteins, and other biochemical activities, thus leading to oxidative stress [4,5]. In addition, ROS influence the expression of a number of genes involved in diverse processes such as growth, cell cycle, programmed cell death, abiotic stress responses, pathogen attack response, systemic signaling, and development [6]. The antioxidant defense system, which includes superoxide dismutase (SOD), peroxidase (POD), catalase (CAT), glutathione reductase (GR), and ascorbate peroxidase (APX), plays a crucial role in normalizing the production of ROS, thereby protecting plants from abiotic stresses [6,7].

5-Aminolevulinic acid (ALA) is an essential biosynthetic precursor and is considered to be a plant growth regulator [8,9]. The compound is a key precursor in the biosynthesis of porphyrin compounds, such as chlorophyll, heme, and plant hormones [10]. In addition, ALA is involved in photosynthesis regulation under abiotic stress. Exogenous ALA application increases chlorophyll accumulation and chlorophyll fluorescence indices in lettuce and oilseed rape [11,12]. It has recently been reported that ALA regulates the expression level of *fructose-1,6-bisphosphatase* (*FBP*), *triose-3-phosphate isomerase* (*TPI*), and *ribulose-1,5-bisphosphate carboxylase/oxygenase small subunit* (*RBCS*), which activate the Calvin cycle of photosynthesis under drought stress [13]. In a previous study we observed that ALA regulates endogenous hormone and nutrient accumulation in cucumber to induce low-temperature stress tolerance [10]. It is also reported that ALA is involved in the chlorophyll biosynthesis pathway under salinity stress [14], and induces antioxidant enzyme activities and endogenous hormone accumulation under low-temperature stress in cucumber seedlings [10]. Previous studies demonstrated that foliar application of ALA may confer plant tolerance to diverse abiotic stresses, such as chilling, high temperature, salinity, drought, weak light, and heavy metals [14–16]. ALA influences a variety of physiological and biochemical activities of plants in response to abiotic stresses, including chlorophyll biosynthesis, nutrient uptake, and antioxidant enzyme activities [14,16]. Furthermore, ALA induces abiotic stress tolerance through activation of numerous types of transcription factors, signal transduction, and chlorophyll and carbohydrate biosynthesis [9]. These findings suggest that ALA can broadly reduce the detrimental effects of environmental influence.

During winter vegetable cultivation, plants are frequently exposed to low temperature and weak light intensity (predominantly clouds or fog), which can negatively influence production. Therefore, this study was designed to investigate the role of ALA in response to a combination (low temperature and weak light) of stresses on cucumber seedling growth, chlorophyll contents, photosynthetic capacity, antioxidant enzyme activity, and ROS accumulation. The information generated from this study will improve our understanding of responses to both stresses and will be useful for security of winter vegetable production.

2. Materials and Methods

2.1. Plant Material and Experimental Setup

Cucumber (*Cucumis sativus* 'Zhongnong 26') seeds were soaked in water at 55 °C for 2–3 h, and then germinated on moist gauze in the dark at 28 °C. The germinated seeds were transplanted into plug trays containing nursery substrate supplemented with ALA (Sigma, St Louis, MO, USA) and incubated at 28/18 °C (light/dark) under 70%–75% humidity and 300–350 $\mu mol \cdot m^{-2} \cdot s^{-1}$ photosynthetically active radiation for 14 h. The experiment consisted of four treatments based on different concentrations of ALA (applied as kg^{-1} substrate):

CK,	Control (no ALA)
T1,	10 mg ALA
T2,	20 mg ALA
T3,	30 mg ALA

The ALA concentrations were mixed with a constant weight of substrate (kg). The substrates were used to fill a 32-cell seedling tray and a germinated seed was sown in each cell. At the first leaf (fully expanded) stage, the seedlings were transferred to a controlled artificial chamber under 16/8 °C (day/night), photosynthetically active radiation of 180 $\mu mol \cdot m^{-2} \cdot s^{-1}$, and a photoperiod of 12 h for 21 d before sampling. The seedlings were irrigated at two-day intervals with Hoagland's solution to fulfil nutritional demand. Each treatment consisted of four replicates.

2.2. Measurement of Plant Growth Parameters

Plant height, stem diameter, and fresh weight were measured with a ruler, vernier caliper, and electronic balance, respectively [10]. Fresh samples were placed in an oven at 105 °C for 30 min, and then dried at 75 °C [1]. Root vitality was determined using the triphenyl tetrazolium chloride method [2]. The seedling vigor index was calculated using the following formula [2].

$$\text{Seedling Vigor Index} = \left(\frac{Hypocotyl\ Diameter}{Plant\ Height} + \frac{Root\ Dry\ Weight}{Shoot\ Dry\ Weight} \right) \times Total\ Dry\ Weight \quad (1)$$

2.2.1. Chlorophyll, Photosynthesis, and Chlorophyll Fluorescence Measurements

Chlorophyll (Chl) contents were determined using an ethanol extraction method, as previously described [2]. Net photosynthetic rate (P_n), stomatal conductance (G_s), transpiration rate (T_r), and intercellular CO_2 concentration (C_i) on the fourth fully expanded leaf from the shoot tip were measured using a portable photosynthesis system (Li-6400XT, LI-COR, Inc., Lincoln, NE, USA).

The portable photosynthesis system was also used for measurement of chlorophyll fluorescence. The fourth fully expanded leaves from the shoot tip were adapted in the dark for 30 min prior to measurement. The maximum photochemical efficiency of photosystem II (F_v/F_m), maximum antenna conversion efficiency (F_v'/F_m'), photochemical quenching (qP), nonphotochemical quenching (NPQ), the actual photochemical efficiency of photosystem II ($\Phi PSII$), and electron transport rate (ETR) were calculated [14].

2.2.2. Determination of Root Activity

The root activity was determined by TTC (Triphenyltetrazolium chloride) reduction method [2]. Briefly, 0.5 g fresh collected roots were cut into 2 cm length and put in 10 mL 0.5 mM PBS buffer containing 0.4% TTC (w/v) and incubate for one hour at 37 °C. The reaction was stopped by using 2 mL H_2SO_4 (1 mol/L) for 15 min, and then remove all solutions, and then add 10 mL 95% ethanol and incubate for 24 h at room temperature (until root turn white). The absorbance was read at 485 nm using spectrophotometer.

Calculation formula:

$$\text{Tetrazole reduction strength (μg/gFW.h)} = (OD + 0.0035)/4*h*W*0.0022\ (h = 4, W = 0.4\sim0.5) \quad (2)$$

2.3. Measurement of $O_2{}^{\cdot-}$, H_2O_2, and Malondialdehyde Contents

Superoxide radical ($O_2{}^{\cdot-}$) and hydrogen peroxide (H_2O_2) contents were determined using assay kits (COMINBIO) with a UV-1800 spectrophotometer (Shimadzu, Kyoto, Japan) in accordance with the manufacturer's instructions. Malondialdehyde (MDA) content was measured using the thiobarbituric acid method [2].

2.4. Activities of Antioxidant Enzymes

Fresh leaves (~0.5 g) were quickly ground with a pestle in an ice-cold mortar with 4 mL of 50 mM phosphate buffer solution (pH 7.8). The homogenate was centrifuged at 10,500 rpm for 20 min. The supernatant was used to determine the activities of antioxidant enzymes. Superoxide dismutase (SOD) activity was measured, with some modifications, based on 50% inhibition of the photochemical reduction of nitro blue tetrazolium at 560 nm. Peroxidase (POD) activity was measured as the increase in absorbance at 470 nm using the method and catalase (CAT) activity was measured as the decline in absorbance at 240 nm [2].

2.5. Statistical Analysis

Each treatment consisted of four independent biological replicates and the entire experiment was repeated three times. The data were statistically analyzed using analysis of variance (ANOVA), and individual treatments were compared using the least significant difference test (LSD; $p = 0.05$) as implemented in Statistix 8.1 software.

3. Results

3.1. Exogenous ALA Application Promoted Cucumber Seedlings Growth

Application of ALA to the substrate significantly increased plant height, stem diameter, leaf area, fresh and dry weight, and seedling vigor index in cucumber seedlings, which were significantly reduced under low-temperature and weak-light stress (Figure 1). Compared with the control treatment (CK), the plant height, stem diameter, leaf area, fresh and dry weight, and seedling vigor index of cucumber seedlings increased by 14.4%, 13.4%, 63.1%, 54.3%, 54.5%, and 53.8%, respectively, in the T2 treatment. Growth parameters in the T1 and T3 treatments were not statistically different, but were significantly higher than those of the CK (Figure 1). The results suggested that ALA application alleviated the detrimental effects of the combined stress of low temperature and weak light, and thus enhanced cucumber seedling growth.

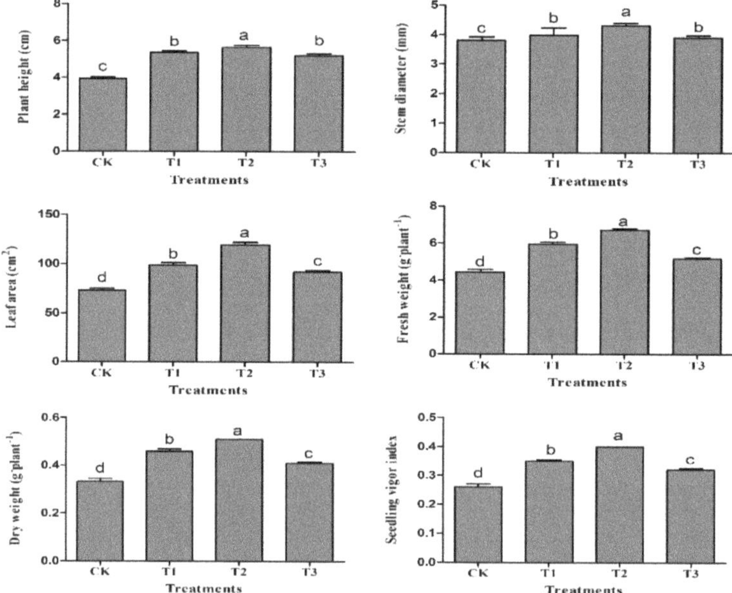

Figure 1. Effect of exogenous ALA application on growth of cucumber seedlings under low-temperature and weak-light stress. Data are the means of four replicates; error bars indicate the standard deviation. Treatments with the same lower-case letter are not significantly different (least significant difference test, $p = 0.05$).

3.2. Exogenous ALA Application Enhanced Root Activity of Cucumber Seedlings

Root activity represents overall vigor, including root metabolic processes, enzyme activities, and water and nutrient absorption and uptake processes, thus it is considered to be an important index for plant response to environmental variables. The present results suggested that root vitality of cucumber seedlings was negatively affected by combined low-temperature and weak-light stress (Figure 2). The ALA-treated seedlings showed significantly enhanced tolerance to low-temperature and weak-light stress, and resulted in a significant increment in root vitality of the cucumber seedlings. Maximum root activity was observed in the T2 treatment and the lowest vitality was recorded in the CK.

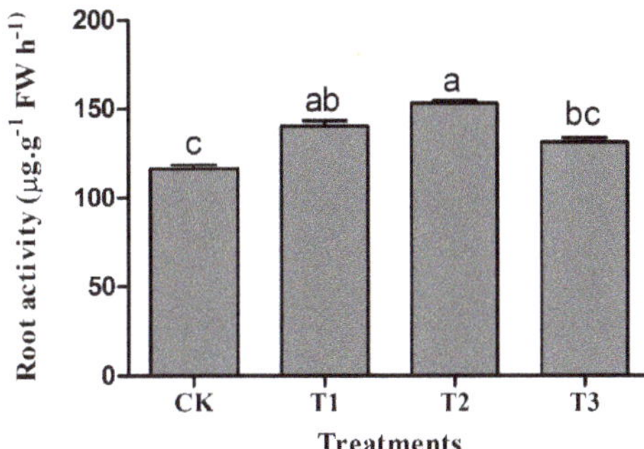

Figure 2. Effect of exogenous ALA application on root activity under low-temperature and weak-light stress. Data are the means of four replicates; error bars indicate the standard deviation. Treatments with the same lower-case letter are not significantly different (least significant difference test, $p = 0.05$).

3.3. Exogenous ALA Application Increased Chlorophyll Content of Cucumber Seedlings

Chlorophylls (Chl) are extremely sensitive to abiotic stress and quickly degrade under an extreme stress intensity, which ultimately reduces photosynthetic capacity. The present results showed that low-temperature and weak-light stress induced a significant decrease in Chl *a*, Chl *b*, and Chl *a+b* contents, whereas the Chl *a/b* ratio was unchanged among the ALA treatments (Figure 3). Compared with the CK, the contents of Chl *a*, Chl *b*, and Chl *a+b* were increased by 22.14%, 28.26%, and 23.59% respectively, in the T2 treatment, by 9.40%, 13.04%, and 10.25% in the T1 and 6.04%, 8.70%, and 6.70% in the T3 treatment (Figure 3). The differences in contents between the T1 and T2 treatments were non-significant, but were higher significantly than those of the CK and lower than those of the T2 treatment. The results showed that exogenous ALA increased Chl contents to reduce the harmful effect of low-temperature and weak-light stress.

The photosynthetic capacity was significantly enhanced by exogenous ALA application and low-temperature and weak-light stress. Significant increases in P_n, G_s, and T_r by 16.50%, 128.57%, and 148.54%, respectively, were observed compared with ALA-treated seedlings (T2; Figure 4). Similarly, the T1 and T3 treatments resulted in a significant increment in photosynthetic parameters compared with those of the CK. The C_i was slightly increased in ALA-treated seedlings, but the difference with the CK was non-significant. These findings indicated that ALA regulated chlorophyll contents and resulted in improved photosynthesis under combined low-temperature and weak-light stress.

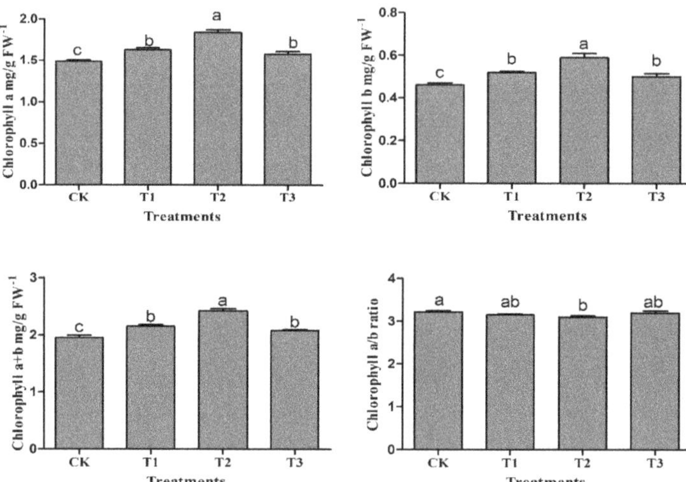

Figure 3. Effect of exogenous ALA application on chlorophyll contents of cucumber seedlings under low-temperature and weak-light stress. Data are the means of four replicates; error bars indicate the standard deviation. Treatments with the same lower-case letter are not significantly different (least significant difference test, $p = 0.05$).

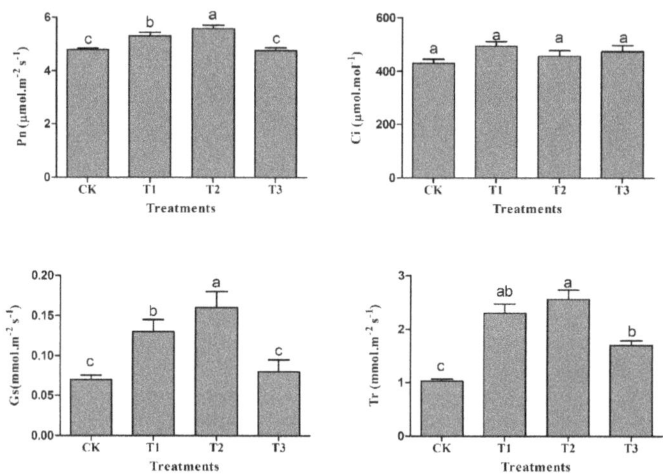

Figure 4. Effect of exogenous ALA on photosynthesis of cucumber seedlings under low-temperature and weak-light stress. Data are the means of four replicates; error bars indicate the standard deviation. Treatments with the same lower-case letter are not significantly different (least significant difference test, $p = 0.05$). P_n, net photosynthetic rate; G_s, stomatal conductance; C_i, intercellular CO_2 concentration; T_r, transpiration rate.

3.4. Effects of exogenous ALA Application on Chlorophyll Fluorescence

Chlorophyll fluorescence analysis is an important and commonly used technique to investigate the plant photosynthetic capacity and response to stress. The present results indicated that F_v/F_m and F_v'/F_m' were non-significantly different among all treatments (Table 1). In general, ALA-treated seedlings showed a significant in qP, ΦPSII, and ETR compared with the CK. However, the opposite trend was observed for NPQ. These findings suggested that ALA played a significant role in abiotic stress tolerance and protected the photosynthetic machinery.

Table 1. Effects of exogenous ALA application in the substrate on chlorophyll fluorescence parameters of cucumber seedlings under low-temperature and weak-light stress.

Treatment	F_v/F_m	F_v'/F_m'	qP	NPQ	ΦPSII	ETR
CK	0.60 ± 0.02a	0.48 ± 0.09a	0.28 ± 0.07d	0.54 ± 0.06a	0.20 ± 0.02c	20.62 ± 2.74bc
T1	0.61 ± 0.03a	0.46 ± 0.04a	0.55 ± 0.08b	0.38 ± 0.06c	0.26 ± 0.03b	21.73 ± 2.76ab
T2	0.61 ± 0.02a	0.44 ± 0.02a	0.63 ± 0.09a	0.37 ± .040c	0.30 ± 0.02a	23.82 ± 1.92a
T3	0.60 ± 0.01a	0.46 ± 0.03a	0.43 ± 0.05c	0.47 ± 0.04b	0.21 ± 0.02c	19.22 ± 1.48c

Data are the means of four replicates ± standard deviation. Treatments with the same lower-case letter within a column are not significantly different (least significant difference test, $p = 0.05$). F_v/F_m, maximum photochemical efficiency of photosystem II; F_v'/F_m', maximum antenna conversion efficiency; qP, photochemical quenching; NPQ, non-photochemical quenching; ΦPSII, actual photochemical efficiency of photosystem II; ETR, electron transport rate.

3.5. Exogenous ALA Application Promoted Antioxidant Enzyme Activities

Overproduction of ROS and accumulation of MDA result in damage to chlorophylls, protein biosynthesis, and DNA, which ultimately results in oxidative stress. Plants have evolved a defense system (antioxidant enzymes), which control ROS overproduction under abiotic stress. In the present study, activities of antioxidant enzyme (SOD, POD, and CAT) were significantly increased in response to ALA treatment compared with those of the CK (Figure 5). The T2 exogenous ALA treatment significantly increased SOD, POD, and CAT activities by 83.91%, 20.27%, and 27.96%, in leaves and 74.58%, 63.97%, and 56.53% in roots, respectively, compared with activities in the CK. The POD activity was significantly higher in T3 leaves and roots compared with those observed in the CK (Figure 5). These findings suggested that exogenous ALA may regulate the plant defense system to reduce the adverse effects of combined low-temperature and weak-light stress.

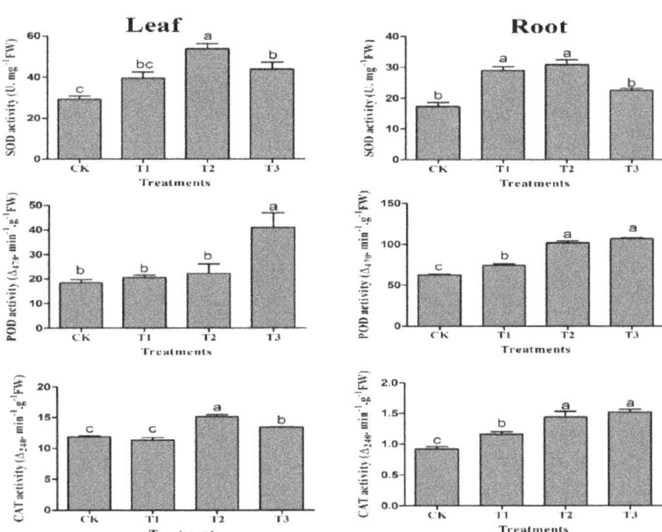

Figure 5. Effect of exogenous ALA application on antioxidant enzyme activities under low-temperature and weak-light stress in cucumber seedlings. Data are the means of four replicates; error bars indicate the standard deviation. Treatments with the same lower-case letter are not significantly different (least significant difference test, $p = 0.05$). SOD, superoxide dismutase; POD, peroxidase; CAT, catalase.

3.6. Exogenous ALA Application Reduced $O_2{}^{\cdot-}$, H_2O_2, and MDA Accumulation

Plant exposure to abiotic stress leads to overproduction of ROS and accumulation of MDA, which are highly reactive and toxic, and affect a variety of physiological and biochemical activities. The ROS and MDA contents were significantly higher in roots and leaves of the CK (Figure 6). The $O_2{}^{\cdot-}$ content in the leaves and roots of CK seedlings was 1.27 and 2.39 µmol g^{-1} FW, respectively, and 0.41 and 1.24 µmol g^{-1} FW in the T2 treatment. The H_2O_2 content in CK leaves and roots were 7.86 and 2.66 µmol g^{-1} FW, respectively, compared with 5.54 and 2.36 µmol g^{-1} FW, respectively, in the T2 treatment. The MDA content was significantly higher in the CK and decreased significantly in the T2 treatment (Figure 6). The ROS and MDA contents were significantly lower in the T1 and T2 treatments compared with those of the CK, but the differences were non-significant for T2. These findings revealed that ALA application plays an important role in stabilizing ROS accumulation and biosynthesis under combined low-temperature and weak-light stress.

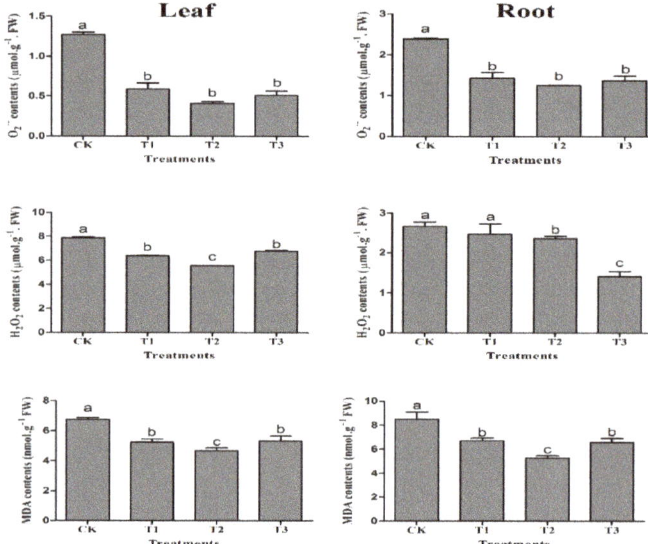

Figure 6. Effect of exogenous ALA application on reactive oxygen species and malondialdehyde contents under low-temperature and weak-light stress in cucumber seedlings. Data are the means of four replicates; error bars indicate the standard deviation. Treatments with the same lower-case letter are not significantly different (least significant difference test, $p = 0.05$). $O_2{}^{\cdot-}$, superoxide radical; H_2O_2, hydrogen peroxide; MDA, malondialdehyde.

4. Discussion

Low-temperature and weak-light stress damage a variety of plant physiological and biochemical metabolic processes, and hence reduce yield [5]. ALA is a vital precursor of the tetrapyrrole biosynthesis pathway, and is considered to be a plant growth regulator and to regulate plant defense mechanisms to mitigate the harmful effects of abiotic stress [9,10]. As reported previously, low temperature has adverse impacts on cucumber seedlings, and results in a significant reduction in Chl accumulation and photosynthetic capacity [12,13]. Exogenous ALA application is involved in regulation of endogenous hormones, chlorophyll and nutrient accumulation, and the plant defense system, and significantly reduces the harmful effects of low temperature and improves cucumber seedling growth [10]. In the present study, the combined stress of low temperature and weak light imposed significant negative effects on cucumber seedling growth (Figure 1). These results were similar to those of previous studies, which reported that ALA stimulates the plant defense system to alleviate the harmful effects of salinity and low-temperature stress and promoted cucumber seedling growth [10,14].

Chlorophyll is highly sensitive to abiotic stress and is quickly degraded, and is also considered to be an indicator of chloroplast development and photosynthesis proficiency. Abiotic stresses increase degradation of Chl [1], which ultimately affects photosynthetic capacity [17–19]. Previous studies have reported that ALA improves Chl accumulation, photosynthetic capacity, and nutrient uptake to reduce the harmful impacts of salinity stress in *Brassica napus* L. [8,20]. In the present study Chl contents (Chl *a*, Chl *b*, and Chl *a+b*) were significantly increased in ALA-treated seedlings (Figure 3). These findings suggested that ALA increased Chl protection under combined low-temperature and weak-light stress. A recent study reported that exogenous ALA application regulates the ALA metabolic pathway and the transcript level of downstream genes (*HEMA1, HEMH, CHLH, POR,* and *CAO*), and ALA accumulation under salinity stress [14]. The activities of glutamyl-tRNA reductase and glutamate-1-semiadelhyde 2,1-aminomutase, which catalyze ALA biosynthesis [20,21], are improved in ALA-treated plants under abiotic stress [22–24]. Transcriptome analysis suggested that ALA regulates thousands of genes that are involved in Chl biosynthesis (*ChlD, ChlH,* and *Chl1-1*), photosynthesis, cell cycle, transaction factors, and defense-related genes [11,25,26]. The results are supported by previous findings that ALA activates chlorophyll biosynthesis and accumulation in bluegrass in response to osmotic stress [25]. These findings help to elucidate the specific role of ALA in the Chl biosynthesis pathway and stimulation of Chl biosynthesis-related gene expression and enzyme activities, thus enhancing Chl accumulation under low-temperature and weak-light stress.

In the present study, photosynthesis capacity and chlorophyll fluorescence are significantly affected by the combined stress of low temperature and weak light (Figures 3 and 4), and were enhanced significantly in ALA-treated seedlings. These findings indicate that ALA reduced the toxic effects of low-temperature and weak-light stress. ALA regulates photosynthesis-related parameters and transcript levels of *RBCS, TPI, FBP, fructose-1,6-bisphosphate aldolase,* and *transketolase* under drought stress in rapeseed plants [13]. In tomato plants, plasma membrane intrinsic proteins (PIPs) genes, such as *PIP1* and *PIP2*, are regulated by exogenous ALA treatment. ALA not only enhances salinity stress tolerance, but also stimulates chlorophyll accumulation, chlorophyll fluorescence, and photosynthetic capacity [13]. In our previous study, we reported that exogenous ALA increases endogenous hormone accumulation, especially of 24-epibrassinolide, which regulates plant defense mechanisms, photosynthesis-related enzymes (such as Rubisco) and increasing the expression level of *rca, rbcS,* and *rbcl* involved in photosynthesis [27–29]. Transcriptome analysis suggested that photosystem II oxygen-evolving enhancer protein, photosystem I subunit, light-harvesting chlorophyll protein complex I and II, ferredoxin, P_n, T_r, ΦPSII, ETR, and qP, are upregulated under ALA treatment [25]. ALA is a crucial precursor in the biosynthesis of all porphyrin compounds, such as chlorophyll, heme, and phytohormones [8,12,13]. These findings indicated that exogenous ALA application stimulated the biosynthesis pathway, that enhanced chlorophyll (Figure 3), photosynthetic capacity (Figure 4), and chlorophyll fluorescence (Table 1), and reduced the detrimental effects of low-temperature and weak-light stress.

Plants increase ROS and MDA accumulation under exposure to abiotic stress, which is highly toxic and causes damaging impacts on chlorophyll, lipid, protein, and carbohydrate biosynthesis [6]. To alleviate these harmful effects, plants have evolved a defense system to scavenge these toxic and reactive species through antioxidation of enzymatic and nonenzymatic systems, which leads to damage and may cause cell death [6,30]. In the present study, low temperature and weak light significantly reduced antioxidant enzyme activities and increased $O_2^{\bullet-}$, H_2O_2 and MDA accumulation, whereas the opposite trend was observed under ALA treatment (Figure 6). Root vitality is an indicator of the overall physiological and biochemical vigor of roots [2], which are extremely sensitive to abiotic stress. Root activity decreased rapidly, and ultimately affected water and nutrient uptake, thus causing negative effects on chlorophyll, photosynthesis, enzyme activities, and growth under abiotic stress [10]. In present study, ALA diminished the detrimental influence of low-temperature and weak-light stress and increase root activity (Figure 2). Previous studies have reported that ALA plays an important role in upregulation of plant defense mechanisms under abiotic stresses [10,14,25]. Antioxidant enzymes (SOD, POD, and CAT) are involved directly in scavenging $O_2^{\bullet-}$ and H_2O_2, and catalyzing their conversion to H_2O and O_2. The current results showed that exogenous ALA enhanced activities of the antioxidant enzymes SOD, POD, and CAT in leaves and roots of cucumber under low temperature and weak light (Figure 5). In cucumber seedlings, significantly enhanced activities of SOD, POD, CAT, APX (Ascorbate peroxidase), and GR (Glutathione reductase), and reduced ROS and MDA accumulation, are observed under ALA treatment combined with low-temperature stress [10]. Previous studies have reported that ALA activates the plant defense system and defense-related genes, such as genes encoding SOD, POD, CAT, and APX, in rice and strawberry under osmotic and photodynamic stresses and reduce overproduction of ROS and MDA [31–33]. ALA is a precursor of heme biosynthesis, and CAT, POD, and APX contain a heme prosthetic group [14], which might be the reason that antioxidant enzyme activities were stimulated in ALA-treated seedlings (Figure 5). A number of defense-related genes, such as those encoding ascorbate/glutathione, CAT, and POD, are upregulated in ALA-treated bluegrass seedlings under osmotic stress. These findings are in line with those of previous studies, in which exogenous ALA upregulated antioxidant enzyme activities and reduced ROS and MDA accumulation in cucumber seedlings under low-temperature stress [8,10,12,33]. Thus, it can be concluded that exogenous ALA application increased tolerance to low-temperature and weak-light stress, and stabilized ROS and MDA accumulation, thus enhancing cucumber seedling growth (Figure 1).

5. Conclusions

The present results have demonstrated that exogenous ALA application to cucumber alleviates growth inhibition by stimulating the plant defense system and stabilizing ROS accumulation, thus enhancing tolerance to low-temperature and weak-light stress. ALA is involved in chlorophyll biosynthesis and accumulation to enhance photosynthetic capacity, and may be involved in carbohydrate and amino acid biosynthesis, which contributes to improved plant growth under low-temperature and weak-light stress. This study provides novel evidence of the potent roles of ALA and provides insight into the ALA regulatory mechanism in conjunction with low-temperature and weak-light stress. ALA was applied through the substrate and induced a distinct response to combined low-temperature and weak-light stress. The results will be helpful for off-seasonal and protected vegetable production in a greenhouse.

Author Contributions: X.Y., A.A. and Y.L. conceived and designed the experiments. A.A. and J.W. performed the experiments. A.A. analyzed the data and wrote the manuscript. X.Y., C.H. and Y.L. contributed in reagents/materials/analysis tools. Y.L. and C.H. review the manuscript. All authors have read and agreed to the published version of the manuscript.

Funding: This work was supported by the National Key Research and Development Program of China (2016YFD0201006), Earmarked Fund for Modern Agro-industry Technology Research System (CARS-25-C-01), Science and Technology Innovation Program of the Chinese Academy of Agricultural Sciences (CAAS-ASTIP-IVFCAAS), and Key Laboratory of Horticultural Crop Biology and Germplasm Innovation, Ministry of Agriculture, China. The funders had no role in study design, data collection and analysis, decision to publish, or preparation of the manuscript.

Conflicts of Interest: The authors declare no conflict of interest.

References

1. Huang, S.; Li, R.; Zhang, Z.; Li, L.; Gu, X.; Fan, W.; Lucas, W.J.; Wang, X.; Xie, B.; Ni, P.; et al. The genome of the cucumber, *Cucumis sativus* L. *Nat. Genet.* **2009**, *41*, 1275–1281. [CrossRef]
2. Anwar, A.; Bai, L.; Miao, L.; Liu, Y.; Li, S.; Yu, X.; Li, Y. 24-Epibrassinolide Ameliorates Endogenous Hormone Levels to Enhance Low-Temperature Stress Tolerance in Cucumber Seedlings. *Int. Mol. Sci.* **2018**, *19*, 2497. [CrossRef] [PubMed]

3. Anwar, A.; Liu, Y.; Dong, R.; Bai, L.; Yu, X.; Li, Y. The physiological and molecular mechanism of brassinosteroid in response to stress: A review. *Biol. Res.* **2018**, *51*, 46. [CrossRef] [PubMed]

4. Xia, X.J.; Wang, Y.J.; Zhou, Y.H.; Tao, Y.; Mao, W.H.; Shi, K.; Asami, T.; Chen, Z.; Yu, J.Q. Reactive oxygen species are involved in brassinosteroid-induced stress tolerance in cucumber. *Plant Physiol.* **2009**, *150*, 801–814. [CrossRef] [PubMed]

5. Shu, S.; Tang, Y.; Yuan, Y.; Sun, J.; Zhong, M.; Guo, S. The role of 24-epibrassinolide in the regulation of photosynthetic characteristics and nitrogen metabolism of tomato seedlings under a combined low temperature and weak light stress. *Plant Physiol. Bioch.* **2016**, *107*, 344–353. [CrossRef] [PubMed]

6. Gill, S.S.; Tuteja, N. Reactive oxygen species and antioxidant machinery in abiotic stress tolerance in crop plants. *Plant Physiol. Bioch.* **2010**, *48*, 909–930. [CrossRef]

7. Xi, Z.; Wang, Z.; Fang, Y.; Hu, Z.; Hu, Y.; Deng, M.; Zhang, Z. Effects of 24-epibrassinolide on antioxidation defense and osmoregulation systems of young grapevines (*V. vinifera* L.) under chilling stress. *Plant Growth Regul.* **2013**, *71*, 57–65. [CrossRef]

8. Naeem, M.S.; Jin, Z.L.; Wan, G.L.; Liu, D.; Liu, H.B.; Yoneyama, K.; Zhou, W.J. 5-Aminolevulinic acid improves photosynthetic gas exchange capacity and ion uptake under salinity stress in oilseed rape (*Brassica napus* L.). *Plant Soil* **2010**, *332*, 405–415. [CrossRef]

9. Wu, Y.; Liao, W.; Dawuda, M.M.; Hu, L.; Yu, J. 5-Aminolevulinic acid (ALA) biosynthetic and metabolic pathways and its role in higher plants: a review. *Plant Growth Regul.* **2019**, *87*, 357–374. [CrossRef]

10. Anwar, A.; Yan, Y.; Liu, Y.; Li, Y.; Yu, X. 5-Aminolevulinic Acid Improves Nutrient Uptake and Endogenous Hormone Accumulation, Enhancing Low-Temperature Stress Tolerance in Cucumbers. *Int. Mol. Sci.* **2018**, *19*, 3379. [CrossRef]

11. Aksakal, O.; Algur, O.; Aksakal, F.; Aysin, F. Exogenous 5-aminolevulinic acid alleviates the detrimental effects of UV-B stress on lettuce (*Lactuca sativa* L) seedlings. *Acta Physiol. Plant.* **2017**, *39*. [CrossRef]

12. Liu, D.; Wu, L.; Naeem, M.S.; Liu, H.; Deng, X.; Xu, L.; Zhang, F.; Zhou, W. 5-Aminolevulinic acid enhances photosynthetic gas exchange, chlorophyll fluorescence and antioxidant system in oilseed rape under drought stress. *Acta Physiol. Plant.* **2013**, *35*, 2747–2759. [CrossRef]

13. Liu, D.; Hu, L.Y.; Ali, B.; Yang, A.G.; Wan, G.L.; Xu, L.; Zhou, W.J. Influence of 5-aminolevulinic acid on photosynthetically related parameters and gene expression in *Brassica napus* L. under drought stress. *Soil Sci. Plant Nutr.* **2016**, *62*, 254–262. [CrossRef]

14. Wu, Y.; Jin, X.; Liao, W.; Hu, L.; Dawuda, M.M.; Zhao, X.; Tang, Z.; Gong, T.; Yu, J. 5-Aminolevulinic Acid (ALA) Alleviated Salinity Stress in Cucumber Seedlings by Enhancing Chlorophyll Synthesis Pathway. *Front. Plant Sci.* **2018**, *9*, 635. [CrossRef] [PubMed]

15. Wang, L.J.; Jiang, W.B.; Huang, B.J. Promotion of 5-aminolevulinic acid on photosynthesis of melon (*Cucumis melo*) seedlings under low light and chilling stress conditions. *Physiol. Plant.* **2004**, *121*, 258–264. [CrossRef]

16. An, Y.; Feng, X.; Liu, L.; Xiong, L.; Wang, L. ALA-Induced Flavonols Accumulation in Guard Cells Is Involved in Scavenging H_2O_2 and Inhibiting Stomatal Closure in Arabidopsis Cotyledons. *Front. Plant Sci.* **2016**, *7*, 1713. [CrossRef]

17. Pandey, S.; Fartyal, D.; Agarwal, A.; Shukla, T.; James, D.; Kaul, T.; Negi, Y.K.; Arora, S.; Reddy, M.K. Abiotic Stress Tolerance in Plants: Myriad Roles of Ascorbate Peroxidase. *Front. Plant Sci.* **2017**, *8*, 581. [CrossRef]

18. Jin, S.H.; Li, X.Q.; Wang, G.G.; Zhu, X.T. Brassinosteroids alleviate high-temperature injury in *Ficus concinna* seedlings via maintaining higher antioxidant defence and glyoxalase systems. *AoB PLANTS* **2015**, *7*. [CrossRef]

19. Ogweno, J.O.; Song, X.S.; Shi, K.; Hu, W.H.; Mao, W.H.; Zhou, Y.H.; Yu, J.Q.; Nogués, S. Brassinosteroids Alleviate Heat-Induced Inhibition of Photosynthesis by Increasing Carboxylation Efficiency and Enhancing Antioxidant Systems in *Lycopersicon esculentum*. *J. Plant Growth Regul.* **2008**, *27*, 49–57. [CrossRef]

20. Tanaka, Y.; Tanaka, A.; Tsuji, H. Effects of 5-Aminolevulinic Acid on the Accumulation of Chlorophyll b and Apoproteins of the Light-Harvesting Chlorophyll a/b-Protein Complex of Photosystem II. *Plant Cell Physiol.* **1993**, *34*, 465–472.

21. Korkmaz, A.; Korkmaz, Y.; Demirkıran, A.R. Enhancing chilling stress tolerance of pepper seedlings by exogenous application of 5-aminolevulinic acid. *Environ. Exp. Bot.* **2010**, *67*, 495–501. [CrossRef]

22. Kwon, S.W.; Sohn, E.J.; Kim, D.W.; Jeong, H.J.; Kim, M.J.; Ahn, E.H.; Kim, Y.N.; Dutta, S.; Kim, D.-S.; Park, J. Anti-inflammatory effect of transduced PEP-1-heme oxygenase-1 in Raw 264.7 cells and a mouse edema model. *Biochem. Bioph. Res. Co.* **2011**, *411*, 354–359. [CrossRef] [PubMed]

23. Nunkaew, T.; Kantachote, D.; Kanzaki, H.; Nitoda, T.; Ritchie, R. Effects of 5-aminolevulinic acid containing supernatants from selected Rhodopseudomonas palustris strains on rice growth under NaCl stress, with mediating effects on chlorophyll, photosynthetic electron transport and antioxidative enzymes. *Electron. J. Biotechn.* **2014**, *17*, 1. [CrossRef]

24. Tsuchiya, T.; Akimoto, S.; Mizoguchi, T.; Watabe, K.; Kindo, H.; Tomo, T.; Tamiaki, H.; Mimuro, M. Artificially produced [7-formyl]-chlorophyll d functions as an antenna pigment in the photosystem II isolated from the chlorophyllide a oxygenase-expressing *Acaryochloris marina*. *BBA-Bioenergetics* **2012**, *1817*, 1285–1291. [CrossRef]

25. Niu, K.; Ma, H. The positive effects of exogenous 5-aminolevulinic acid on the chlorophyll biosynthesis, photosystem and calvin cycle of Kentucky bluegrass seedlings in response to osmotic stress. *Environ. Exp. Bot.* **2018**, *155*, 260–271. [CrossRef]

26. Zhao, Y.Y.; Yan, F.; Hu, L.P.; Zhou, X.T.; Zou, Z.R.; Cui, L.R. Effects of exogenous 5-aminolevulinic acid on photosynthesis, stomatal conductance, transpiration rate, and PIP gene expression of tomato seedlings subject to salinity stress. *Genet. Mol. Res.* **2015**, *14*, 6401–6412. [CrossRef]

27. Wei, L.J.; Deng, X.G.; Zhu, T.; Zheng, T.; Li, P.X.; Wu, J.Q.; Zhang, D.W.; Lin, H.H. Ethylene is Involved in Brassinosteroids Induced Alternative Respiratory Pathway in Cucumber (*Cucumis sativus* L.) Seedlings Response to Abiotic Stress. *Front. Plant Sci.* **2015**, *6*, 982. [CrossRef]

28. Choudhary, S.P.; Yu, J.Q.; Yamaguchi-Shinozaki, K.; Shinozaki, K.; Tran, L.S. Benefits of brassinosteroid crosstalk. *Trends Plant Sci.* **2012**, *17*, 594. [CrossRef]

29. Xia, X.-J.; Huang, L.-F.; Zhou, Y.-H.; Mao, W.-H.; Shi, K.; Wu, J.-X.; Asami, T.; Chen, Z.; Yu, J.-Q. Brassinosteroids promote photosynthesis and growth by enhancing activation of Rubisco and expression of photosynthetic genes in *Cucumis sativus*. *Planta* **2009**, *230*, 1185. [CrossRef]

30. Zhu, T.; Deng, X.; Zhou, X.; Zhu, L.; Zou, L.; Li, P.; Zhang, D.; Lin, H. Ethylene and hydrogen peroxide are involved in brassinosteroid-induced salt tolerance in tomato. *Sci. Rep.* **2016**, *6*, 35392. [CrossRef]

31. Phung, T.H.; Jung, S. Differential antioxidant defense and detoxification mechanisms in photodynamically stressed rice plants treated with the deregulators of porphyrin biosynthesis, 5-aminolevulinic acid and oxyfluorfen. *Biochem. Bioph. Res. Co.* **2015**, *459*, 346–351. [CrossRef] [PubMed]

32. Cai, C.; He, S.; An, Y.; Wang, L. Exogenous 5-aminolevulinic acid improves strawberry tolerance to osmotic stress and its possible mechanisms. *Physiol. Plantarum* **2019**. [CrossRef] [PubMed]

33. Anwar, A.; Li, Y.; He, C.; Yu, X. 24-Epibrassinolide promotes NO_3^- and NH_4^+ ion flux rate and NRT1 gene expression in cucumber under suboptimal root zone temperature. *BMC Plant Biol.* **2019**, *19*, 225.

Article

Differential Aquaporin Response to Distinct Effects of Two Zn Concentrations after Foliar Application in Pak Choi (*Brassica rapa* L.) Plants

Hamideh Fatemi [1], Chokri Zaghdoud [2], Pedro A. Nortes [3], Micaela Carvajal [4] and Maria del Carmen Martínez-Ballesta [4,5,*]

1   Department of Horticulture, University of Mohaghegh Ardabili, Ardabil 56199-13131, Iran; ha.fatemi@yahoo.com
2   Dry Land Farming and Oasis Cropping Laboratory, Institute of Arid Regions of Medenine, University of Gabes, Medenine 4119, Tunisia; chokri_zaghdoud@yahoo.fr
3   Departament of Irrigation, Centro de Edafología y Biología Aplicada del Segura (CEBAS-CSIC), P.O. Box 164, 30100 Espinardo (Murcia), Spain; panortes@cebas.csic.es
4   Department of Plant Nutrition, Centro de Edafología y Biología Aplicada del Segura (CEBAS-CSIC), P.O. Box 164, 30100 Espinardo (Murcia), Spain; mcarvaja@cebas.csic.es
5   Department of Agronomy Engineering, Universidad Politécnica de Cartagena, Paseo Alfonso XIII, 48, 30203 Cartagena (Murcia), Spain
*   Correspondence: mcarmen.ballesta@upct.es; Tel.: +34-968-325457

Received: 26 February 2020; Accepted: 23 March 2020; Published: 24 March 2020

Abstract: Zinc (Zn) is considered an essential element with beneficial effects on plant cells; however, as a heavy metal, it may induce adverse effects on plants if its concentration exceeds a threshold. In this work, the effects of short-term and prolonged application of low (25 µM) and high (500 µM) Zn concentrations on pak choi (*Brassica rapa* L.) plants were evaluated. For this, two experiments were conducted. In the first, the effects of short-term (15 h) and partial foliar application were evaluated, and in the second a long-term (15 day) foliar application was applied. The results indicate that at short-term, Zn may induce a rapid hydraulic signal from the sprayed leaves to the roots, leading to changes in root hydraulic conductance but without effects on the whole-leaf gas exchange parameters. Root accumulation of Zn may prevent leaf damage. The role of different root and leaf aquaporin isoforms in the mediation of this signal is discussed, since significant variations in *PIP1* and *PIP2* gene expression were observed. In the second experiment, low Zn concentration had a beneficial effect on plant growth and specific aquaporin isoforms were differentially regulated at the transcriptional level in the roots. By contrast, the high Zn concentration had a detrimental effect on growth, with reductions in the root hydraulic conductance, leaf photosynthesis rate and Ca^{2+} uptake in the roots. The abundance of the PIP1 isoforms was significantly increased during this response. Therefore, a 25 µM Zn dose resulted in a positive effect in pak choi growth through an increased root hydraulic conductance.

Keywords: aquaporin; *Brassica rapa*; gas exchange parameters; growth; root hydraulic conductance; zinc

---

1. Introduction

Zinc (Zn) is an essential element for plants and animals [1]. In plants, as a co-factor of at least 300 metalloenzymes [2], Zn plays an important role in protein structure [3]. Another important role of Zn is the scavenging, for oxidation, of reactive oxygen species (ROS) in cells under normal and stress conditions [4]. However, when the Zn concentration exceeds a threshold in the cells, it can induce stress in the plants. Thus, at high concentrations, an important decrease in some physiological functions in plants—such as photosynthesis, respiration and reproductive performance—has been observed [5–7].

Zinc fertilization has beneficial effects for crops, since this microelement is involved in the synthesis of tryptophan, a precursor of indole acetic acid (IAA), responsible for growth stimulation [8]. However, shoot growth was significantly reduced at Zn concentrations above 25 mg L^{-1} in nutrient solution or above 170 mg kg^{-1} in the soil. The sensitivity to Zn toxicity differed among other crops, being sensitivity higher in celery > Chinese cabbage > pak choi. But pak choi can accumulate high levels of Zn in their edible parts with negative impact for human health [9]. The threshold between optimal Zn dose for plant development and the amount in which Zn leads to plant toxicity symptoms or human damage needs to be study in this new crop. However, the application of Zn to the soil results less efficient than foliar supply, due to soil and roots limitations and the poor Zn mobility in the phloem [10]. For this reason, repeated foliar sprays of Zn are frequent during vegetables cultivation.

The aquaporins (AQPs) are a family of small (24–30 kDa), pore-forming, integral membrane proteins, and are involved in the transport of water and small solutes such as urea, CO_2, ammonia, silicic acid and boric acid [11–13]. Plant AQPs are subdivided into plasma membrane intrinsic proteins (PIPs), tonoplast intrinsic proteins (TIPs), NOD26-like intrinsic proteins (NIPs), small basic intrinsic proteins (SIPs) and uncharacterised intrinsic proteins (XIPs) [14–16]. AQPs are responsible for 75%–95% of water passage in the plant [17–19] and are the target of many heavy metals and ions within the cells, including Zn. It has been shown that exposure to heavy metals (Zn, Cd or Cu) decreases root and leaf AQPs expression, in order to avoid water loss and maintain the plant water status [20,21]. Furthermore, in *Mesembryanthemum crystallinum*, exposure to 500 µM $ZnSO_4$ caused differential changes in water status, depending on the duration of the treatment [22]. Thus, reductions in the root to stem water flow, leaf water content, relative water content in leaf tissues, transpiration rate, leaf osmotic potential and expression of AQP genes in the root and leaf tissues were observed after 24 and 72 h of Zn treatment. However, the expression of the *PIP1;4* isoform was unaffected, being continuous and stable, while the expression of other AQP genes—such as those of the plasma membrane (*McPIP1;1; McPIP2;1; McPIP2;3*) or tonoplast AQPs (*McTIP1;2 and McTIP2;2*)—was decreased, showing the roles of different AQPs isoforms in the heavy metal response as key elements with specific involvement. Additionally, Gitto and Fricke [21] reported the effects of 0.1 and 1 mM Zn on water relations in barley (*Hordeum vulgare* L.) plants in relation to AQPs expression. They found that the decline in expression of three AQPs (*HvPIP1;2, HvPIP2;4, HvPIP2;5*) was stronger (46%–77%) with the application of 0.1 mM Zn than with 1 mM Zn (20%–50%); the simultaneous reductions in plant transpiration rate and root hydraulic conductivity (Lpr), of 24% (0.1 mM Zn) and 58% (1 mM Zn), respectively, would have limited the transport of Zn to the shoot to avoid major toxicity.

Zinc affects not only AQPs expression, but also their functionality, through the binding of the metal to the thiol groups of the protein, inducing a conformational change in the structure and the closure of the pore [23]. Additionally, heavy metals have been found to diminish water transport in *Actinidia deliciosa* protoplasts [24]. Yukutake et al. [25] showed that the water permeability mediated by AQPs was rapidly and reversibly regulated by dynamic changes in the intracellular Zn^{2+} concentration linked to a disturbed cellular redox state. In addition, there is evidence that Zn may be an integral part of biomembranes, and thus required for the stability and control of the lateral mobility of membrane molecules [26].

It has been postulated that a hydraulic signal sent from the root to the shoot could be responsible for the reduction in leaf turgor after root abiotic stress perception [27]. Similarly, the disruption of the water status during heavy metal stress could be a consequence of a decreased number of stomata and/or their closure for water preservation. Plants must adjust their whole water status to the constant demand of the aerial parts and AQPs may play an important role in the hydraulic balance [21]. The number of stomata was reduced by heavy metals such as copper, cadmium and Zn in *Phaseolus vulgaris* L. [28] and *Beta vulgaris* L. [29]. In Zn-treated plants, the stomata were round in shape and smaller than in control plants [29]. Generally, heavy metals disrupt water flow not only by reductions in stomatal conductance

(Gs), but also by reducing the flow of solutes; this can be a major cause of heavy metals toxicity and the decline in plant biomass at higher concentrations. Interestingly, decreased Gs values have also been reported under Zn deficiency [30], in chickpea plants; their Gs increased when treated with Zn (2.5 μg g^{-1} soil) and they were able to maintain membrane integrity [30]. The values of gas exchange parameters were also increased after Zn application in cotton (*Gossypium hirsutum* L.), but intercellular CO_2 decreased in Zn-treated plants, compared to controls [31].

Pak choi (*Brassica rapa* L. ssp. *chinensis*), also known as Chinese cabbage, is a popular leafy vegetable, grown and consumed worldwide. It is an annual crop that has optimal growth at temperatures ranging from 15 to 20 °C [32]. Although this vegetable is cultivated mainly in Asia, it could be grown in areas of Central Europe from spring to winter, due to its short vegetative growth period and lack of thermal requirements. Pak choi is not frequently found on supermarket shelfs, opening new perspectives for small producers [33].

Pak choi is well known to be tolerant of heavy metals such as Cd [34] and Zn [35], and it accumulates metal ions mainly in the leaves. In addition, the involvement of calcium (Ca) ions in its Zn tolerance has been reported [36]. Related to this, coordinated increases in Zn and Ca accumulation have been described in *Silene maritima* L., the tolerance of the Zn-tolerant population being associated with the tissue Ca levels and its presence in the leaf tissues [37].

Although the short- and long-term effects of heavy metals on plant water transport have been reported in different works [38,39], as well as the effects of low and high Zn concentrations [21], there are no reports describing the short- and long-term, concentration-dependent effects of Zn application in pak choi plants. In spite that pak choi plants may accumulate higher amounts of metals in their leaves, the adverse effects of these elements at molecular level are not well documented and often poorly understood, in particular those related to the changes in root water transport properties.

Thus, because of the importance of the Zn concentration and the treatment duration regarding the effects on plant water relations and their regulation by distinct AQPs isoforms, the specific aim of this work was to study: (1) The effect of the long-term application of Zn at two concentrations on the water status of pak choi plants, in order to identify the target AQPs isoforms and evaluate their differential responses in the beneficial or toxic effects that were induced in an ion-concentration-dependent way; and (2) the effect of the short-term application of Zn at two concentrations, in order to compare the water balance response with that under long-term Zn application and discern the hydraulic signals moving from shoot to root, with the involvement of AQP isoforms in the signalling process.

2. Materials and Methods

2.1. Plant Material and Growth Conditions

Seeds of pak choi (*Brassica rapa* L. ssp. *chinensis*) were pre-hydrated with de-ionised water and aerated continuously for 24 h. After this, the seeds were germinated in vermiculite, in the dark at 28 °C, for 2 days. They were then transferred to a controlled-environment chamber with a 16 h light and 8 h dark cycle, and temperatures of 20 and 15 °C and relative humidity of 60% and 80%, respectively. Photosynthetically active radiation (PAR) of 400 μmol m^{-2} s^{-1} was provided by a combination of fluorescent tubes (Philips TLD 36 W/83, Jena, Germany and Sylvania F36 W/GRO, Manchester, NH, USA) and metal halide lamps (OsramHQI, T 400 W, Berlin, Germany). After 5 days, the seedlings were placed in 15 L containers with continuously-aerated Hoagland nutrient solution [40].

2.2. Experimental Design

Two types of experiment were conducted. In the first, after one week of growth, a fixed amount of solution (18 mL per 6 plants container) of low (25 μM) and high (500 μM) Zn concentration (as $ZnSO_4$) was applied by foliar spraying during 15 days (long-term application), three times in total (1, 7 and 14 days before harvesting). Zn was applied using a 1 mL L^{-1} Tween-20 spraying solution allowing leaf spray retention and an alumni paper section under each spray leaf was used in order to avoid the sprayed Zn to reach the root nutrient solution. Plants were 21 days-old when they were harvested. Distilled water spraying was used as the control.

In the second experiment, after 20 days of plant growth, low (25 μM) and high (500 μM) Zn concentrations (as $ZnSO_4$) were applied by foliar spraying to half of the leaves (+F), leaving the other half without Zn spraying (−F). After 15 h (short-term application) the 21-day-old plants were harvested for measurements. Leaves and roots were separated for determinations in both experimental procedures. The measurements were made in the middle of the photoperiod in order to obtain the highest values for gas exchanges parameters and all samples were collected at this time for the rest of determinations.

2.3. Root Hydraulic Conductance

The root hydraulic conductance (L_0) of the plants was measured by pressurising the roots in a Scholander pressure chamber (UGT: Umwelt-Geräte-Technik GmbH, Freising-Weihenstepha, Germany), as described in Javot et al. [41]. The aerial parts of the plant were removed and the freshly-excised roots were inserted into the pressure chamber, in a plastic tube with the same nutrient solution used for their growth. A gradual increase in pressure (from 0.1 to 0.4 MPa) was applied to the detached roots. The sap that accumulated in this pressure range during a certain time, according to the treatment, was collected in a graduated glass micropipette. The sap flow (Jv) was expressed in mg g (root fresh weight)$^{-1}$ h^{-1} and plotted against pressure (MPa), the slope being the L_0 value in mg g (root fresh weight)$^{-1}$ h^{-1} MPa^{-1}.

2.4. Gas Exchange Measurements

A LI-6400XT photosynthesis system (Li-Cor, Inc., Lincoln, NE, USA), equipped with a LI-6400-40 Leaf Chamber Fluorometer (Li-Cor, Inc., Lincoln, NE, USA) and a LICOR 6400-01 CO_2 injector (Licor Bioscience, Lincoln, NE, U.S.A.), was used to measure the net photosynthetic rate (A), stomatal conductance (Gs) and leaf transpiration (T). Leaf gas exchange was measured in a 2 cm^2 leaf cuvette. During these measurements, the air CO_2 concentration was controlled using the injection system and compressed CO_2-cylinders with a CO_2 concentration of 400 μmol mol^{-1} CO_2. Measurements were made at a PAR of 500 μmol m^{-2} s^{-1}, and at ambient air temperature and relative humidity. The air flow was set to 400 μmol s^{-1}. The third fully-expanded leaf was chosen for the analysis, after 15 h or 15 d of treatment. The measurements were made in the middle of the photoperiod in order to obtain the highest values.

2.5. RNA Extraction and Reverse Transcription

Total RNA was extracted from fully-expanded leaves of both, short and long-term treated plants and they were frozen and ground to a fine powder in liquid nitrogen using a prechilled mortar and pestle. The ground tissues were stored at −80 °C until use. Total RNA was extracted using the RNeasy Plant Mini Kit (Qiagen, Hilden, Germany) according to the manufacturer's protocol. Contaminated DNA from samples was removed with DNase I, using the DNA-free Kit (Ambion, Applied Biosystems, Austin, TX, USA), and the RNA concentration was quantified with a Nanodrop 1000 Spectrophotometer (Thermo Scientific, Waltham, MA USA). The extracted RNA was stored at −80 °C until use.

cDNA was synthesised from 2 mg of total RNA, using M-MLV reverse transcriptase from the RETROscript Kit (Ambion, Applied Biosystems, Austin, TX, USA). Reverse transcription was carried out with heat denaturation of the RNA, according to the manufacturer's instructions.

2.6. Quantitative Real-Time PCR (QRT-PCR) Analyses

To compare the expression of *PIP1* and *PIP2* under different treatments, QRT-PCR analyses were performed as described previously [42], in an Applied Biosystems 7500 Real-Time PCR system (ThermoFisher Scientific S.L, Madrid, Spain).

Gene-specific primers of *Brassica rapa* L. ssp. *pekinensis* genes [43] were used for different PIP isoforms. The primers and the lengths of the amplicons are described in Supplementary Table S1. *B. rapa* var. *italica*, 18s ribosomal RNA [43] was used as the reference gene for standardisation of each sample. After denaturation at 95 °C for 10 min, amplification occurred in a two-step procedure: 15 s of denaturation at 95 °C and 1 min of annealing and extension at 60 °C for 40 cycles, followed by a dissociation stage. Data collection was carried out at the end of each round in step 2. These conditions were used for both target and reference genes and the absence of primer–dimers was checked in controls lacking templates. The amplifications were performed on three independent samples for each treatment (biological replicates) and triplicate reactions were carried out for each sample (technical replicates) in 96-well plates. The transcript levels were calculated using the $2^{-\Delta\Delta Ct}$ method [44]. Standard curves (log of the cDNA dilution vs. Ct) using serial 10-fold dilutions of cDNA were built for each pair of selected primers, obtaining 95% PCR efficiency, corresponding to a slope of −3.44.

2.7. Zinc and Calcium Tissue Analyses

Zinc and Ca were analysed in the oven-dried root and leaf tissues (ca. 100 mg DW). The samples were digested, after HNO_3-H_2O_2 (2:1) addition [45], in a microwave oven (CEM Mars Xpress, North Carolina, USA) and analysed by ICP spectrometry (Iris Intrepid II, Thermo Electron Corporation, Franklin, TN, USA).

2.8. Statistical Analysis

The data were analysed statistically, using the SPSS 13.0 software package (IBM, Armonk, NY, USA), by analysis of variance (ANOVA) and by Tukey's test. Significant differences were determined at $p < 0.05$.

3. Results

3.1. Plant Biomass

A preliminary experiment was carried out in order to determine the effect of different foliar-supplied Zn concentrations on biomass, which allowed the selection of the appropriate Zn concentrations for the subsequent experiments. Plant growth was significantly influenced by long-term (15 days) foliar Zn fertilisation, depending on the concentration (Table 1). Thus, at 25 µM Zn, a marked increase in the total plant fresh (FW) and dry (DW) weights (28.48% and 11.85%, respectively) was observed, compared to the controls. By contrast, significant reductions in these parameters were found at 500 µM Zn (30.62% and 25.69% in plant FW and DW, respectively). At 50 and 100 µM Zn there were no significant changes in the fresh and dry weights with regard to the control and at 1 mM Zn an extreme toxic effect in the plants was observed. Therefore, 25 and 500 µM Zn were chosen as low and high concentrations, respectively, for foliar applications in the short and long term (Figure 1).

Table 1. Effect of long-term exposure to different zinc concentrations on the biomass of pak choi (*Brassica compestris*). Values are means ± SE ($n = 5$).

Zinc Treatments	Shoot Fresh Weight	Root Fresh Weight	Shoot Dry Weight	Root Dry Weight
Control	77.87 ± 5.23b	14.71 ± 1.25b	5.51 ± 0.51b	0.99 ± 0.15a
25 µM	98.83 ± 4.74a	20.12 ± 1.46a	6.02 ± 0.57a	1.25 ± 0.10a
50 µM	66.06 ± 5.09cb	11.08 ± 0.77cb	5.10 ± 0.54cb	0.72 ± 0.08b
100 µM	69.00 ± 5.70cb	11.21 ± 1.38cb	5.28 ± 0.26b	0.72 ± 0.11b
500 µM	54.32 ± 3.08c	9.91 ± 0.47c	4.19 ± 0.20c	0.64 ± 0.08b
1 mM	41.89 ± 4.82c	8.83 ± 1.13c	3.74 ± 0.44d	0.54 ± 0.10c

The different letters indicate significant differences among treatments according to Duncan's test ($p < 0.05$).

Figure 1. Pak choi plants treated with Zn (0, 25 and 500 µM).

3.2. Effect of Zn on Root Hydraulic Conductance (L_0)

After short-term (15 h) partial foliar application of Zn, the behaviour of L_0 at the low (25 µM) concentration was the opposite of that at the high (500 µM) concentration (Figure 2A), the values being significantly increased (35.16%) and decreased (25.13%), respectively, compared to control plants. In the long-term foliar Zn application, a significant effect on L_0 was only recorded at the high concentration, with the values being lower than those of the short-term application (the reduction reached 57.51%, relative to the controls) (Figure 2B).

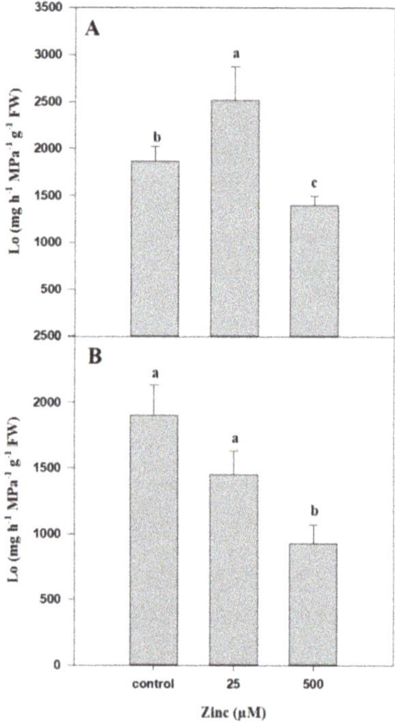

Figure 2. Effect of zinc concentration (0, 25 and 500 µM) on root hydraulic conductance (L_0) of pak choi plants, after foliar application. (**A**) Short-term experiment (15 h) in which half of the leaves were sprayed. (**B**) Long-term experiment (15 days). Values are means ± SE ($n = 5$). Bars with different letters represent significant ($p < 0.05$) differences after ANOVA and an LSD (Least significant difference) test.

3.3. Effect of Zn on Gas Exchange Parameters

At both Zn concentrations, the stomatal conductance (Gs), transpiration rate (E) and net photosynthetic rate per unit leaf area (A) remained without significant changes after the short-term treatment, in both types of leaf (with and without foliar Zn spraying) (Figure 3A–C). After the long-term treatment, Gs was decreased (42.46%) at 500 μM $ZnSO_4$, whereas A was increased and decreased at 25 and 500 μM, respectively, but not E, relative to untreated plants (Figure 3D–F).

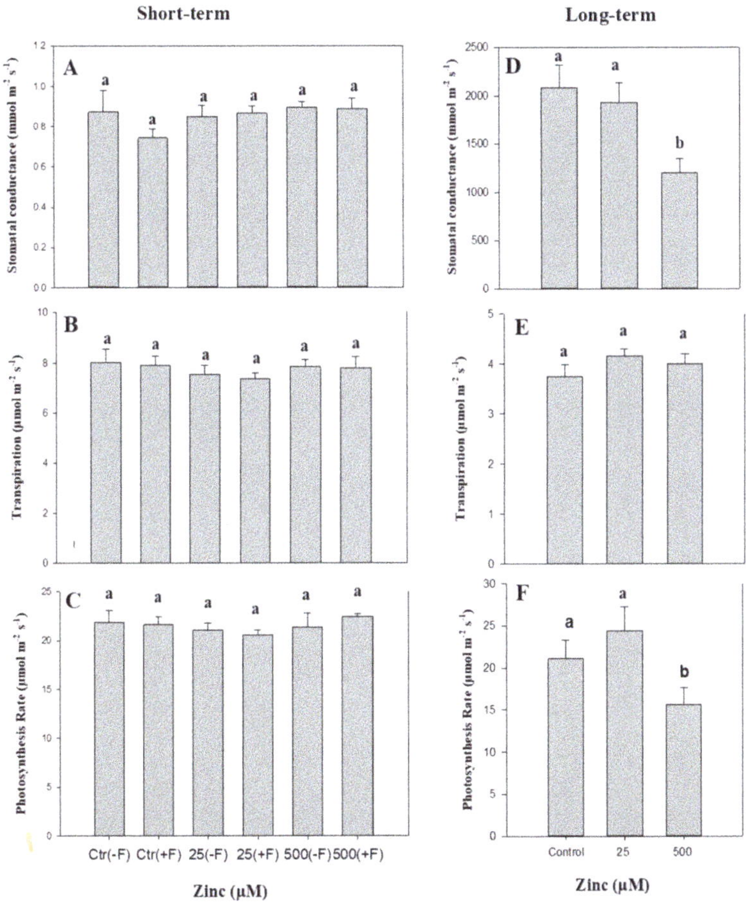

Figure 3. Effect of zinc concentration (0, 25 and 500 μM) on gas exchange parameters of pak choi plants. After short-term application: (**A**) Stomatal conductance, (**B**) transpiration, (**C**) photosynthesis rate, where –F represents the measurements in the leaves that were not sprayed and +F represents the measurements in the leaves that were sprayed. After long-term application: (**D**) Stomatal conductance, (**E**) transpiration, (**F**) carbon assimilation. Values are means ± SE ($n = 5$). Bars with different letters represent significant ($p < 0.05$) differences after ANOVA and an LSD (Least significant difference) test.

3.4. Tissue Contents of Zn and Ca

Macro and micro nutrients were determined in both experiments. However, in addition to Zn, only Ca showed significant differences between treatments.

After 15 h of the partial foliar Zn treatment, the Zn^{2+} content in leaves showed a similar, significant increase at both Zn concentrations, compared to control plants, the values being significantly lower in non-sprayed leaves than in sprayed ones (about 2.2- and 2.90-fold at 25 and 500 μM Zn, respectively) (Figure 4A). However, in roots the Zn^{2+} content significantly increased with the spray concentration of Zn; thus, at 500 μM $ZnSO_4$ this increase reached 2.25-fold relative to the control value (Figure 4B). A pattern of variation similar to that of Zn was observed for the leaf Ca content following the short-term fertilisation (Figure 4C), without differences between the control plant leaves and the non-sprayed leaves of the Zn-treated plants. In roots, the Ca content was significantly increased (by 15.88%) and decreased (by 20.97%) at 25 and 500 μM Zn, respectively, after 15 h of treatment (Figure 4D).

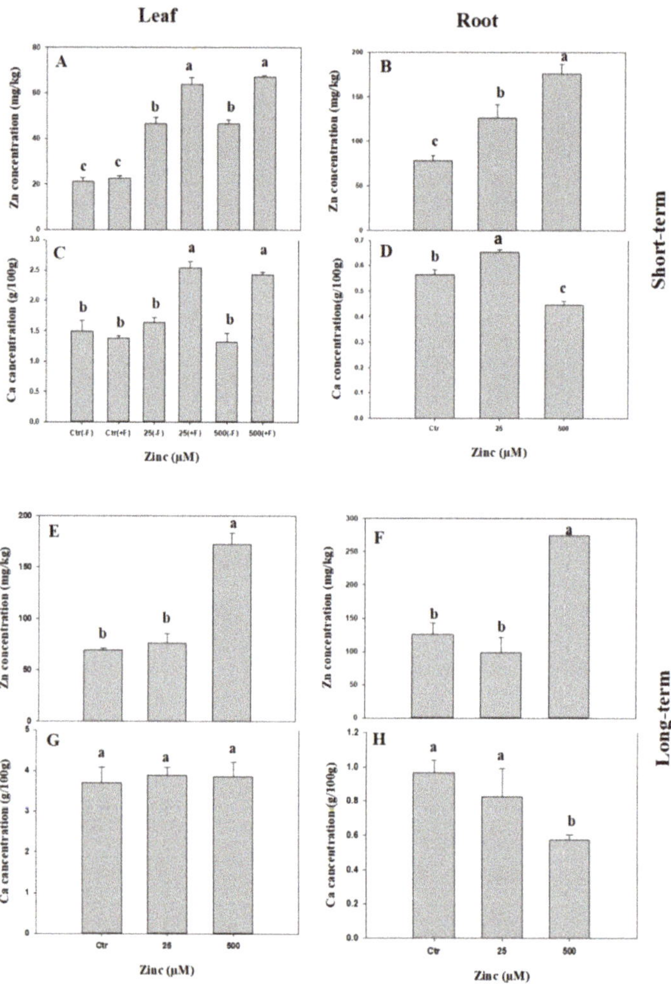

Figure 4. Effect of zinc concentration (0, 25 and 500 μM) on Zn and Ca concentrations of pak choi plants. After short-term application: (**A**) Zn concentration in leaf, (**B**) Zn concentration in root, (**C**) Ca concentration in leaf, (**D**) Ca concentration in root, where −F represents the measurements in the leaves that were not sprayed and +F represents the measurements in the leaves that were sprayed. After long-term application: (**E**) Zn concentration in leaf, (**F**) Zn concentration in root, (**G**) Ca concentration in leaf, (**H**) Ca concentration in root. Values are means ± SE (*n* = 5). Bars with different letters represent significant (*p* < 0.05) differences after ANOVA and an LSD (Least significant difference) test.

In the long-term study the Zn content in the leaves and roots was significantly increased only by the highest Zn concentration, increasing 2.47- and 2.18-fold in leaves and roots, respectively, relative to the controls (Figure 4E,F). Under these conditions, the Ca content was maintained in the leaves, but decreased in the roots (by 40.77%), at 500 μM Zn (Figure 4G,H).

3.5. PIP1 and PIP2 Isoforms Expression in Root and Leaf Tissues

Differences between plant organs in the PIP1 and PIP2 gene expression patterns due to Zn exposure were observed, depending on the experiment duration.

After short-term partial foliar Zn spraying only the highest Zn concentration (500 μM) significantly increased the transcript levels of the aquaporins PIP1;1 and PIP1;2 in leaves (3.71- and 2.42-fold, respectively) (Figure 5A), and of PIP1;4 in roots (3.45-fold), above the control levels (Figure 5B). The changes in these isoforms were specific to the toxic Zn concentration. In addition, while at both Zn concentrations, 25 and 500 μM, the expression levels of PIP2;1 and PIP2;5 in leaves were increased and that of PIP2;2 in roots was decreased, different behaviour of the expression pattern regarding the low and high Zn concentrations was observed for PIP2;2 in leaves and PIP2;1, PIP2;3 and PIP2;7 in roots (Figure 5C,D).

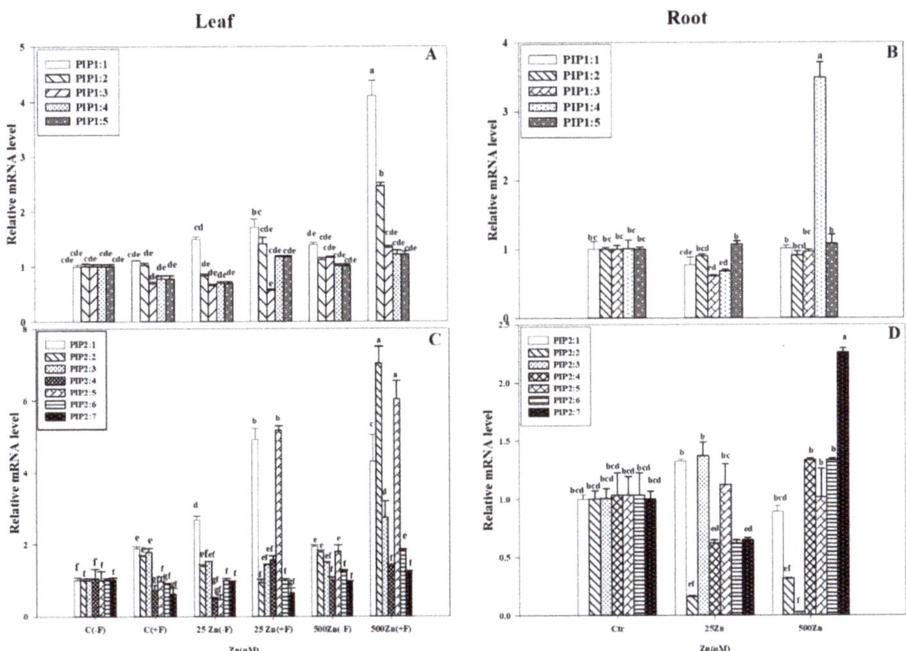

Figure 5. Relative expression level of aquaporin isoforms after short-term application of Zn: (**A**) PIP1 in root, (**B**) PIP1 in leaf, (**C**) PIP2 in root, (**D**) PIP2 in leaf, determined by Q-RT-PCR (quantitative real time-polymerase chain reaction) in control plants and plants treated with 25 or 500 μM Zn. In leaf tissues, gene expression was determined in both non-sprayed (−F) and sprayed (+F) leaves. Mean values and standard errors are shown (*n* = 3). Bars with different letters represent significant (*p* < 0.05) differences after ANOVA and an LSD test.

The plant hydraulic regulation after long-term foliar Zn spraying showed the implication of other aquaporins isoforms. Thus, in roots, while the PIP1;1 isoform remained up-regulated only at 500 μM Zn (1.84-fold), down-regulation of PIP1;3 and PIP2;4 (1.64-fold) was observed only at 25 μM Zn, relative to the levels of control plants (Figure 6A,B). Interestingly, the expression levels of the

PIP2;6 and PIP2;7 isoforms were the ones most significantly affected by long-term Zn exposure, in both plant organs, without differences between the low and high Zn doses (Figure 6C,D). At 500 µM Zn, PIP2;6 and PIP2;7 were up-regulated, by 4.31- and 2.57-fold, respectively, in leaves; whereas, in roots, they were decreased and increased, respectively, 50- and 5.60-fold, compared to the controls.

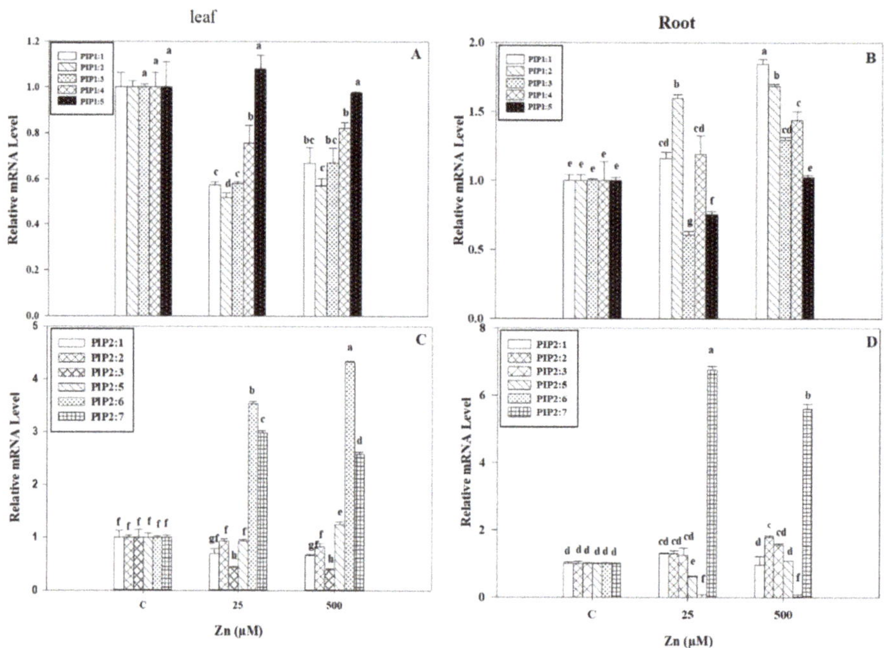

Figure 6. Relative expression level of aquaporin isoforms after long-term application of Zn: (**A**) leaf PIP1, (**B**) leaf PIP2 (**C**)root PIP1 (**D**)root PIP2 genes, determined by Q-RT-PCR in control plants and plants treated with 25 or 500 µM Zn. Mean values and standard errors are shown ($n = 3$). Bars with different letters represent significant ($p < 0.05$) differences after ANOVA and an LSD (Least significant difference) test.

4. Discussion

4.1. Growth, Root Hydraulic Conductance and Gas Exchange Parameters

It has been reported that Zn may promote growth at optimal concentration but at higher or low levels induced a growth reduction by interfering with plant metabolic activities [46]. Similarly, in pak choi plants 25 µM Zn had a beneficial effect on plant growth whereas a 500 µM Zn concentration decreased plant biomass at both, shoot and root level. In addition, water relations and photosynthetic rate may result affected by Zn concentration and duration of Zn exposition.

After short-term Zn spraying, no consistent relationship between L_0 and Gs was observed and whereas L_0 was increased and decreased, respectively, at the low and high Zn concentrations, Gs was maintained under all treatments. Thus, in our plants, the Zn applied by foliar spraying was rapidly (after 15 h) and substantially transported to the roots. This supports previous results obtained in wheat, for which foliar-applied Zn was translocated to leaves, both above and below the treated leaf, as well as to the root tips via the phloem [47,48]. Moreover, the higher content of Zn in root tissues relative to leaves was a mechanism to protect photosynthetic tissues, as no changes in the gas exchange parameters were observed in plants after the short-term foliar treatment, independently of the Zn concentration and zone of spraying.

It has been proposed that the movement of water from root to shoot through the xylem was decreased after metal treatment as consequence of structural responses in the cells [49]. However, in the leaves of *A. rubrum*, the relative water content (RWC) was maintained after metal exposition. The authors indicated that plants were able to compensate the effect of the stress and maintain leaf water status [49]. Similar results were found in pak choi leaf tissues after short-term Zn exposition and root hydraulic conductance did not modulate transpiration. This fact indicated that L_0 and Gs were uncoupled processes under these experimental conditions. In *Beta vulgaris*, an ability to rapidly lose (4 h) and gain (24 h) turgor in plants under abiotic stress (200 mM NaCl) was observed, with reductions in Lp_r consistent with Gs decreases [50], but different stresses and genotypes were described. In any case, it is plausible that L_0 in pak choi plants responded to a rapid shoot to root sensing action of Zn that in addition triggers differential regulation of aquaporin gene expression.

In the long-term Zn experiment, no changes were observed in L_0, Gs or E in plants treated with the low Zn concentration, but growth was correlated to an enhanced photosynthetic rate (A) at the optimal Zn (25 µM) dose. It has been previously reported that photosynthesis, carbonic anhydrase activity and chlorophyll concentrations were correlated with Zn nutrition at low levels [51,52]. However, at high Zn concentration a reduced stomatal conductance was in consonance with L_0 reductions and a photosynthetic rate decrease in pak choi. Similarly, it has been reported that high levels of Zn (200–500 µM) inhibited CO_2 assimilation due to structural and functional disturbances in the photosynthetic process [6,53,54]. Thus, photosynthetic response in our plants to long-term Zn could be attributed to an alteration of the photosystem and pigments rather than to the transpiration change. In pea plants, 1 mM cadmium (Cd) had no statistically significant effect on the transpiration rate but it decreased this parameter a 23.1% regarding control in barley plants [55], pointing out the genotype-dependence of the effect of metals on transpiration.

4.2. Calcium and Zinc Concentrations in Pak Choi Plants after Treatments

The short-term Zn accumulation in our plants was accompanied by significant increases in the Ca^{2+} contents only in sprayed leaves—to a similar extent at both Zn concentrations—and in roots at the low dose. Calcium is a crucial intracellular messenger and its homeostasis can be modified rapidly by hormonal and environmental stimuli, including metal [56]. The correlation of Zn and Ca in Zn-tolerant and non-tolerant populations of *Silene maritima* was demonstrated by Baker [37]. The important role of Ca in the alleviation of heavy metals toxicity in plants occurs through the prevention of a decrease in the negative charge on the plasma membrane, decreasing the activity of heavy metal ions on the plasma membrane surface, and the maintenance of Ca-related signalling pathways in the presence of the toxicant(s) [57–59]. Davis and Parker [36] also reported that Zn toxicity was highly correlated with the Ca:Zn ratio and reduced stem biomass. According to this, in our pak choi plants the root Ca increment at the low Zn concentration reflects the lack of toxicity of this heavy metal at low doses after a short-term foliar application. By contrast, the decrease in the root Ca content in response to the short-term application of a high Zn concentration is indicative of Zn toxicity and is related to the antagonistic relationship between Zn and Ca in plants [60].

In the long-term experiment the Zn and Ca contents in both organs of plants sprayed with low Zn returned towards control values, compared to the short-term treatment. These results suggest the involvement of an efficient Zn-detoxification mechanism, maintained over time after the low Zn-dose application. Other brassicas, such as *Brassica campestris* L. and *Raphanus sativus* L., have been found to act as hyper-accumulator plants, with higher Zn concentrations in their leaves—relative to the levels of other heavy metals—after irrigation with sewage water [61].

4.3. PIP Aquaporin Expression Remove

With the short-term Zn application, no aquaporin expression changes were found in non-spraying leaves and only in the sprayed ones there were significant increases in *PIP1* and *PIP2* expression after 15 h and depending on the Zn dose. This was concomitant with a similar Ca accumulation at both Zn doses. The maintenance of critical Ca levels in the apoplast may be required to keep membrane stability and the water channels open through phosphorylation [62–64], reinforcing previous view that AQP phosphorylation may be among the initial targets of Zn toxicity [65]. In fact, Przedpelska-Wasowicz and Wierzbicka [66] showed that, in *Allium cepa* L., the fast (10 min) toxic effects of heavy metals at the cellular level involved an AQP gating mechanism and this response cannot be ruled out in pak choi plants, being gene expression regulation a compensatory mechanism of the water channel gating.

Our results showed a short-term modulation of leaf *PIP2;1* and *PIP2;5*, which were up-regulated at both Zn concentrations. Exposure of plants to heavy metals (Cd, Ni and Zn) is known to induce water deficit in plant organs [67]. Thus, the leaf *PIP2;1* and *PIP2;5* isoforms may play a key role in water dynamics as part of a rapid Zn-induced sensing response and osmotic stress prevention. The over-expression of these two AQPs isoforms was previously observed to favour water transport into the inner leaf tissues, preventing a fall in leaf water potentials and reducing xylem tensions [68,69]. By contrast, in this study, an up-regulation of *PIP1;1*, *PIP1;2* and *PIP2;2* in pak choi leaves was only observed after short-term application of 500 µM Zn. This fact suggests the implication of these isoforms in the regulation of water homeostasis under a rapid and toxic Zn effect, as reported in several studies for different abiotic stresses [70–73]. Thus, co-expression of *PIP1;2* and *PIP2;5* resulted in increased water-channel activity in *Zea mays* plants [74]. In addition, transgenic rice plants over-expressing *PIP1;1* or *PIP2;2* developed enhanced tolerance to 200 mM mannitol and mild salt stress [70]. In our plants, upregulation of a few individual PIPs in the leaves may redirect water flow into specific cells—which is crucial for plant survival, maintaining gas exchange mechanisms, as it was observed from our data and according to those observed by Alexandersson et al. [75] under water deficit.

Regarding root AQPs expression, in our short-term experiment, among the root *PIP1* genes, only *PIP1;4* showed overexpression at the highest Zn dose. In *Arabidopsis thaliana*, the transcription level of *PIP1;4* isoforms increased more than five-fold over the first 48 h of drought stress (250 mM mannitol), in leaves and roots, as well as in response to salt (150 mM NaCl) and cold stresses [14]. However, expression of this gene was unaffected during *M. crystallinum* adaptation to 1 d of Cu, Zn [22] or salinity [76] stress. After rapid application of a high Zn dose to pak choi plants, *PIP1;4* may function as promotor of water transport from roots to the aerial parts, to maintain water relations and gas exchange in the plants, even at low L_0. The fact that only AQPs located in the plasma membrane of sprayed leaves and roots had modified gene expression suggests the involvement of these proteins in the rapid shoot-to-root hydraulic communication, previous to the Zn effect on leaf transpiration [77].

Additionally, in short-term experiment, root *PIP2;2* and *PIP2;3* showed significant sensitivity to Zn application, especially at high Zn concentration. However, the functional redundancy within AQPs [78] represents a challenge to the determination of the overall role of any given type of AQP in the response to Zn stress. In fact, the decline of both *PIP2;2* and *PIP2;3*—which had high homology, showing 96.8% amino acid identity—could have been compensated by the enhancement of *PIP2;7*. In other reports, a transient up-regulation of *PIP2;3* gene expression upon short-term (2 to 96 h) salt stress has been also reported, implying a possible influence of *PIP2;3* on the short-term response to ionic or purely-osmotic stress [14,79,80]. In addition, rapid (1 to 24 h) up-regulation of *PIP2;7* was observed to be involved in the osmoregulation in plants only under high-toxicity stress [81,82], and the efflux of metalloids from roots to avoid toxicity under excessive levels [83,84], which could explain its overexpression in our pak choi roots only at 500 µM Zn, when root accumulation of Zn^{2+} ions was high enough. In any case, similar to that which occurs with other stresses and genotypes, *PIP2;2*, *PIP2;3* and *PIP2;7* showed a coordinated expression in pak choi plants exposed to Zn.

In the long-term application, by contrast to the short-term experiment, *PIP2;7* was up-regulated in both plant organs, independently of the Zn dose. Apart from its role in plant growth, facilitating water transport into the rapidly-elongating root cells [85], *PIP2;7* facilitates H_2O_2 diffusion across the plasma membrane [86,87], which is an important signal in the regulation of multiple genes associated with abiotic stress tolerance [88,89]. Thus, in pak choi plants *PIP2;7* acts as a sensitive target element for Zn in prolonged heavy metal exposure and its role as part of this signalling response to Zn must be elucidated.

In addition, reductions in root *PIP1;3*, *PIP1;5*, *PIP2;5* and *PIP2;6* expression in plants treated with 25 μM Zn were not correlated with the maintenance of L_0, but other significant increments in the AQP mRNA levels were observed such as for *PIP1;1*; *PIP1;2*, *PIP1;4* and *PIP2;7* which may contribute to L_0. After the long-term 500 μM -Zn treatment, when Zn^{2+} ions were largely accumulated in leaves and to a greater extent in roots, the majority of *PIP1* isoforms were upregulated in roots. However, L_0 was significantly decreased in these conditions, concomitant with a reduction in the root Ca^{2+} content. This fact pointing out the role of PIP2 subfamily to water transport contribution and the aquaporin regulation through other mechanisms alternative to a transcriptional control as Ca^{2+} dependent-protein phosphorylation. There were also concomitant reductions in A and Gs after long-term exposure to high Zn, indicating severe injury to the photosynthetic apparatus, which finally induced a reduction in plant FW and DW. These results were in consonance with the similar effects of excess of Zn in *Jatropha* plants [90].

5. Conclusions

Pak choi plants have a differential response to Zn depending on the treatment intensity and duration. A rapid shoot-to-root hydraulic signal was involved in the response to short-term partial foliar Zn application, affecting root hydraulic conductance, that it was increased or decreased depending on the Zn dose applied. However, in these plants, L_0 was a no-coordinated process with leaf transpiration and the closure of the stoma under these conditions. Similar root patterns of variation for L_0 and Ca ions with distinct Zn concentrations pointed out the effect of Zn on Ca availability and the importance that this element could have in processes such as aquaporin gating. Leaf *PIP2;1* and *PIP2;5* aquaporin isoforms play a key role sensing the rapid Zn-induced response at leaf level and may act in controlling water dynamics as part of this rapid Zn-induced sensing response preventing osmotic stress. By contrast, *PIP1;1* and *PIP2;2* could be involved in the regulation of leaf water homeostasis, specifically under a toxic Zn effect.

Long-term exposition to low-Zn dose had a beneficial effect on plant growth through an increased leaf photosynthesis rate and maintaining plant water balance. However, high Zn concentrations induced a stomatal closure that together with a decrease in the photosynthesis rate and water transport uptake led to a reduced plant biomass. Differential pattern of aquaporin isoforms reveals that almost all *PIP1* isoforms and *PIP2;6* and *PIP2;7* were involved in sensing the long-term response to Zn treatments reflecting a major Zn dose dependence the PIP1 subfamily. Based on the transcriptional response, other mechanisms of aquaporin regulation at protein level need to be elucidated for the isoforms involved in the response to low and high Zn concentrations and particular function of each isoform in the Zn response have to be deeper studied.

Supplementary Materials: The following are available online at http://www.mdpi.com/2073-4395/10/3/450/s1, Table S1: Primer sequence used for real time and RT-PCR amplification of Aquaporin genes of *B. rapa*.

Author Contributions: Experimental design, M.C. and M.d.C.M.-B.; funding acquisition, M.C.; experimental measurements and investigation, H.F., P.A.N. and M.d.C.M.-B.; statistical analysis and investigation, C.Z.; interpretation and validation of the results, H.F., C.Z., P.A.N., M.C. and M.d.C.M.-B.; writing—review and editing, H.F., C.Z., P.A.N., M.C. and M.d.C.M.-B.; All authors have read and agreed to the published version of the manuscript.

Funding: This work was funded by the Spanish Ministerio de Economía, Industria y Competitividad (AGL2016-80247-C2-1-R). H.F. was supported by the Iranian Ministry of Science, Research and Technology.

Acknowledgments: The authors thank David Walker for correction of the English language and style.

Conflicts of Interest: The authors declare no conflicts of interest.

References

1. Prasad, A.S. Zinc: The biology and therapeutics of an ion. *Ann. Intern. Med.* **1996**, *125*, 142–144. [CrossRef] [PubMed]
2. Coleman, J.E. Zinc enzymes. *Curr. Opin. Chem. Biol.* **1998**, *2*, 222–234. [CrossRef]
3. Andreini, C.; Banci, L.; Bertini, I.; Rosato, A. Zinc through the three domains of life. *J. Proteome Res.* **2006**, *5*, 3173–3178. [CrossRef]
4. Broadley, M.R.; White, P.J.; Hammond, J.P.; Zelko, I.; Lux, A. Zinc in plants. *New Phytol.* **2007**, *173*, 677–702. [CrossRef]
5. Ali, G.; Srivastava, P.S.; Iqbal, M. Influence of cadmium and zinc on growth and photosynthesis of *Bacopa monniera* cultivated in vitro. *Biol. Plant.* **2000**, *43*, 599–601. [CrossRef]
6. Khudsar, T.; Uzzafar, M.; Iqbal, M.; Sairam, R.K. Zinc-induced changes in morpho-physiological and biochemical parameters in *Artemisia annua*. *Biol. Plant.* **2004**, *48*, 255–260. [CrossRef]
7. Kholodova, V.P.; Volkov, K.S.; Kuznetsov, V.V. Adaptation of the common ice plant to high copper and zinc concentrations and their potential using for phytoremediation. *Russ. J. Plant Physiol.* **2005**, *52*, 748–757. [CrossRef]
8. Mallick, M.F.R.; Muthukrishnan, C.R. Effect of micronutrients on tomato (*Lycopersicon esculentum* Mill). II. Effect on flowering, fruit-set and yield. *South Indian Hort.* **1980**, *28*, 14–20.
9. Long, X.X.; Yang, X.E.; Ni, W.Z.; Ye, Z.Q.; He, Z.L.; Calvert, D.V.; Stoffella, J.P. Assessing zinc thresholds for phytotoxicity and potential dietary toxicity in selected vegetable crops. *Commn. Soil Sci. Plant. Anal.* **2003**, *34*, 1421–1434. [CrossRef]
10. Swietlik, D. Zinc nutrition of fruit trees by foliar sprays. *Acta Hort.* **2002**, *93*, 123–129. [CrossRef]
11. Deshmukh, R.; Bélanger, R.R. Molecular evolution of aquaporins and silicon influx in plants. *Funct. Ecol.* **2016**, *30*, 1277–1285. [CrossRef]
12. Kitchen, P.; Day, R.E.; Salman, M.M.; Conner, M.T.; Bill, R.M.; Conner, A.C. Beyond water homeostasis: Diverse functional roles of mammalian aquaporins. *Biochim. Biophys. Acta* **2015**, *1850*, 2410–2421. [CrossRef] [PubMed]
13. Kreida, S.; Törnroth-Horsefield, S. Structural insights into aquaporin selectivity and regulation. *Curr. Opin. Struct. Biol.* **2015**, *33*, 126–134. [CrossRef] [PubMed]
14. Jang, J.Y.; Kim, Y.D.; Kim, J.S.; Kang, H. An expression analysis of a gene family encoding plasma membrane aquaporins in response to abiotic stresses in *Arabidopsis thaliana*. *Plant Mol. Biol.* **2004**, *54*, 713–725. [CrossRef] [PubMed]
15. Danielson, J.Å.H.; Johanson, U. Unexpected complexity of the aquaporin gene family in the moss *Physcomitrella patens*. *BMC Plant Biol.* **2008**, *8*, 45. [CrossRef]
16. Ishibashi, K.; Hara, S.; Kondo, S. Aquaporin water channels in mammals. *Clin. Exp. Nephrol.* **2009**, *13*, 107–117. [CrossRef]
17. Henzler, T.; Ye, Q.; Steudle, E. Oxidative gating of water channels (aquaporins) in Chara by hydroxyl radicals. *Plant Cell Environ.* **2004**, *27*, 1184–1195. [CrossRef]
18. Maurel, C. Aquaporins and water permeability of plant membranes. *Annu. Rev. Plant Physiol. Plant Mol. Biol.* **1997**, *48*, 399–429. [CrossRef]
19. Tyerman, S.D.; Bohnert, H.J.; Maurel, C.; Steudle, E.; Smith, J.A.C. Plant aquaporins: Their molecular biology, biophysics and significance for plant water relations. *J. Exp. Bot.* **1999**, *50*, 1055–1071. [CrossRef]
20. Yamaguchi, H.; Fukuoka, H.; Arao, T.; Ohyama, A.; Nunome, T.; Miyatake, K.; Negoro, S. Gene expression analysis in cadmium-stressed roots of a low cadmium-accumulating solanaceous plant, *Solanum torvum*. *J. Exp. Bot.* **2010**, *61*, 423–437. [CrossRef] [PubMed]
21. Gitto, A.; Fricke, W. Zinc treatment of hydroponically grown barley plants causes a reduction in root and cell hydraulic conductivity and isoform-dependent decrease in aquaporin gene expression. *Physiol. Plant.* **2018**, *164*, 176–190. [CrossRef]
22. Kholodova, V.; Volkov, K.; Abdeyeva, A.; Kuznetsov, V. Water status in *Mesembryanthemum crystallinum* under heavy metal stress. *Environ. Exp. Bot.* **2011**, *71*, 382–389. [CrossRef]
23. Tazawa, M.; Asai, K.; Iwasaki, N. Characteristics of Hg- and Zn-sensitive water channels in the plasma membrane of Chara cells. *Bot. Acta* **1996**, *109*, 388–396. [CrossRef]

24. Qiu, Q.S.; Wang, Z.Z.; Cai, Q.G.; Jiang, R.X. Characterization of aquaporins at the plasma membrane of leaf callus protoplasts from *Actinidia deliciosa* cv. Hayward. *Acta Bot. Sin.* **2000**, *42*, 143–147.

25. Yukutake, Y.; Hirano, Y.; Suematsu, M.; Yasui, M. Rapid and reversible inhibition of aquaporin-4 by zinc. *Biochemistry* **2009**, *48*, 12059–12061. [CrossRef]

26. Rygol, J.; Arnold, W.M.; Zimmermann, U. Zinc and salinity effects on membrane transport in *Characonnivens*. *Plant Cell Environ.* **1992**, *15*, 11–23. [CrossRef]

27. Muries, B.; Carvajal, M.; Martínez-Ballesta, M.C. Response of three broccoli cultivars to salt stress, in relation to water status and expression of two leaf aquaporins. *Planta* **2013**, *237*, 1297–1310. [CrossRef]

28. Kasim, W.A. Changes induced by copper and cadmium stress in the anatomy and grain yield of *Sorghum bicolor* (L.) Moench. *Int. J. Agri. Biol.* **2006**, *15*, 123–128.

29. Sagardoy, R.; Vázquez, S.; Florez-Sarasa, I.D.; Albacete, A.; Ribas-Carbó, M.; Flexas, J.; Abadía, J.; Morales, F. Stomatal and mesophyll conductances to CO_2 are the main limitations to photosynthesis in sugar beet (*Beta vulgaris*) plants grown with excess zinc. *New Phytol.* **2010**, *187*, 145–158. [CrossRef]

30. Khan, H.R.; McDonald, G.K.; Rengel, Z. Zinc fertilization and water stress affects plant water relations, stomatal conductance and osmotic adjustment in chickpea (*Cicer arientinum* L.). *Plant Soil* **2004**, *267*, 271–284. [CrossRef]

31. Ahmed, N.; Ahmad, F.; Abid, M.; Aman Ullah, M. Impact of zinc fertilisation on gas exchange characteristics and water use efficiency of cotton crop under arid environment. *Pak. J. Bot.* **2009**, *41*, 2189–2197.

32. Acikgoz, F.E. Seasonal variations on quality parameters of Pak Choi (*Brassica rapa* L. subsp. chinensis L.). *Adv. Crop Sci. Technol.* **2016**, *4*, 4. [CrossRef]

33. Dixon, G.R. Vegetable brassicas and related crucifers. In *Crop Production Science in Horticulture*, 14th ed.; CABI: Walllingford, Oxfordshire, 2007; p. 327.

34. Fang, X.Z.; Zhu, Z.J.; Sun, G.W. Effects of different concentrations of cadmium on growth and antioxidant system in *Brassica campestris* ssp. *Chinensis*. *J. Agro-Environ. Sci.* **2004**, *23*, 877–880.

35. Chen, X.L.; Xu, Y.R.; Cui, X.M.; Wu, X.B. Studies on zinc tolerance and accumulation characteristic of pakchoi (*Brassica campestris* L. ssp. *chinensis* (L.) Makino var. communis Tsen et Lee). *China Vegetables* **2010**, *14*, 19–25.

36. Davis, J.G.; Parker, M.B. Zinc toxicity symptom development and partioning of biomass and zinc in peanut plants. *J. Plant Nutr.* **1993**, *16*, 2353–2369. [CrossRef]

37. Baker, A.J.M. Ecophysiological aspects of zinc tolerance in *Silene maritime*. *New Phytol.* **1978**, *80*, 635–642. [CrossRef]

38. Salah, S.A.; Barrington, S.F. Effect of soil fertility and transpiration rate on young wheat plants (*Triticum aestivum*) Cd/Zn uptake and yield. *Agr. Water Manag.* **2006**, *82*, 177–192. [CrossRef]

39. Llamas, A.; Ullrich, C.I.; Sanz, A. Ni^{2+} toxicity in rice: Effect on membrane functionality and plant water content. *Plant Physiol. Biochem.* **2008**, *46*, 905–910. [CrossRef]

40. Hoagland, D.T.; Arnon, D.I. The water culture method for growing plants without soil. *Calif. Agr. Expt. Sta. Circ.* **1938**, *347*, 1–39.

41. Javot, H.; Lauvergeat, V.; Santoni, V.; Martin-Laurent, F.; Güclü, J.; Vinh, J.; Heyes, J.; Franck, K.I.; Schäffner, A.R.; Bouchez, D.; et al. Role of a single aquaporin isoform in root water uptake. *Plant Cell* **2003**, *15*, 509–522. [CrossRef]

42. Muries, B.; Faize, M.; Carvajal, M.; Martínez-Ballesta, M.C. Identification and differential induction of the expression of aquaporins by salinity in broccoli plants. *Mol. BioSyst.* **2011**, *7*, 1322–1335. [CrossRef] [PubMed]

43. Kayum, M.A.; Park, J.I.; Nath, U.K.; Biswas, M.K.; Kim, H.-T.; Nou, I.-S. Genome-wide expression profiling of aquaporin genes confer responses to abiotic and biotic stresses in *Brassica rapa*. *BMC Plant Biol.* **2017**, *17*, 23. [CrossRef]

44. Livak, K.J.; Schmittgen, T.D. Analysis of relative gene expression data using real-time quantitative PCR and the 2(-Delta Delta C(T)) Method. *Methods* **2001**, *25*, 402–408. [CrossRef]

45. Hansen, T.H.; Laursen, K.H.; Persson, D.P.; Pedas, P.; Husted, S.; Schjoerring, J.K. Micro-scaled high-throughput digestion of plant tissue samples for multi-elemental analysis. *Plant Method.* **2009**, *5*, 1–11. [CrossRef] [PubMed]

46. Mukhopadhyay, M.; Das, A.; Subba, P.; Bantawa, P.; Sarkar, B.; Ghosh, P.D.; Mondal, T.K. Structural, physiological and biochemical profiling of tea plants (*Camellia sinensis* (L.) O. Kuntze) under zinc stress. *Biol. Plant.* **2013**, *57*, 474–480. [CrossRef]

47. Haslett, B.S.; Reid, R.J.; Rengel, Z. Zinc mobility in wheat: Uptake and distribution of zinc applied to leaves or roots. *Ann. Bot.* **2001**, *87*, 379–386. [CrossRef]

48. Rengel, Z. Genotypic difference in micronutrient use efficiency in crops. *Commn. Soil Sci. Plant. Anal.* **2001**, *323*, 1163–1186. [CrossRef]

49. De Silva, N.D.G.; Cholewa, E.; Ryser, P. Effects of combined drought and heavy metal stresses on xylem structure and hydraulic conductivity in red maple (*Acer rubrum* L.). *J. Exp. Bot.* **2012**, *63*, 5957–5966. [CrossRef]

50. Vitali, V.; Bellati, J.; Soto, G.; Ayub, N.D.; Amodeo, G. Root hydraulic conductivity and adjustments in stomatal conductance: Hydraulic strategy in response to salt stress in a halotolerant species. *AoB Plants* **2015**, *7*, plv136. [CrossRef]

51. Ohki, K. Effect of Zinc nutrition on photosynthesis and carbonic anhydrase activity in cotton. *Physiol. Plant.* **1976**, *38*, 300–304. [CrossRef]

52. Arif, N.; Yadav, V.; Singh, S.; Singh, S.; Ahmad, P.; Mishra, R.K.; Sharma, S.; Tripathi, D.K.; Dubey, N.K.; Chauhan, D.K. Influence of high and low levels of plant-beneficial heavy metal ions on plant growth and development. *Front. Environ. Sci.* **2016**, *4*, 69. [CrossRef]

53. Sagardoy, R.; Morales, F.; Lopez-Millan, A.F.; Abadıa, A.; Abadıa, J. Effects of zinc toxicity on sugar beet (*Beta vulgaris* L.) plants grown in hydroponics. *Plant Biol.* **2009**, *11*, 339–350. [CrossRef]

54. Vassilev, A.; Nikolova, A.; Koleva, L.; Lidon, F. Effects of excess Zn on growth and photosynthetic performance of young bean plants. *J. Phytol.* **2011**, *3*, 58–62.

55. Januškaitienė, I. Impact of low concentration of cadmium on photosynthesis and growth of pea and barley. *Environ. Res. Eng. Manag.* **2010**, *3*, 24–29.

56. White, P.; Broadley, M.R. Calcium in plants. *Ann. Bot.* **2003**, *92*, 487–511. [CrossRef]

57. Kinraide, T.B.; Pedler, J.F.; Parker, D.R. Relative effectiveness of calcium and magnesium in the alleviation of rhizotoxicity in wheat induced by copper, zinc, aluminum, sodium, and low pH. *Plant Soil* **2004**, *259*, 201–208. [CrossRef]

58. Kobayashi, Y.; Kobayashi, Y.; Watanabe, T.; Shaff, J.F.; Ohta, H.; Kochian, L.V.; Wagatsuma, T.; Kinraide, T.B.; Koyama, H. Molecular and physiological analysis of Al^{3+} and H^+ rhizotoxicities at moderately acidic conditions. *Plant Physiol.* **2013**, *163*, 180–192. [CrossRef]

59. Pandey, P.; Srivastava, R.K.; Dubey, R.S. Salicylic acid alleviates aluminum toxicity in rice seedlings better than magnesium and calcium by reducing aluminum uptake, suppressing oxidative damage and increasing antioxidative defense. *Ecotoxicology* **2013**, *22*, 656–670. [CrossRef]

60. Prasad, R.; Shivay, Y.S.; Kumar, D. Interactions of zinc with other nutrients in soils and plants—A review. *Indian J. Fertil.* **2016**, *12*, 16–26.

61. Khan, I.; Ghani, A.; Rehman, A.U.; Awan, S.A.; Noreen, A.; Khalid, I. Comparative analysis of heavy metal profile of *Brassica campestris* (L.) and *Raphanus sativus* (L.) irrigated with municipal waste water of Sargodha city. *J. Clin. Toxicol.* **2016**, *6*, 307. [CrossRef]

62. Azad, A.K.; Sawa, Y.; Ishikawa, T.; Shibata, H. Phosphorylation of plasma membrane aquaporin regulates temperature-dependent opening of tulip petals. *Plant Cell Physiol.* **2004**, *45*, 608–617. [CrossRef]

63. Jia, X.Y.; Xu, C.Y.; Jing, R.L.; Li, R.Z.; Mao, X.G.; Wang, J.P.; Chang, X.P. Molecular cloning and characterization of wheat calreticulin (CRT) gene involved in drought stressed responses. *J. Exp. Bot.* **2008**, *59*, 739–751. [CrossRef]

64. Martínez-Ballesta, M.C.; Cabañero, F.; Olmos, E.; Periago, P.M.; Maurel, C.; Carvajal, M. Two different effects of calcium on aquaporins in salinity-stressed pepper plants. *Planta* **2008**, *228*, 15–25. [CrossRef]

65. Ariani, A.; Barozzi, F.; Sebastiani, L.; Di Toppi, L.S.; Di Sansebastiano, G.P.; Andreucci, A. AQUA1 is a mercury sensitive poplar aquaporin regulated at transcriptional and post-translational levels by Zn stress. *Plant Physiol. Biochem.* **2019**, *135*, 588–600. [CrossRef]

66. Przedpelska-Wasowicz, E.M.; Wierzbicka, M. Gating of aquaporins by heavy metals in *Allium cepa* L. epidermal cells. *Protoplasma* **2011**, *248*, 663–671. [CrossRef]

67. Rucińska-Sobkowiak, R. Water relations in plants subjected to heavy metal. *Acta Physiol. Plant.* **2016**, *38*, 257. [CrossRef]

68. Aroca, R.; Ferrante, A.; Vernieri, P.; Chrispeels, M.J. Drought, abscisic acid and transpiration rate effects on the regulation of PIP aquaporin gene expression and abundance in *Phaseolus vulgaris* plants. *Ann. Bot.* **2006**, *98*, 1301–1310. [CrossRef]

69. Maurel, C.; Verdoucq, L.; Luu, D.T.; Santoni, V. Plant aquaporins: Membrane channels with multiple integrated functions. *Annu. Rev. Plant Biol.* **2008**, *59*, 595–624. [CrossRef]

70. Guo, L.; Wang, Z.Y.; Lin, H.; Cui, W.E.; Chen, J.; Liu, M.; Chen, Z.L.; Qu, L.J.; Gu, H. Expression and functional analysis of the rice plasma-membrane intrinsic protein gene family. *Cell Res.* **2006**, *16*, 277–286. [CrossRef]

71. Sreedharan, S.; Shekhawat, U.K.S.; Ganapathi, T.R. Transgenic banana plants overexpressing a native plasma membrane aquaporin MusaPIP1;2 display high tolerance levels to different abiotic stresses. *Plant Biotechnol. J.* **2013**, *11*, 942–952. [CrossRef]

72. Xu, Y.; Hu, W.; Liu, J.; Zhang, J.; Jia, C.; Miao, H.; Xu, B.; Jin, Z. A banana aquaporin gene, MaPIP1;1, is involved in tolerance to drought and salt stresses. *BMC Plant Biol.* **2014**, *14*, 59. [CrossRef]

73. Wang, L.; Liu, Y.; Feng, S.; Yang, J.; Li, D.; Zhang, J. Roles of plasmalemma aquaporin gene *StPIP1* in enhancing drought tolerance in potato. *Front. Plant Sci.* **2017**, *8*, 616. [CrossRef]

74. Fetter, K.; Van, V.W.; Moshelion, M.; Chaumont, F. Interactions between plasma membrane aquaporins modulate their water channel activity. *Plant Cell* **2004**, *16*, 215–228. [CrossRef]

75. Alexandersson, E.; Fraysse, L.; Sjövall-Larsen, S.; Gustavsson, S.; Fellert, M.; Karlsson, M.; Johanson, U.; Kjellbom, P. Whole gene family expression and drought stress regulation of aquaporins. *Plant Mol. Biol.* **2005**, *59*, 469–484. [CrossRef]

76. Abdeeva, A.R.; Kholodova, V.P.; Kuznetsov, V.V. Expression of aquaporin genes in the common ice plant during induction of the water-saving mechanism of CAM photosynthesis under salt stress conditions. *Dokl. Biol. Sci.* **2008**, *418*, 30–33. [CrossRef]

77. Vandeleur, R.K.; Sullivan, W.; Athman, A.; Jordans, C.; Gilliham, M.; Kaiser, B.N.; Tyerman, S.D. Rapid shoot-to-root signalling regulates root hydraulic conductance via aquaporins. *Plant Cell Environ.* **2014**, *37*, 520–538. [CrossRef]

78. Da Ines, O. Functional Analysis of PIP2 Aquaporins in *Arabidopsis thaliana*. Ph.D. Thesis, Ludwig Maximilians Universität, Munich, Germany, 2008.

79. Boursiac, Y.; Chen, S.; Luu, D.T.; Sorieul, M.; van den Dries, N.; Maurel, C. Early effects of salinity on water transport in Arabidopsis roots. Molecular and cellular features of aquaporin expression. *Plant Physiol.* **2005**, *139*, 790–805. [CrossRef]

80. Maathuis, F.J.; Filatov, V.; Herzyk, P.; Krijger, G.C.; Axelsen, K.B.; Chen, S.; Green, B.J.; Li, Y.; Madagan, K.L.; Sánchez-Fernández, R.; et al. Transcriptome analysis of root transporters reveals participation of multiple gene families in the response to cation stress. *Plant J.* **2003**, *35*, 675–692. [CrossRef]

81. Weig, A.; Deswarte, C.; Chrispeels, M.J. The major intrinsic protein family of *Arabidopsis* has 23 members that form three distinct groups with functional aquaporins in each group. *Plant Physiol.* **1997**, *114*, 1347–1357. [CrossRef]

82. Pih, K.T.; Kabilan, V.; Lim, J.H.; Kang, S.G.; Piao, H.L.; Jin, J.B.; Hwang, I. Characterization of two new channel protein genes in *Arabidopsis*. *Mol. Cells* **1999**, *9*, 84–90.

83. Mosa, K.A.; Kumar, K.; Chhikara, S.; Mcdermott, J.; Liu, Z.; Musante, C.; White, J.C.; Dhankher, O.P. Members of rice plasma membrane intrinsic proteins subfamily are involved in arsenite permeability and tolerance in plants. *Transgenic Res.* **2012**, *21*, 1265–1277. [CrossRef] [PubMed]

84. Kumar, K.; Mosa, K.; Chhikara, S.; Musante, C.; White, J.C.; Dhankher, O.P. Two rice plasma membrane intrinsic proteins, OsPIP2;4 and OsPIP2;7, are involved in transport and providing tolerance to boron toxicity. *Planta* **2014**, *239*, 187–198. [CrossRef] [PubMed]

85. Markakis, M.N.; De Cnodder, T.; Lewandowski, M.; Simon, D.; Boron, A.; Balcerowicz, D.; Doubbo, T.; Taconnat, L.; Renou, J.P.; Höfte, H.; et al. Identification of genes involved in the ACC-mediated control of root cell elongation in *Arabidopsis thaliana*. *BMC Plant Biol.* **2012**, *12*, 208. [CrossRef] [PubMed]

86. Dynowski, M.; Schaaf, G.; Loque, D.; Moran, O.; Ludewig, U. Plant plasma membrane water channels conduct the signalling molecule H_2O_2. *Biochem. J.* **2008**, *414*, 53–61. [CrossRef]

87. Bienert, G.P.; Møller, A.L.B.; Kristiansen, K.A.; Schulz, A.; Møller, I.M.; Schjoerring, J.K.; Jahn, T.P. Specific aquaporins facilitate the diffusion of hydrogen peroxide across membranes. *J. Biol. Chem.* **2007**, *282*, 1183–1192. [CrossRef]

88. Neill, S.J.; Desikan, R.; Hancock, J.T. Hydrogen peroxide signalling. *Curr. Opin. Plant Biol.* **2002**, *5*, 388–395. [CrossRef]

89. Ben Rejeb, K.; Benzarti, M.; Debez, A.; Bailly, C.; Savouré, A.; Abdelly, C. NADPH oxidase-dependent H_2O_2 production is required for salt-induced antioxidant defense in *Arabidopsis thaliana*. *J. Plant Physiol.* **2015**, *174*, 5–15. [CrossRef]
90. Luo, Z.B.; He, X.J.; Chen, L.; Tang, L.; Gao, S.; Chen, F. Effects of zinc on growth and antioxidant responses in *Jatropha curcas* seedlings. *Int. J. Agric. Biol.* **2010**, *12*, 119–124.

Article

New Eco-Friendly Polymeric-Coated Urea Fertilizers Enhanced Crop Yield in Wheat

Ricardo Gil-Ortiz [1,*], Miguel Ángel Naranjo [1,2], Antonio Ruiz-Navarro [3], Marcos Caballero-Molada [1,2], Sergio Atares [2], Carlos García [3] and Oscar Vicente [4]

[1] Institute for Plant Molecular and Cell Biology (UPV-CSIC), Universitat Politècnica de València, 46022 Valencia, Spain; mnaranjo@ibmcp.upv.es (M.Á.N.); marcamo2@ibmcp.upv.es (M.C.-M.)

[2] Fertinagro Biotech S.L., Polígono de la Paz, C/ Berlín s/n, 44195 Teruel, Spain; sergio.atares@tervalis.com

[3] Centre for Soil and Applied Biology Science of Segura (CEBAS-CSIC), Espinardo University Campus, 30100 Murcia, Spain; ruiznavarro@cebas.csic.es (A.R.-N.); cgarizq@cebas.csic.es (C.G.)

[4] Institute for the Preservation and Improvement of Valencian Agrodiversity (COMAV), Universitat Politècnica de València, 46022 Valencia, Spain; ovicente@upvnet.upv.es

* Correspondence: rigilor@ibmcp.upv.es

Received: 5 March 2020; Accepted: 19 March 2020; Published: 23 March 2020

Abstract: Presently, there is a growing interest in developing new controlled-release fertilizers based on ecological raw materials. The present study aims to compare the efficacy of two new ureic-based controlled-release fertilizers formulated with water-soluble polymeric coatings enriched with humic acids or seaweed extracts. To this end, an experimental approach was designed under controlled greenhouse conditions by carrying out its subsequent field scaling. Different physiological parameters and crop yield were measured by comparing the new fertilizers with another non polymeric-coated fertilizer, ammonium nitrate, and an untreated 'Control'. As a result, on the microscale the fertilizer enriched with humic acids favored a better global response in the photosynthetic parameters and nutritional status of wheat plants. A significant 1.2-fold increase in grain weight yield and grain number was obtained with the humic acid polymeric fertilizer versus that enriched with seaweed extracts; and also, in average, higher in respect to the uncoated one. At the field level, similar results were confirmed by lowering N doses by 20% when applying the humic acid polymeric-coated produce compared to ammonium nitrate. Our results showed that the new humic acid polymeric fertilizer facilitated crop management and reduced the environmental impact generated by N losses, which are usually produced by traditional fertilizers.

Keywords: coated-urea fertilizer; humic acid; lignosulfonate; natural polymers; seaweed extract; wheat

---

1. Introduction

According to the Food and Agriculture Organization of the United Nations (FAO), wheat is the world's largest cultivated crop per hectare and the third largest cereal to be produced [1]. In the European Union, 144.5 million tons were harvested in 2016 and production is estimated to increase by 3.5% each year. In fact, the world's production in 2017 was expected to come to 744.5 million tons, an increase that comes close to 1000% since 1990/1991. Current cereal production demand and gradual soil impoverishment mean that it is increasingly necessary to apply more fertilizers, mainly nitrogenous ones [2]. High quantities of nitrogen (N) per hectare need to be applied to soil to produce optimum wheat grain yields [3]. N-organic mineralization in soil is a slow process that requires the action of soil microorganisms and must also be given the necessary environmental conditions [4,5]. Plants absorb N in the form of exchangeable ammonium (NH_4^+) or nitrates (NO_3^-), which are highly mobile compounds in soil that can be easily lost by volatilization or leaching, which leads to environmental problems and/or toxicity for plants [6,7]. The legal restrictions associated with pollution limitations have been set to preserve the environment [8]. Traditional N fertilizers, such as sulfates, nitrates, or urea, are characterized by high constant N-release kinetics [9–11]. Fertilizer granules rapidly decompose and a strong N release occurs when applied to soil. Later emissions slow down and additional covert applications are usually necessary for crops to achieve expected yields [12]. This lack of efficiency implies figures like 90% for N loss of the total N applied [8,13,14]. N fertilization efficiency depends on different variables, such as environmental conditions, coupling soil/plant or management practices [9,15].

Fertilizer manufacturers have concentrated in recent decades to produce slow-release and controlled-release fertilizers (SRFs/CRFs) as enhanced-efficiency nitrogen fertilizers (EENF) [16]. SRFs are long-chain molecules of lower solubility than traditional fertilizers like formaldehyde, isobutylene diurea, or methylene urea, for which the biodegradability is proportional to the microbiological activity of the soil. CRFs action, on the other hand, depends mainly on diffusion through coatings and not directly on biodegradation, thus being more efficient in controlling release of nutrients [17,18]. The main advantages of these slow- or controlled-release fertilizer generations are summarized in numerous reviews [8,9,18]: (1) extending the durability of fertilizers by providing small amounts for a longer time; (2) lowering the number of fertilizer applications, generally to a single background application, by prolonging their time of action; (3) cutting costs by eliminating the typical covert applications of traditional fertilizers; (4) reducing environmental pollution by limiting the amount of fertilizer released being assimilated in soil/the plant system. Urea is the major N source used in plant nutrition [19]. Synthetic SRF urea formulations are, for example, urea formaldehyde, isobutylidene diurea, crotonylidene diurea, or sulfur-coated urea fertilizers [20]. Today, controlled-released coated urea (CRCU) is the most important technology being developed in the fertilizer industry [13,18]. To manufacture CRCU, urea is usually coated with polymers that control N release by diffusion based on the permeability of polymer coatings [13,18,21]. Release of N from polymeric CRCU is not significantly influenced by microorganisms of soils because nutrient release can be better controlled compared to sulfur-based coatings [18]. In fact, emissions are influenced mainly by environmental factors like temperature or humidity [17,22] and also by intrinsic factors of fertilizers, such as nutrient composition, coating thickness, granular shape, and diameter [17,18].

In recent decades, the use of synthetic polymers, like those based on sulfur, resins, or thermoplastic materials, has been hampered by legal restrictions that limit pollution due to these materials' difficult degradation [13]. Such products are used in many other industries to manufacture pesticides, herbicides, pheromones, fungicides or growth regulators [8,23,24]. Their marketed forms are encapsulations, reservoir laminate structures, or monolithic systems [8]. Carbohydrate and lignin-based polymer-coated urea has been indicated as an alternative to solve problems related to N emissions of traditional fertilizers and to avoid environmental problems concerning synthetic polymers [18,25,26]. In fact, they are ecologically friendly and easily available at cheap prices, which are the main restrictions of using CRFs. For example, coatings based on starch, ethyl cellulose, or lignin have been successfully employed to slow down N release from urea [13,26,27]. Including bio-inhibitors as urease or nitrification inhibitors in fertilizers is a commonplace practice to slow down N releases [2,9,28]. Biostimulants like amino acids, humic/fulvic acids or seaweed extracts offer beneficial properties for crops, such as biofortification and resistance to different abiotic stresses, e.g., drought or salinity [29–33].

At the beginning of their development, SCRFs were limited mainly to horticultural and ornamental crops, and actually are not well established in extensive cropping as more research is necessary to perform cheaper and more ecological fertilizers [34–36]. The objective of this research was to compare the efficacy of new ecological CRCU with traditional fertilizers in physiological terms, and also grain yield and quality in wheat. The novelty of this research lies in the combination of eco-friendly polymers as byproducts from the production of wood pulp, urease inhibitors, and natural biostimulants in the same fertilizer.

2. Materials and Methods

2.1. Experimental Design

A comparison of the effectiveness of the different polymeric-coated formulations was made under greenhouse conditions and then scaled to field essays. Experiment 1 (*Triticum aestivum*). Wheat was grown in a greenhouse at ambient temperature and humidity at the facilities of the Valencian Institute of Agrarian Research (IVIA) (Moncada, Spain) from autumn 2014 to spring 2015. Sowing was carried out in pots (22 cm high × 16 cm Ø) placed in watertight trays, sized 54 × 39 × 9 cm (6 pots/tray) with three repetitions per treatment on 7 November. The pots were filled with a non-fertilized soil from a fallow area close to the greenhouse in IVIA's grounds. Culture was developed with extra lighting for a 12:12 photoperiod and irrigated with distilled water. High temperatures were reduced and partly controlled by means of automatic systems for shading, ventilation, and water cooling. Temperature and humidity were measured every hour with a digital thermo-hygrometer. The average temperatures during the whole culture cycle, from November to May were 18.8 ± 7.2 °C (39 °C max. and 4.9 °C min.), and average relative humidity 55.3 ± 12.8% (88.0% max. and 20.4% min.). Experiment 2 (*Triticum spelta*). Three separated grids of 400 m^2, divided into 16 individual surfaces of 25 m^2 each, were designed at the field level in a plot located in Teruel (Spain) at GPS coordinates 40°22′17.4″ N, 1°05′58.9″ W. Four treatments, including an untreated Control, were placed by quadruplicate. Sowing was carried out mechanically in the winter at a dose of 250,000 seeds/ha. Culture was surface-irrigated on a bi-weekly basis with well water.

2.2. Fertilizer Treatments

Different N fertilizers developed by Fertinagro Biotech S.L. (Teruel, Spain) were tested and their efficacy was compared. As these fertilizers are patented, the exact composition is not herein presented. To study the influence of the different coating compositions and thicknesses, the following fertilizers were tested in Experiment 1: (1) DURAMON® (Fertinagro Biotech S.L., Teruel, Spain), composed of urea, including a urease inhibitor (monocarbamida dihidrogenosulfate—MCDHS) with no coating (ES 2 204 307 patent); (2) a new controlled-released urea fertilizer based on DURAMON® technology, but also 3% lignosulfonate-coated with humic acids (hereafter CRF_A); (3) the same as (2), but 5% lignosulfonate-coated with seaweed extracts (CRF_B). The three formulates had the same N composition (24–0–0), but a different coverture percentage. Fertilizers were applied in wheat at doses of 150 kg ha^{-1} (nitrogen fertilizer units—NFU) as a basal dressing for maximum yields based on theoretical extraction by crops. Experiment 2: Based on the physiological responses observed in the greenhouse experiment, the best CRF was selected and applied to the field at 100% doses and with a reduction to 80%. Both doses were compared with ammonium nitrosulfate (NSA, 26–0–0) (Fertinagro Biotech S.L., Teruel, Spain) as the traditional fertilizer. Maximum doses of 80 kg ha^{-1} were applied in the phenological state of tillering (27 April), based on the historical average yields obtained in the area. In both experiments, the same repetitions with untreated plants were included (CONTROL).

2.3. Soil Fertility Characterization

Several soil properties were measured to characterize soil fertility in both experiments. pH and EC were determined in a 1/5 (*w/v*) aqueous soil extract by shaking for 2 h, followed by centrifugation at 26916 g for 15 min and filtration. pH was measured by a pH meter (Crison mod.2001, Barcelona, Spain) and EC with a Conductivity meter (Crison micro CM2200, Barcelona, Spain). Total and organic soil C (SOC) and total N (N) were determined by combustion gas chromatography in a Flash EA 1112 Thermo Finnigan (Franklin, MA, USA) elemental analyzer after eliminating carbonate by acid digestion with HCl. The total nutrient contents (P, K, Ca, Mg, Cu, Fe, K, Mg, Mn, and Zn) were extracted by aqua regia digestion (3:1, *v/v*, HCl/HNO_3) and determined by ICP-AES (Thermo Elemental Iris Intrepid II XDL, Franklin, MA, USA). Analysis showed that both cultures grew on N-poor soil (Table 1).

Table 1. Fertility of the soil used in the experimental analysis from the first 15 cm of soil surface. Data on total nitrogen (N), total carbon (C) and organic carbon (CO) and other macro- and micronutrients are shown. Values are means ± SD ($n = 5$) at the beginning of the experiment.

Parameters	Mean ± SD (%)	
	Microscale	Field
Total nitrogen (g 100 g^{-1})	0.09 ± 0.02	0.19 ± 0.03
Total carbon (g 100 g^{-1})	2.06 ± 0.28	6.76 ± 0.34
Organic carbon (g 100 g^{-1})	0.66 ± 0.07	1.84 ± 0.28
pH	8.75 ± 0.095	8.26 ± 0.11
EC (µS cm^{-1})	120.7± 27.99	109.12 ± 47.40
P (g 100 g^{-1})	0.064 ± 0.001	0.03 ± 0.01
K (g 100 g^{-1})	0.339 ± 0.14	0.74 ± 0.08
Mg (g 100 g^{-1})	0.285 ± 0.04	0.26 ± 0.04
Ca (g 100 g^{-1})	3.561 ± 0.42	1.35 ± 0.61
Fe (g 100 g^{-1})	11.113 ± 1.77	15.94 ± 6.27
Cu (mg kg^{-1})	15.929 ± 1.45	7.46 ± 1.81
Mn (mg kg^{-1})	191.99 ± 11.68	216.75 ± 73.57
Zn (mg kg^{-1})	26.951 ± 2.50	24.66 ± 4.61

2.4. Plant Growth

Photosynthetically active flag leaves (PAFL) were characterized in the phenological state of panicles swelling (booting stage) by fresh weight (g) and foliar surface (cm^2) using a LI-3100C area meter (LI-COR®, Lincoln, Nebraska, USA). Some plant material was weighed before being dried at 65 °C until a constant mass was obtained to determine dry weight (g). Differences in the total dry weight, length (cm), primary stem length (cm), and tillers number were determined at the end of the culture.

2.5. Leaf Greenness and Effective Quantum Yield of Photosystem II

Leaf greenness was measured in the booting stage using an SPAD-502 Chlorophyll meter (Konica-Minolta, Osaka, Japan). The effective quantum yield of photosystem II electron transport (ΦPSII), which represents the electron transport efficiency between photosystems within light-adapted leaves, was checked with a leaf fluorometer (Fluorpen FP100, Photos System Instrument, Drásov, Czech Republic). Both parameters were measured in a minimum of 25 PAFL.

2.6. Gas Exchange Analysis

Gas exchange measurements were taken at noon in five plants per treatment using a portable infrared gas analyzer LCpro-SD, equipped with a PLU5 LED light unit (ADC BioScientific Ltd., Hoddesdon, UK). The selected flag leaves in wheat (booting stage) of Experiment 1 were analyzed in a leaf chamber (6.25 cm^2) to determine the following parameters: stomatal conductance (*gs*) (expressed as mmol m^{-2} s^{-1}), net photosynthetic rate (*A*) (µmol m^{-2} s^{-1}), transpiration (*E*) (mol m^{-2} s^{-1}), and intercellular CO_2 concentration (*Ci*) (µmol mol^{-1}) under ambient CO_2, temperature, and relative humidity conditions. They were recorded by programming increasing photosynthetically active radiations (PAR) of 348, 522, 696, 870, 1218, and 1566 µmol m^{-2} s^{-1}. Water-use efficiency (WUE) and intrinsic WUE were calculated as the ratio between *A/gs* and *A/E*, expressed as µmol (CO_2 assimilated) mol^{-1} (H_2O transpired).

2.7. Foliar Nutrient Analysis

Foliar analyses were performed from the fresh samples collected in the phenological state of panicles swelling (booting stage) 70 days after plants emerged. Samples were composed of a pool with a minimum of four flag leaves taken from different plants in the same treatment. Four replicates per treatment and culture were collected and kept at −80 °C until they were biochemically analyzed. The compositions in macro- (N, P, K, Ca, Mg and S) and micronutrients (Fe, Cu, Mn, Zn, B, and Mo) were determined by Inductively Coupled Plasma Optical Emission Spectrometry (ICP-OES). N content was estimated by an N-Pen N 100 apparatus (Photon System Instruments, Drásov, Czech Republic).

2.8. Growth, Yield, and Cereal Grain Composition

Once grain ripening had been completed at 138 days in greenhouses after emergence, the remaining plants per culture were harvested and characterized in growth and grain yield terms. The growth parameters of the total dry weight of aerial parts, primary stem length, tillers number, ears number, and ear length and weight were measured. Yield was determined by measuring the total dry grain weight, one-hundred grain weight and grain number. Nitrogen Use Efficiency (NUE) for each fertilizer treatment was calculated as agronomic efficiency according to [37]:

$$\text{NUE} \left(\text{kg kg}^{-1} \right) = \frac{\text{Grain yield of the fertilized area } - \text{ Grain yield of the unfertilized area}}{\text{Quantity of N applied as N fertilizer}} \tag{1}$$

At the field level and 90 days after applying fertilizers, the biomass of aerial parts, ears weight, and grain weight was studied. The Harvest Index (HI) was calculated as the grain weight/biomass of aerial parts. Different quality parameters were also measured in the grain. A representative composite sample was prepared separately for each treatment, pooling fractions of plant material for each replication. Subsequently, each composite mixture was ground and analyzed based on food quality analysis methods (Comission Regulation EC N° 152/2009 of 27 January): humidity (gravimetric by drying in an oven at 130 °C), ashes (gravimetric by incineration at 550 °C), lipids (extraction without hydrolisis in Soxtec Avanti—Foss), protein (Kjeldahl method using Foss automatic distillation equipment, Foss, Hillerød, Denmark), crude fiber (gravimetric), and total carbohydrates (volumetric using Luff Schoorl reagent). Analysis were carried out by the Valencia's Agrifood Laboratory (Burjassot, Spain).

2.9. Statistics

The differences between fertilizers treatments were tested by analysis of variance (ANOVA) at 95% confidence. Prior to the ANOVA, the data requirements of normality and homogeneity of variances were checked according to Levene's and Shapiro–Wilk tests. When the null ANOVA hypothesis was rejected, post hoc comparisons were made to establish the possible statistical differences among the different treatments applied using the Fisher's LSD test. The statistical Statgraphics Centurion XV, version 15.2.05 software program (Statpoint Technologies, Inc., Warrenton, VA, USA) was used to perform the analysis.

3. Results

3.1. Plant Growth, Leaf Greenness and Effective Quantum Yield of Photosystem II

No significant differences were observed for the various treatments performed for chlorophyll content and ΦPSII, measured by nondestructive techniques in the phenological state of panicles swelling (booting stage), although they were significant compared to the CONTROL (Table 2). Similar results were obtained when studying PAFL fresh weight content, dry weight and area. The total fresh weight of aerial parts was 1.4- and 1.7-fold significantly higher for the plants fertilized with DURAMON® compared to CRF$_A$ and CRF$_B$, but no differences were observed for dry weight. The responses of fertilizer treatments on plant growth, foliar area, and root development are shown in Figure 1.

Table 2. Effects of fertilizer treatments CRF$_A$, CRF$_B$, and DURAMON® on photosynthetic parameters (effective quantum yield of photosystem II—ΦPSII, leaf greenness, nitrogen content, total fresh weight of aerial parts (g), dry weight of aerial parts (%), photosynthetically active flag leaves—PAFL fresh weight, PAFL area, and leaf area index—LAI) in *Triticum aestivum* leaves compared to the Control in the phenological state of panicles swelling (booting state). Values are means ± SD (*n* = 18 for ΦPSII, leaf greenness, and N content; *n* = 8 for the other growth parameters).

Parameters	CRF$_A$	CRF$_B$	DURAMON®	CONTROL
ΦPSII	0.69 ± 0.03 b	0.69 ± 0.03 b	0.68 ± 0.03 b	0.58 ± 0.07 a
Leaf greenness content (SPAD units)	54.5 ± 2.3 b	52.5 ± 2.3 b	54 ± 1.5 b	40.2 ± 6.8 a
N content (%)	5.5 ± 0.4 b	5 ± 0.3 ab	5.3 ± 0.4 ab	3.5 ± 0.9 a
Total fresh weight (aerial part) (g)	58.9 ± 9.3 b	46.9 ± 20.9 ab	79.6 ± 15.9 c	38.5 ± 10.5 a
Dry weight (aerial part) (%)	29 ± 8 a	33.3 ± 7.6 ab	35.4 ± 3.2 ab	32.9 ± 3.5 b
PAFL fresh weight (g)	24.5 ± 12.7 b	26.6 ± 25.5 b	27.2 ± 9.2 b	2.8 ± 3.8 a
PAFL dry weight (%)	6.13 ± 3.02 b	6.67 ± 5.81 b	7.23 ± 2.09 b	0.89 ± 1.13 a
PALF area (cm^2)	200.2 ± 69.8 b	156.9 ± 96.7 b	229.5 ± 72.7 b	67.6 ± 43.1 a
LAI	1 ± 0.3 b	0.8 ± 0.5 b	1.1 ± 0.4 b	0.3 ± 0.2 a

[1] Different letters in the same row indicate significant statistical differences (Fisher's LSD test, $P < 0.05$).

Figure 1. Responses of fertilizer treatments on plant growth (**A,B**), foliar area (**C,D**) and root development (**E**) of *Triticum aestivum* in the phenological state of panicles swelling (**A,C,D,E**) and at the end of the experiment (**B**). Treatments from left to right: CRF$_A$, CRF$_B$, DURAMON®, and CONTROL. Bars correspond to 10 cm.

3.2. Gas Exchange Analysis

Significant increases in *gs*, *A*, and *E* were noted in the plants treated with the different fertilizers as increasing PAR levels were applied from 348 to 1566 μmol m^{-2} s^{-1} (Figure 2A–C). Conversely, *Ci* showed a decreasing tendency in all the applied treatments (Figure 2D).

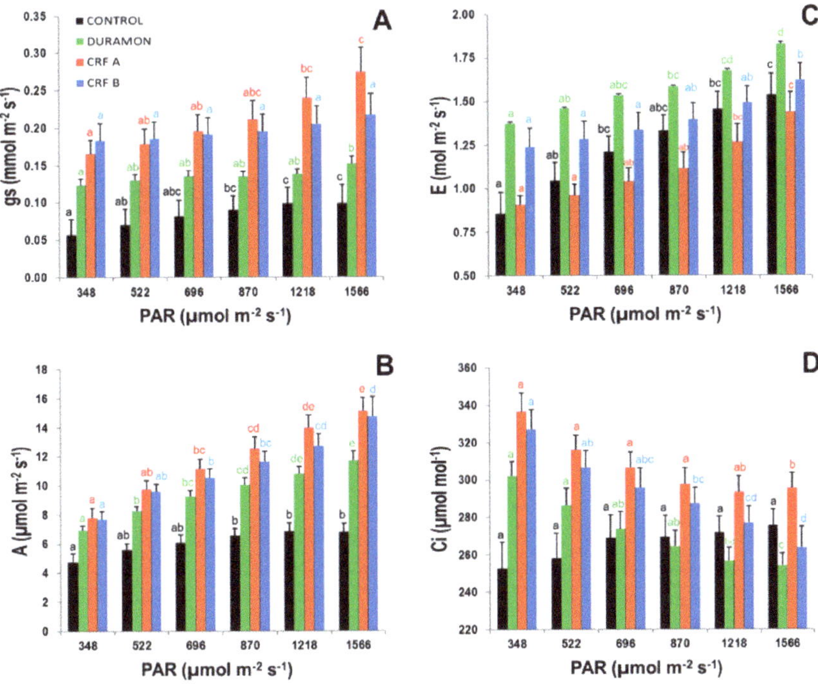

Figure 2. Gas exchange responses to different photosynthetically active radiation rates of *Triticum aestivum* treated with fertilizers CRF$_A$, CRF$_B$, DURAMON®, and CONTROL in the phenological state of panicles swelling (booting stage). (**A**) Stomatal conductance, (**B**) net photosynthetic rate, (**C**) transpiration rate, and (**D**) substomatal CO$_2$ concentration. Values represent means $\pm$ SE ($n = 6$). Different letters for each treatment (same color) indicate statistically significant differences (ANOVA, $P < 0.05$).

The levels of *gs*, *A* and *Ci* for CRF$_A$ were higher than those found for CRF$_B$ and DURAMON® but were lower for *E*. Significant differences were found in all the studied gas exchange parameters between CRFs and DURAMON®. After the maximum PAR application of 1566 μmol m^{-2} s^{-1}, the levels of *A* did not significantly differ for both CRFs, but *gs* and *Ci* were 1.3- and 1.1-fold significantly higher for CRF$_A$ than for CRF$_B$. At the same PAR, the *A* levels were 1.3-fold significantly higher for CRFs than for DURAMON®. The *gs* and *A* levels for the CONTROL plants were significantly lower at all the studied PAR compared to those of the different treatments. However, the *E* and *Ci* levels for the CONTROL plants were only significantly different for DURAMON®, as was *Ci* with CRF$_A$. *A/gs* significantly differed when globally comparing CRFs with DURAMON® and the CONTROL (Figure 3A), but the *A/E* levels were significantly higher for CRF$_A$ than for CRF$_B$, DURAMON® and the CONTROL, which also significantly differed from one another. The maximum *A/E* levels were produced within the 870 to 1218 μmol m^{-2} s^{-1} range (Figure 3B). The *A/E* levels in the CRF$_A$ leaves were 1.3- and 1.8-fold significantly higher than CRF$_B$ and DURAMON® at a PAR of 870 μmol m^{-2} s^{-1}.

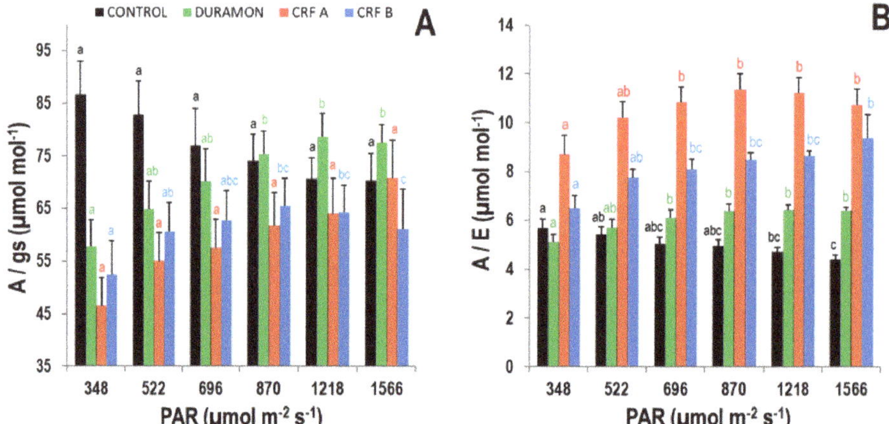

Figure 3. Water-use efficiency responses to different photosynthetically active radiation rates of *Triticum aestivum* treated with fertilizers CRF$_A$, CRF$_B$, DURAMON®, and the CONTROL in the phenological state of panicles swelling (booting stage). (**A**) Water-use efficiency and (**B**) intrinsic water-use efficiency. Values represent means ± SE (n = 6). Different letters for each treatment (same color) indicate statistically significant differences (ANOVA, $P < 0.05$).

3.3. Foliar Nutrient Content

No significant differences appeared in the PAFL macronutrient concentrations of N, K, Ca, Mg, and S at the beginning of ears formation (Figure 4A). The plants fertilized with CRF$_A$ presented 1.1- and 1.2-fold higher foliar N average levels than CRF$_B$ and DURAMON®. The foliar P concentrations in the plants fertilized with CRF$_A$ were 1.2-fold significantly higher than for CRF$_B$ and DURAMON®. On average, the plants fertilized with CRF$_B$ presented slightly higher contents of K, Ca, Mg, and S. Regarding micronutrient foliar contents, the plants fertilized with CRF$_A$ presented 1.7- and 1.5-fold higher Fe levels than CRF$_B$ and DURAMON®, respectively (Figure 4B). Cu and Zn contents were 1.4- and 1.2-fold significantly higher in the plants fertilized with CRFs than in those fertilized with DURAMON®, respectively. Foliar Mn concentrations did not differ significantly among the distinct fertilizer treatments and the CONTROL levels came close to the critical thresholds. The foliar B levels were significantly higher in the plants treated with CRF$_B$ compared to those treated with DURAMON®, but the quantitative CONTROL levels came close to CRF$_B$. Mo in PAFL content was < 2 mg kg^{-1} DW (dry weight) in all the treatments and the CONTROL.

No clear correspondence was obtained for the macro- and micronutrient concentrations quantified at the foliar level compared to those found in roots (Figure 4C, 4D). Quantitatively, the root N concentrations were around half of those obtained at the foliar level. The remaining macronutrient contents were slightly lower in roots, except for Ca. Micronutrients almost doubled in roots compared to foliar content but were 30-fold higher in Fe content. No significant differences were obtained for the three compared treatments in the macro- and micronutrient contents at the root level, except for Ca and Mg, which were higher in the plants treated with CRFs compared to DURAMON®.

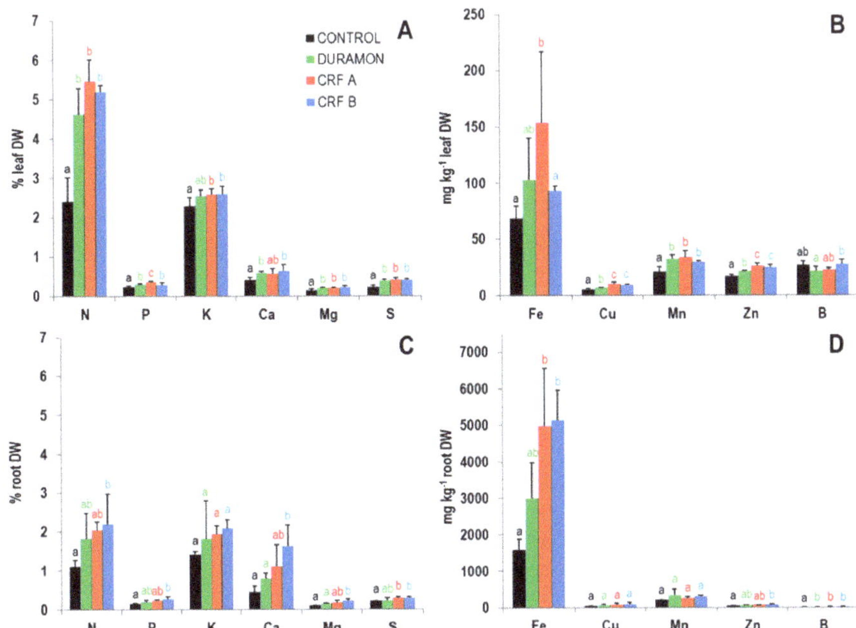

Figure 4. Macro- (**A**,**C**) and micronutrient (**B**,**D**) contents in leaves (**A**,**B**) and roots (**C**,**D**), for the different treatments with the polymeric-coated fertilizers (CRF$_A$, CRF$_B$), DURAMON®, and the CONTROL in the phenological state (booting stage). Results of macronutrients (N, P, K, Ca, Mg, and S) are expressed in percentage of DW, and micronutrients (Fe, Cu, Mn, Zn, B) in mg kg^{-1} DW. Values are means ± SD (n = 4). Different letters for a specific macro- or micronutrient in each panel indicate statistically significant differences between treatments (ANOVA, P < 0.05).

3.4. Growth, Yield, and Cereal Grain Composition

The results about the measured growth and yield parameters in greenhouses are shown in Table 3. No significant differences were found in the different treatments for the measured growth parameters. Significant differences were observed in the CONTROL with the CRF$_A$- and DURAMON®-treated plants for the total dry weight of aerial parts, and with DURAMON® for ear weight and number. Regarding the yield parameters, total dry grain weight was significantly higher for CRF$_A$ than for CRF$_B$. No significant differences were observed in grain weight among treatments, but differences were significant compared to the CONTROL. Total grain number was 1.2-fold significantly higher for CRF$_A$ than for CRF$_B$ but was not significant compared to DURAMON®. At the field level, no significant differences were found in the various applied treatments when comparing growth and grain yield parameters (Table 4). However, NUE was 28 and 38% higher for CRF$_A$ 80%, compared to CRF$_A$ 100% and NSA; these differences were statistically significant. No significant differences were observed among treatments and the CONTROL for the analyzed grain parameters. As a result, on average the values were 1.6% CRF$_A$ ash, 12.9% humidity, 1.7% lipids, 11.4% protein, 2.5% crude fiber, and 65.8% total carbohydrates. No significant differences were obtained comparing the studied quality parameters between CRF$_A$ and the rest of treatments and CONTROL.

Table 3. Comparison of growth and grain yield parameters among the applied fertilizer treatments CRF_A, CRF_B, and DURAMON® compared to the Control in *Triticum aestivum*. Values are means ± SD ($n = 10$) at the end of the culture—138 days after plant emergence.

Parameters	CRF_A	CRF_B	DURAMON®	CONTROL
Total dry weight (aerial part) (g)	40.2 ± 3.4 b	36.7 ± 6.0 ab	41.4 ± 7.7 b	31.7 ± 3.9 a
Primary stem length (cm)	60.8 ± 2.1 a	61.1 ± 3.0 a	53.3 ± 7.9 a	53.6 ± 10.0 a
Tillers number	10.0 ± 0.9 a	10.2 ± 1.6 a	9.8 ± 2.8 a	11.5 ± 2.4 a
Ears number	10.7 ± 1.3 a	9.8 ± 1.1 a	9.7 ± 2.3 a	9.2 ± 2.1 a
Ear weight (g)	2.7 ± 0.2 ab	2.4 ± 0.2 ab	3.0 ± 0.3 b	2.0 ± 0.2 a
Ear length (cm)	13.4 ± 0.4 ab	13.2 ± 0.9 ab	13.9 ± 0.3 b	12.9 ± 0.9 a
Total dry grain weight (g)	22.6 ± 2.0 c	18.6 ± 2.7 b	20.4 ± 3.2 bc	15.9 ± 3.0 a
Grain weight ($n = 100$)	4.8 ± 0.2 b	4.8 ± 0.4 b	4.8 ± 0.3 b	4.4 ± 0.3 a
Total grain number	473.7 ± 46.3 c	392.7 ± 67.4 ab	428.4 ± 79.8 bc	364.7 ± 52.8 a

[1] Different letters in the same row indicate statistically significant differences (Fisher's LSD test, $P < 0.05$).

Table 4. Field harvest comparison of the growth parameters and grain yield of *Triticum spelta* among the different applied fertilizer treatments CRF_A 100%, CRF_A 80%, and NSA compared to the Control. Values are means ± SD ($n = 10$) at the end of the culture—90 days after applying fertilizers.

Parameters	CRF_A 100%	CRF_A 80%	NSA	CONTROL
Biomass of the aerial part (t ha^{-1})	9.32 ± 1.34 b	9.61 ± 1.86 b	8.78 ± 1.41 b	4.91 ± 1.82 a
Ear weight (t ha^{-1})	3.58 ± 0.54 b	3.67 ± 0.54 b	3.76 ± 0.67 b	1.91 ± 0.46 a
Grain weight (t ha^{-1})	2.35 ± 0.45 b	2.49 ± 0.12 b	2.19 ± 0.37 b	1.16 ± 0.43 a
Nitrogen Use Efficiency (kg kg^{-1} N)	14.93 ± 5.58 a	20.76 ± 1.88 b	12.89 ± 4.66 a	-
Harvest Index	0.31 ± 0.05 a	0.33 ± 0.08 a	0.3 ± 0.01 a	0.29 ± 0.02 a

[1] Different letters in the same row indicate statistically significant differences (Fisher's LSD test, $P < 0.05$).

4. Discussion

Nutrient release contained in fertilizers depends on many factors, including environmental conditions, crop management, and the chemical composition of fertilizers [2,38,39]. Nowadays, the slowing down of nutrient emissions in fertilization is a challenge that has already been overcome [11,18,24]. However, high production costs and contamination linked to synthetic fertilizers and waste materials, together with increasingly restrictive environmental policies, have forced new ecological materials to be sought to allow sustainable fertilization [40,41]. The use of water-soluble synthetic products or natural polymers based on lignin has been indicated as an alternative to these problems because they can be obtained in large quantities and at cheap prices from the waste generated in the paper industry, from wood and other sources [42,43].

Research conducted with CRFs has shown that their efficiency is generally higher to that of traditional fertilizers and SRFs [9]. In fact, SRFs are more sensitive to high temperature and sandy soils [44]. Nevertheless, most research works conducted to date with CRFs have focused mainly on crops with a high added value, such as horticultural, ornamental or wood products, and have obtained different results [45–52]. It is important to point out that the main challenge of CRFs application to crops is to successfully provide the amount of nutrients that plants need and in a fractional manner. Moreover, there are also the important advantages that CRFs offer 'per se' in both crop management and the environment. In fact, CRF applications are usually unique, which means savings in crop handling from avoiding successive top-dressing fertilizer applications. Finally, nutrient doses are usually lower in CRFs than those applied with traditional fertilizers, and N losses by evaporation or leaching consistently lower.

In two experiments, this research compares the effectiveness of two lignosulfonate-based polymer-coated urea fertilizers: an analogous non-coated urea, and ammonium nitrosulfate as a traditional fertilizer. Based on the experimental design, the CRF with the best behavior was selected based on the responses of wheat to different physiological and yield parameters on the microscale. In a second stage, the selected CRF was compared with ammonium nitrosulfate in the field. In physiological terms, significant differences were found in growth, chlorophyll content and ΦPSII among CRF_A, CRF_B and DURAMON® compared with the untreated plants in early crop development stages. Lower values of E and higher of Ci were detected in plants treated with DURAMON®, as compared to the CRFs treatments. Intrinsic efficiency in the water use of CRF_A was significantly higher compared to CRF_B and DURAMON®. To explain these results, the enhanced effects of CRFs were produced by a combination of the individual effects of lignosulfonates and biostimulants on nutrient supply. In fact, lignosulfonates or sulfonated lignin have a variety of functional groups that provide unique colloidal properties and act as chelating agents [53]. The humic substances contained in CRF_A may promote plant development by stimulating root and shoot growth as they can enhance nutrient use efficiency by facilitating the assimilation of macro- and microelements [54,55]. Seaweed extracts can enhance chlorophyll and carotenoid contents in plant shoots, root thickness, and biomass [56]. However, the effectiveness of marketed biostimulants depends very much on their origin because their composition and proportions usually change [57].

Better physiological responses in the state of panicles formation suggest that plants would increase yield and biomass parameters at the end of the crop. A significant correlation was also found between higher levels of photosynthesis during grain formation and increased crop yields obtained in wheat [58,59]. In fact, on a microscale, the total yield expressed in dry grain weight was significantly higher in CRF_A than in CRF_B; and was, on average, also higher than DURAMON®. The observed differences were due to a large number of grains harvested by the production of 1.1-fold more spikes in the plants fertilized with CRF_A compared to CRF_B and DURAMON®. On average the plants fertilized with DURAMON® produced larger sized and heavier spikes with more grains. This could be explained by the faster N-release kinetics of DURAMON® compared to CRFs, as lacking lignosulfonate-polymeric coverage. A bigger N supply in the first crop stages could explain why the plants fertilized with DURAMON® seemed to be slightly more advanced in their phenological status compared to CRFs. Physiological requirements of wheat are established as 3 kg N Qm^{-1} grain; therefore, the theoretical yield that should have been obtained at a dose of 150 kg N ha^{-1} was 5 t ha^{-1}. Our results showed exceeded yields of 6.4 grain t ha^{-1} with CRF_A, 5.3 grain t ha^{-1} with CRF_B and 5.8 grain t ha^{-1} with DURAMON®. Despite DURAMON® not being a CRF because it lacks polymer-coating, it could be considered an SRF for being formulated with urease inhibitor MCDHS, which is also contained in CRFs. This would explain why the DURAMON®-treated plants gave yields close to CRFs. In fact, the minor differences between DURAMON® and CRFs might indicate that DURAMON® could be also used successfully to maintain N availability for plants over time in wheat. As examples, when using nitrification and urease inhibitors in wheat, maize and barley, it was obtained better crop yields and N_2O mitigation than SRCFs [2]. Further, better performance for CRFs and those formulated with nitrification inhibitors compared to traditional ones in maize with a reduction in N_2O emissions up to 21% that did not affect yields [39]. The lower yields obtained with CRF_B could be explained by excessive N emission slowdown by having formulated with a 2% thicker polymeric coverage. The best results obtained with CRF_A were confirmed at field level when comparing the NUE between the applied fertilizer treatments. It was possible to obtain yields close to those observed with CRF_A 100% and NSA, by applying CRF_A with a 20% less N content, but significantly increasing the NUE by more than four times. Even though it has been reported that NUE can vary depending on factors like the doses applied or climatic conditions [2], no reductions in grain yield were observed when applying different CRF formulations by reducing N content in a similar proportion [60,61].

The macronutrients analysis showed that wheat plants had a good NPK nutritional status in phenological state at the beginning of ear formation. The N concentrations in the treated plants fell within the N leaf DW 4–6% range, which is considered suitable for obtaining high yields [62], but no statistically significant differences appeared in the applied treatments. On the contrary, P concentrations were 1.2-fold significantly higher in CRF_A than in CRF_B and DURAMON®. In all cases, P levels fell within the range considered optimal for good plant development (0.2–0.5% leaf DW). Although the applied N fertilizer was not mixed with P, it is known that soil N applications can stimulate root growth and increase cation exchange capacity to favor Ca uptake, and P uptake indirectly [63]. In cereals, increased yields and improvements in the content of macro- and micronutrients of crops have been achieved in barley, maize, rice, or wheat using SCRFs [15,64–69]. No significant differences were found in the K, Ca, Mg, and S contents in leaves among treatments, and their levels were medium to high. Regarding micronutrients, the plants fertilized with CRF$_A$ presented 1.7- and 1.5-fold higher Fe levels than CRF$_B$ and DURAMON®, respectively. However, the Fe levels were optimum for maximum yields (21–200 mg kg^{-1} leaf DW). Cu and Zn contents were 1.4- and 1.2-fold significantly higher in the plants fertilized with CRFs compared to those fertilized with DURAMON®, but concentrations were at the lowest levels within the range considered normal for Cu and Zn (5–50 and 20–70 mg kg^{-1} foliar DW, respectively). The Mn content did not differ significantly for the different fertilizer treatments and presented lower levels (16–200 mg kg^{-1} leaf DW). Fertilizer treatments did not significantly affect B content as the CONTROL plants had similar levels. The Mo levels were very low and were lower than 2 mg kg^{-1} leaf DW in all the treatments.

5. Conclusions

In the present study, we have carried out a comparison of two lignin-coated controlled release fertilizers enriched with humic substances (CRF$_A$) or seaweed extracts (CRF$_B$) with a similar non polymeric-coated fertilizer (DURAMON®) and with an ammonium nitrosulfate one (NSA). Our results showed that plants performed better when they were fertilized with CRFs coated with humic substances, although the improvement in crop yield was not excessive compared to the seaweed-coated one and that the uncoated one. However, a significant improvement in crop yield and the measured physiological parameters of wheat plants was achieved with Fertinagro's controlled release fertilizers compared to the traditional NSA. Fertilization with these new technified CRFs greatly favored the wheat crop management by making it possible to carry out one single application as a basal dressing, which simplified crop handling. Smaller amounts of N in formulations gave important advantages, such as reduced costs and minimized N losses, which thus avoids the common contamination problems that usually occur when applying traditional fertilizers. We conclude that it is possible to use this technology in extensive cropping, as lignin-based polymers are economically feasible and environmentally friendly.

Author Contributions: Conceptualization, R.G.-O.; data curation, R.G.-O., and A.R.-N.; formal analysis, R.G.-O. and A.R.-N.; funding acquisition, M.Á.N., O.V. and S.A.; investigation, R.G.-O., A.R.-N., and M.C.-M.; methodology, R.G.-O., M.Á.N., A.R.-N., C.G., and O.V.; project administration, M.Á.N.; supervision, C.G. and O.V.; validation, O.V., A.R.-N.; visualization, M.Á.N., M.C.-M., and S.A.; writing–original draft, R.G.-O.; writing–review and editing, R.G.-O., O.V., and A.R.-N. All authors have read and agreed to the published version of the manuscript.

Funding: This research was funded by the Spanish Ministry of Economy and Competitiveness, grant number RTC-2014-1457-5, with the project entitled "Los CRFs como alternativa a los fertilizantes tradicionales: buscando una mayor protección del medio ambiente".

Acknowledgments: The authors are grateful to Manuel Talón (Centre of Genomics–IVIA) to provide the facilities to develop the greenhouse experiments and Ángel Boix for his support in the plants maintenance.

Conflicts of Interest: The authors declare no conflict of interest.

References

1. FAOSTAT. Available online: http://www.fao.org/faostat/en/#data/QC (accessed on 23 February 2020).
2. Feng, J.F.; Li, F.B.; Deng, A.X.; Feng, X.M.; Fang, F.P.; Zhang, W.J. Integrated assessment of the impact of enhanced-efficiency nitrogen fertilizer on N$_2$O emission and crop yield. *Agric. Ecosyst. Environ.* **2016**, *231*, 218–228. [CrossRef]
3. Zuk-Golaszewska, K.; Zeranskal, A.; Krukowska, A.; Bojarczuk, J. Biofortification of the nutritional value of foods from the grain of *Triticum durum* desf. *by an agrotechnical method: A scientific review. J. Elem.* **2016**, *21*, 963–975.

4. Barakat, M.; Cheviron, B.; Angulo-Jaramillo, R. Influence of the irrigation technique and strategies on the nitrogen cycle and budget: A review. *Agric. Water Manage.* **2016**, *178*, 225–238. [CrossRef]

5. Zak, D.R.; Holmes, W.E.; MacDonald, N.W.; Pregitzer, K.S. Soil temperature, matric potential, and the kinetics of microbial respiration and nitrogen mineralization. *Soil Sci. Soc. Am. J.* **1999**, *63*, 575–584. [CrossRef]

6. Achat, D.L.; Augusto, L.; Gallet-Budynek, A.; Loustau, D. Future challenges in coupled C–N–P cycle models for terrestrial ecosystems under global change: A review. *Biogeochemistry* **2016**, *131*, 173–202. [CrossRef]

7. Di, H.J.; Cameron, K.C. Inhibition of nitrification to mitigate nitrate leaching and nitrous oxide emissions in grazed grassland: A review. *J. Soils Sediments* **2016**, *16*, 1401–1420. [CrossRef]

8. Akelah, A. Novel utilizations of conventional agrochemicals by controlled release formulations. *Mater. Sci. Eng. C-Biomimetic Mater. Sens. Syst.* **1996**, *4*, 83–98. [CrossRef]

9. Shaviv, A.; Mikkelsen, R.L. Controlled-release fertilizers to increase efficiency of nutrient use and minimize environmental degradation - A review. *Fertil. Res.* **1993**, *35*, 1–12. [CrossRef]

10. Ni, X.Y.; Wu, Y.J.; Wu, Z.Y.; Wu, L.; Qiu, G.N.; Yu, L.X. A novel slow-release urea fertiliser: Physical and chemical analysis of its structure and study of its release mechanism. *Biosyst. Eng.* **2013**, *115*, 274–282.

11. Prasad R, R.G.B.; Lakhdive, B.A. Nitrification retarders and slow-release nitrogen fertilizers. *Adv. Agron.* **1971**, *23*, 337–383.

12. Wang, Z.H.; Miao, Y.F.; Li, S.X. Wheat responses to ammonium and nitrate N applied at different sown and input times. *Field Crops Res.* **2016**, *199*, 10–20. [CrossRef]

13. Naz, M.Y.; Sulaiman, S.A. Slow release coating remedy for nitrogen loss from conventional urea: A review. *J. Control. Release* **2016**, *225*, 109–120. [CrossRef] [PubMed]

14. Chien, S.H.; Prochnow, L.I.; Cantarella, H. Recent developments of fertilizer production and use to improve nutrient efficiency and minimize environmental impacts. In *Advances in Agronomy*; Sparks, D.L., Ed.; Elsevier Academic Press Inc: San Diego, CA, USA, 2009; Volume 102, pp. 267–322.

15. Diez, J.A.; Caballero, R.; Bustos, A.; Roman, R.; Cartagena, M.C.; Vallejo, A. Control of nitrate pollution by application of controlled release fertilizer (CRF), compost and an optimized irrigation system. *Fertil. Res.* **1996**, *43*, 191–195. [CrossRef]

16. Halvorson, A.D.; Snyder, C.S.; Blaylock, A.D.; Del Grosso, S.J. Enhanced-Efficiency Nitrogen Fertilizers: Potential role in nitrous oxide emission mitigation. *Agron. J.* **2014**, *106*, 715–722. [CrossRef]

17. Carson, L.C.; Ozores-Hampton, M. Factors affecting nutrient availability, placement, rate, and application timing of controlled-release fertilizers for Florida vegetable production using seepage irrigation. *Horttechnology* **2013**, *23*, 553–562. [CrossRef]

18. Azeem, B.; KuShaari, K.; Man, Z.B.; Basit, A.; Thanh, T.H. Review on materials & methods to produce controlled release coated urea fertilizer. *J. Control. Release* **2014**, *181*, 11–21. [PubMed]

19. Herrera, J.M.; Rubio, G.; Haner, L.L.; Delgado, J.A.; Lucho-Constantino, C.A.; Islas-Valdez, S.; Pellet, D. Emerging and Established Technologies to Increase Nitrogen Use Efficiency of Cereals. *Agronomy-Basel* **2016**, *6*, 25. [CrossRef]

20. Dou, H.; Alva, A.K. Nitrogen uptake and growth of two citrus rootstock seedlings in a sandy soil receiving different controlled-release fertilizer sources. *Biol. Fert. Soils* **1998**, *26*, 169–172. [CrossRef]

21. Razali, R.; Daud, H.; Nor, S.M. Modelling and Simulation of Nutrient Dispersion from Coated Fertilizer Granules. In Proceedings of the 3rd International Conference on Fundamental and Applied Sciences, Kuala Lumpur, Malaysia, 3–5 June 2014; pp. 442–448.

22. Feng, C.; Lu, S.Y.; Gao, C.M.; Wang, X.G.; Xu, X.B.; Bai, X.; Gao, N.N.; Liu, M.Z.; Wu, L. "Smart" fertilizer with temperature- and pH-responsive behavior via surface-initiated polymerization for controlled release of nutrients. *ACS Sustain. Chem. Eng.* **2015**, *3*, 3157–3166. [CrossRef]

23. Kenawy, E.R. Recent advances in controlled release of argochemicals. *J. Macromol. Sci.-Rev. Macromol. Chem. Phys.* **1998**, *C38*, 365–390. [CrossRef]

24. Dubey, S.; Jhelum, V.; Patanjali, P.K. Controlled release agrochemicals formulations: A review. *J. Sci. Ind. Res. India* **2011**, *70*, 105–112.

25. Majeed, Z.; Ramli, N.K.; Mansor, N.; Man, Z. A comprehensive review on biodegradable polymers and their blends used in controlled-release fertilizer processes. *Rev. Chem. Eng.* **2015**, *31*, 69–96. [CrossRef]

26. Chowdhury, M.A. The controlled release of bioactive compounds from lignin and lignin-based biopolymer matrices. *Int. J. Biol. Macromol.* **2014**, *65*, 136–147. [CrossRef] [PubMed]

27. Fernandez-Perez, M.; Garrido-Herrera, F.J.; Gonzalez-Pradas, E.; Villafranca-Sanchez, M.; Flores-Cespedes, F. Lignin and ethylcellulose as polymers in controlled release formulations of urea. *J. Appl. Polym. Sci.* **2008**, *108*, 3796–3803. [CrossRef]

28. Abalos, D.; Jeffery, S.; Sanz-Cobena, A.; Guardia, G.; Vallejo, A. Meta-analysis of the effect of urease and nitrification inhibitors on crop productivity and nitrogen use efficiency. *Agric. Ecosyst. Environ.* **2014**, *189*, 136–144. [CrossRef]

29. Battacharyya, D.; Babgohari, M.Z.; Rathor, P.; Prithiviraj, B. Seaweed extracts as biostimulants in horticulture. *Sci. Hortic.* **2015**, *196*, 39–48. [CrossRef]

30. Calvo, P.; Nelson, L.; Kloepper, J.W. Agricultural uses of plant biostimulants. *Plant Soil* **2014**, *383*, 3–41. [CrossRef]

31. Canellas, L.P.; Olivares, F.L.; Aguiar, N.O.; Jones, D.L.; Nebbioso, A.; Mazzei, P.; Piccolo, A. Humic and fulvic acids as biostimulants in horticulture. *Sci. Hortic.* **2015**, *196*, 15–27. [CrossRef]

32. Colla, G.; Nardi, S.; Cardarelli, M.; Ertani, A.; Lucini, L.; Canaguier, R.; Rouphael, Y. Protein hydrolysates as biostimulants in horticulture. *Sci. Hortic.* **2015**, *196*, 28–38. [CrossRef]

33. du Jardin, P. Plant biostimulants: Definition, concept, main categories and regulation. *Sci. Hortic.* **2015**, *196*, 3–14. [CrossRef]

34. Birrenkott, B.A.; Craig, J.L.; McVey, G.R. A leach collection system to track the release of nitrogen from controlled-release fertilizers in container ornamentals. *Hortscience* **2005**, *40*, 1887–1891. [CrossRef]

35. Clark, M.J.; Zheng, Y.B. Species-specific fertilization can benefit container nursery crop production. *Can. J. Plant Sci.* **2015**, *95*, 251–262. [CrossRef]

36. Cox, D.A. Reducing nitrogen leaching-losses from containerized plants - the effectiveness of controlled-release fertilizers. *J. Plant Nutr.* **1993**, *16*, 533–545. [CrossRef]

37. Agegnehu, G.; Nelson, P.N.; Bird, M.I. The effects of biochar, compost and their mixture and nitrogen fertilizer on yield and nitrogen use efficiency of barley grown on a Nitisol in the highlands of Ethiopia. *Sci. Total Environ.* **2016**, *569–570*, 869–879. [CrossRef] [PubMed]

38. Huett, D.O.; Gogel, B.J. Longevities and nitrogen, phosphorus, and potassium release patterns of polymer-coated controlled-release fertilizers at 30 degrees C and 40 degrees C. *Commun. Soil Sci. Plan.* **2000**, *31*, 959–973. [CrossRef]

39. Yang, L.; Wang, L.G.; Li, H.; Qiu, J.J.; Liu, H.Y. Impacts of fertilization alternatives and crop straw incorporation on N_2O emissions from a spring maize field in Northeastern China. *J. Integr. Agric.* **2014**, *13*, 881–892. [CrossRef]

40. Harrison, R.; Webb, J. A review of the effect of N fertilizer type on gaseous emissions. In *Advances in Agronomy*; Sparks, D.L., Ed.; Elsevier Academic Press Inc: San Diego, CA, USA, 2001; Volume 73, pp. 65–108.

41. Obreza, T.A.; Rouse, R.E.; Sherrod, J.B. Economics of controlled-release fertilizer use on young citrus trees. *J. Prod. Agric.* **1999**, *12*, 69–73. [CrossRef]

42. Garcia, C.; Vallejo, A.; Diez, J.A.; Garcia, L.; Cartagena, M.C. Nitrogen use efficiency with the application of controlled release fertilizers coated with kraft pine lignin. *Soil Sci. Plant Nutr.* **1997**, *43*, 443–449. [CrossRef]

43. Treinyte, J.; Grazuleviciene, V.; Ostrauskaite, J. Biodegradable polymer composites with nitrogen- and phosphorus-containing waste materials as the fillers. *Ecol. Chem. Eng. S.* **2014**, *21*, 515–528.

44. Medina, L.C.; Sartain, J.B.; Obreza, T.A.; Hall, W.L.; Thiex, N.J. Evaluation of a soil incubation method to characterize nitrogen release patterns of slow- and controlled-release fertilizers. *J. AOAC Int.* **2014**, *97*, 643–660. [CrossRef]

45. Gasparin, E.; Araujo, M.M.; Saldanha, C.W.; Tolfo, C.V. Controlled release fertilizer and container volumes in the production of Parapiptadenia rigida (Benth.) Brenan seedlings. *Acta Sci.-Agron.* **2015**, *37*, 473–481. [CrossRef]

46. Haver, D.L.; Schuch, U.K. Production and postproduction performance of two New Guinea Impatiens cultivars grown with controlled-release fertilizer and no leaching. *J. Am. Soc. Hortic. Sci.* **1996**, *121*, 820–825. [CrossRef]

47. Jacobs, D.F.; Salifu, K.F.; Seifert, J.R. Growth and nutritional response of hardwood seedlings to controlled-release fertilization at outplanting. *For. Ecol. Manage.* **2005**, *214*, 28–39. [CrossRef]

48. Kaplan, L.; Tlustos, P.; Szakova, J.; Najmanova, J. The influence of slow-release fertilizers on potted chrysanthemum growth and nutrient consumption. *Plant Soil Environ.* **2013**, *59*, 385–391. [CrossRef]

49. Kinoshita, T.; Yano, T.; Sugiura, M.; Nagasaki, Y. Effects of controlled-release fertilizer on leaf area index and fruit yield in high-density soilless tomato culture using low node-order pinching. *PLoS ONE* **2014**, *9*, 10. [CrossRef]

50. Kinoshita, T.; Yamazaki, H.; Inamoto, K.; Yamazaki, H. Analysis of yield components and dry matter production in a simplified soilless tomato culture system by using controlled-release fertilizers during summer-winter greenhouse production. *Sci. Hortic.* **2016**, *202*, 17–24. [CrossRef]

51. Oliet, J.; Planelles, R.; Segura, M.L.; Artero, F.; Jacobs, D.F. Mineral nutrition and growth of containerized *Pinus halepensis* seedlings under controlled-release fertilizer. *Sci. Hortic.* **2004**, *103*, 113–129. [CrossRef]

52. Pack, J.E.; Hutchinson, C.M.; Simonne, E.H. Evaluation of controlled-release fertilizers for northeast Florida chip potato production. *J. Plant Nutr.* **2006**, *29*, 1301–1313. [CrossRef]

53. Vishtal, A.; Kraslawski, A. Challenges in industrial applications of technical lignins. *BioResources* **2011**, *6*, 3547–3568.

54. Cacco, G.; Attina, E.; Gelsomino, A.; Sidari, M. Effect of nitrate and humic substances of different molecular size on kinetic parameters of nitrate uptake in wheat seedlings. *J. Plant Nutr. Soil Sci.* **2000**, *163*, 313–320. [CrossRef]

55. Nardi, S.; Ertani, A.; Francioso, O. Soil-root cross-talking: The role of humic substances. *J. Plant Nutr. Soil Sci.* **2017**, *180*, 5–13. [CrossRef]

56. Michalak, I.; Gorka, B.; Wieczorek, P.P.; Roj, E.; Lipok, J.; Leska, B.; Messyasz, B.; Wilk, R.; Schroeder, G.; Dobrzynska-Inger, A.; et al. Supercritical fluid extraction of algae enhances levels of biologically active compounds promoting plant growth. *Eur. J. Phycol.* **2016**, *51*, 243–252. [CrossRef]

57. Lotze, E.; Hoffman, E.W. Nutrient composition and content of various biological active compounds of three South African-based commercial seaweed biostimulants. *J. Appl. Phycol.* **2016**, *28*, 1379–1386. [CrossRef]

58. Gaju, O.; DeSilva, J.; Carvalho, P.; Hawkesford, M.J.; Griffiths, S.; Greenland, A.; Foulkes, M.J. Leaf photosynthesis and associations with grain yield, biomass and nitrogen-use efficiency in landraces, synthetic-derived lines and cultivars in wheat. *Field Crops Res.* **2016**, *193*, 1–15. [CrossRef]

59. Richards, R.A. Selectable traits to increase crop photosynthesis and yield of grain crops. *J. Exp. Bot.* **2000**, *51*, 447–458. [CrossRef] [PubMed]

60. Ji, Y.; Liu, G.; Ma, J.; Xu, H.; Yagi, K. Effect of controlled-release fertilizer on nitrous oxide emission from a winter wheat field. *Nutr. Cycl. Agroecosyst.* **2012**, *94*, 111–122. [CrossRef]

61. Zhang, J.S.; Wang, C.Q.; Li, B.; Liang, J.Y.; He, J.; Xiang, H.; Yin, B.; Luo, J. Effects of controlled release blend bulk urea on soil nitrogen and soil enzyme activity in wheat and rice fields. *Chin. J. Appl. Ecol.* **2017**, *28*, 1899–1908.

62. Jones, J.B. Plant tissue analysis in micronutrients. In *Micronutrients in Agriculture*, 2nd ed.; Mortvedt, J.J., Ed.; Soil Science Society of America, Inc.: Madison, WI, USA, 1991; pp. 477–521.

63. Grunes, D.L. Effect of nitrogen on the availability of soil and fertilizer phosphorus to plants. *Adv. Agron.* **1959**, *11*, 369–396.

64. Zhao, B.; Dong, S.T.; Zhang, J.W.; Liu, P. Effects of controlled-release fertiliser on nitrogen use efficiency in summer maize. *PLoS ONE* **2013**, *8*, 8. [CrossRef]

65. Dong, Y.J.; He, M.R.; Wang, Z.L.; Chen, W.F.; Hou, J.; Qiu, X.K.; Zhang, J.W. Effects of new coated release fertilizer on the growth of maize. *J. Soil Sci. Plant Nutr.* **2016**, *16*, 637–649. [CrossRef]

66. Mi, W.H.; Yang, X.; Wu, L.H.; Ma, Q.X.; Liu, Y.L.; Zhang, X. Evaluation of nitrogen fertilizer and cultivation methods for agronomic performance of rice. *Agron. J.* **2016**, *108*, 1907–1916. [CrossRef]

67. Roshanravan, B.; Soltani, S.M.; Mahdavi, F.; Rashid, S.A.; Yusop, M.K. Preparation of encapsulated urea-kaolinite controlled release fertiliser and their effect on rice productivity. *Chem. Speciation Bioavailab.* **2014**, *26*, 249–256. [CrossRef]

68. Morikawa, C.K.; Saigusa, M.; Nakanishi, H.; Nishizawa, N.K.; Hasegawa, K.; Mori, S. Co-situs application of controlled-release fertilizers to alleviate iron chlorosis of Paddy rice grown in calcareous soil. *Soil Sci. Plant Nutr.* **2004**, *50*, 1013–1021. [CrossRef]

69. Morikawa, C.K.; Saigusa, M.; Nishizawa, N.K.; Mori, S. Importance of contact between rice roots and co-situs applied fertilizer granules on iron absorption by paddy rice in a calcareous paddy soil. *Soil Sci. Plant Nutr.* **2008**, *54*, 467–472. [CrossRef]

Article

Assessing the Potential of Extra-Early Maturing Landraces for Improving Tolerance to Drought, Heat, and Both Combined Stresses in Maize

Charles Nelimor [1,2,3], Baffour Badu-Apraku [2,*], Antonia Yarney Tetteh [4], Ana Luísa Garcia-Oliveira [2] and Assanvo Simon-Pierre N'guetta [3]

[1] West African Science Service Centre on Climate Change and Adapted Land-use (WASCAL), Graduate Research Program on Climate Change and Biodiversity, Université Felix Houphouët Boigny, Abidjan 31 B.P. 165, Cote d'Ivoire; nelimor.c@edu.wascal.org

[2] International Institute of Tropical Agriculture, Ibadan 200001, Nigeria; o.oliveira@cgiar.org

[3] Department of Bioscience, Université Felix Houphouët Boigny, Abidjan 22 B.P. 461, Cote d'Ivoire; nguettaewatty@gmail.com

[4] Department of Biochemistry and Biotechnology, Kwame Nkrumah University of Science and Technology, University Post Office Box PMP, Kumasi 00233, Ghana; aytetteh@gmail.com

* Correspondence: b.badu-apraku@cgiar.com; Tel.: +234-8108-482590

Received: 17 December 2019; Accepted: 4 February 2020; Published: 25 February 2020

Abstract: Maize landrace accessions constitute an invaluable gene pool of unexplored alleles that can be harnessed to mitigate the challenges of the narrowing genetic base, declined genetic gains, and reduced resilience to abiotic stress in modern varieties developed from repeated recycling of few superior breeding lines. The objective of this study was to identify extra-early maize landraces that express tolerance to drought and/or heat stress and maintain high grain yield (GY) with other desirable agronomic/morpho-physiological traits. Field experiments were carried out over two years on 66 extra-early maturing maize landraces and six drought and/or heat-tolerant populations under drought stress (DS), heat stress (HS), combined both stresses (DSHS), and non-stress (NS) conditions as a control. Wide variations were observed across the accessions for measured traits under each stress, demonstrating the existence of substantial natural variation for tolerance to the abiotic stresses in the maize accessions. Performance under DS was predictive of yield potential under DSHS, but tolerance to HS was independent of tolerance to DS and DSHS. The accessions displayed greater tolerance to HS (23% yield loss) relative to DS (49% yield loss) and DSHS (yield loss = 58%). Accessions TZm-1162, TZm-1167, TZm-1472, and TZm-1508 showed particularly good adaptation to the three stresses. These landrace accessions should be further explored to identify the genes underlying their high tolerance and they could be exploited in maize breeding as a resource for broadening the genetic base and increasing the abiotic stress resilience of elite maize varieties.

Keywords: Abiotic stress; climate change; combined drought and heat stress; drought; heat stress; genetic resources; landrace accessions; maize

---

1. Introduction

Cultivation of maize (*Zea mays* L.) across the different agro-ecological zones of Africa dates to precolonial times after its introduction by Portuguese sailors in the late fifteenth century [1]. Hybridization between different populations, natural and artificial selection, and cultivation in diverse edaphic and climatic conditions, led to a plethora landraces/local varieties adapted to different agro-ecological zones, cultivation practices, and uses [2]. Whereas about 45% of these landraces/local varieties are still being grown across sub-Saharan Africa (SSA), many of these genotypes were collected and preserved in germplasm banks for avoiding the threat of extinction due to adoption of modern varieties or hybrids [3]. Historically, maize breeders identified and composited the most productive of these landraces to generate genetically diverse populations, constituting the foundation of hybrid maize breeding and developing open-pollinated varieties (OPVs) that displayed high yielding with tolerance to biotic and abiotic stresses [4,5].

Nonetheless, nowadays breeders are generally reluctant to use landraces because of the long-term commitment required to identify useful, novel diversity and introgress it into well-adapted elite cultivars while reducing the effects of undesired linked genes [6]. Breeders often resort to their own working collection consisting of elite breeding lines and some germplasm lines as parents in crossing, resulting in recirculating of the same germplasm. As a direct consequence, some newly developed maize varieties, including hybrids exhibit reduced genetic diversity [7], which may limit genetic gains and resilience to abiotic stresses. For example, by assessing the genetic diversity among selected elite CIMMYT Maize Hybrids in East and Southern Africa (ESA), Masuka et al. [7] observed that repeated use of four inbred parents resulted in narrowing of the genetic base of 29 to 58% of the hybrids, which could pose major risk in case of pest or disease outbreaks that are most likely under the prevailing climate changes.

The global climate over the past decade has changed rapidly, and temperatures are predicted to increase [8]. Similarly, precipitation patterns are expected to change significantly, which could adversely affect crop productivity either through drought or waterlogging [8]. Drought stress (DS) and heat (HS) stress are the two most critical and frequently co-occurring abiotic factors on farmers' fields. The tropical location, socio-economic, demographic, policy and farming characteristics of SSA heighten the risk of these stresses [9,10]. Thus, climate change represents a major impediment to African economy and subsistence.

Predicted impacts of climate change on major staple crops including maize in SSA are significant and primarily negative [11,12]. Drought and HS impede maize performance at all stages of plant growth and development. At flowering and early grain-filling stages of maize, DS and HS result in longer anthesis-silking interval (ASI) owing to the delay of silk extrusion, premature lodging, and reduced rates of net photosynthesis arising from oxidative damage to chloroplasts. Damage to chloroplasts is initiated by other detrimental effects of stomatal closure, leaf firing, tassel blasting, and senescence [13–18]. Consequently, reduced pollination efficiency, abnormal development of the embryo sac, as well as zygotic and early kernel abortion occur [14,15]. For these reasons, DS and HS occurring just before and shortly after pollination have the most profound negative effects on maize grain yield (GY) [16]. Yield losses attributable to DS and HS at flowering and the early grain-filling stages of maize were estimated at between 46–90% [17–19] and 45–55%, respectively [19,20]. Worse of all, DS and HS often co-occur under natural field conditions and their combined effects on growth, GY and related traits can either be synergistic, antagonistic or hypo-additive [21–24]. Recently, we screened 36 early maturing maize accessions for tolerance to abiotic stresses and found that on average, GY was reduced by 46%, 55% and 66% under DS, HS and DSHS conditions, respectively [19]. The high GY loss under DSHS reflects hypo-additive effect of DS and HS, which under extreme conditions could force farmers to abandon their farmlands [18,25].

To meet the food and feed needs of the projected population by 2050, it is essential for maize improvement programs in SSA to focus on the development of maize varieties with climate-adaptive traits, specifically, enhanced DS and/or HS tolerance either through conventional or biotechnology means. Genetic resources of maize are crucial to this goal. In particular, the landraces of maize, given their millennia evolutionary history and adaptation to low-input agricultural systems, harbor wide genetic diversity for improved productivity, climate adaptation, nutritional value, and quality attributes [26]. Despite their potential, most genetic diversity studies on maize landraces have focused on the analysis of variation in genetic parameters without much consideration for traits that confer tolerance to abiotic stress conditions. Only few studies have assessed the amount and nature of variation in landrace accessions of maize under individual and combined DS and HS [19,27].

Besides high yield potential, an appropriate cultivar for large areas of West and Central Africa (WCA), especially the Sudan savanna agro-ecological zone where growing cycles are short, must be DS and/or HS-tolerant at an extra-early stage (requiring 80–85 days to maturity) or early stage (90–95 days to maturity) [28]. The success of maize improvement for DS tolerance in tropical maize is well acknowledged [29]. Nonetheless, new DS tolerant source populations are needed to sustain increased genetic gains for food security [29]. Moreover, in comparison to DS, research on HS tolerance in tropical maize is still at the infant stage [29,30], and more studies are needed to complement these breeding efforts. Therefore, performing phenotypic screens to capture the genetic diversity that exist in extra-early landrace accessions of maize resting idle in germplasm banks under DS, HS and DSHS conditions is a promising strategy to uncover new genetic sources, which when introgressed into breeding stocks could contribute to the broadening of the genetic base and development of 'next generation' maize varieties with enhanced quality and other end-user preferred traits.

The present study aimed at identifying DS, HS and DSHS tolerant extra-early maturing maize landraces by exploring agronomic and morpho-physiological traits along with grain yield.

2. Materials and Methods

2.1. Plant Materials

One hundred and ninety-six (196) landrace accessions of maize from Burkina Faso, Ghana and Togo, were randomly sampled from gene banks at the International Institute of Tropical Agriculture (IITA), Nigeria and the Plant Genetics Resources Institute of Ghana in January 2017. Records on collection dates and geographical co-ordinates of the collection sites of these landraces were not available. However, the agro-ecologies/environments in these countries differ in a number of ways in terms of temperature and precipitation, and the adaptation of the accessions to these contrasting agro-ecologies may have been different. We characterized the maize collections under non-stress (NS) conditions for two years, and identified traits related to adaptation to local environmental conditions [31]. A total of 66 landraces were selected for the present study based on variation in standability, plant architecture, agronomic traits (lodging, earlier flowering, anthesis-silking interval, ear husk cover) and grain yield potential under optimal conditions. Six extra-early maturing drought and /or heat-tolerant populations developed by the Maize Improvement Program at the International Institute of Tropical Agriculture (IITA-MIP), Ibadan, Nigeria were included as checks. Details on the 72 maize accessions assessed in this study are available in Supplementary Table S1.

2.2. Agronomic Management

In the present study, each location-year combination was regarded as an environment. At all environments, the experiment was arranged in a 9 × 8 alpha lattice design with two replications. Each plot consisted of one-row plots each 3-m long with plants within and between rows spaced 0.40 m and 0.75 m apart, respectively. Three seeds were sown per hill and thinned to two, two weeks after emergence, resulting in a final plant population density of 66,666 plants per hectare. All stress trials were conducted during the dry season to allow DS, HS and DSHS to be imposed at the period considered most critical for maize growth and development [16].

As control, trials conducted under NS conditions were planted between June and July at the IITA experimental stations at Ikenne (6°53′ N, 3°42′ E, 60 m altitude, 1200 mm annual rainfall) and Mokwa (9°18′ N, 5°185 4′ E, altitude 457 m, 1100 mm annual rainfall), all in Nigeria, in 2017 and 2018. At 2 weeks after planting (WAP), 60 kg ha^{-1} each of nitrogen (N), phosphorus (P) and potassium (K) was applied. Four weeks later, the trials were top-dressed with 30 kg ha^{-1} of urea. In the stress trials, all plots received 60 kg N ha^{-1}, 60 kg P ha^{-1} and 60 kg K ha^{-1} as NPK 15–15–15 at sowing. A second application of N (30 kg N ha^{-1}) was applied as urea two weeks later. Weeds were controlled with herbicides and/or manually.

2.2.1. Drought Stress Trials

Drought experiments were planted at Ikenne during the last week of November in 2017 and 2018, so that flowering and early grain filling stages occurred in January when the incidence of rainfall was negligible. In both years, irrigation was applied using a sprinkler system (flow rate 17 mm day^{-1}) for the first 25 days after planting (DAP). Irrigation was suspended two weeks before anthesis until physiological maturity, so that the maize plants depended on stored water in the soil for growth and development.

2.2.2. Heat and Combined Drought and Heat Stress Trials

HS and DSHS trials were carried out at Kadawa (11°39′ N, 8°27′ E, 500 m altitude), Nigeria, where extreme DS at elevated temperatures occur between February and June each year. During this period, air temperature often varies between 33 and 45 °C [18], allowing for the establishment of trials under HS and DSHS. The HS and DSHS trials were laid in adjacent blocks, 15 m apart and planting was done on the same day in mid-February 2018 and 2019. Temperature and rainfall patterns during the experiment were measured by an automatic weather station installed at Kano, Nigeria. The average temperature at the experimental sites during the study period ranged from 36 °C/17 °C (day/night) to 40 °C/24 °C (day/night) in 2018, and from 31 °C/18 °C (day/night) to 39 °C/26 °C (day/night) in 2019 (Figure 1). There were minor incidences of rainfall after grain filling in May and June in both years (Figure 1). Flowering and grain-filling stages occurred in April, coinciding with extreme DS at elevated temperatures, which resulted in irreversible tissue injuries (leaf firing and tassel blasting) to susceptible genotypes (Supplementary Figure S1). During the reproductive stages in April, the minimum temperature was 36 °C/18 °C (day/night) in 2018, and 36 °C/23 °C (day/night) in 2019. The maximum temperature observed was 41 °C/27 °C (day/night) in 2018, and 41 °C/29 °C (day/night) in 2019, with a mean of 39 °C/23 °C (day/night) and 39 °C/26 °C (day/night) in 2018 and 2019, respectively (Figure 2). Irrigation was applied twice weekly on both HS and DSHS blocks using a furrow irrigation system. Irrigation was suspended on the DSHS block at 32 DAP but was resumed after grain-filling and applied once in order to avoid complete loss of trials. The HS block on the hand continued to receive irrigation every four days until physiological maturity.

2.3. Trait Measurement

At flowering, days to anthesis (AD) and silking (SD) were recorded when 50% of the plants in a plot had shed pollen and extruded silks, respectively. ASI was computed as SD minus AD. Plants with leaf firing (LF) and tassel blasting (TB) were counted on HS and DSHS plots and converted to percentages. At physiological maturity, plant and ear heights (PLHT and EHT) were measured on ten representative plants per plot as the length from the base of the plant to the height of the first tassel branch and the node bearing the upper ear, respectively. Plant aspect (PASP) was visually scored based on the general appeal of plants in a plot (standability, vigour, plant, and ear height, uniformity of plants, ear placement and size, as well as disease damage and lodging) using a scale of 1 to 9, where 1 = excellent overall phenotypic appeal; 2 = very good overall phenotypic appeal; 3 = good overall phenotypic appeal; 4 = satisfactory overall phenotypic appeal; 5 = acceptable phenotypic appeal; 6 = undesirable phenotypic appeal, 7 = poor overall phenotypic appeal, 8 = very poor phenotypic appeal and 9 = completely undesirable phenotypic appeal. Similarly, husk cover (HC) was rated on a scale of 1 to 9, where 1 = husks tightly arranged and extended beyond the ear tip and 9 = exposed ears. Stay green characteristic (SG) was recorded on all stressed plots at 70 DAP using a scale of 1 to 9, where 1 = 10% dead leaf area; 2 = 20% dead leaf area; 3 = 30% dead leaf area, 4 = 40% dead leaf area; 5 = 50% dead leaf area; 6 = 60% dead leaf area; 7 = 70% dead leaf area; 8 = 80% dead leaf area and 9 represented 90–100% dead leaf area. Few days before harvest, root and stalk lodging (RL and SL) were recorded as the percentage of plants leaning more than 30° from the vertical, and percentage plants broken at or below the highest ear node, respectively then all plants were hand harvested. At harvest, ear aspect (EASP) was rated based on the general appeal of the ears without the husks (ear size and number; uniformity of size, colour, and texture; extent of grain filling, insect and disease damage) using a scale of 1 to 9, where 1 = excellent (clean, uniform, large, and well-filled and disease-free ears); 2 = very good ears with no disease damage and fully filled grains; 3 = good ears with no disease damage and fully filled grains; 4 = no disease, fully filled grains, one or two irregularity in cob size poor; 5 = mild disease damage and fully filled grains, one or two irregularity in cob size, 6 = severe disease damage and fully filled grains, smaller cobs, non-uniform cob size; 7 = severe disease, scanty grain filling, few ears, non-uniformity of cobs; 8 = severe disease damage, very scanty grain filling, few ears, and 9 = only one or no ears produced (Supplementary Figure S2). Ears with rot (EAROT) were counted on plot basis and converted to percentage. Grain yield (kg ha^{-1}) was estimated based on 80% shelling percentage for NS plots. In contrast, total shelled grain weight was obtained for stress plots and GY was adjusted to 15% moisture content.

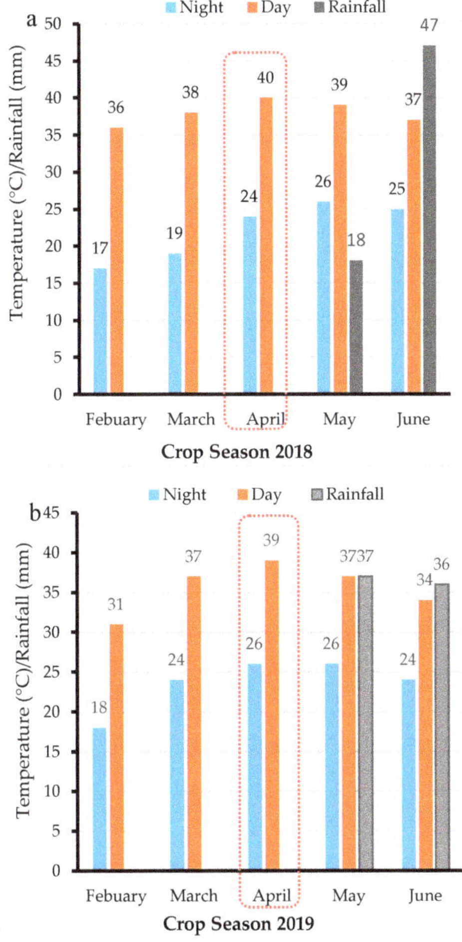

Figure 1. Monthly average temperature and rainfall during the experimental period at Kadawa, Nigeria. Note: Dotted rectangles represent flowering and early grain filling stages. (**a**) Trial period in 2018 (**b**) Trial period in 2019.

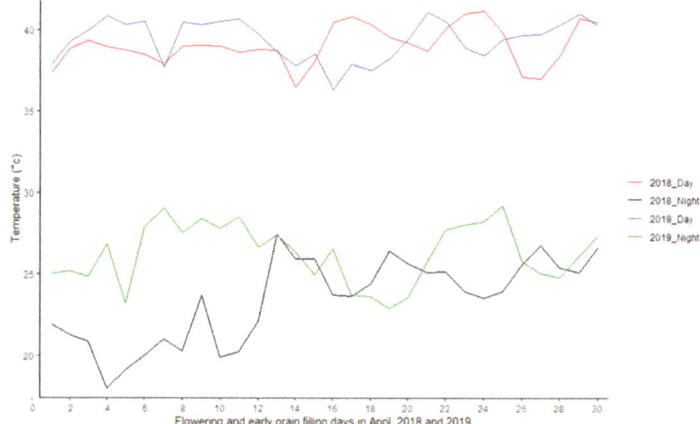

Figure 2. Average temperature (night and day) at Kadawa during flowering and early grain filling in 2018 and 2019.

2.4. Statistical Analysis

Variance components were estimated for each trial/environment by restricted maximum likelihood (Reml) using PROC Mixed of SAS version 9.4 [32]. The corresponding linear mixed model for the response variable was represented by:

$$Y_{ikl} = \mu + R_k + IB(R)_{lk} + G_i + e_{ikl} \tag{1}$$

where Y is the trait of interest, μ is the overall mean, R_k is the effect of the *kth* replicate, $IB(R)_{lk}$ is the effect of the incomplete block within the *kth* replicate, G_i is the effect of the *ith* genotype, and e_{ikl} is the experimental error. All effects were considered random except replicates. Broad-sense heritability (H) of grain yield was estimated for each environment as:

$$H = \frac{\sigma_g^2}{\left(\sigma_g^2 + \frac{\sigma_e^2}{r}\right)} \tag{2}$$

where σ_g^2 and σ_e^2 represented genotypic and residual variance, respectively and r denote the number of replicates. Trials with H of GY less than 0.30 were highly influenced by the environment and were removed from the analyses.

Combined analysis of variance (ANOVA) across environments were performed for each treatment (NS, HS, DS, and DSHS) with PROC GLM in SAS 9.4 using a RANDOM statement with the TEST option [32]. In the ANOVA, test environments, replicates, blocks and their interactions were considered as random factors while accessions were regarded as fixed effects.

Repeatability of the traits [33] under each treatment were computed on accession-mean basis using the following formula:

$$R = \frac{\sigma_g^2}{\sigma_g^2 + \frac{\sigma_{ge}^2}{e} + \frac{\sigma_e}{re}} \tag{3}$$

where σ_g^2 is the genotypic variance, σ_{ge}^2 is the variance of genotype × environment, σ_e is the residual variance; e is the number of environments, and r is the number of replicates per environment.

Phenotypic correlations among grain yield of the treatments were computed to determine the mechanism of abiotic stress tolerance using R library Performance Analytics [34]. Genetic correlations between phenology (AD and SD) and other secondary traits were estimated following the procedure of Cooper et al. [35] using Multi Environment Trial Analysis (META) [36].

Sequential path coefficient analysis [37] was used to identify traits with significant contributions to GY of the maize accessions under the evaluation conditions. A stepwise regression analysis was used to categorize the predictor traits into first, second and third order based on their individual contributions to total variation in GY with minimized multicollinearity [38]. The first step involved the regression of all the traits on GY and those with significant contributions to GY at $p < 0.05$ were identified as first order traits. Subsequently, traits that were not identified as first order traits were regressed on each of the first order traits to identify those with significant contributions to GY through each of the first order traits and were categorized as second-order traits. The procedure was repeated to identify traits in subsequent orders. The path coefficients were represented by the standardized b-values obtained from the regression analysis [38–40]. The sequential path coefficient including the stepwise regression analyses was done using the Statistical Package for Social Sciences, SPSS version 17.0 [41]

A base index (BI) that integrated superior grain yield, EPP, ASI, PASP, EASP, and SG was used to select the best and worst performing genotypes under each treatment [40]. Each trait was first standardized with standard deviation of 1 and a mean of zero to minimize the effect of the different scales prior to integrating into the BI. The BI was computed using the equation:

$$BI = [(2 \times YLD_S) + EPP - ASI - PASP - EASP - SG] \tag{4}$$

where YLD_S is GY under stress, $PASP$ is plant aspect, $EASP$ is ear aspect, EPP is ears per plant, ASI is anthesis-silking interval and SG is the stay-green characteristic. A positive BI value indicated tolerance to the applied stress while negative values indicated susceptibility [40].

Principal components analysis (PCA) was performed and the results were graphically visualized in a biplot that displayed different genetic groups (highlighted with different colours) and the association between accessions and measured traits. Grain yield was regressed on TB scores, plotted on a graph to show the effect of TB on GY of the maize accessions under HS and DSHS. Standardized data of the traits included in the BI selection were subjected to cluster analysis, in which phylogenetic constellation plots were generated, depicting the genetic relationships among the accessions under each stress treatment. The PCA, regression and cluster analysis were performed using JMP pro 14.10 [32].

3. Results

3.1. Analysis of Variance and Broad-Sense Heritability

Detailed values for variance components are presented in Tables 1 and 2. There was a large variation among the accessions for GY and other traits under NS, DS, HS and DSHS, which facilitated the grouping of abiotic stress-tolerant genotypes from their susceptible counterparts, as well as identification of traits that maximize variance in tolerance to the imposed stresses. Broad-sense heritability estimates of GY of individual trials ranged from 0.88 to 0.96 under NS, 0.51 to 0.61 under DS, 0.48 to 0.55 under HS, and 0.51 to 0.68 under DSHS (Supplementary Table S1). Hence, no trial was eliminated from the combined analysis. Within-treatments, repeatability measured across environmental conditions ranged from 0.41 for HC to 0.94 for GY under NS, 0.30 for EPP to 0.95 for EHT under DS, 0.21 for TB to 0.94 for AD under HS, and from 0.25 for EPP to 0.95 for AD under DSHS (Tables 1 and 2).

Table 1. Mean squares, and repeatability of grain yield and other traits of 72 extra-early maturing maize accessions evaluated under non-stress and drought stress conditions between 2017 and 2019 at Kadawa, Nigeria.

SV	df	GY	AD	SD	ASI	PLHT	EHT	HC	EPP	PASP	EASP	SG	RL	SL	EAROT
Non-Stress Conditions (NS)															
Env	2	5,746,960.3**	1.3	108.8*	133.4**	83913.4**	32981.7**	86.7**	0.5**	24.5**	3.3*	-	-	-	-
Rep (Env)	3	169,368.8	9.5*	0.8	5.4	1485.9*	1260.4**	7.6**	0	1.4*	0	-	-	-	-
Block (Env × Rep)	42	530,887.1*	3.4	8	5.2	690.1**	139	1.3**	0	0.9*	1.2*	-	-	-	-
Genotype	71	2,466,309.9**	47.7**	75.6**	13.2**	686.7**	344.2**	0.8*	0.1**	1.7**	3.8**	-	-	-	-
Env × Genotype	142	84,207.7	5.5**	13.9*	6.8*	180.3	85.9	0.5	0	0.4	0.4	-	-	-	-
Error	171	204,531.2	2.5	7.9	5.1	214.4	95	0.4	0	0.4	0.7	-	-	-	-
Repeatability		0.94	0.89	0.81	0.5	0.75	0.77	0.41	0.7	0.82	0.86	-	-	-	-
Grand Mean		2992.38	47	50	3	152.72	68.81	4	0.92	5	5	-	-	-	-
Drought Stress (DS)															
Env	1	3,976,484	75.4**	130.7**	19.4**	10811.4**	852.3**	54.2**	1.6**	24.3**	34.9**	9.4**	0.9**	0.5**	-
Rep (Env)	2	2,258,791	4.9*	6.5*	0.5	280.5	206.4*	0.9	0	0.6	0.7	4.8**	0.2**	0	-
Block (Env × Rep)	28	1,039,093	3.2*	3.4*	0.6	402.9**	177.5**	0.7	0	0.5	0.5	0.94*	0	0	-
Genotype	71	6,056,025.5 **	72.4**	76.4**	2.6**	1531.6**	946.2**	3.0**	0.1**	4.2**	4.2**	1.6**	0*	0**	-
Env × Genotype	71	1,840,190	3.9**	5.2**	1.5**	105.2	43	1.0*	0.0**	0.6*	0.8**	0.37	0	0	-
Error	114	1,638,119	1.6	2	0.8	109.5	60	0.7	0	0.4	0.4	0.39	0	0	-
Repeatability		0.69	0.94	0.93	0.45	0.94	0.95	0.68	0.3	0.86	0.82	0.79	0.33	0.68	-
Grand Mean		1523.92	46	50	4	116.73	54.95	4	0.68	5	5	4	0.1	0.22	-
Reduction (%)		0.49	1	-	1	0.24	0.20	-	0.26	-	-	-	-	-	-

*, ** Significance at 0.01 and 0.001, respectively; GY: Grain yield (kg/ha); AD: Days to 50% anthesis; SD: Days to 50% silking; ASI: Anthesis-silking interval; PLHT: Plant height (cm); EHT: Ear height (cm); HC: Husk cover (scale:1-9); EPP: Ears per plant; PASP: Plant aspect (scale:1-9); EASP: Ear aspect (scale:1-9); SG: Stay green characteristic (scale:1-9); RL: Root lodging (%); SL: Stalk lodging (%); EAROT: Ear rot.

Table 2. Mean squares and repeatability of grain yield and other traits of 72 extra-early maturing maize accessions evaluated under heat stress and combined drought and heat stress conditions between 2017 and 2019 at Kadawa, Nigeria.

SV	df	GY	AD	SD	ASI	PLHT	EHT	HC	EPP	PASP	EASP	SG	RL	SL	TB	LF	EAROT
						Heat Stress (HS)											
Env	1	16,037,142.5**	9.8	20.1*	1.8	40110.6**	23562.6**	103.9**	1.8**	0.2	8.7*	48.3**	0.2**	0.1*	0.2**	0.0*	-
Rep (Env)	2	8,533,488.5**	11.9	11.6	1	287.9	652.8*	1.7	0.6**	0.8	2.9*	3.5*	0.1**	0	0.0*	0	-
Block (Env × Rep)	28	1,970,909.7*	9.1*	9.2*	1.1	1363.7**	501.8**	1	0.1	0.6	1.3*	0.7	0	0	0	0	-
Genotype	71	2,330,918.4**	59.9**	51. 0**	3.4**	1181.9**	716.6**	1.5*	0.1**	2.0**	2.0**	1.4**	0	0.0**	0.0*	0.0*	-
Env × Genotype	71	753,004.1	4.5	3.2	1.3	286.3	168.4	0.9	0	0.3	0.6	0.6	0.0*	0	0.0*	0	-
Error	114	841,366	4.3	4.6	1.5	336.8	135.6	0.7	0	0.4	0.7	0.5	0	0	0	0	-
Repeatability		0.67	0.94	0.93	0.63	0.76	0.79	0.34	0.63	0.85	0.69	0.56	0.37	0.64	0.21	0.38	-
Mean		2301.39	57	60	3	144.12	59.65	4	0.82	5	5	3	7.33	12.71	3.91	4.19	-
Reduction (%)		0.23	−10	−10	-	0.06	0.13	-	0.11	-	-	-	-	-	-	-	-
						Combined Drought and Heat Stress (DSHS)											
Env	1	613,509,9.7*	1.9	0.1	2.5	155177.3**	58356.3**	38.9**	2.3*	0.1	5.7*	23.0**	0.1*	0	0	0.1*	-
Rep (Env)	2	843,164.7	14.4*	17.2*	0.3	522.3*	21.3	1.4	0.1	0.5	2.9*	0.4	0	0	0	0.1*	-
Block (Env × Rep)	28	1,522,778.5*	7.7*	13.0**	2.8*	913.3**	263.3**	2.0**	0.3	0.7	2.8**	0.9	0	0.0*	0	0.0*	-
Genotype	71	2,575,212.4**	47.3**	46.6**	4.1**	1010.2**	719.7**	1.6**	0.2*	1.8**	3.5**	2.1**	0.0*	0.0**	0.0**	0.0*	-
Env × Genotype	71	4,923,88.2	2.3	3.1	0.7	254.1*	113.4**	0.6	0.2	0.3	0.8	0.6	0	0	0	0	-
Error	114	595,224.5	3.2	4.7	1.7	161.7	46.8	0.7	0.2	0.5	0.8	0.6	0	0	0	0	-
Repeatability		0.79	0.95	0.93	0.66	0.77	0.86	0.62	0.25	0.78	0.77	0.69	0.57	0.62	0.64	0.55	-
Grand Mean		1258.51	56	58	3	142.68	58.59	4	0.68	5	5	4	9.75	15.71	9.94	11.97	-
Reduction (%)		0.58	−9	−8	-	0.07	0.15	-	0.26	-	-	-	-	-	-	-	-

*, ** Significance at 0.01 and 0.001, respectively; GY: Grain yield (kg/ha); AD: Days to 50% anthesis; SD: Days to 50% silking; ASI: Anthesis-silking interval; PLHT: Plant height (cm); EHT: Ear height (cm); HC: Husk cover (scale:1-9); EPP: Ears per plant; PASP: Plant aspect (scale:1-9); EASP: Ear aspect (scale:1-9); SG: Stay green characteristic (scale:1-9); RL: Root lodging (%); SL: Stalk lodging (%); TB: Tassel blast (scale:1-9); LF: Leaf firing (scale:1-9); EAROT: Ear rot.

Mean GY under NS was ~2992 kg/ha, ~2301 kg/ha under HS, ~1524 kg/ha under DS, and ~1259 kg/ha under DSHS (Tables 1 and 2). Compared with the NS environment, DS reduced GY by between 2.5% to ~85% (mean = 49%), 1.1% to 68.4% with an average of 23% under HS and, between 10% to 94% with a mean of 58% under DSHS. Anthesis was, on average, reached after 47 and 46 days under NS and DS conditions, respectively. Under HS, anthesis was delayed by 10 days and by 9 days under DSHS. Similarly, silking was on the average reached after 50 days under both NS and DS conditions but delayed by 10 days under HS and 8 days under DSHS. Despite the stress, ASI remained largely unchanged under all the treatments (averaging 3 days under NS, HS and DSHS, and 4 days under DS). Plant height was reduced by ~24% under DS and by 6% and 7% under HS and DSHS, respectively. Reduction in EPP was highest under DS (26%) and DSHS (26%) and lowest under HS (11%). Traits such as HC, and PASP and EASP were unaffected by the stresses as indicated by the average score of 4 and 5 under all the contrasting conditions. Leaf senescence, RL, SL, TB, and LF were on average, higher under DSHS relative to HS conditions.

3.2. Genetic Correlations and Sequential Regression Analysis

Phenotypic correlations between grain yields under NS, DS, HS and DSHS are presented in Table 3. Grain yield observed under NS was strongly and positively correlated with GY under HS ($r = 0.75$; $p < 0.0001$) while the correlation between GY under NS and those under DS, and DSHS were moderate and positive. Similarly, the correlation between GY under DS and DSHS was positive and moderate ($r = 0.60$; $p < 0.0001$). However, weak and positive phenotypic correlations were observed between GY under DS and HS ($r = 0.48$; $p < 0.001$) as well as HS and DSHS ($r = 0.37$; $p < 0.001$). Strong and positive correlations ($r \geq 0.73$; $p < 0.0001$) were observed for both AD and SD under NS and those under DS, and DSHS as well as between DS and DSHS. No significant correlation was observed between EPP under the different treatments except between NS and DS ($r = 0.37$; $p < 0.001$). Weak to moderate correlations were recorded for PASP under the different treatments except those between HS and DS ($r = 0.19$) and, HS and DSHS ($r = 0.05$). A similar trend was observed for EASP under the treatments. Negative genetic correlations were observed between phenology (AD and SD) and GY, HC, PASP, EASP, SG, LF, and TB under the applied stresses whereas, genetic correlations between flowering traits (AD and SD) and ASI, PLHT and EHT were positive (Table 4).

Table 3. Coefficient of phenotypic correlations between treatments for traits of 72 extra-early maturing maize accessions evaluated between 2017 and 2019 in Nigeria.

Trait	NS vs. DS	NS vs. HS	NS vs. DSHS	HS vs. DS	DS vs. DSHS	HS vs. DSHS
Grain yield	0.66***	0.75***	0.62***	0.48**	0.60***	0.37***
Anthesis days	0.90***	−0.05	0.87***	−0.04	0.82***	−0.13
Silking	0.80***	−0.07	0.84***	−0.08	0.73***	−0.08
Anthesis-Silking Interval	0.23*	−0.09	0.42***	−0.13	0.13	−0.28*
Plant height	0.48***	0.30*	0.64***	0.18	0.29*	0.23*
Ear height	0.60***	0.21	0.77***	0.24*	0.14	0.55***
Ears per plant	0.37**	0.11	0.06	0.03	0.22	−0.07
Ear aspect	0.54***	0.32**	0.33**	0.19	0.52***	0.05
Plant aspect	0.46***	0.14	0.63***	0.18	0.46***	0.04
Stay green	-	-	-	0.08	0.38***	0.03
Leaf firing	-	-	-	-	-	−0.20
Tassel blast	-	-	-	-	-	−0.10

*, **, *** Significant at 0.05, 0.01 and 0.001 probability levels, respectively.

Table 4. Genetic correlation between phenology (days to anthesis and silking) and other secondary traits of 72 extra-early maize accessions evaluated under contrasting environments between 2017 and 2019 in Nigeria.

Trait	NS	DS	HS	DSHS
Days to anthesis				
Grain yield	0.34	−0.71	−0.42	−0.23
Silking days	0.95	0.99	0.98	0.97
Anthesis-silking interval	0.42	0.12	0.51	0.20
Husk cover	−0.48	−0.71	−0.92	−0.63
Plant height	0.58	0.80	0.91	0.79
Ear height	0.65	0.87	0.98	0.87
Plant aspect	−0.31	−0.76	−0.87	−0.62
Ear aspect	−0.32	−0.83	−0.41	−0.12
Ears per plant	0.06	−0.89	0.19	−0.34
Stay green	-	−0.49	−0.88	−0.49
Leaf firing	-	-	−0.79	−0.26
Tassel blast	-	-	−0.98	−0.12
Days to Silking				
Grain yield	0.14	-0.65	-0.31	-0.14
Anthesis-silking interval	0.68	0.25	0.34	0.01
Husk cover	−0.24	−0.64	−0.82	−0.60
Plant height	0.35	0.80	0.87	0.81
Ear height	0.40	0.86	0.95	0.85
Plant aspect	−0.06	−0.71	−0.81	−0.59
Ear aspect	−0.09	−0.78	−0.26	−0.04
Ears per plant	−0.33	−0.77	0.08	−0.53
Stay green	−0.28	-	−0.76	−0.45
Leaf firing	-	-	−0.79	−0.20
Tassel blast	-	-	−0.98	-

NS: Non stress; DS: Drought stress; HS: Heat stress; DSHS: Combined drought and heat stress.

Under DS, stepwise multiple regression analysis identified EASP, PASP, and SL as the first order traits with significant contributions to GY (explaining 96 % of the total variation in GY) of the maize accessions (Figure 3). Of these first order traits, PASP had the highest negative direct effect on GY (−0.87) while SL contributed the least (−0.42) to GY. Traits, which contributed indirectly to GY through one or two of the first order traits, included EPP, EHT, RL, HC, SG, and AD. These traits were thus, classified into the second order. The traits classified into the third order were ASI, PLHT, SD, and EAROT, each contributing to variation in GY of the maize accessions through one or more of the second order traits.

Across the HS treatments, EASP, EPP, SG, SL, and LF were identified as first order traits responsible for 83% of the total variation in GY (Figure 4). Of these first order traits, EASP had the highest negative direct effect on GY (−0.54), while LF recorded the least direct negative effect on GY (−0.12). Only EPP had direct positive contribution to GY (0.25). Six traits, namely PASP, SD, EAROT, ASI, HC and TB contributed indirectly to GY through one or more of the first order traits and were thus categorized into the second order. Among the second order traits, ASI and TB contributed indirectly to GY through SG, and LF and SL, respectively. Plant aspect had the highest negative (−0.43) indirect contribution to GY through EPP, while the highest positive indirect contribution to GY was also observed for PASP through EASP (0.85). Four traits (EHT, AD, RL and PLHT) contributed indirectly to GY through one or two of the second order traits and were categorized into third order traits.

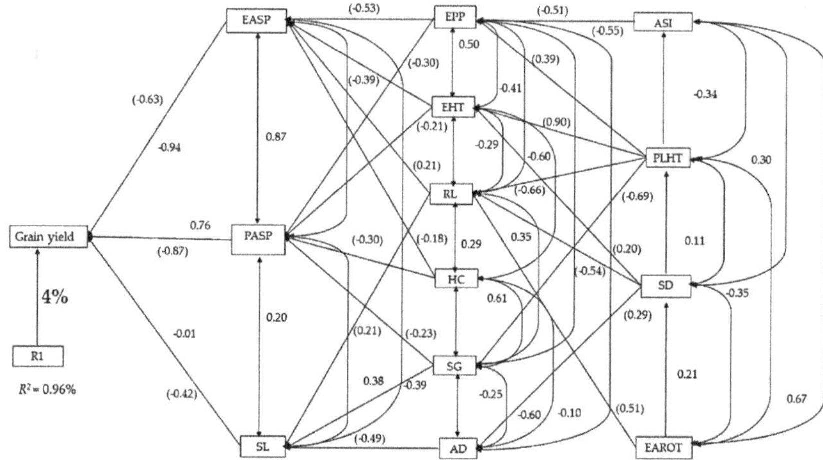

Figure 3. Path analysis diagram depicting the causal relationship of measured traits of the 72 maize accessions under drought-stressed conditions. Note: Value written in bold is the error effect; the direct path coefficients are values in parenthesis and other values are correlation coefficients. R1 is error effects, R^2 = coefficient of determination. AD: Days to 50% anthesis; SD: Days to 50% silking; ASI: Anthesis-silking interval; PLHT: Plant height; EHT: Ear height; HC: Husk cover; EPP: Ears per plant; PASP: Plant aspect; EASP: Ear aspect; SG: Stay green characteristic; RL: Root lodging; SL: Stalk lodging; TB: Tassel blast; LF: Leaf firing; EAROT: Ear rot.

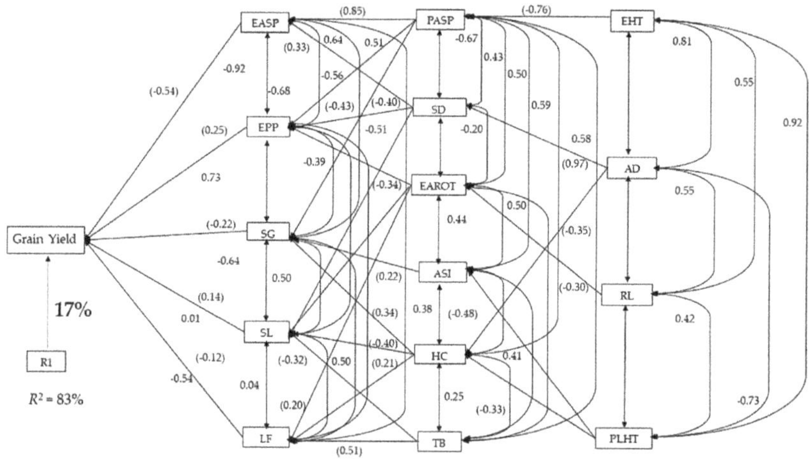

Figure 4. Path analysis diagram depicting the causal relationship of measured traits of the 72 maize accessions under heat stress conditions. Note: Values in bold are the error effect; the direct path coefficients are values in parenthesis and other values are correlation coefficients. R1 is error effects, R^2 = coefficient of determination. AD: Days to 50% anthesis; SD: Days to 50% silking; ASI: Anthesis-silking interval; PLHT: Plant height; EHT: Ear height; HC: Husk cover; EPP: Ears per plant; PASP: Plant aspect; EASP: Ear aspect; SG: Stay green characteristic; RL: Root lodging; SL: Stalk lodging; TB: Tassel blast; LF: Leaf firing; EAROT: Ear rot.

Under DSHS, traits classified in the first order (EASP, SG, PASP and ASI) explained 88% of the total variation in GY of the maize accessions (Figure 5). Each of these traits had direct negative contribution to GY with EASP being the highest contributor (−0.66). Seven traits, namely EPP, LF, AD, RL, HC, EHT and EAROT were categorized into the second order traits, each contributing to GY through one or two of the first order traits. Of the second order traits, EPP was found to have the highest negative indirect effect on GY through EASP (−0.69). Traits contributing to variation in GY through two or more of the second order traits included TB, SL, SD, and PLHT. Of these, TB contributed negatively to variation in GY through EPP (−0.35), AD (−0.26) while PLHT contributed positively to GY through EHT (0.62).

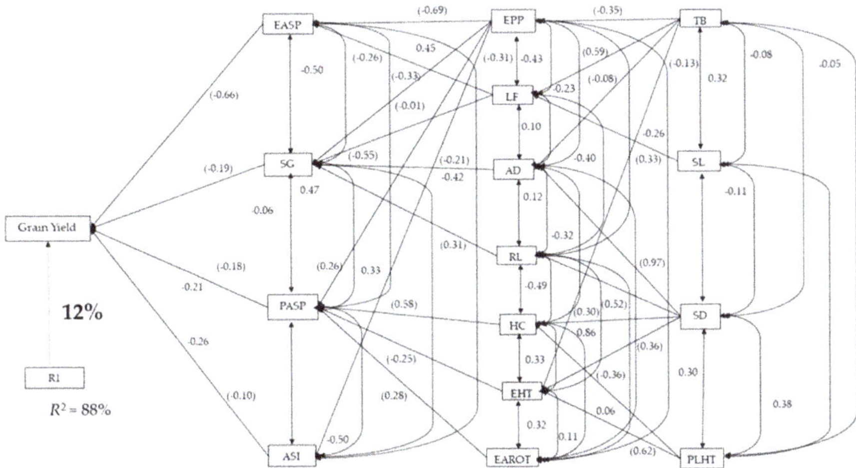

Figure 5. Path analysis diagram depicting the causal relationship of measured traits of the 72 maize accessions under combined drought and heat stressed conditions. Note: Values in bold is the error effects; the direct path coefficients are values in parenthesis while other values are correlation coefficients. R1 is error effects, and R^2 = coefficient of determination. AD: Days to 50% anthesis; SD: Days to 50% silking; ASI: Anthesis-silking interval; PLHT: Plant height; EHT: Ear height; HC: Husk cover; EPP: Ears per plant; PASP: Plant aspect; EASP: Ear aspect; SG: Stay green characteristic; RL: Root lodging; SL: Stalk lodging; TB: Tassel blast; LF: Leaf firing; EAROT: Ear rot.

3.3. Performance of Accessions under the Contrasting Environment

The distribution of the accessions in terms of GY performance under the abiotic stresses is presented in Figure 6. Under DS, DSHS and HS, about 38%, 29% and 7% of the accessions yielded below 1000 kg/ha, respectively while about 47%, 42% and 31% of the accessions produced yields between 1001 to 2000 kg/ha under DSHS, DS, and HS, respectively. Whereas under HS, 42% of the maize accessions yielded between 2001–3000 kg/ha, only a small proportion of them produced yields between 2001–3000 kg/ha (9% under DS and 18% under DSHS). None of the landrace accessions yielded >4000kg/ha under any of the stresses.

Given that selection solely for GY potential under abiotic stress condition is considered inefficient for accelerating genetic gain [42], a base index that integrated GY with other important secondary traits (ASI, EPP, PASP, EASP, and SG) was used as criterion to select accessions tolerant to each of the stresses (accessions with positive BI values) as well as across the contrasting environments. Summary of the top accessions (best check, and 15 landraces) and worse five landraces identified by the base index under each research condition is presented in Table 5. Under DS, the BI values ranged from −10.6 for TZm−1510 (with GY of ~543 kg/ha) to 13.4 for the check 3-TZEE-W HDT C3 STR C5 (with GY of ~3863 kg/ha). Of the top 15 landrace accessions based on the BI values, four (TZm-1440, TZm-1163, TZm-1162, TZm-1500) yielded between 3000 and 3487 kg/ha, six (TZm-1486, TZm-1160, TZm-1508, TZm-1449, TZm-1472 and TZm-1159) yielded between 2000 kg/ha and 3000 kg/ha while the remaining yielded below 2000 kg/ha. All accessions with negative BI values yielded below 1000 kg/ha (Table 5).

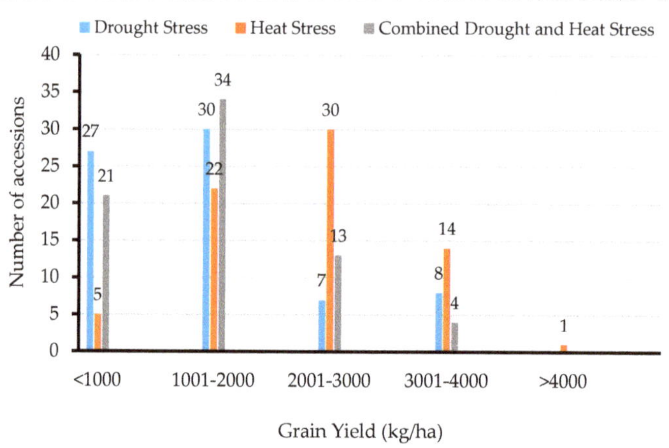

Figure 6. Distribution of grain yield of 72 extra-early maize accessions evaluated under drought, heat and combined drought and heat stress conditions during 2017 and 2019 at Ikenne and Kadawa, Nigeria.

Table 5. Grain yield and base index values of the best check and the top 15, and worse five landrace accessions evaluated under drought, heat, and combined drought and heat stress environments at Ikenne and Kadawa, Nigeria between 2017 and 2019.

Drought Stress (DS)			Heat Stress (HS)			Combined Drought and Heat Stress (DSHS)		
Accession	GY (kg/ha)	BI	Accession	GY (kg/ha)	BI	Accession	GY (kg/ha)	BI
Check 3	3863	13.4	Check 5	4723	14.5	Check 4	3899	10.9
TZm-1440	3287	11.4	TZm-1167	3895	12.3	TZm-1486	3167	8.2
TZm-1163	3487	11.3	TZm-1157	3614	9.1	TZm-1162	3174	7.7
TZm-1500	3086	11.1	TZm-1178	3896	8.0	TZm-1472	2334	4.8
TZm-1162	3256	10.5	TZm-1472	3238	7.8	TZm-1171	2070	4.7
TZm-1486	2815	9.2	TZm-1163	3673	6.4	TZm-1440	2373	4.5
TZm-1160	2215	7.4	TZm-1158	3247	6.3	TZm-1470	2391	4.2
TZm-1174	1886	6.5	TZm-1352	2524	6.1	TZm-1160	2187	3.9
TZm-1349	1932	5.1	TZm-1162	3137	4.7	TZm-1481	2039	3.8
TZm-1496	1685	4.9	TZm-1179	3956	4.5	TZm-1508	1797	3.1
TZm-1449	2404	4.9	TZm-1508	3350	4.2	TZm-1483	2042	2.6
TZm-1508	2521	4.5	TZm-1329	3415	3.9	TZm-1485	1735	2.5
TZm-1472	2015	4.3	TZm-1443	2984	3.9	TZm-1167	1700	2.5
TZm-1159	2026	4.3	TZm-1561	3277	3.9	TZm-1496	1802	2.4
TZm-1511	1861	4.1	TZm-1511	2793	3.9	TZm-1506	2210	2.2
TZm-1167	1926	3.7	TZm-1454	3194	3.4	TZm-1448	1853	2.0
TZm-1169	530	−8.8	TZm-1497	907	−9.36	TZm-1510	336	−6.8
TZm-1493	571	−9.0	TZm-1493	907	−9.37	TZm-1509	281	−7.3
GH-4863	430	−9.6	TZm-1177	936	−9.41	TZm-1176	366	−7.3
TZm-1165	501	−10.4	TZm-1170	932	−12.08	TZm-1480	467	−8.6
TZm-1510	543	−10.6	TZm-1498	501	−12.29	TZm-1173	152	−11.3

GY = Grain yield; BI = Base index; Check 3 = TZEE-W HDT C3 STR C5; Check 4 = TZEE-Y HDT C3 STR C5; Check 5 = 2014 TZEE-Y DTH STR.

Under HS, the BI values ranged from −12.29 for TZm-1498 (GY = 501 kg/ha) to 14.5 for the check5 -2014 TZEE-Y DTH STR (GY = 4723 kg/ha). All the top 15 landrace accessions based on the BI values yielded above 2500 kg/ha with TZm-1167, TZm-1157, TZm-1178, TZm-1163, and TZm-1179 yielding above 3500 kg/ha. As with DS, all accessions with negative BI values under HS yielded below 1000 kg/ha (Table 5).

Similarly, under DSHS, the BI values ranged from −11.3 for TZm-1173 (GY = 152 kg/ha) to 10.9 for check 4-TZEE-Y HDT C3 STR C5 (GY = 3899 kg/ha). Two of the top-yielding 15 landrace accessions identified by the BI (TZm-1486 and TZm-1162) yielded approximately 3000 kg/ha while TZm-1472, TZm-1440 and TZm-1470 yielded above 2300 kg/ha. The worse five landrace accessions under DSHS yielded below 500 kg/ha (Table 5).

Based on the BI values, seven landrace accessions (TZm-1159, TZm-1162, TZm-1163, TZm-1167, TZm-1472, TZm-1500 and TZm-1508) were tolerant to both DS and HS, eight (TZm-1160, TZm-1162, TZm-1167, TZm-1440, TZm-1472, TZm-1486, TZm-1496 and TZm-1508) were tolerant to both DS and DSHS. Only five landrace accessions (TZm-1167, TZm-1162, TZm-1472, TZm-1508 and TZm-1506) were tolerant to both HS and DSHS, while four (TZm-1162, TZm-1167, TZm-1472 and TZm-1508) showed good performance across all the individual and combined stresses.

3.4. Principal Component Biplot and Cluster Analysis

The biplot of principal components 1 and 2 under the different treatment conditions are presented in Figure 7. Under DS, the PCA biplot explained ~68% of the total variability among the genotypes (Figure 7a), 63% under HS (Figure 7b) and 55% under DSHS (Figure 7c). Under each stress condition, tolerant to very tolerant accessions were largely located in the lower left of the plot and were mainly associated with increased GY and EPP whiles their susceptible to very susceptible counterparts, which were characterized by increased ASI, LF, TB, SG, RL, and SL, poor HC, EASP, and PASP, and to some extent increased AD, and SD were found sparingly distributed in the upper and lower right sides of the biplot. The strong positive correlation between GY and EPP was evident by the acute angle between their respective vectors of similar length. Similarly, the negative correlation between GY and ASI, LF, TB, SG, RL, SL, HC, EASP, and PASP can be seen from the opposite direction of the variable vectors.

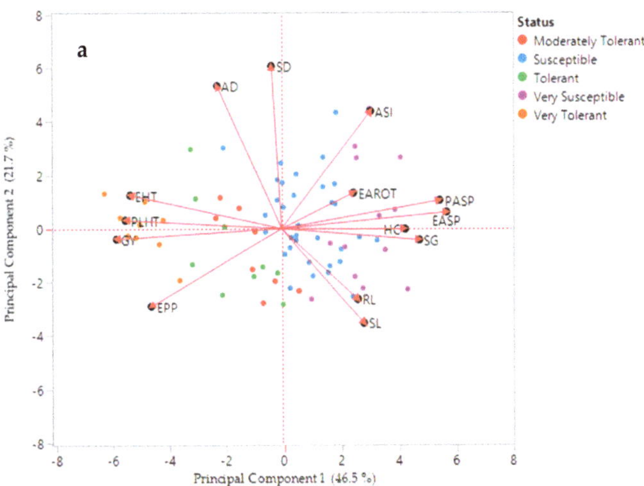

Figure 7. *Cont.*

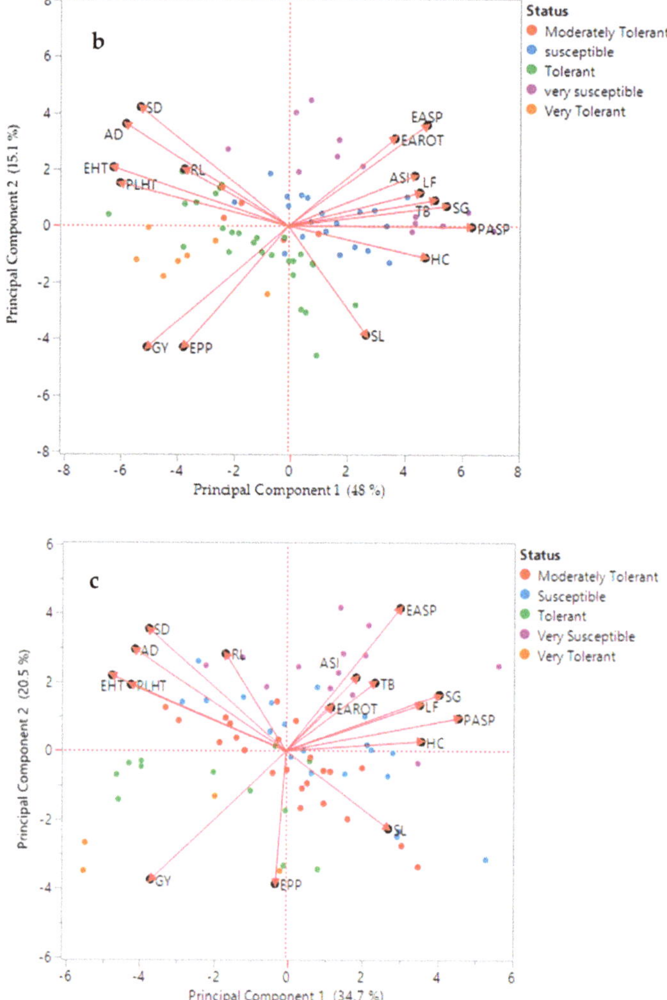

Figure 7. Biplot of 72 extra-early maize accessions and variables under drought (**a**), heat (**b**), and combined drought and heat stress (**c**) conditions. NB: The length of each variable vector is proportional to its contribution to total variation of the accessions, and the direction of the vector indicates its relative contribution to the principal components. AD: Days to 50% anthesis; SD: Days to 50% silking; ASI: Anthesis-silking interval; PLHT: Plant height; EHT: Ear height; HC: Husk cover; EPP: Ears per plant; PASP: Plant aspect; EASP: Ear aspect; SG: Stay green characteristic; RL: Root lodging; SL: Stalk lodging; TB: Tassel blast; LF: Leaf firing; EAROT: Ear rot.

Results of the regression of GY on TB showed that, largely accessions with high TB (%) had low grain yields (Figure 8). TB accounted for 15% and 28% of the yield reduction of the extra-early maize accessions under DSHS and HS, respectively.

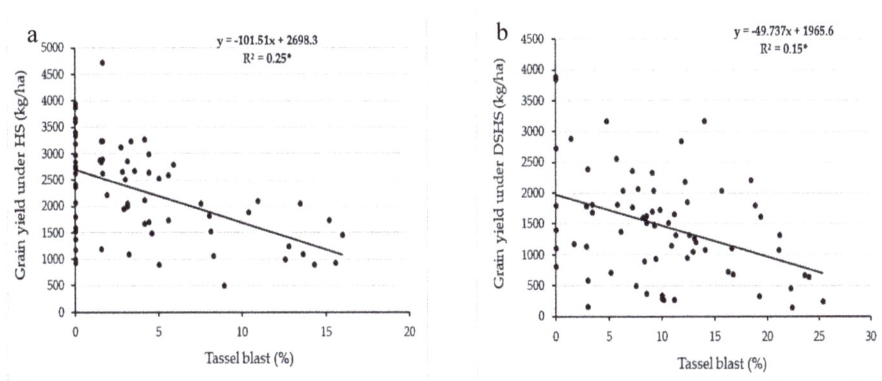

Figure 8. Regression of grain yield (kg/ ha) of 72 extra-early maize landraces including six drought and/or heat-tolerant populations/varieties, on tassel blast under heat stress (**a**) and combined drought and heat stress (**b**).

Under DS, the phylogenetic constellation plot generated using GY and the secondary traits included in the index selection classified the 72 extra-early maize accessions into five major groups, each further divided into subgroups (Figure 9). The number of accessions in the major clusters ranged from six in cluster IV to 31 in cluster III. Accessions in cluster I and II were characterized by high GY (>2000 kg/ha), increased EPP (≥0.80), delayed senescence (average rating of 3), desirable PASP and EASP (4 to 5). Consequently, the average BI values of these groups were high (averaging, 11.72 for cluster I and 5.0 for cluster II) (Supplementary Table S3).

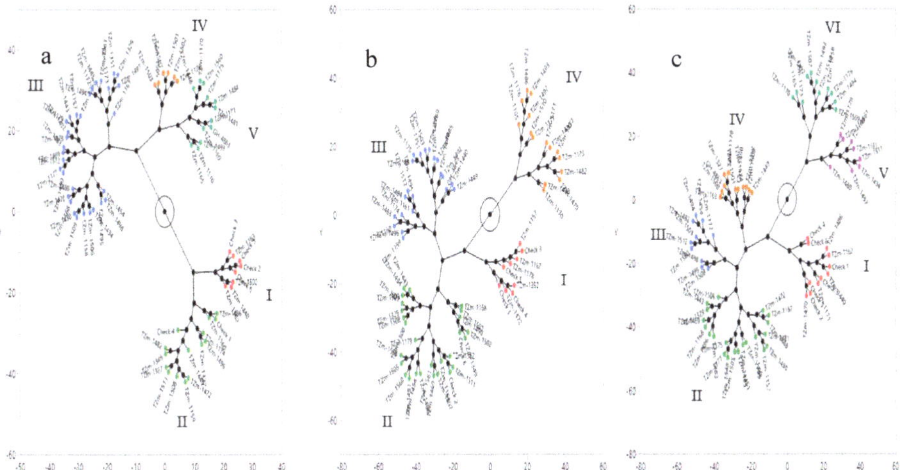

Figure 9. Phylogenetic constellation plots displaying the relationships between 66 extra-early maize landraces and six improved populations/varieties evaluated under managed drought stress (**a**), heat stress (**b**) and combined drought and heat stress (**c**). For each treatment, clusters I, and II were represented largely by tolerant accessions, while the remaining clusters consisted of susceptible accessions.

Similarly, under HS, the phylogenetic constellation plot separated the 72 accessions into five major clusters, with each further separated into sub-clusters that ranged from two in cluster I to four in cluster II (Figure 9). The first major cluster contained nine genotypes while the second consisted of 27 individuals. Individuals of the two clusters (I and II) were characterized by short ASI (2 to 3 days), good plant and ear aspect scores (4 to 5), and increased GY (>2800 kg/ha), hence the positive average BI values (9.73 for cluster I and 2.72 for II). Of the nine accessions in cluster I, five were landraces (TZm-1157, TZm-1167, TZm-1178, TZm-1352, and TZm-1472) and the other four were improved population checks. Accessions of the remaining clusters had very poor plant and ear aspect scores, reduced EPP, low GY and hence, negative BI values (Supplementary Table S3).

Under DSHS, six major groups of accessions were revealed by the phylogenetic constellation plot (Figure 9). The number of individuals in the clusters varied from eight in cluster IV to 21 in cluster II. The first major group consisted of nine accessions, including four checks, and five landrace accessions (TZm-1486, TZm-1162, TZm-1440, TZm-1171, and TZm-1470). The accessions of this group displayed desirable plant and ear aspect scores, and had good ears per plant and GY, which was reflected in their high positive index values (Supplementary Table S3). The second major group comprised accessions with good plant and ear aspect scores and had relatively good ears per plant, moderate grain yield as well as positive index values. The remaining four clusters contained accessions that recorded poor plant and ear aspect scores, reduced ears per plant, low GY, and negative BI values (Supplementary Table S3).

4. Discussion

Frequent occurrence of extreme weather conditions owing to global climate change has heightened the need for genetic improvement of major staple crops for tolerance to abiotic stresses. However, as selection pressure increases for a specific trait, genetic variability inevitably decreases [43], leading to reduced breeding gains. This is particularly true for tropical maize in which little additional gain in maximum drought-tolerance has been achieved in the last decade [29]. Addressing this issue and hence, ensuring progress in genetic improvement of maize under abiotic stress conditions requires identification of donor lines with beneficial traits that will bring new genetic variation [29,44]. Landrace gene pools of maize from areas that frequently experience DS at elevated temperatures may provide a useful source of novel alleles for abiotic stress tolerance [27,45]. In the present study, we evaluated 66 extra-early landrace accessions of maize representing gene pools from Burkina Faso, Ghana, and Togo, together with six abiotic stress-tolerant populations from the IITA-MIP under field conditions of DS, HS and DSHS imposed at the reproductive stages of crop growth and development for two years.

As shown in Figure 1, the HS trials were performed under high temperatures, while the DSHS trials were exposed to prolonged DS at elevated temperatures. In particular, temperatures during flowering and early grain-filling stages substantially exceeded the optimal threshold for lowland tropical maize (34 °C during the day and 23 °C at night; Figure 2) with no incidence of rainfall, indicating that the sites selected for this study were appropriate for screening the maize accessions for tolerance to HS and DSHS. These sites were used for screening maize for high levels of tolerance to DS and/or HS in earlier studies [13,18,19].

The presence of significant genotypic differences for all measured traits of the maize accessions under DS, HS, and DSHS conditions suggested that superior genotypes and traits conferring tolerance to the stresses could be identified and selected. Even though highly significant statistical differences were detected among genotypes, genotype × environment interaction was not significant for most traits indicating that the environments were similar in stress severity. This observation could be attributed to the coincidence of the imposed stresses at stages most critical for growth and development of the maize accessions. Moreover, the high repeatability estimates observed for majority of the measured traits including GY indicated that most of the variances observed in the present study can be attributed to differences among the studied accessions. These observations largely provided credibility in the performance of the accessions for breeding purposes. Low repeatability estimates across sites for some

measured traits were probably the result of inconsistent expressions as indicated by high genotype ×
environment variance. Similar findings were reported in maize under multiple stresses [19,29].

In agreement with previous studies [19], the wide range of variation in GY losses observed in
this study indicated that indeed the applied stresses were severe and that the yield levels observed
could be attributed to stress tolerance. Compared with the NS environment, HS, DS, and DSHS, on the
averaged reduced GY by ~23%, 49%, and 58%, respectively (Tables 1 and 2), suggesting that the effect
of the combined stresses on GY of the extra-early maize accessions was higher than the individual
effects but lower than their sum (hypo-additive effect). These results corroborated the findings of
earlier workers who reported higher yield losses from the combined effects of DS and HS than DS and
HS applied alone in cereals including maize [17,19,29,46]. The high GY loss under DSHS compared to
HS and DS could be attributed to the interaction effects of HS and DS on stomatal movements [46].
Stress-induced changes in morpho-physiological properties of the maize accessions might have caused
osmotic imbalances under DSHS, resulting in the high yield losses. Under DS at elevated temperatures,
plants either close their stomata to prevent water loss or keep stomata opened to cool the leaves
through transpiration [46]. The long delay in anthesis and silking (by ≥8 days) under HS and DSHS
was most likely the result of severe cold stress due to harmattan at the time of planting, that might
have delayed seed emergence and extended pre-flowering developmental stages. Reduction in plant
and ear heights were on the average, higher under DS (26%) compared to HS (6%) and DSHS (7%).
These results suggested the occurrence of DS during the early stages of growth and development of
the maize accessions and that the plants were only affected by HS and DSHS towards the end of the
vegetative phase. Indeed, in the present study, DS was imposed at the early growth stages (25 DAP)
compared to the HS and DSHS, which had the treatments imposed at 32 DAP. Under DS and DSHS,
the number of ears per plant were strongly reduced (26%) relative to HS conditions, and this might
have contributed to the high yield losses observed under DS (49%) and DSHS (58%) compared to HS
(23%). This was further evident from the strong association of EPP and GY with the high yielding
accessions as revealed by the PCA biplot analysis (Figure 7).

Grain yield measured under NS conditions was to some extent predictive of the performance
under the applied stresses as indicated by the correlations across the treatments, suggesting that
similar physiological mechanisms may be conditioning yield potential under NS and DS, HS and
DSHS conditions. These observations, in part agreed with the results of previous studies [18,29].
The correlation between GY under NS and DS (0.66) observed in this study is similar to that (0.63)
reported by Cairns et al. [29] but slightly lower than that (0.75) recorded by Meseka et al. [18]. In the
present study, the moderately strong positive correlation between GY under DS and DSHS ($r = 0.60$;
$p < 0.0001$) relative to DS and HS ($r = 0.48$; $p < 0.001$) as well as HS and DSHS ($r = 0.37$; $p < 0.001$)
suggested that while yield performance under DS was predictive of attainable yield under DSHS,
tolerance to HS was independent of tolerance to DS and DSHS in the extra-early maize accessions.
Cairns et al. [29] suggested that tolerance to combined DS and HS in maize was genetically distinct
from tolerance to the individual stresses, and tolerance to either stress alone did not confer tolerance to
DSHS conditions. Similarly, Meseka et al. [18] found that tolerance to DS was independent of tolerance
to DSHS. The discrepancy between our results and those of Cairns et al. [29] and Meseka et al. [18]
particularly for the mechanism of DS and DSHS tolerance could be attributed to differences in maturity
classes of the genetic materials investigated. Cairns et al. [29] evaluated early, intermediate and late
maturing inbred lines while Meseka et al. [18] assessed drought-tolerant three-way cross maize hybrids,
and a local variety that were intermediate to late maturing as compared to the present study where
extra-early maturing maize accessions were studied. The strong correlations observed between the
measured traits under the different treatments implied the presence of common genetic elements
regulating the expression of these traits under the research conditions. The negative genetic correlations
observed between phenology and GY, PASP, EASP and HC under DS, HS and DSHS indicated that
early flowering was associated with higher GY and desirable PASP, EASP, and HC, which might have
contributed resilience to the stresses.

It is widely acknowledged that selection for increased GY together with highly heritable secondary traits can lead to remarkable progress in genetic gains under abiotic stress conditions [42]. For example, significant genetic gains were reported under low nitrogen stress and DS by complementing selection for GY potential with key secondary traits [47]. In the present study, sequential multiple regression analysis identified EASP, PASP, SL, and SG, and to some extent, EPP, and LF as the principal determinants of GY, explaining more than 80% of the differences in GY levels observed under the different stresses (Figures 3–5). This together with the moderate to high repeatability estimates of these traits indicated their potential to improve selection efficiency for GY under the abiotic stresses. Regression of TB on GY revealed that TB accounted for 15% and 25% of the yield reduction of the maize accessions under DSHS and HS conditions, respectively (Figure 8). Reduction in GY by 28% was attributed to TB under DSHS [18].

To allow efficient identification and selection of accessions tolerant to the stresses, we used a base index that integrated superior grain yield, EPP, anthesis-silking interval, plant and ear aspects, and the stay-green characteristic under each applied stress [38]. Promising accessions were identified, most of them tolerant to the individual stresses (Table 4). The tolerant landraces might contain novel resistance genes or combinations of resistance gene that would be valuable for 'climate smart' maize breeding efforts. In particular, eight landrace accessions were tolerant to DS and DSHS. Only five landrace accessions performed well across HS and DSHS, while four more accessions namely, TZm-1162, TZm-1167, TZm-1472, and TZm-1508 had outstanding performance across the three stresses. Therefore, development of lines from accessions that performed well across the stresses might be a successful strategy for abiotic stress tolerance maize breeding since hybrids resulting from the combination of the parents can have high performance for both DS and HS. The fact that only four accessions were tolerant across all the stresses was most likely the result of different physiological and morphological mechanisms conditioning tolerance to the three stresses applied in this study. Therefore, it is important to evaluate maize germplasm under each abiotic stress separately.

The genetic relationships/relatedness among germplasm under abiotic stress conditions are extremely important in determining specific groups of accessions that have good levels of tolerance to a specific stress. Herein, GY and stress adaptive secondary traits included in the base index selection were used to examine these relationships. The aggregation of the genotypes into four or more clusters under each stress further highlighted the potential of the landrace accessions to provide new genetic variation for abiotic stress tolerance in maize. In agreement with previous studies [18,19,48], cluster analyses showed a clear distinction between tolerant accessions and their susceptible counterparts. Under each stress condition, the highly tolerant landrace accessions clustered together with the most tolerant drought and/or heat-tolerant checks. This result suggested that the outstanding landraces and the resistant check cultivars were genetically similar. Majority of the outstanding landraces originated from Burkina Faso (a Sahel country). Thus, natural and artificial selection under the drier and hotter climatic conditions of the Sahel might have resulted in the excellent levels of abiotic stress resistance in these landraces. This observation provided further evidence that the superior landrace accessions identified in this study might contain novel resistance genes or combinations of resistance gene that could be valuable for expanding genetic base and thus, enhancing genetic gains in maize under the abiotic stress conditions.

5. Conclusions

This study uncovered striking levels of abiotic stress tolerance diversity among the extra-early landrace maize accessions, and identified traits potentially associated with tolerance to the stresses. Drought and heat stress, either individually or combined had significant negative effect on grain yield and other morpho-physiological traits of the maize accessions. Plant and ear aspects, stay green, lodging, leaf firing, and ears per plant were key to increased yield potential under the applied stresses. The performance pattern of the superior landrace accessions was similar to those of the best drought and/or heat-tolerant populations under DS, HS, and DSHS conditions. Therefore, they may be interesting for the development of germplasm tolerant to the stresses in SSA. In particular, accessions TZm-1162, TZm-1167, TZm-1472, and TZm-1508 showed good adaptation to the three stress. These landraces should be prioritised for further improvement of key adaptive traits and their introgression into maize breeding programs in SSA can play a considerable role in addressing the effects of drought and heat stress on maize. Extensive screening for abiotic stress tolerance in extra-early landrace accessions of maize should be undertaken to unearth further sources of tolerance.

Supplementary Materials: The following are available online at http://www.mdpi.com/2073-4395/10/3/318/s1, Table S1: List of the 72 extra-early maize accessions evaluated for tolerance to drought, heat, and combined drought and heat stress between 2017 and 2019 at Ikenne and Kadawa, Nigeria. Table S2: Genotypic and residual variance, and broad-sense heritability estimates of grain yield (kg/ha) of individual trials. Table S3: Cluster means (base index values and other secondary traits) of 72 extra-early maize accessions evaluated under non stress, drought, heat stress, and combined drought and heat stress conditions between 2017 and 2019, in Nigeria. Table S4: List of abbreviations and their explanations. Figure S1: Susceptible genotype (**left**) showing symptoms of leaf firing and tassel blasting under drought stress at elevated temperature at Kadawa, Nigeria. Figure S2: Ear aspect rating (plot basis) of extra-early maize accessions evaluated under heat stress conditions between 2017 and 2019 at Kadawa, Nigeria.

Author Contributions: Conceptualization: A.S.-P.N. and A.Y.T.; Methodology: C.N., B.B.-A. and A.L.G.-O.; Supervision: B.B.-A., A.S.-P.N., A.Y.T. and A.L.G.-O.; Data analysis and manuscript draft: C.N.; Manuscript review and editing, B.B.-A., A.Y.T., A.S.-P.N. and A.L.G.-O. All authors have read and agreed to the published version of the manuscript.

Funding: This research was financed by the German Federal Ministry of Education, through the West African Science Service Centre on Climate Change and Adapted Land-use (WASCAL) and, in part by the Bill & Melinda Gates Foundation [OPP1134248] through the funding support to the Stress Tolerant Maize for Africa (STMA) Project.

Acknowledgments: The first author is grateful to WASCAL for the PhD fellowship. We are grateful to the Genetic Resource Centre at IITA, Ibadan, Nigeria and the Plant Genetics Resources Institute at Bunso, Ghana for providing the maize accessions used in this study.

Conflicts of Interest: The authors declare no conflicting interest.

References

1.   McCann, J.C. *Maize and Grace: Africa's Encounter with a New World Crop, 1500–2000*; Harvard University Press: Cambridge, MA, USA, 2005.

2.   Warburton, M.L.; Reif, J.C.; Frisch, M.; Bohn, M.; Bedoya, C.; Xia, X.C.; Crossa, J.; Franco, J.; Hoisington, D.; Pixley, K.; et al. Genetic diversity in CIMMYT non-temperate maize germplasm: Landraces, open pollinated varieties, and inbred lines. *Crop Sci.* **2008**, *48*, 617–624. [CrossRef]

3.   Abate, T.; Fisher, M.; Abdoulaye, T.; Kassie, G.T.; Luduka, R.; Marenya, P.; Asnake, W. Characteristics of maize cultivars in Africa: How modern are they and how many do smallholder farmers grow? *Agric. Food Secur.* **2017**, *6*, 30. [CrossRef]

4.   Edmeades, G.O.; Trevisan, W.; Prasanna, B.M.; Campos, H. Tropical maize (*Zea mays* L.). In *Genetic Improvement of Tropical Crops*; Campos, H., Caligari, P.D.S., Eds.; Springer: Cham, Switzerland, 2017; pp. 57–109.

5.   Wu, X.; Wang, A.; Guo, X.; Liu, P.; Zhu, Y.; Li, X.; Chen, Z. Genetic characterization of maize germplasm derived from Suwan population and temperate resources. *Hereditas* **2019**, *156*, 2. [CrossRef]

6.   Singh, S.; Vikram, P.; Sehgal, D.; Burgueño, J.; Sharma, A.; Singh, S.K.; Sansaloni, C.P.; Joynson, R.; Brabbs, T.; Ortiz, C.; et al. Harnessing genetic potential of wheat germplasm banks through impact-oriented-prebreeding for future food and nutritional security. *Sci. Rep.* **2018**, *8*, 12527. [CrossRef]

7.   Masuka, B.P.; van Biljon, A.; Cairns, J.E.; Das, B.; Labuschagne, M.; MacRobert, J.; Olsen, M. Genetic diversity among selected elite CIMMYT maize hybrids in East and Southern Africa. *Crop Sci.* **2017**, *57*, 2395–2404. [CrossRef]

8.   Edenhofer, O. (Ed.) *Climate Change 2014: Mitigation of Climate Change*; Cambridge University Press: Cambridge, UK, 2015; p. 3.

9.   Muller, C.; Cramer, W.; Hare, W.L.; Lotze-Campen, H. Climate change risks for African agriculture. *Proc. Natl. Acad. Sci. USA* **2011**, *108*, 4313–4315. [CrossRef]

10.   Kurukulasuriya, P.; Mendelsohn, R.; Hassan, R.; Benhin, J.; Deressa, T.; Diop, M.; Eid, H.M.; Fosu, K.Y.; Gbetibouo, G.; Jain, S.; et al. Will African agriculture survive climate change? *World Bank Econ. Rev.* **2006**, *20*, 367–388. [CrossRef]

11. Cairns, J.E.; Hellin, J.; Sonder, K.; Araus, J.L.; MacRobert, J.F.; Thierfelder, C.; Prasanna, B.M. Adapting maize production to climate change in Sub-Saharan Africa. *Food Secur.* **2012**, *5*, 345–360. [CrossRef]

12. Knox, J.; Hess, T.; Daccache, A.; Wheeler, T. Climate change impacts on crop productivity in Africa and South Asia. *Environ. Res.* **2012**, *7*, 034032. [CrossRef]

13. Badu-Apraku, B.; Fakorede, M.A.B. Improvement of Early and Extra-Early Maize for Combined Tolerance to Drought and Heat Stress in Sub-Saharan Africa. In *Advances in Genetic Enhancement of Early and Extra-Early Maize for Sub-Saharan Africa*; Springer: Cham, Switzerland, 2017; pp. 311–358.

14. Cicchino, M.; Rattalino-Edreria, J.I.; Uribelarrea, M.; Otegui, M.E. Heat stress in field-grown maize: Response of physiological determinants of grain yield. *Crop Sci.* **2011**, *50*, 1438–1448. [CrossRef]

15. Zaidi, P.H.; Zaman-Allah, M.; Trachsel, S.; Seetharam, K.; Cairns, J.E.; Vinayan, M.T. *Phenotyping for Abiotic Stress Tolerance in Maize Heat Stress: A Field Manual*; CIMMYT: Mexico City, Mexico, 2016.

16. Araus, J.L.; Serret, M.D.; Edmeades, G.O. Phenotyping maize for adaptation to drought. *Front. Physiol.* **2012**, *3*, 305. [CrossRef]

17. NeSmith, D.S.; Ritchie, J.T. Effects of soil water-deficits during tassel emergence on development and yield components of maize (*Zea mays* L.). *Field Crop Res.* **1992**, *28*, 251–256. [CrossRef]

18. Meseka, S.; Menkir, A.; Bossey, B.; Mengesha, W. Performance Assessment of Drought Tolerant Maize Hybrids under Combined Drought and Heat Stress. *Agronomy* **2018**, *8*, 274. [CrossRef]

19. Nelimor, C.; Badu-Apraku, B.; Tetteh, A.Y.; N'guetta, A.S.P. Assessment of Genetic Diversity for Drought, Heat and Combined Drought and Heat Stress Tolerance in Early Maturing Maize Landraces. *Plants* **2019**, *8*, 518. [CrossRef]

20. Bassu, S.; Brisson, N.; Durand, J.L.; Boote, K.; Lizaso, J.; Jones, J.W.; Basso, B. How do various maize crop models vary in their responses to climate change factors? *Glob. Chang. Biol.* **2014**, *20*, 2301–2320. [CrossRef]

21. Wardlaw, I.F. Interaction between drought and chronic high temperature during kernel filling in wheat in a controlled environment. *Ann. Bot.* **2002**, *90*, 469–476. [CrossRef]

22. Shah, N.H.; Paulsen, G.M. Interaction of drought and high temperature on photosynthesis and grain-filling of wheat. *Plant Soil* **2003**, *257*, 219–226. [CrossRef]

23. Prasad, P.V.V.; Pisipati, S.R.; Momicilovic, I.; Ristic, Z. Independent and combined effects of high temperature and drought stress during grain filling on plant yield and chloroplast EF-Tu expression in spring wheat. *J. Agron. Crop Sci.* **2011**, *197*, 430–441. [CrossRef]

24. Pradhan, G.P.; Prasad, P.V.V.; Fritz, A.K.; Kirkham, M.B.; Gil, B.S. Effects of drought and high temperature stress on synthetic hexaploid wheat. *Funct. Plant Biol.* **2012**, *39*, 190–198. [CrossRef]

25. Heiniger, R.W. *The Impact of Early Drought on Corn Yield*; North Carolina State University: Raleigh, NC, USA, 2001.

26. Dwivedi, S.L.; Ceccarelli, S.; Blair, M.W.; Upadhyaya, H.D.; Are, A.K.; Ortiz, R. Landrace germplasm for improving yield and abiotic stress adaptation. *Trends Plant Sci.* **2016**, *21*, 31–42. [CrossRef]

27. Castro-Nava, S.; Ramos-Ortiz, V.H.; Reyes-Méndez, C.A.; Briones-Encinia, C.A.F.; López-Santillán, J.A. Preliminary field screening of maize landrace germplasm from northeastern Mexico under high temperatures. *Maydica* **2011**, *56*, 77–82.

28. Gedil, M.; Menkir, A. An integrated molecular and conventional breeding scheme for enhancing genetic gain in maize in Africa. *Front. Plant Sci.* **2019**, *10*. [CrossRef] [PubMed]

29. Cairns, J.E.; Crossa, J.; Zaidi, P.H.; Grudloyma, P.; Sanchez, C.; Araus, J.L.; Menkir, A. Identification of drought, heat, and combined drought and heat tolerant donors in maize. *Crop Sci.* **2013**, *53*, 1335–1346. [CrossRef]

30. Alam, M.S.; Seetharam, K.; Zaidi, P.H.; Dinesh, A.; Vinayan, M.T.; Nath, U.K. Dissecting heat stress tolerance in tropical maize (*Zea mays* L.). *Field Crop. Res.* **2017**, *204*, 110–119. [CrossRef]

31. Nelimor, C.; Badu-Apraku, B.; Nguetta, S.P.A.; Tetteh, A.Y.; Garcia-Oliveira, A.L. Phenotypic characterization of maize landraces from Sahel and Coastal West Africa reveals marked diversity and potential for genetic improvement. *J. Crop Improv.* **2019**. [CrossRef]

32. SAS Institute Inc. *SAS User's Guide: Statistics*; Version 9.4; SAS Institute Inc.: Cary, NC, USA, 2017.

33. Falconer, D.S.; Mackay, T.F.C. *Introduction to Quantitative Genetics*, 4th ed.; Longman: New York, NY, USA, 1996.

34. Peterson, B.G.; Carl, P.; Boudt, K.; Bennett, R.; Ulrich, J.; Zivot, E.; Lestel, M.; Balkissoon, K.; Wuertz, D.; Peterson, M.B.G. Performance Analytics: Econometric Tools for Performance and Risk Analysis. R Package Version 1. Available online: https://cran.r-project.org/web/packages/PerformanceAnalytics/index.html (accessed on 15 August 2014).

35. Cooper, M.; DeLacy, I.H.; Basford, K.E. Relationship among analytical methods used to analyse genotypic adaptation in multi-environment trials. In *Plant Adaptation and Crop Improvement*; Cooper, M., Hammer, G.L., Eds.; Cambridge University Press: Cambridge, UK, 1996; pp. 193–224.

36. Vargas, M.; Combs, E.; Alvarado, G.; Atlin, G.; Mathews, K.; Crossa, J. META: A suit of SAS programs to analyse multi-environment breeding trials. *Agron. J.* **2013**, *105*, 11–19. [CrossRef]

37. Mohammadi, S.A.; Prasanna, B.M.; Singh, N.N. Sequential path model for determining interrelationships among grain yield and related characters in maize. *Crop Sci.* **2003**, *43*, 1690–1697. [CrossRef]

38. Badu-Apraku, B.; Akinwale, R.O.; Oyekunle, M. Efficiency of secondary traits in selecting for improved grain yield in extra-early maize under Striga-infested and Striga-free environments. *Plant Breed.* **2014**, *133*, 373–380. [CrossRef]

39. Talabi, A.O.; Badu-Apraku, B.; Fakorede, M.A.B. Genetic variances and relationship among traits of an early-maturing maize population under drought-stress and low-N environments. *Crop Sci.* **2016**, *57*, 681–692. [CrossRef]

40. Badu-Apraku, B.; Fakorede, M.A.B.; Talabi, A.O.; Oyekunle, M.; Akaogu, I.C.; Akinwale, R.O.; Annor, B.; Melaku, G.; Fasanmade, Y.; Aderounmu, M. Gene action and heterotic groups of early white quality protein maize inbreds under multiple stress environments. *Crop Sci.* **2015**, *56*, 183–199. [CrossRef]

41. Statistical Package for Social Sciences (SPSS) Inc. *SPSS Base 17.0 for Windows User's Guide*; SPSS Inc.: Chicago, IL, USA, 2007.

42. Bolaños, J.; Edmeades, G.O. Eight cycles of selection for drought tolerance in lowland tropical maize. I. Responses in grain yield, biomass, and radiation utilization. *Field Crop. Res.* **1993**, *31*, 233–252. [CrossRef]

43. Araus, J.L.; Kefauver, S.C.; Zaman-Allah, M.; Olsen, M.S.; Cairns, J.E. Translating high-throughput phenotyping into genetic gain. *Trends Plant Sci.* **2018**, *23*, 451–466. [CrossRef] [PubMed]

44. Das, B.; Atlin, G.N.; Olsen, M.; Burgueño, J.; Tarekegne, A.; Babu, R.; Ndou, E.N.; Mashingaidze, K.; Moremoholo, L.; Ligeyo, D.; et al. Identification of donors for low-nitrogen stress with maize lethal necrosis (MLN) tolerance for maize breeding in sub-Saharan Africa. *Euphytica* **2019**, *215*, 80. [CrossRef]

45. Gouesnard, B.; Zanetto, B.; Welcker, C. Identification of adaptation traits to drought in collections of maize landraces from southern Europe and temperate regions. *Euphytica* **2015**, *209*, 565–584. [CrossRef]

46. Mahrookashani, A.; Siebert, S.; Hüging, H.; Ewert, F. Independent and combined effects of high temperature and drought stress around anthesis on wheat. *J. Agron. Crop Sci.* **2017**, *203*, 453–463. [CrossRef]

47. Edmeades, G.O.; Bolaños, J.; Chapman, S.C.; Lafiite, H.R.; Bänziger, M. Selection improves drought tolerance in tropical maize populations. I. Gains in biomass, grain yield and harvest index. *Crop Sci.* **1999**, *39*, 1306–1315. [CrossRef]

48. Tandzi, L.N.; Bradley, G.; Mutengwa, C. Morphological responses of maize to drought, heat and combined drought and heat stresses. *J. Biol. Sci.* **2019**, *19*, 7–16. [CrossRef]

Article

Transcriptomic Analysis of Female Panicles Reveals Gene Expression Responses to Drought Stress in Maize (*Zea mays* L.)

Shuangjie Jia [1,†], Hongwei Li [1,†], Yanping Jiang [1], Yulou Tang [1], Guoqiang Zhao [1], Yinglei Zhang [1], Shenjiao Yang [2], Husen Qiu [2], Yongchao Wang [1], Jiameng Guo [1], Qinghua Yang [1,*] and Ruixin Shao [1,*]

[1] The Collaborative Center Innovation of Henan Food Crops, National Key Laboratory of Wheat and Maize Crop Science, Henan Agricultural University, Zhengzhou 450046, China; jiasj2017@126.com (S.J.); L_hongwei@126.com (H.L.); jiangyanping.up@gmail.com (Y.J.); tyl134679@163.com (Y.T.); z1013468268@163.com (G.Z.); yinglei609@163.com (Y.Z.); wangyongchao723@163.com (Y.W.); guojiameng@hotmail.com (J.G.)

[2] Farmland Irrigation Research Institute, CAAS/National Agro-ecological System Observation and Research Station of Shangqiu, Xinxiang 453002, China; shenjiao@gmail.com (S.Y.); qiuhusen2008@163.com (H.Q.)

[*] Correspondence: shaoruixin@henau.edu.cn (R.S.); yangqinghua@henau.edu.cn (Q.Y.)

[†] These authors are equal contribution to this study.

Received: 8 January 2020; Accepted: 14 February 2020; Published: 24 February 2020

Abstract: Female panicles (FPs) play an important role in the formation of yields in maize. From 40 days after sowing to the tasseling stage for summer maize, FPs are developing and sensitive to drought. However, it remains unclear how FPs respond to drought stress during FP development. In this study, FP differentiation was observed at 20 and 30 days after drought (DAD) and agronomic trait changes of maize ears were determined across three treatments, including well-watered (CK), light drought (LD), and moderate drought (MD) treatments at 20, 25, and 30 DAD. RNA-sequencing was then used to identify differentially expressed genes (DEGs) in FPs at 30 DAD. Spikelets and florets were suppressed in LD and MD treatments, suggesting that drought slows FP development and thus decreases yields. Transcriptome analysis indicated that 40, 876, and 887 DEGs were detected in LD/CK, MD/CK, and MD/LD comparisons. KEGG pathway analysis showed that 'biosynthesis of other secondary metabolites' and 'carbohydrate metabolism' were involved in the LD response, whereas 'starch and sucrose metabolism' and 'plant hormone signal transduction' played important roles in the MD response. In addition, a series of molecular cues related to development and growth were screened for their drought stress responses.

Keywords: transcriptome analysis; summer maize; drought; female panicle

1. Introduction

Under the influence of global warming, changes in climatic conditions are creating unusual weather phenomena worldwide, often imposing drought stress on crops [1,2]. From the agricultural perspective, drought often results in decreased crop productivity and growth [3–5], especially for cereal crops. Maize (*Zea mays* L.) is one of the most important cereal crops and has the most extensive planting area globally [6,7]. One of the most important factors limiting maize growth and development around the world is a lack of water [8–11]. Accordingly, improving tolerance of maize to drought stress is essential for achieving high and stable yields in cereal crops.

As a multidimensional stress, water limitation triggers a wide variety of plant responses; these range from responses at the physiological and biochemical levels to the molecular level [12–16]. When external drought stimuli are perceived and captured by sensors on cell membranes, the signals are transmitted through multiple signal transduction pathways. Then, plant can regulate the expression of drought-responsive genes to protect themselves from the harmful effects of external stimuli [17,18]. The expressed products of drought-responsive genes are mainly proteins involved in signaling cascades and transcriptional regulation (such as protein kinase, protein phosphatase, and transcription factors) and functional proteins [19]. With the rapid development of high-throughput sequencing technologies, transcriptome analyses have been conducted to identify stress-mediated differences at the level of gene expression. Previous research has shown that many significantly differentially regulated genes that were associated with drought tolerance are induced in different organs of maize plants [20–26]. For example, 249 and 3000 differentially expressed genes (DEGs) were involved in root tissues after 6 h of light and severe drought stress, respectively [23]. In leaves, a total of 619 DEGs and 126 transcripts had their expression levels altered by drought stress at flowering time [24]. In tassels, 1902 DEGs were found after 5–7 days of drought stress [25]. In young ears, a total of 1825 DEGs were identified on the 5th day of drought stress at the V9 stage [26].

The panicle stage (from jointing to flowering) is the key stage for panicle differentiation and development in maize, the number of rows per ear and the number of grains per row are dependent on spikelet and floret differentiation at this time [27]. Therefore, the growth and development of female panicles (FPs) play an important role in the formation of maize yields. Although great advances in understanding differentiation of FPs and how drought stress affects genes transcription in FPs have been achieved in the past few decades [26,28–31], so far, progress in understanding the general molecular basis of FP development in response to long-term drought stress across the panicle stage has not been reported.

Accordingly, in this study, maize inbred line PH6WC (6WC) was used as drought-sensitive experimental material [32], soil water was controlled by means of drip irrigation for 30 days, and the gene expression dynamics of developing FPs at 30 days after drought (DAD) were investigated using transcriptomic analysis. The DEGs were identified and assigned to functional categories to reveal the various metabolic pathways in FPs that are involved in responses to long-term drought stress at different levels. Furthermore, differences in transcription factors between treatments were also analyzed. Overall, the exploration and function prediction of drought-response genes in maize FPs represent an efficient approach to improving the molecular breeding of drought-resistance maize cultivars.

2. Materials and Methods

2.1. Plant Material and Growth Conditions

Field experiments were conducted at field experiment stations (34°31′ N, 115°35′ E, 50.7 m above sea level) during the maize growing season (June–October, 2018) in Shangqiu (Henan, China). The maize inbred line 6WC was grown in 9 experimental plots (each plot was 2 m wide, 3.3 m long, 1.8 m deep) which were under a movable awning and filled with luvo-aquic soils, with a 20 cm sand filter layer at the bottom. Maize were planted into four rows, with 40 cm between rows in each plot. Two to three seeds were sown at each acupoint, with subsequent thinning to one seedling conducted at the trifoliate stage (V3). The final stand density was 8 plants m^{-2}. During the jointing and tasseling stages, topdressing fertilizer was applied, and weeds, insects, and diseases were controlled throughout the experiment. The top soil (0–40 cm layer) had a pH (water) of 7.3, mean mineral P content of 3.24 g/kg, and inorganic N at sowing of 3.60 g/kg. The average daily maximum and minimum temperatures of the field experiment during the trial were 32.98 °C and 20.71 °C, respectively.

2.2. Drought Stress Treatments

During FP development, soil moisture was controlled by means of drip irrigation at 80 ± 5% of the field water capacity (FWC) (well-watered, CK), 60 ± 5% of FWC (light drought, LD), and 45 ± 5% of FWC (moderate drought, MD). The meter was checked every morning and evening throughout the growth period to guide adjustments of the soil moisture. When the treated soil moisture dropped towards its lower limit, moderate drip irrigation was carried out until the upper limit level was reached, and the irrigation volume was measured by a water meter. After 30 DAD, the drought treatment plots were rehydrated to the CK level. Other field management measures reflected standard field management practices.

2.3. Measurement of Morphology and Microscopic Observation of Female Panicles

Plant height was measured from the ground to the top of the leaves in their natural growth state at 20, 25, and 30 DAD. The length and width of all the green leaves were measured by ruler in order to calculate leaf area, and the leaf area index (LAI) was determined according to this method [33]. Dry matter accumulation in stalks, leaves, tassels and ears of maize plants were measured at 20, 25, and 30 DAD. After 30 min of defoliation at 105 °C, dry weight was determined after being dried at 75 °C until a constant weight was reached. The percentage of drought limitation was calculated as (T2−T1)/T1. Here, T2 was shoot dry matter under the MD or LD treatment, while T1 was shoot dry matter under the control or LD treatment [34]. FPs were dissected with a dissecting needle at 20 and 27 DAD and analyzed under a stereomicroscope (Guanpujia, SMZ-B2, Beijing, China) to observe the differentiation of developing female inflorescences. There were three biological replicates in each group.

2.4. RNA Isolation and Illumina Sequencing

Three plants with consistent growth were selected from each treatment at 30 DAD, FP bracts were sampled, and the upper, middle, and lower parts of the ears were mixed evenly and then frozen at −80 °C prior to RNA-sequencing (RNA-seq) analysis. Total RNA was extracted using the mirVana miRNA Isolation Kit (Ambion, Inc., Austin, TX, USA). RNA integrity was evaluated using the Agilent 2100 Bioanalyzer (Agilent Technologies, Santa Clara, CA, USA). Shanghai OE Biotech (Shanghai, China) conducted RNA-Seq library construction and high-throughput sequencing based on total RNA from the female inflorescence. The libraries were sequenced on the Illumina sequencing platform (HiSeq 2500l, Illumina, San Diego, CA, USA) with 125 bp paired-end reads. The raw RNA-seq data have been uploaded to NCBI SRA (BioProject ID: PRJNA604094).

2.5. Read Mapping and Differential Expression Analysis

Base calling was conducted using the raw image data generated by sequencing to obtain sequence data, and the called raw data (raw reads) were stored in fastq format. Raw data (raw reads) were processed using Trimmomatic [35]. The reads containing poly-N runs and low-quality reads were removed to obtain the clean reads. Then, the clean reads were mapped to the reference (NCBI_B73_v4) genome [36] using HISAT2 (version 2.2.1.0) [37]. The Fragments Per kb Per Million Reads (FPKM) values were calculated using cufflinks (version 2.2.1) [38,39], followed by differential expression analysis using DESeq (version1.18.0) [40]. Genes with |fold change| > 2 and $p < 0.05$ were identified as differentially expressed genes with p presented as raw p-values rather than FDR adjusted p-values.

2.6. GO and KEGG Enrichment Analysis

The significantly expressed Gene Ontology (GO) terms were selected by GO enrichment analysis according to the GO database (http://geneontology.org/). The differences in the frequency of assignment of GO terms in the DEG set were compared with the expressed genes in the CK, LD, and MD samples ($p < 0.05$). Functional groups encompassing DEGs were identified based on GO analysis, and pathway analysis was conducted according to the Kyoto Encyclopedia of Genes and Genomes (KEGG) database (http://www.genome.jp/kegg/), with manual reannotation based on several databases and a literature search.

2.7. Differential Expression Verification by Quantitative Real-Time PCR (qRT-PCR)

Transcriptome sequencing data were validated by qRT-PCR. Total RNA was reverse transcribed using EasyScript One-Step gDNA Removal and cDNA Synthesis SuperMix (TRANS, Beijing, China). The qRT-PCR experimental methods used HiScript Q RT SuperMix for qPCR (Vazyme, Nanjing, China). The primer sequences were designed using Primer 5 and are listed in Supplementary Materials Table S1. The relative quantification $2^{-\Delta\Delta Ct}$ method was used to calculate the expression level of target genes in different treatments.

2.8. Statistical Analysis

Data collation and graphic rendering were conducted with SIGMAPLOT 14.0 (Systat Software Inc., San Jose, CA, USA), Microsoft Excel, and Microsoft PowerPoint 2016 software (Microsoft Corp., Redmond, WA, USA). All data are expressed as the mean ± SD value from three independent experiments unless otherwise stated. Data were analyzed by one-way ANOVA using Duncan's multiple range test at a $p < 0.05$ significance threshold in SPSS (IBM Corp., Armonk, NY, USA).

3. Results

3.1. Female Panicle Development, Phenotypic Change and Yield Components

Plant height, LAI, dry matter accumulation, and the percentage of drought limitation were measured at 20, 25, and 30 DAD. For example, plant height was reduced by 17.63%, 17.01%, and 16.44% under the LD treatment compared with CK, respectively; furthermore, MD significantly decreased plant height by 25.86%, 26.34%, and 29.73% (Figure 1a), respectively. Leaf area index was significantly decreased under MD at 20, 25, and 30 DAD compared with CK plants, but was not significantly changed at 20 and 30 DAD under LD (Figure 1b). For dry matter accumulation, MD significantly decreased the shoot dry matter at 20, 25, and 30 DAD compared with CK plants, but it was not significantly reduced in the LD vs. CK comparison at 20 and 30 DAD (Figure 1c). In addition, the percentage of drought limitation had the highest absolute values at 20, 25, and 30 DAD after MD treatment. However, the percentage of drought limitation was not significantly affected by LD treatment at 20 and 30 DAD compared with CK (Figure 1d).

From 20 to 30 DAD, according to the book *Corn Growth and Development* [41], FPs may be in the process of differentiation. To explore the responses of FPs to drought stress, maize plants were dissected. The length and diameter of FPs were significantly decreased under the MD treatment at 20 and at 30 DAD; silk was seen in CK and LD plants, but there was no floral differentiation MD plants (Figure 2a). Proportion of dry matter in FP was decreased under LD and MD treatments (Figure 2b). To investigate the effect of drought stress on development and number of mature ears, ear size, and dry matter, yield components were determined (Figure 2c, Supplementary Materials Table S2), which showed that LD and MD treatments significantly decreased ear size and increased the bald tip length. For this reason, the numbers of rows and kernels were reduced by 14.00% and 29.00% under the LD treatment and 19.00% and 43.00%, respectively, under the MD treatment. Therefore, drought resulted in great losses in grain yield of 32.00% and 35.00% under the LD and MD treatments, respectively (Figure 2c, Supplementary Materials Table S2).

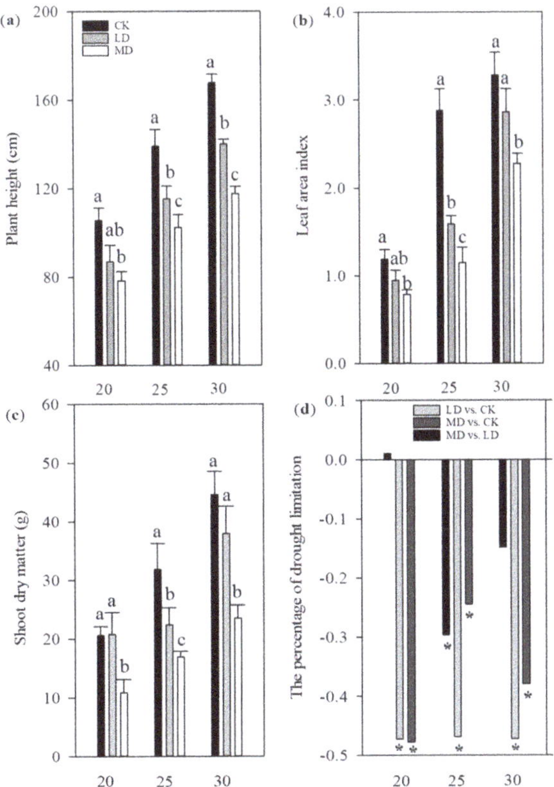

Figure 1. Agronomic traits changes of maize plants in response to different drought stress. Effect of drought stress on (**a**) plant height; (**b**) leaf area index; (**c**) shoot dry matter; and (**d**) the percentage of drought limitation at 20, 25, and 30 days after drought (DAD). Values are the means of the replicates ± sd. Different lowercase letters and * symbols indicate statistical significance of differences at a $p < 0.05$ level. There were three treatments, including well-watered (CK), light drought (LD), and moderate drought (MD), and five biological replicates were sampled for each treatment.

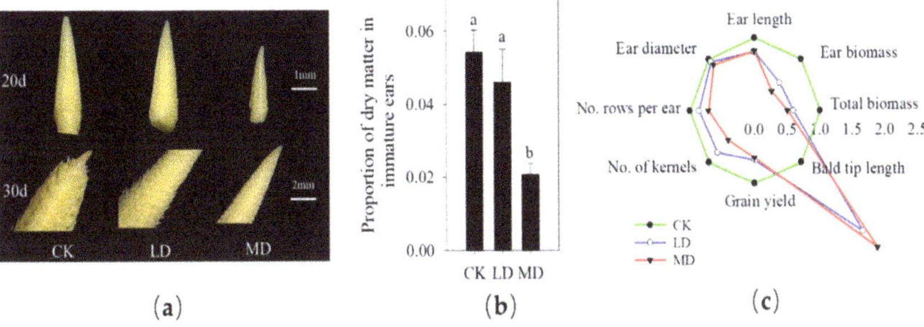

Figure 2. Development of female inflorescence and agronomic traits changes of ears in response to drought stress. (**a**) Contrasting sizes and differentiation of the control and drought-treated plants at 20 and 30 DAD. (**b**) Proportion of dry matter in immature ears at 30 DAD. (**c**) Radar chart showing changes in yield traits for mature maize plants grown under well-watered (CK), light drought (LD), and moderate drought (MD) conditions. (**b**) and (**c**) were calculated with 15 and 10 biological replicates sampled for each treatment.

3.2. Overview of RNA Sequencing and Mapping

A total of 49.42 million raw reads were obtained from PH6WC transcriptome libraries (Supplementary Materials Table S3). More than 96.72% of them (47.80 million clean reads) remained after discarding low-quality reads and reads containing adaptor sequences, which were then used for downstream analyses. The clean reads were mapped to the B73 reference genome (ZmB73_RefGen_v4). Overall, 94.34–94.52% of clean reads from nine samples were mapped onto the reference genome (Supplementary Materials Table S3). On average, approximately 43.93 (91.29%), 43.73 (91.13%), and 43.70 (91.17%) million reads from the CK, LD, and MD treatments, respectively, were uniquely mapped onto the reference genome.

Compared with the CK treatment, only 40 genes (log_2 foldchange > 1 and $p < 0.05$), including nine up-regulated and 31 down-regulated genes, showed significantly differential expression in the LD treatment, and a total of 212 up-regulated and 664 down-regulated genes were identified in the MD treatment. A total of 887 DEGs, including 208 up-regulated and 679 down-regulated genes, were identified in the MD versus LD comparison (Figure 3a). A Venn diagram of the DEGs illustrated that there were 10 common genes that appeared in the LD vs. CK and MD vs. CK comparisons, five genes shared between the LD vs. CK and MD vs. LD comparisons, and 565 genes shared between the MD vs. LD and MD vs. CK comparisons. However, there were no DEGs commonly expressed in all three comparisons (Figure 3b).

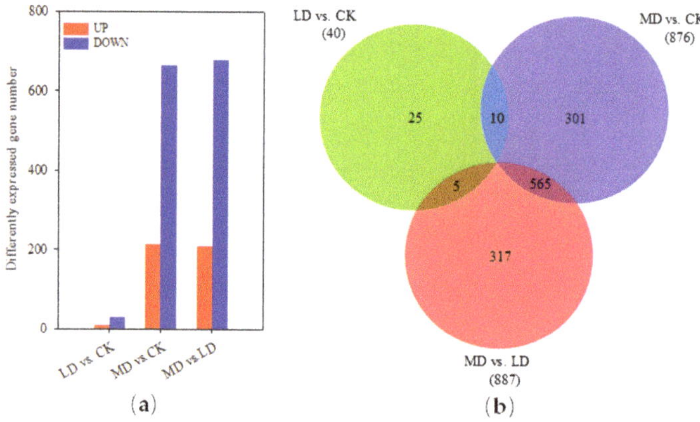

Figure 3. Identification and characterization of differentially expressed genes (DEGs) between the drought treatment and control plants. (**a**) The number of DEGs in three comparison groups. (**b**) A Venn diagram comparison summarizing overlaps in differentially expressed genes among the three comparisons.

3.3. Gene Expression Validation by qRT-PCR

To investigate the changes in gene expression at the mRNA level, eight randomly selected genes and three specific genes *cuc2* (LOC103629107), *TE1* (LOC541683), *DLF1* (LOC100037791) were analyzed using quantitative real-time RT-PCR for validation of RNA-seq. The level of expression of the genes amplified is shown in Figure 4.

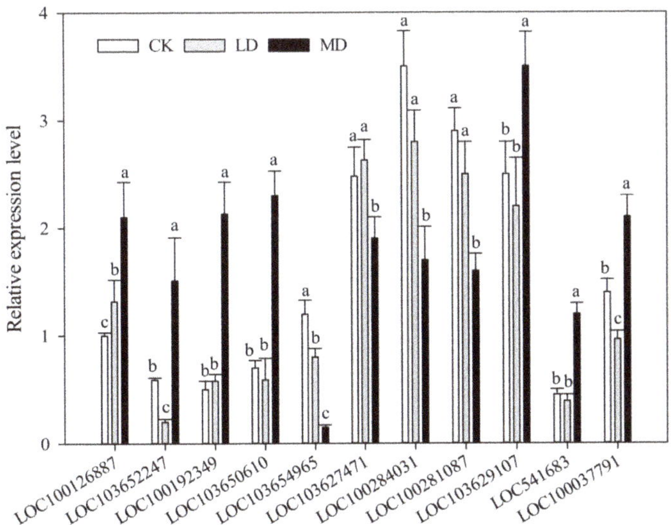

Figure 4. The expression patterns of eleven genes in female panicle tissues under well-watered (CK), light drought (LD), and moderate drought (MD) conditions by qRT-PCR. Values are the mean ± SD of three independent experiments. Maize β-actin expression was used as a control.

Raw data were compared to transcriptomics data (Supplementary Materials Table S4), which closely resembled each other, validating the differential expression of the genes identified as being under drought stress.

3.4. GO Annotation and Enrichment

A total of 26, 629, and 630 DEGs were assigned by GO analysis conducted based on the genes from the LD vs. CK, MD vs. LD, and MD vs. LD comparisons, respectively. The most significantly regulated 20 terms among biological processes from the MD vs. CK and MD vs. LD comparison genes are shown in Figure 5a,b, but there were no significant terms (i.e., terms with gene number > 2 and $p < 0.05$) resulting from the LD vs. CK comparison. The up-regulated terms from the MD vs. CK comparison are involved in "regulation of timing of plant organ formation," "developmental process," and "regulation of cell proliferation." The down-regulated terms "response to water deprivation" and "post-embryonic plant morphogenesis" were also enriched. The up-regulated terms from the MD vs. LD comparison are "regulation of timing of plant organ formation," "regulation of cell proliferation," and "gibberellin biosynthetic process." The down-regulated terms "reductive pentose-phosphate cycle," "phosphate ion homeostasis," and "ethylene-activated signaling pathway" were enriched among genes from the MD vs. LD comparison. The genes associated with GO terms related to development, growth, and responses to stimulus were also significantly different among the three comparisons. These genes that changed in their levels of transcriptional expression were basically the same in the MD vs. CK and MD vs. LD comparison groups, but lower in expression differences in the LD vs. CK comparison group (Figure 5c,d, Supplementary Materials Table S5).

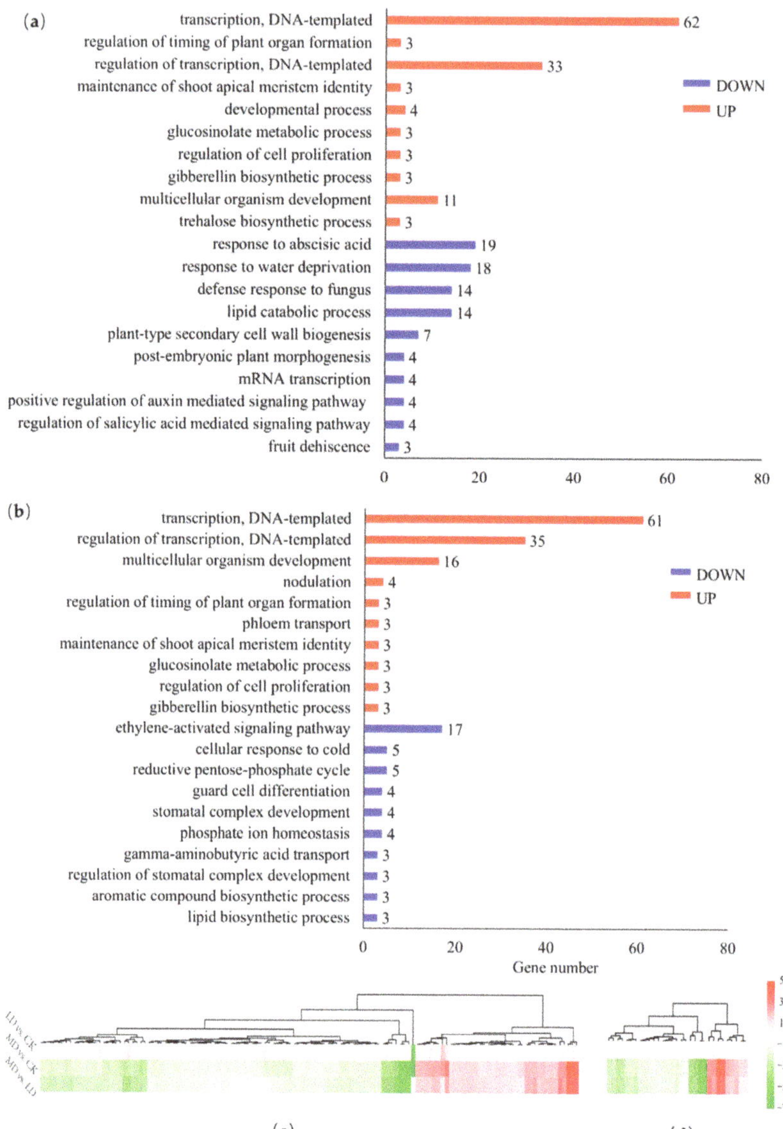

Figure 5. Top 20 biological processes enriched by the up-and down-regulated genes in the (**a**) MD vs. CK and (**b**) MD vs. LD comparisons. Expression pattern of the differentially expressed genes associated with (**c**) development progress and (**d**) growth in the three comparisons. Colors indicate the $\log_2$ fold change values. Red indicates up-regulation, and green indicates down-regulation in that comparison.

3.5. Metabolic Pathways Related to Soil Drought Stress

To further characterize genes affected by drought stress, we performed a KEGG pathway classification analysis to identify functional enrichment of DEGs. Thus, 8, 72, and 74 terms were significantly enriched in the transcriptome profile comparisons of LD vs. CK, MD vs. CK, and MD vs. LD groups (Supplementary Materials Table S6). The significant differences in the top 20 enriched KEGG pathways in the MD vs. CK and MD vs. LD comparisons are shown in Figure 6. In the

MD vs. CK comparison, genes associated with the pathway "Starch and sucrose metabolism" were most enriched followed by those associated with "plant hormone signal transduction" (Figure 6a, Supplementary Materials Table S7). In the MD vs. LD group, "Plant hormone signal transduction," "Starch and sucrose metabolism," and "Glycosphingolipid biosynthesis" were the three most enriched terms (Figure 6a, Supplementary Materials Table S7).

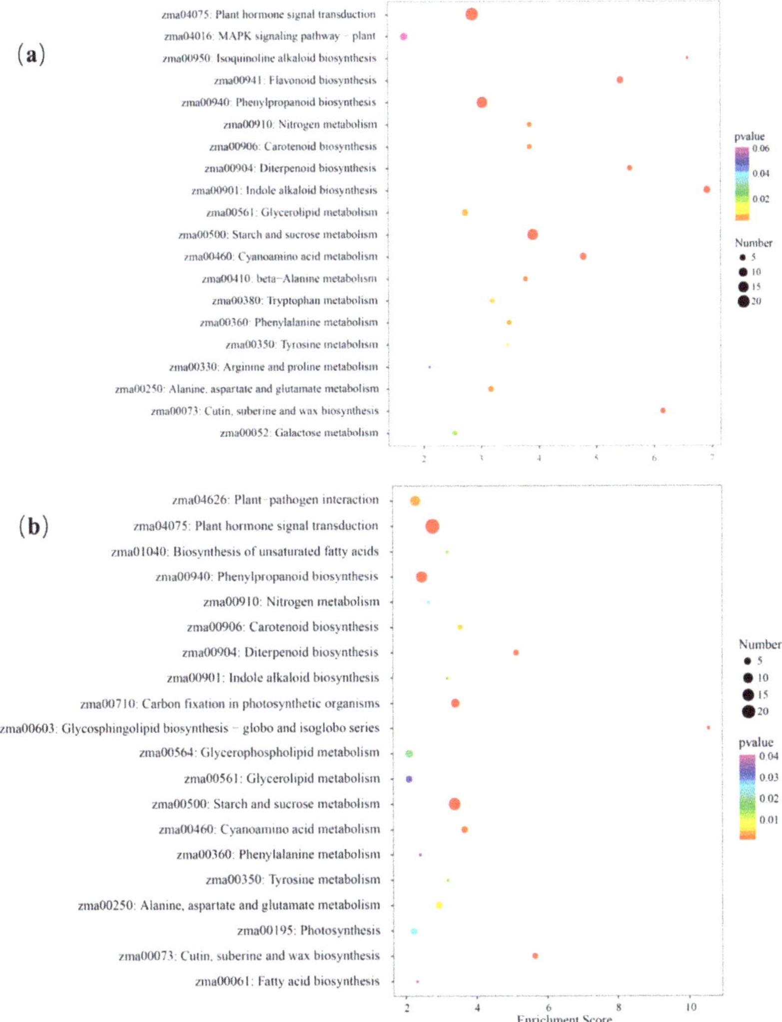

Figure 6. Top 20 enriched KEGG pathways in the (**a**) "MD vs. CK" and (**b**) "MD vs. LD" comparisons. Pathway entries with the corresponding number of genes (among those pathways with more than two genes) are shown, and the corresponding -log$_{10}$ *p*-value of each entry is sorted in descending order. The number of DEGs in each pathway is positively related to the size of plot, and the *p*-values shown in red are more significant.

4. Discussion

4.1. Responses of Plant Growth and Female Panicle Differentiation to Soil Drought Stress

The panicle stage is the most important productive stage in corn development, and soil drought stress in this stage can affect the plant growth rate, prolong the growth processes of the panicle stage, hinder the normal differentiation and development of ears, and ultimately lead to decreased crop seed setting rates and yields [42–46]. Further, the drought response depends on the time and intensity of water loss as well as the developmental stage [47,48]. In this study, LD and MD treatments compared with the CK treatment decreased green leaf area and significantly suppressed shoot dry matter accumulation over the prolonged drought treatment, and relative to the LD treatment, the MD treatment affected plant growth much more (Figure 1), which is consistent with previous research by Boonjung et al. [49].

FPs are the precursor to maize ears, and the differentiation and development of FPs mainly occurs from the V9 to VT stages, which include growth cone extension (V9), spikelet differentiation (V11), floret differentiation (V12), and formation of the sexual organs (VT). Developing organs are sensitive to drought, especially during their early phases [50]. When drought occurs between the V9 and VT stages, how does the degree of drought affect the formation of ears? In our study, spikelet and floral differentiation, as observed under stereomicroscope, were significantly inhibited and thus delayed by soil drought (Figure 2). Some studies have shown that the number of kernel rows is determined at the spikelet differentiation stage, and the floret differentiation period is the key period that affects grain number [51–53]. Here, mature ears in the MD and LD treatments were much shorter and thinner than those in the CK treatment; in addition, the bald tip length and number of unfilled grains were both increased under drought treatments. As indicated above, drought affected grain yields as well (Figure 2c).

4.2. Genes Involved in Development and Growth in Response to Soil Drought Stress

Drought treatments affected the expression of genes associated with development and growth of the inflorescence (Figure 4). *Terminal ear 1* (*te1*) maize mutants have shortened internodes, abnormal phyllotaxy, leaf pattern defects, and partial feminization of tassels [54]. Similarly, *cup-shaped cotyledon 2* (*cuc2*) mutants have been reported to have abnormalities in the regulation of the shoot meristem boundary and formation and subsequent development [55,56]. Here, *cuc2* were up-regulated under moderate drought stress (Figure 4, Supplementary Table S5), combined with developmental change (Figure 2a), implying that the differential expression of the gene under MD treatments may be related to the mature delay of the FPs tissue. *DLF1* was also up-regulated under MD stress (Figure 4, Supplementary Materials Table S5), suggesting that the trans-activator protein encoded by, this gene plays an important role in the signal transduction pathway and the regulation of plant growth at the FP development stage [57].

4.3. Genes Involved in Auxin Signaling in Response to Soil Drought Stress

Auxin is an important phytohormone that is closely related to plant resistance to adverse environmental conditions, and it can induce rapid and transient expression of some genes, including auxin response factor genes (ARF) and primary auxin response genes (Aux/IAA, GH3, SAUR and LBD); the protein products of these genes can specifically bind to ARFs to activate or inhibit downstream gene expression under drought [58–61]. In the current study, auxin signaling genes were involved in the response to drought, as the expression of IAA-conjugating genes (GH3) was up-regulated, and the expression levels of auxin biosynthesis genes were down-regulated after MD stress (Figure 7), leading to the reduction of auxin levels (Supplementary Materials Figure S1a). This implies that drought improves GH3 transcription to help maintain endogenous auxin at an appropriate level in maize [62,63]. However, when the concentration of auxin increases, auxin combines with transport inhibitor response 1 (TIR1), causing Aux/IAA ubiquitination and degradation; then, ARF is released, which further activates the expression of small auxin-up RNA (SAUR) genes [64]. SAUR genes are early auxin-responsive genes involved in plant growth, and the SAUR family regulates a series of cellular, physiological, and developmental processes in response to environmental signals [65–67]. In our study, three SAUR genes were down-regulated under the MD treatment (Figure 7), which may explain MD-induced inhibition of maize growth.

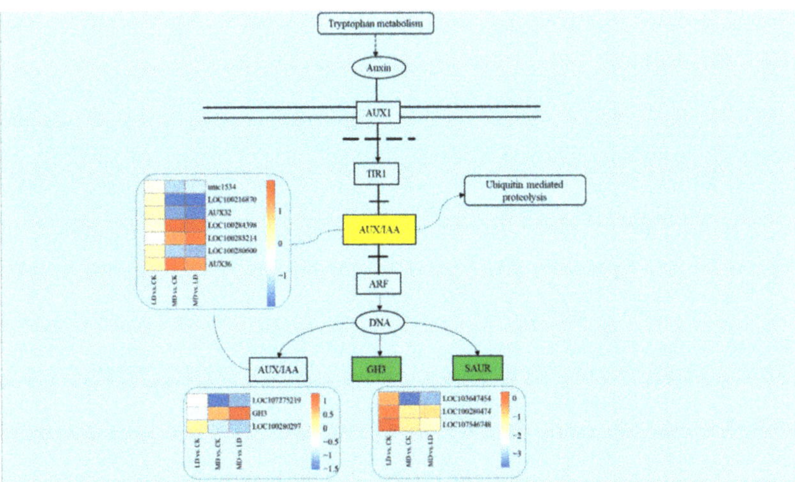

Figure 7. Genes involved in auxin plant hormone signal transduction pathway in the Kyoto Encyclopedia of Genes and Genomes (KEGG). Differentially expressed genes involved in AUX/IAA, GH3, and SAUR were shown by heat-maps, and the number was calculated with log2foldchange in three comparisons. Color of heat-maps represented different fold change. Yellow box means involved significantly differentially expressed genes were mainly up-regulated, and green box means down-regulated.

4.4. Reactive Oxygen Scavenging System and Ion Channel in Response to Soil Drought Stress

Limited water supply enhances the production of reactive oxygen species (ROS) [68,69], and plants are protected by glutathione *S*-transferase (GST) and other antioxidant enzymes scavenging excessive ROS from the damage caused by ROS [70]. This is because GST comprises a large superfamily of multifunctional protein and participates in ascorbic acid (AsA)/glutathione (GSH) cycling pathways [68]. Here, probable glutathione *S*-transferase *GST12* was significantly down-regulated under the MD treatment compared with the CK treatment. However, *GSTU6* (LOC103637303) and GST activity was up-regulated under MD stress (Supplementary Materials Table S8 and Figure S1b), which may scavenge ROS and protect both plant cell membrane structure and protein activity [71,72], implying that GST is involved in responding to drought stress [73,74].

Supplementary Materials: The following are available online at http://www.mdpi.com/2073-4395/10/2/313/s1, Table S1: The primer sequences for qRT-PCR, Table S2: Ear and grain yield traits at harvest time under CK, LD and MD treatments, Table S3: Number of reads based on RNA-Seq data of CK, LD and MD treatments, Table S4: RNA-Seq expression levels of the eight genes for qRT-PCR, Table S5: Expression pattern of the differently expressed genes about development progress (A) and growth (B) under CK, LD and MD treatments, Table S6: Number of enriched KEGG pathways terms and DEGs in three comparisons (LD vs.CK, MD vs. CK; MD vs. LD), Table S7: Differentially expressed genes (DEGs) in the top 20 enriched KEGG pathways in two comparisons (MD vs. CK; MD vs. LD), Table S8: Genes significantly enriched in Glutathione metabolism in three comparisons (LD vs.CK, MD vs. CK; MD vs. LD), Figure S1: Effect of drought stress treatments on the content of IAA (a) and the activities of GST (b) in female panicles. CK, well-watered; LD, light drought; MD, moderate drought, five biological replicates were sampled for each treatment.

Author Contributions: S.J. did the experiment and wrote the manuscript. H.L. analyzed the data. Y.J., Y.T., G.Z. and Y.Z. did a part of experiment. S.Y., H.Q., Y.W. and J.G. gave some good suggestions on the manuscript. Q.Y.

revised the manuscript. R.S. designed this experiment, revised the manuscript and provided the funding. All authors have read and agreed to the published version of the manuscript.

Funding: This research was funded by Open Foundation of CAAS/Key Laboratory of Crop Water Use and Regulation, Ministry of Agriculture (FIRI2019-02-0103), Open Foundation of State Key Laboratory of Crop Biology (2019KF03), Scientific and Technological Innovation Talents in Colleges and Universities in Henan (20HASTIT036) and Central Public-interest Scientific Institution Basal Research Fund (Farmland Irrigation Research Institute, CAAS) (FIRI2017-19).

Conflicts of Interest: The authors declare no conflict of interest.

References

1. Lisar, S.Y.S.; Motafakkerazad, R.; Hossain, M.M.; Rahman, I.M.M. Water Stress in Plants: Causes, Effects and Responses. In *Water Stress*; Rahman, M.M., Hasegawa, H., Eds.; Intech: Rijeka, Croatia, 2012; pp. 1–14. ISBN 978-953-307-963-9.
2. Dai, A. Drought under global warming: A review. *WIREs Clim. Chang.* **2011**, *2*, 45–65. [CrossRef]
3. Rollins, J.A.; Habte, E.; Templer, S.E.; Colby, T.; Schmidt, J.; Von Korff, M. Leaf proteome alterations in the context of physiological and morphological responses to drought and heat stress in barley (*Hordeum vulgare* L.). *J. Exp. Bot.* **2013**, *64*, 3201–3212. [CrossRef] [PubMed]
4. Potopová, V.; Boroneanţ, C.; Boincean, B.; Soukup, J. Impact of agricultural drought on main crop yields in the Republic of Moldova. *Int. J. Climatol.* **2016**, *36*, 2063–2082. [CrossRef]
5. Daryanto, S.; Wang, L.X.; Jacinthe, P.A. Global synthesis of drought effects on food legume production. *PLoS ONE* **2015**, *10*, e0127401. [CrossRef] [PubMed]
6. Cao, L.R.; Lu, X.M.; Zhang, P.Y.; Wang, G.R.; Wei, L.; Wang, T.C. Systematic Analysis of Differentially Expressed Maize ZmbZIP Genes between Drought and Rewatering Transcriptome Reveals bZIP Family Members Involved in Abiotic Stress Responses. *Int. J. Mol. Sci.* **2019**, *20*, 4103. [CrossRef]
7. Hussain, S.; Rao, M.J.; Anjum, M.A.; Ejaz, S.; Zakir, I.; Ali, M.A.; Ahmad, N.; Ahmad, S. Oxidative stress and antioxidant defense in plants under drought conditions. In *Plant Abiotic Stress Tolerance*; Hasanuzzaman, M., Hakeem, K.R., Nahar, K., Alharby, H.F., Eds.; Springer: Cham, Switzerland, 2019; pp. 207–219. ISBN 978-3-030-06117-3.
8. Meeks, M.; Murray, S.; Hague, S.; Hays, D. Measuring maize seedling drought response in search of tolerant germplasm. *Agronomy* **2013**, *3*, 135–147. [CrossRef]
9. Vaughan, M.M.; Block, A.; Christensen, S.A.; Allen, L.H.; Schmelz, E.A. The effects of climate change associated abiotic stresses on maize phytochemical defenses. *Phytochem. Rev.* **2018**, *17*, 37–49. [CrossRef]
10. Calanca, P.P. Effects of abiotic stress in crop production. In *Quantification of Climate Variability, Adaptation and Mitigation for Agricultural Sustainability*; Ahmed, M., Stockle, C.O., Eds.; Springer: Cham, Switzerland, 2017; pp. 165–180. ISBN 978-3-319-32057-1.
11. EL Sabagh, A.; Hossain, A.; Islam, M.S.; Barutcular, S.; Fahad, S.; Ratnasekera, D.; Kumar, N.; Meena, R.S.; Vera, P.; Saneoka, H. Role of osmoprotectants and soil amendments for sustainable soybean (*Glycine max* L.) production under drought condition: A review. *J. Exp. Biol. Agric. Sci.* **2018**, *6*, 32–41. [CrossRef]
12. Yang, M.; Geng, M.Y.; Shen, P.F.; Chen, X.H.; Li, Y.J.; Wen, X.X. Effect of post-silking drought stress on the expression profiles of genes involved in carbon and nitrogen metabolism during leaf senescence in maize (*Zea mays* L.). *Plant Physiol. Biochem.* **2019**, *135*, 304–309. [CrossRef]
13. Chiuta, N.; Mutengwa, C. Response of yellow quality protein maize inbred lines to drought stress at seedling stage. *Agronomy* **2018**, *8*, 287. [CrossRef]
14. Zhan, J.P.; Li, G.S.; Ryu, C.H.; Ma, C.; Zhang, S.S.; Lloyd, A.; Hunter, B.G.; Larkins, B.A.; Drews, G.N.; Wang, X.F.; et al. Opaque-2 regulates a complex gene network associated with cell differentiation and storage functions of maize endosperm. *Plant Cell* **2018**, *30*, 2425–2446. [CrossRef] [PubMed]
15. Jain, D.; Ashraf, N.; Khurana, J.P.; Kameshwari, M.N. The 'Omics' Approach for Crop Improvement Against Drought Stress. In *Genetic Enhancement of Crops for Tolerance to Abiotic Stress: Mechanisms and Approaches*; Rajpal, V.R., Sehgal, D., Kumar, A., Raina, S.N., Eds.; Springer: Cham, Switzerland, 2019; pp. 183–204. ISBN 978-3-319-91955-3.
16. Dastogeer, K.M.G.; Li, H.; Sivasithamparam, K.; Jones, M.G.K.; Wylie, S.J. Fungal endophytes and a virus confer drought tolerance to Nicotiana benthamiana plants through modulating osmolytes, antioxidant enzymes and expression of host drought responsive genes. *Environ. Exp. Bot.* **2018**, *149*, 95–108. [CrossRef]

17. Kaur, G.; Asthir, B. Molecular responses to drought stress in plants. *Biol. Plant* **2017**, *61*, 201–209. [CrossRef]

18. Fang, Y.J.; Xiong, L.Z. General mechanisms of drought response and their application in drought resistance improvement in plants. *Cell. Mol. Life Sci.* **2015**, *72*, 673–689. [CrossRef] [PubMed]

19. Nakashima, K.; Yamaguchi-Shinozaki, K.; Shinozaki, K. The transcriptional regulatory network in the drought response and its crosstalk in abiotic stress responses including drought, cold, and heat. *Front. Plant Sci.* **2014**, *5*, 170. [CrossRef] [PubMed]

20. Feng, F.; Qi, W.W.; Lv, Y.D.; Yan, S.M.; Xu, L.M.; Yang, W.Y.; Yuan, Y.; Chen, Y.H.; Zhao, H.; Song, R.T. OPAQUE11 is a central hub of the regulatory network for maize endosperm development and nutrient metabolism. *Plant Cell* **2018**, *30*, 375–396. [CrossRef]

21. An, Y.X.; Chen, L.; Li, Y.X.; Li, C.H.; Shi, Y.S.; Song, Y.C.; Zhang, D.F.; Li, Y.; Wang, T.Y. Candidate loci for the kernel row number in maize revealed by a combination of transcriptome analysis and regional association mapping. *BMC Plant Biol.* **2019**, *19*, 201. [CrossRef]

22. Wilson, J. *Control of Maize Development by MicroRNA and Auxin Regulated Pathways*; East Carolina University: Greenville, NC, USA, 2018.

23. Opitz, N.; Paschold, A.; Marcon, C.; Malik, W.A.; Lanz, C.; Piepho, H.P.; Hochholdinger, F. Transcriptomic complexity in young maize primary roots in response to low water potentials. *BMC Genomics* **2014**, *15*, 741. [CrossRef]

24. Song, K.; Kim, H.C.; Shin, S.; Kim, K.H.; Moon, J.C.; Kim, J.Y.; Lee, B.M. Transcriptome analysis of flowering time genes under drought stress in maize leaves. *Front. Plant Sci.* **2017**, *8*, 267. [CrossRef]

25. Li, L. *The Major Metabolic Pathways at Maize Developing Young Tassel in Response to Drought Stress and Identification of Drought-Tolerant Candidate SNAC Genes*; Xinjiang Agricultural University: Wulumuqi, Xinjiang, China, 2015.

26. Wang, B.M.; Liu, C.; Zhang, D.F.; He, C.M.; Zhang, J.R.; Li, Z.X. Effects of maize organ-specific drought stress response on yields from transcriptome analysis. *BMC Plant Biol.* **2019**, *19*, 335. [CrossRef]

27. Zhao, M.; Wang, Q.X.; Wang, K.J.; Li, C.H.; Hao, J.P. Maize. In *Crop Cultivation Science: North*; Yu, Z.W., Ed.; China Agriculture Press: Beijing, China, 2003; pp. 69–111. ISBN 9787109179363.

28. Chen, D.Q.; Wang, S.W.; Cao, B.B.; Cao, D.; Leng, G.H.; Li, H.B.; Yin, L.N.; Shan, L.; Deng, X.P. Genotypic variation in growth and physiological response to drought stress and re-watering reveals the critical role of recovery in drought adaptation in maize seedlings. *Front. Plant Sci.* **2016**, *6*, 1241. [CrossRef] [PubMed]

29. Oury, V.; Caldeira, C.F.; Prodhomme, D.; Pichon, J.P.; Gibon, Y.; Turc, O. Is change in ovary carbon status a cause or a consequence of maize ovary abortion in water deficit during flowering? *Plant Physiol.* **2016**, *171*, 997–1008. [CrossRef] [PubMed]

30. Khandagale, S.G.; Dubey, R.B.; Sharma, V.; Khan, R. Response of physiological traits of maize to moisture stress induced at different developmental stages. *Int. J. Chem. Stud.* **2018**, *6*, 2757–2761.

31. Ma, C.Y.; Li, B.; Wang, L.N.; Xu, M.L.; Zhu, L.E.; Jin, H.Y.; Wang, Z.C.; Ye, J.R. Characterization of phytohormone and transcriptome reprogramming profiles during maize early kernel development. *BMC Plant Biol.* **2019**, *19*, 197. [CrossRef]

32. Jia, S.J.; Li, H.W.; Jiang, Y.P.; Zhao, G.Q.; Wang, H.Z.; Yang, S.J.; Yang, Q.H.; Guo, J.M.; Shao, R.X. Effects of drought on photosynthesis and ear development characteristics of maize. *Acta Ecol. Sin.* **2020**, *3*, 1–9.

33. Li, Y.B.; Song, H.; Zhou, L.; Xu, Z.Z.; Zhou, G.S. Tracking chlorophyll fluorescence as an indicator of drought and rewatering across the entire leaf lifespan in a maize field. *Agric. Water Manag.* **2019**, *211*, 190–201. [CrossRef]

34. Xu, Z.Z.; Zhou, G.S.; Shimizu, H. Are plant growth and photosynthesis limited by pre-drought following rewatering in grass? *J. Exp. Bot.* **2009**, *60*, 3737–3749. [CrossRef]

35. Bolger, A.M.; Lohse, M.; Usadel, B. Trimmomatic: A flexible trimmer for Illumina sequence data. *Bioinformatics* **2014**, *30*, 2114–2120. [CrossRef]

36. Jiao, Y.P.; Peluso, P.; Shi, J.H.; Liang, T.; Stitzer, M.C.; Wang, B.; Campbell, M.S.; Stein, J.C.; Wei, X.H.; Chin, C.S.; et al. Improved maize reference genome with single-molecule technologies. *Nature* **2017**, *546*, 524–527. [CrossRef]

37. Kim, D.; Langmead, B.; Salzberg, S.L. HISAT: A fast spliced aligner with low memory requirements. *Nat. Methods* **2015**, *12*, 357–360. [CrossRef]

38. Roberts, A.; Pimentel, H.; Trapnell, C.; Pachter, L. Identification of novel transcripts in annotated genomes using RNA-Seq. *Bioinformatics* **2011**, *27*, 2325–2329. [CrossRef] [PubMed]

39. Roberts, A.; Trapnell, C.; Donaghey, J.; Rinn, J.L.; Pachter, L. Improving RNA-Seq expression estimates by correcting for fragment bias. *Genome Biol.* **2011**, *12*, R22. [CrossRef] [PubMed]

40. Anders, S.; Huber, W. *Differential Expression of RNA-Seq Data at the Gene Level–the DESeq Package*; European Molecular Biology Laboratory: Heidelberg, Germany, 2012.

41. Abendroth, L.J.; Elmore, R.W.; Boyer, M.J.; Marlay, S.K. Vegetative Stages (VE to VT). In *Corn Growth and Development*; Iowa State University: Ames, IA, USA, 2011; pp. 13–27.

42. Mueller, N.D.; Gerber, J.S.; Johnston, M.; Ray, D.K.; Ramankutty, N.; Foley, J.A. Closing yield gaps through nutrient and water management. *Nature* **2012**, *490*, 254–257. [CrossRef]

43. Zhang, X.B.; Lei, L.; Lai, J.S.; Zhao, H.M.; Song, W.B. Effects of drought stress and water recovery on physiological responses and gene expression in maize seedlings. *BMC Plant Biol.* **2018**, *18*, 68. [CrossRef] [PubMed]

44. Hayano-Kanashiro, C.; Calderón-Vázquez, C.; Ibarra-Laclette, E.; Herrera-Estrella, L.; Simpson, J. Analysis of gene expression and physiological responses in three Mexican maize landraces under drought stress and recovery irrigation. *PLoS ONE* **2009**, *4*, e7531. [CrossRef] [PubMed]

45. Huo, Y.J.; Wang, M.P.; Wei, Y.Y.; Xia, Z.L. Overexpression of the maize psbA gene enhances drought tolerance through regulating antioxidant system, photosynthetic capability, and stress defense gene expression in tobacco. *Front. Plant Sci.* **2016**, *6*, 1223. [CrossRef] [PubMed]

46. Avramova, V.; AbdElgawad, H.; Vasileva, I.; Petrova, A.S.; Holek, A.; Mariën, J.; Asard, H.; Beemster, G.T.S. High antioxidant activity facilitates maintenance of cell division in leaves of drought tolerant maize hybrids. *Front. Plant Sci.* **2017**, *8*, 84. [CrossRef]

47. Witt, S.; Galicia, L.; Lisec, J.; Cairns, J.; Tiessen, A.; Araus, J.L.; Palacios-Rojas, N.; Fernie, A.R. Metabolic and phenotypic responses of greenhouse-grown maize hybrids to experimentally controlled drought stress. *Mol. Plant* **2012**, *5*, 401–417. [CrossRef]

48. Bartels, D.; Souer, E. Molecular responses of higher plants to dehydration. In *Plant Responses to Abiotic Stress*; Hirt, H., Shinozaki, K., Eds.; Springer: Berlin/Heidelberg, Germany, 2004; pp. 9–38. ISBN 9783-540-20037-6.

49. Boonjung, H.; Fukai, S. Effects of soil water deficit at different growth stages on rice growth and yield under upland conditions. 2. Phenology, biomass production and yield. *Field Crops Res.* **1996**, *48*, 47–55. [CrossRef]

50. Su, Z.; Ma, X.; Guo, H.H.; Sukiran, N.L.; Guo, B.; Assmann, S.M.; Ma, H. Flower development under drought stress: Morphological and transcriptomic analyses reveal acute responses and long-term acclimation in Arabidopsis. *Plant Cell* **2013**, *25*, 3785–3807. [CrossRef]

51. Gonzalez, V.H.; Lee, E.A.; Lukens, L.L.; Swanton, C.J. The relationship between floret number and plant dry matter accumulation varies with early season stress in maize (*Zea mays* L.). *Field Crops Res.* **2019**, *238*, 129–138. [CrossRef]

52. Nielsen, R.L. *Ear Size Determination in Corn*; Corny News Network Articles; Purdue University: West Lafayette, IN, USA, 2007.

53. Hu, X.J.; Wang, H.W.; Diao, X.Z.; Liu, Z.F.; Li, K.; Wu, Y.J.; Liang, Q.J.; Wang, H.; Huang, C.L. Transcriptome profiling and comparison of maize ear heterosis during the spikelet and floret differentiation stages. *BMC Genomics* **2016**, *17*, 959. [CrossRef]

54. Jeffares, D.C. *Molecular Genetic Analysis of the Maize Terminal Ear1 Gene and in Silico Analysis of Related Genes*; Massey University: Palmerston North, New Zealand, 2001.

55. Vroemen, C.W.; Mordhorst, A.P.; Albrecht, C.; Kwaaitaal, M.A.C.J.; Vries, S.C. The CUP-SHAPED COTYLEDON3 gene is required for boundary and shoot meristem formation in Arabidopsis. *Plant Cell* **2003**, *15*, 1563–1577. [CrossRef]

56. Hibara, K.; Karim, M.R.; Takada, S.; Taoka, K.; Furutani, M.; Aida, M.; Tasaka, M. Arabidopsis CUP-SHAPED COTYLEDON3 regulates postembryonic shoot meristem and organ boundary formation. *Plant Cell* **2006**, *18*, 2946–2957. [CrossRef]

57. Cai, Y.H.; Chen, X.J.; Xie, K.; Xing, Q.K.; Wu, Y.W.; Li, J.; Du, C.H.; Sun, Z.X.; Guo, Z.J. Dlf1, a WRKY transcription factor, is involved in the control of flowering time and plant height in rice. *PLoS ONE* **2014**, *9*, e102529. [CrossRef]

58. Chen, Z.H.; Yuan, Y.; Fu, D.; Shen, C.J.; Yang, Y.J. Identification and expression profiling of the auxin response factors in Dendrobium officinale under abiotic stresses. *Int. J. Mol. Sci.* **2017**, *18*, 927. [CrossRef] [PubMed]

59. Porco, S.; Larrieu, A.; Du, Y.J.; Gaudinier, A.; Goh, T.; Swarup, K.; Swarup, R.; Kuempers, B.; Bishopp, A.; Lavenus, J.; et al. Lateral root emergence in Arabidopsis is dependent on transcription factor LBD29 regulation of auxin influx carrier LAX3. *Development* **2016**, *143*, 3340–3349. [CrossRef] [PubMed]

60. Dinesh, D.C.; Villalobos, L.I.A.C.; Abel, S. Structural biology of nuclear auxin action. *Trends Plant Sci.* **2016**, *21*, 302–316. [CrossRef] [PubMed]

61. Jung, H.R.; Lee, D.K.; Do Choi, Y.; Kim, J.K. OsIAA6, a member of the rice Aux/IAA gene family, is involved in drought tolerance and tiller outgrowth. *Plant Sci.* **2015**, *236*, 304–312. [CrossRef]

62. Asghar, M.A.; Li, Y.; Jiang, H.K.; Sun, X.; Ahmad, B.; Imran, S.; Yu, L.; Liu, C.Y.; Yang, W.Y.; Du, J.B. Crosstalk between Abscisic Acid and Auxin under Osmotic Stress. *Agron. J.* **2019**, *111*, 2157–2162. [CrossRef]

63. Kelley, K.B.; Riechers, D.E. Recent developments in auxin biology and new opportunities for auxinic herbicide research. *Pestic. Biochem. Phys.* **2007**, *89*, 1–11. [CrossRef]

64. Han, S.; Hwang, I. Integration of multiple signaling pathways shapes the auxin response. *J. Exp. Bot.* **2017**, *69*, 189–200. [CrossRef] [PubMed]

65. Luo, J.; Zhou, J.J.; Zhang, J.Z. Aux/IAA gene family in plants: Molecular structure, regulation, and function. *Int. J. Mol. Sci.* **2018**, *19*, 259. [CrossRef] [PubMed]

66. Bai, Q.S.; Hou, D.; Li, L.; Cheng, Z.C.; Ge, W.; Liu, J.; Li, X.P.; Mu, S.H.; Gao, J. Genome-wide analysis and expression characteristics of small auxin-up RNA (SAUR) genes in moso bamboo (*Phyllostachys edulis*). *Genome* **2016**, *60*, 325–336. [CrossRef]

67. Ren, H.; Gray, W.M. SAUR proteins as effectors of hormonal and environmental signals in plant growth. *Mol. Plant* **2015**, *8*, 1153–1164. [CrossRef]

68. Han, D.G.; Zhang, Z.Y.; Ding, H.B.; Chai, L.J.; Liu, W.; Li, H.X.; Yang, G.H. Isolation and characterization of MbWRKY2 gene involved in enhanced drought tolerance in transgenic tobacco. *J. Plant Interact.* **2018**, *13*, 163–172. [CrossRef]

69. Ahmad, N.; Malagoli, M.; Wirtz, M.; Hell, R. Drought Stress in Maize Causes Differential Acclimation Responses of Glutathione and Sulfur Metabolism in Leaves and Roots. *BMC Plant Biol.* **2016**, *16*, 247. [CrossRef]

70. Zhang, X.; Tao, L.; Qiao, S.; Du, B.H.; Guo, C.H. Roles of Glutathione S-transferase in Plant Tolerance to Abiotic Stresses. *J. Chin. Biotechnol.* **2017**, *37*, 92–98.

71. Kong, X.X.; Li, B.Z.; Yang, J.S. Research progress in microalgae resistance to cadmium stress. *Microbiol. Chin.* **2017**, *44*, 1980–1987.

72. George, S.; Venkataraman, G.; Parida, A.A. chloroplast-localized and auxin-induced glutathione S-transferase from phreatophyte Prosopis juliflora confer drought tolerance on tobacco. *J. Plant Physiol.* **2010**, *167*, 311–318. [CrossRef]

73. Nakamura, A.; Umemura, I.; Gomi, K.; Hasegawa, Y.; Kitano, H.; Sazuka, T.; Matsuoka, M. Production and characterization of auxin-insensitive rice by overexpression of a mutagenized rice IAA protein. *Plant J.* **2006**, *46*, 297–306. [CrossRef] [PubMed]

74. Gu, H.H.; Yang, Y.; Xing, M.H.; Yue, C.P.; Wei, F.; Zhang, Y.J.; Zhao, W.E.; Huang, J.Y. Physiological and transcriptome analyses of *Opisthopappus taihangensis* in response to drought stress. *Cell Biosci.* **2019**, *9*, 56. [CrossRef] [PubMed]

 *agronomy*

Article

Comparison of Biochemical, Anatomical, Morphological, and Physiological Responses to Salinity Stress in Wheat and Barley Genotypes Deferring in Salinity Tolerance

Muhammad Zeeshan [1], Meiqin Lu [3], Shafaque Sehar [1], Paul Holford [4] and Feibo Wu [1,2,*]

[1] Institute of Crop Science, Department of Agronomy, College of Agriculture and Biotechnology, Zijingang Campus, Zhejiang University, Hangzhou 310058, China; 11616102@zju.edu.cn (M.Z.); 21516206@zju.edu.cn (S.S.)

[2] Jiangsu Co-Innovation Center for Modern Production Technology of Grain Crops, Yangzhou University, Yangzhou 225009, China

[3] Australian Grain Technologies, Narrabri, NSW 2390, Australia; meiqin.lu@ausgraintech.com

[4] School of Science and Health, Western Sydney University, Penrith, NSW 2751, Australia; p.holford@westernsydney.edu.au

* Correspondence: wufeibo@zju.edu.cn; Tel./Fax: +86-571-88982827

Received: 27 November 2019; Accepted: 9 January 2020; Published: 15 January 2020

Abstract: A greenhouse hydroponic experiment was performed using salt-tolerant (cv. Suntop) and -sensitive (Sunmate) wheat cultivars and a salt-tolerant barley cv. CM72 to evaluate how cultivar and species differ in response to salinity stress. Results showed that wheat cv. Suntop performed high tolerance to salinity, being similar tolerance to salinity with CM72, compared with cv. Sunmate. Similar to CM72, Suntop recorded less salinity induced increase in malondialdehyde (MDA) accumulation and less reduction in plant height, net photosynthetic rate (Pn), chlorophyll content, and biomass than in sensitive wheat cv. Sunmate. Significant time-course and cultivar-dependent changes were observed in the activities of antioxidant enzymes such as superoxide dismutase (SOD), peroxidase (POD), catalase (CAT), ascorbate peroxidase (APX), and glutathione reductase (GR) in roots and leaves after salinity treatment. Higher activities were found in CM72 and Suntop compared to Sunmate. Furthermore, a clear modification was observed in leaf and root ultrastructure after NaCl treatment with more obvious changes in the sensitive wheat cv. Sunmate, rather than in CM72 and Suntop. Although differences were observed between CM72 and Suntop in the growth and biochemical traits assessed and modified by salt stress, the differences were negligible in comparison with the general response to the salt stress of sensitive wheat cv. Sunmate. In addition, salinity stress induced an increase in the Na^+ and Na^+/K^+ ratio but a reduction in K^+ concentrations, most prominently in Sunmate and followed by Suntop and CM72.

Keywords: antioxidants; ultrastructure; osmotic stress; salinity; wheat; barley

---

1. Introduction

Saline soils are a major problem in many countries with the Environment Program of United Nations estimating that of the 9–34% of the world's irrigated land is adversely affected by salinity [1]. Salinity can kill plants and other soil organisms and is referred to as a "silent killer" in some regions or as "white death" in others as it invokes images of a lifeless, shining land studded with dead trees. Approximately 32 million ha of dry lands [2] and 60 million ha of irrigated land [3] are affected by human-induced soil salinization, and it is well documented that salinity is one of the most severe environmental stresses hampering crop production [4,5]. At high electrical conductivity (EC) resulting from salinization, crop yields can decline drastically rendering crop cultivation no longer profitable and making soil amendments inevitable [6]. World agriculture needs to feed about 2.3 billion people globally by 2050 [7]; thus, it is imperative to understand the mechanisms associated with tolerance to salinity so that breeding programs and agronomic practices can be put in place that will allow production to meet this increasing demand [8].

Saline soils limit plant growth due to osmotic stress, ionic toxicity, and a reduced ability to take up essential minerals [9,10]. In extreme cases, root cells may lose water instead of absorbing it due to the hyperosmotic pressure of the soil solution. Water deficits affect a cascade of physical, signaling, gene expression, biochemical, and physiological pathways and processes, resulting in decreased cell elongation, wilting, and, ultimately, plant death; these harmful effects of salinity can be considered as water-deficit effects [3,11,12]. In saline soils, NaCl comprises 50–80% of the total soluble salts [13] causing elevated, and potentially toxic, concentrations of Na^+ and/or Cl^- in the plant. These ions affect many enzymes or cellular functions such as photosynthesis signaling systems [14–16]. In addition, because of their physicochemical similarities and a shared transport system, the Na^+ in the soil solution of saline soils competes for uptake with K^+ [17] and can lead to K^+ deficiency [18,19]. The induced K^+ deficiency inhibits growth because it plays a critical role in maintaining cell turgor, membrane potentials, and enzyme activities.

As a consequence of the primary effects of salinity described above, secondary stresses such as oxidative stress often occurs due to an overproduction of reactive oxygen species (ROS) [20]. These ROS cause lipid peroxidation leading to increased membrane fluidity and permeability [21,22], the denaturation of functional and structural proteins [23], and can affect nucleic acids through base modifications, induce inter- and intra-strand crosslinks, crosslinks with proteins as well as creating strand breaks [24]. However, plants have developed comprehensive internal resistance systems to combat the outcomes of ROS that are comprised of enzymatic as well as non-enzymatic antioxidants [25]. ROS-scavenging enzymes include those that are playing a direct role in the processing of ROS such as superoxide dismutase (SOD), peroxidase (POD), and catalase (CAT), and those glutathione reductase (GR) and ascorbate peroxidases (APX) that mediate in the reaction cycle of antioxidant chemicals such as glutathione (GSH) and ascorbic acid (AsA) [26–30]. The other half of the antioxidant machinery includes nonenzymatic antioxidants comprising of ascorbic acid, phenolic compounds (flavonoids, anthocyanins), α-tocopherol, carotenoids, and amino acid cum osmolyte proline. Besides the synthesis and modulation of osmolytes some phytohormones and regulatory molecules also play prominent role in triggering salinity tolerance effector molecules [31].

Barley and wheat have different salt tolerances capacities and are grown as major grain crops in both saline and non-saline soils [32]. Previous studies have focused on salinity stress in either barley or wheat alone, with little inter-specific comparison. Thus, this study is the first to compare the mechanisms that confer salinity tolerance in these two species. We aimed to explore the similarities or differences in their physiological mechanisms upon exposure to salt stress. We also hypothesized that there may be species-specific mechanisms that can be co-related with the salt sensitivity of wheat or the tolerance of barley. Thus, this research can enhance our understanding of holistic salinity tolerance mechanisms and will aid in the breeding of salt-tolerant crops.

2. Material and Methods

2.1. Plant Material and Growth Conditions

A shade house hydroponic experiment was carried out on the Zijingang Campus, Zhejiang University, China. Two wheat (*Triticum aestivum* L.) cv. Suntop (salt-tolerant) and Sunmate (salt-sensitive) and a salt-tolerant barley (*Hordeum vulgare* L.) cv. CM72, were used in the experiment. Suntop and Sunmate are two high yielding Australian Prime Hard varieties bred by Australian Grain Technologies (AGT, Narrabri, Australia. Although they were derived from the same cross, Suntop and Sunmate differ significantly in salt tolerance. Seeds of each cultivar were disinfected in 2% (*v/v*)

H_2O_2 then washed thoroughly using double distilled water (ddH_2O). The seeds were germinated on filter paper in germination boxes in a plant growth chamber (23/18 °C, day/night) in darkness for 3 days and incubated for further 4 days in the light. The uniform 7-day-old seedlings of each cultivar were selected and transplanted into 5 L pots in a hydroponic solution containing 4.5 L basic nutrient solution (BNS) as described by [33,34], with continuous aeration using air pumps. Each container was covered with a polystyrene plate with 7 evenly spaced holes (2 plants per hole) and placed in a greenhouse with natural light and a temperature of 20 ± 2 °C/day and 15 ± 2 °C/night. At the two-leaf stage, plants were treated with 100 mM NaCl; non-NaCl treated plants were used as controls (BNS). The solution pH was adjusted to 5.8 with NaOH or HCl, as required. The experiment was arranged in a randomized block design with four replications. Plants were sampled at 1, 5, 10, and 15 days after treatment (DAT) for time-course analysis of the salt treatments. For morphological and physiological analyses, plants were harvested 25 DAT and either analyzed immediately or frozen in liquid nitrogen and stored at −80 °C for further analysis.

2.2. Measurement of Growth Traits and Mineral Concentrations

Shoot height, root length, and fresh weights were determined 25 DAT (days after treatment) and then the samples were separated into shoot and root and perfectly washed with ddH_2O to eliminate any foreign material. Samples from each treatment with 4 biological replicates were oven dried at 75 °C for three days and subsequently, the dry weight of the roots and shoots were determined in gram. Later each dried sample was weighed (about 0.2 g), ground, and made into ashes by heating the samples at 550 °C for half a day. Before dilution with ddH_2O, the ashes were digested in 30% HNO_3 and then Na^+ and K^+ concentration were quantified using flame atomic absorption spectrometry (SHIMADZU AA-6300, Columbia, Maryland, USA) [35].

2.3. Measurement of Photosynthesis Parameters, Chlorophyll Contents, and Chlorophyll Fluorescence

Intact, second fully-expanded leaves from the apex were used to measure relative chlorophyll content with the help of a handheld chlorophyll meter (Minolta SPAD-502, Tokyo, Japan) according to Wu et al. [36]. Three measurements were recorded from each leaf and averaged. The gas exchange parameters (i.e., photosynthetic rates (Pn), intercellular CO_2 concentrations (Ci), stomatal conductance (Gs), and transpiration rates (Tr)) were measured on a bright sunny day between 9 a.m. to 11 a.m. using a Li-Cor-6400 portable photosynthesis system (Li-Cor, Lincoln, NE, USA).

Chlorophyll fluorescence (*Fv/Fm*) was measured at 25 DAT according to Genty et al. [37]. Both treated and control plants were shifted to an experimental room, kept in the dark for 25 min, flag leaf were cut for the determination of chlorophyll fluorescence using a pulse-modulated chlorophyll fluorimeter using IMAGING-PAM (Walz; Effeltrich, Germany) image processing software. Fluorescence values observed comprised of *Fo*, initial/minimal fluorescence, *Fm*, the maximal fluorescence value, and *Fv/Fm*, the maximum quantum yield of PSII photochemistry. The data were noted at five different points at 40, 70, 120, 150, and 180 mm from leaf tips.

2.4. Lipid Peroxidation and Antioxidative Enzyme Activity Assay

The roots and second leaf from the apex were sampled at 1, 5, 10, and 15 DAT. Lipid peroxidation was measured in the tissues and expressed as malondialdehyde (MDA) content using the TBA (thiobarbituric acid) method according to the Wu et al. [34]. The activity-specific and non-specific absorbance was determined at 532 and 600 nm, respectively. Enzymatic antioxidants activity was determined as described by Leul and Zhou [38]. Briefly, 0.2 g of frozen leaf and root plant tissue were ground in a pre-chilled mortar and pestle and homogenized in 2 mL of 1 M Tris buffer (pH 8). Later, the samples were briefly centrifuged at 10,000× g at 4 °C for about 15 min and the supernatants were used for the following assays. The activity assay of superoxide dismutase (SOD, EC 1.15.1.1), peroxidase (POD, EC 1.11.1.7), and catalysis (CAT, EC 1.11.1.6) were recorded according to Wu et al. [34]. Ascorbate peroxidase (EC 1.11.1.11) activity was determined at 290 nm using ascorbate (AsA) as a substrate and 2.8 mM cm^{-1} as an extinction co-efficient [39], while Jiang and Zhang [40] methods were used to determine the activity of glutathione reductase (GR, EC. 1.6.4.2).

2.5. Cell Ultrastructure

For transmission electron-microscopy, fresh roots (about 2–3 mm in length) and leaf pieces (about 1 mm^2) without veins were hand sectioned and treated with 100 mM sodium phosphate buffer (PBS; pH 7.0) containing 2.5% (*v/v*) glutaraldehyde and placed overnight at 4 °C, then briefly washed; this step was performed 3 times with the same buffer. Each sample was treated for 60 min with 1% osmium tetroxide OsO$_4$ (*v/v*) followed by washing with PBS (sodium phosphate buffer) further 3 times. Thereafter, the leaf and root samples were dehydrated with a diluted series of ethanol (50%, 60%, 70%, 80%, 90%, 95%, and 100%) for about 20 min in each solution, later all the samples were dried for 20 min in concentrated acetone. Finally, ultrathin sections (80 nm) were cut and affixed to copper grids for study using transmission electron microscopy (TEM 1230EX, JEOL, Akishima, Tokyo, Japan) at 60 kV.

2.6. Statistical Analysis

Statistical analysis was performed with the Data Processing System statistical software package [41] using ANOVA followed by Duncan's Multiple Range Tests (DMRT) to evaluate significant treatment effect at a significance level of $p < 0.05$. Origin Pro (Version 8.0, Origin lab corporation, Wellesley Hills, Wellesley, MA, USA) was used to prepare graphs.

3. Results

3.1. Plant Growth Parameters

Salt inhibited the growth of the barley and wheat plants, with the treated plants showing wilting, necrosis and chlorosis (Figure 1A). Salt damage was most severe in the wheat cv., Sunmate, while in the other cultivars, the damage was less pronounced. Salinity stress significantly ($p < 0.05$) reduced plant biomass in the wheat and barley cultivars (Figure 1B). In comparison to the other two cultivars, the effects of salt stress on plant growth was much more noticeable in Sunmate; it had the least effect on shoot height and the biggest effect on shoot weight. Shoot height was reduced under salinity stress by 29%, 12%, and 13% in the Sunmate, Suntop and CM72 cultivars, respectively. The fresh shoot weight was reduced by 68%, 55%, and 59%, while shoot dry weight was reduced 68%, 53%, and 49% in Sunmate, Suntop and CM72 in salinity stress plants respectively. Similarly, compared to the control plants, the root length was reduced under salinity stress by 37%, 8%, and 24% in Sunmate, Suntop, and CM72, respectively, while the reductions in fresh root weight were 42%, 33%, and 11% and dry root weight were 65%, 39%, and 30% in Sunmate, Suntop, and CM72, respectively in salinity treated plants (Table S1).

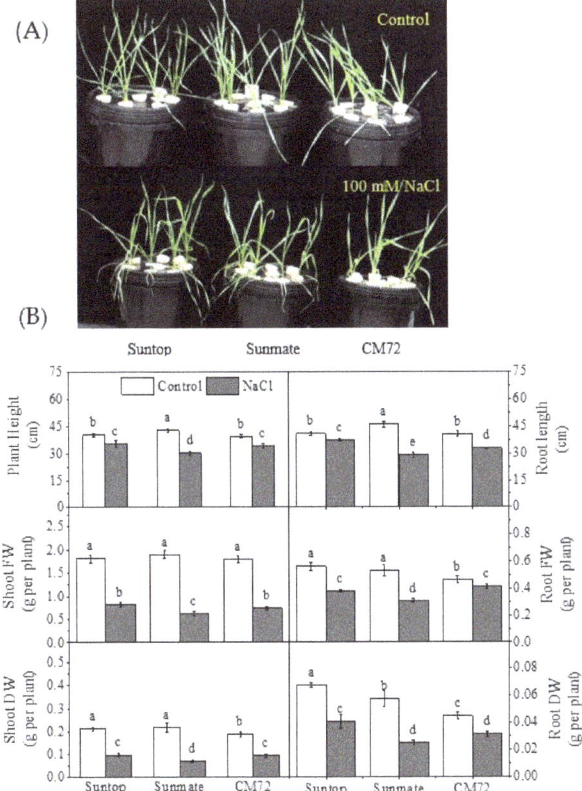

Figure 1. Morphology (**A**) and growth parameters (**B**) of seedlings of two wheat cv., Suntop and Sunmate, and one barley cv., CM72, 25 days after treatment with NaCl. Control and NaCl represent 0 and 100 mM NaCl, respectively. Values are means + SE ($n = 4$). For each parameter, means annotated with the same letter are not significantly different from each other according to Duncan's Multiple Range Tests at $p \leq 0.05$.

3.2. Chlorophyll and Photosynthetic Parameters

Gas exchange parameters were recorded 25 DAT (days after treatment) and significant ($p < 0.05$) decreases in net photosynthetic rate (Pn), stomatal conductance (Gs), intercellular CO_2 concentration (Ci), and transpiration rate (Tr) were detected in both wheat and barley in comparison to their respective controls (Figure 2A–F). No significant changes were observed in Gs, chlorophyll contents and Fv to Fm ratios, however, significant differences were observed in Pn, Ci, and Tr among all the cultivars in the salinity treated plants. Interestingly, the two-salt tolerant cultivars, Suntop and CM72, showed no significant difference in regard to the photosynthetic parameters, but differences were noted in Sunmate, which is salt-sensitive.

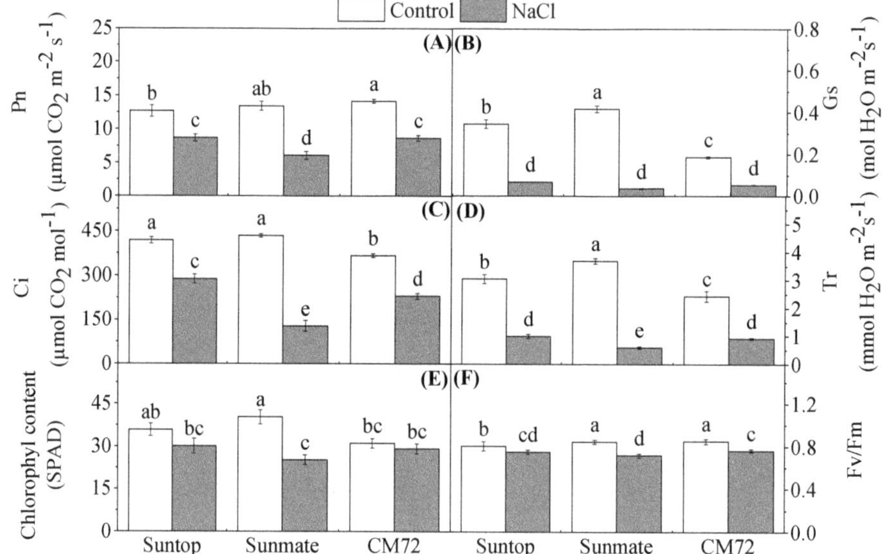

Figure 2. Effect of salinity stress on photosynthetic traits in two wheat cv., Suntop and Sunmate, and the barley cv., CM72, 25 days after treatment with 100 mM NaCl. Pn (**A**), Gs (**B**), Ci (**C**), Tr (**D**) and *Fv/Fm* (**F**), represent net photosynthetic rate, stomatal conductance, intercellular CO_2 concentration, transpiration rate, and a maximum quantum yield of photosystem II photochemistry of the second fully expanded leaves, respectively. The chlorophyll content was measured as SPAD (**E**) (Soil Plant analysis Development). Values are means + SE ($n = 4$). For each parameter, means annotated with the same letter are not significantly different from each other according to Duncan's Multiple Range Tests at $p \leq 0.05$.

3.3. Shoots and Roots Na^+, K^+ Concentration, and Na^+:K^+ Ratio

Internal Na^+ and K^+ concentrations were determined, salinity significantly ($p < 0.05$) increased Na^+, decreased K^+ content, and increased Na^+:K^+ ratio in both shoots and roots of all cultivars in the saline-treated plants relative to the control plants (Figure 3A–F). In general, roots contained more Na^+ and K^+ compared to shoots, regardless of the cultivar or treatment. With regard to the plants given the salt treatment, in both shoots and roots, the increase in the Na^+ content followed the trend CM72 < Suntop < Sunmate in both organs, while the K^+ content decreased in the following trend CM72 > Suntop > Sunmate. Interestingly, the increase in Na^+ content among the cultivars was inversely proportional to the decrease in K^+ content. As a consequence of the changes in both minerals, the Na^+:K^+ ratio increased under salt stress. The greatest Na^+:K^+ ratios were observed in Sunmate (0.339) and Suntop (0.127), while the smallest were observed in CM72 (0.075) in shoots. The same trend in Na^+:K^+ values were also observed in the roots (Table S2).

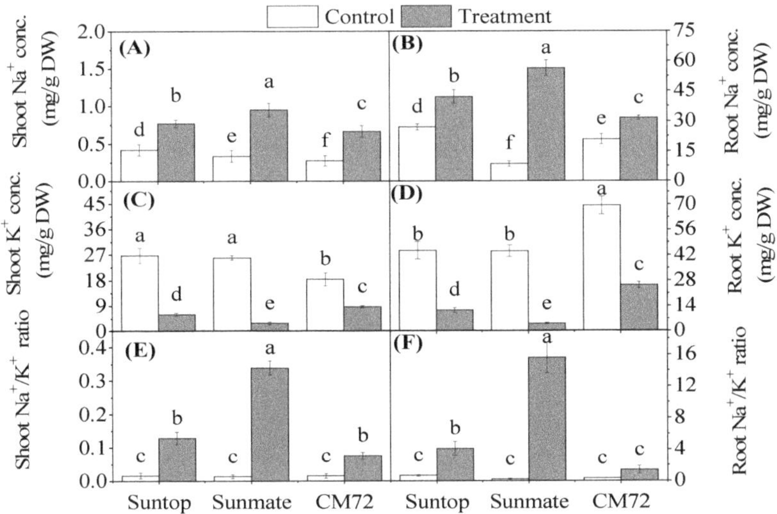

Figure 3. Effect of salinity stress on Na$^+$ and K$^+$ concentrations and Na$^+$:K$^+$ ratio in shoots (**A,C,E**) and roots (**B,D,F**) of two wheat cv., Suntop and Sunmate, and the barley cv., CM72, 25 days after treatment with 100 mM NaCl. Error bars represent SE (n = 4). Different letters indicate significant differences ($p \leq 0.05$) among the 3 cultivars.

3.4. Lipid Peroxidation Assay and Antioxidative Enzyme Activities

Lipid peroxidation measurements at 1, 5, 10, and 15 DAT showed that salt stress induced significant changes among the cultivars and treatments (Figure 4A,B and Supplementary Tables S3 and S4). Regardless of the cultivar, MDA contents were significantly increased by the salt treatment in both leaves and roots, indicating enhanced lipid peroxidation. In general, in plants given salinity treatments, the MDA content was highest in Sunmate followed by Suntop then CM72, with the highest increase observed 15 and 10 d after treatment in leaves and roots, respectively.

Significant differences ($p < 0.05$) were detected among the cultivars and between the treatment in both roots and leaves for all measured antioxidant enzymes (Supplementary Tables S3 and S4, and expression relative to control activities in Figure 5A–J). In general, in the leaves, the relative activities of all enzymes were highest in Suntop, followed by CM72 and then Sunmate; however, in the roots, there was little difference between CM72 and Suntop. For SOD in the leaves, the activities of this enzyme were similar on Days 1 and 5, rose on Day 10, and then dropped below those measured on Day 1. In the roots, SOD activities rose on Day 10 and remained high on Day 15. For POD in the leaves, activities rose on Day 5 and then declined during the remainder of the assessment period, whilst in the roots POD activities did not rise until Day 10 and then declined. For APX in the leaves, activities started to rise on Day 10 and were highest on Day 15 whilst in the root's activities rose on Day 10 and remained high. For CAT in leaves, activities were highest on Day 5 and then declined, except in Suntop, where a decline was observed on Day 10. In the roots, CAT activities tended to stay on similar levels throughout the treatment period. Finally, for GR for the two wheat cv., there was little change in activities in the leaves during the assessment period; however, for CM72, GR activities were highest on Days 5 and 10 and then declined. In the roots, for Suntop, GR activities increased on Day 5 and then remained high, for Sunmate, activities increased on Day 10, and then reduced and for CM72, GR activities were high throughout the experiment. Peaks of antioxidant enzyme activity were observed generally earlier in shoots than in roots, while the earliest peak was observed for CAT and the latest for APX.

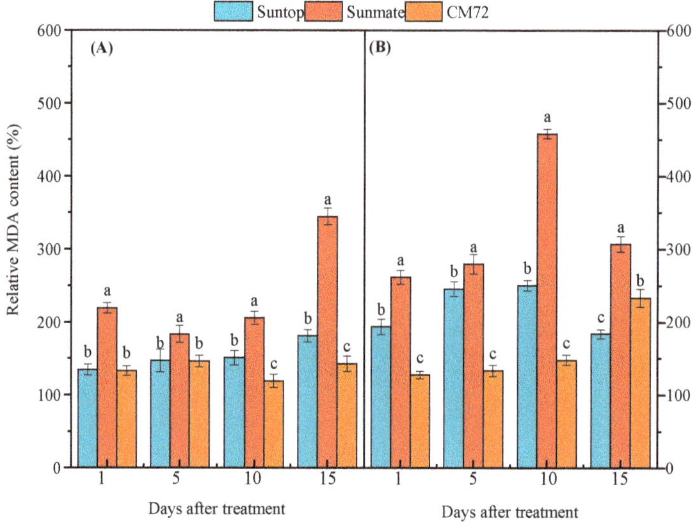

Figure 4. Effect of salinity stress on malondialdehyde contents (MDA, nmol^{-1} FW) in leaves (**A**) and roots (**B**) of Suntop (Aqua), Sunmate (Red), and CM72 (Orange) 1, 5, 10, and 15 days after treatment with 100 mM NaCl. The data are expressed as a percentage of control content. Different letters indicate significant differences ($p \leq 0.05$) among the 3 cultivars within each sampling day. Error bars represent SE ($n = 4$).

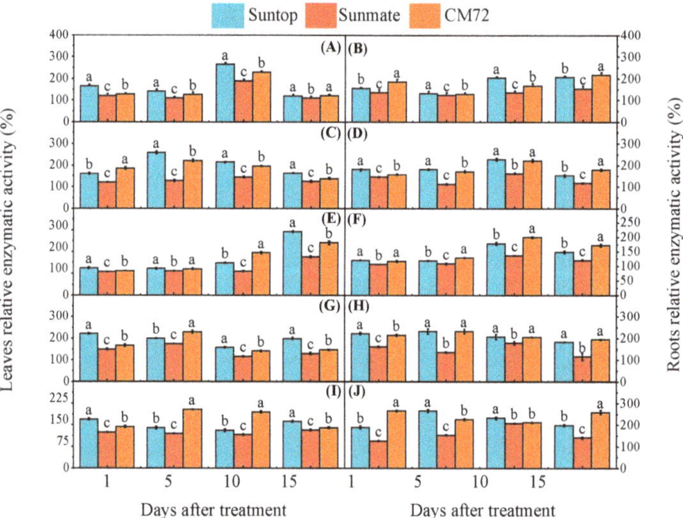

Figure 5. Effect of salinity stress on (**A,B**) superoxide dismutase (SOD) (U g^{-1} FW); (**C,D**) peroxidase (POD) (OD 470 g^{-1} min^{-1}); (**E,F**) ascorbate peroxidases (APX) (mmol g^{-1} FW min^{-1}); (**G,H**) catalase (CAT) (mmol^{-1} FW min^{-1}; (**I,J**) glutathione reductase (GR) (mmol g^{-1} FW min^{-1}) activities in leaves (**A,C,E,G,I**) and roots (**B,D,F,H,J**) of Suntop (Aqua), Sunmate (Red), and CM72 (Orange) 1, 5, 10, and 15 days after treatment with 100 mM NaCl. The data are expressed as a percentage of control activities. Different letters indicate significant differences ($p \leq 0.05$) among the 3 cultivars within each sampling day. Error bars represent SE ($n = 4$).

3.5. Leaf and Root Ultrastructure

The chloroplast ultrastructure of Sunmate was more severely affected by salt stress relative to controls and also to Suntop and CM72. Under control conditions, the chloroplasts of Suntop mesophyll cells usually had normal morphology with distinct grana and stroma lamellae, large starch grains with numerous plastoglobuli and well-organized, round mitochondria (Figure 6A); after the salt treatment, there were fewer plastoglobuli, no starch grains were apparent, and the grana and stroma lamellae were diffuse (Figure 6B). In contrast, chloroplasts of Sunmate were severely damaged by salinity stress, i.e., the chloroplast envelope showed disintegration with reduced grana stacks and less distinct thylakoids membranes, swollen oval-shaped mitochondria and larger osmophilic plastoglobuli (Figure 6C,D). As with Suntop, the chloroplasts of CM72 remained relatively normal in response to the salt treatment except for the disappearance of starch grains and thinner lamellae (Figure 6E,F).

When viewed using transmission electron microscopy, the root cells of all cultivars grown without salt treatment (control) had dense cytoplasm and organelles, and organized and large nuclei and nucleoli (Suntop, Figure 7A; Sunmate, Figure 7C; CM72, Figure 7E). Treatment with salt induced a number of ultrastructural changes from mild to severe, with the most visible alteration being the disappearance of nucleoli and vacuoles in Sunmate (Suntop, Figure 7B; Sunmate, Figure 7D; CM72, Figure 7F). Suntop and CM72 had clear nucleoli and larger and several vacuoles in comparison with Sunmate. However, the size of the nucleoli increased in CM72, and to a lesser extent in Suntop, upon exposure to salt.

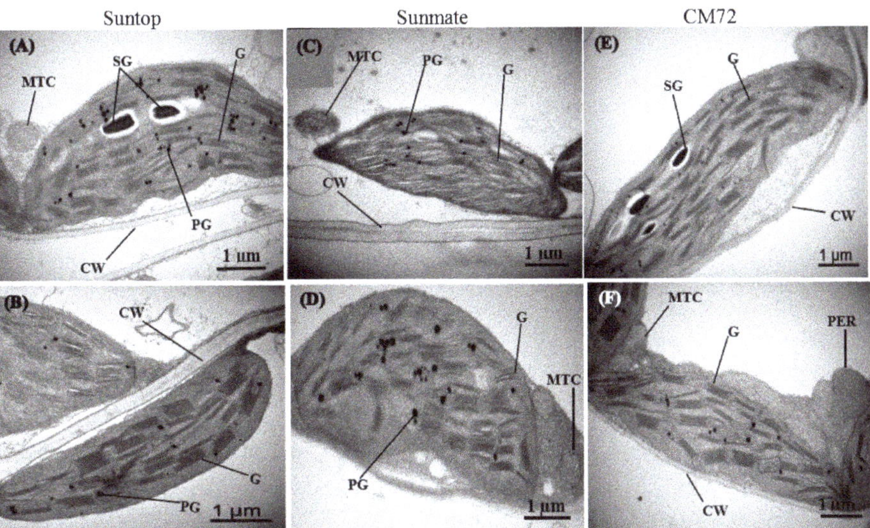

Figure 6. Transmission electron micrographs of chloroplasts of leaves of Suntop (**A,B**), Sunmate (**C,D**), and CM72 (**E,F**) under control (top panel) and 100 mM NaCl (bottom panel). CW, cell wall; G, grana; MTC, mitochondria; PG, plastoglobuli; SG, starch grains.

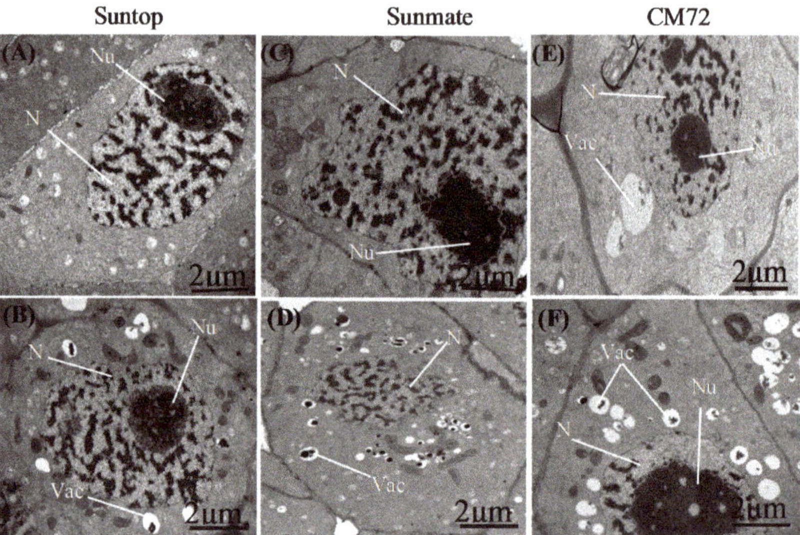

Figure 7. Electron micrographs of roots of Suntop (**A**,**B**), Sunmate (**C**,**D**), and CM72 (**E**,**F**) under control (top panel) and 100 mM NaCl (bottom panel). CW, cell wall; Nu, nucleolus; N, nucleus; Vac, vacuole.

4. Discussion

The effects of the treatments differed for different plant organs; therefore, the effects on shoots and roots of both species are considered separately.

Reduced biomass, a marked perturbation in photosynthetic parameters along with reduced chlorophyll contents resulting from salinity stress were observed in the wheat and barley cultivars. These effects are possibly due to either single or combined effects of reduced stomatal conductance, inhibition of metabolic phenomena, and increased ROS generation which can increase oxygen-induced cellular damage [42]. The reductions in stomatal conductance (Gs), photosynthesis rates (Pn), and leaf chlorophyll contents due to salinity were greater in Sunmate than in Suntop and CM72 (Figure 2). In a study conducted using sorghum, Netondo et al. [43] found that changes in stomatal conductance (Gs) and intercellular CO_2 concentration (Ci) were positively correlated under salt stress, concluding that stomatal conductance (Gs) was the key factor arresting net photosynthesis rates (Pn) under saline stress. The lower stomatal conductance (Gs) accompanied by low chlorophyll contents in Sunmate could contribute to the higher inhibition of net photosynthesis rates (Pn). Usually, plants close their stomata upon the onset of stressful conditions to save water, consequently reducing stomatal conductance (Gs) and photosynthesis [44]. The effect of salinity might be a secondary influence, arbitrated by the lower partial pressure of CO_2 in the green parts of the plant due to the stomata closure on the photosynthesis-related enzyme activities [45,46].

The *Fv/Fm* ratio reflects the photochemical efficiency of PSII [47]. Results in this study show that even a small but significant reduction in *Fv/Fm* with the greatest decline in Sunmate followed by CM72 then Suntop was recorded (Figure 2); these results are consistent with that presented by Ahmad et al. [48] and Ibrahim et al. [47]. NaCl stress can disturb the photosynthesis biochemistry, limiting the efficiency of two photosystems due to the disordering of chloroplast integrity [47]. In our study, salinity altered leaf chloroplast ultrastructure causing swelling of thylakoids, diffuse granular and stroma lamellae, a larger number of large-sized plastoglobuli and a reduction in leaf chlorophyll content in the sensitive wheat cultivar, Sunmate; these changes were not seen to the same extent in Suntop and CM72 (Figure 6). There may be several reasons for the disruptions to thylakoids and the chloroplast envelope. These include higher accumulation of lipids in chloroplasts, ion toxicity, or imbalance [49], and osmotic imbalance between chloroplast and stroma [50], which in turn cause a reduction of photosynthetic efficiency and the electron transport activity of chloroplasts [51].

Additionally, upon exposure to salinization, severe disruption of nuclei and nuclear membranes of roots were detected in Sunmate but to a lesser extent in Suntop and CM72 (Figure 7). Salinity largely affects roots because of their direct contact with the soil. Therefore, to protect the whole plant from the adverse effects of salinity, roots should better tolerate salinity stress [47]. Zhang and Blumwald [52] noted that in tolerant plants, Na^+ is kept away from the cytosol by compartmentalizing it into the vacuole, and due to the lack of this ability in sensitive plants, dehydration and ionic imbalance disturbed the metabolic process in sensitive plants [53]. We observed that the nucleolus disappeared in Sunmate. A common consequence of this type of alteration inside the nucleus would be a loss of function and/or even cell death [53].

It is important to determine Na^+ and K^+ concentrations and Na^+:K^+ ratios in shoots and roots to understand mechanisms of salinity tolerance [54]. In our study, under the salinity stress, significant differences were found in shoot Na^+ and K^+ along with Na^+:K^+ ratios in both species relative to controls, with the most severe effect in Sunmate (Figure 3A–F, Table S2). In general, CM72 accumulated less Na^+ and more K^+ in shoots followed by Suntop and then Sunmate. Hence, the low Na^+:K^+ ratios in CM72 and Suntop may explain the tolerance of these cultivars. Root to shoot Na^+ and K^+ translocation is limited, as all genotypes accumulated more Na^+ and K^+ in roots than in shoots. These results are consistent with the idea of differences in translocation restricting Na^+ movement to the shoot being one of the mechanisms of salinity tolerance. Na^+ and K^+ are interdependent under salinity stress. Previous studies have found a decrease in K^+ content in several plant species resulting from high salinity [35,55]. Increased Na^+ concentrations in root zones have an antagonistic effect on K^+ uptake. Consequently, a deficiency of K^+ has created stunting growth and reducing yields [56].

The most general consequence of salinization is the accumulation of hazardous substances in plant cells especially ROS such as singlet oxygen (O_2), superoxide radicals (O_2^-), and hydrogen peroxide (H_2O_2); these species cause damage to proteins, lipids, and nucleic acids thereby promoting rapid plant death [57]. Malondialdehyde (MDA), a product of polyunsaturated fatty acid peroxidation [58], is commonly considered as a sign of the extent of oxidation damage under stress [27,59]. The hostile influences of NaCl stress on lipid peroxidation have been reported in other plants, for example in *Brassica juncea* [60] and *Vicia faba* [61], and MDA has been widely recognized as a good salinity tolerance marker in plant species [62]. We found significantly lower MDA contents in the shoots and roots of CM72 and Suntop compared to the Sunmate (Figure 4A,B). These data suggest that CM72 and Suntop were better protected than Sunmate against oxidative damage under salinity stress.

After salt treatment, tolerant plants eventually develop an enhanced antioxidant enzyme system to handle the effects of ROS. In our study, significantly increased SOD, POD, APX, CAT, and GR activities were found in roots and leaf tissues of both species in the NaCl treated plants (Figure 5A–J). However, the relative activities of these enzymes were recorded higher in Suntop and CM72 than in cultivar Sunmate in both tissues. SOD provides the first line of defense against ROS and protects plants from severe damage generated by O_2^- and H_2O_2 in the presence of metal ions [63]. Many studies have found that salinity positively promotes SOD activity in tolerant cultivars in both roots [64] and leaves [65]. Subsequent reactions are required to convert the H_2O_2 produced by SOD to H_2O because H_2O_2 is still toxic to plants and reactions involving POD, CAT, and APX are important. Our research corroborates previous studies [47,48] had measured an enhanced activity of SOD, POD, and CAT in plants treated with a high NaCl dose and the activities of these enzymes were again higher in the two tolerant genotypes. Feki et al. [66] and Koca et al. [67] also demonstrated that tolerance to salinity in wheat and sesame genotypes was associated with lower MDA contents and higher activities of antioxidant enzymes. Thus, it is evident from our results and the results of others [10,68] that the higher POD, CAT, and APX activities coordinate with SOD activity to deal with the undesirable effect of O_2^- and H_2O_2 and the activities of these enzymes are strongly correlated with tolerance to salt-induced oxidative stress in wheat and barley.

The activity of GR in the leaves and roots was higher in CM72 compared to Suntop and Sunmate (Figure 5I,J). Other studies working with salt-sensitive and tolerant genotypes suggested that higher GR activities relate to salt tolerance [46,65]. The higher GR activity might be able to elevate $NADP^+$ concentrations to gain electrons from the photosynthetic electron transport chain thereby reducing the production of ROS [69]. Our results also suggest that the salt-tolerant cultivar may exhibit a more active ascorbate-glutathione cycle.

5. Conclusions

Although differences were observed between CM72 and Suntop in the growth and biochemical traits assessed and modified by salt stress, the differences are negligible in comparison with the response to the salt stress of sensitive wheat cv. Sunmate. The distinct differences between wheat and barley were lower MDA content, lower Na^+/K^+ ratio and a higher level of APX and GR content in the roots of barley cultivar CM72. These results lead us to infer that differences in response to salinity may be just as great within a species as between species. The most obvious mechanisms for salt tolerance in the tolerant barley and wheat cultivars are the increased activities of ROS-scavenging enzymes and a more balanced $Na^+:K^+$ ratio. Our results indicated that Suntop is highly tolerant against salinity, which is quite similar to barley CM72. Novel salt-tolerant related genes may be identified in Suntop for improving the salt tolerance of wheat cultivars, except for commercial application in saline-alkali soils.

Supplementary Materials: The following are available online at http://www.mdpi.com/2073-4395/10/1/127/s1, Table S1: Effect of salinity on plant growth and biomass of Suntop, Sunmate and CM72, 25 days after treatment with 100 mM NaCl, Table S2: Shoot and root Na^+ and K^+ concentration and Na^+/K^+ ratio of two wheat cv. Suntop and Sunmate, and one barley cv. CM72, 25 days after treatment with 100 mM NaCl, Table S3: Effect of salinity stress on SOD, POD, CAT, APX and GR activities and MDA contents in the shoots of Suntop, Sunmate and CM72, after 1, 5, 10 and 15 days 100 mM NaCl treatment, Table S4: Effect of salinity stress on SOD, POD, CAT, APX, and GR activities and MDA contents in the roots of Suntop, Sunmate, and CM72, after 1, 5, 10, and 15 days 100 mM NaCl treatment.

Author Contributions: Conceptualization, F.W. and M.L.; data curation, M.Z. and F.W.; formal analysis, M.Z.; funding acquisition, F.W.; investigation, M.Z. and S.S.; methodology, F.W. and M.Z.; project administration, F.W.; supervision, F.W.; validation, M.Z., S.S., P.H., and F.W.B.; writing—original draft, M.Z.; writing—review and editing, M.Z., F.W., and P.H. All authors have read and agreed to the published version of the manuscript.

Funding: This project was supported by the Key Research Foundation of Science and Technology Department of Zhejiang Province of China (2016C02050-9-7).

Conflicts of Interest: The authors declare no conflict of interest.

References

1. Ghassemi, F.; Jakeman, A.J.; Nix, H.A. *Salinisation of Land and Water Resources: Human Causes, Extent, Management and Case Studies*; CAB International Publishing: Wallingford, UK, 1995; p. 526.

2. FAO. Global Network on Integrated Soil Management for Sustainable Use of Salt-Affected Soils. 2000. Available online: http://www.fao.org/ag/AGL/agll/spush/intro.htm (accessed on 10 May 2004).

3. Zhang, H.X.; Hodson, J.N.; Williams, J.P.; Blumwald, E. Engineering salt-tolerant Brassica plants: characterization of yield and seed oil quality in transgenic plants with increased vacuolar sodium accumulation. *Proc. Natl. Acad. Sci. USA* **2001**, *98*, 12832–12836. [CrossRef] [PubMed]

4. Flowers, T.J. Improving crop salt tolerance. *J. Exp. Bot.* **2004**, *55*, 307–319. [CrossRef] [PubMed]

5. Munns, R.; Tester, M. Mechanisms of salinity tolerance. *Annu. Rev. Plant Biol.* **2008**, *59*, 651–681. [CrossRef] [PubMed]

6. Parida, A.K.; Das, A.B. Salt tolerance and salinity effects on plants: A review. *Ecotoxicol. Environ. Saf.* **2005**, *60*, 324–349. [CrossRef]

7. FAO. *The Special Challenge for Sub-Saharan Africa. How to Feed the World 2050*; FAO: Rome, Italy, 2009.

8. Ashraf, M.; Harris, P.J.C. Potential biochemical indicators of salinity tolerance in plants. *Plant Sci.* **2004**, *166*, 3–16. [CrossRef]

9. Ferguson, L.; Grattan, S.R. How salinity damages citrus: osmotic effects and specific ion toxicities. *Hort. Technol.* **2005**, *15*, 95–99. [CrossRef]

10. Munns, R.; James, R.A.; Lauchli, A. Approaches to increasing the salt tolerance of wheat and other cereals. *J. Exp. Bot.* **2006**, *57*, 1025–1043. [CrossRef]

11. Apse, M.P.; Blumwald, E. Engineering salt tolerance in plants. *Curr. Opin. Biotechol.* **2002**, *13*, 146–150. [CrossRef]

12. Machado, R.M.; Serralheiro, R.P. Soil salinity: effect on vegetable crop growth. Management practices to prevent and mitigate soil salinization. *Horticulturae* **2017**, *3*, 30. [CrossRef]

13. Rengasamy, P. World salinization with emphasis on Australia. *J. Exp. Bot.* **2006**, *57*, 1017–1023. [CrossRef]

14. Slabu, C.; Zörb, C.; Steffens, D.; Schubert, S. Is salt stress of faba bean (*Vicia faba*) caused by Na+ or Cl–toxicity? *J. Plant Nutr. Soil Sci.* **2009**, *172*, 644–651. [CrossRef]

15. Tavakkoli, E.; Rengasamy, P.; McDonald, G.K. High concentrations of Na^+ and Cl^- ions in soil solution have simultaneous detrimental effects on growth of faba bean under salinity stress. *J. Exp. Bot.* **2010**, *61*, 4449–4459. [CrossRef] [PubMed]

16. Cheeseman, J.M. The integration of activity in saline environments: Problems and perspectives. *Funct. Plant Biol.* **2013**, *40*, 759–774. [CrossRef]

17. Schachtman, D.P.; Liu, W.H. Molecular pieces to the puzzle of the interaction between potassium and sodium uptake in plants. *Trends Plant Sci.* **1999**, *4*, 281–287. [CrossRef]

18. Ball, M.C.; Chow, W.S.; Anderson, J.M. Salinity-induced potassium deficiency causes loss of functional photosystem II in leaves of the grey mangrove, *Avicennia marina*, through depletion of the atrazine-binding polypeptide. *Funct. Plant Biol.* **1987**, *14*, 351–361. [CrossRef]

19. Botella, M.A.; Martinez, V.; Pardines, J.; Cerda, A. Salinity induced potassium deficiency in maize plants. *J. Plant Physiol.* **1997**, *150*, 200–205. [CrossRef]

20. Pang, C.H.; Wang, B.S. Oxidative stress and salt tolerance in plants. In *Progress in Botany*; Lüttge, U., Beyschlag, W., Murata, J., Eds.; Springer: Berlin/Heidelberg, Germany, 2008; Volume 69, pp. 231–245.

21. Wong-Ekkabut, J.; Xu, Z.; Triampo, W.; Tang, I.M.; Tieleman, D.P.; Monticelli, L. Effect of lipid peroxidation on the properties of lipid bilayers: A molecular dynamics study. *Biophys. J.* **2007**, *93*, 4225–4236. [CrossRef]

22. Sharma, P.; Jha, A.B.; Dubey, R.S.; Pessarakli, M. Reactive oxygen species, oxidative damage, and antioxidative defense mechanism in plants under stressful conditions. *J. Bot.* **2012**, *2012*, 1–26. [CrossRef]

23. Smirnoff, N. Plant resistance to environmental stress. *Curr. Opin. Biotech.* **1998**, *9*, 214–219. [CrossRef]

24. Jena, N.R. DNA damage by reactive species: Mechanisms, mutation and repair. *J. Biosci.* **2012**, *37*, 503–517. [CrossRef]

25. Noctor, G.; Foyer, C.H. Ascorbate and glutathione: Keeping active oxygen under control. *Annu. Rev. Plant Biol.* **1998**, *49*, 249–279. [CrossRef] [PubMed]

26. Takahashi, M.; Asada, K. Superoxide production in aprotic interior of chloroplast thylakoids. *Arch. Biochem. Biophys.* **1988**, *267*, 714–722. [CrossRef]

27. Apel, K.; Hirt, H. Reactive oxygen species: Metabolism, oxidative stress, and signal transduction. *Annu. Rev. Plant Biol.* **2004**, *55*, 373–399. [CrossRef] [PubMed]

28. Mittler, R.; Vanderauwera, S.; Gollery, M.; Van Breusegem, F. Reactive oxygen gene network of plants. *Trends Plant Sci.* **2004**, *9*, 490–498. [CrossRef] [PubMed]

29. Dietz, K.J.; Jacob, S.; Oelze, M.L.; Laxa, M.; Tognetti, V.; de Miranda, S.M.; Baier, M.; Finkemeier, I. The function of peroxiredoxins in plant organelle redox metabolism. *J. Exp. Bot.* **2006**, *57*, 1697–1709. [CrossRef]

30. Türkan, I.; Demiral, T. Recent developments in understanding salinity tolerance. *Environ. Exp. Bot.* **2009**, *67*, 2–9. [CrossRef]

31. Gupta, B.; Huang, B. Mechanism of salinity tolerance in plants: Physiological, biochemical, and molecular characterization. *Int. J. Genom.* **2014**, *2014*, 1–18. [CrossRef]

32. Pessarakli, M. Dry matter yield, nitrogen-15 absorption, and water uptake by green bean under sodium chloride stress. *Crop Sci.* **1991**, *31*, 1633–1640. [CrossRef]

33. Wu, F.B.; Zhang, G.P.; Yu, J.S. Interaction of cadmium and four microelements for uptake and translocation in different barley genotypes. *Comm. Soil Sci. Plant Analysis* **2003**, *34*, 2003–2020. [CrossRef]

34. Wu, F.B.; Zhang, G.P.; Dominy, P. Four barley genotypes respond differently to cadmium: Lipid peroxidation and activities of antioxidant capacity. *Environ. Exp. Bot.* **2003**, *50*, 67–78. [CrossRef]

35. Chen, Z.; Zhou, M.; Newman, I.A.; Mendham, N.J.; Zhang, G.; Shabala, S. Potassium and sodium relations in salinized barley tissues as a basis of differential salt tolerance. *Funct. Plant Biol.* **2007**, *34*, 150–162. [CrossRef]

36. Wu, F.B.; Lianghuan, W.; Fuhua, X. Chlorophyll meter to predict nitrogen sidedress requirements for short-season cotton (*Gossypium hirsutum* L.). *Field Crop Res.* **1998**, *56*, 309–314.

37. Genty, B.; Briantais, J.M.; Baker, N.R. The relationship between the quantum yield of photosynthetic electron transport and quenching of chlorophyll fluorescence. *Biochim. Biophys. Acta-Gen. Subj.* **1989**, *990*, 87–92. [CrossRef]

38. Leul, M.; Zhou, W.J. Alleviation of waterlogging damage in winter rape by uniconazole application: Effects on enzyme activity, lipid peroxidation, and membrane integrity. *J. Plant Growth Regul.* **1999**, *18*, 9–14. [CrossRef]

39. Chen, F.; Wang, F.; Wu, F.B.; Mao, W.H.; Zhang, G.P.; Zhou, M.X. Modulation of exogenous glutathione in antioxidant defense system against Cd stress in the two barley genotypes differing in Cd tolerance. *Plant Physiol. Biochem.* **2010**, *48*, 663–672. [CrossRef]

40. Jiang, M.; Zhang, J. Effect of abscisic acid on active oxygen species, antioxidative defence system and oxidative damage in leaves of maize seedlings. *Plant Cell. Physiol.* **2001**, *42*, 1265–1273. [CrossRef]

41. Tang, Q.; Feng, M. *Practical Statistics and Its DPS Statistics Software Package*; China Agriculture Press: Beijing, China, 1997.

42. Neill, S.J.; Desikan, R.; Clarke, A.; Hurst, R.D.; Hancock, J.T. Hydrogen peroxide and nitric oxide as signaling molecules in plants. *J. Exp. Bot.* **2002**, *53*, 1237–1247. [CrossRef]

43. Netondo, G.W.; Onyango, J.C.; Beck, E. Sorghum and salinity. II. Gas exchange and chlorophyll fluorescence of sorghum under salt stress. *Crop Sci.* **2004**, *44*, 806–811.

44. Taiz, L.; Zeiger, E. *Plant Physiology*; Sinauer Associates Inc.: Sunderland, MA, USA, 2002.

45. Lawlor, D.W.; Cornic, G. Photosynthetic carbon assimilation and associated metabolism in relation to water deficits in higher plants. *Plant Cell Environ.* **2002**, *25*, 275–294. [CrossRef]

46. Meloni, D.A.; Oliva, M.A.; Martinez, C.A.; Cambraia, J. Photosynthesis and activity of superoxide dismutase, peroxidase and glutathione reductase in cotton under salt stress. *Environ. Exp. Bot.* **2003**, *49*, 69–76. [CrossRef]

47. Ibrahim, W.; Ahmed, I.M.; Chen, X.; Cao, F.; Zhu, S.; Wu, F. Genotypic differences in photosynthetic performance, antioxidant capacity, ultrastructure and nutrients in response to combined stress of salinity and Cd in cotton. *BioMetals* **2015**, *28*, 1063–1078. [CrossRef] [PubMed]

48. Ahmed, I.M.; Cao, F.; Zhang, M.; Chen, X.; Zhang, G.; Wu, F. Difference in yield and physiological features in response to drought and salinity combined stress during anthesis in Tibetan wild and cultivated barleys. *PLoS ONE* **2013**, *8*, e77869. [CrossRef] [PubMed]

49. Yamane, K.; Kawasaki, M.; Taniguchi, M.; Miyake, H. Differential effect of NaCl and polyethylene glycol on the ultrastructure of chloroplasts in rice seedlings. *J. Plant Physiol.* **2003**, *160*, 573–575. [CrossRef] [PubMed]

50. Naeem, M.S.; Warusawitharana, H.; Liu, H.; Liu, D.; Ahmad, R.; Waraich, E.A.; Xua, L.; Zhou, W.J. 5-aminolevulinic acid alleviates the salinity-induced changes in *Brassica napus* as revealed by the ultrastructural study of chloroplast. *Plant Physiol. Biochem.* **2012**, *57*, 84–92. [CrossRef] [PubMed]

51. Parida, A.K.; Das, A.B.; Mittra, B. Effects of NaCl stress on the structure, pigment complex composition, and photosynthetic activity of mangrove *Bruguiera parviflora* chloroplasts. *Photosynthetica* **2003**, *41*, 191–200. [CrossRef]

52. Zhang, H.X.; Blumwald, E. Transgenic salt-tolerant tomato plants accumulate salt in foliage but not in fruit. *Nat. Biotechnol.* **2001**, *19*, 765–768. [CrossRef] [PubMed]

53. Katsuhara, M.; Kawasaki, T. Salt stress induced nuclear and DNA degradation in meristematic cells of barley roots. *Plant Cell Physiol.* **1996**, *l37*, 169–173. [CrossRef]

54. Kader, M.A.; Lindberg, S. Uptake of sodium in protoplasts of salt-sensitive and salt-tolerant cultivars of rice, *Oryza sativa* L. determined by the fluorescent dye SBFI. *J. Exp. Bot.* **2005**, *56*, 3149–3158. [CrossRef]

55. Genc, Y.; McDonald, G.K.; Tester, M. Reassessment of tissue Na^+ concentration as a criterion for salinity tolerance in bread wheat. *Plant Cell. Environ.* **2007**, *30*, 1486–1498. [CrossRef]

56. Reuter, D.; Robinson, J.B. *Plant Analysis: An Interpretation Manual*, 2nd ed.; CSIRO Publishing: Melbourne, Australia, 1997.

57. Ozfidan-Konakci, C.; Yildiztugay, E.; Kucukoduk, M. Upregulation of antioxidant enzymes by exogenous gallic acid contributes to the amelioration in *Oryza sativa* roots exposed to salt and osmotic stress. *Environ. Sci. Pollut. Res. Int.* **2015**, *22*, 1487–1498. [CrossRef]

58. Del, R.D.; Stewart, A.J.; Pellegrini, N. A review of recent studies on malondialdehyde as toxic molecule and biological marker of oxidative stress. *Nutr. Metab. Cardiovasc Dis.* **2005**, *15*, 316–328. [CrossRef] [PubMed]

59. Davey, M.W.; Stals, E.; Panis, B.; Keulemans, J.; Swennen, R.L. High-throughput determination of malondialdehyde in plant tissues. *Anal. Biochem.* **2005**, *347*, 201–207. [CrossRef] [PubMed]

60. Ahmad, P.; Hakeem, K.R.; Kumar, A.; Ashraf, M.; Akram, N.A. Salt-induced changes in photosynthetic activity and oxidative defense system of three cultivars of mustard (*Brassica juncea* L.). *African J. Biotechnol.* **2012**, *11*, 2694–2703.

61. Azooz, M.M.; Youssef, A.M.; Ahmad, P. Evaluation of salicylic acid (SA) application on growth, osmotic solutes and antioxidant enzyme activities on broad bean seedlings grown under diluted seawater. *Int. J. Plant Physiol. Biochem.* **2011**, *3*, 253–264.

62. Katsuhara, M.; Otsuka, T.; Ezaki, B. Salt stress-induced lipid peroxidation is reduced by glutathione S-transferase, but this reduction of lipid peroxides is not enough for a recovery of root growth in Arabidopsis. *Plant Sci.* **2005**, *169*, 369–373. [CrossRef]

63. Bowler, C.; Montagu, M.V.; Inze, D. Superoxide dismutase and stress tolerance. *Annu. Rev. Plant Physiol. Plant Mol. Biol.* **1992**, *43*, 83–116. [CrossRef]

64. Shalata, A.; Mittova, V.; Volokita, M.; Guy, M.; Tal, M. Response of the cultivated tomato and its wild salt-tolerant relative *Lycopersicon pennellii* to salt-dependent oxidative stress: The root antioxidative system. *Physiol. Plant* **2001**, *112*, 487–494. [CrossRef]

65. Hernandez, J.A.; Jimenez, A.; Mullineaux, P.; Sevilla, F. Tolerance of pea (*Pisum sativum* L.) to long-term salt stress is associated with induction of antioxidant defences. *Plant Cell Environ.* **2000**, *23*, 853–862. [CrossRef]

66. Feki, K.; Tounsi, S.; Brini, F. Comparison of an antioxidant system in tolerant and susceptible wheat seedlings in response to salt stress. *Spanish J. Agric. Res.* **2017**, *15*, e0805. [CrossRef]

67. Koca, H.; Bor, M.; Özdemir, F.; Türkan, I. The effect of salt stress on lipid peroxidation, antioxidative enzymes and proline content of sesame cultivars. *Environ. Exp. Bot.* **2007**, *60*, 344–351. [CrossRef]

68. Temel, A.; Gozukirmizi, N. Physiological and molecular changes in barley and wheat under salinity. *Appl. Biochem. Biotechol.* **2015**, *175*, 2950–2960. [CrossRef] [PubMed]

69. Reddy, R.A.; Chaitanya, K.V.; Vivekanandan, M. Drought-induced responses of photosynthesis and antioxidant metabolism in higher plants. *J. Plant Physiol.* **2004**, *161*, 1189–1202. [CrossRef] [PubMed]

Article

Transcriptomic Profiling of Pomegranate Provides Insights into Salt Tolerance

Cuiyu Liu [1,2], Yujie Zhao [1,2], Xueqing Zhao [1,2], Jinping Wang [1,2], Mengmeng Gu [3,*] and Zhaohe Yuan [1,2,*]

1   Co-Innovation Center for Sustainable Forestry in Southern China, Nanjing Forestry University, Nanjing 210037, China; liucuiyu88@gmail.com (C.L.); z1184985369@163.com (Y.Z.); zhaoxq402@163.com (X.Z.); wangjp0304@163.com (J.W.)
2   College of Forestry, Nanjing Forestry University, Nanjing 210037, China
3   Department of Horticultural Sciences, Texas A&M AgriLife Extension Service, College Station, TX 77843-2134, USA
*   Correspondence: mgu@tamu.edu (M.G.); zhyuan88@hotmail.com (Z.Y.); Tel.: +1-979-845-8545 (M.G.); +86-025-8542-7056 (Z.Y.)

Received: 17 October 2019; Accepted: 24 December 2019; Published: 27 December 2019

Abstract: Pomegranate (*Punica granatum* L.) is widely grown in arid and semi-arid soils, with constant soil salinization. To elucidate its molecular responses to salt stress on mRNA levels, we constructed 18 cDNA libraries of pomegranate roots and leaves from 0 (controls), 3, and 6 days after 200 mM NaCl treatment. In total, we obtained 34,047 genes by mapping to genome, and then identified 2255 DEGs (differentially expressed genes), including 1080 up-regulated and 1175 down-regulated genes. We found that the expression pattern of most DEGs were tissue-specific and time-specific. Among root DEGs, genes associated with cell wall organization and transmembrane transport were suppressed, and most of metabolism-related genes were over-represented. In leaves, 41.29% of DEGs were first suppressed and then recovered, including ions/metal ions binding-related genes. Also, ion transport and oxidation-reduction process were restricted. We found many DEGs involved in ABA, Ca^{2+}-related and MAPK signal transduction pathways, such as ABA-receptors, Ca^{2+}-sensors, MAPK cascades, TFs, and downstream functional genes coding for HSPs, LEAs, AQPs and PODs. Fifteen genes were selected to confirm the RNA-seq data using qRT-PCR. Our study not only illuminated pomegranate molecular responses to salinity, but also provided references for selecting salt-tolerant genes in pomegranate breeding processes.

Keywords: pomegranate; salt stress; transcriptome; tissue-specific; signaling transduction pathways; transcription factors

1. Introduction

Soil salinization is defined as the excess or deposition of salt ions in land, which may interfere with plant growth. With the aggravation of soil salinization, it has become a considerable threat to healthy and sustainable development of worldwide agriculture [1]. Approximate 20% of the global cultivated lands and 50% of the irrigated lands are affected by salinity [2]. Plants exposed to saline conditions mainly suffer from osmotic stress, ion toxicity, and nutrient deficiency [3,4]. Consequently, all of the significant processes involved in plant growth and development, such as photosynthesis, protein synthesis, energy conversion and ion balance, could be affected by salinity [5]. The effects of salinity depend, not only on species, genotypes, and the age of plants, but also on the duration and intensity of stress [6]. Meanwhile, plants can adapt to saline environment with following strategies: (1) Efficiently controlling the uptake, transport, and compartmentalization of toxic ions;

(2) synthesis of osmoregulation substances and activation of antioxidant enzymes; (3) formation of unique morphological structures, such as succulent leaves, salt glands and bladders [7,8].

Plant salt-tolerance is a quantitative trait controlled by multiple genes [4]. Using transcriptomic analysis, researchers can identify the salt-related genes with differential and temporal expression patterns in plants, and provide a clear picture of transcripts in response to salt stress. These salt-related genes are identified and categorized into two types according to the functions of proteins [4]. The first type is the effector code the functional proteins, which are directly involved in the physiological and biochemical responses to salt stress in plants. These proteins include the superoxide dismutase (SOD) [9], ascorbate peroxidase (APX) [10], high-affinity potassium transporter (HKT) [11], Na^+/H^+ antiporter (NHX) [12], aquaporin (AQP) [13], late embryogenesis abundant (LEA) [14], and H^+-ATPase (VHA) [15], etc. The second type is regulator that involved in regulating the expressions of genes and the signal transduction pathways, such as transcription factors (TFs) and various kinases [16]. These well-characterized TFs include Apetala2/ethylene response factor (AP2/ERF), dehydration-responsive element-binding protein/C-repeat binding factor (DREB/CBF), WRKY, NAC, basic leucine zipper (bZIP), MYB, and basic helix-loop-helix (bHLH) family members [4]. These genes regulate the expressions of downstream genes via various ways, then may influence the plants salt-tolerance eventually [17]. Under salt stress, many signaling transduction pathways are stimulated in plants, such as Ca^{2+}-related, abscisic acid (ABA), and mitogen-activated protein kinase (MAPKs), as well as the crosstalk networks among them, which play crucial roles in responses to salt stress [4,18].

Pomegranate (*Punica granatum* L.) is an emerging commercial fruit tree of the Lythraceae family [19]. In recent years, pomegranate is increasingly popular with extensive usage of its fruits and products, and it is considered a 'super fruit' with high nutritional and medicinal values [20]. The species is widely grown in arid and semiarid regions where the availability and irrigation of saline water are significant issues [21]. Thus, it is important to explore pomegranate potential salt tolerance. Considerable research has been done on pomegranate physiological and biochemical responses to salinity, especially on its growth, ion balance, osmoregulation, and the scavenging of reactive oxygen species (ROS) [22–24]. In this study, we used roots and leaves of a pomegranate cultivar 'Taishanhong' with whole genomic sequence released to perform deep transcriptome sequencing and then demonstrate a global representation of potential candidate genes under salt stress. We aim to unravel the fundamental molecular mechanisms in pomegranate underlying the responses to salt stress.

2. Materials and Methods

2.1. Plant Growth and Stress Treatments

Uniform rooted cuttings of pomegranate ('Taishanhong') were obtained from the Pomegranate Repository of Nanjing Forestry University, China. They were grown in plastic pots (2.5 L) containing medium (1:1 by volume of perlite and peat, 1.5 ± 0.1 Kg) in a climate controlled chamber for six months (14 h light 26 °C/10 h dark 22 °C), and fertigated weekly with $\frac{1}{2}$ Hoagland's solution [25].

At the beginning of experiment, forty-five plants were randomly selected and divided into 3 groups, fifteen plants per group. The roots and leaves of the first group of plants were harvested as control samples (T0R and T0L). The The other two groups of plants were fertigated once with 500 mL of $\frac{1}{2}$ Hoagland's solution containing 200 mM NaCl. A saucer was placed under each container to collect leachate solution during the experiment period. The roots and leaves of the second group of plants on day 3 (T1R and T1L) and the third group of plants on day 6 (T2R and T2L) were harvested later. There was not visible difference between the treated and untreated plant (Figure S1). Samples from five plants in the same group were pooled together as a replicate due to the small amount of biomass. All samples were immediately frozen in liquid nitrogen and then stored at −80 °C.

2.2. RNA Preparation, cDNA Library Construction and Sequencing

Total RNA was extracted from pomegranate roots and leaves using the Total RNA Kit (Tiangen, Beijing, China) according to the manufacturer's instructions. RNA purity, concentration and integrity were checked using the NanoPhotometer® spectrophotometer (IMPLEN, Westlake Village, CA, USA), Qubit® RNA Assay Kit in Qubit® 2.0 Flurometer (Life Technologies, Carlsbad, CA, USA) and the RNA Nano 6000 Assay Kit of the Agilent Bioanalyzer 2100 (Agilent Technologies, Santa Clara, CA, USA), respectively.

A total amount of 3 µg RNA per sample was used as input material for the RNA sample preparations. Sequencing libraries were generated using NEB Next® Ultra™ RNA Library Prep Kit for Illumina® (NEB, Ipswich, MA, USA) following manufacturer's recommendations and index codes were added to attribute sequences to each sample. The RNA samples were concentrated using magnetic oligo (dT) beads and then broken into short fragments using an RNA Fragmentation buffer (NEB Next First Strand Synthesis Reaction Buffer (5X), Ambion, Austin, TX, USA). The first-strand cDNA was synthesized using random hexamer primer and M-MuLV Reverse Transcriptase (RNase H-). Second-strand cDNA was synthesized subsequently using DNA Polymerase I and RNase H. After adenylation of 3'ends of DNA fragments, NEB Next Adaptor with hairpin loop structure were ligated to prepare for hybridization. The cDNA fragments with suitable lengths (150~200 bp) were purified with AMPure XP system (Beckman Coulter, Beverly, MA, USA) to construct the final cDNA libraries. Then 3 µL USER Enzyme (NEB, USA) was used with size-selected, adaptor-ligated cDNA at 37 °C for 15 min followed by 5 min at 95 °C before PCR. Next, the selected cDNA fragments were enriched via PCR with Phusion High-Fidelity DNA polymerase, Universal PCR primers and Index (X) Primer. PCR products were purified with AMPure XP system and the cDNA libraries were assessed on the Agilent Bioanalyzer 2100 system. Finally, the clustering of the index-coded samples was performed on a cBot Cluster Generation System using TruSeq PE Cluster Kit v3-cBot-HS (Illumina, Inc., San Diego, CA, USA) according to the manufacturer's instructions. After cluster generation, the library preparations were sequenced on an Illumina HiSeq 4000 platform (Illumina, Inc., San Diego, CA, USA) and 150 bp paired-end reads were generated.

2.3. Sequence Assembly

Low quality reads and reads containing adapter and ploy-N were filtered by NGS QC ToolKit [26] from raw reads. At the same time, Q20, Q30, GC-content and sequence duplication level of the clean data were calculated. All the downstream analyses were based on clean data with high quality. The clean data from all 18 libraries were separately mapped to the pomegranate genome assembly (ASM220158v1) using HISAT2 software [27]. StringTie [28] was used to construct and identify both known and novel transcripts from HISAT2 alignment results. StringTie was also used to count the reads numbers mapped to each gene. After that, FPKM (Fragments Per Kilobase of exon per million fragments Mapped) were calculated based on the length of the gene and reads count mapped to this gene [27].

2.4. Identification and Functional Annotation of DEGs

All genes were compared against various protein databases by BLASTX, including the Nr (NCBI non-redundant protein sequences); Pfam (Protein family); KOG/COG (Clusters of Orthologous Groups of proteins), and Swiss-Prot (A manually annotated and reviewed protein sequence database) with an E-value cut-off of 10^{-5}. Then genes with the best BLAST hit (the highest score) were chosen along with their protein functional annotations. DESeq and Q-value were employed to evaluate the differential expression genes (DEGs) between controls and treatments. The false discovery rate (FDR) and $\log_2$FC (log of fold change) were calculated for all genes, and only transcripts with FDR < 0.05 and$|\log_2$(fold change)$|\geq1$ were considered as DEGs. To annotate the DEGs with gene ontology (GO) terms, the Nr BLAST results were imported into the Blast2GO program [29]. To examine the expression patterns of DEGs, the expression data from roots and leaves (T0, T1 and T2) were normalized to 0, $\log_2$(T1/T0), $\log_2$(T2/T0), and then clustered by Short Time-series Expression Miner software (STEM) separately, using a FDR correction method and *p*-value $\leq$ 0.05 as the cutoff [30]. The results of GO annotations in each pattern were enriched and refined using TBtools v0.6652 (Toolbox for Biologists, https://github.com/CJ-Chen/TBtools). We used KOBAS software [31] to test the enrichment of DEGs in KEGG pathways.

2.5. Validation of RNA-Seq by qRT-PCR

We selected 15 DEGs as experimental validation by quantitative real-time PCR (qRT-PCR), which were performed with three biological and three technical replicates for each cDNA sample. The primers for these genes were designed with NCBI primer-BLAST (Table S1). Reverse transcription was conducted with the GoScriptTM Reverse Transcription System (Promega, Madison, WI, USA). We conducted qRT-PCR in a 7500 fast Real-Time PCR system (Applied Biosystems, CA, MA, USA), and analyzed results with the $\Delta\Delta$Ct method, and *Pgactin* (F: ATCCTCCGTCTTGACCTTG, R: TGTCCGTCAGGCAACTCAT) gene was used as a reference gene. Each reaction was carried out in a final volume of 20 µL, containing 7.5 µL of ddH$_2$O, 10 µL of SYBR Green PCR master mix, 0.5 µL of each gene-specific primer and 2 µL of diluted cDNA. The PCR thermal cycling conditions were as follows: 95 °C for 10 min; 40 cycles of 95 °C for 5 s, 60 °C for 30 s and 72 °C for 30 s. Data were collected during the extension step: 95 °C for 15 s, 60 °C for 1 min, 95 °C for 30 s and 60 °C for 15 s.

3. Results

3.1. Overview of Sequencing and Mapping

To obtain a global overview of the salt-induced changes at whole-transcriptomic scale in pomegranate, we constructed 18 cDNA libraries from the roots and leaves of 0 (controls), 3 and 6 days after 200 mM NaCl treatment. A high correlation (R^2 > 0.80) between biological replicates was observed for all treatments (Figure S2), which indicated that the biological replicates were reliable in this study. In totally, sequencing libraries yielded 519.47 million reads with 150 bp for both paired ends. After adapter removal and refining, we obtained 155.84 Gb of clean data, and Q30 percentage were all above 90.08% (Table 1). The ratios of reads mapping to the pomegranate genome were high, with values ranging from 93.61% to 95.33%. Then, the unique-mapped reads were 91.89–93.79% and the mapped sequences in genome exon regions were 62.88–69.32% (Table 1). Finally, we assembled 29,226 transcripts from the 18 cDNA libraries of pomegranate roots and leaves under salt stress. To splice the genes more completely and accurately, the StringTie software was employed to reconstruct the transcripts, and then retrieved 34,047 assemblies, including 5396 novel transcripts. Finally, 26,444 genes (including 2421 new genes) were assembled from annotated genes using 7 different databases (see methods), and can be used as a reference of pomegranate genome annotation (Table S2).

Table 1. Summary of RNA-Seq results and the alignments in pomegranate genome.

Tissues	Samples ID	Read Number (M)	Base Number (M)	GC Content (%)	Q30 (%)	Mapped Reads (%)	Unique Alignments (%)	Mapped to Exonic (%)
Root (CK)	T01	31,493,153	9,447,945,900	50.43	91.10	94.06	92.27	66.54
	T02	28,438,357	8,531,507,100	50.10	91.78	94.53	92.76	66.84
	T03	26,939,697	8,081,909,100	50.03	91.27	94.99	93.33	62.88
Leaf (CK)	T04	33,023,098	9,906,929,400	52.08	91.22	95.19	93.59	69.32
	T05	27,023,343	8,107,002,900	51.60	90.70	94.88	93.34	68.26
	T06	25,709,176	7,712,752,800	51.48	91.84	95.21	93.54	68.57
Root (3 d)	T07	31,433,566	9,430,069,800	49.95	91.16	94.46	92.67	66.13
	T08	28,144,078	8,443,223,400	50.16	91.23	94.12	92.32	67.30
	T09	26,542,339	7,962,701,700	50.51	91.32	94.63	92.88	67.69

Table 1. *Cont.*

Tissues	Samples ID	Read Number (M)	Base Number (M)	GC Content (%)	Q30 (%)	Mapped Reads (%)	Unique Alignments (%)	Mapped to Exonic (%)
Leaf (3 d)	T10	28,771,523	8,631,456,900	51.83	91.24	94.97	93.36	68.38
	T11	29,759,918	8,927,975,400	51.32	91.46	95.33	93.79	68.04
	T12	31,628,291	9,488,487,300	51.18	90.08	94.91	93.03	67.80
Root (6 d)	T13	31,761,056	9,528,316,800	50.56	91.91	93.73	91.89	66.87
	T14	29,807,151	8,942,145,300	50.07	91.40	93.89	92.01	66.11
	T15	25,298,243	7,589,472,900	50.29	91.46	93.61	91.89	66.33
Leaf (6 d)	T16	30,075,932	9,022,779,600	51.41	90.84	95.35	93.79	68.50
	T17	27,643,746	8,293,123,800	51.37	91.18	95.18	93.63	68.38
	T18	25,980,948	7,794,284,400	50.88	91.38	94.32	92.66	67.57
Total	—	519,473,615	155,842,084,500	—	—	—	—	—

Reads Number: total Number of paid-end reads in Clean Data; Base Number: total number of bases in Clean Data; Q30%: the ratio of nucleotides with quality value ≥ 30; GC content: The ratio of guanidine and cytosine nucleotides; Mapped ratio: The percentage of Mapped Reads in proportion of Clean reads.

3.2. Identification and Annotation of DEGs

A total of 2255 DEGs were identified from salt treatments with FDR < 0.05 and $|\log_2(\text{fold change})| \geq 1$, including 1080 up-regulated and 1175 down-regulated genes, and more DEGs were found in roots (1623 genes) than in leaves (632 genes) (Table S2). To provide insights into the underlying functions of pomegranate transcripts under salt stress, DEGs were annotated after they were compared to well-studied sequences in the GO, Nr, Swiss-Prot, KEGG, and KOG databases (Table S2). In this study, most of DEGs in T1R were distributed in T subgroup (Signal transduction mechanisms), E subgroup (Amino acid transport and metabolism) and G subgroup (Carbohydrate transport and metabolism) (Figure 1a). DEGs of T2R and T1L samples were mostly concentrated in R subgroup (General function prediction only), Q subgroup (Secondary metabolites biosynthesis, transport and cabalisms), and K subgroup (Transcription) (Figure 1b,c). The DEGs in T2L were distributed in O subgroup (Posttranslational modification, protein turnover and chaperones), P (Inorganic ion transport and metabolisms) and Q subgroup (Secondary metabolites biosynthesis, transport and cabalisms) (Figure 1d). These subgroups were close to the metabolism, such as DNA transcription, cell division, protein modification and energy conversion.

The DEGs of roots and leaves were also analyzed by KEGG pathways enrichment. These pathways were mainly classified into 5 categories (Figure S3) and showed the top 20 pathways (Figure 2). The majority of pathways in pomegranate roots and leaves were associated with metabolisms, such as global and overview maps, amino acid metabolism, carbohydrate metabolism, lipid metabolism, and energy metabolism, etc. (Figure S3). The metabolism-related genes were significantly enriched in metabolic pathways (ko01100), biosynthesis of secondary metabolites (ko01110), cysteine and methionine metabolism (ko00270), phenylpropanoid biosynthesis (ko00940), starch and sucrose metabolism (ko00500), amino sugar and nucleotide sugar metabolism (ko005200) and protein processing in endoplasmic reticulum (ko04141) pathways. Also, there were some DEGs involved in environmental adaptation (plant-pathogen interaction, ko04626) and signal transduction (plant hormone signal transduction, ko04075) pathways (Figure 2; Figure S3). Genes clustered in genetic information processing were enriched in folding, sorting and degradation, replication and repair, transcription and translation, including protein processing in endoplasmic reticulum (ko04141), DNA replication (ko03030), RNA polymerase (ko03020), RNA transport (ko03013) and RNA degradation (ko03018). In addition, genes involved in the photosynthetic pathways, such as carotenoid biosynthesis (ko00906), porphyrin and chlorophyll metabolism (ko00860) and photosynthesis-antenna proteins (ko00196) were suppressed by salinity.

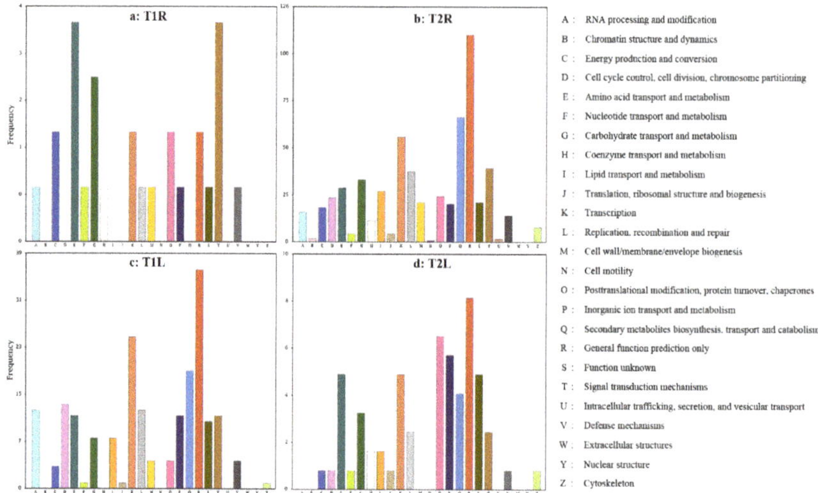

Figure 1. COG classification of differentially expressed genes (DEGs) in pomegranate roots and leaves. DEGs enrichments in T1R (**a**), T2R (**b**), T1L (**c**) and T2L (**d**).

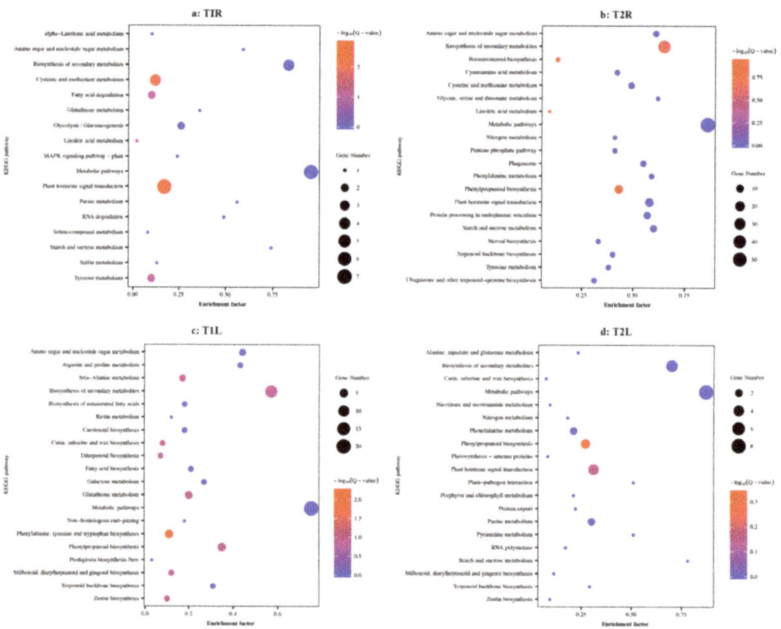

Figure 2. The top 20 KEGG pathway of DEGs in pomegranate root and leaf. The horizontal axis showed an enrichment factor, the smaller the enrichment factor is the more significant enrichment of DEGs in this pathway. While the vertical axis illustrated the KEGG pathway, and the color illustrated the -log(Q-value), red color is more reliable and significance enrichment in this pathway. The size of black spots showed the number of DEGs enriched into each pathway. DEGs enrichments in T1R (**a**), T2R (**b**), T1L (**c**), and T2L (**d**).

3.3. The Expressional Patterns of DEGs

Numbers of DEGs increased in roots with the duration of salt-stress, which were opposite in leaves. The genes were both with little overlap in roots and leaves at two points of stress time (Figure 3a). Only a small portion (2.0%) of DEGs (20 up-regulated and 24 down-regulated) shared the common expression tendency between roots and leaves, and even less (1.2%) DEGs showed an utterly opposite tendency between two tissues (3) genes were up-regulated in leaves and down-regulated in roots, 23 genes were down-regulated in leaves but up-regulated in roots) (Figure 3b). The remaining majority of DEGs were exclusively up-regulated or down-regulated in either tissue. Almost 85.9% of up-regulated and 53.4% of down-regulated DEGs were activated in roots after 6-days salt stress, most of down-regulated genes in T1L recovered in T2L (Figure 3c,d), and only 22 (2.0%) genes were suppressed in leaves at two points of time.

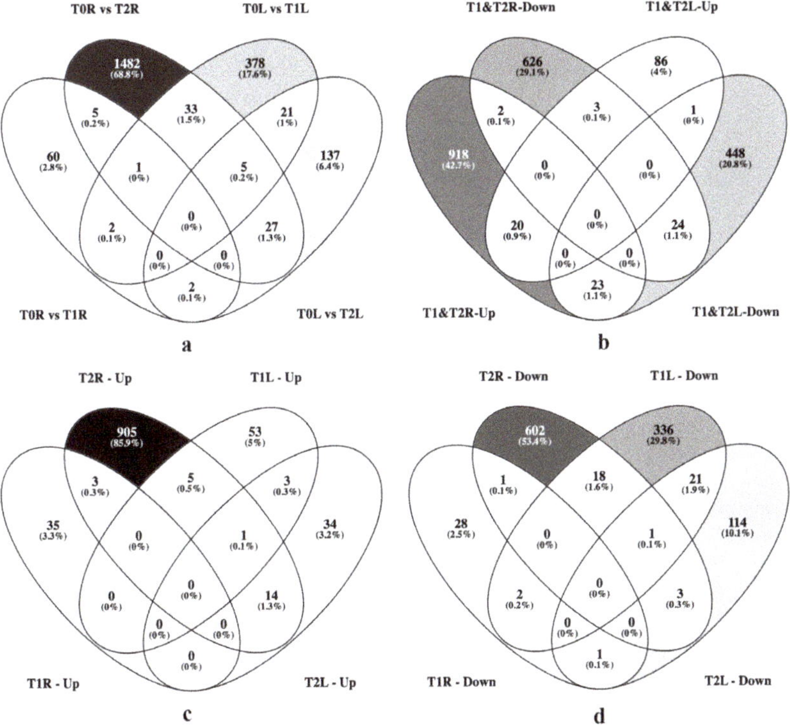

Figure 3. The numbers of up-regulated and down-regulated genes in treatments (T1R, T2R, T1L, and T2L), comparing to the controls (T0R and T0L). (**a**) All DEGs in T1R, T2R, T1L, and T2L; (**b**) Up- or down-regulated genes in leaves and roots; (**c**) Up-regulated genes in T1R, T2R, T1L, and T2L; and (**d**) Down-regulated genes in T1R, T2R, T1L, and T2L.

The STEM was employed to analyze the expressional patterns of DEGs. All the 1,623 DEGs in roots were clustered into 8 profiles, whereby 1,340 DEGs were significantly clustered into 4 profiles with p-value $\leq 2 \times 10^{-5}$ (Figure 4a). The two up-regulated patterns were Profile 4 (31.55%, 512 DEGs) and Profile 7 (15.16%, 246 DEGs), and two down-regulated patterns were Profile 3 (28.59%, 464 DEGs) and Profiles 0 (7.27%, 118 DEGs). All the 632 DEGs in leaf samples were also clustered into 8 profiles and then 2 enriched profiles, including one first down-regulated and then the up-regulated pattern of Profile 2 (41.29%, 261 DEGs, p-value = 3×10^{-32}), and one down-regulated pattern of Profile 1 (16.61%, 105 DEGs, p-value = 6×10^{-17}) (Figure 4b).

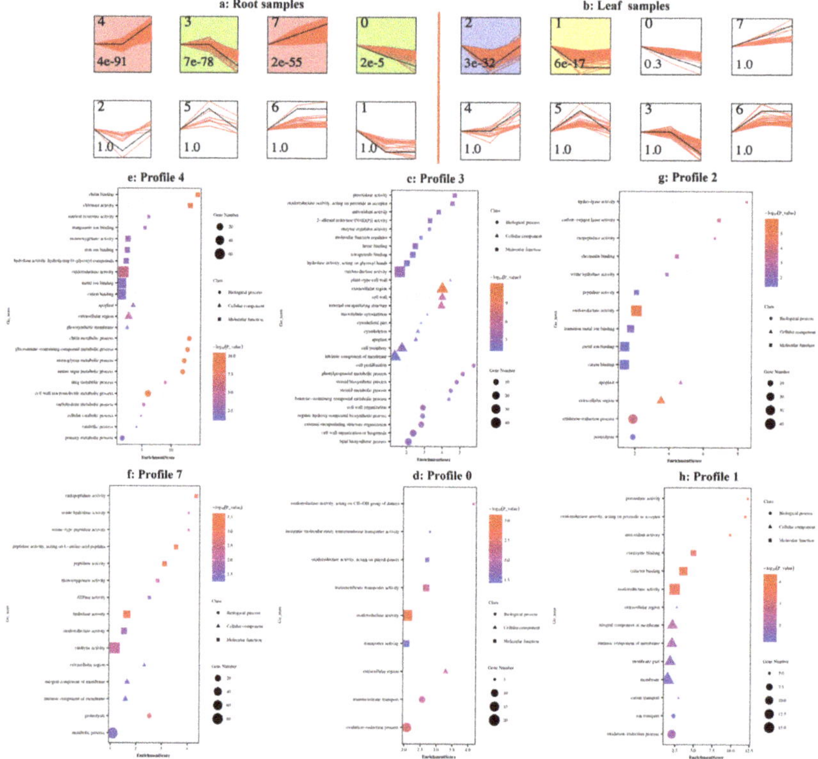

Figure 4. The expression patterns of DEGs in pomegranate roots and leaves under NaCl stress using STEM. The P-value in roots; (**a**) and leaves (**b**) are shown. The trends of each pattern showed in black lines and the clustered genes are shown in red lines. GO enrichments of Profile 4 (**c**), Profile 7 (**d**) Profile 3 (**e**) and Profile 0 (**f**) in root libraries (T0R, T1R and T2R); and Profile 2 (**g**) and Profile 1 (**h**) in leaf libraries (T0L, T1Land T2L).

Next, DEGs within these profiles were subjected to GO-term enrichment analysis. They were classified into three main categories, including cellular component, biological process, and molecular function, and the top 30 sub-categories (if more than 30, *p*-value < 0.05) of each category were shown (Figure 4). In roots, genes in up-regulated pattern of Profile 4 were just inducted in T2R. They were enriched in chitinase activity, chitin binding, oxidoreductase activity, metal ion, and cation binding under molecular function; chitin, glucosamine-containing compound, aminoglycan, amino sugar and cell wall macromolecule metabolic process under biological process (Figure 4c; Table S3). Gens were continuously suppressed with the extension of stress in Profile 7, and they were enriched in endopeptidase activity, peptidase activity, hydrolase activity, and catalytic activity under molecular function and proteolysis and metabolic process under biological process (Figure 4d; Table S3). The top subcategories of down-regulated pattern (Profile 3) were extracellular region (GO:0005576), cell wall (GO:0005618) and external encapsulating structure (GO:0030312) under cellular component, and lipid biosynthetic process (GO:0008610), cell wall organization (GO:0071555), and external encapsulating structure organization (GO:0045229) under biological process. (Figure 4e; Table S4). Many DEGs coding cell wall components were mostly suppressed in T2R samples, including four extensins, two pectin acetylesterase, three polygalacturonase, and three xyloglucan endotransglucosylase/hydrolase proteins. At the same time, transmembrane transport (GO:0055085) and oxidation-reduction process (GO:0055114) in Profile 0 were inhibited under salt stress. Many genes were identified in these biological process, such as adenine/guanine permease, aquaporin, cationic amino acid transporter, sugar carrier, and zinc transporter (Figure 4f; Table S4). Profile 2 showed a first down- and then up-regulated pattern in leaves, there were many subcategories, such as proteolysis and oxidation-reduction process under biological process, oxidoreductase activity, metal ion, and cation binding under molecular function. The top subcategories in down-regulated pattern of Profile 1 were coenzyme binding, cofactor binding and oxidoreductase activity under molecular function, and oxidation-reduction process, cation and ion transport under biological process. We identified genes coding ion transporters, such as cation/calcium exchanger, zinc transporter, ammonium transporter 1 member, copper-transporting ATPase, potassium transporter (Figure 4h; Table S5).

3.4. ABA Signaling Pathway

The ABA signaling pathway mainly consists of three protein classes: ABA receptors (PYR/PYL/RCAR), type 2C protein phosphatase (PP2Cs) and sucrose non-fermenting1-related protein kinase 2 (SnRK2s). In our study, three ABA receptors, *PYLs*, were significantly down-regulated (Figure 5), while 13 *PP2Cs* have different expressional patterns in pomegranate roots and leaves under salt stress. Among these genes, seven *PP2Cs* were up-regulated in roots or leaves, five *PP2Cs* down-regulated, and another gene up-regulated in roots and down-regulated in leaves (Figure 5). In contrast, just one *SnRK2* was identified among the DEGs, and it was up-regulated in leaves, which allowed the accumulation of phosphorylated downstream ABFs and activation of ABA-response genes.

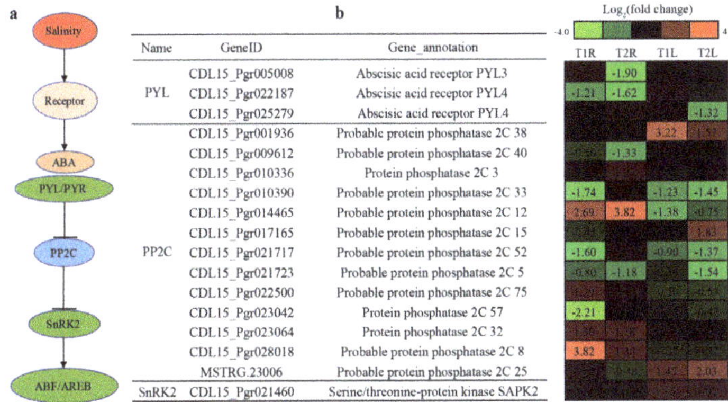

Name	GeneID	Gene_annotation
PYL	CDL15_Pgr005008	Abscisic acid receptor PYL3
	CDL15_Pgr022187	Abscisic acid receptor PYL4
	CDL15_Pgr025279	Abscisic acid receptor PYL4
PP2C	CDL15_Pgr001936	Probable protein phosphatase 2C 38
	CDL15_Pgr009612	Probable protein phosphatase 2C 40
	CDL15_Pgr010336	Protein phosphatase 2C 3
	CDL15_Pgr010390	Probable protein phosphatase 2C 33
	CDL15_Pgr014465	Probable protein phosphatase 2C 12
	CDL15_Pgr017165	Probable protein phosphatase 2C 15
	CDL15_Pgr021717	Probable protein phosphatase 2C 52
	CDL15_Pgr021723	Probable protein phosphatase 2C 5
	CDL15_Pgr022500	Probable protein phosphatase 2C 75
	CDL15_Pgr023042	Protein phosphatase 2C 57
	CDL15_Pgr023064	Protein phosphatase 2C 32
	CDL15_Pgr028018	Probable protein phosphatase 2C 8
	MSTRG.23006	Probable protein phosphatase 2C 25
SnRK2	CDL15_Pgr021460	Serine/threonine-protein kinase SAPK2

Figure 5. DEGs involved in ABA signal transduction pathway in pomegranate under salt stress. (**a**) ABA signal transduction pathway; (**b**) DEGs annotations and expressional levels. The color represents with $\log_2$(fold change), the DEGs with $\log_2$FC $\geq$ 1 were up-regulated (red) and $\log_2$FC $\leq$ 1 were down-regulated (green) (FDR < 0.05).

3.5. Ca^{2+}-Related Signaling Pathways

Thirty DEGs involved in Ca^{2+}-related signaling pathway were identified, these transcripts coded function proteins included three Ca^{2+}-ATPases (ACAs), four cation/H^+ antiporters (CAXs), two cation/calcium exchangers (CCXs), four glutamate receptor (GLRs), ten calcium-binding proteins (CaM/CMLs), two CBL-interacting protein kinases (CIPKs) and five calcium-dependent protein kinases (CDPKs) (Figure 6). Expectedly, two *ACAs* were up-regulated in roots (T2R), and one was down-regulated in leaves (Figure 6). Ten *CAXs* expressed differently, including two *CAXs* located on vacuoles were both up-regulated in T2R. Two *CIPKs* were significantly up-regulated in leaves, but no obvious change was observed in roots. The majority of *CaM/CMLs* were up-regulated in roots and down-regulated in leaves. The *CDPKs* were up- or down-regulated in roots and/or leaves. Also, two DEGs, that were involved in the MAPK signaling pathway were identified, and both were down-regulated in leaves (Figure 6).

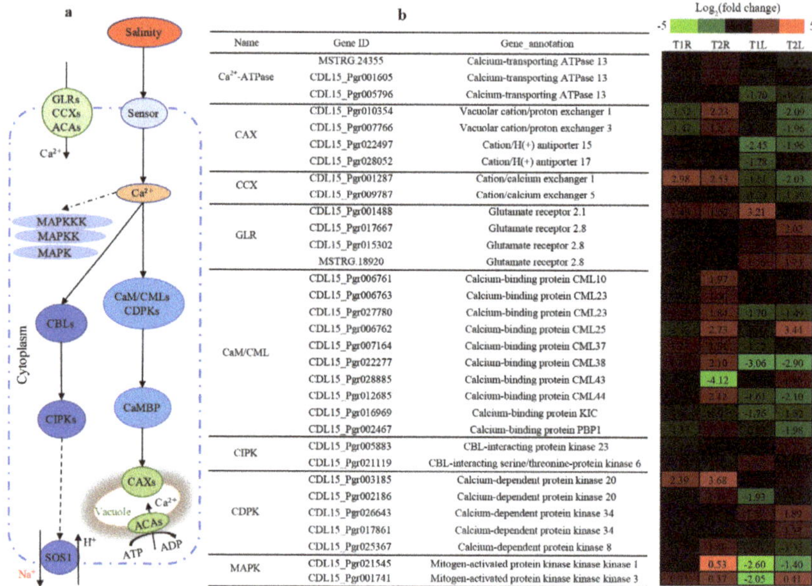

Name	Gene ID	Gene_annotation	T1R	T2R	T1L	T2L
Ca²⁺-ATPase	MSTRG.24355	Calcium-transporting ATPase 13				
	CDL15_Pgr001605	Calcium-transporting ATPase 13				
	CDL15_Pgr005796	Calcium-transporting ATPase 13			1.70	
CAX	CDL15_Pgr010354	Vacuolar cation/proton exchanger 1		2.23		-2.09
	CDL15_Pgr007766	Vacuolar cation/proton exchanger 3				-1.96
	CDL15_Pgr022497	Cation/H(+) antiporter 15			-2.45	-1.96
	CDL15_Pgr028052	Cation/H(+) antiporter 17			1.78	
CCX	CDL15_Pgr001287	Cation/calcium exchanger 1	2.98	2.53		-2.03
	CDL15_Pgr009787	Cation/calcium exchanger 5				
GLR	CDL15_Pgr001488	Glutamate receptor 2.1		3.21		
	CDL15_Pgr017667	Glutamate receptor 2.8				-2.02
	CDL15_Pgr015302	Glutamate receptor 2.8				
	MSTRG.18920	Glutamate receptor 2.8				
CaM/CML	CDL15_Pgr006761	Calcium-binding protein CML10		-1.97		
	CDL15_Pgr006763	Calcium-binding protein CML23				
	CDL15_Pgr027780	Calcium-binding protein CML23	1.84	-1.70	-1.49	
	CDL15_Pgr006762	Calcium-binding protein CML25		2.73		3.44
	CDL15_Pgr007164	Calcium-binding protein CML37				
	CDL15_Pgr022277	Calcium-binding protein CML38		2.10	-3.06	-2.90
	CDL15_Pgr028885	Calcium-binding protein CML43	-4.12			
	CDL15_Pgr012685	Calcium-binding protein CML44	2.42	-1.61	-2.10	
	CDL15_Pgr016969	Calcium-binding protein KIC			-1.75	1.92
	CDL15_Pgr002467	Calcium-binding protein PBP1				-1.98
CIPK	CDL15_Pgr005883	CBL-interacting protein kinase 23				
	CDL15_Pgr021119	CBL-interacting serine/threonine-protein kinase 6				
CDPK	CDL15_Pgr003185	Calcium-dependent protein kinase 20	2.39	3.68		
	CDL15_Pgr002186	Calcium-dependent protein kinase 20			-1.93	
	CDL15_Pgr026643	Calcium-dependent protein kinase 34				1.92
	CDL15_Pgr017861	Calcium-dependent protein kinase 34				
	CDL15_Pgr025367	Calcium-dependent protein kinase 8				
MAPK	CDL15_Pgr021545	Mitogen-activated protein kinase kinase kinase 1		0.53	-2.60	1.40
	CDL15_Pgr001741	Mitogen-activated protein kinase kinase kinase 3		0.37	-2.05	0.41

Figure 6. The DEGs involved in the Ca²⁺-related signal transduction pathways. (**a**) Ca²⁺-related signal transduction pathway; (**b**) DEGs annotations and expressional levels. The color represents with log₂(fold change), the DEGs with log₂FC ≥ 1 were up-regulated (red) and log₂FC ≤ 1 were down-regulated (green) (FDR < 0.05).

3.6. The Transcription Factors (TFs)

For identification of the TFs in pomegranate transcriptome, all the mapped genes were analyzed by BLAST against the Plant Transcription Factor Database (PlantTFDB, http://planttfdb.cbi.pku.edu.cn). A total of 1346 TFs that were classified into 55 putative families (Table S6). Among these TFs, twenty-seven TF families, including 151 genes expressed differently under salt stress compared to controls. The most abundant of differential expression TFs included NAC, ERF, MYB-related, C2H2, MYB, bHLH, GRAS, WRKY, LBD, B3 and bZIP family genes (Table 2; Table S6).

Under NaCl stress, we found that 12 of 19 *NAC* genes were up-regulated, and 10 of 19 genes were down-regulated in leaves or roots of pomegranate plants. Interestingly, we also observed that 18 *PgNACs* significantly changed in T2R when compared to controls, and 4 genes were up-regulated or down-regulated in roots, but reversed in leaves (Table 2; Table S6). These results suggested that many *NACs* were involved in salt stress, but there were different potential responding mechanisms of NAC domain genes to salinity. Thirteen *MYB* genes were induced by salt treatment in pomegranate (Table 2). Among these genes, six in roots and four in leaves were up-regulated, and four in roots and eight in leaves were down-regulated. Moreover, sixteen *MYB-related* genes were detected in our RNA-Seq analyses, thirteen genes were down-regulated, and seven genes were up-regulated by salt stress (Table 2; Table S6). But so far, little was reported that the MYB-related type proteins are related with the responses to salt stress.

The AP2/ERF superfamily is divided into three families: the AP2 family proteins containing two repeated AP2/ERF domains, the ERF family proteins, containing a single AP2/ERF domain, and the RAV family proteins containing a B3 domain. There were 19 *PgERFs*, and 3 *PgAP2s* significantly expressed in treatments when compared to controls (Table 2). Fifteen *ERFs* were down-regulated, and all of *AP2* were repressed by salinity. There were 9 up-regulated and 8 down-regulated *C2H2*, three up-regulated and nine down-regulated *bHLH*, five up-regulated and three down-regulated *WRKY* identified in pomegranate roots and leaves. Remarkably, nine *GRASs* were up-regulated in the root, but the no-significant change in leaves (Table 2; Table S6).

Table 2. The different expressional TFs in pomegranate under NaCl stress.

TFs	Total No.	DEGs	T1R		T2R		T1L		T2L	
			Up	Down	Up	Down	Up	Down	Up	Down
NAC	97	19	1	1	12	6	1	2	1	4
ERF	114	19	1	1	8	5	1	7		9
MYB_related	91	16	-	-	4	3	1	11	3	2
C2H2	91	13	-	-	8	3	-	1		2
MYB	79	13	-	1	6	4	-	8	4	2
bHLH	102	11	-	1	2	5	1	4	1	1
GRAS	50	9	1	-	9	-	-	-	-	-
LBD	38	7	2	2	2	2	-	2	-	1
WRKY	66	7	-	-	5		-	2	1	3
B3	49	3	-	-	1	1	-	1		1
bZIP	45	3	-	-	1	2	-	1	1	1
AP2	8	3	-	-	-	3	-	-	-	-
G2-like	40	3	-	-	1	1	-		-	1
HD-ZIP	36	3	-	-	-	2	-	-	-	1
HSF	21	3	-	-	3	-	-	-	-	-
Others	419	19	-	-	8	3	-	6	1	3
Total	1346	151	5	6	70	40	6	46	13	31

3.7. qRT-PCR Validation

To confirm the reliability of the expression levels obtained from the RNA-Seq, fifteen DEGs, including five *PgERFs*, three *PgMYBs* and seven *PgNACs* were selected for qRT-PCR assays. The results showed that the expression level of each transcript closely corresponded to the transcript level estimated from the sequence data with $R^2 \geq 0.85$ (Figure 7), which implies reproducibility and accuracy of the RNA-Seq results.

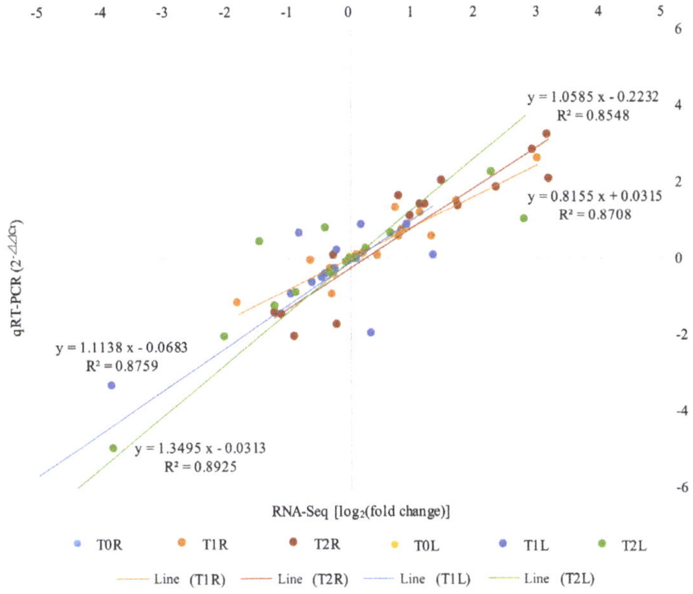

Figure 7. The relationship between qRT-PCR and RNA-Seq date. The color circle is gene expression level. The lines represent co-relationship with dependent linear equations.

4. Discussion

RNA-sequencing techniques have proven to be beneficial and economical for scanning the transcribed genes, both in model and non-model plant species. In this study, we used the RNA-Seq approach as a powerful tool to elucidate the molecular responses to salt stress in pomegranate. Finally, we reconstructed the transcripts and identified 5396 novel genes from the 18 cDNA libraries. There were significant differences between the percentage of total mapped reads and the mapped reads in exonic regions. Theoretically, the ratio of sequencing reads that produced from mature mRNA mapped into exonic regions could be 100%. However, a portion of 30.68–37.12% reads were also mapped into intronic and intergenic regions for the following reasons: (1) The coding and non-coding sequences in the reference genome might have been annotated incorrectly; (2) the new alternative splicing (SNPs) and novel genes resulted from the alternative mRNA splicing and transcripts reconstructing; (3) variations exist between sequencing and reference genome genotypes, which are attributed to the gaps and diversity of their sequences [32,33].

Under salt stress, plants firstly experience osmotic stress due to a disorder of water uptake. Then a process of gradual recovery due to the partially or completely re-established uptake of water occurs within days after the salt treatment [34,35]. We identified 2255 DEGs through the 18 mRNA libraries. From the third to sixth day after treatment, there were significant differences between pomegranate roots and leaves, with little overlap of DEGs responding salinity (Figure 3). More salt-related genes were up-regulated in roots, while the majority of suppressed genes recovered with salt-treating process in leaves. In contrast, the osmotic adjustment of leaves started only after roots had reached new water equilibrium [36]. These tissue-specific and time-specific differences between roots and leaves have also been reported in *Arabidopsis thaliana* (L.) [37], *Populus euphratica* (Oliv.) [36], and *Millettia pinnata* (L.) [38]. Many DEGs were involved in down-regulated patterns in roots, such as cell wall organization, transmembrane transport and oxidation-reduction process, but the proteolysis and metabolic process were over-expressed. Otherwise, most of DEGs involved in oxidation-reduction process and ion transport in leaves were suppressed (Figure 4). The cell wall plays an important role in protecting plant from salt toxicity [39]. In our study, extensin, pectin acetylesterase, polygalacturonase, and xyloglucan endotransglucosylase/hydrolase proteins were identified in roots (Table S4). These proteins are important components of the cell wall, which affect plant growth by mediating cell enlargement and expansion [39]. Many DEGs involved in transmembrane transport code for various channels, carriers and pumps, and ion transporters, such as aquaporin, cationic amino acid transporter, sugar carrier, and zinc transporter (Table S4). They worked together to transport coenzymes, amino acids, carbohydrates, and ions under salt stress. The suppressed genes coding proteins, such as dehydrogenase, cytochrome P450s and peroxidase involved in oxidation-reduction process, play essential roles in responding to salinity [40,41] (Tables S4 and S5). These results indicated that salinity affected pomegranate growth and development by inhibiting the cell division and transmembrane transport, as well as slowing down the redox reactions.

On the other hand, over-represented genes in roots were enriched in metabolic process, carbohydrate metabolic process, and catabolic process, etc. These metabolism-related genes in roots were also enriched in KEGG pathways, such as metabolic pathways (ko01100), biosynthesis of secondary metabolites (ko01110), phenylpropanoid biosynthesis (ko00940), starch and sucrose metabolism (ko00500), amino sugar and nucleotide sugar metabolism (ko005200) (Figure 2; Table S3). The carbohydrates such as glycan, starch, sucrose, amino sugar have been demonstrated as osmolytes and energy resouces in plant responses to salt stress, especially in halophyte species [42]. The accelerated metabolisms in pomegranate might contribute to its adaption responses to salt stress. Similar expression patterns of metabolism-related genes were observed in *Oryza sativa* (L.) [43], *Thellungiella halophila* (C. A. Mey.) [42] and *Helianthus tuberosus* (L.) [44] under salt stress. Metal ions/cations, such as K^+, Ca^{2+}, Mg^{2+}, Fe^{2+}, Cu^{2+}, and Zn^{2+}, act as cofactors participating in the catalytic activity of the enzymes they bind to [44]. The up-regulated DEGs involved in metal ion/cation binding might accelerate the catalytic activity. At the same time, the ion transport process was inhibited in roots and leaves (Tables

S3 and S5). The restriction of ion transport indicated that the excess of Na$^+$ inhibited the uptake of mineral ions [4]. The results correspond to the physiological responses in other pomegranate cultivars under salt stress [45]. We suspected that pomegranate could cope with the toxic ions to some extent via decreasing uptake of ions from soil and increasing utilization of ions in cells. The toxic ions access into the leaf cells with the transpiration, but the detrimental effect is a slow process taking weeks or months, eventually cause salt toxicity in leaves [46]. Plants were only exposed to salinity stress just for six days in our research, so further study is needed to reveal the long-term effect of salinity.

ABA is a critical hormone, which regulates plant growth, development, as well as responses to environmental stresses such as salinity, drought, heat, cold, wound, and pathogen [47,48]. Once ABA is produced, ABA-bound receptors bind ABA, inhibit *PP2Cs*, such as *ABI1* and *ABI2*, thereby activate *SnRK2s* [49]. The *SnRK2s* are involved in plant responses to abiotic stresses, which can phosphorylate the ABA-responsive element binding factor (ABF) and trigger the expression of ABA-responsive genes [50]. In our study, three ABA receptors, *PYLs*, were significantly down-regulated in roots or leaves (Figure 5), which is consistent with results in tea (*Camellia sinensis* L.) [51] and grape (*Vitis vinifera* L.) [52]. The down-regulation of *PYLs* can reduce plant sensitivity to ABA in response to salt stress and help plants adapt to salinity. The expressional differences of *PP2Cs* between roots and leaves in pomegranate are similar to other plants, which indicate that *PP2Cs* may participate in response to salinity via different strategies [52,53]. *PYR/PYLs* may regulate *SnRK2s* directly or indirectly, however, whether *SnRK2s* responding to various abiotic stresses in an ABA-independent or ABA-dependent pathway needs further investigation.

Ca^{2+} referred as a second messenger plays a very important role in many stress-related signaling transduction pathways. Under various abiotic stresses, over-accumulation of ions induce a temporary fluctuations in plant cytosolic ([Ca^{2+}]$_{cyt}$) levels [54]. The genes involved in CIPKs and CDPKs pathways, such as *ACAs*, *CAXs*, *GLRs*, *CaM/CMLs*, *CBLs* etc., participate in transporting and binding Ca^{2+}, sensing and relaying signals in plant cell under salinity stress [55,56]. In our study, we found two *ACAs* and two *CAXs* located on vacuoles were up-regulated in T2R, and the majority of *CaM/CMLs* were up-regulated in roots but down-regulated in leaves (Figure 6). Previous reports revealed a central role for CaMs in the regulation of Ca^{2+} channels and pumps, like CNGCs and ACAs [57,58]. The negative effect of CaM on CNGC activity provides a direct feedback pathway to restrict the influx of Ca^{2+} into plant cells [58]. The CaM stimulates the activity of these *ACAs* by preventing their auto-inhibition [58]. Collectively, under salt stress, pomegranate plants restrain the excess influx of Ca^{2+} into root cells under salinity. Meanwhile, it coped with the excess Ca^{2+} via removing Ca^{2+} from the cytosol by *ACAs* and *CAXs* in plasma membrane, including efflux of excess Ca^{2+} into the outer rhizosphere and/or the influx into vacuoles [59,60].

Specifically, transcription factors, such as NAC, MYB, AP2/ERF, bHLH, WRKY, GRAS family genes, regulated the expression of downstream genes to cope with various stresses [17]. Our study indicated that the TFs of pomegranate participated in the response to salinity with different patterns. For example, most of *NAC* and *C2H2* genes were up-regulated in roots, while most of *MYB* and *MYB-related* genes were down-regulated in leaves (Table 2; Table S6). Interestingly, nine *GRASs* were all up-regulated in T2R, which is mainly due to that these proteins play roles in plant development, including root development, axillary shoot development, and maintenance of the shoot apical meristem [61]. We also found 57 DEGs coding for salt-inducted effectors, including two *SODs*, two *APXs*, nineteen *PODs*, five *LEAs*, five *AQPs*, ten *HSFs*, three *AKTs,* and one *HKT* (Table S7). Two *APXs* were up-regulated and two *SODs* were down-regulated in leaves. Fourteen and nine *PODs* were suppressed in roots and leaves, respectively. Most of *AQPs* (4 of 5) were down-regulated, while all of *LEAs* (5 of 5) and majority of *HSFs* (9 of 10) were up-regulated in pomegranate tissues under salt stress. *LEAs* and *HSPs* were reported to prevent protein denaturation and maintain cell membrane fluidity under stress [62]. Then the DEGs of *LEA* and *HSF* in pomegranate contributed to mitigating salt stress. The expressions of three *AKTs* and one *HKT* were significantly down-regulated, which suggested that the influxes of K$^+$ and Na$^+$ in the cell were restricted [63].

Briefly, when pomegranate plants are exposed to high salinity, Na^+ enters the cell and simultaneously generates stress signals. The messengers Ca^{2+}, ROSs (reactive oxygen species), and ABA transduced the signals, and then the cascades, including Ca^{2+}-sensors, MAPK cascades and TFs, involved in secondary and tertiary regulatory networks were activated (Figure S4). TFs regulated the expression of downstream functional genes, such as *HSPs*, *LEAs*, *AQPs*, *SODs*, and *APXs*, to cope with salt stress.

5. Conclusions

In summary, our research sheds light on the pomegranate molecular response mechanisms to salinity. The differentially expressed genes could also provide references for selecting salt-tolerant breeding materials in pomegranate breeding processes.

Accession Codes: The raw data were deposited in NCBI Sequence Read Archive (SRA) database (http://www.ncbi.nlm.nih.gov/Traces/sra) under accession number SRP148507.

Supplementary Materials: The following are available online at http://www.mdpi.com/2073-4395/10/1/44/s1, Figure S1: The images of untreated and treated plants. Figure S2: Pearson correlation of all samples, Figure S3: Pathways classification into cellular process, environmental information processing, genetic processing, metabolism and organismal system based on the KEGG analysis, Figure S4: Signaling networks in pomegranate under salt stress, Table S1: The primers used for qPCR validation of 15 DEGs, Table S2: Total numbers of DEGs assigned to databases, Table S3: DEGs involved in up-regulated patterns of pomegranate roots, Table S4: DEGs involved in down-regulated patterns of pomegranate roots, Table S5: DEGs involved in expressional patterns of pomegranate leaves, Table S6: The numbers of differently expressed TFs in pomegranate under NaCl stress, Table S7: The salt-inducted DEGs in pomegranate under NaCl stress.

Author Contributions: Conceptualization, C.L. and Z.Y.; methodology, C.L. and Y.Z.; formal analysis, C.L., Y.Z., J.W., and X.Z.; investigation, C.L., Y.Z., X.Z., and J.W.; writing—original draft preparation, C.L.; writing—review and editing, C.L., X.Z., M.G., and Z.Y.; supervision, M.G. and Z.Y.; Funding acquisition, Z.Y. All authors have read and agreed to the published version of the manuscript.

Funding: This work was supported by the Initiative Project for Talents of Nanjing Forestry University (GXL2014070, GXL2018032), the Natural Science Foundation of Jiangsu Province (BK20180768), the Doctorate Fellowship Foundation of Nanjing Forestry University, and the Priority Academic Program Development of Jiangsu High Education Institutions (PAPD).

Conflicts of Interest: The authors declare no conflict of interest.

References

1. Li, J.; Pu, L.; Zhu, M.; Zhang, R. The present situation and hot issues in the salt-affected soil research. *Acta Geogr. Sin.* **2012**, *67*, 1233–1245.
2. Zhu, J.K. Plant salt tolerance. *Trends Plant Sci.* **2001**, *6*, 66–71. [CrossRef]
3. Pandey, P.; Ramegowda, V.; Senthil-Kumar, M. Shared and unique responses of plants to multiple individual stresses and stress combinations: Physiological and molecular mechanisms. *Front. Plant Sci.* **2015**, *6*, 723. [CrossRef]
4. Munns, R.; Tester, M. Mechanisms of salinity tolerance. *Annu. Rev. Plant Biol.* **2008**, *59*, 651–681. [CrossRef]
5. Zhang, H.; Han, B.; Wang, T.; Chen, S.; Li, H.; Zhang, Y.; Dai, S. Mechanisms of plant salt response: Insights from proteomics. *J. Proteome Res.* **2011**, *11*, 49–67. [CrossRef]
6. Bui, E.N. Soil salinity: A neglected factor in plant ecology and biogeography. *J. Arid Environ.* **2013**, *92*, 14–25. [CrossRef]
7. Apse, M.P.; Blumwald, E. Na^+ transport in plants. *FEBS Lett.* **2007**, *581*, 2247–2254. [CrossRef]
8. Munns, R.; James, R.A.; Gilliham, M.; Flowers, T.J.; Colmer, T.D. Tissue tolerance: An essential but elusive trait for salt-tolerant crops. *Funct. Plant Biol.* **2016**, *43*, 1103–1113. [CrossRef]
9. Attia, H.; Karray, N.; Msilini, N.; Lachaâl, M. Effect of salt stress on gene expression of superoxide dismutases and copper chaperone in *Arabidopsis thaliana*. *Biol. Plant.* **2011**, *55*, 159–163. [CrossRef]
10. Shafi, A.; Chauhan, R.; Gill, T.; Swarnkar, M.K.; Sreenivasulu, Y.; Kumar, S.; Kumar, N.; Shankar, R.; Ahuja, P.S.; Singh, A.K. Expression of *SOD* and *APX* genes positively regulates secondary cell wall biosynthesis and promotes plant growth and yield in *Arabidopsis* under salt stress. *Plant Mol. Biol.* **2015**, *87*, 615–631. [CrossRef]

11. Kumar, S.; Beena, A.S.; Awana, M.; Singh, A. Salt-induced tissue-specific cytosine methylation downregulates expression of *HKT* genes in contrasting wheat (*Triticum aestivum* L.) genotypes. *DNA Cell Biol.* **2017**, *36*, 283–294. [CrossRef] [PubMed]

12. Yokoi, S.; Quintero, F.J.; Cubero, B.; Ruiz, M.T.; Bressan, R.A.; Hasegawa, P.M.; Pardo, J.M. Differential expression and function of *Arabidopsis thaliana NHX* Na$^+$/H$^+$ antiporters in the salt stress response. *Plant J.* **2010**, *30*, 529–539. [CrossRef] [PubMed]

13. Hu, W.; Yuan, Q.; Wang, Y.; Cai, R.; Deng, X.; Wang, J.; Zhou, S.; Chen, M.; Chen, L.; Huang, C. Overexpression of a wheat aquaporin gene, *TaAQP8*, enhances salt stress tolerance in transgenic tobacco. *Plant Cell Physiol.* **2012**, *53*, 2127–2141. [CrossRef] [PubMed]

14. Duan, J.; Cai, W. *OsLEA3-2*, an abiotic stress induced gene of rice plays a key role in salt and drought tolerance. *PLoS ONE* **2012**, *7*, e45117. [CrossRef]

15. Zhou, A.; Liu, E.; Ma, H.; Feng, S.; Gong, S.; Wang, J. NaCl-induced expression of *AtVHA-c5* gene in the roots plays a role in response of Arabidopsis to salt stress. *Plant Cell Rep.* **2018**, *37*, 443–452. [CrossRef]

16. Hasegawa, P.M.; Bressan, R.A.; Zhu, J.K.; Bohnert, H.J. Plant cellular and molecular responses to high salinity. *Annu. Rev. Plant Physiol. Plant Mol. Biol.* **2000**, *51*, 463–499. [CrossRef]

17. Deinlein, U.; Stephan, A.B.; Horie, T.; Luo, W.; Xu, G.; Schroeder, J.I. Plant salt-tolerance mechanisms. *Trends Plant Sci.* **2014**, *19*, 371–379. [CrossRef]

18. Guo, S.M.; Tan, Y.; Chu, H.J.; Sun, M.X.; Xing, J.C. Transcriptome sequencing revealed molecular mechanisms underlying tolerance of *Suaeda salsa* to saline stress. *PLoS ONE* **2019**, *14*, e0219979. [CrossRef]

19. Yuan, Z.; Fang, Y.; Zhang, T.; Fei, Z.; Han, F.; Liu, C.; Liu, M.; Xiao, W.; Zhang, W.; Wu, S. The pomegranate (*Punica granatum* L.) genome provides insights into fruit quality and ovule developmental biology. *Plant Biotechnol. J.* **2018**, *16*, 1363–1374. [CrossRef]

20. Silva, J.A.T.D.; Rana, T.S.; Narzary, D.; Verma, N.; Meshram, D.T.; Ranade, S.A. Pomegranate biology and biotechnology: A review. *Sci. Hortic.* **2013**, *160*, 85–107. [CrossRef]

21. Holl, D.; Hatib, K.; Bar-Ya'Akov, I. Pomegranate: Botany, horticulture, breeding. *Hortic. Rev.* **2009**, *35*, 127–191.

22. Karimi, H.R.; Hasanpour, Z. Effects of salinity and water stress on growth and macro nutrients concentration of pomegranate (*Punica granatum* L.). *J. Plant Nutr.* **2014**, *37*, 1937–1951. [CrossRef]

23. Okhovatianardakani, A.R.; Mehrabanian, M.; Dehghani, F.; Akbarzadeh, A. Salt tolerance evaluation and relative comparison in cuttings of different pomegranate cultivars. *Plant Soil Environ.* **2010**, *56*, 176–185. [CrossRef]

24. Liu, C.; Yan, M.; Huang, X.; Yuan, Z. Effects of salt stress on growth and physiological characteristics of pomegranate (*Punica granatum* L.) cuttings. *Pak. J. Bot.* **2018**, *50*, 457–464.

25. Feng, Z.T.; Deng, Y.Q.; Fan, H.; Sun, Q.J.; Sui, N.; Wang, B.S. Effects of NaCl stress on the growth and photosynthetic characteristics of *Ulmus pumila* L. seedlings in sand culture. *Photosynthetica* **2014**, *52*, 313–320. [CrossRef]

26. Patel, R.K.; Jain, M. NGS QC Toolkit: A toolkit for quality control of next generation sequencing data. *PLoS ONE* **2012**, *7*, e30619. [CrossRef]

27. Kim, D.; Pertea, G.; Trapnell, C.; Pimentel, H.; Kelley, R.; Salzberg, S.L. TopHat2: Accurate alignment of transcriptomes in the presence of insertions, deletions and gene fusions. *Genome Biol.* **2013**, *14*, R36. [CrossRef]

28. Pertea, M.; Kim, D.; Pertea, G.M.; Leek, J.T.; Salzberg, S.L. Transcript-level expression analysis of RNA-seq experiments with HISAT, StringTie and Ballgown. *Nat. Protoc.* **2016**, *11*, 1650–1667. [CrossRef]

29. Götz, S.; Garcíagómez, J.M.; Terol, J.; Williams, T.D.; Nagaraj, S.H.; Nueda, M.J.; Robles, M.; Talón, M.; Dopazo, J.; Conesa, A. High-throughput functional annotation and data mining with the Blast2GO suite. *Nucleic Acids Res.* **2008**, *36*, 3420–3435. [CrossRef]

30. Ernst, J.; Nau, G.J.; Bar-Joseph, Z. Clustering short time series gene expression data. *Bioinformatics* **2005**, *21*, i159–i168. [CrossRef]

31. Mao, X.; Cai, T.; Olyarchuk, J.G.; Wei, L. Automated genome annotation and pathway identification using the KEGG Orthology (KO) as a controlled vocabulary. *Bioinformatics* **2005**, *21*, 3787–3793. [CrossRef] [PubMed]

32. Wang, J.; Zhu, J.; Zhang, Y.; Fan, F.; Li, W.; Wang, F.; Zhong, W.; Wang, C.; Yang, J. Comparative transcriptome analysis reveals molecular response to salinity stress of salt-tolerant and sensitive genotypes of indica rice at seedling stage. *Sci. Rep.* **2018**, *8*, 2085. [CrossRef] [PubMed]

33. Zhou, Y.; Yang, P.; Cui, F.; Zhang, F.; Luo, X.; Xie, J. Transcriptome analysis of salt stress responsiveness in the seedlings of Dongxiang wild rice (*Oryza rufipogon* Griff.). *PLoS ONE* **2016**, *11*, e0146242. [CrossRef]

34. Munns, R.; James, R.A.; Läuchli, A. Approaches to increasing the salt tolerance of wheat and other cereals. *J. Exp. Bot.* **2006**, *57*, 1025–1043. [CrossRef]

35. Munns, R. Genes and salt tolerance: Bringing them together. *New Phytol.* **2005**, *167*, 645–663. [CrossRef]

36. Monika, B.; Mikael, B.; Basia, V.; Atef, A.O.; Payam, F.; Dennis, J.; Ottow, E.A.; Cullmann, A.D.; Joachim, S.; Jaakko, K.R. Linking the salt transcriptome with physiological responses of a salt-resistant Populus species as a strategy to identify genes important for stress acclimation. *Plant Physiol.* **2010**, *154*, 1697–1709.

37. Ma, S.; Gong, Q.; Bohnert, H.J. Dissecting salt stress pathways. *J. Exp. Bot.* **2006**, *57*, 1097–1107. [CrossRef]

38. Huang, J.; Xiang, L.; Hao, Y.; Chen, S.; Zhang, W.; Huang, R.; Zheng, Y. Transcriptome characterization and sequencing-based identification of salt-responsive genes in *Millettia pinnata*, a semi-mangrove plant. *DNA Res.* **2012**, *19*, 195–207. [CrossRef]

39. Le Gall, H.; Philippe, F.; Domon, J.-M.; Gillet, F.; Pelloux, J.; Rayon, C. Cell wall metabolism in response to abiotic stress. *Plants* **2015**, *4*, 112–166. [CrossRef]

40. Fatehi, F.; Hosseinzadeh, A.; Alizadeh, H.; Brimavandi, T.; Struik, P.C. The proteome response of salt-resistant and salt-sensitive barley genotypes to long-term salinity stress. *Mol. Biol. Rep.* **2012**, *39*, 6387–6397. [CrossRef]

41. Bushman, B.S.; Amundsen, K.L.; Warnke, S.E.; Robins, J.G.; Johnson, P.G. Transcriptome profiling of Kentucky bluegrass (*Poa pratensis* L.) accessions in response to salt stress. *BMC Genom.* **2016**, *17*, 48. [CrossRef] [PubMed]

42. Wang, X.; Chang, L.; Wang, B.; Wang, D.; Li, P.; Wang, L.; Yi, X.; Huang, Q.; Peng, M.; Guo, A. Comparative proteomics of *Thellungiella halophila* leaves from plants subjected to salinity reveals the importance of chloroplastic starch and soluble sugars in halophyte salt tolerance. *Mol. Cell. Proteom.* **2013**, *12*, 2174–2195. [CrossRef] [PubMed]

43. Boriboonkaset, T.; Theerawitaya, C.; Yamada, N.; Pichakum, A.; Supaibulwatana, K.; Cha-um, S.; Takabe, T.; Kirdmanee, C. Regulation of some carbohydrate metabolism-related genes, starch and soluble sugar contents, photosynthetic activities and yield attributes of two contrasting rice genotypes subjected to salt stress. *Protoplasma* **2013**, *250*, 1157–1167. [CrossRef] [PubMed]

44. Zhang, A.; Han, D.; Wang, Y.; Mu, H.; Zhang, T.; Yan, X.; Pang, Q. Transcriptomic and proteomic feature of salt stress-regulated network in Jerusalem artichoke (*Helianthus tuberosus* L.) root based on de novo assembly sequencing analysis. *Planta* **2018**, *247*, 715–732. [CrossRef] [PubMed]

45. Sun, Y.; Niu, G.; Masabni, J.G.; Ganjegunte, G. Relative salt tolerance of 22 pomegranate (*Punica granatum*) cultivars. *HortScience* **2018**, *53*, 1513–1519. [CrossRef]

46. Aroca, R.; Porcel, R.; Ruiz-Lozano, J.M. Regulation of root water uptake under abiotic stress conditions. *J. Exp. Bot.* **2011**, *63*, 43–57. [CrossRef] [PubMed]

47. Finkelstein, R.R.; Gampala, S.S.; Rock, C.D. Abscisic acid signaling in seeds and seedlings. *Plant Cell* **2002**, *14*, S15–S45. [CrossRef]

48. Shinozaki, K.; Yamaguchi-Shinozaki, K. Molecular responses to dehydration and low temperature: Differences and cross-talk between two stress signaling pathways. *Curr. Opin. Plant Biol.* **2000**, *3*, 217–223. [CrossRef]

49. Fan, W.; Zhao, M.; Li, S.; Bai, X.; Li, J.; Meng, H.; Mu, Z. Contrasting transcriptional responses of *PYR1/PYL/RCAR* ABA receptors to ABA or dehydration stress between maize seedling leaves and roots. *BMC Plant Biol.* **2016**, *16*, 99. [CrossRef]

50. Zhang, H.; Li, W.; Mao, X.; Jing, R.; Jia, H. Differential activation of the wheat SnRK2 family by abiotic stresses. *Front. Plant Sci.* **2016**, *7*, 420. [CrossRef]

51. Wan, S.; Wang, W.; Zhou, T.; Zhang, Y.; Chen, J.; Xiao, B.; Yang, Y.; Yu, Y. Transcriptomic analysis reveals the molecular mechanisms of *Camellia sinensis* in response to salt stress. *Plant Growth Regul.* **2018**, *84*, 481–492. [CrossRef]

52. Boneh, U.; Biton, I.; Zheng, C.; Schwartz, A.; Ben-Ari, G. Characterization of potential ABA receptors in *Vitis vinifera*. *Plant Cell Rep.* **2012**, *31*, 311–321. [CrossRef] [PubMed]

53. Cao, J.; Min, J.; Peng, L.; Chu, Z. Genome-wide identification and evolutionary analyses of the PP2C gene family with their expression profiling in response to multiple stresses in *Brachypodium distachyon*. *BMC Genom.* **2016**, *17*, 175. [CrossRef] [PubMed]

54. McCormack, E.; Tsai, Y.-C.; Braam, J. Handling calcium signaling: Arabidopsis *CaMs* and *CMLs*. *Trends Plant Sci.* **2005**, *10*, 383–389. [CrossRef] [PubMed]

55. Dang, Z.H.; Zheng, L.L.; Jia, W.; Zhe, G.; Wu, S.B.; Zhi, Q.; Wang, Y.C. Transcriptomic profiling of the salt-stress response in the wild recretohalophyte *Reaumuria trigyna*. *BMC Genom.* **2013**, *14*, 29. [CrossRef] [PubMed]

56. Yang, Y.; Zhang, C.; Tang, R.; Xu, H.; Lan, W.; Zhao, F.; Luan, S. Calcineurin B-Like proteins CBL4 and CBL10 mediate two independent salt tolerance pathways in Arabidopsis. *Int. J. Mol. Sci.* **2019**, *20*, 2421. [CrossRef]

57. Virdi, A.S.; Singh, S.; Singh, P. Abiotic stress responses in plants: Roles of calmodulin-regulated proteins. *Front. Plant Sci.* **2015**, *6*, 809. [CrossRef]

58. Cheval, C.; Aldon, D.; Galaud, J.-P.; Ranty, B. Calcium/calmodulin-mediated regulation of plant immunity. *Biochim. Biophys. Acta Mol. Cell Res.* **2013**, *1833*, 1766–1771. [CrossRef]

59. Apse, M.P.; Sottosanto, J.B.; Blumwald, E. Vacuolar cation/H^+ exchange, ion homeostasis, and leaf development are altered in a T-DNA insertional mutant of *AtNHX1*, the Arabidopsis vacuolar Na^+/H^+ antiporter. *Plant J.* **2010**, *36*, 229–239. [CrossRef]

60. Huda, K.M.; Banu, M.S.; Tuteja, R.; Tuteja, N. Global calcium transducer P-type Ca^{2+}-ATPases open new avenues for agriculture by regulating stress signalling. *J. Exp. Bot.* **2013**, *64*, 3099–3109. [CrossRef]

61. Bolle, C. The role of GRAS proteins in plant signal transduction and development. *Planta* **2004**, *218*, 683–692. [CrossRef] [PubMed]

62. Hoekstra, F.A.; Golovina, E.A.; Buitink, J. Mechanisms of plant desiccation tolerance. *Trends Plant Sci.* **2001**, *6*, 431–438. [CrossRef]

63. Xu, M.; Chen, C.; Cai, H.; Wu, L. Overexpression of *PeHKT1*; 1 Improves Salt Tolerance in Populus. *Genes* **2018**, *9*, 475. [CrossRef] [PubMed]

Article

Treatment of Sweet Pepper with Stress Tolerance-Inducing Compounds Alleviates Salinity Stress Oxidative Damage by Mediating the Physio-Biochemical Activities and Antioxidant Systems

Khaled A. Abdelaal [1], **Lamiaa M. EL-Maghraby** [2], **Hosam Elansary** [3,4], **Yaser M. Hafez** [1], **Eid I. Ibrahim** [5], **Mostafa El-Banna** [6], **Mohamed El-Esawi** [7,8] and **Amr Elkelish** [9,*]

[1] Plant Pathology and Biotechnology Lab., Excellence Center (EPCRS), Faculty of Agriculture, Kafrelsheikh University, Kafrelsheikh 33516, Egypt; khaled_elhaies@yahoo.com (K.A.A.); hafezyasser@gmail.com (Y.M.H.)

[2] Agricultural Biochemistry Department, Faculty of Agriculture, Zagazig University, Zagazig 44511, Egypt; dr_lamiaa222@yahoo.com

[3] Plant Production Department, College of Food and Agricultural Sciences, King Saud University, P.O. Box 2455, Riyadh 11451, Saudi Arabia; helansary@ksu.edu.sa

[4] Floriculture, Ornamental Horticulture and Garden Design Department, Faculty of Agriculture, Alexandria University, Alexandria 21526, Egypt

[5] Rice Biotechnology Lab., Rice Research Dep., Field Crops Research Institute, Sakha, Kafr El-Sheikh 33717, ARC, Egypt; Eid.ibrahim@gmail.com

[6] Agricultural Botany Department, Faculty of Agriculture, Mansoura University, Mansoura 35516, Egypt; el-banna@mans.edu.eg

[7] Botany Department, Faculty of Science, Tanta University, Tanta 31527, Egypt; mohamed.elesawi@science.tanta.edu.eg

[8] Sainsbury Laboratory, University of Cambridge, Cambridge CB2 1LR, UK

[9] Botany Department, Faculty of Science, Suez Canal University, Ismailia 41522, Egypt

* Correspondence: amr.elkelish@science.suez.edu.eg; Tel.: +20-1005145454

Received: 28 November 2019; Accepted: 19 December 2019; Published: 23 December 2019

Abstract: Salinity stress occurs due to the accumulation of high levels of salts in soil, which ultimately leads to the impairment of plant growth and crop loss. Stress tolerance-inducing compounds have a remarkable ability to improve growth and minimize the effects of salinity stress without negatively affecting the environment by controlling the physiological and molecular activities in plants. Two pot experiments were carried out in 2017 and 2018 to study the influence of salicylic acid (1 mM), yeast extract (6 g L^{-1}), and proline (10 mM) on the physiological and biochemical parameters of sweet pepper plants under saline conditions (2000 and 4000 ppm). The results showed that salt stress led to decreasing the chlorophyll content, relative water content, and fruit yields, whereas electrolyte leakage, malondialdehyde (MDA), proline concentration, reactive oxygen species (ROS), and the activities of antioxidant enzymes increased in salt-stressed plants. The application of salicylic acid (1 mM), yeast extract (6 g L^{-1}), and proline (10 mM) markedly improved the physiological characteristics and fruit yields of salt-stressed plants compared with untreated stressed plants. A significant reduction in electrolyte leakage, MDA, and ROS was also recorded for all treatments. In conclusion, our results reveal the important role of proline, SA, and yeast extracts in enhancing sweet pepper growth and tolerance to salinity stress via modulation of the physiological parameters and antioxidants machinery. Interestingly, proline proved to be the best treatment.

Keywords: *Capsicum annuum* L.; salt stress; salicylic acid; yeast; proline

---

1. Introduction

Global food safety is seriously dependent on crops and their supplies, which require considerable increases for servicing the gap between production and demand [1]. The necessity of improving crop production has been much more emergent in the last few years due to the expanding population, which will exceed to 9.7 billion by 2050. Undoubtedly, increases in the population will exert pressure on crops and food resources [1]. Simultaneously, global warming, as well as various biotic and abiotic stresses, hinder the growth and yields of agricultural crops [2]. Among abiotic stresses, salinity is recognized as one of the main restricting factors affecting the growth and productivity of agricultural crops, especially in arid and semiarid regions [3]. Salinity stress causes a reduction in growth and biomass, chlorophyll degradation, water status modification, malfunctions in stomatal functions, modifications in transpiration and respiration, and disequilibria in ion ratios [4,5]. Furthermore, plants develop cytotoxic-activated oxygen under saline conditions, which might seriously interfere with healthy metabolisms as a result of the oxidative damage of lipids, proteins, and nucleic acids [6,7]. Salinization may additionally lead to the excessive intracellular generation of reactive oxygen species (ROS) such as hydroxyl radicals (OH) and superoxide radicals (O_2^-) [8]. Plants confront these sorts of oxidants by developing several defensive mechanisms, including antioxidant enzymes and molecules that eliminate potentially cytotoxic types of activated oxygen [9,10].

Sweet pepper (*Capsicum annuum* L.) is an important vegetable crop that is grown for local consumption, and which has a high economic value in the Egyptian agricultural market. Farmers started to utilize saline water to partially fulfil crop water demands. The pepper plant is not a salt-tolerant vegetable, and about 14% of fruit yield loss occurs as a result of each increase in salt level of 1.0 dS/m [11]. Previous investigations have been conducted to mitigate the harmful impact of salt stress on sweet pepper, but most have not been sufficient or broadly applicable. As a result, the search for cheaper, ecologically-friendly strategies for salinity amelioration which enhance the growth and productivity of sweet pepper has been very important to the agriculture sector [12].

Numerous studies have found that implementing exogenous chemicals improves salt stress tolerance in plants [13]; examples of such chemicals are phytohormones such as salicylic acid, sterols, and methyl jasmonate [2,14]. Other chemicals such as polyamines, melatonin, and sodium nitroprusside have also been used to enhance the tolerance of various crop plants to saline conditions [15].

Salicylic acid is an essential phenolic compound that regulates plant growth processes and responses to different environmental factors [16]. It is a stress tolerance inducer and an important signal in many physiological processes, such as proline metabolism and photosynthesis. It reduces oxidative stress in plants under environmental stress and enhances plant growth and productivity under salt- [17] and drought-stress conditions [18]. Foliar application of SA-enhanced growth characteristics of sweet pepper plants [6] has increased the chlorophyll concentrations and enzyme activitives in barley plants, as well as counteracting the deleterious impacts of salinity on faba beans [19]. Yeast extracts are the main source of various important compounds, such as amino acids, phytohormones, and vitamins [20,21]. The use of active yeast extracts has been shown to decrease the damaging impact of drought conditions on pea plants, and enhanced the growth performance and yield of stressed plants [22]. Yeast extract applications have led to improvements in the growth characteristics of bean and corn plants, such as the dry weight of leaves, the leaf area, and the number of leaves under drought conditions [23]. The application of yeast and NPK fertilizers has significantly enhanced chlorophyll concentrations and root yields in sugar beet plants [22]. Seaweed extracts have also improved plant tolerance to abiotic stresses. For example, the application of *Ecklonia maxima* seaweed extract has been shown to enhance the tolerance of zucchini squash plants to salinity stress by improving plant performance, shoot biomass yield, fruit quality, leaf gas exchange rate, SPAD index, and leaf nutritional status under saline conditions [24]. Furthermore, proline has a positive impact on the activity of enzymes and osmotic adjustment under stress conditions, while protecting enzyme denaturation and modulating osmoregulation [25]. The application of proline-modulated antioxidant enzymes such as peroxidase (POX) and catalase (CAT) in tobacco plants under salinity conditions plays a significant role in protein synthesis and accumulation in plants under stress conditions like drought and salinity in order to enhance the growth characteristics and yield [26–30].

Considering the variable effectiveness levels of salicylic acid, proline, and yeast extract on plants, as well as the harmful impact of salinity stress on the growth and productivity of important crops, the present study aims to evaluate and compare the levels of effectiveness of the three stress tolerance inducers, i.e., salicylic acid (1 mM), yeast extract (6 g L^{-1}), and proline (10 mM), on the growth characteristics, antioxidants, physiological and biochemical parameters, and yield of sweet pepper plants (*Capsicum annuum* L.) grown under the same saline conditions in order to determine which stress tolerance inducer should be recommended for further enhancements of crop performance and tolerance.

2. Materials and Methods

2.1. Experiments Design and Treatments

Pot experiments were performed at Agricultural Botany Department, Faculty of Agriculture, Kafrelsheikh University, Egypt during the growing seasons of 2017 and 2018. Laboratory analyses were carried out at the Plant Pathology & Biotechnology Lab, and the EPECRS Excellence Center Kafrelsheikh University, Egypt. This research was conducted to study the impacts of salicylic acid (1 mM), yeast extract (6 g L^{-1}), and proline (10 mM) on the growth characteristics and biochemical and yield parameters of salt-stressed sweet pepper plants (*Capsicum annuum* L.). Irrigation water was artificially salinized by applying NaCl at concentrations of 2000 and 4000 ppm. The seeds of sweet pepper cv. California Wonder were obtained from Sun Seed Company in USA. Ten seeds were sown in the nursery using foam trays. Forty-two days after sowing, seedlings were transplanted into pots (30 cm diameter); each pot contained 8 kg soil and 2 plants. The physical and chemical soil characteristics were recorded, according to the methods described by Abdelaal et al. [21], as follows. pH: 8.2; N: 32.4 ppm; P: 10.5 ppm; K: 289 ppm; electrical conductivity: 1.8 dS m^{-1}, soil organic matter: 1.9%; sand: 17.3%; silt: 35.5%; and clay: 47.2%. Fertilizers were added in two equal doses as recommended (NPK, 135:40:35 kg/ha), plus essential micronutrients, whereas the first dose was added 15 days after transplanting and the second at the beginning of flowering stage [31]. The plants were treated twice (20 and 40 days after transplanting) with salicylic acid (1 mM), yeast (6 g L^{-1}) and proline (10 mM). The experiment was done in a completely randomized design with five replicates (five pots with two plants each), and the following measurements were recorded after collecting the plant samples.

2.2. Physiological and Biochemical Analysis

For physiological and biochemical analyses, the samples were collected at 90 days after transplantation for use in the following assays.

2.2.1. Chlorophyll a and b Determination

For chlorophyll a and b determination, 5 mL N-N Dimethyl formamid was added to 1 g sweet pepper fresh leaves and placed in a refrigerator for 24 h. Following the centrifugation at 4000 g for 15 min, the optical density was calculated using spectrophotometer at 647 and 664 nm, according to Moran [32].

2.2.2. Calculation of Leaves Relative Water Content (RWC %) and Electrolyte Leakage (EL %)

The relative water content (RWC) in leaves was recorded according to the formula of Sanchez et al. [33] as follows: RWC = (FW − DW)/ (TW − DW) × 100, where FW is fresh weight, DW is dry weight, and TW is turgid weight. Electrolyte leakage (EL %) was estimated using the formula of Dionisio-Sese and Tobita [34] as follows: EL (%) = Initial electrical conductivity/final electrical conductivity × 100.

2.2.3. Proline Content Determination

Proline was assayed according to the method described by Bates et al. [35] with minor modifications. In brief, a plant sample (0.6 g) was extracted in sulfosalicylic acid (5%) followed by centrifugation at 10000 g for 7 min. The supernatants were diluted with water, mixed with 2% ninhydrin, heated at 94 °C for 30 min, and then cooled. Toluene was then added to the mixture, and the upper aqueous phase was spectrophotometrically assayed at 520 nm.

2.2.4. Calculation of Lipid Peroxidation and Reactive Oxygen Species (Superoxide and Hydrogen Peroxide)

The lipid peroxidation as malondialdehyde (MDA) in plant samples was calculated according to the method described by Heath and Packer [36] with minor modifications. In brief, 0.6 g of plant sample was extracted in TCA (0.1%), followed by centrifugation at 13,000 g for 8 min. The supernatants were mixed with thiobarbituric acid (0.5%) and TCA, and heated at 92 °C for 35 min, followed by cooling and centrifugation at 12,000 g for 8 min. Next, the supernatants' absorbance was measured at 532 and 660 nm. Superoxide and hydrogen peroxide levels were also determined according to the method described by Badiani et al. [37].

2.2.5. Antioxidant Enzymes Activity (CAT and POX)

Plant samples (1.5 g) were extracted in Tris-HCl (100 mM, pH 7.5) containing Dithiothreitol (5 mM), $MgCl_2$ (10 mM), EDTA (1 mM), magnesium acetate (5 mm), PVP-40 (1.6%), aphenylmethanesulfonyl fluoride (1 mM), and aproptinin (1 μg mL^{-1}). The mixed solutions were filtered and centrifuged for 8 min at 13,000 rpm. The supernatants were utilized to record enzymes activities. The activity of CAT and POX of leafy samples was determined according to the method described by Aebi [38] and Hammerschmidt et al. [39]. The supernatant absorbance was shown spectrophotometrically to be 470 nm.

2.2.6. Fruit yields

At 120 days after transplanting, the number of fruits per plant, the fruit fresh weight per plant (g), and the total fruit yield (ton hectare^{-1}) were recorded.

2.3. Statistical Analysis

Data represent the mean ± SD (standard deviation). Two-way analysis of variance was performed using SPSS ver. 19 (SPSS Inc., Chicago, IL, USA). A Tukey's test was also carried out to determine whether a significant difference ($p < 0.05$) existed between mean values.

3. Results

3.1. Chlorophyll a and b Concentrations

According to our results in Figure 1, the concentrations of chlorophyll a and b were significantly decreased in sweet pepper plants under salt-stress conditions; the lowest values were recorded with 4000 ppm compared with 2000 ppm and control plants in the two growing seasons. However, the salt stressed plants treated with salicylic acid, yeast extract, and proline showed significant increases in chlorophyll a and chlorophyll b concentrations compared with stressed untreated plants in both seasons. Under salt stresses of 2000 and 4000 ppm, the maximum concentrations of chlorophyll a and b were recorded with proline treatment in both seasons.

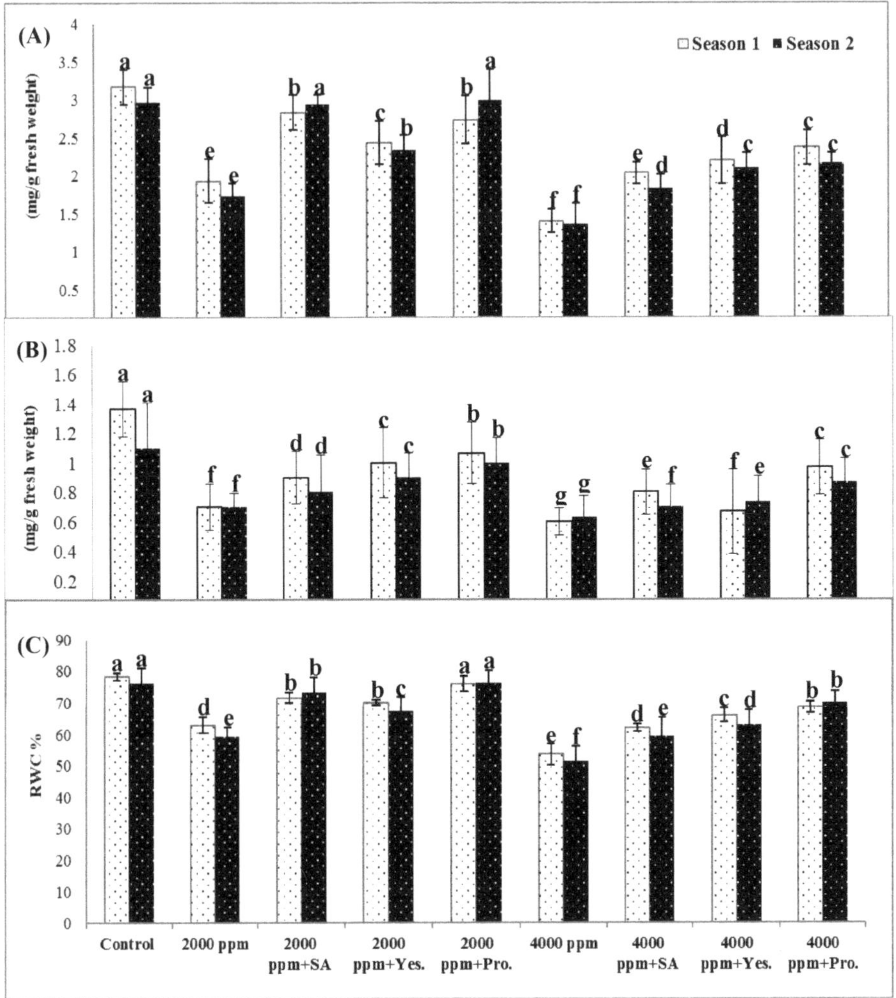

Figure 1. Effect of salinity stress (2000 and 4000 ppm NaCl) and supplementation of SA, yeast, and proline on the contents of (**A**) chlorophyll a, (**B**) chlorophyll b, (**C**) relative water content (RWC) in sweet pepper in the seasons of 2017 and 2018. Data is mean (±SE) of five replicates. Different letters in each Figure represent significant differences at $p < 0.05$.

3.2. Relative Water Content (RWC %)

Data obtained in Figure 1 showed that RWC decreased considerably in salt stressed plants; the greatest reduction was recorded in the plants exposed to salinity at 4000 ppm compared with control plants. The exogenous application of salicylic acid (1 mM), yeast (6 g L^{-1}), and proline (10 mM) caused a significant increase in RWC in salt stressed plants (2000 and 4000 ppm) compared with salt stressed untreated plants. Furthermore, the best treatments under salinity of 2000 ppm were salicylic acid and proline. Under salt treatment at 4000 ppm, the application of yeast extract (6 g L^{-1}) and proline (10 mM) showed the highest RWC in sweet pepper plants compared with SA treatment in stressed untreated plants in both seasons.

3.3. Electrolyte Leakage (EL %)

It may be noted from Figure 2 that salt stress at 2000 and 4000 ppm caused a significant increase in electrolyte leakage (EL); the maximum increase was recorded with a salinity level of 4000 ppm in both seasons. Interestingly, electrolyte leakage was significantly decreased upon the foliar application of salicylic acid (1 mM), yeast extract (6 g L^{-1}), and proline (10 mM) compared with control plants in both seasons. The best treatment was proline under a salt stress of 2000 ppm in both seasons (Figure 2).

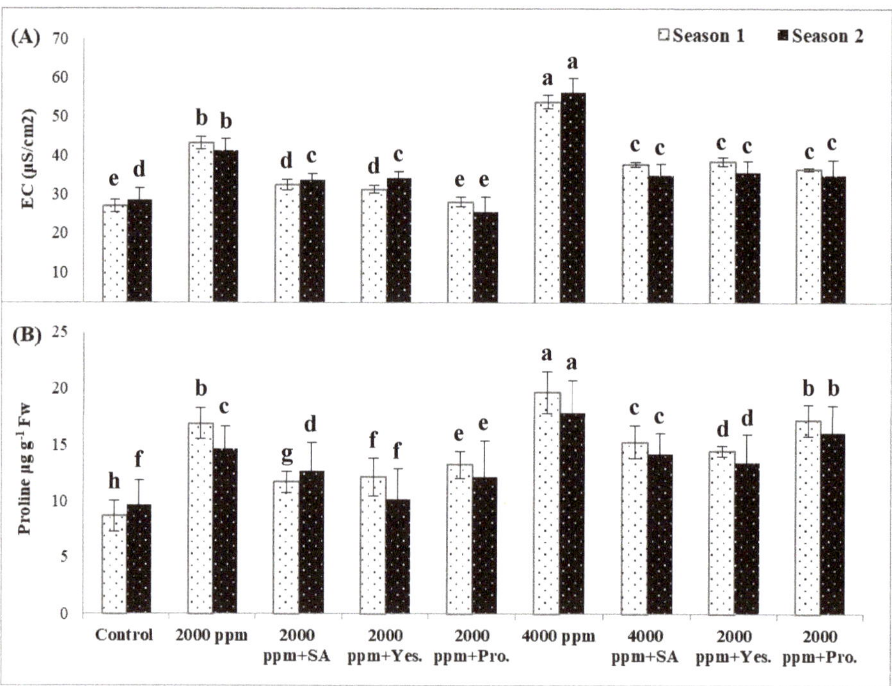

Figure 2. Effect of salinity stress (2000 and 4000 ppm NaCl) and supplementation of SA, yeast, and proline on the contents of (**A**) electrolyte leakage, (**B**) proline in sweet pepper in the seasons of 2017 and 2018. Data is mean (±SE) of five replicates. Different letters in each Figure represent significant differences at $p < 0.05$.

3.4. Proline Concentration

It is evident that proline had markedly accumulated in sweet pepper plants; the highest concentration was recorded with a salinity at 4000 ppm in comparison to the control plants (Figure 2). Intriguingly, the application of salicylic acid, yeast extract, and proline resulted in enhanced proline concentration under all salinity levels; the greatest result was observed with proline (10 mM).

3.5. Lipid Peroxidation (MDA) and Reactive Oxygen Species (Superoxide and Hydrogen Peroxide).

The results showed that lipid peroxidation (i.e., malondialdehyde or MDA), superoxide, and hydrogen peroxide were significantly increased under salt conditions compared with control plants in both seasons (Figure 3). The maximum levels of MDA, superoxide, and hydrogen peroxide were recorded at a salinity level of 4000 ppm, followed by 2000 ppm, in both seasons. On the other hand, the application of salicylic acid, yeast extract, and proline significantly reduced MDA, O$_2^-$, and H$_2$O$_2$ concentrations under all salinity levels compared to the stressed untreated plants. The best results were obtained with SA and proline.

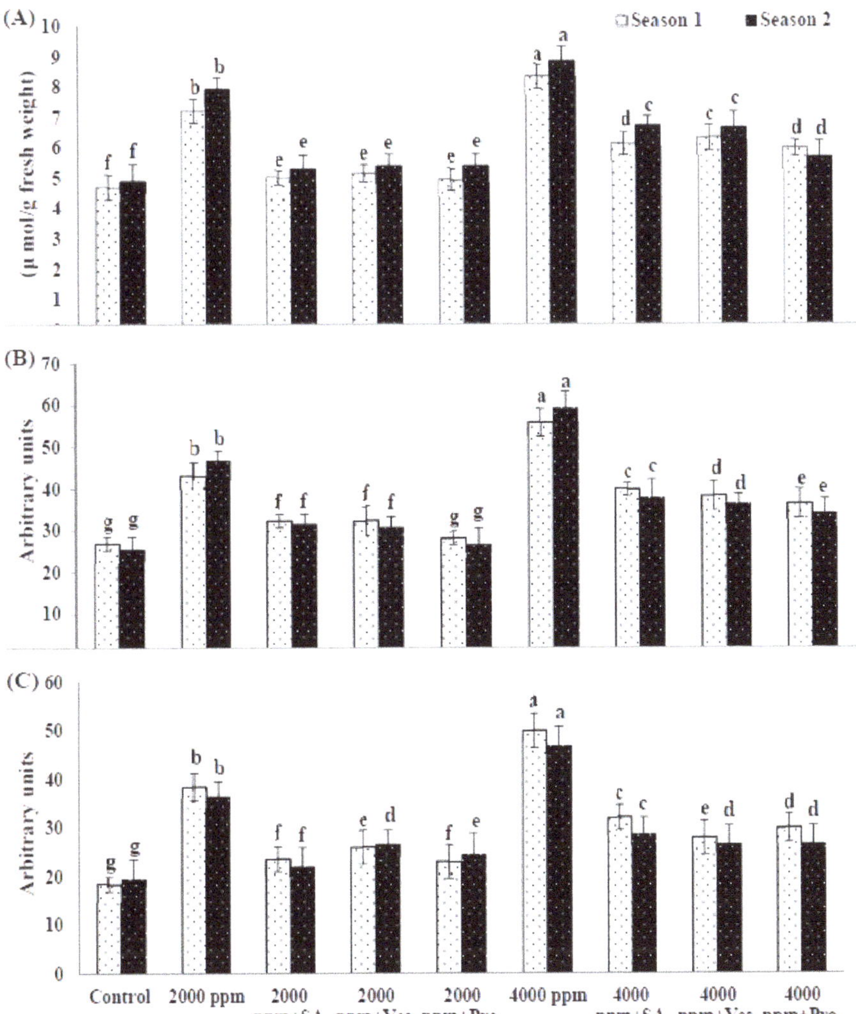

Figure 3. Effect of salinity stress (2000 and 4000 ppm NaCl) and supplementation of SA, yeast, and proline on the contents of (**A**) lipid peroxidation, (**B**) superoxide, (**C**) hydrogen peroxide in sweet pepper in the seasons of 2017 and 2018. Data is mean (±SE) of five replicates. Different letters in each Figure represent significant differences at $p < 0.05$.

3.6. Antioxidant Enzymes Activity

Antioxidant enzyme activities (CAT and POX) were assayed. The data presented in Figure 4 shows that the plants exposed to salt stress (2000 and 4000 ppm) had higher CAT and POX activity compared with control plants in both seasons. On the other hand, applications of salicylic acid (1 mM), yeast extract (6 g L^{-1}), and proline (10 mM) led to reductions in the activities of CAT and POX in the salt-stressed plants.

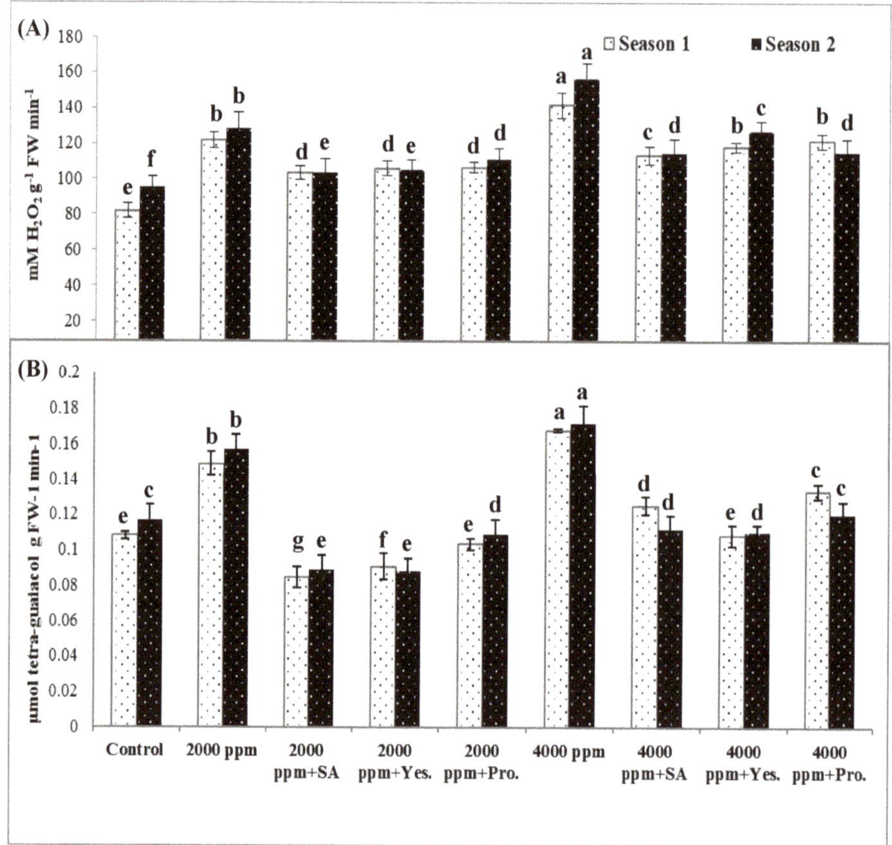

Figure 4. Effect of salinity stress (2000 and 4000 ppm NaCl) and supplementation of SA, yeast, and proline on the activity of (**A**) Catalase (CAT), (**B**) peroxides (POX) in sweet pepper in the seasons of 2017 and 2018. Data is mean (±SE) of five replicates. Different letters in each Figure represent significant differences at $p < 0.05$.

3.7. Number of Fruits per Plant, Fruit Fresh Weight, and Total Fruit Yield (Ton Hectare^{-1}).

According to our findings in Figure 5, salt stress at 2000 and 4000 ppm caused significant decreases in fruit number per plant, fruit fresh weight, and the total fruit yield (ton hectare^{-1}) in both seasons. The lowest values of these traits were recorded with salt stressed plants at 4000 ppm concentration, followed by 2000 ppm. Nevertheless, the exogenous application of salicylic acid, yeast, and proline significantly improved the number of fruits per plant, fruit fresh weight, and total fruit yield (ton hectare^{-1}) in the stressed treated plants compared with stressed untreated plants.

Interestingly, SA and proline treatments gave the maximum values of the three studied characteristics at a salinity concentration of 2000 ppm in the two seasons (Figure 5). Under salinity stress of 4000 ppm, the best results of fruit number per plant, fruit fresh weight, and total fruit yield (ton hectare^{-1}) were recorded with proline, followed by SA and yeast extract in both seasons.

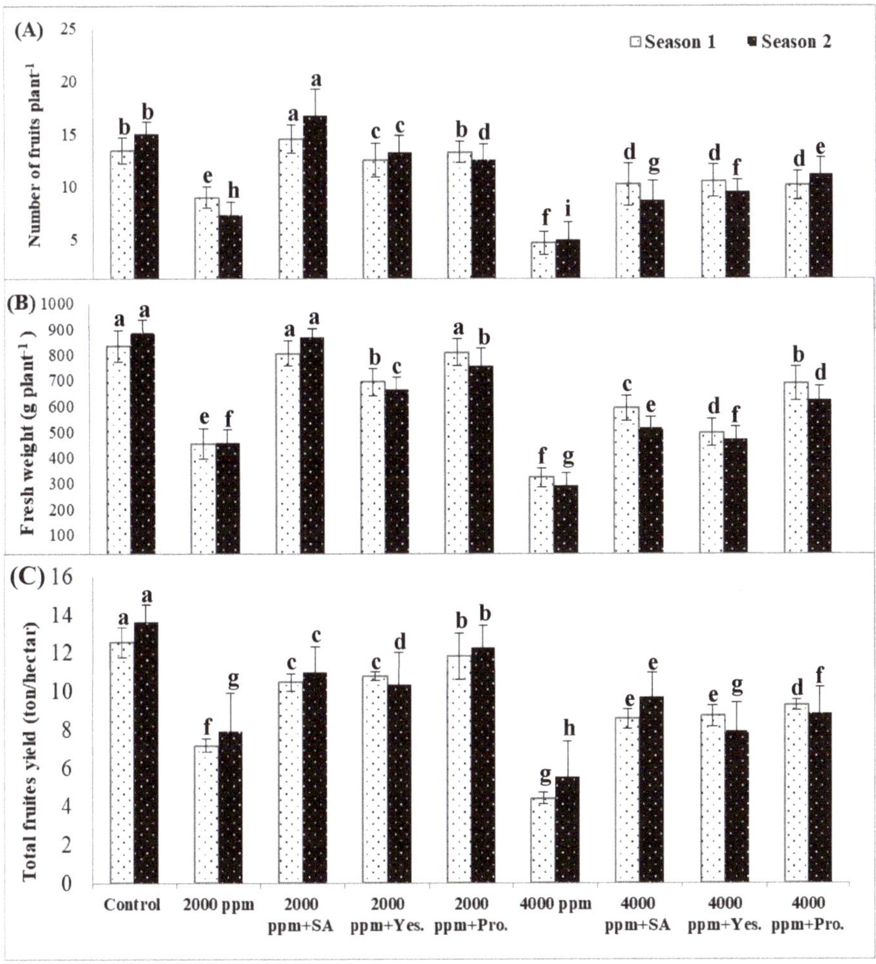

Figure 5. Effect of salinity stress (2000 and 4000 ppm NaCl) and supplementation of SA, yeast, and proline on (**A**) number of fruits plant^{-1}, (**B**) fresh weight plant^{-1}, (**C**) total fruit yield (ton hectare^{-1}) in sweet pepper in the seasons of 2017 and 2018. Data is mean (±SE) of five replicates. Different letters in each Figure represent significant differences at $p < 0.05$.

4. Discussion

The exogenous application of salicylic acid, proline, and yeast extract previously exhibited variable effectiveness levels on plant performance and tolerance to the harmful impact of salinity stress; therefore, the present study assessed and compared the effectiveness levels of these stress tolerance inducers on the growth characteristics, antioxidant levels, physiological and biochemical parameters, and yield of sweet pepper plants (*Capsicum annuum* L.) grown under the same saline conditions in order to determine which stress tolerance inducer should be recommended for the enhancement of crop performance and tolerance. In the present study, salt stress significantly decreased the aforementioned physiological parameters of sweet pepper plants. This reduction in chlorophyll a and b concentrations could be due to the effect of salinity on chlorophyll-degrading enzyme (chlorophyllase) activity, which reduces the chlorophyll synthesis level or negatively affects the structure and number of chloroplasts [40–42]. The chloroplast is one of the most vital organelles for photosynthesis and plant production, and is dramatically affected by abiotic stresses [43,44]. The obtained results indicated that foliar applications of SA (1 mM), yeast extract (6 g L^{-1}), and proline (10 mM) led to increased chlorophyll a and b contents in salt-stressed plants. These findings are in harmony with those obtained by Saleh et al. [25], Abdelaal et al. [21], and Soliman et al. [2]. These results might be due to the antioxidant scavenging influence of SA on chlorophyll degradation under saline conditions [45,46]. Correspondingly, the role of yeast extract in chlorophyll concentration enhancement might be due to the fact that yeast is rich in many essential elements, vitamins, and amino acids, which improve chlorophyll concentrations under stress conditions [21]. Moreover, our results showed that proline application minimized the harmful effects of all salinity levels on chlorophyll a and b concentrations due to its ability to function as a scavenger for ROS. Thus, proline plays a pivotal role in enzyme activation and protects chlorophyll from degradation under salt-stress conditions [47,48].

Additionally, relative water content (RWC) was decreased under salt stress. This decrease may be due to the reduction in water uptake [49] and/or its harmful effect on cell wall structure [50,51]. In contrast, RWC was significantly increased in stressed plants treated with SA, yeast, and proline. The ameliorative effects of these treatments on RWC could be due to the increase in osmoregulators, as well as to osmotic adjustment in plant cells [23,52,53].

Salt stress causes adverse effects on sweet pepper plants, including increased electrolyte leakage percentage. This increase may be due to the damaging effects on plasma membrane and selective permeability resulting in an increase in electrolyte leakage. This result is similar to that obtained in [54,55]. Conversely, the foliar application of SA, yeast, and proline led to decreased electrolyte leakage levels in all treatments. This beneficial effect could be due to the protective role of SA, yeast, and proline in plasma membrane stability and increasing soluble metabolite accumulation. A similar result was indicated by Ishikawa and Evans [56] and Huang et al. [57], who reported that osmoregulators improve plant growth and yield under various stress conditions.

Proline concentration was significantly increased in response to salt-stress conditions. This increment represents an important mechanism to minimize the deleterious impact of salinity stress and enhance plant growth [58]. The foliar application of SA, yeast, and proline under salt conditions may minimize the destructive effect of salinity on plant growth and improve proline accumulation. Similarly, SA application led to improved plant growth characteristics in maize plants under salt conditions [59]. Our results are in agreement with those of Huang et al. [60], Li et al. [61], and Gharsallah et al. [62]. Lipid peroxidation as MDA is an important factors indicating oxidative damage induced by salt stress. Lipid peroxidation was significantly boosted in salt-stressed (2000 and 4000 ppm) sweet pepper plants. Nonetheless, lipid peroxidation content was significantly decreased upon the foliar application of SA, yeast, and proline. These results may be attributed to the pivotal role of these treatments in decreasing oxidative stress damage, and consequently, in causing MDA reduction [25,48,53].

In the current study, superoxide and hydrogen peroxide, which are indicators of oxidative stress, were significantly produced in sweet pepper plants treated with NaCl at 2000 and 4000 ppm. This increase in superoxide and hydrogen peroxide production may be due to the fact that reactive oxygen species have a critical role under stress conditions in adjusting development, differentiation, redox levels, and stress signaling in the chloroplasts, mitochondria, and peroxisomes of plant cells [63,64]. Moreover, the high levels of hydrogen peroxide and superoxide are the main reasons for oxidative stress in the plant cells exposed to various stresses. Our results are supported by the findings of previous studies [65–67]. The application of SA, yeast, and proline on salt-stressed sweet pepper plants led to reductions in the formation of superoxide and hydrogen peroxide. This effect may be due to the role of these treatments in stabilizing protein structures and maintaining the redox states of plant cells, as well as stimulating antioxidant enzymes system [6,21,48]. Under salt stress (2000 and 4000 ppm), antioxidant enzyme activities were significantly increased in sweet pepper plants in order to combat the harmful impact of salt by adjusting osmotic balance. In agreement with our findings, similar results were noted in various plants under saline and drought conditions [68,69].

The activation of CAT and POX enzymes under salt conditions plays a key role in the improvement of plant defense systems. In the current study, the exogenous foliar application of SA, yeast, and proline led to improved antioxidant enzymes activity, as well as guarding the plant cells against oxidative stress and dehydration of the plasma membrane under salt-stress conditions. These results were supported by the findings reported in various plants [70–72].

The reductions of fruit numbers per plant, fruit fresh weight, and total fruit yield (ton hectare^{-1}) under salt conditions are possibly due to the adverse impacts of salinity on the growth characteristics and physiological processes such as water uptake, photosynthesis, flowering, and fruit formation, which led to diminished yields. Accordingly, the highest level of salt (4000 ppm) was adversely more effective than the lowest one (2000 ppm). The same trends of salt stress were previously described in faba bean [73] and strawberry plants [74]. Our results indicate that proline treatment was the best, followed by SA and yeast treatments. This useful effect of proline may be due to its pivotal role in osmotic regulation, enzyme activation, and protein synthesis, which consequently enhances the growth and yield characteristics of stressed plants [47,75,76]. Also, SA plays an essential role as a stress tolerance inducer via reducing the oxidative damage and enhancing plant productivity under salt stress. These results are in harmony with previous findings of Gupta and Huang [77], Ahanger et al. [78], and Husen et al. [79].

5. Conclusions

According to our findings, salt stress caused significant decreases in chlorophyll concentrations, relative water content, and fruit yields. However, lipid peroxidation, proline, electrolyte leakage, and reactive oxygen species were increased. Based on the results, the foliar application of salicylic acid (1 mM), yeast extract (6 g L^{-1}), and proline (10 mM) was an effective method by which to overcome the injurious effects of salt stress on sweet pepper plants. It may be concluded that relative water content and chlorophyll concentration, as well as antioxidant enzyme activity, were significantly modulated in the stressed treated sweet pepper plants. In contrast, electrolyte leakage and lipid peroxidation were decreased in treated sweet pepper plants under salt conditions. Thus, the application of salicylic acid, yeast, and proline led to a decrease in the harmful effects of salt stress by regulating osmolytes and antioxidants, which ultimately enhances the growth characteristics and fruit yields of sweet pepper plants. Interestingly, proline proved to be the best treatment for the further enhancement of plant performance and tolerance to salinity stress.

Author Contributions: K.A.A., L.M.E.-M., H.E., Y.M.H., E.I.I., M.E.-B., M.E.-E. and A.E. performed the experiments, analyzed the data, and wrote the manuscript. All authors have read and agreed to the published version of the manuscript.

Funding: This work was funded by king Saud University funding through Researchers Supporting Project number (RSP-2019/118).

Acknowledgments: The authors extend their appreciation to king Saud University funding through Researchers Supporting Project number (RSP-2019/118) and also to all members of PPBL and EPCRS excellence center, Fac. of Agric., Kafrelsheikh University.

Conflicts of Interest: The authors have declared no conflict of interest.

References

1. Majeed, A.; Muhammad, Z. Salinity: A Major Agricultural Problem—Causes, Impacts on Crop Productivity and Management Strategies. In *Plant Abiotic Stress Tolerance*; Hasanuzzaman, M., Hakeem, K.R., Nahar, K., Alharby, H.F., Eds.; Springer International Publishing: Cham, Switzerland, 2019; pp. 83–99, ISBN 978-3-030-06117-3.
2. Soliman, M.H.; Alayafi, A.A.M.; El Kelish, A.A.; Abu-Elsaoud, A.M. Acetylsalicylic acid enhance tolerance of Phaseolus vulgaris L. to chilling stress, improving photosynthesis, antioxidants and expression of cold stress responsive genes. *Bot. Stud.* **2018**, *59*, 6. [CrossRef] [PubMed]

3. Elkeilsh, A.; Awad, Y.M.; Soliman, M.H.; Abu-Elsaoud, A.; Abdelhamid, M.T.; El-Metwally, I.M. Exogenous application of β-sitosterol mediated growth and yield improvement in water-stressed wheat (Triticum aestivum) involves up-regulated antioxidant system. *J. Plant Res.* **2019**, *132*, 881–901. [CrossRef] [PubMed]

4. Hasan, M.K.; El Sabagh, A.; Sikdar, M.S.; Alam, M.J.; Ratnasekera, D.; Barutcular, C.; Abdelaal, K.A.; Islam, M.S. Comparative adaptable agronomic traits of blackgram and mungbean for saline lands. *Plant Arch.* **2017**, *17*, 589–593.

5. El-Esawi, M.A.; Alayafi, A.A. Overexpression of Rice *Rab7* Gene Improves Drought and Heat Tolerance and Increases Grain Yield in Rice (*Oryza sativa* L.). *Genes (Basel)* **2019**, *10*, 56. [CrossRef]

6. Abdelaal, K.A. Effect of salicylic acid and abscisic acid on morpho-physiological and anatomical characters of faba bean plants (*Vicia faba* L.) under drought stress. *J. Plant Prod.* **2015**, *6*, 1771–1788. [CrossRef]

7. Elkelish, A.A.; Alnusaire, T.S.; Soliman, M.H.; Gowayed, S.; Senousy, H.H.; Fahad, S. Calcium availability regulates antioxidant system, physio-biochemical activities and alleviates salinity stress mediated oxidative damage in soybean seedlings. *J. Appl. Bot. Food Qual.* **2019**, *92*, 258–266.

8. Al Hassan, M.; Chaura, J.; Donat-Torres, M.P.; Boscaiu, M.; Vicente, O. Antioxidant responses under salinity and drought in three closely related wild monocots with different ecological optima. *AoB Plants* **2017**, *9*. [CrossRef]

9. Al Mahmud, J.; Bhuyan, M.H.M.B.; Anee, T.I.; Nahar, K.; Fujita, M.; Hasanuzzaman, M. Reactive Oxygen Species Metabolism and Antioxidant Defense in Plants Under Metal/Metalloid Stress. In *Plant Abiotic Stress Tolerance*; Hasanuzzaman, M., Hakeem, K.R., Nahar, K., Alharby, H.F., Eds.; Springer International Publishing: Cham, Switzerland, 2019; pp. 221–257, ISBN 978-3-030-06117-3.

10. Elansary, H.O.; Szopa, A.; Kubica, P.; Ekiert, H.; Ali, H.M.; Elshikh, M.S.; Abdel-Salam, E.M.; El-Esawi, M.; El-Ansary, D.O. Bioactivities of traditional medicinal plants in Alexandria. *Evid.-Based. Complement. Altern. Med.* **2018**, *2018*. [CrossRef]

11. El-Hifny, I.M.; El-Sayed, M.A. Response of Sweet Pepper plant Growth and Productivity to Application of Ascorbic Acid and Biofertilizers under Saline Conditions. *Aust. J. Basic Appl. Sci.* **2011**, *5*, 1273–1283.

12. Hernández, J.A. Salinity Tolerance in Plants: Trends and Perspectives. *Int. J. Mol. Sci.* **2019**, *20*, 2408. [CrossRef]

13. Nguyen, H.M.; Sako, K.; Matsui, A.; Suzuki, Y.; Mostofa, M.G.; Ha, C.V.; Tanaka, M.; Tran, L.-S.P.; Habu, Y.; Seki, M. Ethanol Enhances High-Salinity Stress Tolerance by Detoxifying Reactive Oxygen Species in Arabidopsis thaliana and Rice. *Front. Plant Sci.* **2017**, *8*. [CrossRef] [PubMed]

14. Yoon, J.Y.; Hamayun, M.; Lee, S.-K.; Lee, I.-J. Methyl jasmonate alleviated salinity stress in soybean. *J. Crop. Sci. Biotechnol.* **2009**, *12*, 63–68. [CrossRef]

15. Savvides, A.; Ali, S.; Tester, M.; Fotopoulos, V. Chemical Priming of Plants Against Multiple Abiotic Stresses: Mission Possible? *Trends Plant Sci.* **2016**, *21*, 329–340. [CrossRef]

16. An, C.; Mou, Z. Salicylic Acid and its Function in Plant ImmunityF. *J. Integr. Plant Biol.* **2011**, *53*, 412–428. [CrossRef] [PubMed]

17. Rao, S.; Du, C.; Li, A.; Xia, X.; Yin, W.; Chen, J. Salicylic Acid Alleviated Salt Damage of Populus euphratica: A Physiological and Transcriptomic Analysis. *Forests* **2019**, *10*, 423. [CrossRef]

18. Brito, C.; Dinis, L.-T.; Moutinho-Pereira, J.; Correia, C.M. Drought Stress Effects and Olive Tree Acclimation under a Changing Climate. *Plants* **2019**, *8*, 232. [CrossRef]

19. Hernández-Ruiz, J.; Arnao, M. Relationship of Melatonin and Salicylic Acid in Biotic/Abiotic Plant Stress Responses. *Agronomy* **2018**, *8*, 33. [CrossRef]

20. Barnett, J.A.; Yarrow, D.; Payne, R.W.; Barnett, L. *Yeasts: Characteristics and Identification*, 3rd ed.; Cambridge University Press: Cambridge, UK; New York, NY, USA, 2000; ISBN 978-0-521-57396-2.

21. Abdelaal, K.A.; Hafez, Y.M.; El Sabagh, A.; Saneoka, H. Ameliorative effects of Abscisic acid and yeast on morpho-physiological and yield characteristics of maize plant (*Zea mays* L.) under water deficit conditions. *Fresenius Environ. Bull.* **2017**, *26*, 7372–7383.

22. Xi, Q.; Lai, W.; Cui, Y.; Wu, H.; Zhao, T. Effect of Yeast Extract on Seedling Growth Promotion and Soil Improvement in Afforestation in a Semiarid Chestnut Soil Area. *Forests* **2019**, *10*, 76. [CrossRef]

23. Kasim, W.; AboKassem, E.; Ragab, G. Ameliorative effect of Yeast Extract, IAA and Green-synthesized Nano Zinc Oxide on the Growth of Cu-stressed *Vicia faba* Seedlings. *Egypt. J. Bot.* **2017**, *57*, 1–16. [CrossRef]

24. Rouphael, Y.; De Micco, V.; Arena, C.; Raimondi, G.; Colla, G.; Pascale, S. Effect of *Ecklonia maxima* seaweed extract on yield, mineral composition, gas exchange, and leaf anatomy of zucchini squash grown under saline conditions. *J. Appl. Phycol.* **2017**, *29*, 459–470. [CrossRef]

25. Saleh, A.A.H.; Abu-Elsaoud, A.M.; Elkelish, A.A.; Sahadad, M.A.; Abdelrazek, E.M. Role of External Proline on Enhancing Defence Mechanisms of Vicia Faba L. Against Ultraviolet Radiation. *Am.-Eurasian J. Sustain. Agric.* **2015**, *9*, 13.

26. Ali, Q.; Anwar, F.; Ashraf, M.; Saari, N.; Perveen, R. Ameliorating effects of exogenously applied proline on seed composition, seed oil quality and oil antioxidant activity of maize (*Zea mays* L.) under drought stress. *Int. J. Mol. Sci.* **2013**, *14*, 818–835. [CrossRef] [PubMed]

27. Hasanuzzaman, M.; Alam, M.M.; Rahman, A.; Hasanuzzaman, M.; Nahar, K.; Fujita, M. Exogenous Proline and Glycine Betaine Mediated Upregulation of Antioxidant Defense and Glyoxalase Systems Provides Better Protection against Salt-Induced Oxidative Stress in Two Rice (*Oryza sativa* L.) Varieties. *BioMed Res. Int.* **2014**, *2014*. [CrossRef]

28. El-Amier, Y.; Elhindi, K.; El-Hendawy, S.; Al-Rashed, S.; Abd-ElGawad, A. Antioxidant System and Biomolecules Alteration in Pisum sativum under Heavy Metal Stress and Possible Alleviation by 5-Aminolevulinic Acid. *Molecules* **2019**, *24*, 4194. [CrossRef] [PubMed]

29. Qadeer, U.; Ahmed, M.; Hassan, F.; Akmal, M. Impact of Nitrogen Addition on Physiological, Crop Total Nitrogen, Efficiencies and Agronomic Traits of the Wheat Crop under Rainfed Conditions. *Sustainability* **2019**, *11*, 6486. [CrossRef]

30. Kaundun, S.S.; Jackson, L.V.; Hutchings, S.-J.; Galloway, J.; Marchegiani, E.; Howell, A.; Carlin, R.; Mcindoe, E.; Tuesca, D.; Moreno, R. Evolution of Target-Site Resistance to Glyphosate in an Amaranthus palmeri Population from Argentina and Its Expression at Different Plant Growth Temperatures. *Plants* **2019**, *8*, 512. [CrossRef]

31. Abdelaal, K.A. Pivotal Role of Bio and Mineral Fertilizer Combinations on Morphological, Anatomical and Yield Characters of Sugar Beet Plant (*Beta vulgaris* L.). *Middle East. J. Agric.* **2015**, *4*, 717–734.

32. Moran, R. Formulae for Determination of Chlorophyllous Pigments Extracted with N,N-Dimethylformamide 1. *Plant Physiol.* **1982**, *69*, 1376–1381. [CrossRef]

33. Sánchez, F.J.; de Andrés, E.F.; Tenorio, J.L.; Ayerbe, L. Growth of epicotyls, turgor maintenance and osmotic adjustment in pea plants (*Pisum sativum* L.) subjected to water stress. *Field Crop. Res.* **2004**, *86*, 81–90. [CrossRef]

34. Dionisio-Sese, M.L.; Tobita, S. Antioxidant responses of rice seedlings to salinity stress. *Plant Sci.* **1998**, *135*, 1–9. [CrossRef]

35. Bates, L.S.; Waldren, R.P.; Teare, I.D. Rapid determination of free proline for water-stress studies. *Plant Soil* **1973**, *39*, 205–207. [CrossRef]

36. Heath, R.L.; Packer, L. Photoperoxidation in isolated chloroplasts. I. Kinetics and stoichiometry of fatty acid peroxidation. *Arch. Biochem. Biophys.* **1968**, *125*, 189–198. [CrossRef]

37. Badiani, M.; De Biasi, M.G.; Colognola, M.; Artemi, F. Catalase, peroxidase and superoxide dismutase activities in seedlings submitted to increasing water deficit. *Agrochimica* **1990**, *34*, 90–102.

38. Aebi, H. Catalase in vitro. In *Methods in Enzymology*; Oxygen Radicals in Biological Systems; Academic Press: New York, NY, USA, 1984; Volume 105, pp. 121–126.

39. Hammerschmidt, R.; Nuckles, E.M.; Kuć, J. Association of enhanced peroxidase activity with induced systemic resistance of cucumber to Colletotrichum lagenarium. *Physiol. Plant Pathol.* **1982**, *20*, 73–82. [CrossRef]

40. El-Esawi, M.A.; Alaraidh, I.A.; Alsahli, A.A.; Alamri, S.A.; Ali, H.M.; Alayafi, A.A. *Bacillus firmus* (SW5) augments salt tolerance in soybean (*Glycine max* L.) by modulating root system architecture, antioxidant defense systems and stress-responsive genes expression. *Plant Physiol. Biochem.* **2018**, *132*, 375–384. [CrossRef]

41. El-Esawi, M.A.; Al-Ghamdi, A.A.; Ali, H.M.; Alayafi, A.A. *Azospirillum lipoferum* FK1 confers improved salt tolerance in chickpea (*Cicer arietinum* L.) by modulating osmolytes, antioxidant machinery and stress-related genes expression. *Environ. Exp. Bot.* **2019**, *159*, 55–65. [CrossRef]

42. El-Esawi, M.A.; Al-Ghamdi, A.A.; Ali, H.M.; Alayafi, A.A.; Witczak, J.; Ahmad, M. Analysis of genetic variation and enhancement of salt tolerance in French pea. *Int. J. Mol. Sci.* **2018**, *19*, 2433. [CrossRef]

43. Suo, J.; Zhao, Q.; David, L.; Chen, S.; Dai, S. Salinity Response in Chloroplasts: Insights from Gene Characterization. *IJMS* **2017**, *18*, 1011. [CrossRef]

44. Yang, X.; Li, Y.; Qi, M.; Liu, Y.; Li, T. Targeted Control of Chloroplast Quality to Improve Plant Acclimation: From Protein Import to Degradation. *Front. Plant Sci.* **2019**, *10*, 958. [CrossRef] [PubMed]

45. Shah, S.; Houborg, R.; McCabe, M. Response of Chlorophyll, Carotenoid and SPAD-502 Measurement to Salinity and Nutrient Stress in Wheat (*Triticum aestivum* L.). *Agronomy* **2017**, *7*, 61.

46. Bulgari, R.; Franzoni, G.; Ferrante, A. Biostimulants Application in Horticultural Crops under Abiotic Stress Conditions. *Agronomy* **2019**, *9*, 306. [CrossRef]

47. Hayat, S.; Hayat, Q.; Alyemeni, M.N.; Wani, A.S.; Pichtel, J.; Ahmad, A. Role of proline under changing environments. *Plant Signal. Behav.* **2012**, *7*, 1456–1466. [CrossRef] [PubMed]

48. Dawood, M.G.; Taie, H.A.A.; Nassar, R.M.A.; Abdelhamid, M.T.; Schmidhalter, U. The changes induced in the physiological, biochemical and anatomical characteristics of *Vicia faba* by the exogenous application of proline under seawater stress. *S. Afr. J. Bot.* **2014**, *93*, 54–63. [CrossRef]

49. Parvin, K.; Hasanuzzaman, M.; Bhuyan, M.H.M.B.; Nahar, K.; Mohsin, S.M.; Fujita, M. Comparative Physiological and Biochemical Changes in Tomato (*Solanum lycopersicum* L.) under Salt Stress and Recovery: Role of Antioxidant Defense and Glyoxalase Systems. *Antioxidants* **2019**, *8*, 350. [CrossRef]

50. Acosta-Motos, J.; Ortu?o, M.; Bernal-Vicente, A.; Diaz-Vivancos, P.; Sanchez-Blanco, M.; Hernandez, J. Plant Responses to Salt Stress: Adaptive Mechanisms. *Agronomy* **2017**, *7*, 18. [CrossRef]

51. Abdelaal, K.A.A.; Hafez, Y.M.; El-Afry, M.M.; Tantawy, D.S.; Alshaal, T. Effect of some osmoregulators on photosynthesis, lipid peroxidation, antioxidative capacity, and productivity of barley (Hordeum vulgare L.) under water deficit stress. *Environ. Sci. Pollut. Res.* **2018**, *25*, 30199–30211. [CrossRef]

52. Gholami Zali, A.; Ehsanzadeh, P. Exogenous proline improves osmoregulation, physiological functions, essential oil, and seed yield of fennel. *Ind. Crop. Prod.* **2018**, *111*, 133–140. [CrossRef]

53. Hafez, E.; Omara, A.E.D.; Ahmed, A. The Coupling Effects of Plant Growth Promoting Rhizobacteria and Salicylic Acid on Physiological Modifications, Yield Traits, and Productivity of Wheat under Water Deficient Conditions. *Agronomy* **2019**, *9*, 524. [CrossRef]

54. El-Esawi, M.A.; Alaraidh, I.A.; Alsahli, A.A.; Ali, H.M.; Alayafi, A.A.; Witczak, J.; Ahmad, M. Genetic Variation and Alleviation of Salinity Stress in Barley (*Hordeum vulgare* L.). *Molecules* **2018**, *23*, 2488. [CrossRef] [PubMed]

55. El-Esawi, M.A.; Alaraidh, I.A.; Alsahli, A.A.; Alzahrani, S.M.; Ali, H.M.; Alayafi, A.A.; Ahmad, M. *Serratia liquefaciens* KM4 Improves Salt Stress Tolerance in Maize by Regulating Redox Potential, Ion Homeostasis, Leaf Gas Exchange and Stress-Related Gene Expression. *Int. J. Mol. Sci.* **2018**, *19*, 3310. [CrossRef] [PubMed]

56. Ishikawa, H.; Evans, M.L. Electrotropism of Maize Roots: Role of the Root Cap and Relationship to Gravitropism. *Plant Physiol.* **1990**, *94*, 913–918. [CrossRef] [PubMed]

57. Huang, D.; Sun, Y.; Ma, Z.; Ke, M.; Cui, Y.; Chen, Z.; Chen, C.; Ji, C.; Tran, T.M.; Yang, L.; et al. Salicylic acid-mediated plasmodesmal closure via Remorin-dependent lipid organization. *Proc. Natl. Acad. Sci. USA* **2019**, *116*, 21274–21284. [CrossRef]

58. Verbruggen, N.; Hermans, C. Proline accumulation in plants: A review. *Amino Acids* **2008**, *35*, 753–759. [CrossRef]

59. El-Katony, T.M.; El-Bastawisy, Z.M.; El-Ghareeb, S.S. Timing of salicylic acid application affects the response of maize (*Zea mays* L.) hybrids to salinity stress. *Heliyon* **2019**, *5*, e01547. [CrossRef]

60. Huang, Z.; Zhao, L.; Chen, D.; Liang, M.; Liu, Z.; Shao, H.; Long, X. Salt Stress Encourages Proline Accumulation by Regulating Proline Biosynthesis and Degradation in Jerusalem Artichoke Plantlets. *PLoS ONE* **2013**, *8*, e62085. [CrossRef]

61. Li, T.; Hu, Y.; Du, X.; Tang, H.; Shen, C.; Wu, J. Salicylic acid alleviates the adverse effects of salt stress in Torreya grandis cv. Merrillii seedlings by activating photosynthesis and enhancing antioxidant systems. *PLoS ONE* **2014**, *9*, e109492. [CrossRef]

62. Gharsallah, C.; Fakhfakh, H.; Grubb, D.; Gorsane, F. Effect of salt stress on ion concentration, proline content, antioxidant enzyme activities and gene expression in tomato cultivars. *AoB Plants* **2016**, *8*, plw055. [CrossRef]

63. Wang, Y.; Li, X.; Li, J.; Bao, Q.; Zhang, F.; Tulaxi, G.; Wang, Z. Salt-induced hydrogen peroxide is involved in modulation of antioxidant enzymes in cotton. *Crop J.* **2016**, *4*, 490–498. [CrossRef]

64. El-Esawi, M.A.; Elkelish, A.; Elansary, H.O.; Ali, H.M.; Elshikh, M.; Witczak, J.; Ahmad, M. Genetic Transformation and Hairy Root Induction Enhance the Antioxidant Potential of *Lactuca serriola* L. *Oxid. Med. Cell. Longev.* **2017**, *2017*. [CrossRef] [PubMed]

65. Lin, C.C.; Kao, C.H. Effect of NaCl stress on H_2O_2 metabolism in rice leaves. *Plant Growth Regul.* **2000**, *30*, 151–155. [CrossRef]

66. Hernandez, M.; Fernandez-Garcia, N.; Diaz-Vivancos, P.; Olmos, E. A different role for hydrogen peroxide and the antioxidative system under short and long salt stress in *Brassica oleracea* roots. *J. Exp. Bot.* **2010**, *61*, 521–535. [CrossRef] [PubMed]

67. Li, Q.; Lv, L.R.; Teng, Y.J.; Si, L.B.; Ma, T.; Yang, Y.L. Apoplastic hydrogen peroxide and superoxide anion exhibited different regulatory functions in salt-induced oxidative stress in wheat leaves. *Biol. Plant.* **2018**, *62*, 750–762. [CrossRef]

68. Vighi, I.L.; Benitez, L.C.; Amaral, M.N.; Moraes, G.P.; Auler, P.A.; Rodrigues, G.S.; Deuner, S.; Maia, L.C.; Braga, E.J.B. Functional characterization of the antioxidant enzymes in rice plants exposed to salinity stress. *Biol. Plant.* **2017**, *61*, 540–550. [CrossRef]

69. Pérez-Labrada, F.; López-Vargas, E.R.; Ortega-Ortiz, H.; Cadenas-Pliego, G.; Benavides-Mendoza, A.; Juárez-Maldonado, A. Responses of Tomato Plants under Saline Stress to Foliar Application of Copper Nanoparticles. *Plants* **2019**, *8*, 151. [CrossRef]

70. El-Esawi, M.A.; Elansary, H.O.; El-Shanhorey, N.A.; Abdel-Hamid, A.M.E.; Ali, H.M.; Elshikh, M.S. Salicylic Acid-Regulated Antioxidant Mechanisms and Gene Expression Enhance Rosemary Performance under Saline Conditions. *Front. Physiol.* **2017**, *8*, 716. [CrossRef]

71. Đorđević, N.O.; Todorović, N.; Novaković, I.T.; Pezo, L.L.; Pejin, B.; Maraš, V.; Tešević, V.V.; Pajović, S.B. Antioxidant Activity of Selected Polyphenolics in Yeast Cells: The Case Study of Montenegrin Merlot Wine. *Molecules* **2018**, *23*, 1971. [CrossRef]

72. Mohammadrezakhani, S.; Hajilou, J.; Rezanejad, F.; Zaare-Nahandi, F. Assessment of exogenous application of proline on antioxidant compounds in three Citrus species under low temperature stress. *J. Plant Inter.* **2019**, *14*, 347–358. [CrossRef]

73. Abdul Qados, A.M.S. Effect of salt stress on plant growth and metabolism of bean plant (*Vicia faba* L.). *J. Saudi Soc. Agric. Sci.* **2011**, *10*, 7–15. [CrossRef]

74. Yildirim, E.; Karlidag, H.; Turan, M. Mitigation of salt stress in strawberry by foliar K, Ca and Mg nutrient supply. *Plant Soil Environ.* **2009**, *55*, 213–221. [CrossRef]

75. Huang, Y.; Bie, Z.; Liu, Z.; Zhen, A.; Wang, W. Protective role of proline against salt stress is partially related to the improvement of water status and peroxidase enzyme activity in cucumber. *Soil Sci. Plant Nutr.* **2009**, *55*, 698–704. [CrossRef]

76. Sharma, A.; Shahzad, B.; Kumar, V.; Kohli, S.K.; Sidhu, G.P.S.; Bali, A.S.; Handa, N.; Kapoor, D.; Bhardwaj, R.; Zheng, B. Phytohormones Regulate Accumulation of Osmolytes Under Abiotic Stress. *Biomolecules* **2019**, *9*, 285. [CrossRef] [PubMed]

77. Gupta, B.; Huang, B. Mechanism of Salinity Tolerance in Plants: Physiological, Biochemical, and Molecular Characterization. *Int. J. Genomics* **2014**, *2014*. [CrossRef] [PubMed]

78. Ahanger, M.A.; Tomar, N.S.; Tittal, M.; Argal, S.; Agarwal, R.M. Plant growth under water/salt stress: ROS production; antioxidants and significance of added potassium under such conditions. *Physiol. Mol. Biol. Plant* **2017**, *23*, 731–744. [CrossRef]

79. Husen, A.; Iqbal, M.; Sohrab, S.S.; Ansari, M.K.A. Salicylic acid alleviates salinity-caused damage to foliar functions, plant growth and antioxidant system in Ethiopian mustard (*Brassica carinata* A. Br.). *Agric. Food Secur.* **2018**, *7*, 44. [CrossRef]

Article

Screening of Provitamin-A Maize Inbred Lines for Drought Tolerance: Beta-Carotene Content and Secondary Traits

Aleck Kondwakwenda *, Julia Sibiya, Rebecca Zengeni, Cousin Musvosvi and Samson Tesfay

School of Agriculture, Earth and Environmental Sciences, University of KwaZulu-Natal, Private Bag X01, Scottsville 3209, Pietermaritzburg, South Africa; sibiyaj@ukzn.ac.za (J.S.); zengeni@ukzn.ac.za (R.Z.); cousinmusvosvi@gmail.com (C.M.); tesfay@ukzn.ac.za (S.T.)
* Correspondence: alekondwa@gmail.com; Tel.: +27-332-606-246

Received: 17 September 2019; Accepted: 21 October 2019; Published: 29 October 2019

Abstract: Provitamin A maize (*Zea mays* L.) biofortification is an ideal complementary means of combating vitamin A deficiency (VAD) in sub-Saharan Africa where maize consumption is high coupled by high VAD incidences. However, drought remains a major abiotic constraint to maize productivity in this region. Comprehensive drought screening of initial breeding materials before advancing them is important to achieve genetic gain. In this study, 46 provitamin-A inbred lines were screened for drought tolerance in the greenhouse and field under drought and optimum conditions using β-carotene content (BCC), grain yield (GY), and selected morphophysiological and biochemical traits. The results revealed that BCC, morphophysiological and biochemical traits were effective in discriminating among genotypes. Number of ears per plant (EPP), stomatal conductance (Gs), delayed leaf senescence (SEN), leaf rolling (RL), chlorophyll content (CC) and free proline content (PC) proved to be ideal traits to use when indirectly selecting for GY by virtue of having relative efficiency of indirect selection values that are greater than unity and considerable genetic variances under either or both conditions. The findings of this study form the basis of initial germplasm selection when improving provitamin A maize for drought tolerance.

Keywords: Provitamin A; maize; drought; morphological; physiological; biochemical; β-carotene

---

1. Introduction

Provitamin A maize has orange or yellow endosperm, which contains precursors of vitamin A in the form of carotenoids, hence the name "provitamin A maize". Provitamin A carotenoids include β-carotene, α-carotene and β-cryptoxanthin. Beta-carotene is the most important of the three carotenoids because it has higher provitamin A activity owing to its unique double ring molecular structure [1]. However, the ordinary (not improved) yellow maize grown and consumed throughout the world has β-carotene content of less than 1.5 µg g^{-1} [2], which is too low given that the required target of total provitamin A carotenoid is 15 µg g^{-1} [3]. Developing provitamin A maize cultivars with higher levels of provitamin A carotenoids through biofortification is a sustainable, cheap and effective complementary solution to VAD challenges faced by many developing nations [4].

An online Biofortification Priority Index (BPI) tool, developed and managed by HarvestPlus, shows that maize provitamin A biofortification as VAD intervention is most suitable for maize consuming developing countries, particularly the Southern Africa region (www.harvestplus.org/knowledge-market/BPI). However, maize production in this region is vulnerable to drought due to recurring low annual precipitation coupled by poor coping capacity of most farmers. For instance, in 1992 and 2002, most of the southern African countries experienced the worst droughts resulting in over 60% maize yield loss in the whole region [5]. In 2013, 770 million people in the Southern African Development

Community (SADC) were at risk of food insecurity due to severe mid-season dry spells [6]. In 2016, eight of South Africa's nine provinces were declared food-insecure due to drought [7].

Drought stress affects maize at almost all growth stages, but the flowering and grain filling stages are the most susceptible, with yield losses of over 90% reported when drought coincides with these growth stages [8]. Genetic improvement of maize for drought tolerance through breeding is a sustainable solution to reduce the impacts of drought. However, breeding for drought tolerance is a complex task because the trait is controlled by many genes, and is highly affected by genotype and environment interaction. Furthermore, grain yield, which is the trait of interest, has very low variation and heritability under drought conditions, which makes selection difficult [9]. Comprehensive screening forms the foundation of any successful drought tolerance breeding program [10]. Additionally, screening materials for drought tolerance at the initial stages of breeding often imparts tolerance to other related stresses such as low nitrogen stress [11].

To increase the chances of selecting the appropriate genotypes, breeders should meticulously consider all the available information at screening phase. Screening maize for drought tolerance entails the selection of high yielding genotypes under water deficit stress and/or optimum conditions. Indirect selection for grain yield via related secondary traits helps to circumvent the challenge of poor grain yield variation and low heritability under drought conditions [12]. Stress-tolerant indices and multivariate statistics are often useful when selecting best genotypes.

Indirect selection for drought tolerance involves selecting for multi-secondary traits that are highly correlated to grain yield and have high heritability values. Furthermore, an ideal secondary trait should have efficiency of indirect selection relative to direct selection of greater than a unity [13]. Morphological and physiological (morphophysiological) traits that are associated with maize drought tolerance include anthesis-silking interval (ASI), leaf rolling (RL), chlorophyll content (CC), leaf senescence (SEN), and number of ears per plant (EPP) [12,14]. Stomatal conductance (Gs) analysis as a physiological response to drought stress has not been widely applied as a screening criterion for drought tolerance in maize and therefore, information about its correlation with grain yield and heritability under drought stress and optimum conditions is not well established.

Biochemical changes that are drought-induced in plants include increase in stress signaling hormones and proteins regulators such as proline and abscisic acid (ABA) among others [15,16]. Proline is an amino acid, which plays an osmoregulatory role in plants under drought conditions [17]. Despite the presence of genetic variation of proline content in plants under drought stress and wide application of proline analysis in understanding drought tolerance of other crops such as wheat [18], cowpea [19], and peanut [20], it has not been applied in large-scale maize drought tolerance screening.

Given the importance of maize provitamin A biofortification in maize-consuming developing countries and the prevailing devastating impacts of drought to maize productivity, it is important to investigate the effectiveness of integrated application of morphophysiological and biochemical traits in screening provitamin A maize inbred lines for drought tolerance. The objectives of the study were to: (i) determine the level of genotypic variation for drought tolerance among tropical provitamin A maize inbred lines in regards to secondary traits, (ii) screen candidate lines for drought tolerance based on grain yield and β-carotene content, and (iii) identify secondary traits that can be effectively used for indirect selection of grain yield under drought stressed and non-stressed conditions. The study forms the basis of germplasm selection for use in provitamin A maize drought tolerance breeding programs.

2. Materials and Methods

2.1. Plant Materials and Study Sites

Fifty inbred lines were screened for drought stress tolerance under managed drought conditions. Inbred lines consisted of 46 provitamin A (orange endosperm) and four drought tolerant white endosperm (non-provitamin A) checks. The 46 provitamin-A inbred lines were sourced from the provitamin A biofortification nurseries of International Maize and Wheat Improvement Center

(CIMMYT) (33) and International Institute of Tropical Agriculture (IITA) (13). The four drought tolerant checks were obtained from the Agricultural Research Council (ARC), Grain Crops Institute, Potchefstroom, South Africa. Names, codes and other information about the inbred lines are given in supplementary Table S1. The respective institutions could not provide pedigree information. The study was carried out across four environments (Env), which were two greenhouse and two field in KwaZulu-Natal Province of South Africa. Greenhouse trials were carried out at the University of KwaZulu-Natal (UKZN), Agriculture Pietermaritzburg campus (29°46′ S, 30°58′ E) from January to April 2017 (Env1) and May to September 2017 (Env2). The field trials were carried out at Ukulinga Research Farm in Pietermaritzburg (29°40′ S, 30°24′ E) from April to August 2017 (Env3) and at Makhathini Research station, Jozini, South Africa (27°39′ S, 32°17′ E) from May to September 2018 (Env4). Supplementary Table S2 shows monthly weather data for the four environments during the respective growing periods.

2.2. Experimental Design and Crop Establishment

An alpha lattice design was used to screen the 50 genotypes with two replications containing five incomplete blocks with ten genotypes each and two water regimes (water stress, S and optimum conditions, W) across all the four environments. In the field, the plot size was two rows of 5 m with 0.75 m between the rows and intra row spacing of 0.30 m. Plots were planted with two seeds per station and thinned to one plant 2 weeks after crop emergence. In the greenhouse, a plot was made of four 5 L perforated plastic pots with two plants in each pot, which were thinned to one plant per pot 2 weeks after crop emergence. Pine-bark growing media mixed with loam soil at a ratio of 3:1 was used in the greenhouse. In the field, the soil was predominantly black clay loam soil at both sites. The water stress treatment (S) for all the experiments was implemented in accordance with CIMMYT protocols of withholding irrigation at two weeks prior to expected anthesis date [21]. The water stress condition was maintained until 5 weeks after the flowering of 50% of the genotypes then a single irrigation was applied at grain filling stage. In the field, the optimum treatment (W) involved a 10-day interval sprinkler irrigation throughout the growing period. In the greenhouse, W consisted of drip irrigation for 3 min, four times per day. Across all the environments, compound fertilizer was applied at the rate of 150 kg N, 65 kg P and 65 kg K ha^{-1} at the time of planting and top-dressing fertilizer was applied at five weeks after emergence at a rate of 60 kg N ha^{-1}. In the field, weeds were controlled using Gramoxone® (Syngenta, Greensboro, NC, USA) at a rate of 5 L ha^{-1} and manual weeding whilst hand weeding was done in the greenhouse. Coragen® (Dupont, Washington, DC, USA) and Karate® (Syngenta, Greensboro, NC, USA) insecticides were used to control insects at a rate of 1 L ha^{-1} across all environments when it was necessary.

2.3. Plant Characteristics

2.3.1. Morphophysiological Traits

Data for the following morphophysiological and biochemical traits were collected under both water regimes across all the four environments. Most traits were measured following a CIMMYT protocol described by [21]. Both field and greenhouse grain yield (GY) was estimated per plot area bases and converted to tonnes per hectare. The plot area in the greenhouse was determined by multiplying the cylinder-shaped pot area by 4 since a plot was made up of four pots. Similarly, the plot area for the field experiments was determined by calculating the area of a rectangular shaped plot. Number of ears per plant (EPP) was computed as the number of ears with at least one fully developed grain divided by the number of harvested plants. Days to anthesis (DA) is the number of days after planting to when 50% of the plants in a plot start to shed pollen. Days to silking (DS) is the number of days after planting when 50% of plant in a plot produces silks. Anthesis-silking interval (ASI) was calculated as DS minus DA. Leaf rolling (RL) was scored using a scale of 0 to 10 where 0 = unrolled leaf and 10 = leaf rolled like an onion and scores were converted to percentage (%) with measurements taken twice after imposing drought treatment. Leaf senescence (SEN) was scored using a scale from 1 to 10 (1 = 10%; 2 = 20%; 3 = 30%; 4 = 40%; 5 = 50%; 6 = 60%; 7 = 70%; 8 = 80%; 9 = 90%; and 10 = 100% dead leaf area) at 3, 5 and 7 weeks after 50% of the plants reached anthesis. Chlorophyll content (CC) was measured from the adaxial surface of the second top fully expanded leaf of five plants per plot at 3, 5 and 7 weeks after 50% of the plants reached anthesis using SPAD-502-Plus chlorophyll meter (Konica Minolta, Osaka, Japan). Stomatal conductance (Gs) was measured from the abaxial surface of the second top fully expanded leaf using a SC–1 leaf porometer (Decagon Devices®, Pullman, WA, USA) at 3, 5 and 7 weeks after 50% of the plants reached anthesis. Chlorophyll content (CC) and stomatal conductance were measured at midday periods (1200–1400 h). Beta carotene content (BCC) and proline (PC) content were determined and measured as described in the following section.

2.3.2. Biochemical Traits

Beta-carotene content was measured from kernels harvested from the self pollinated plants of each of the 46 provitamin inbred lines. A sample of 20 g, which contained about 30 to 50 kernels, was randomly collected from each of the 46 provitamin-A inbred lines and dispatched to Agricultural Research Council (ARC), Science Analytical Laboratory, Pretoria, South Africa (http://www.arc.agric.za) for β-carotene analysis. High-performance liquid chromatography (HPLC) was used for analysis following a protocol for dried maize kernels as described by [22]. The β-carotene analysis was done three times per sample, giving three data points per each genotype.

Proline analysis was performed at UKZN, Crop Science laboratory following a protocol by [23]. Fresh leaf samples were collected from the second top fully expanded leaves for the S and W treatments of both field and greenhouse experiments at 3 weeks after imposing the S treatment. The leaf samples were freeze-dried at very low temperature (−74 °C) using liquid nitrogen before grinding them into fine powder. A 0.5 g ground leaf sample was homogenized in 10 mL of 3% aqueous sulfosalicylic acid and the homogenate was filtered. Two ml of filtrate was mixed with 2 mL acid-ninhydrin and 2 mL of glacial acetic acid for 1 h in a water bath at 100 °C. After cooling, 4 mL of toluene were added and then mixed vigorously using a rotor. The top mixture containing proline within toluene was decanted from the aqueous phase then taken for UV visible spectrophotometer analysis for the absorbance measurement at a wavelength of 520 nm using a model UV-1800 spectrophotometer (Shimadzu Corporation, Kyoto, Japan). The proline concentration was calculated using the formula shown in Equation (1) [23].

$$
\begin{aligned}
\text{Proline Content} \quad &(\mu g \text{ per gram of dry leaf tissue})\\
&= [(\mu g \text{proline}/mL) \times mL \text{ toluene})/115.5 \, \mu g \\
&/\mu mole]/[(g \text{ sample})/5]
\end{aligned}
\tag{1}
$$

where 115.5 is the molecular weight of proline.

2.4. Data Analysis

2.4.1. Analysis of Variance, Mean Performance and Stress-Tolerant Index

Analysis of variance (ANOVA) for all morphophysiological and biochemical traits was done after carrying out a test of homogeneity of variances. A lattice procedure of R software version 3.5.1 [24] was used to carry out the ANOVA following a mixed model (Equation (2)). Genotypes were considered as fixed effects and environments as random:

$$
Y_{ijklm} = \mu + Re_i + B(Re)_{ij} + G_k + E_l + W_m + GE_{kl} + GW_{km} + EW_{lm} + GEW_{klm} + \varepsilon_{ijklm}
\tag{2}
$$

where Y_{ijklm} is the trait of interest, μ is the mean effect, Re_i is the effect of the ith replicate, $B(Re)_{ij}$ is the effect of the jth incomplete block within the ith replicate, G_k is the effect of the kth genotype, E_l is the effect of the lth environment, W_m is effect of the mth water regime while GE_{kl}, GW_{km}, EW_{lm} and GEW_{klm} are the respective interactions and ε_{ijklm} is the residual error term.

Beta-carotene was analysed separately using the general linear model given in Equation (3).

$$Yij = \mu + Gj + S(G)ij \tag{3}$$

where Yij—the performance of ith sample of the jth genotype; Gj—the effect of jth genotype; $S(G)ij$—sample within genotype, which is the error term.

Least significant difference (LSD) test was carried out at 0.5 α level to separate the means.

Stress tolerance index (STI) described by [25] was used to select high yielding genotypes under both conditions as shown in Equation (4):

$$STI = \frac{Ypi \times Ysi}{Yp^2} \tag{4}$$

where Ys—grain yield of a genotype i under drought-stressed condition; Yp—grain yield of i genotype under non-stressed condition, and Xp—mean yield of genotypes under non-stressed condition. Genotypes with high STI value and β-carotene content greater than 1.5 µg g^{-1} were selected.

2.4.2. Variance Components and Heritability

Variance components and broad sense heritability (H) were computed for morphophysiological and biochemical traits in R software following a procedure described by [26] as shown in Equations (5) and (6).

$$\delta^2 p = \delta^2 g + \frac{\delta^2 ge}{e} + \frac{\delta^2}{re} \tag{5}$$

$$H = \frac{\delta^2 g}{\delta^2 p} \tag{6}$$

where $\delta^2 p$—phenotypic variance, $\delta^2 g$—genotypic variances, $\delta^2 ge$—genotype by environment interaction variance, δ^2—error variance, r—number of replications, e—number of environments and H—broad sense heritability.

Heritability classifications guidelines described by [27] were used to describe the heritability levels exhibited by the measured traits in this study in which values from 0 to 0.3 was low, 0.3 to 0.6 was moderate and >0.6 was high.

2.4.3. Principal Component Biplot, Trait Correlations and Relative Selection Efficiency

Phenotypic and genotypic correlations among morphophysiological and biochemical traits were computed separately for each water regime using the META-R software version 6.04 [28]. Correlation classifications guidelines described by [29] were used to explain the correlations among the traits in this study in which correlation coefficients values from ±0.9 to ±1.00 were considered very high correlations, ±0.7 to ±0.9 were high, ±0.5 to ±0.7 were moderate, ±0.3 to ±0.5 were low and ±0.00 to ±0.3 were negligible. A combined principal component PCA biplot was computed to graphically show the traits associated with each water regime. Equation (7) was used to test the efficiency of indirect selection of grain yield via secondary traits relative to direct selection as outlined by [13].

$$\text{Relative efficiency of indirect selection} = \frac{|r_g| h_X}{h_{GY}} \tag{7}$$

where $|r_g|$ is the value of the genotypic correlation between GY and a secondary trait, h_X is the square root of the broad sense heritability of trait, and h_{GY} is the square root of the broad sense heritability of grain yield. According to [13] the most desirable secondary traits should have an efficiency of indirect selection relative to direct selection of greater than unity.

3. Results

3.1. Analysis of Variance, Mean Performances and Stress-Tolerant Index

A combined ANOVA was carried out after separate ANOVA had shown significant ($p < 0.05$) effects of genotype, water regime and their interaction for the studied traits (Table 1). The combined ANOVA revealed that the genotype, water regime, environment and their respective interactions had significant ($p < 0.001$ and $p < 0.05$) effects on GY and other traits except DA, EPP and PC, which were not significantly affected by the interaction of the environment and water regime. Genotypes exhibited significantly ($p < 0.001$) different mean BCC (Table 1).

Mean performance of the top ten and bottom five genotypes with >1.5 µg g^{-1} BCC and ranked in descending order of STI values are presented in Table 2. Thirty-one genotypes had >1.5 µg g^{-1} BCC. Mean BCC was 2.05 µg g^{-1} with genotype 24 (CLHP0022) ranked first in the STI ranking of genotypes with >1.5 µg g^{-1} BCC. Genotypes 1 (CLHP00306) and 11 (CLHP00378) had the highest BCC of 4.22 µg g^{-1} whilst 14 (CLHP0343) had the lowest value of 0.64 µg g^{-1}. The mean performance of all the 50 genotypes under the two water regimes are given in supplementary Table S3. Mean STI was 0.53. Fifty percent of genotypes surpassed the mean STI value. The highest STI value was 0.94 exhibited by entry 27 (CLHP0005) whilst entry 20 (CLHP0364) had the least STI value of 0.23. Genotype 50 (CML569) had the highest STI value of 0.71 among the four drought tolerant checks. The proportion of test genotypes that were ranked higher than the best check on the STI ranking was 10.86%.

The mean GY under optimum and stress conditions were 1.72 t ha^{-1} and 0.88 t ha^{-1}, respectively. This resulted in 51.2% mean yield loss due to drought stress with the highest and lowest percentage yield losses of 64.17% and 29.21%, respectively. Genotype 50 (CML569) was the best yielding check under both drought-stressed and non-stressed conditions with 0.94 t ha^{-1} and 2.26 t ha^{-1}, respectively. The proportion of test genotypes that yielded higher than the best check was 45.57% and 8.70% under drought stressed and non-stressed conditions, respectively. Mean DA was reduced from 75.03 days under optimum conditions to 69.40 days under water stress conditions. Mean ASI increased from a mean of 1.97 days under optimum conditions to a mean of 8.56 days under drought stress conditions. Mean EPP was reduced by drought stress from 2.24 under optimum to 1.68 under drought stress conditions.

Mean RL increased from 3.04% under optimum conditions to 49.78% under water stress. Stomatal conductance was severely reduced from a mean of 368.94 mmol m^{-2} s^{-1} under optimum conditions to a mean of 49.78 mmol m^{-2} s^{-1} under drought stress conditions. Leaf senescence increased due to drought stress from a mean of 11.60% under optimum conditions to 50.18% under water stress conditions. Proline content increased from a mean of 31.83 µg g^{-1} under optimum conditions to a mean of 149.23 µg g^{-1} under water stress conditions with genotype 31 (CML486) having the highest PC of 230.63 µg g^{-1} under water stress conditions.

Table 1. Mean squares and significant tests after combined analysis of variance for nine morphophysiological traits of 50 inbred lines.

SOV	DF	GY	ASI	DA	EPP	RL	Gs	SEN	CC	PC	SOV	DF	BCC
Rep	1	0.33 ***	0.02 *	1.62	0.24	5.28	3370.51	28.13	11.05	570.12	Gen	45	2.758 ***
Rep.Bloc	8	0.37 ***	9.20 ns	156.74 ***	0.19	346.60 ***	10,845.54 ***	239.00 ***	35.05 **	2076.57	Error	92	0.006
Gen	49	1.70 ***	124.35 ***	487.55 ***	2.23 ***	1737.56 ***	17,231.90 ***	1993.14 ***	131.57 ***	6212.13 ***			
Env	3	24.22 ***	10.71 ***	54.82 ***	5.04 ***	516.00 ***	13,750.30 *	1472.46 ***	495.93 ***	889.32			
WR	1	142.85 ***	8685.62 ***	6339.38 ***	63.85 ***	287,471.53 ***	20,372,674.06 ***	297,606.12 ***	88,914.06 ***	2,756,772.93 ***			
Gen.Env	147	0.53 ***	23.21 ***	117.80 ***	0.48 ***	357.14 ***	10,469.86 ***	231.03 ***	31.12 ***	1517.61 ***			
Gen.WR	49	0.28 ***	102.51 ***	281.83 ***	0.92 ***	1715.20 ***	21,625.12 ***	1847.35 ***	94.69 ***	6224.34 ***			
Env.WR	3	13.07 ***	97.87 ***	89.21	0.47	670.90 ***	11,839.37 ***	1964.46 ***	284.66 ***	106.78			
Gen.Env.WR	147	0.22 ***	18.13 ***	119.03 ***	0.43 ***	341.42 ***	10,995.21 ***	215.09 ***	33.01 ***	1506.32 ***			
Error	391	0.01	1.99	9.36	0.20	10.23	1789.44	29.28	6.92	468.59			

SOV—Source of Variation, Rep—Replication, Bloc—incomplete block, Gen—genotype, Env—environment, WR—water regime, DF—degrees of freedom, GY—grain yield, ASI—anthesis-silking interval, DA—days to anthesis, EPP—ears per plant, RL—leaf rolling, SEN—leaf senescence, Gs—stomatal conductance, CCI—chlorophyll content index, PC—proline content, BCC—β-carotene content, * $p < 0.05$, ** $p < 0.01$, *** $p < 0.001$ and ns, not significant.

Table 2. Mean performance of the top 15 and bottom 5 provitamin A maize inbred lines with >1.5 µg g^{-1} BCC and ranked in descending order of STI values.

Gen	STI	BCC	GY		ASI		DA		EPP		RL		Gs		SG		CC		PC	
			S	W	S	W	S	W	S	W	S	W	S	W	S	W	S	W	S	W
									Top 10											
24	0.77	2.73	1.00	2.27	6.88	0.75	77.63	79.00	1.88	2.75	27.50	2.00	26.03	420.63	31.25	11.25	19.18	44.45	148.93	33.20
1	0.75	4.22	1.00	2.23	12.75	2.38	74.88	71.25	1.63	2.50	27.50	1.63	26.60	330.11	37.50	10.00	22.21	41.93	175.82	24.88
8	0.70	2.26	1.00	2.08	4.75	2.13	73.63	78.38	2.00	2.13	12.50	1.88	35.53	353.66	33.75	13.75	20.99	37.17	193.53	25.57
17	0.69	1.59	0.99	2.07	4.00	2.25	80.13	74.50	1.88	2.25	26.25	3.25	40.64	413.58	26.25	10.00	19.23	44.53	178.11	40.96
42	0.68	2.64	0.95	2.12	5.25	0.25	53.88	71.75	2.50	2.63	13.75	3.50	24.12	406.85	27.50	13.75	24.33	39.06	165.51	34.52
38	0.64	2.76	0.91	2.08	0.50	1.38	63.50	71.38	2.38	2.75	30.00	2.50	20.14	347.01	47.50	12.50	16.73	43.74	140.98	28.30
44	0.60	2.53	0.94	1.88	5.50	1.38	58.00	64.88	2.13	2.00	16.25	3.00	37.85	361.02	30.00	10.00	25.19	35.24	213.15	25.96
7	0.58	2.61	1.00	1.73	4.63	2.25	79.25	80.75	2.00	2.38	18.75	3.88	34.79	376.43	25.00	10.00	19.79	40.66	205.98	35.33
39	0.58	1.69	0.91	1.90	6.88	1.75	62.00	64.63	1.75	2.25	21.25	4.00	17.52	292.60	43.75	10.00	21.73	39.95	175.97	23.09
25	0.58	1.67	0.85	2.01	4.75	2.63	79.50	75.13	1.75	2.13	22.50	2.88	44.85	273.27	27.50	11.25	23.01	38.75	184.89	28.27
									Bottom 5											
21	0.34	1.97	0.75	1.32	11.38	1.13	78.63	81.25	1.25	2.13	78.75	4.00	66.66	384.13	71.25	12.50	16.32	41.94	103.21	31.60
32	0.31	2.92	0.81	1.15	12.13	2.88	77.50	74.13	1.00	2.50	66.25	3.88	92.69	376.40	85.00	12.50	12.19	36.62	89.56	35.81
19	0.31	1.81	0.75	1.21	12.38	3.50	74.88	78.00	1.13	2.13	58.75	2.50	73.76	349.29	81.25	11.25	15.14	38.61	100.77	27.59
12	0.27	2.88	0.70	1.16	10.63	1.63	75.63	77.88	1.38	1.63	58.75	2.13	66.85	314.55	66.25	13.75	16.04	38.04	139.47	28.92
20	0.23	3.39	0.69	0.97	-9.88	0.88	80.88	72.00	1.13	1.75	87.50	3.50	68.86	338.51	88.75	12.50	10.07	34.20	95.98	28.24
Mean		2.05	0.88	1.72	8.56	1.97	69.40	75.03	1.68	2.24	40.95	3.04	49.78	368.94	50.18	11.60	18.80	39.88	149.23	31.83
CV		3.65	7.87	5.33	17.96	63.75	1.93	5.51	20.69	23.40	10.85	27.36	1.38	16.11	13.26	26.78	4.28	9.09	20.24	13.45
LSD		0.12	0.07	0.09	1.52	1.24	1.32	4.08	0.34	0.52	4.38	0.82	0.68	58.60	6.56	3.06	0.79	3.58	29.79	4.22

Gen—Genotype/inbred line, STI—Stress-Tolerant Index, GY—grain yield, BCC—β-carotene content; ASI—anthesis-silking interval, DA—days to anthesis, EPP—number of ears per plant, SEN—leaf senescence, RL—leaf rolling, Gs—stomatal conductance, CC—chlorophyll content, PC—proline content, CV—coefficient of variation, LSD—least significant difference, S—Water stress conditions and W—well-watered condition.

3.2. Heritability and Variance Components

Grain yield, DA, EPP, Gs and CC exhibited high heritability and genotypic variance values under non-stressed conditions whilst ASI, RL, SEN and PC had high values under drought conditions (Table 3). Heritability estimates ranged from 0.250 to 0.998 under non-drought conditions and 0.127 to 0.987 under drought conditions. Most traits, except Gs, were characterized by sharp differences in heritability values between the two water regimes. For instance, heritability estimates for GY yield dropped from 0.621 under non-stressed conditions to 0.398 under drought conditions whilst PC had heritability values of 0.263 and 0.777 under non-stressed and drought conditions, respectively. Generally, variance due to genotype by environment interaction was higher under drought conditions than under non-drought conditions.

Table 3. Variance components and heritability of measured morphophysiological traits.

Variance Components	Traits								
	GY	ASI	DA	EPP	RL	Gs	SEN	CC	PC
Non-stressed conditions (W)									
Gen ($\delta^2 g$)	0.098	0.189	73.122	0.189	0.074	506.778	0.656	18.235	7.753
Env ($\delta^2 e$)	0.238	0.025	0.000	0.025	0.214	0.385	0.323	1.471	0.000
Gen.Env ($\delta^2 ge$)	0.232	1.052	51.808	0.138	0.910	3.887	3.094	16.008	78.695
Error (δ^2)	0.012	0.683	2.580	0.123	1.618	0.525	9.500	0.714	16.873
Phenotypic ($\delta^2 p$)	0.157	0.537	86.396	0.239	0.504	507.815	2.617	22.326	29.536
Heritability (H)	0.621	0.352	0.846	0.792	0.147	0.998	0.251	0.817	0.263
Stressed conditions (S)									
Gen ($\delta^2 g$)	0.091	23.487	12.852	0.055	345.544	405.282	424.332	2.265	1200.237
Env ($\delta^2 e$)	0.021	0.463	2.244	0.127	4.748	0.036	29.510	5.685	0.000
Gen.Env ($\delta^2 ge$)	0.417	17.578	69.455	0.143	343.666	7.164	194.567	9.653	977.106
Error (δ^2)	0.273	2.328	30.552	0.005	19.500	26.971	48.751	12.963	917.053
Phenotypic ($\delta^2 p$)	0.193	28.173	34.034	0.091	433.898	410.444	479.067	6.298	1559.145
Heritability (H)	0.398	0.834	0.378	0.602	0.796	0.987	0.886	0.360	0.770

3.3. Principal Component Biplot Analysis

A combined principal components biplot was constructed to show traits that were more outstanding in discriminating among genotypes per each water regime (Figure 1). First and second principal components explained 80.4% of the total variation. Performance of genotypes with respect to the measured nine traits under water-stressed conditions (prefixed with SG) were located on the negative side of the biplot whilst majority of genotypes under optimum conditions (prefixed with WG) were on the positive side of the biplot. Traits ASI, RL, SEN and PC were more discriminating among genotypes under drought conditions whilst GY, Gs, CC, DA and EPP were more discriminating among genotypes under non-stressed conditions. Vector length is relative to the discriminating power of the respective trait. Hence, the trait discriminating power within the non-stressed condition in descending order was EPP, DA, GS, CC and GY. On the other hand, PC, RL, SEN and ASI, was the respective order of discriminating under drought stress conditions. Genotypes positioned at or close to a vector of a traits are more associated with that trait. Also genotypes that are located at the tip end of trait vector excelled in the respective trait. For instance, genotype 39 (TZM1224) excelled in proline content (PC) under drought conditions whilst genotypes 2 (CLHP00306) and 25 (CLHP0113) were further on the GY vector under well-watered conditions. Very few genotypes were associated and excelled in EPP and DA under non-stress conditions.

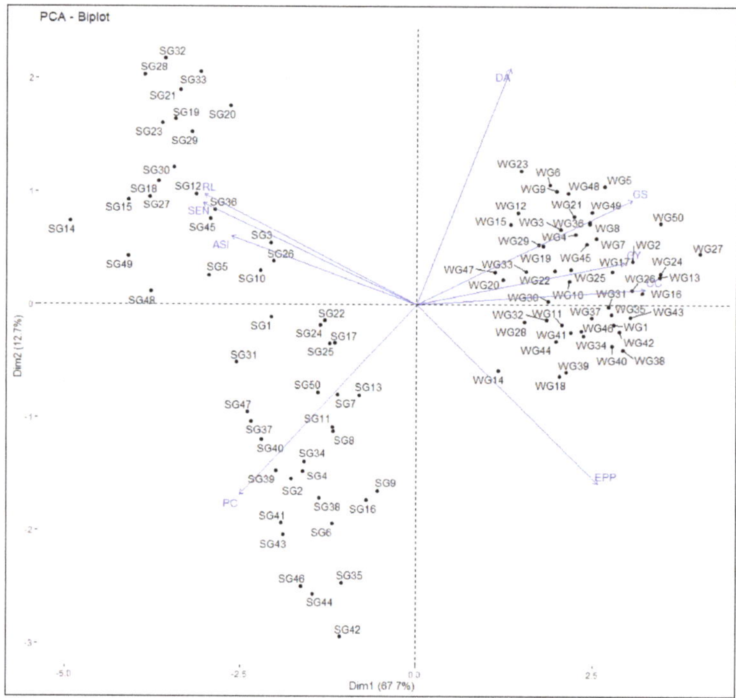

Figure 1. A combined principal component biplot showing genotypes clustering under stress conditions and well-watered conditions. ASI—anthesis-silking interval, CC—chlorophyll content, DA—days to anthesis, EPP—ears per plant, GY—grain yield, Gs—stomatal conductance, RL—leaf rolling, SEN—leaf senescence, PC—proline content, SG—genotypes under water stress conditions, WG -genotypes grown under optimum conditions.

3.4. Phenotypic Correlation Analysis

Phenotypic correlation coefficients (*r*) among measured morphophysiological and biochemical traits under non-stressed conditions (upper diagonal) and drought conditions (lower diagonal) (Table 4). Under drought stress, GY had significant ($p < 0.001$) and positive correlations that were high with EPP, and low with RL, CC and PC. It also had a significant and negative correlation that was moderate with ASI and Gs, and low with DA, and SEN. Number of ears per plant had significant ($p < 0.001$) and negative correlations which were high with ASI, moderate with DA and Gs, and low with SEN, CC and PC. Stomatal conductance had significant ($p < 0.001$) and moderate positive correlations with RL and SEN. In addition, PC had a significant ($p < 0.001$,) low positive correlation with RL and SEN. It also had a significant ($p < 0.05$), low negative correlation with CC. Under non-stressed conditions, GY had significant ($p < 0.001$) and positive correlation, which was moderate with CC and EPP, low with Gs and DA, and negligible with RL. It also had significant and negative correlation, which was moderate with ASI, low with SEN and negligible with PC. Number of ears per plant had a significant ($p < 0.001$) negative correlation which was moderate with ASI and low with SEN. On the other hand, EPP had a significant ($p < 0.001$) positive correlation, which was moderate with Gs and low with CC. Proline content had negligible correlation with SEN and CC under non-stressed conditions.

Table 4. Phenotypic correlation coefficients describing association of traits under S (lower diagonal) and W (upper diagonal) conditions.

		Non-Stressed Condition (W)								
		GY	**ASI**	**DA**	**EPP**	**RL**	**Gs**	**SEN**	**CC**	**PC**
	GY	1	−0.593 ***	0.462 ***	0.502 ***	0.187 ***	0.496 ***	−0.205 **	0.612 ***	0.109 *
	ASI	−0.694 ***	1	0.446 ***	−0.520 ***	−0.013	−0.307	0.166	−0.004	−0.067
	DA	−0.444 **	0.510 **	1	0.363 ***	−0.308	0.046	0.106	0.267	0.068
Water Stressed (S)	EPP	0.774 ***	−0.711 ***	−0.563 ***	1	0.096 ***	0.626 ***	−0.338 ***	0.324 ***	0.088
	RL	0.446 *	0.458	0.207	0.513 **	1	0.152	0.128	−0.061	0.262
	Gs	−0.566 *	0.333 *	0.232	−0.573 ***	0.538***	1	−0.120 ***	0.246 *	0.492
	SEN	−0.423 ***	0.365	0.041	−0.486 **	0.334 **	0.584 **	1	−0.250 **	0.038 *
	CC	0.406 ***	−0.209	−0.24	0.455 ***	−0.466 *	−0.390 *	−0.396 ***	1	−0.188 *
	PC	0.317 ***	0.374	−0.076	0.415 ***	0.472 ***	−0.235	0.489 ***	−0.356 *	1

GY—grain yield, ASI—anthesis-silking interval, CC—chlorophyll content, DA—days to anthesis, EPP—ears per plant, Gs—stomatal conductance, RL—leaf rolling, PC -proline content, SEN—leaf senescence, *,**,*** indicate level of significance of the correlation at $p < 0.05$, $p < 0.01$ and $p < 0.001$, respectively.

Across environments phenotypic correlation coefficients for grain yield only were computed in order to compare greenhouse and field environments (supplementary Table S4). Environments were further subdivided into greenhouse non-stressed, greenhouse-stressed, field-non-stressed and field-stressed. Correlations among all the four environmental subdivisions ranged from moderate to high ($r = 0.456$ to $r = 0.751$). Significant ($p < 0.001$) and high correlation ($r = 0.751$) was observed between greenhouse and field-non-stressed whilst greenhouse and field-stressed had moderate correlations ($r = 0.643$). Significant ($p < 0.001$) moderate correlations were also observed between greenhouse non-stressed and stressed ($r = 0.553$), and field-stressed and non-stressed ($r = 0.585$). Greenhouse non-stressed and field-stressed had a significant ($p < 0.05$) moderate correlation ($r = 0.502$) whilst greenhouse stressed and field-non-stressed had a significant ($p < 0.05$) moderate correlation ($r = 0.456$).

3.5. Relative Efficiency of Indirect Selection

Relative efficiency of indirect selection through secondary traits for grain yield ranged from 0.142 for DA to 1.370 for Gs under drought conditions and from 0.012 for PC under to 1.235 for Gs under non-drought conditions (Table 5). Traits that exhibited relative selection efficiency of greater than unity are EPP, RL, Gs, SEN and PC under drought conditions, and EPP, Gs and CC under optimum conditions. All the genetic correlations coefficients used in calculating the relative efficiency of indirect selection are given in supplementary Table S5.

Table 5. Genetic correlations and the relative efficiency of indirect selection through secondary traits for grain yield improvement under drought stress and optimum conditions.

Secondary Traits	Genetic Correlation (rg) with GY		Relative Efficiency of Indirect Selection	
	Stressed	Non-Stressed	Stressed	Non-Stressed
Anthesis silking interval	0.352 *	0.19 ns	0.510	0.143
Days to anthesis	0.146 ***	0.163 ns	0.142	0.190
Ears per plant	0.909 ***	0.912 ***	1.118	1.030
Leaf Rolling	0.934 ***	0.334 ns	1.321	0.163
Stomatal conductance	0.715 ***	0.974 ***	1.126	1.235
Leaf senescence	0.918 ***	0.842 ***	1.370	0.535
Chlorophyll content	0.738 ***	0.996 ***	0.702	1.142
Proline content	0.829 ***	0.018 ns	1.153	0.012

$^{*} p < 0.05$, $^{***} p < 0.001$, ^{ns}, not significant.

4. Discussion

Comprehensive screening of germplasm forms the basis of selecting the right materials, which in turn increases the chances of achieving genetic gain in plant breeding. In this study, the ANOVA showed highly significant genotypic differences with respect to all the studied traits. This indicates the feasibility of genetic improvement for drought tolerance and β-carotene content given that genetic variation is a prerequisite for genetic advance [30]. In a related study, [31] also reported a significant genetic variation among inbred lines obtained from CIMMYT and IITA stress-tolerant breeding programs. The significance of the water regime and its interaction with the genotype for all the traits indicate that the imposed drought stress was effective in discriminating among the genotypes. Furthermore, the highly significant genotype by environment interaction effects observed for most of the traits showed that there were performance differences by inbred lines across the environments. This confirms the complexity of drought tolerance breeding as this makes performance ranking difficult [12]. However, the significant high and moderate positive correlations observed between greenhouse and field environments after subdividing them along water regimes suggest that the two screening environments had almost similar discriminating abilities. This validates the ranking of genotypes within water regimes across the four environments.

Trait performance inconsistence was observed between the two water regimes. That is, combined principal component biplot revealed that RL, SEN, PC and ASI were more discriminating under drought conditions while EPP, GY, CC, Gs and DA were more discriminating under non-stressed conditions. Similarly, variables Gs, GY, EPP, DA and CC had higher heritability and genotypic variance estimates under optimum conditions than under drought conditions whilst the opposite pattern was observed for ASI, SEN, PC and RL (Table 2). In this regard, our findings concur with [32], who reported a decline in the genotypic variances for GY and EPP whilst that of ASI increased under drought conditions. This inconsistency justifies the importance of using a combination of multi-traits when screening germplasm for contrasting growing conditions [33]. This also suggests the need of separating the breeding programs to target different growing conditions in which different secondary traits would be utilized for selection. South Africa, like most SSA countries, has mixed growing conditions; therefore, separating the breeding target environments would be ideal.

The low heritability and genotypic variance estimates exhibited by GY under drought stress agree with findings by some of the previous researchers [34,35]. The observed yield loss (51.2%) due to drought stress was lower than the 81% reported by [36] who attributed greater variation under drought stress to kernel number. In this study, greater variation can be attributed to EPP, RL, Gs, SEN and PC under drought conditions, and EPP, Gs and CC under non-stressed conditions. Stomatal conductance was one of the traits that consistently accounted for greater variation under both conditions. This coupled with having a relative efficiency greater than unity suggests that Gs can be effectively used as selection proxy for GY [13]. This concurs with [37], who reported Gs as the major physiological trait that can effectively discriminate among genotypes between drought tolerant and susceptible plant genotypes. However, its low correlation with GY under non-stressed conditions reduces its effective utilization as an indirect selection criterion for production under optimum conditions.

The EPP and ASI are by far the most applied traits in maize drought tolerance studies [21,33,38]. In the current study, EPP was one of the largest contributors to the total genetic variation as demonstrated by high heritability and genotypic variance estimates under both drought stress and optimum conditions. Furthermore, by having higher discriminatory power as indicated by a longer PCA vector and high correlation with GY under both conditions confirms that EPP is an important trait in maize drought tolerance studies as reported by other researchers [9,36]. This, in addition to having a relative efficiency value greater than a unity justifies the use of EPP in indirect selection for GY. Despite having a moderate correlation with GY and moderate to high heritability estimates under both conditions, the effective use of ASI in indirect selection for GY would be limited by having a relative efficiency value of less than a unity. This is contrary to the findings by [32,39], who reported a relative efficiency of indirect selection for ASI, which was greater than a unity. However, it should be noted that although ASI did not demonstrate to be an ideal trait for indirect selection for GY, it could still be utilized in drought tolerant maize breeding by virtue of having moderate correlation with GY and moderate to high heritability estimates.

The observed significant correlations between GY and photosynthesis related traits such as SEN, RL, CC and Gs confirms the importance of photosynthesis for maize yield [40,41]. Leaf senescence and RL had relative efficiency values of greater than unity under drought and therefore can be used in indirect selection for GY. This concurs with [42] who found delayed leaf senescence to be useful in indirectly selecting for GY in maize under drought conditions. Moderate correlation between PC and GY coupled with high heritability estimate and relative efficiency of greater than unity under drought conditions, infers that genotypes that exhibited high PC under drought stress can be selected as drought tolerant. This supports the claim that under drought conditions proline is released to effect plant cell osmotic adjustments which helps to conserve cell turgor [17,43]. In a related study, [44] reported an increase in PC of 47% and 114% after exposing wheat genotypes to reproductive and grain filling drought stresses, respectively. However, lack of high correlation between PC and GY under non-stressed conditions hinders its effective use as a selection proxy for GY for optimum production.

Although this study did not investigate the effects of water regime, environment and the interaction of genotype and environment on BCC, there is enough evidence in literature that provitamin A content cannot be influenced by genotype and environment interaction but can be affected by the environment [45–47]. This makes selection for provitamin A an easy task. Furthermore, in their studies, [47] and [48] reported no significant correlation between GY and BCC, indicating that the two key traits can be improved simultaneously. The maximum BCC of 4.22 μg g^{-1} observed in this study was lower than 13.22 μg g^{-1} reported by [2]. This difference could be attributed to the natural superiority of temperate maize, which was studied by the later, over the its tropical counterpart [49].

5. Conclusions

This study showed that BCC, and the morphophysiological and biochemical traits applied in the screening of provitamin-A inbred lines for drought tolerance were effective in discriminating among the evaluated genotypes. There was considerably high genetic variation among the provitamin A genotypes under study that can be utilized when breeding for drought tolerance. The study also demonstrated that EPP, Gs, RL, SEN and PC can be effectively used in indirect selection of GY under drought-stressed conditions whilst EPP, Gs, and CC were ideal traits for GY indirect selection under non-stressed conditions. By having more secondary traits with a relative efficiency greater than unity under drought stress than under non-stressed, the study confirmed that indirect selection would be more useful than direct selection under drought conditions where GY exhibited low heritability and genotypic variance estimates. Consequently, based on BCC >1.5 μg g^{-1}, which was greater than that of ordinary maize and the ranking for STI values, 30 inbred lines were selected to be utilized in developing drought tolerant provitamin A maize varieties.

Supplementary Materials: The following are available online at http://www.mdpi.com/2073-4395/9/11/692/s1, Table S1: List of maize inbred lines used in the study, Table S2: Monthly weather data during the greenhouse trials (Env & Env2) at UKZN and field trials (Env3 and Env4) at Ukulinga and Makhatini research stations in South Africa, Table S3: Mean performance of all the genotypes evaluated under drought (S) and non-stressed (W) conditions across four environments ranked by the selection index (SI) values, Table S4: Pearson's correlation coefficients (r) describing association of four environments grouped according to water regimes, Table S5: Genetic correlation describing association of traits under S (lower diagonal) and W (upper diagonal) conditions.

Author Contributions: Conceptualization, A.K. and J.S.; Data curation, A.K. and C.M.; Formal analysis, A.K., C.M., and S.T.; Funding acquisition, J.S.; Investigation, A.K.; Methodology, A.K., J.S., C.M. and S.T.; Project administration, J.S.; Supervision, J.S. and R.Z.; Validation, A.K., J.S., R.Z., C.M. and S.T.; Writing—original draft, A.K. Writing—review & editing, A.K., J.S., R.Z., C.M. and S.T.

Funding: This research was funded by Alliance for a Green Revolution in Africa (AGRA) grant number 2014 PASS 013, The World Academy of Science (TWAS) and the National Research Foundation (NRF) of South Africa grant number SFH150914142525.

Acknowledgments: The authors would like to thank Andile Mshengu for administrative duties. CIMMYT and IITA are also acknowledged for providing the germplasm.

Conflicts of Interest: The authors declare no conflict of interest.

References

1. Thurnham, D. Vitamin A and carotenoids. In *Essentials of Human Nutrition*; Mann, J., Truswell, A.S., Eds.; Oxfrd University Press: Oxford, UK, 2012; pp. 191–216.
2. Harjes, C.E.; Rocheford, T.R.; Bai, L.; Brutnell, T.P.; Kandianis, C.B.; Sowinski, S.G.; Stapleton, A.E.; Vallabhaneni, R.; Williams, M.; Wurtzel, E.T.; et al. Natural Genetic Variation in Lycopene Epsilon Cyclase Tapped for Maize Biofortification. *Science* **2008**, *319*, 330–333. [CrossRef]
3. Andersson, M.; Saltzman, A.; Singh Virk, P.; Pfeiffer, W. Progress update: Crop development of biofortified staple food crops under HarvestPlus. *Afr. J. Food Agric. Nutr. Dev.* **2017**, *17*, 11905–11935. [CrossRef]
4. Bouis, H.E.; Saltzman, A. Improving nutrition through biofortification: A review of evidence from HarvestPlus, 2003 through 2016. *Glob. Food Secur.* **2017**, *12*, 49–58. [CrossRef]
5. Magorokosho, C.; Tongooona, P. Selection for Drought Tolerance in Two Tropical Maize Populations. *Afr. Crop Sci. J.* **2003**, *11*, 151–161. [CrossRef]

6. DAFF. *Trends in the Agricultural Sector*; Department of Agriculture, Forestry and Fisheries: Pretoria, South Africa, 2013.

7. FAO. *Southern Africa el Niño Response Plan (2016/2017)*; Food and Agriculture Organization of the United Nations: Rome, Italy, 2017.

8. Lu, Y.; Hao, Z.; Xie, C.; Crossa, J.; Araus, J.-L.; Gao, S.; Vivek, B.S.; Magorokosho, C.; Mugo, S.; Makumbi, D.; et al. Large-scale screening for maize drought resistance using multiple selection criteria evaluated under water-stressed and well-watered environments. *Field Crop. Res.* **2011**, *124*, 37–45. [CrossRef]

9. Almeida, G.D.; Nair, S.; Borém, A.; Cairns, J.; Trachsel, S.; Ribaut, J.-M.; Bänziger, M.; Prasanna, B.M.; Crossa, J.; Babu, R. Molecular mapping across three populations reveals a QTL hotspot region on chromosome 3 for secondary traits associated with drought tolerance in tropical maize. *Mol. Breed.* **2014**, *34*, 701–715. [CrossRef] [PubMed]

10. Blum, A. Phenotyping and Selection. In *Plant Breeding for Water-Limited Environments*; Springer: New York, NY, USA, 2011; pp. 153–216.

11. Bänziger, M.; Edmeades, G.; Lafitte, H. Selection for drought tolerance increases maize yields across a range of nitrogen levels. *Crop Sci.* **1999**, *39*, 1035–1040. [CrossRef]

12. Betrán, F.J.; Beck, D.; Bänziger, M.; Edmeades, G.O. Secondary traits in parental inbreds and hybrids under stress and non-stress environments in tropical maize. *Field Crop. Res.* **2003**, *83*, 51–65. [CrossRef]

13. Falconer, D.S. *Introduction to Quantitative Genetics, 2nd edn London*; Longman: London, UK, 1981.

14. Edmeades, G. *Progress in Achieving and Delivering Drought Tolerance in Maize—An Update*; ISAAA: Ithaca, NY, USA, 2013; p. 130.

15. Yang, S.; Vanderbeld, B.; Wan, J.; Huang, Y. Narrowing Down the Targets: Towards Successful Genetic Engineering of Drought-Tolerant Crops. *Mol. Plant* **2010**, *3*, 469–490. [CrossRef] [PubMed]

16. Helander, J.D.M.; Vaidya, A.S.; Cutler, S.R. Chemical manipulation of plant water use. *Bioorganic Med. Chem.* **2016**, *24*, 493–500. [CrossRef]

17. Shao, H.-B.; Chen, X.-Y.; Chu, L.-Y.; Zhao, X.-N.; Wu, G.; Yuan, Y.-B.; Zhao, C.-X.; Hu, Z.-M. Investigation on the relationship of proline with wheat anti-drought under soil water deficits. *Colloids Surf. B Biointerfaces* **2006**, *53*, 113–119. [CrossRef]

18. Vendruscolo, E.C.G.; Schuster, I.; Pileggi, M.; Scapim, C.A.; Molinari, H.B.C.; Marur, C.J.; Vieira, L.G.E. Stress-induced synthesis of proline confers tolerance to water deficit in transgenic wheat. *J. Plant Physiol.* **2007**, *164*, 1367–1376. [CrossRef] [PubMed]

19. Zegaoui, Z.; Planchais, S.; Cabassa, C.; Djebbar, R.; Belbachir, O.A.; Carol, P. Variation in relative water content, proline accumulation and stress gene expression in two cowpea landraces under drought. *J. Plant Physiol.* **2017**, *218*, 26–34. [CrossRef] [PubMed]

20. Zhang, M.; Wang, L.-F.; Zhang, K.; Liu, F.-Z.; Wan, Y.-S. Drought-induced responses of organic osmolytes and proline metabolism during pre-flowering stage in leaves of peanut (*Arachis hypogaea* L.). *J. Integr. Agric.* **2017**, *16*, 2197–2205. [CrossRef]

21. Bänziger, M.; Edmeades, G.O.; Beck, D.; Bellon, M. *Breeding for Drought and Nitrogen Stress Tolerance in Maize: From Theory to Practice*; CIMMYT: Mexico City, Mexico, 2000.

22. Menkir, A.; Liu, W.; White, W.S.; Maziya-Dixon, B.; Rocheford, T. Carotenoid diversity in tropical-adapted yellow maize inbred lines. *Food Chem.* **2008**, *109*, 521–529. [CrossRef]

23. Bates, L.S.; Waldren, R.P.; Teare, I.D. Rapid determination of free proline for water-stress studies. *Plant Soil* **1973**, *39*, 205–207. [CrossRef]

24. R Development Core Team. *R: A Language and Environment for Statistical Computing*; R Foundation for Statistical Computing: Vienna, Austria, 2018.

25. Fernandez, G.C. Effective Selection Criteria for Assessing Plant Stress Tolerance. In Proceedings of the International Symposium on Adaptation of Vegetables and Other Food Crops in Temperature and Water Stress, Shanhua, Taiwan, 13–16 August 1992; pp. 257–270.

26. Allard, R. *Principles of Plant Breeding*; Wiley: New York, NY, USA, 1999.

27. Robinson, H.F.; Comstock, R.E.; Harvey, P.H. Estimates of heritability and the degree of dominance in corn. *Agron. J.* **1949**, *41*, 353–359. [CrossRef]

28. Alvarado, G.; López, M.; Vargas, M.; Pacheco, Á; Rodríguez, F.; Burgueño, J.; Crossa, J. META-R (Multi Environment Trail Analysis with R for Windows) Version 6.04 CIMMYT Research Data & Software Repository Network: 2019. Available online: https://data.cimmyt.org/dataset.xhtml?persistentId=hdl:11529/10201 (accessed on 20 August 2019).

29. Mukaka, M.M. Statistics Corner: A guide to appropriate use of Correlation coefficient in medical research. *Malawi Med. J.* **2012**, *24*, 69–71.

30. Pixley, K.; Rojas, N.P.; Babu, R.; Mutale, R.; Surles, R.; Simpungwe, E. Biofortification of maize with provitamin A carotenoids. In *Carotenoids in Human Nutrition and Health*; Tanumihardjo, S.A., Ed.; Springer Science and Business Media: New York, NY, USA, 2013.

31. Wen, W.; Araus, J.L.; Shah, T.; Cairns, J.; Mahuku, G.; Bänziger, M.; Torres, J.L.; Sánchez, C.; Yan, J. Molecular Characterization of a Diverse Maize Inbred Line Collection and its Potential Utilization for Stress Tolerance Improvement. *Crop Sci.* **2011**, *51*, 2569. [CrossRef]

32. Bolaños, J.; Edmeades, G.O. The importance of the anthesis-silking interval in breeding for drought tolerance in tropical maize. *Field Crop. Res.* **1996**, *48*, 65–80. [CrossRef]

33. Araus, J.L.; Serret, M.D.; Edmeades, G. Phenotyping maize for adaptation to drought. *Front. Physiol.* **2012**, 3. [CrossRef] [PubMed]

34. Bänziger, M.; Cooper, M. Breeding for low input conditions and consequences for participatory plant breeding examples from tropical maize and wheat. *Euphytica* **2001**, *122*, 503–519. [CrossRef]

35. Betrán, F.J.; Beck, D.; Bänziger, M.; Edmeades, G.O. Genetic Analysis of Inbred and Hybrid Grain Yield under Stress and Nonstress Environments in Tropical Maize. *Crop Sci.* **2003**, *43*, 807–817. [CrossRef]

36. Cairns, J.E.; Sanchez, C.; Vargas, M.; Ordoñez, R.; Araus, J.L. Dissecting Maize Productivity: Ideotypes Associated with Grain Yield under Drought Stress and Well-watered Conditions. *J. Integr. Plant Biol.* **2012**, *54*, 1007–1020. [CrossRef] [PubMed]

37. Grzesiak, M.T.; Grzesiak, S.; Skoczowski, A. Changes of leaf water potential and gas exchange during and after drought in triticale and maize genotypes differing in drought tolerance. *Photosynthetica* **2006**, *44*, 561–568. [CrossRef]

38. Campos, H.; Cooper, M.; Edmeades, G.O.; Löffler, C.; Schussler, J.R.; Ibañez, M. Changes in drought tolerance in maize associated with fifty years of breeding for yield in the U.S. corn belt. *Maydica* **2006**, *51*, 369–381.

39. Ziyomo, C.; Bernardo, R. Drought tolerance in maize: Indirect selection through secondary traits versus genomewide selection. *Crop Sci.* **2013**, *53*, 1269–1275. [CrossRef]

40. Cabrera-Bosquet, L.; Molero, G.; Nogués, S.; Araus, J.L. Water and nitrogen conditions affect the relationships of Δ13C and Δ18O to gas exchange and growth in durum wheat. *J. Exp. Bot.* **2009**, *60*, 1633–1644. [CrossRef]

41. Vadez, V. Root hydraulics: The forgotten side of roots in drought adaptation. *Field Crop. Res.* **2014**, *165*, 15–24. [CrossRef]

42. Zheng, H.J.; Wu, A.Z.; Zheng, C.C.; Wang, Y.F.; Cai, R.; Shen, X.F.; Xu, R.R.; Liu, P.; Kong, L.J.; Dong, S.T. QTL mapping of maize (*Zea mays*) stay-green traits and their relationship to yield. *Plant Breed.* **2009**, *128*, 54–62. [CrossRef]

43. Marcińska, I.; Czyczyło-Mysza, I.; Skrzypek, E.; Filek, M.; Grzesiak, S.; Grzesiak, M.T.; Janowiak, F.; Hura, T.; Dziurka, M.; Dziurka, K.; et al. Impact of osmotic stress on physiological and biochemical characteristics in drought-susceptible and drought-resistant wheat genotypes. *Acta Physiol. Plant.* **2013**, *35*, 451–461. [CrossRef]

44. Moayedi, A.A.; Nasrulhaq-Boyce, A.; Tavakoli, H. Application of physiological and biochemical indices for screening and assessment of drought tolerance in durum wheat genotypes. *Aust. J. Crop Sci.* **2011**, *5*, 1014–1018.

45. Menkir, A.; Maziya-Dixon, B. Influence of genotype and environment on beta-carotene content of tropical yellow-endosperm maize genotypes [*Zea mays* L.; Nigeria]. *Maydica* **2004**, *49*, 313–318.

46. Pfeiffer, W.H.; McClafferty, B. HarvestPlus: Breeding Crops for Better Nutrition. *Crop Sci.* **2007**, *47*, S-88–S-105. [CrossRef]

47. Suwarno, W.B.; Pixley, K.V.; Palacios-Rojas, N.; Kaeppler, S.M.; Babu, R. Formation of Heterotic Groups and Understanding Genetic Effects in a Provitamin A Biofortified Maize Breeding Program. *Crop Sci.* **2014**, *54*, 14. [CrossRef]

Agronomy **2019**, *9*, 692

48. Egesel, C.O.; Wong, J.C.; Lambert, R.J.; Rocheford, T.R. Combining Ability of Maize Inbreds for Carotenoids and Tocopherols. *Crop Sci.* **2003**, *43*, 818–823. [CrossRef]

49. Menkir, A.; Rocheford, O.; Maziya-Dixon, B.; Tanumihardjo, S. Exploiting natural variation in exotic germplasm for increasing provitamin-A carotenoids in tropical maize. *Euphytica* **2015**, *205*, 203–217. [CrossRef]

Article

Sodium Azide Priming Enhances Waterlogging Stress Tolerance in Okra (*Abelmoschus esculentus* L.)

Emuejevoke D. Vwioko [1], Mohamed A. El-Esawi [2,3,*], Marcus E. Imoni [1], Abdullah A. Al-Ghamdi [4], Hayssam M. Ali [4], Mostafa M. El-Sheekh [2], Emad A. Abdeldaym [5] and Monerah A. Al-Dosary [4]

[1]   Department of Plant Biotechnology, Faculty of Life Sciences, University of Benin, P.O. Box 1154, Benin City, Nigeria; emuejevoke.vwioko@yahoo.com (E.D.V.); marcus.imoni@yahoo.com (M.E.I.)
[2]   Botany Department, Faculty of Science, Tanta University, Tanta 31527, Egypt; mostafaelsheikh@science.tanta.edu.eg
[3]   Sainsbury Laboratory, University of Cambridge, Cambridge CB2 1LR, UK
[4]   Botany and Microbiology Department, College of Science, King Saud University, P.O. Box 2455, Riyadh 11451, Saudi Arabia; abdaalghamdi@ksu.edu.sa (A.A.A.-G.); hayhassan@ksu.edu.sa (H.M.A.); almonerah@ksu.edu.sa (M.A.A.-D.)
[5]   Vegetable Crops Department, Faculty of Agriculture, Cairo University, Giza P.O. Box 12613, Egypt; emad.abdeldaym@agr.cu.edu.eg
*   Correspondence: mohamed.elesawi@science.tanta.edu.eg; Tel.: +20-102-4824-643

Received: 6 September 2019; Accepted: 24 October 2019; Published: 25 October 2019

Abstract: Waterlogging stress adversely affects crop growth and yield worldwide. Effect of sodium azide priming on waterlogging stress tolerance of okra plants was investigated. The study was conducted as a field experiment using two weeks old plants grown from 0%, 0.02%, and 0.05% sodium azide (NaN_3)-treated seeds. The waterlogging conditions applied were categorized into control, one week, and two weeks. Different growth and reproductive parameters were investigated. Activity and expression of antioxidant enzymes, root anatomy, and soil chemical analysis were also studied. Results showed that sodium azide priming inhibited germination. The germination percentages recorded were 92.50, 85.00, and 65.00 for 0%, 0.02%, and 0.05% NaN_3-treated seeds, respectively, nine days after planting. Waterlogging conditions depressed plant height ten weeks after planting. Under waterlogging conditions, NaN_3 promoted plant height and number of leaves formed. NaN_3 also supported the survival of plants and formation of adventitious roots under waterlogging conditions. Waterlogging conditions negatively affected the redox potential, organic C, N, and P concentrations in the soil but enhanced Soil pH, Fe, Mn, Zn, and SO_4. Under waterlogging conditions, NaN_3 increased the average number of flower buds, flowers, and fruits produced in comparison to control. Moreover, NaN_3 highly stimulated the development of aerenchyma which in turn enhanced the survival of okra plants under waterlogging conditions. NaN_3 priming also enhanced the activities and gene expression level of antioxidant enzymes (ascorbate peroxidase, APX; catalase, CAT) under waterlogging conditions. In conclusion, this study demonstrated that NaN_3 priming could improve waterlogging stress tolerance in okra.

Keywords: sodium azide; okra; waterlogging stress; antioxidants; gene expression

---

1. Introduction

Okra (*Abelmoschus esculentus* L.) is one of the economically important vegetable crops grown in tropical and sub-tropical regions of the world [1]. Okra originated in Ethiopia and was then reproduced in the Mediterranean area, North Africa, and India [1]. Environmental stresses negatively affect the growth, yield, and biological activities of plants worldwide [2–6]. In particular, waterlogging conditions influence the growth and yield of okra plants through causing hypoxic or anoxic conditions, which in turn affect various physiological processes in roots, including carbohydrate metabolism, gas exchanges, and water relations [7–9]. The oxygen-deficient soil environments may lead to changes in the composition and decomposition activities of microbes. Waterlogging conditions also affect soil factors such as EC, pH, soil structure, hydraulic conductance, porosity, and organics [10,11]. Plants could adapt to waterlogging conditions via activating their self-defense mechanisms and developing adventitious roots and hypertrophied stem bases with lenticels and aerenchyma cells [7,12]. Such aerenchyma cells could enhance organ porosity and root aeration [13,14]. These morphological features help plants to manage the low oxygen tension within the tissues, prevent anoxia, and maintain root functions and plant survival.

Applications of chemicals to plants, either as foliar or seed treatments, may induce their physiological mechanisms, leading to plant growth stimulation and stress tolerance [7,15,16]. For instance, seed pretreatment with salicylic acid enhances plant growth, antioxidant activities, and tolerance to harsh environmental factors such as heavy metal, herbicides, low temperature and salt stress [17,18]. Ethylene is also described as a signaling molecule in plants and has been projected as capable of inducing survival traits and tolerance under waterlogging conditions via up-regulating the activity of antioxidant enzymes and genes linked to aerenchyma formation, leaf senescence, adventitious roots, and epinasty [7,19–21]. However, ethylene application as a proactive measure for ameliorating envisaged waterlogging condition on a wide scale may not be appreciated. Hence, seed priming techniques may be easier to enhance growth and yield. Sodium azide (NaN_3) has been successfully used for creating genetic variability and enhancing agronomic traits of crop plants. It affects crops based on the concentration applied. Gnanamurthy et al. [22] and Shagufta et al. [23] reported that NaN_3 priming delayed and inhibited the germination of maize and fenugreek, respectively. However, Vwioko and Onobun [24] reported that NaN_3 enhanced the germination percentage and height of okra plants. Al-Qurainy [25] and Zuzana et al. [26] also stated that NaN_3 stimulated the plant height of *Eruca sativa* and *Diospyros lotus*, respectively. On the other hand, Adamu and Aliyu [27] and Gnanamurthy et al. [22] revealed that NaN_3 priming inhibited plant height. NaN_3 priming also regulates various physiological and molecular mechanisms in plants and modulates the activities of catalase, peroxidase, and cytochrome oxidase [28]. Molecular changes induced by NaN_3 treatments produce mutations by base substitution, leading to changes in amino acid sequences. NaN_3 is reckoned to be an efficient reagent that induces a broad and high variation of morphological and yield parameters in cultivated species. However, it is not popularly used to initiate plants tolerance to environmental factors. Environmental stresses such as salinity and water stress [29,30] increase production of free radical in plants, and resistance to the unfavorable conditions often involves stimulation of the antioxidant response. Haq et al. [31], El Kaaby et al. [32], and Kuasha et al. [33] carried out in vitro studies on the ability of NaN_3 to confer salt tolerance in plants. Haq et al. [31] stated that one of the three cultivars of sugarcane studied regenerated plantlets that were salt tolerant, while El Kaaby et al. [32] and Kuasha et al. [33] stated that NaN_3 depressed the responses of the explants of tomato and sugarcane to salinity stress. Salim et al. [34] also studied the effect of NaN_3 on various plant traits, including disease resistance, yield, antioxidant activities, pigmentations, and salinity and drought stress tolerance. However, the role of NaN_3 in regulating waterlogging stress responses has not been studied yet. Therefore, the main aim of the present study was to assess the ability of NaN_3 to induce waterlogging stress tolerance in okra plants.

2. Materials and Methods

2.1. Plant Material and Application of Sodium Azide Treatments

Seeds of okra variety Clemson spineless produced by Technism (Longué-Jumelles, France) were obtained and used in this study. Okra seeds were soaked in sodium azide treatments, i.e., 0%, 0.02%, and 0.05% (*w/v*), at room temperature (27 °C) for 5 h with a continuous gentle stirring. After 5 h, the seeds were removed and washed 5 times with deionized water to remove all traces of NaN_3. NaN_3 treatments were classed as mild (0.02%) and severe (0.05%).

2.2. Soil Preparation for Potted Field Experiment

Top soil (0–15 cm deep) was collected from the Demonstration Farm, Faculty of Agriculture, University of Benin, Nigeria. The soil type is categorized as ultisol. The composite soil sample was air-dried for three weeks and sieved to remove gravel and other particles. Each experimental pot was filled with 5 kg of soil. Thirty-six (36) pots were prepared to make twelve pots for each NaN_3 treatment. The undersides of the experimental pots were not perforated so that they could retain water.

2.3. Sowing of Seeds in Nursery Beds, Transplanting into Experimental Pots, and Acclimatization

Twelve soil nursery beds (measuring 2 feet by 2 feet) were prepared for the sowing seeds. The beds were allocated to the treated seeds, i.e., 0%, 0.02%, and 0.05% NaN_3. The seeds were sown at a depth of 2–3 cm. Germination records were collected every day for two weeks. After two weeks in the nursery beds, four plants were transferred into each experimental pot and taken to the open field. The plants were allowed to acclimatize for another two weeks in the field before flooding condition was introduced.

2.4. Application of Flooding or Waterlogging Conditions

When the plants were four weeks old, flooding of experimental pots with tap water was carried out. Three conditions of flooding or waterlogging were set up; no flooding (NF), one-week flooding (1 WF), and two weeks flooding (2 WF). Flooding of the pots was done up to 2 cm mark above the soil level. The water level was maintained in each pot by topping daily after inspection during the period.

2.5. Growth Parameters Measured

The field data collected were germination percentage, stem girth, plant height, number of leaves formed, survival percentage of plants, number of adventitious roots formed, number of flower buds formed, number of flowers, and number of fruits produced.

2.6. Soil Chemical Analyses

Soil chemical factors like pH, electrolyte conductivity (EC), redox potential (Eh), nitrogen, phosphorus, sulphate, organic carbon, iron, manganese, zinc, and total soluble phenolics were determined using standard methods. The soil analysis was carried out for the soil samples collected after plant harvested. pH, EC, and Eh were estimated in a soil-water slurry (ratio 1:3) [35]. Total nitrogen was estimated following Kjeldahl method [36]. Total soluble phenolic analysis was done based on the modified citrate extraction protocol followed by Folin–Ciocalteau colorimetric methodology [37]. The methodologies of Appiah and Ahenkorah [38] and Ben Mussa et al. [39] were used to determine sulphate content. Phosphorus measurement was conducted following the methodology of Bray and Kurtz [40]. Walkley–Black chromic acid wet oxidation methodology [41] was used to estimate the organic carbon. Iron content was determined following the hydroxylamine and 1,10- phenanthroline protocol [42]. Manganese was determined following the permanganate oxidation procedures [42]. The determination of zinc was carried using atomic absorption spectrophotometer (Shimadzu Europa GmbH, Duisburg, Germany).

2.7. Soil Microflora Counts

Presence of bacteria and fungi in the soil samples was investigated after plant harvest. Serial dilution processes were used in the analysis of soil microflora. Ten grams of the samples were dispensed into sterile beakers and mixed thoroughly with 90 mL sterile distilled water. Each sample was serially diluted from the stock sample and then transferred to the first tube 9 mL of sterilized water to give 10^{-1} dilution, from which further dilution up to 10^{-4} was made. The pour plate method was utilized for inoculation on a sterilized nutrient agar (NA) or potato dextrose agar (PDA), impregnated with antifungal or antibacterial agents for the growth of bacterial or fungal isolates, respectively. Nutrient agar plates were kept for 24–48 hrs at 37 °C for bacterial growth. Potato dextrose agar was incubated at room temperature (30 ± 2 °C) for 3–5 days. Total viable colonies were then counted for the microbial isolates and represented in terms of colony forming units (cfu/g). Viable counts obtained were recorded with reference to the serial dilution used [43,44].

2.8. Root Anatomy

Harvested plant roots were washed and used to make microscopic slides to examine internal tissues. Root sections were immersed in paraffin wax and left to solidify. Sections were cut and dewaxed by clamping in the microtome. Aniline blue stain was applied to the sections to show a clear contrast of air spaces (aerenchyma) formed. Excess stains were removed by ethanol before oven-drying. Following oven-drying, slides were viewed and then photographed using the microscope IRMECO model IM-660 T1 (IRMECO GmbH & Co. KG, Geesthacht, Germany) with a camera connected to PC. Observations were done under X10 objective lens.

2.9. Antioxidant Enzyme Assays

Activities of catalase (CAT) and ascorbate peroxidase (APX) were determined in the leafy tissue of the NF, 1 WF, and 2 WF plants treated with 0%, 0.02%, and 0.05% NaN_3 collected at the tenth week after planting following the method of Zhang and Kirkham [45]. In brief, 0.25g of leafy tissue was homogenized in 3 mL of solution, composed of PBS (50 mM), EDTA (0.2 mM), and 1% PVP, and centrifuged. Supernatants were assayed to detect the absorbance at 290 nm (for APX) and 240 nm (for CAT).

2.10. RNA Isolation, cDNA Synthesis, and Quantitative RT-PCR

Quantitative real-time PCR (qRT-PCR) assay was conducted to evaluate the expression level of antioxidant enzyme-encoding genes (*APX*, *CAT*) in the leafy tissue of the NF, 1 WF, and 2 WF plants treated with 0%, 0.02%, and 0.05% NaN_3 collected at the tenth week after planting. Total RNA samples were isolated from the tissue following Qiagen RNeasy Plant Mini kit. DNA removal and cDNA synthesis were performed using Qiagen RNase-Free DNase Set and Qiagen Reverse Transcription kit, respectively. Quantitative RT-PCR was performed following Qiagen QuantiTect SYBR Green PCR kit protocol. PCR conditions, housekeeping gene, and gene-specific primers were used as reported by Vwioko et al. [7]. The primer pair 5′-TGCCCTTCTATTGTGGTTCC-3′ and 5′-GATGAGCACACTTTGGAGGA-3′ was used for *CAT* amplification, whereas the primer pair 5′-ACCAATTGGCTGGTGTTGTT-3′ and 5′-TCACAAACACGTCCCTCAAA-3′ was used for *APX* amplification. The primer pair 5′-TTCCTTGATGATGCTTGCTC-3′ and 5′-TTGACAGCTCTTGGGTGAAG-3′ was used for the housekeeping gene (*UBQ1*) amplification.

2.11. Statistical Analysis

Mean and standard deviation were measured for the data obtained for the different traits measured. Two-way analysis of variance was conducted using NaN_3 treatments and flooding conditions as factors. Tukey's test was conducted to determine the significance of values. Statistical analyses were performed using SPSS ver. 19 (SPSS Inc., Chicago, IL, USA).

3. Results

3.1. Germination of NaN₃-Treated Seeds

The germination was first recorded for okra seeds given control (0%) treatments 2 days after planting (2 DAP). Germination was recorded for 0.02 and 0.05% NaN_3-treated seeds three days after planting (3 DAP). Eight days after planting (8 DAP), the highest and least percentage of germination were recorded for 0% and 0.05% NaN$_3$-treated seeds, respectively (Figure 1). Twenty-four hours delay in germination was recorded for the NaN$_3$-treated seeds.

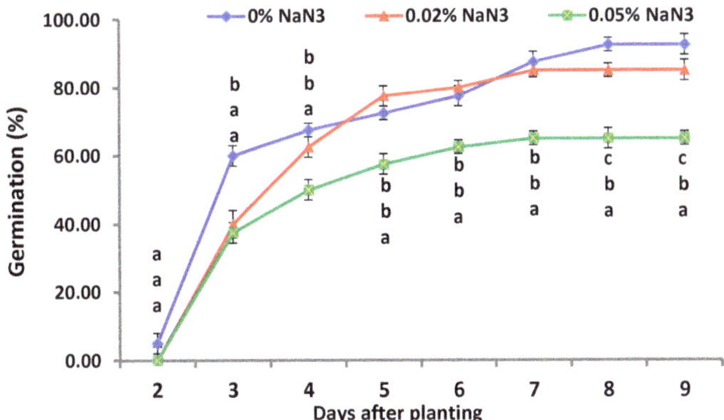

Figure 1. Percentage of germination of NaN$_3$-treated okra seeds sown in nursery. Values = mean ± SD, *n* = 4. Mean values with similar letters at the same day after planting (DAP) are not significantly different at $p \leq 0.05$.

3.2. Plant Height

Values obtained for plant height showed that non-waterlogged plants produced the highest values irrespective of the NaN$_3$ treatment given to the seeds ten weeks after planting (10 WAP). For example, mean values obtained for plant height were 31.5, 29.5, and 31.1 cm for 0%, 0.02%, and 0.05%, respectively, under non-waterlogging condition, 10 WAP (Table 1). Under one-week waterlogging condition, the values recorded for 0%, 0.02%, and 0.05% were 15.2, 21.8, and 19.4 cm, respectively, 10 WAP. Similarly, under two weeks waterlogging conditions, the values recorded for 0%, 0.02%, and 0.05% were 16.3, 22.4, and 19.9 cm, respectively, 10 WAP, indicating growth stimulations for plants grown from 0.02% and 0.05% NaN$_3$-treated seeds.

Table 1. Height (cm) of okra plants grown from NaN$_3$-treated seeds subjected to different waterlogging conditions four weeks after planting (WAP).

NaN$_3$ Treatment	Waterlogging Conditions	2 WAP	4 WAP	6 WAP	8 WAP	10 WAP
0%	Non-waterlogging	8.6 [b] ± 0.45	14.7 [e] ± 0.78	18.8 [a] ± 0.28	23.7 [a] ± 1.19	31.5 [a] ± 1.28
	One-week waterlogging	7.7 [c] ± 0.47	17.2 [ab] ± 0.68	17.9 [abc] ± 0.43	18.3 [c] ± 0.25	15.2 [c] ± 0.62
	Two weeks waterlogging	8.3 [b] ± 0.30	15.8 [cd] ± 0.24	16.8 [d] ± 0.94	18.9 [bc] ± 1.05	19.3 [b] ± 0.29
0.02%	Non-waterlogging	9.5 [a] ± 0.62	12.0 [f] ± 0.30	18.0 [abc] ± 0.60	24.5 [a] ± 1.31	29.5 [ab] ± 0.68
	One-week waterlogging	8.3 [bc] ± 0.45	17.9 [ab] ± 0.66	18.4 [ab] ± 0.42	18.8 [bc] ± 0.47	21.8 [d] ± 0.35
	Two weeks waterlogging	7.9 [bc] ± 0.09	17.7 [ab] ± 0.91	18.5 [ab] ± 1.23	19.7 [b] ± 1.30	22.4 [bc] ± 0.45
0.05%	Non-waterlogging	8.1 [bc] ± 0.78	11.9 [f] ± 0.83	17.6 [bcd] ± 0.49	23.5 [a] ± 0.88	31.1 [a] ± 0.98
	One-week waterlogging	7.9 [bc] ± 0.21	17.2 [ab] ± 0.48	17.7 [bcd] ± 0.63	18.0 [c] ± 0.72	19.4 [c] ± 1.60
	Two weeks waterlogging	7.8 [bc] ± 0.22	16.1 [bc] ± 0.28	17.1 [cd] ± 0.20	18.1 [c] ± 0.71	19.9 [c] ± 0.34

Values = mean ± S.D., *n* = 4, WAP = weeks after planting. Mean values with similar letters as superscript in one column are not significantly different at $p \leq 0.05$.

3.3. Stem Girth

The highest stem girth values were obtained for okra plants grown under non-waterlogging conditions (Table 2). Ten weeks after planting, the values recorded for the stem girth of okra plants grown under two-week waterlogging conditions were statistically significant compared to those recorded for plants grown under and non-waterlogging conditions (Table 2).

Table 2. Stem girth (cm) of okra plants grown from NaN_3-treated seeds subjected to different waterlogging conditions four weeks after planting (WAP).

NaN_3 Treatment	Waterlogging Conditions	2 WAP	4 WAP	6 WAP	8 WAP	10 WAP
0%	Non-waterlogging	0.81 [a] ± 0.02	0.95 [c] ± 0.05	1.10 [b] ± 0.08	1.25 [a] ± 0.05	1.35 [a] ± 0.05
	One-week waterlogging	0.80 [a] ± 0.01	1.02 [b] ± 0.09	1.07 [b] ± 0.09	1.15 [b] ± 0.05	1.27 [b] ± 0.05
	Two weeks waterlogging	0.85 [a] ± 0.05	1.17 [a] ± 0.05	1.27 [a] ± 0.05	1.27 [a] ± 0.05	1.27 [b] ± 0.05
0.02%	Non-waterlogging	0.76 [b] ± 0.05	1.02 [b] ± 0.09	1.10 [c] ± 0.08	1.25 [a] ± 0.05	1.37 [a] ± 0.05
	One-week waterlogging	0.89 [a] ± 0.09	1.12 [a] ± 0.05	1.17 [b] ± 0.05	1.25 [a] ± 0.05	1.37 [a] ± 0.05
	Two weeks waterlogging	0.80 [b] ± 0.08	1.15 [a] ± 0.05	1.27 [a] ± 0.05	1.27 [a] ± 0.05	1.30 [b] ± 0.00
0.05%	Non-waterlogging	0.75 [b] ± 0.05	1.05 [b] ± 0.05	1.17 [b] ± 0.05	1.30 [a] ± 0.08	1.45 [a] ± 0.05
	One-week waterlogging	0.82 [a] ± 0.07	1.07 [b] ± 0.05	1.20 [b] ± 0.08	1.20 [b] ± 0.08	1.32 [b] ± 0.09
	Two weeks waterlogging	0.85 [a] ± 0.05	1.17 [a] ± 0.09	1.27 [a] ± 0.05	1.27 [a] ± 0.05	1.32 [b] ± 0.05

Values = mean ± S.D., $n = 4$, WAP = weeks after planting. Mean values with similar letters as superscript in one column are not significantly different at $p \leq 0.05$.

3.4. Number of Leaves Formed, Number of Adventitious Roots Produced, and Percentage of Survival of Plants

The total number of leaves formed per plant recorded indicated that the plants grown under non-waterlogging condition produced the highest number of leaves 10 WAP. The combination of waterlogging conditions and NaN_3 treatments gave higher values for number of leaves formed than when the waterlogging condition is applied only (Table 3). For example, total number of leaves under non-waterlogging conditions were 16, 16.5, and 16.5 for 0%, 0.02%, and 0.05%, respectively. Whereas in one-week waterlogging conditions, values were 13, 14, and 15 for plants grown from 0%, 0.02%, and 0.05% NaN_3-treated seeds.

Table 3. Number of leaves, average number of adventitious roots produced, and survival percentage of okra plants grown from NaN_3-treated seeds under waterlogging conditions 10 WAP.

NaN_3 Treatment	Waterlogging Conditions	No. Leaves per Plant	No. Adventitious Roots per Plant	Survival Percentage
0%	Non-waterlogging	16.0 [a] ± 2.30	0 [c]	100.0 [a] ± 0.00
	One-week waterlogging	13.0 [b] ± 1.10	10.7 [b] ± 7.18	33.3 [b] ± 27.22
	Two weeks waterlogging	12.0 [b] ± 0.00	13.0 [a] ± 8.67	25.0 [b] ± 16.67
0.02%	Non- waterlogging	16.5 [a] ± 1.00	0 [c]	100.0 [a] ± 0.00
	One-week waterlogging	14.0 [b] ± 1.60	15.5 [b] ± 1.29	50.0 [b] ± 19.25
	Two weeks waterlogging	13.0 [b] ± 1.15	21.0 [a] ± 0.81	33.3 [c] ± 0.00
0.05%	Non- waterlogging	16.5 [a] ± 1.00	0 [c]	100.0 [a] ± 0.00
	One-week waterlogging	15.0 [b] ± 1.15	18.0 [b] ± 1.63	50.0 [b] ± 19.25
	Two weeks waterlogging	13.0 [c] ± 1.15	22.2 [a] ± 1.25	50.0 [b] ± 19.25

Values = mean ± S.D., $n = 4$, WAP = weeks after planting. Mean values with similar letters as superscript in one column are not significant different at $p \leq 0.05$.

Plants did not form adventitious roots under non-waterlogging conditions. However, the production of adventitious roots was observed in plants subjected to waterlogging condition. Plants subjected to two weeks of waterlogging condition initiated higher numbers of adventitious roots (Table 3). Furthermore, plants grown from 0.05% NaN_3-treated seeds produced the highest number of adventitious roots recorded. The combination of NaN_3 concentration and waterlogging condition supported the greater production of adventitious roots in okra.

Ten weeks after planting, the number of plants that survived the waterlogging conditions is shown in Table 3. Higher percentage of survival was recorded with the combination of sodium azide and waterlogging condition. For example, under two weeks waterlogging condition, the percentage of survival of okra plants were 25, 33.3, and 50 for 0%, 0.02%, and 0.05% NaN_3-treated seeds, respectively. Similarly, for one-week waterlogging condition, percentage of survival of okra plants were 33.3, 50, and 50 for 0%, 0.02%, and 0.05% NaN_3-treated seeds, respectively.

3.5. Number of Flower Buds, Flowers, and Fruits Produced

The number of flower buds, flowers, and fruits are shown in Table 4. The waterlogging condition caused a decrease in all the reproductive parameters considered. For example, the average number of flower buds recorded for plants grown from control seeds (0.00% NaN_3 treatment) were 5.5, 2.75, and 1.75 for NF, 1 WF, and 2 WF conditions, respectively. Similarly, average number of flowers recorded for the same plants were 5, 2, and 1, respectively. Moreover, the average number of fruits recorded for the same plants were 4.5, 1.25, and 0.5, respectively. The average number of flower buds, flowers, and fruits recorded for plants grown from 0.05% NaN_3-treated seeds and subjected to waterlogging conditions were higher than those recorded for non-treated plants.

Table 4. Average number of flower buds, flowers, and fruits formed per plant of okra grown from NaN3 treated seeds subjected to waterlogging conditions ten weeks after planting.

NaN_3 Treatment	Waterlogging Conditions	Number of Flower Buds	Number of Flowers	Number of Fruits
0%	Non-waterlogging	5.5 [a] ± 0.57	5.0 [a] ± 0.81	4.5 [a] ± 0.57
	One-week waterlogging	2.7 [b] ± 1.25	2.0 [b] ± 1.41	1.2 [b] ± 0.95
	Two weeks waterlogging	1.7 [b] ± 1.25	1.0 [b] ± 0.81	0.5 [b] ± 0.57
0.02%	Non-waterlogging	5.0 [a] ± 0.81	5.0 [a] ± 0.81	3.5 [a] ± 1.29
	One-week waterlogging	2.5 [b] ± 0.57	1.7 [b] ± 0.50	1.5 [b] ± 1.00
	Two weeks waterlogging	1.7 [b] ± 0.5	1.2 [b] ± 0.95	1.2 [b] ± 0.95
0.05%	Non-waterlogging	5.0 [a] ± 0.81	4.5 [a] ± 1.29	3.7 [a] ± 1.89
	One-week waterlogging	3.2 [b] ± 0.95	2.5 [b] ± 0.57	2.2 [b] ± 0.91
	Two weeks waterlogging	2.2 [b] ± 0.95	1.5 [b] ± 0.57	1.5 [b] ± 0.57

Values = mean ± S.D., $n = 4$. Values with similar letters as superscript are not significantly different.

3.6. Soil Microflora Counts

The average values obtained for bacteria and fungi counts are shown in Table S1. The bacterial counts were higher than fungal counts in all soil samples analyzed. The bacterial count values were higher in soils collected from waterlogging condition, while the fungal count values were higher in soils collected from non-waterlogging condition. Soils collected from two-week waterlogging conditions gave the least fungal counts.

3.7. Soil Chemical Analysis

There were clear differences in many of the soil chemical parameters analyzed between soil samples collected from non-waterlogging and waterlogging experimental pots (Table S2). The differences in values obtained shows a regular pattern. For example, pH values for NF were 6.0–6.1 while higher values were recorded for 1 WF and 2 WF. Redox potential (Eh) values were consistently higher for NF than 1 WF and 2 WF. Soil Eh ranged from 23.60–24.10 for NF and 7.2–7.4 for 1 WF and 2 WF. The highest values of sulphate ion (SO_4) concentrations and electrolyte conductivity (EC) readings were observed in 1 WF soil samples. Mean values for non-treated soil EC were 228, 413, and 125 µS/cm for NF, 1 WF, and 2 WF, respectively. Similarly, mean values for SO_4 concentration in non-treated soil were 0.52, 1.13, and 0.80 mg/Kg for NF, 1 WF, and 2 WF, respectively. Organic carbon, total nitrogen and available phosphorus contents in soil followed the same reduction pattern under one- and two-week waterlogging conditions. Approximately, 10-fold reductions in organic carbon and total nitrogen contents were observed under waterlogging conditions. The records for soil metallic factors like Fe, Zn, and Mn showed the same pattern where the values were higher in soil samples collected from one- and two-week waterlogging conditions. Mean values obtained for Fe were 116.3, 242.1, and 243.3 mg/kg for NF, 1 WF, and 2 WF, respectively, for soil samples collected from pots where 0% NaN_3 plants were grown. The mean values recorded for Zn in soil samples collected from pots containing 0% NaN_3 plants were 14.2, 22.7, and 35.4 mg/kg for NF, 1 WF, and 2 WF, respectively. The mean values of Mn in the same soil samples were 1.34, 9.68, and 12.9 mg/kg for NF, 1 WF, and 2 WF, respectively. The mean values of total phenol content show low variation.

3.8. Anatomy of Okra Roots

There were structural differences in the anatomy of okra root sections obtained from non-waterlogged and waterlogged plants (Figures 2–4). The presence of air channels (lacunae) was conspicuously absent in non-waterlogged root sections (Figure 2). The development of aerenchyma in the cortex and stele were very conspicuous in root sections of plants subjected to waterlogging conditions (Figures 3 and 4). Furthermore, the aerenchyma cells observed in root sections of waterlogged plants were larger in plants grown from 0.05% NaN$_3$-treated seeds than those from 0.02% NaN$_3$-treated seeds (Figures 3 and 4). This suggests an explanation for the higher percentage of survival recorded for plants grown from 0.05% NaN$_3$-treated seeds. The walls of the aerenchyma cells are thick to prevent their collapse.

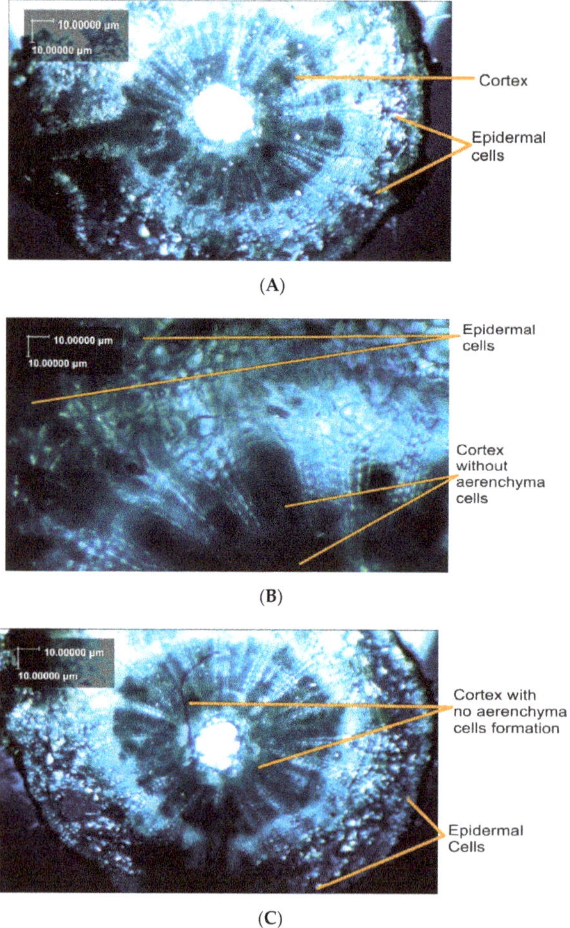

Figure 2. Root sections of okra plants grown from different concentrations of NaN$_3$-treated seeds show no aerenchyma cells formed under non-waterlogging conditions. (**A**) 0% NaN$_3$, (**B**) 0.02% NaN$_3$, (**C**) 0.05% NaN$_3$.

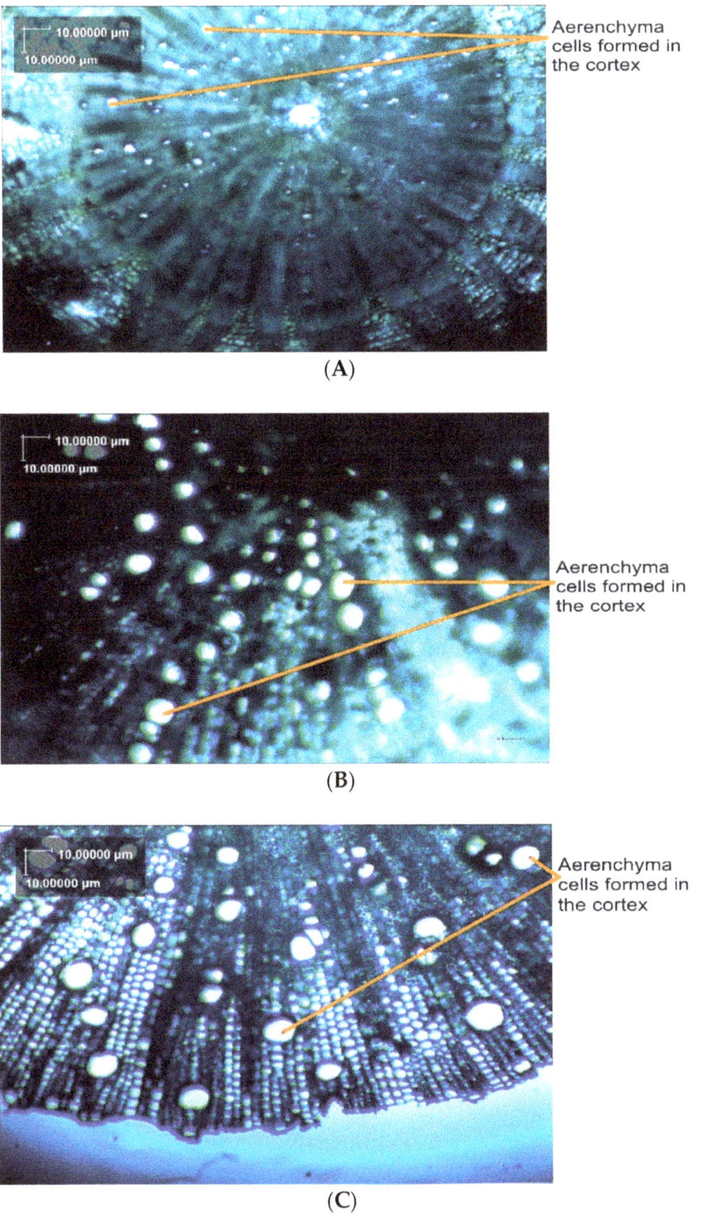

Figure 3. Root sections of okra plants grown from different concentrations of NaN_3-treated seeds show aerenchyma cells formed under one-week waterlogging conditions. (**A**) 0% NaN_3, (**B**) 0.02% NaN_3, (**C**) 0.05% NaN_3.

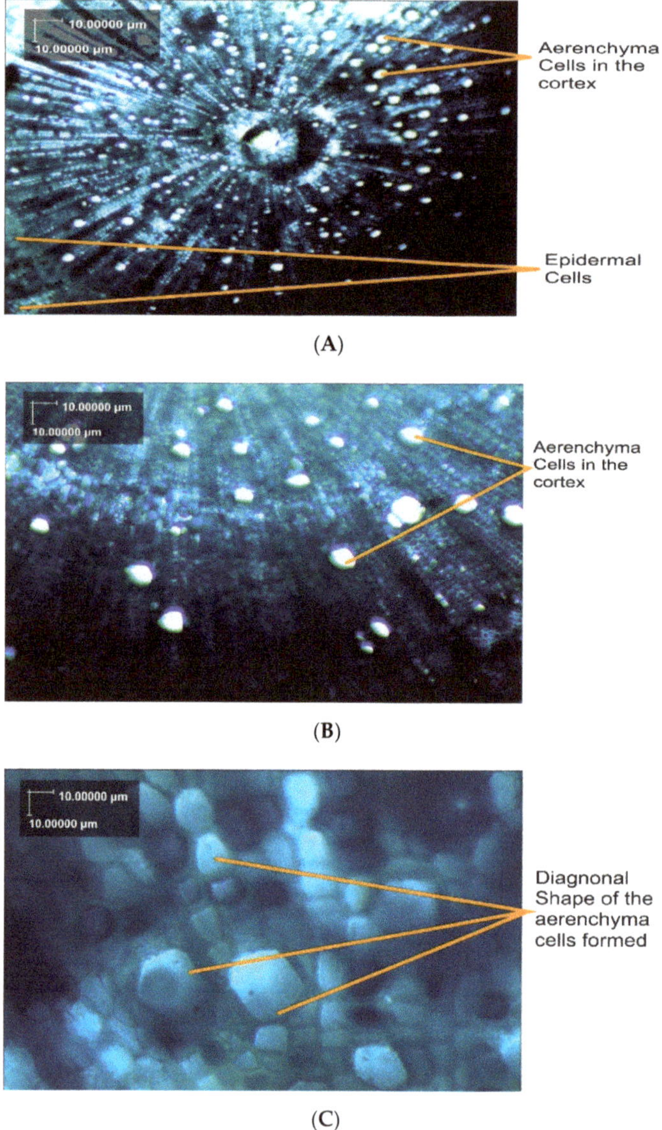

Figure 4. Root sections of okra plants grown from different concentrations of NaN$_3$-treated seeds show aerenchyma cells formed under two-week waterlogging conditions. (**A**) 0% NaN$_3$, (**B**) 0.02% NaN$_3$, (**C**) 0.05% NaN$_3$.

3.9. Antioxidant Enzymes Activity and Gene Expression Analyses

The effects of the waterlogging condition and NaN$_3$ treatments on the activities and expression levels of antioxidant enzymes (APX, CAT) in the leaf tissues were investigated. The activity and expression level of APX enzyme were significantly enhanced in plants exposed to waterlogging and sodium azide treatments with respect to non-treated (control) plants (Figure 5). Additionally, under waterlogging conditions, the activity and expression level of CAT enzyme were slightly enhanced in plants treated with sodium azide, as compared to non-treated plants (Figure 5).

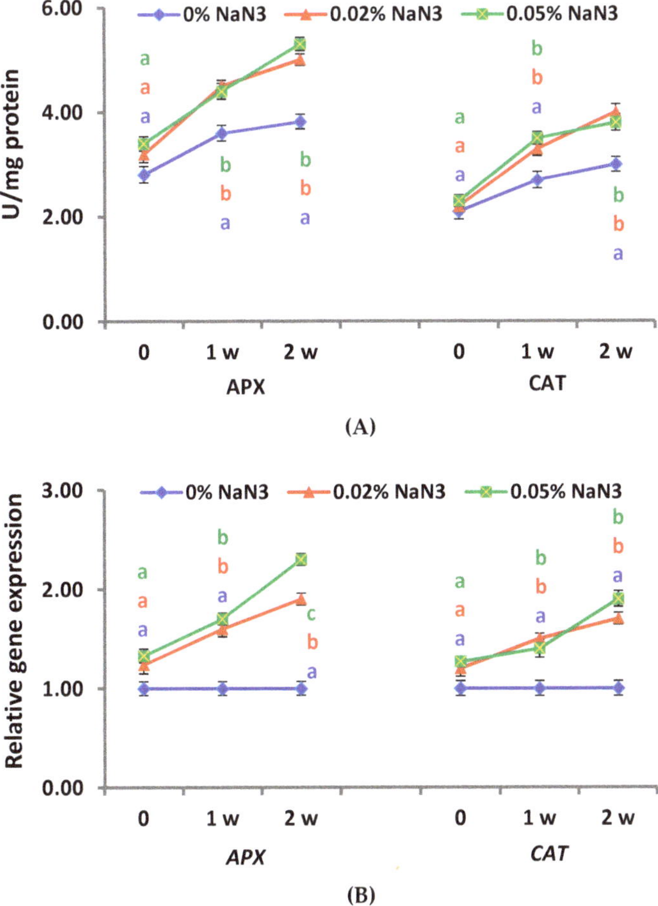

Figure 5. Activity (**A**) and gene expression levels (**B**) of APX and CAT enzymes in okra plants grown from NaN3-treated seeds under waterlogging conditions ten weeks after planting (WAP). Values = mean ± SD, $n = 4$. Mean values with similar letters at the same WAP are not significantly different at $p \leq 0.05$.

4. Discussion

Waterlogging stress has adverse impacts on crop development and productivity. Waterlogging-induced oxygen depletion results in changes in plant morphology and metabolism. Waterlogging conditions also cause inhibition of photosynthesis, leaf chlorophyll degradation, and early leaf senescence [46]. Negative impacts of flooding might be due to the reduced level of gas diffusion in water, which does not allow terrestrial plants to survive for a long period. Plants develop specific traits to improve gas exchange and cope with waterlogging conditions. These traits include formation of adventitious roots and aerenchyma cells, as well as elongation of stem root juncture above the water surface. These efficiently ameliorate the stress-induced hypoxic or anoxic conditions. The presence of aerenchyma cells facilitates exchange of gases between aerial and submerged plant parts [47]. Kawai et al. [48] proposed that the development of aerenchyma in tissues and organs decreases the number of cells requiring oxygen for respiration. However, the development of adaptive traits to waterlogging stress is species-dependent [7,49,50]. Enhanced formation of aerenchyma was observed upon treating rice plants with exogenous ethylene [14].

In the present study, NaN$_3$ treatments enhanced waterlogging stress tolerance and aerenchyma formation in okra. The results also showed that NaN$_3$ treatments affected okra germination. NaN$_3$-caused seed germination inhibition has also been reported in different plant species [22,23,51,52]. However, NaN$_3$ stimulated the germination of okra plants [24]. This germination inhibition was dependent on the concentration of NaN$_3$ used as seed treatment. Three days after planting (3 DAP), germination has been recorded in all NaN$_3$ treatments applied. Under waterlogging conditions, NaN$_3$ promoted okra growth 10 WAP, indicating that plants grown from 0.02% NaN$_3$-seed treatments exhibited better performance than those grown from 0.05% NaN$_3$-seed treatments. These findings were in a harmony with that reported by Al-Qurainy [25] and Zuzana et al. [26] who stated that NaN$_3$ could stimulate the plant growth and height of *Eruca sativa* and *Diospyros lotus*, respectively. Moreover, the difference in the number of leaves formed under waterlogging and non-waterlogging conditions was significant. Plants grown from NaN$_3$-treated seeds formed more leaves than those from non-treated seeds. Additionally, plants that were grown from 0.02% and 0.05% NaN$_3$-treated seeds produced a greater number of adventitious roots under waterlogging conditions. The emergence of adventitious roots is preceded by epidermal cell death at the nodes of submerged rice plants [47]. The activities leading to epidermal cell death for the emergence of adventitious roots occurred more in plants grown from NaN$_3$-treated seeds. Waterlogging conditions negatively affected the reproductive parameters recorded for okra plants in the current study. These findings are in harmony with that reported by Vwioko et al. [7]. Plants grown from 0.05% NaN$_3$-treated seeds formed a higher number of buds than plants produced from 0% NaN$_3$-treated seeds subjected to two-week waterlogging conditions. Plants grown from 0.05% NaN$_3$-treated seeds also produced more fruits than the control plants under two-week waterlogging conditions.

Waterlogging conditions cause depletion of soil oxygen due to microbial respiration. The reduction of soil oxygen urges anaerobic microorganisms to shift to alternative electron acceptors for their metabolic requirements [53]. Bacteria and fungi ratio in soil community are altered whenever there are soil inundations. Soil bacteria and fungi have a critical role in decomposition and nutrient cycling [54]. In the current investigation, microbial count results exhibited an increase in the bacteria populations and reduction in the fungi populations. The decrease in fungi populations has been previously reported [53,55–57]. Therefore, under waterlogging conditions, fungi presence is less prevalent than bacteria. Fungi require aerobic conditions to thrive but are inhibited by the scarcity of oxygen in the flooded soil environments. Fungi germinate from spores under flooding slowly, resulting in a decreased colonization. Unger et al. [53] suggested that some microbial groups may thrive well under flooded conditions. Gram-positive bacteria showed higher levels compared to Gram-negative bacteria under waterlogging conditions. Mentzer et al. [57] reported that flooding exhibited greater effect than nutrient loading on the microbial community and profoundly altered the composition and functional components.

Water copiously influences several physicochemical processes in soil, particularly under flooded conditions. This begins with the cutoff of oxygen supply to soil environments under waterlogging stress. The lack of oxygen promotes anaerobic metabolism by microbes through utilizing a decomposable organic matter. A reduction in soil redox potential and an increase in pH are recorded [58]. The soil Eh data recorded in a soil-water suspension rightly predicts the level of transformations present in the waterlogged soil [59]. Other important chemical changes in flooded soils indicate the prevalence of reduced forms of nitrogen, oxygen, iron, manganese, or sulphur in soil [53]. There are changes in phase or solubility because of redox reactions. For example, nitrate-nitrogen is transformed into gaseous forms (N_2, NO_2, N_2O) and lost, resulting in nitrogen depletion of soil [60]. In the present study, the soil chemical analysis showed that waterlogging conditions increased pH towards neutral, reduced soil Eh, organic carbon, total nitrogen and available phosphorus. These soil factors indicate higher reduction-oxidation reactions in soils under waterlogging conditions. These patterns of chemical environments and transformations are suspected to favor the tolerant bacteria for their higher counts recorded in waterlogged soil samples. The chemical environments attained under waterlogging soil conditions met the metabolic needs of tolerant bacteria. The decomposition of complex organic compounds is slow under anoxic conditions and in some cases leads to detection of higher amounts of phenolics [53] in waterlogged than in non-waterlogged soils. The present study does not reveal changes in the total phenolics of soil samples, suggesting that either the soil is devoid of complex organics for microbes to degrade under waterlogging conditions, or the microbes utilized readily available forms of carbons that are root exudates. Carbon enters the soil profile via the decomposition of plant residue on the surface or via root exudates in the upper soil horizon [53].

In the current study, root anatomy showed some peculiar features with waterlogged plants. Plants did not develop air-chambers in the cortex and stele regions under non-waterlogging conditions. However, plants subjected to waterlogging conditions formed aerenchyma cells. Further examination of the micrographs showed that plants grown from NaN_3-treated seeds produce more aerenchyma cells than those grown from untreated seeds. It was evident that 0.05% NaN_3-treated seeds produce plants with the highest aerenchyma development and increased with increasing the duration of waterlogging conditions. The formation of aerenchyma in the root as an adaptive trait contributed to the survival of okra plants exposed to waterlogging conditions. Furthermore, under waterlogging conditions, the activities and expression levels of APX and CAT enzymes were enhanced in plants treated with NaN_3 compared to non-treated plants in the present study. The survival of plants in stressed environments might be attributed to the induction of expression levels of antioxidant compounds. Salim et al. [34] reported that NaN_3-treated seeds produce mutant plants that showed higher antioxidation capacities than the normal plants. Moreover, Jeng et al. [61] revealed that these mutants induced increased antioxidant capacities through the generation of scavenging metabolics (DPPH, LPI ability, FRAP, and ABTS radical scavenging activities) than the wild type. In addition, the antioxidant enhancements could be linked to the accumulation of phenolics, anthocyanin, and proanthocyanidins at higher levels in the seed coats. These results are in harmony with that reported by Elfeky et al. [62] who stated that *Helianthus annus* plants grown from NaN_3-treated seeds initiated and induced higher antioxidant capacities than those grown from untreated seeds via increasing carotenoids, peroxidase activity, and protein content. In conclusion, sodium azide priming could enhance waterlogging stress tolerance in okra plants through enhancing the growth and reproductive parameters, inducing the formation of adventitious roots and aerenchyma cells, and increasing the activities and gene expression level of antioxidant enzymes.

Supplementary Materials: The following are available online at http://www.mdpi.com/2073-4395/9/11/679/s1, Table S1: Total bacteria and fungi count of soil samples analyzed after plant harvest, Table S2: Values obtained for soil factors in soil samples collected from different experimental pots after plant growth under waterlogging conditions.

Author Contributions: M.A.E.-E. and E.D.V. designed and performed the experiments, analyzed the data, and wrote and revised the manuscript. M.E.I., A.A.A.-G., H.M.A., E.A.A., and M.A.A.-D. helped with analysis and revision of the manuscript. M.M.E.-S. revised the manuscript. All the authors approved the final version of the manuscript.

Funding: The authors would like to extend their sincere appreciation to the Deanship of Scientific Research at King Saud University for funding this Research group no. RG 1440-054. The authors would also like to thank University of Benin in Nigeria and Tanta University in Egypt for supporting this work.

Conflicts of Interest: The authors declare no conflict of interest.

References

1. Gemede, H.F.; Ratta, N.; Haki, G.D.; Woldegiorgis, A.Z.; Bey, F. Nutritional Quality and Health Benefits of Okra (*Abelmoschus esculentus*): A Review. *Int. J. Nut. Food Sci.* **2015**, *4*, 208–215. [CrossRef]
2. El-Esawi, M.A.; Al-Ghamdi, A.A.; Ali, H.M.; Alayafi, A.A.; Witczak, J.; Ahmad, M. Analysis of Genetic Variation and Enhancement of Salt Tolerance in French Pea. *Int. J. Mol. Sci.* **2018**, *19*, 2433. [CrossRef] [PubMed]
3. El-Esawi, M.A.; Al-Ghamdi, A.A.; Ali, H.M.; Ahmad, M. Overexpression of *AtWRKY30* Transcription Factor Enhances Heat and Drought Stress Tolerance in Wheat (*Triticum aestivum* L.). *Genes* **2019**, *10*, 163. [CrossRef] [PubMed]

4. El-Esawi, M.A.; Elkelish, A.; Elansary, H.O.; Ali, H.M.; Elshikh, M.; Witczak, J.; Ahmad, M. Genetic transformation and hairy root induction enhance the antioxidant potential of *Lactuca serriola* L. *Oxid. Med. Cell. Longev.* **2017**, *2017*. [CrossRef]

5. El-Esawi, M.A.; Alayafi, A.A. Overexpression of Rice *Rab7* Gene Improves Drought and Heat Tolerance and Increases Grain Yield in Rice (*Oryza sativa* L.). *Genes* **2019**, *10*, 56. [CrossRef] [PubMed]

6. El-Esawi, M.A.; Alayafi, A.A. Overexpression of *StDREB2* Transcription Factor Enhances Drought Stress Tolerance in Cotton (*Gossypium barbadense* L.). *Genes* **2019**, *10*, 142. [CrossRef]

7. Vwioko, E.; Adinkwu, O.; El-Esawi, M.A. Comparative physiological, biochemical and genetic responses to prolonged waterlogging stress in okra and maize given exogenous ethylene priming. *Front. Physiol.* **2017**, *8*, 632. [CrossRef]

8. Heschbach, C.; Mult, S.; Kreuzwieser, J.; Kopriva, S. Influence of anoxia on whole plant sulphur nutrition of flooding tolerant poplar (*Populus tremula* × *P. alba*). *Plant Cell Environ.* **2005**, *28*, 167–175. [CrossRef]

9. Herrera, A.; Tezara, W.; Marin, O.; Rengifo, E. Stomatal and non-stomatal limitations of photosynthesis in trees of a tropical seasonally flooded forest. *Physiol. Plant.* **2008**, *134*, 41–48. [CrossRef]

10. Syversten, J.P.; Zablotowicz, R.M.; Smith, M.L. Soil-temperature and flooding effects on two species of citrus. 1. Plant growth and hydraulic conductivity. *Plant Soil* **1983**, *72*, 3–12.

11. Setter, T.L.; Waters, I.; Sharma, S.K.; Singh, K.N.; Kulshreshtha, N.; Yaduvanshi, N.P.S.; Ram, P.C.; Singh, B.N.; Rane, J.; McDonald, G.; et al. Review of wheat improvement for waterlogging tolerance in Australia and India: The importance of anaerobiosis and element toxicities associated with different soils. *Ann. Bot.* **2009**, *103*, 221–235. [CrossRef] [PubMed]

12. Calvo-Polanco, M.; Senorans, J.; Zwiazek, J.J. Role of adventitious roots in water relations of tamarack (*Larix* laricina) seedlings exposed to flooding. *BMC Plant Biol.* **2012**, *12*, 99–107. [CrossRef] [PubMed]

13. Sauter, M. Root responses to flooding. *Curr. Opin. Plant Biol.* **2013**, *16*, 282–286. [CrossRef] [PubMed]

14. Takahashi, H.; Yamauchi, T.; Colmer, T.; Nakazono, M. Aerenchyma Formation in Plants. In *Low-Oxygen Stress in Plants, Oxygen Sensing and Adaptive Responses to Hypoxia*, 1st ed.; Van Dongen, J.T., Licausi, F, Eds.; Plant Cell Monographs; Springer: New York, NY, USA, 2014; Volume 21, pp. 247–265.

15. Janda, T.; Szalai, G.; Tari, I.; Paldi, E. Hydroponic treatments with salicylic acid decreases the effects of chilling injury in maize (*Zea mays* L.) plants. *Planta* **1999**, *208*, 175–180. [CrossRef]

16. Rajasekaran, L.R.; Blake, T.J. New plant growth regulators protect photosynthesis and enhance growth under drought of jack pine seedlings. *J. Plant Growth Reg.* **1999**, *18*, 171–181. [CrossRef]

17. Gondor, O.K.; Pál, M.; Darkó, É.; Janda, T.; Szalai, G. Salicylic Acid and Sodium Salicylate Alleviate Cadmium Toxicity to Different Extents in Maize (*Zea mays* L.). *PLoS ONE* **2016**, *11*, e0160157. [CrossRef]

18. Vwioko, E.D. Performance of soybean (*Glycine max* L.) in salt-treated soil environment following salicylic acid mitigation. *NISEB J.* **2013**, *13*, 44–49.

19. Jackson, M.B. Ethylene-promoted elongation: An adaptation to submergence stress. *Ann. Bot.* **2008**, *101*, 229–248. [CrossRef]

20. Vidoz, M.L.; Loreti, E.; Mensuali, A.; Alpi, A.; Perata, P. Hormonal interplay during adventitious root formation in flooded tomato plants. *Plant J.* **2011**, *63*, 551–562. [CrossRef]

21. Sasidharan, R.; Voesenek, L.A.C.J. Ethylene-mediated acclimations to flooding stress. *Plant Physiol.* **2015**, *169*, 3–12. [CrossRef]

22. Gnanamurthy, S.; Dhanavel, D.; Girija, M.; Pavadai, P.; Bharathi, T. Effect of chemical mutagenesis on quantitative traits of maize (*Zea mays* (L.). *Int. J. Res. Bot.* **2012**, *2*, 34–36.

23. Shagufta, B.; Aijaz, A.W.; Irshad, A.N. Mutagenic sensitivity of gamma rays, EMS and sodium azide in *Trigonella foenumgraecum* L. *Sci. Res. Rep.* **2013**, *3*, 20–26.

24. Vwioko, D.E.; Onobun, E. Vegetative response of ten accessions of *Abelmoschus esculentus* (L) Moench. treated with sodium azide. *J. Life Sci. Res. Dis.* **2015**, *2*, 13–24.

25. Al-Qurainy, F. Effects of sodium azide on growth and yield traits of *Eruca sativa* (L.). *World Appl. Sci. J.* **2009**, *7*, 220–226.

26. Zuzana, K.; Katarína, R.; Elena, Z.; Maria, L.B.; Ján, B. Sodium azide induced morphological and molecular changes in persimmon (*Diospyros lotus* L.). *Agriculture* **2012**, *58*, 57–64.

27. Adamu, A.K.; Aliyu, H. Morphological effects of sodium azide on tomato (*Lycopersicon esculentum* Mill.). *Sci. World J.* **2007**, *2*, 9–12.

28. Gruszka, D.; Szarejko, L.; Maluszynski, M. Sodium azide as a mutagen. In *Plant Mutation Breeding and Biotechnology*; Shu, Q., Forster, B.P., Nakagawa, H., Eds.; CABI Publishing Company: Wallingford, UK, 2012; pp. 159–166.

29. Kravchik, M.; Bernstein, N. Effects of salinity on the transcriptome of growing maize leaf cells points at differential involvement of the antioxidative response in cell growth restriction. *BMC Genom.* **2013**, *16*, 14–24.

30. Mittler, R. Oxidative stress, antioxidant and stress tolerance: A review. *Trends Plant Sci.* **2002**, *7*, 405–410. [CrossRef]

31. Haq, I.U.; Memon, S.; Gill, N.P.; Rajput, M.T. Regeneration of plantlets under NaCl-stress from NaN$_3$ treated sugarcane explants. *Afr. J. Biotechnol.* **2011**, *10*, 16152–16156.

32. El Kaaby, E.A.J.; Al-Ajeel, S.A.; Al-Anny, J.A.; Al-Aubaidy, A.A.; Ammar, K. Effect of the chemical mutagen sodium azide on plant regeneration of two tomato cultivars under salinity stress condition in vitro. *J. Life Sci.* **2015**, *9*, 25–31. [CrossRef]

33. Kuasha, M.; Nasiruddin, K.M.; Hassan, L. Effects of sodium azide on callus in sugarcane. *Discovery* **2016**, *52*, 1683–1688.

34. Salim, K.; Fahad, A.-Q.; Firoz, A. Sodium azide: A chemical mutagen for enhancement of agronomic traits of crop plants. *Int. J. Sci. Tech.* **2009**, *4*, 1–21.

35. Ademoroti, C.A. *Standard Methods for Water and Effluent Analysis*, 1st ed.; Foludex Press Ltd.: Ibadan, Nigeria, 1996.

36. Bremner, J.M. Determination of nitrogen in soil by the Kjeldahl method. *J. Agric. Sci.* **1960**, *55*, 11–33. [CrossRef]

37. Blum, U. Benefits of citrate over EDTA for extracting phenolics from soils and plant debris. *J. Chem. Ecol.* **1997**, *23*, 347–362. [CrossRef]

38. Appiah, M.R.; Ahenkorah, Y. Determination of available sulphate in some soils of Ghana considering five extraction methods. *Biol. Fertil. Soils* **1989**, *8*, 80–86. [CrossRef]

39. Ben Mussa, S.A.; Elferjani, H.S.; Haroun, F.A.; Abdelnabi, F.F. Determination of available nitrate, phosphate and sulphate in soil samples. *Int. J. PharmTech Res.* **2009**, *1*, 598–604.

40. Bray, R.H.; Kurtz, L.T. Determination of total organic carbon and available phosphorus in soils. *Soil Sci.* **1945**, *59*, 39–48. [CrossRef]

41. Bremner, J.M.; Jenkinson, D.S. Determination of organic carbon in soil. I. oxidation by dichromate of organic matter in soil and plant materials. *J. Soil Sci.* **1960**, *11*, 394–402. [CrossRef]

42. Islam, M.S.; Halim, M.A.; Safiullah, S.; Islam, M.S.; Islam, M.M. Analysis of organic matter, ion and manganese in soil of arsenic affected Singair Area, Bangladesh. *Res. J. Environ. Toxicol.* **2009**, *3*, 31–35.

43. Harrigan, W.F.; McCance, M.E. *Laboratory Methods in Foods and Dairy Microbiology*, 8th ed.; Academic Press: London, UK, 1990.

44. Holt, J.G.; Sneath, P.H.; Krieg, N.R. *Bergey's Manual of Determinative Bacteriology*, 9th ed.; Lippincott, Williams and Wilkins Publishers: Baltimore, MD, USA, 2002; p. 787.

45. Zhang, J.; Kirkham, M.B. Enzymatic Responses of the Ascorbate-Gluta-thione Cycle to Drought in Sorghum and Sunflower Plants. *Plant Sci.* **1996**, *113*, 139–147. [CrossRef]

46. Zou, X.; Hu, C.; Zeng, L.; Xu, M.; Zhang, X. A comparison of screening methods to identify waterlogging tolerance in the field in *Brassica napus* (L.) during plant ontogeny. *PLoS ONE* **2014**, *9*, e89731. [CrossRef] [PubMed]

47. Steffens, B.; Geske, T.; Sauter, M. Aerenchyma formation in the rice stem and its promotion by H$_2$O$_2$. *New Phytol.* **2011**, *190*, 369–378. [CrossRef] [PubMed]

48. Kawai, M.; Samarajeewa, P.K.; Barrero, R.A.; Nishigushi, M.; Uchimiya, H. Cellular dissection of the degradation pattern of cortical cell death during aerenchyma formation of rice roots. *Planta* **1998**, *204*, 277–287. [CrossRef]

49. Fukao, T.; Xu, K.; Ronald, P.C.; Bailey-Serres, J. A variable cluster of ethylene response factor-like genes regulates metabolic and developmental acclimation responses to submergence in rice. *Plant Cell* **2006**, *18*, 2021–2034. [CrossRef] [PubMed]

50. Hattori, Y.; Nagai, K.; Furukawa, S.; Song, X.-J.; Kawano, R.; Sakakibara, H.; Wu, J.; Matsumoto, T.; Yoshimura, A.; Kitano, H.; et al. The ethylene response factors SNORKEL 1 and SNORKEL 2 allow rice to adapt to deep water. *Nature* **2009**, *460*, 1026–1030. [CrossRef]

51. Mensah, J.K.; Obadoni, B. Effects of sodium azide on yield parameters of groundnut (*Arachis hypogaea* L.). *Afr. J. Biotechnol.* **2007**, *6*, 668–671.

52. Nakweti, R.K.; Franche, C.; Ndiku, S.L. Effects of sodium azide (NaN₃) on seeds germination, plantlets growth and in vitro antimalarial activities of *Phyllantus odontadenius* Mull. *Arg. Amer. J. Exp. Agric.* **2015**, *5*, 226–238.

53. Unger, I.M.; Kennedy, A.C.; Muzika, R.-M. Flooding effects on soil microbial communities. *Appl. Soil Ecol.* **2009**, *42*, 1–8. [CrossRef]

54. Suzuki, C.; Kunito, T.; Aono, T.; Liu, C.-T.; Oyaizu, H. Microbial indices of soil fertility. *J. Appl. Microbiol.* **2005**, *98*, 1062–1074. [CrossRef]

55. Bossio, D.A.; Scow, K.M. Impacts of carbon and flooding on soil microbial communities: Phospholipid fatty acid profiles and substrate utilization patterns. *Microb. Ecol.* **1998**, *35*, 265–378. [CrossRef]

56. Drenovsky, R.E.; Vo, D.; Graham, K.J.; Scow, K.M. Soil water content and organic carbon availability are major determinants of soil microbial community composition. *Microb. Ecol.* **2004**, *48*, 424–430. [CrossRef] [PubMed]

57. Mentzer, J.L.; Goodman, R.M.; Balser, T.C. Microbial responses over time to hydrologic and fertilization treatments in a simulated wet prairie. *Plant Soil* **2006**, *284*, 85–100. [CrossRef]

58. Stover, R.H. Flooding of soil for disease control. In *Soil Disinfection*; Mulder, D., Ed.; Elsevier: Amsterdam, The Netherlands, 1979.

59. Labuda, S.Z.; Vetchinnikov, A.A. Soil susceptibility on reduction as an index of soil properties applied in the investigation upon soil devastation. *Ecol. Chem. Eng.* **2011**, *18*, 333–344.

60. Vepraskas, M.J.; Faulkner, S.P. Redox chemistry of hydric soils. In *Wetlands Soils: Genesis, Hydrology, Landscapes and Classification*; Richardson, J.L., Vepraskas, M.J., Eds.; Lewis Publishers: Boca Raton, FL, USA, 2001.

61. Jeng, T.L.; Tseng, T.H.; Lai, C.C.; Wu, M.T.; Sung, J.M. Antioxidative charactrisation of NaN₃- induced common bean mutants. *Food Chemistry.* **2010**, *119*, 1006–1011. [CrossRef]

62. Elfeky, S.; Abo-Hamad, S.; Saad-Allah, K.M. Physiological impact of sodium azide on *Helianthus annus* seedlings. *Int. J. Agron. Agric. Res.* **2014**, *4*, 102–109.

Article

Effect of Heat Stress on Growth and Physiological Traits of Alfalfa (*Medicago sativa* L.) and a Comprehensive Evaluation for Heat Tolerance

Misganaw Wassie [1,2], Weihong Zhang [1], Qiang Zhang [3], Kang Ji [1,2] and Liang Chen [1,*]

[1] Key Laboratory of Plant Germplasm Enhancement and Specialty Agriculture, Wuhan Botanical Garden, Chinese Academy of Sciences, Wuhan 430074, China; misgie2010@yahoo.com (M.W.); zwh2019wbgcas@163.com (W.Z.); jikang17@mails.ucas.ac.cn (K.J.)

[2] University of Chinese Academy of Sciences, Beijing 100049, China

[3] College of Agronomy, Hunan Agricultural University, Changshan 410000, China; zhangqhunau@163.com

* Correspondence: chenliang1034@126.com; Tel.: +152-0716-6216; Fax: +86-27-8751-0251

Received: 28 August 2019; Accepted: 27 September 2019; Published: 28 September 2019

Abstract: Alfalfa (*Medicago sativa* L.) is a valuable forage legume, but its production is largely affected by high temperature. In this study, we investigated the effect of heat stress on 15 alfalfa cultivars to identify heat-tolerant and -sensitive cultivars. Seedlings were exposed to 38/35 °C day/night temperature for 7 days and various parameters were measured. Heat stress significantly reduced the biomass, relative water content (RWC), chlorophyll content, and increased the electrolyte leakage (EL) and malondialdehyde (MDA) content of heat-sensitive alfalfa cultivars. However, heat-tolerant cultivars showed higher soluble sugar (SS) and soluble protein (SP) content. The heat tolerance of each cultivar was comprehensively evaluated based on membership function value. Cultivars with higher mean membership function value of 0.86 (Bara310SC) and 0.80 (Magna995) were heat tolerant, and Gibraltar and WL712 with lower membership function value (0.24) were heat sensitive. The heat tolerance of the above four cultivars were further evaluated by chlorophyll *a* fluorescence analysis. Heat stress significantly affected the photosynthetic activity of heat-sensitive cultivars. The overall results indicate that Bara310SC and WL712 are heat-tolerant and heat-sensitive cultivars, respectively. This study provides basic information for understanding the effect of heat stress on growth and productivity of alfalfa.

Keywords: alfalfa; evaluation; growth; heat stress; physiological traits

---

1. Introduction

Heat stress is one of the major abiotic stresses limiting plant growth and development. When plants are exposed to high temperature, several cellular injuries, including cell death, may occur within minutes, which then leads to an appalling failure of cellular organization [1,2]. In addition, the rapid closure of stomata, reduction in cell size, an increase in stomatal, trachomatous densities, and xylem vessels of both root and shoot were reported to occur in response to heat stress [3]. However, different plant species may show different responses to heat stress [4]. Generally, heat stress triggers various morphological, physiological, biochemical, and molecular changes to inhibit plant growth and development. Heat stress inhibits seed germination; causes scorching; twigs and burning of leaves, branches, and stems; leaf senescence and abscission; shoot and root growth inhibition; fruit discoloration and damage; reduced yield; and finally plant death [5,6]. High temperature stress also affects shoot net assimilation and decreases the overall dry weight of the plant [4].

It is well established that heat stress has detrimental impacts on various key physiological, biochemical, and metabolic processes of plants, and disrupts normal cellular homeostasis [2]. It promotes the overproduction and accumulation of reactive oxygen species (ROS), malondialdehyde (MDA) production due to lipid peroxidation, photoinhibition, protein denaturation, and accumulation of compatible solutes [7–9]. The oxidative stress caused by heat stress further leads to cellular injury, membrane proteins breakage, lipid peroxidation, photosynthetic pigment degradation, and enzymes and nucleic acid denaturation [4,10,11]. Furthermore, heat stress influences plant photosynthesis and respiration processes to curtail the life cycle and reduce plant productivity [12]. The heat stress sensitivity of plants varies with the plant genotype and the stage of plant development, but the effect is highly dependent on genotype and species, as well as with abundant inter- and intraspecific variations [12].

Alfalfa (*Medicago sativa* L.) is one of the most important perennial forage legume species. Due to its outstanding nutritional quality, alfalfa is an excellent source of feed nutrient for animals. High temperature stress is a limiting factor for alfalfa cultivation [13,14]. Previous studies have shown that increasing temperature above optimal level markedly affects alfalfa's morphological, physiological, and proteomic processes, and reduced photosynthetic rate, destroyed plasma membrane structure, and accelerated the process of aging [15–18]. Therefore, it is of a great significance to develop heat stress-tolerant alfalfa cultivar that withstand heat stress-induced growth inhibition and biomass reduction. Studying plants' physiology in response to heat stress could be helpful to further understand the molecular tolerance traits [3] and will provide fundamental knowledge to develop heat-tolerant cultivars. It is well reported that different genotypes of a single plant species demonstrate high degrees of variation for heat tolerance; therefore, the selection of varieties with high thermotolerance potential is crucial to further improve thermotolerance. The genetic variability present in alfalfa could be exploited to evaluate and screen for high-temperature tolerance. Moreover, compared with other abiotic stresses reports in alfalfa, studies about the effect of heat stress on alfalfa growth and physiology are very limited. Thus, the objectives of this study were to investigate the effect of heat stress on the growth and physiological traits of alfalfa, to evaluate for heat tolerance, and to identify heat-tolerant and heat-sensitive alfalfa cultivars.

2. Materials and Methods

2.1. Plant Materials, Growth Conditions, and Heat Treatment

In this study, fifteen alfalfa cultivars (*Medicago sativa* L.) were used and the details of the cultivars are presented in Table 1. Ten seeds of each cultivar were sown in each plastic pot filled with clay, sand, and loamy soil (1:1:2, v/v). The pots were then kept in a greenhouse with a temperature of 25 °C, relative humidity of approximately 60%, light intensity of 500–550 μmol m^{-2} s^{-1}, and a photoperiod of 14 h/10 h light/dark [19]. There were five replications for each cultivar and treatment. The seedlings were watered daily to field capacity level and fertilized once a week with a half-strength Hoagland nutrient solution. The four-week-old seedlings were divided into two groups for heat treatment. The control group was kept in a greenhouse at 25 °C and the treatment group was transferred into a growth incubator and treated at 38/35 °C light/dark and light intensity of 500–550 μmol m^{-2} s^{-1} for 7 days.

Table 1. Alfalfa cultivars used in the study.

S. No	Cultivars	Origin	Dormancy Rate	S. No	Cultivars	Origin	Dormancy Rate
1	Gibraltar	USA	2	9	WL440HQ	USA	6
2	Golden Queen	USA	2.5	10	WL525HQ	USA	8
3	SK3010	Canada	2.5	11	Magna995	USA	9
4	Bara310SC	USA	3	12	Siriver	Australia	9
5	WL354HQ	USA	3.9	13	WL656HQ	USA	9.3
6	55V48	USA	5	14	Nofollow	China	9.6
7	WL363HQ	USA	5	15	WL712	USA	10.2
8	Sanditi	France	5				

2.2. Plant Biomass Measurement

After seven days of heat stress treatment, five plants for each of the treatment and control group were randomly harvested from each pot (25 plants for each cultivar and treatment). The roots of selected plants were washed with distilled water and separated from the shoot. The fresh weight of roots and shoots were measured separately using analytical balance (precision 0.0001 g). Dry weight was measured after drying the shoot and root in an oven at 80 °C for 24 h. Total fresh and dry biomasses were calculated using the following formula:

$$TFB = SFW + RFW$$
$$TDB = SDW + RDW \tag{1}$$

where TFW is the total fresh biomass, SFW is the shoot fresh weight, RFW is the root fresh weight, TDB is the total dry biomass, SDW is the shoot dry weight, and RDW is the root dry weight.

2.3. Physiological Traits Analysis

Fresh leaves were used to measure relative water content, electrolyte leakage, chlorophyll content, and soluble sugar content, while MDA and soluble protein content were determined from liquid nitrogen dried leaves which were stored at −80 °C freezer.

2.4. Relative Water Content (RWC) Measurement

For RWC determination, fresh weight (FW) was measured from the fully expanded top leaves, and Turgid weight (TW) was recorded after dipping the leaves in distilled water for 12 h. Dry weight (DW) was then measured after oven drying the leaves at 80 °C for 24 h and RWC was calculated using the following formula:

$$RWC\% = ((FW - DW)/(TW - DW)) \times 100 \tag{2}$$

where FW is the fresh weight of leaves, DW is the dry weight of leaves, and TW is the turgid weight of leaves.

2.5. Chlorophyll Content Determination

For chlorophyll content determination, 0.1 g fresh alfalfa leaves were placed into a centrifuge tube containing 5 mL of 95% alcohol. The test tubes were wrapped with aluminum foil and incubated for 48 h at room temperature in the dark. The absorbance of the chlorophyll extract was read at 665 and 649 nm. The chlorophyll content was calculated according to the following formula:

$$Chl\ a\ (mg \cdot g^{-1}\ FW) = (13.95 \times D665 - 6.88 \times D649) \times 0.005 \div W \tag{3}$$

$$Chl\ b\ (mg \cdot g^{-1}\ FW) = (24.96 \times D649 - 7.32 \times D665) \times 0.005 \div W \tag{4}$$

$$Total\ Chl\ (mg \cdot g^{-1}\ FW) = (18.08 \times D649 + 6.63 \times D665) \times 0.005 \div W \tag{5}$$

where D is the absorbance of the chlorophyll extract and W is the fresh weight leaves (g).

2.6. Electrolyte Linkage (EL) Measurement

The electrolyte leakage was measured using the method described by Huang et al. [20]. Briefly, 0.5 g of fresh alfalfa leaves were collected and washed with deionized water three times and then transferred into a 50 mL plastic centrifuge tube that was filled with 15 mL of deionized water. The tubes were incubated at room temperature for 12 h on a conical shaker and initial conductivity (EL1) using a conductivity meter (JENCO-3173, Jenco Instruments, Inc., San Diego, CA, USA). To release all electrolytes, the leaves were killed by autoclaving at 121 °C for 30 min. The tubes were then cooled at room temperature and the second conductivity (EL2) was measured. The relative EL was calculated by using the formula:

$$\text{Relative EL (\%)} = (EL1/EL2) \times 100. \tag{6}$$

2.7. Soluble Sugar Content Determination

For soluble sugar content determination, 0.1 g fresh leaves were placed into a test tube containing 5 mL distilled water. The tubes were then sealed with plastic film and soaked in the boiling water bath for 30 min. The supernatant was collected and extraction continued for the second time using the residue. The supernatant from the primary and secondary extraction was mixed together, and 0.5 mL extract was taken and added to each tube containing 1.5 mL distilled water. Then 0.5 mL anthrone reagent (1 g anthrone dissolved in 50 mL ethyl acetate) and 5 mL 98% (W/V) H_2SO_4 (NA) was added to the test tubes and mixed gently, then placed in a boiling water bath (100 °C) for 10 min. The tubes were then cooled rapidly under running cold water and the absorbance was measured at 620 nm against the blank reagent. The concentration of total soluble sugar was obtained from the glucose standard curve. Finally, the total soluble sugar content was calculated using the equation:

$$\text{Soluble sugar content (\%)} = (C \times 7.5)/(W \times 104) \tag{7}$$

where C is soluble sugar concentration from the standard curve (µg), and W is the fresh weight of leaves (g).

2.8. Preparation of Crude Enzyme Extract

For crude enzyme extraction, about 0.3 g of alfalfa leaves, which were dried with liquid nitrogen, were ground into powder in liquid nitrogen using prechilled mortar and pestle (4 °C). Then, 5 mL of 150 mM sodium phosphate buffer (PBS), pH 7.4, was added to the powder and the homogenate was centrifuged at 12,000 rpm for 20 min at 4 °C. The supernatant was collected and used as a crude extract to determine malondialdehyde (MDA) and soluble protein content.

2.9. Determination of MDA Content

The MDA content was determined by the thiobarbituric acid (TBA) method according to a previous report [21]. Briefly, previously prepared 1 mL crude enzyme extract was mixed with 2 mL of reaction mixture containing 20% (v/v) trichloroacetic acid and 0.5% (v/v) thiobarbituric acid. The mixture was then heated for 30 min in a 95 °C water bath and directly cooled to room temperature. The mixture was then centrifuged at 12,000 g for 10 min at 20 °C, and the absorbance of the supernatant was read at 450, 532, and 600 nm with a spectrophotometer (UV2600, UNIC, Shanghai, China). The MDA content was calculated using the following formula:

$$\text{MDA (mmol·g}^{-1}\text{ FW)} = [6.425 \times (OD532 - OD600) - 0.559 \times OD450] \times Vt/(Vs \times FW) \tag{8}$$

where Vt is the volume of extraction liquid (mL), Vs is the volume of extraction solution (mL), and FW is the fresh weight of samples (g).

2.10. Soluble Protein Content Determination

Soluble protein content was determined following the Bradford assay method [22]. Briefly, 100 µL crude enzyme extract was added to a tube containing 3 mL Bradford working solution. After 10 min, the absorbance was measured at 595 nm using a spectrophotometer (UV2600, UNIC, Shanghai, China). The protein concentration of the crude extract was determined from the standard curve established by the reference solution of bovine serum albumin (BSA) and the OD595 value of the sample.

2.11. A Comprehensive Evaluation of Alfalfa for Heat Tolerance

The comprehensive evaluation was performed according to the previous report [23] based on membership function value, which were calculated from the heat tolerance coefficients. The correlation of each biomass and physiological trait for all alfalfa cultivars was analyzed using heat tolerance coefficients. The heat tolerance coefficient of all biomass and physiological traits was calculated using the following equation:

$$\text{Heat tolerance coefficient} = (\text{HT}/\text{CK}) \times 100 \tag{9}$$

where CK is the mean value of a single trait under the control treatment and HT is the mean value of a single trait under heat treatment.

The membership function values of all biomass and physiological traits were calculated from the heat tolerance coefficient using the formula (F1 and F2). Formula F1 was used for the traits that are directly related to heat tolerance, and F2 was used for the traits that are inversely related traits.

$$\text{F1 (Xi)} = (\text{Xi} - \text{Xmin})/(\text{Xmax} - \text{Xmin}) \tag{10}$$

$$\text{F2 (Xi)} = 1 - (\text{Xi} - \text{Xmin})/(\text{Xmax} - \text{Xmin}) \tag{11}$$

where X is the *i* heat tolerance coefficient, max is the maximum heat tolerance coefficient value from all cultivars of the *i* trait, and Xmin is the minimum heat tolerance coefficient from all cultivars of the *i* trait.

2.12. Chlorophyll a (Chl a) Fluorescence Analysis

Based on the comprehensive heat tolerance evaluation, four alfalfa cultivars—two relatively heat tolerant (Bara310SC and Magna995) and two relatively heat sensitive (Gibraltar and WL712)—were selected and further evaluated by Chl *a* fluorescence analysis to select the most heat-tolerant and -sensitive alfalfa. Chl a fluorescence transient was measured using a pulse–amplitude modulation (PAM) fluorometer (PAM2500 Heinz Walz GmbH, Eichenring, Germany) according to the previous report [21]. Briefly, the leaves were kept in the dark for 30 min, and all measurements were taken using a saturating light intensity of 2000 μmol photons m^{-2} s^{-1}. Five measurements were taken for each cultivar and treatment. The strong light pulses inducted Chl a fluorescence emission, which was subsequently measured and digitized between 10 μs and 300 ms. The chlorophyll fluorescence induction curve (OJIP) transients were analyzed using the JIP-test.

2.13. Statistical Analysis

Data were subjected to analysis of variance (ANOVA) using SPSS software version 22.0 (IBM Corporation, Chicago, IL, USA). All values were shown as mean ± SD (Standard Deviation) ($n = 5$). The independent sample *t*-test was employed to compare the control and the treatment groups using the least significant difference (LSD) test. All statistical results were considered significant at $p \leq 0.05$. All figures were created by Origin 9.0 (Origin Lab, Inc., Hampton, MA, USA).

3. Results

3.1. Effect of Heat Stress on Alfalfa Plant Biomass

Heat stress affected the biomass of all alfalfa cultivars (Table 2). Compared with the control, heat stress significantly reduced the shoot fresh weight of Golden Queen (45.74%), 55V48 (34.07%), WL354HQ (33.47%), Gibraltar (31.55%), and WL712 (28.57%). In addition, significantly higher shoot dry weight reduction was noticed in WL354HQ (55.10%), Golden Queen (45.45%), WL712 (40.43%), and Gibraltar (35.14%) ($p < 0.05$).

Table 2. Effect of heat stress on alfalfa plant biomass (g).

Cultivars	SFW		SDW		RFW		RDW		TFB		TDB	
	CK	HT	CK	HT	CK	HT	CK	HT	CK	HT	CK	HT
Gibraltar	1.87	1.28 *	0.37	0.24 *	0.88	0.66 *	0.16	0.08 *	2.75	1.94 *	0.53	0.32 *
Golden queen	1.88	1.02 *	0.44	0.24 *	0.92	0.63 *	0.19	0.11	2.8	1.64 *	0.62	0.35
SK3010	2.09	2.07	0.66	0.54 *	0.94	0.65 *	0.18	0.12 *	3.03	2.72 *	0.84	0.77
Bara310SC	1.85	1.71	0.46	0.4	0.88	0.84	0.16	0.15	2.73	2.55	0.62	0.55
WL354HQ	2.39	1.59 *	0.49	0.22 *	0.87	0.69 *	0.17	0.12 *	3.26	2.28 *	0.66	0.34 *
55V48	2.26	1.49	0.52	0.35 *	0.9	0.74 *	0.18	0.10 *	3.16	2.22 *	0.7	0.45 *
WL363HQ	2	1.56	0.6	0.5 *	0.86	0.71	0.16	0.13	2.86	2.27	0.76	0.63
Sanditi	2.13	1.97	0.59	0.37 *	0.94	0.63 *	0.19	0.10 *	3.06	2.59	0.78	0.47 *
WL440HQ	2.01	1.66	0.54	0.49	0.91	0.81 *	0.18	0.10 *	2.92	2.47	0.72	0.59
WL525HQ	1.92	1.45	0.55	0.34	0.94	0.77 *	0.18	0.12 *	2.86	2.21	0.73	0.46
Magna995	2.02	1.61 *	0.54	0.44	0.93	0.91	0.19	0.18	2.96	2.52 *	0.72	0.63
Siriver	1.93	1.58 *	0.52	0.37	0.87	0.72 *	0.14	0.11	2.8	2.31 *	0.66	0.48
WL656HQ	2.07	1.69 *	0.52	0.41 *	0.91	0.69	0.15	0.09 *	2.98	2.38 *	0.67	0.5 *
Nofollow	1.93	1.85	0.53	0.49	0.88	0.83	0.18	0.12 *	2.81	2.69 *	0.71	0.61
WL712	1.82	1.3 *	0.47	0.28 *	0.88	0.75 *	0.19	0.09 *	2.7	2.05 *	0.65	0.38 *

The data in the table are mean ($n = 5$). Asterisk indicates statistical significance difference between the control (CK) and heat treatment (HT) at $p < 0.05$, independent sample t-test. SFW, shoot fresh weight; SDW, shoot dry weight; RFW, root fresh weight; RDW, root dry weight; TFB, total fresh biomass; TDB, total dry biomass.

Furthermore, heat stress remarkably affected the fresh and dry root weights of most cultivars, with significant reductions recorded in SK3010, Sanditi, and Gibraltar (Table 2). Relative to the control, heat stress significantly decreased the total fresh biomass of Golden Queen, WL354HQ, 55V48, Gibraltar, and WL712 cultivars by 41.43%, 30.06%, 29.75%, 29.45%, and 24.04%, respectively ($p < 0.05$). Similarly, total dry biomass was significantly reduced by 48.48%, 43.55%, 41.54%, 39.74%, and 39.62% in WL354HQ, Golden Queen, WL712, Sanditi, and Gibraltar cultivars, respectively (Table 2). The biomass of Gibraltar, Golden Queen, and WL712 were highly affected by heat stress, which may indicate the sensitivity of the cultivars. By contrast, Bara310SC, Magna995, and WL363HQ cultivars were able to maintain their biomass under the heat stress condition, which may indicate their tolerance.

3.2. Heat Stress Reduced the Relative Water Content (RWC) of Alfalfa

Heat stress obviously decreased the leaf relative water content of all alfalfa cultivars compared to the control (Figure 1A). Of which, significantly higher decrement was noted in WL354HQ, WL712, Sanditi, WL440HQ, 55V48, and Siriver cultivars by 15.29%, 13.28%, 12.54%, 12.42%, 11.43%, and 10.01%, respectively ($p < 0.05$) (Figure 1A). Despite the reduction of RWC in Gibraltar, Golden Queen, and SK3010 cultivars being small (<10%), it was still significant compared to the control (Figure 1A). The result revealed that heat stress had a higher impact on the RWC of some alfalfa cultivars, especially on WL354HQ and WL712 cultivars, which showed higher water loss under heat stress. Bara310SC and Magna995 cultivars sustained their relative water content under heat stress and could be heat tolerant, but cultivar WL712 could be heat sensitive, as evidenced by higher water loss.

3.3. Heat Stress Increased the Electrolyte Leakage (EL) of Alfalfa

Heat stress affected the membrane integrity and stability and increased the EL of all alfalfa cultivars, as shown in Figure 1B, and was significant for the majority of the cultivars. Heat stress significantly increased the EL of Golden Queen, 55V48, Siriver, WL712, SK3010, WL354HQ, Nofollow, Gibraltar, WL525HQ, Sanditi, WL440HQ, and WL656HQ by 61.36%, 56.72%, 53.45%, 53.20%, 52.86%, 52.08%, 49.56%, 45.78%, 42.12%, 36.84%, and 36.45%, respectively ($p < 0.05$) (Figure 1B). However, Bara310SC cultivar had a lower increase in EL of 24.07%. The result showed that heat stress had a huge impact on the membrane stability of the majority of alfalfa cultivars, as shown by significantly higher increment in EL.

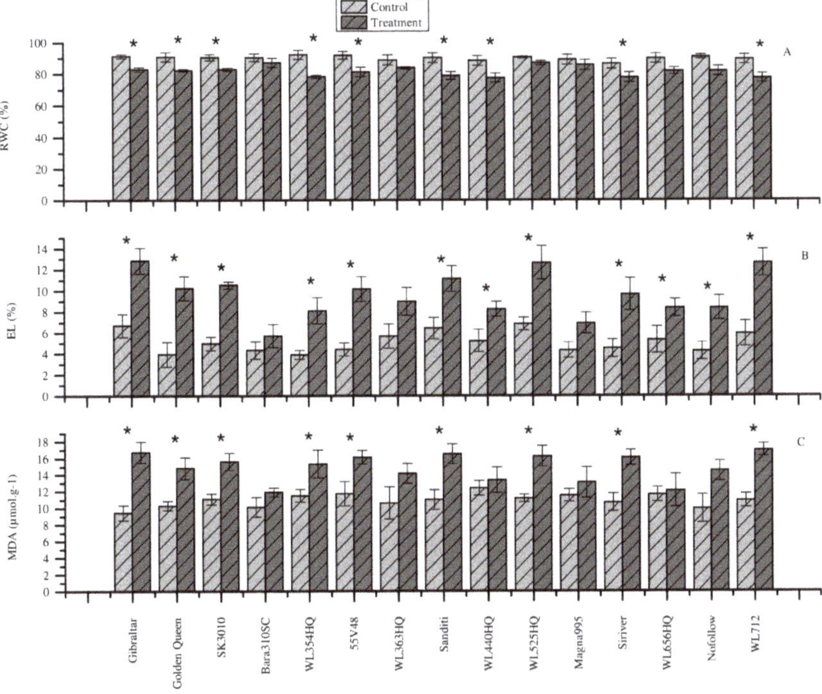

Figure 1. Effects of heat stress on physiological parameters. (**A**) Relative water content. (**B**) Electrolyte leakage. (**C**) MDA content. Each bar represents the mean ($n = 5$) and the error bar indicates the standard deviation. Asterisks indicate statistically significant differences between control and treatment group for each cultivar ($p < 0.05$), independent sample t-test. EL is electrolyte leakage, MDA is malondialdehyde.

3.4. Effect of Heat Stress on Lipid Peroxidation

As indicated in EL, heat stress damaged the membrane of alfalfa and caused lipid peroxidation, which was manifested by higher MDA content in all cultivars (Figure 1C). However, heat stress had no significant effect on some alfalfa cultivars such as Bara310SC, WL363HQ, WL440HQ, Magna995, WL656HQ, and Nofollow (Figure 1C). Meanwhile, under heat stress, Gibraltar, WL712, Siriver, Sanditi, WL525HQ, and Golden Queen cultivars showed significantly higher MDA content compared to the control ($p < 0.05$). The results showed that different alfalfa cultivars had a different level of sensitivity to heat stress. The higher the MDA content, the higher the lipid peroxidation and the greater the membrane damage.

3.5. Heat Stress Decreased the Chlorophyll Content of Alfalfa

It is obvious that heat stress affects the chlorophyll content of plant leaves, and the same was true for all alfalfa cultivars (Figure 2). The Chl content of some cultivars was more significantly and highly affected by heat stress than others compared to the control. In particular, heat stress significantly reduced the chlorophyll content of Gibraltar, Golden Queen, SK310, WL354HQ, WL363HQ, Sanditi, WL440HQ, WL525HQ, Siriver, and WL712 compared to control. Meanwhile, higher reduction in Chl a was observed in Gibraltar, WL354HQ, Golden Queen, Siriver, WL712, and Sanditi cultivars by 40.30%, 36.06%, 35.06%, 33.41%, 31.28%, and 31.11%, respectively (Figure 2A). Similar reduction in Chl b content was observed in WL712 (44.14%), WL354HQ (34.44%), WL440HQ, (31.53%), Golden Queen (24.69%), and Gibraltar (22.59%) (Figure 2B). In addition, higher reduction in total Chl content was noted in WL712, Gibraltar, 13, and Golden Queen (36.57%, 33.11%, 31.29%, and 31.05%, respectively)

(Figure 2C). The results suggested that some cultivars might be more sensitive to heat stress, as indicated by higher Chl content reduction under heat stress.

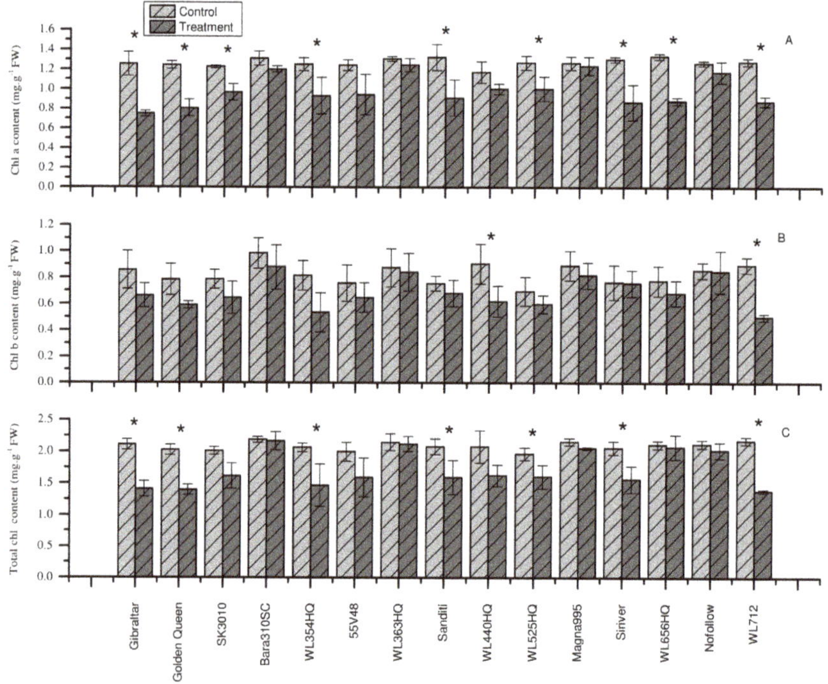

Figure 2. Effect of heat stress on the chlorophyll content of alfalfa cultivars. (**A**) Chlorophyll a content. (**B**) Chlorophyll b content. (**C**) Total chlorophyll content. Each bar represents the mean ($n = 5$) and the error bar indicates the standard deviation. Asterisks indicate statistical significant differences between control and treatment group for each cultivar ($p < 0.05$), independent sample t-test.

3.6. Effect of Heat Stress on Soluble Sugar Content

The soluble sugar content of all alfalfa cultivars showed an increment under heat stress relative to the control (Figure 3A). Heat stress significantly increased the soluble sugar content of Bara310SC, Magna995, WL363HQ, Nofollow, WL525HQ, and WL354HQ by 20.20%, 15.57%, 11.53%, 7.13%, 4.68%, and 2.22%, respectively (Figure 3A). The results showed that these cultivars produce higher soluble sugar to maintain their osmotic potential and organize proteins and cellular structures under heat stress, which increases heat tolerance.

3.7. Effect of Heat Stress on Soluble Protein Content

Similar to soluble sugar content, heat stress increased the soluble protein content of all alfalfa cultivars as shown in Figure 3B. Compared to the control, significantly higher soluble protein content was noted in Bara310SC, SK3010, WL363HQ, 55V48, WL440HQ, Magna995, and WL354HQ cultivars under heat stress ($p < 0.05$). Heat stress increased the soluble protein content of these cultivars by 36.18%, 26.73%, 25.52%, 24.91%, 24.44%, 21.01%, 18.91%, and 18.63%, respectively (Figure 3B). These results indicate that soluble protein plays an important role in alfalfa heat stress response.

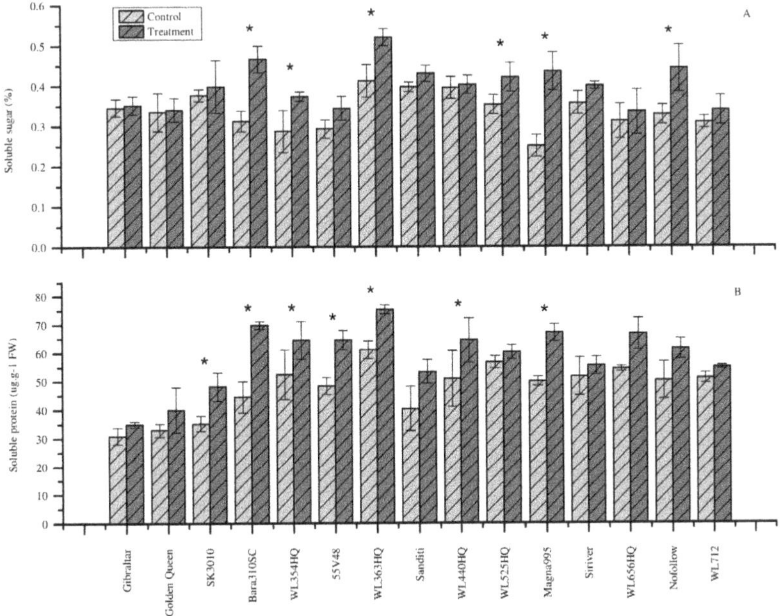

Figure 3. Effect of heat stress on soluble sugar and soluble protein content. (**A**) Soluble sugar content. (**B**) Soluble protein content. Each bar represents the mean ($n = 5$) and the error bar indicates the standard deviation. Asterisks indicate statistical significant differences between control and treatment plant ($p < 0.05$), independent sample t-test.

3.8. A Comprehensive Evaluation of the Heat Tolerance of Alfalfa Cultivars

All biomass and physiological traits were standardized for the comprehensive heat tolerance evaluation using the heat tolerance coefficient. The heat tolerance coefficient of each biomass and physiological traits (indexes) are presented in Table 3. The heat tolerance coefficients were further used to calculate the membership function values of alfalfa cultivars.

Table 3. Heat tolerance coefficients of biomass and physiological traits (indexes) of alfalfa cultivars.

Cultivars	SFW	SDW	RFW	RDW	TFB	TDB	RWC	EL	MDA	Chla	Chlb	Tchl	SS	SP
Gibraltar	68.76	66.16	74.29	47.05	70.54	60.31	86.97	236.10	155.76	59.70	77.41	66.89	111.38	112.94
Golden Queen	61.11	71.02	68.11	64.66	63.20	69.12	90.56	328.84	134.05	64.94	75.31	68.95	101.66	121.71
SK3010	76.67	63.64	68.60	70.24	74.27	64.99	91.55	186.63	176.98	78.96	82.25	80.25	105.62	136.49
Bara310SC	92.45	87.84	94.75	92.48	93.19	89.04	96.45	131.69	102.06	91.66	89.28	98.91	149.40	156.68
WL354HQ	66.53	44.90	79.13	70.10	69.88	51.31	84.71	208.66	133.16	74.50	65.56	70.98	130.37	122.89
55V48	65.71	68.58	82.10	54.55	70.36	64.98	88.57	299.32	132.07	76.00	85.49	79.60	117.63	133.17
WL363HQ	77.93	83.33	82.76	82.45	79.38	83.15	94.50	122.85	105.07	95.16	95.91	98.72	126.29	123.32
Sanditi	92.43	63.66	67.36	53.47	84.77	61.15	87.46	219.36	150.00	68.89	90.13	76.60	108.49	132.35
WL440HQ	82.50	89.59	89.08	55.13	84.55	80.96	87.58	254.19	107.80	85.60	68.47	78.13	102.01	126.59
WL525HQ	75.57	61.38	81.18	68.15	77.42	63.07	96.13	184.42	123.20	79.23	86.02	81.64	119.07	106.21
Magna995	79.58	82.14	97.58	101.63	85.27	87.12	96.28	137.04	113.47	97.79	91.58	95.22	174.57	134.26
Siriver	81.89	72.23	83.15	75.71	82.29	73.00	89.99	214.81	166.06	66.59	98.87	75.87	112.05	107.56
WL656HQ	81.74	78.14	76.08	61.50	80.01	74.39	90.99	157.36	103.87	66.04	87.96	98.17	107.40	122.42
Nofollow	84.02	73.24	91.47	69.00	83.68	71.75	90.08	150.95	103.87	92.76	98.88	95.23	135.21	122.09
WL712	71.23	60.85	84.97	50.25	75.73	57.81	86.72	213.66	128.23	68.72	55.86	63.43	110.16	107.78

SFW, shoot fresh weight; SDW, shoot dry weight; RFW, root fresh weight; RDW, root dry weight; TFB, total fresh biomass; TDB, total dry biomass; RWC, relative water content; EL, electrolyte leakage; MDA, malnodialdehyde content; SS, soluble sugar content; SP, soluble protein content.

Furthermore, correlation analysis was performed to investigate the relationship between traits. The results revealed that shoot fresh weight and shoot dry weight were strongly positively correlated with total fresh biomass and total dry biomass, respectively ($r = 0.94$) ($p < 0.01$) (Table 4). Root fresh weight was strongly positively correlated with Chl b and SS, and root dry weight was strongly positively correlated with SS, RWC, and total dry biomass (Table 4). Total fresh biomass was significantly negatively correlated with EL. Total fresh biomass was significantly positively correlated with total dry biomass, and total dry biomass was strongly positively correlated with Chl b ($r = 0.80$) and total Chl content ($r = 0.80$) (Table 4). These results indicated that electrolyte leakage and MDA content were negatively correlated with the rest of traits, which were significant for Chl b, total Chl, and SS for electrolyte leakage, and Chlb and total Chl for MDA. However, electrolyte leakage was positively correlated with MDA content ($r = 0.40$). Chl a ($r = 0.79$) and Chl b ($r = 0.94$) were strongly positively correlated with total Chl content. In addition, there was a positive correlation between soluble sugar and soluble protein content (Table 4). Hence, all biomass and physiological traits were used to evaluate the heat tolerance of all alfalfa cultivars.

The membership function value calculated from heat tolerance coefficient was used to evaluate the heat tolerance of all alfalfa cultivars. As shown in Table 5, Bara310SC and Magna995 cultivars had higher mean membership function value (0.86 and 0.80), respectively. By contrast, Gibraltar (0.26) and WL712 (0.26) showed lower mean membership function value. Based on this result, Bara310SC and Magna995 were ranked "one" and "two", respectively whereas Gibraltar and WL712 had the same rank, "14" (Table 5). Furthermore, the mean membership function value was used for Euclidean distance cluster analysis. The results showed that Bara310SC and Magna995 cultivars were clustered into one group, and Gibraltar and WL712 were clustered in another group (data are not shown). Finally, the rank and cluster results were combined to evaluate the heat tolerance of the cultivars. Thus, Bara310SC and Magna995 were found to be heat tolerant, whereas Gibraltar and WL712 were found to be heat-sensitive alfalfa cultivars. To screen the most heat-tolerant and heat-sensitive alfalfa, the four cultivars (Bara310SC, Magna995, Gibraltar, and WL712) were further evaluated by chlorophyll a fluorescence analysis.

3.9. Alteration of Chlorophyll a Fluorescence under Heat Stress

Chlorophyll a fluorescence was measured to further screen the most heat-tolerant and heat-sensitive alfalfa cultivars based on the photosynthesis behavior under heat stress. OJIP curve was constructed from fluorescence transient measurement. The effect of heat stress on basic photosynthetic parameters, specific energy fluxes, quantum yield and efficiency, and performance indexes were investigated.

Table 4. Correlation analysis of biomass and physiological traits.

	SFW	SDW	RFW	RDW	TFB	TDB	RWC	EL	MDA	CHLa	CHLb	Tchl	SS	SP
SFW	1.00	0.49	0.33	0.33	0.94 **	0.52 *	0.34	−0.61 *	−0.24	0.58 *	0.34	0.41	0.26	0.42
SDW		1.00	0.51 *	0.40	0.60 *	0.94 **	0.53 *	−0.29	−0.55 *	0.70 **	0.72 **	0.73 **	0.26	0.41
RFW			1.00	0.56 *	0.62 *	0.61 *	0.40	−0.48	−0.64 *	0.26	0.68 **	0.58 *	0.74 **	0.21
RDW				1.00	0.48	0.68 **	0.77 **	−0.64 *	−0.31	0.37	0.64 *	0.61 *	0.81 **	0.45
TFB					1.00	0.66 **	0.44	−0.67 **	−0.42	0.55 *	0.51	0.53 *	0.47	0.42
TDB						1.00	0.70 **	−0.46	−0.56 *	0.70 **	0.80 **	0.80 **	0.49	0.49
RWC							1.00	−0.57 *	−0.37	0.41	0.56 *	0.57 *	0.55 *	0.32
EL								1.00	0.40	−0.38	−0.62 *	−0.58 *	−0.61 *	−0.23
MDA									1.00	−0.38	−0.53 *	−0.61 *	−0.44	−0.22
CHLa										1.00	0.63 *	0.79 **	0.33	0.65 **
CHLb											1.00	0.94 **	0.57 *	0.42
Tchl												1.00	0.54 *	0.59 *
SS													1.00	0.43
SP														1.00

** Correlation is significant at the 0.01 level (2-tailed). * Correlation is significant at the 0.05 level (2-tailed). SFW, shoot fresh weight; SDW, shoot dry weight; RFW, root fresh weight; RDW, root dry weight; TFB, total fresh biomass; TDB, total dry biomass; RWC, relative water content; EL, electrolyte leakage; MDA, malnodialdehyde content; SS, soluble sugar content; SP, soluble protein content.

Table 5. Membership function value of alfalfa cultivars.

Cultivars	SFW	SDW	RFW	RDW	TFB	TDB	RWC	EL	MDA	Chla	Chlb	Tchl	SS	TSP	Mean	Rank
Gibraltar	0.24	0.48	0.23	0.00	0.24	0.24	0.19	0.81	0.28	0.00	0.50	0.10	0.13	0.13	0.26	14
Golden Queen	0.00	0.58	0.02	0.32	0.00	0.47	0.50	0.50	0.57	0.14	0.45	0.16	0.00	0.31	0.29	13
SK3010	0.50	0.42	0.04	0.42	0.37	0.36	0.58	0.42	0.00	0.51	0.61	0.47	0.05	0.60	0.38	9
Bara310SC	1.00	0.96	0.91	0.83	1.00	1.00	1.00	0.00	1.00	0.84	0.78	1.00	0.65	1.00	0.86	1
WL354HQ	0.17	0.00	0.39	0.42	0.22	0.00	0.00	1.00	0.58	0.39	0.23	0.21	0.39	0.33	0.31	12
55V48	0.15	0.53	0.49	0.14	0.24	0.36	0.33	0.67	0.60	0.43	0.69	0.46	0.22	0.53	0.42	7
WL363HQ	0.54	0.86	0.51	0.65	0.54	0.84	0.83	0.17	0.96	0.93	0.93	0.99	0.34	0.34	0.67	3
Sanditi	1.00	0.42	0.00	0.12	0.72	0.26	0.23	0.77	0.36	0.24	0.80	0.37	0.09	0.52	0.42	10
WL440HQ	0.68	1.00	0.72	0.15	0.71	0.79	0.24	0.76	0.92	0.68	0.29	0.41	0.00	0.40	0.55	5
WL525HQ	0.46	0.37	0.46	0.39	0.47	0.31	0.97	0.03	0.72	0.51	0.70	0.51	0.24	0.00	0.44	11
Magna995	0.59	0.83	1.00	1.00	0.74	0.95	0.99	0.01	0.85	1.00	0.83	0.90	1.00	0.56	0.80	2
Siriver	0.66	0.61	0.52	0.53	0.64	0.57	0.45	0.55	0.15	0.18	1.00	0.35	0.14	0.03	0.46	8
WL656HQ	0.66	0.74	0.29	0.26	0.56	0.61	0.53	0.47	0.98	0.17	0.75	0.98	0.08	0.32	0.53	6
Nofollow	0.73	0.63	0.80	0.40	0.68	0.54	0.46	0.54	0.98	0.87	1.00	0.90	0.46	0.31	0.67	4
WL712	0.32	0.36	0.58	0.06	0.42	0.17	0.17	0.73	0.45	0.24	0.00	0.00	0.11	0.03	0.26	14

SFW, shoot fresh weight; SDW, shoot dry weight; RFW, root fresh weight; RDW, root dry weight; TFB, total fresh biomass; TDB, total dry biomass; RWC, relative water content; EL, electrolyte leakage; MDA, malnodialdehyde content; SS, soluble sugar content; SP, soluble protein content.

3.9.1. OJIP Transient Curve

OJIP transient curve was constructed based on the fluorescence measurement relative to the time (Figure 4). OJIP transient curves of control groups were higher than those of heat treatment groups for all cultivars. The JIP-test was applied to further investigate the structural alteration, functional parameters, and photosynthetic behaviors under heat stress treatment. Basic fluorescence parameters, specific energy fluxes, quantum yield efficiency, and performance index were extracted and analyzed.

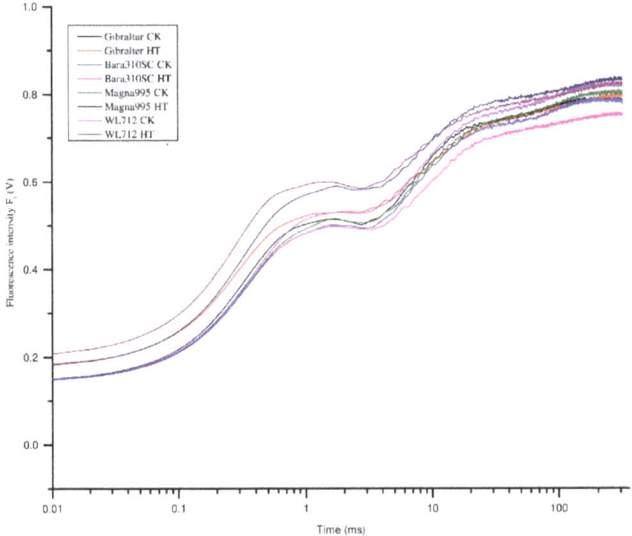

Figure 4. Effect of heat stress on the chlorophyll *a* fluorescence transient (OJIP curves) of 4 alfalfa cultivars after seven days of heat treatment. CK represents control treatment and HT represents heat treatment (*n* = 5).

3.9.2. Basic Photosynthetic Parameters (F0, Fj, Fi, Fm, F300 μs, and Fv/Fm)

The basic fluorescence parameters were extracted from the OJIP transient curve (Table 6). Heat stress increased the F0 of Gibraltar, Magna99,5 and WL712 alfalfa cultivars, and was significant for Gibraltar and WL712 cultivars when compared with the control (Table 6). Heat stress also increased the Fm (Maximal fluorescence) of alfalfa. Relative to the control, heat stress decreased the Fv/Fm of Gibraltar, Bara310SC, Magna995, and WL712 by 6.10%, 1.25%, 2.50%, and 6.25%, respectively, and was significant in WL712 cultivar (Table 6). The results indicated that WL712 was highly affected by heat stress, which revealed high heat sensitivity.

3.9.3. Specific Energy Fluxes (TP0/RC, ETO/RC, RE0/RC, and ABS/RC)

Heat stress affected the specific energy fluxes of alfalfa. All heat-treated plants showed higher TPO/RC and were significant for WL712 (Table 6). Cultivars showed different responses in ETO/RC under heat stress. Heat stress significantly decreased the ETO/RC of WL712 but increased in Bara310SC and Magna995 (relatively heat tolerant) (Table 6). Unlike other cultivars, heat stress reduced the RE0/RC of Bara310SC (Table 6). On the other hand, heat-treated Gibraltar, Magna995, and WL712 plants showed higher ABS/RC compared to the control and was significant for WL712 (Table 6).

Table 6. Photosynthetic parameters extracted from OJIP fluorescence transients.

	Gibraltar		Bara310SC		Magna995		WL712	
	CK	**HT**	**CK**	**HT**	**CK**	**HT**	**CK**	**HT**
	Data extracted from OJIP fluorescence transient curves							
F0	0.15	0.19 *	0.16	0.18	0.18	0.19	0.18	0.22 *
Fj	0.50	0.53	0.49	0.50	0.51	0.58 *	0.53	0.60 *
Fi	0.72	0.73	0.72	0.70	0.74	0.77	0.76	0.77
Fm	0.84	0.85	0.81	0.84	0.86	0.89	0.87	0.87
F300µs	0.37	0.43	0.37	0.36	0.36	0.43	0.35	0.50 *
Fv/Fm	0.82	0.77	0.80	0.79	0.80	0.78	0.80	0.75 *
	Specific energy fluxes (per active PS II reaction center)							
TPo/RC	2.53	2.75	2.41	2.44	2.22	2.47	1.92	2.95
ETo/RC	1.25	1.31	1.27	1.17	1.14	1.08	0.95	1.24
REo/RC	0.42	0.50	0.43	0.39	0.40	0.41	0.32	0.50
ABS/RC	3.11	3.55	3.04	3.04	2.79	3.16	2.41	3.93
	Quantum yields and efficiencies							
ϕpo	0.82	0.77	0.79	0.80	0.80	0.78	0.80	0.75 *
φEo	0.41	0.37	0.41	0.38	0.41	0.34 *	0.39	0.32 *
δRo	0.34	0.38	0.34	0.34	0.35	0.38	0.34	0.39
RC/ABS	0.33	0.28	0.33	0.33	0.36	0.32	0.42	0.26 *
	Performance indexes							
PI$_{ABS}$	1.08	0.60 *	0.90	0.86	1.01	0.65 *	1.10	0.37 *
PItotal	0.63	0.36	0.48	0.44	0.55	0.38	0.57	0.25 *

F0, minimal fluorescence; Fj, fluorescence intensity at the J-step (2 ms) of OJIP; Fi, fluorescence intensity at the I-step (30 ms) of OJIP; Fm, maximal fluorescence; F300 µs, fluorescence intensity at 300 µs; Fv/Fm, maximum quantum yield of photosystem; ABS/RC, absorption flux (of antenna Chls) per RC; TR0/RC, trapping flux (leading to QA reduction) per RC; ETO/RC, electron transport flux (further than QA$^-$) per RC; REO/RC, electron flux reducing end electron acceptors at the PS I acceptor side, per RC; φpo, maximum quantum yield for primary photochemistry, namely FV/FM; φEo, quantum yield of the electron transport flux from QA to QB; δRo, efficiency/probability with which an electron from the intersystem electron carriers moves to reduce end electron acceptors at the PSI acceptor side (RE); RC/ABS, QA-reducing RCs per PSII antenna Chl (reciprocal of ABS/RC). The data in the table are mean ($n = 5$), and asterisks indicate statistical significance difference between control (CK) and heat treatment (HT) at $p < 0.05$, independent sample t-test.

3.9.4. Quantum Yield and Efficiency (φpo, φEo, δRo, and RC/ABS)

Heat stress affected all quantum yield efficiency components (Table 6). Heat stress noticeably decreased the φpo and δRo of all alfalfa cultivars except Bara310SC. Relative to control, WL712 showed significant reductions in φpo, φEo, and RC/ABS under heat stress. Meanwhile, heat stress significantly decreased the φEo of Magna995 (Table 6).

3.9.5. Performance Indexes (PI$_{ABS}$ and PItotal)

Heat stress markedly decreased the photosynthetic performance indexes (PIABS and PItotal) of all alfalfa cultivars (Table 6). Meanwhile, heat stress significantly reduced the PIABS of Gibralter, Magna995, and WL712 by 44.44%, 35.64%, and 66.36%, respectively, compared to control ($p < 0.05$). In addition, WL712 showed significantly lower PItotal under heat stress, indicating its more heat sensitivity than others. However, heat stress had no significant effect on the performance indexes of Bara310SC (Table 6).

4. Discussion

High-temperature stress changes morphological, biochemical, and physiological processes of plants [4,24]. In this study, heat stress had an obvious negative effect on alfalfa plant biomass and most of the cultivars showed significant reductions in biomass following heat stress treatment. It has been reported that heat stress causes leaf wilting, leaf curling, leaf yellowing, reduction in shoot growth, root growth, root number, root diameter, plant height, and biomass [1,25]. Thus, the decrease in alfalfa plant biomass could be associated with the reduction in plant height, wilting, and falling off of leaves caused by heat stress. Our results were consistent with previous reports in maize [26], sugarcane [27], and wheat [28]. The biomass of most alfalfa cultivars was significantly affected, indicating that heat stress has a huge impact on alfalfa productivity. Like biomass, heat stress caused a significant reduction in the RWC of most alfalfa cultivars. Our results are in agreement with previous reports in rice [29] and wheat [28]. Similar with the findings of Sita et al. [30] heat-tolerant alfalfa cultivars had higher RWC than heat-sensitive ones. The decrease in leaf water content might affect plant metabolism and decrease plant growth and biomass. The reduction in leaf relative water content could be associated with the reduction in the number, mass, and growth of the roots under heat stress, which ultimately limits the supply of water and nutrients to the above-ground parts of the plant [3]. Taken together, significant reductions in biomass and RWC could be indicators of alfalfa heat stress sensitivity.

It is well documented that the plant membrane is sensitive to various abiotic stresses, and stress condition increased lipid peroxidation and impaired membrane selectivity [31]. In this study, both EL and MDA content, which are indicators of stress sensitivity, were higher in heat-treated alfalfa plants compared to the control. The membrane stability of most alfalfa cultivars was significantly affected by heat stress, which may reveal the heat sensitivity of the cultivars. Our results were in agreement with the findings of Kumar et al. [32] in chickpea, Sita et al. [30] in lentil, and Hu et al. [21] in tall fescue, who reported higher EL under heat stress. The increase in lipid peroxidation might be as a result of the overproduction and accumulation of ROS, which then causes membrane peroxidation, protein degradation, and DNA damage to severely inhibit growth [33–35]. Consequently, our results revealed that heat stress highly damaged the membrane integrity and stability of alfalfa, especially heat-sensitive cultivars. However, some cultivars like Bara310SC and Magna995 had lower EL and MDA content than others, indicating that these cultivars could maintain their membrane integrity and stability under heat stress. It has been reported that the maintenance of membrane integrity and stability under stress conditions is a major component of tolerance [36] and is essential to sustained photosynthetic and respiratory performance [37]. Thus, Bara310SC and Magna995 with lower EL and MDA content after heat stress could be heat tolerant.

It is well established that photosynthetic pigments such as chlorophyll a and b are sensitive to high-temperature stress. Heat stress results in plant leaf pigment loss and significantly damages photosynthetic activities [38]. In the present study, heat treatment decreased the chlorophyll content (Chl a, Chl b, and total Chl) of all alfalfa cultivars, and a more pronounced effect was observed in heat-sensitive cultivars. The decrease in chlorophyll content might be attributed to the chlorophyll degradation or inhibition of chlorophyll biosynthesis [39]. In addition, the effect of high temperature on the pigments and other photosynthetic apparatus is due to the production of toxic oxygen species (oxidative damage) and reduction in antioxidative defense [32]. Thus, significant increases in EL and MDA content and a decrease in chlorophyll content might be interconnected, indicating the heat sensitivity of the cultivars. Our result revealed that cultivars with higher EL and MDA content had lower chlorophyll content, biomass, and RWC, suggesting that heat stress had a greater effect on some cultivars including WL712, Gibraltar, and Golden Queen.

When exposed to heat stress, plants accumulate compatible solutes, such as soluble sugar, to protect the plant from stress-induced damage by maintaining membrane stability and cell water balance, and by buffering the cellular redox potential and homeostasis [40]. The accumulation of compatible solute is an important adaptive mechanism, directly participating in osmotic adjustment [41]. Similarly, soluble proteins, which are induced by stress, play a role in stress tolerance, presumably via hydration of cellular structures [27]. In the current study, heat stress increased the soluble sugar and protein content of alfalfa plants. We found significant and higher soluble sugar content in Bara310SC, Magna995, WL363HQ. These cultivars could adjust their osmotic balance and cellular homeostasis, which is one of the tolerance mechanisms. Similar results were reported in lettuce [42] and moth bean [43]. In this study, we found higher relative water content in heat-tolerant alfalfa, which was in agreement with soluble sugar and protein content, thus entailing great implications for heat tolerance [27]. The result revealed that soluble sugar and protein could play a considerable role for alfalfa heat tolerance by maintaining the water balance and cellular homeostasis. In this study, cultivars with higher soluble sugar and protein content had lower lipid peroxidation and membrane damage under heat stress, which implies that soluble sugar and protein ameliorates heat-induced damage in those cultivars. Our results are also supported by those of Lang-Mladek et al. [44], who stated that osmolyte production under heat stress is thought to increase protein stability and stabilize the structure of the membrane bilayer. Similarly, Khan et al. [45] and Kumar et al. [46] found significantly higher soluble protein content in wheat under heat stress, and maximum accumulation of soluble protein content was observed in thermotolerant wheat genotypes, which was similar with our result. The overall results indicate that soluble sugar and protein play a remarkable role in alleviating heat-induced damage of alfalfa, while increasing heat tolerance.

The heat tolerance of alfalfa was evaluated comprehensively by membership function value, and the results showed that alfalfa cultivars had different sensitivities to heat stress. Cultivars with higher membership function value were considered as heat tolerant, whereas cultivars with lower membership function value were heat sensitive. As shown, Bara310SC and Magna995 had higher membership function value, 0.86 and 0.80, respectively, and were considered as relatively heat tolerant. By contrast, with lower membership function value (0.24), Gibraltar and WL712 were relatively heat-sensitive alfalfa. Furthermore, chlorophyll a fluorescence analysis was performed on the above four cultivars to identify the most heat sensitive and heat tolerant alfalfa cultivars. Chlorophyll a fluorescence analysis is a powerful and nondestructive method to study the photosynthetic behavior of plants and is widely applied to screen tolerant species [21,47]. In this study, heat stress affected the chlorophyll a fluorescence and altered the OJIP fluorescence transient curve of all alfalfa cultivars. The change in the OJIP fluorescence transient curve might have been caused by the oxidation of electron transport chains, which results from the reduction in the electron donor of PSII reaction centers under high temperature. Basic photosynthetic parameters were extracted from fluorescence transient and the results showed that heat stress increased the F0 and decreased Fv/Fm of all alfalfa cultivars, which were significant for sensitive cultivars (WL712). Higher F0 indicates the elevated damage of chloroplast by heat stress, resulting in blocked energy transfer to the PS II traps and a decrease of the quantum efficiency of PS II [48]. Fv/Fm is commonly used to analyze heat-induced damage to PSII [47], and heat stress decreases Fv/Fm in a range of plant species [49]. For many plant species, the approximate optimal Fv/Fm value is in the range of 0.79 to 0.84, with lowered values indicating plant stress [50]. The heat-tolerant cultivar (Bara310SC) had 0.79 and 0.80 Fv/Fm under heat stress and control condition, respectively.

Heat stress altered the specific energy flux parameters (TP0/RC, ETO/RC, RE0/RC, and ABS/RC) of all alfalfa cultivars. ABS/RC and TR0/RC were higher under heat stress, which indicates the inactivation of absorption and trapping reaction centers. Similar results have been reported by Zushi et al. [51] in tomato leaf and fruit under heat stress. In addition, heat stress markedly altered the quantum yield and efficiency parameters (φpo, φEo, and δRo). The result revealed that the behaviors of PS II on both the electron donor and acceptor side were blocked due to heat stress and PSI was less damaged than PSII [51]. The energy fluxes such as φpo, φEo, and RC/ABS of PS II were lower, whereas δRo was higher under heat treatment. Similarly, the decrease in RC/ABS was observed in heat-stressed Spirulina [52]. On the other hand, Stefanov et al. [53] reported an increase of δRo in bean plants immediately after heat treatment. This result suggested a difference in energy flux between PS I and PS II in response to heat. It was reported that PS II is the most temperature-sensitive component of the photosynthetic apparatus [54,55]. Our results also confirmed that PS II is more sensitive to heat stress than PS I, which could be as a result of thylakoid membrane fluidity caused by heat stress.

Alteration of specific energy fluxes and quantum yield efficiency of the photosystem could affect the overall photosynthetic performance of alfalfa. Performance indexes (PI_{ABS} and PItotal) were measured to investigate the changes in leaf photosynthetic performance. The performance index (PIABS) is a parameter sensitive to various types of stress. PItotal reveals the changes in intersystem electrons and the energy conservation from exciton to the reduction of PSI end acceptors [56]. Heat stress noticeably decreased the performance indexes of all heat-treated alfalfa cultivars, and a significant decrease was observed in sensitive cultivar (WL712). Another heat sensitive cultivar (Gibraltar) showed a significant reduction in PI_{ABS}. A similar result was reported in tall fescue under heat stress [21]. In addition, Fahad et al. [29] reported a significant reduction in the photosynthetic activities of two rice cultivars under high day and night temperatures. The decrease in performance indexes could indicate the lower photochemistry of PSII [57]. However, heat stress had no significant effect on the performance indexes of heat-tolerant cultivars (Bara310SC and Magna995), and Bara310SC showed higher performance indexes under heat stress than others. The result revealed that Bara310SC was more tolerant to heat stress compared to others under heat stress. The overall Chl a fluorescence analysis results suggested that Bara310SC and WL712 are the most heat-tolerant and heat-sensitive alfalfa cultivars, respectively. Further studies will be done to understand the molecular mechanisms of the heat tolerance of alfalfa.

5. Conclusions

Heat stress affected the biomass and physiological characteristics of alfalfa cultivars. The more pronounced effect was observed in sensitive cultivars such as WL712, Gibraltar, and Golden Queen, as evidenced by a significant decrease in biomass, RWC, chlorophyll content, and photosynthetic performance, and significant increases in EL and MDA under heat stress. Heat-tolerant cultivars showed significantly higher soluble sugar and protein content and performed better under heat stress. Among the fifteen cultivars evaluated, Bara310SC was the most heat tolerant and WL712 was the most heat-sensitive one. These cultivars can be used to explore the molecular mechanisms of heat tolerance in alfalfa plants.

Author Contributions: Conceptualization, M.W. and L.C.; methodology, M.W., W.Z. and Q.Z.; software, M.W. and K.J.; validation, M.W.; W.Z. and L.C.; formal analysis, M.W.; investigation, M.W.; resources, L.C.; data curation, M.W., Q.Z. and W.Z.; writing—original draft preparation, M.W.; writing—review and editing, M.W. and L.C.; visualization, M.W.; supervision, L.C. and K.J.; project administration, L.C.; funding acquisition, L.C.

Funding: This work was funded by the National Natural Science Foundation of China (NSFC) (Grant Nos. 31672482 and 31401915).

Conflicts of Interest: The authors declare no conflict of interest.

References

1. Ahuja, I.; de Vos, R.C.H.; Bones, A.M.; Hall, R.D. Plant molecular stress responses face climate change. *Trends Plant Sci.* **2010**, *15*, 664–674. [CrossRef] [PubMed]

2. Hasanuzzaman, M.; Nahar, K.; Alam, M.; Roychowdhury, R.; Fujita, M. Physiological, biochemical, and molecular mechanisms of heat stress tolerance in plants. *Int. J. Mol. Sci.* **2013**, *14*, 9643–9684. [CrossRef] [PubMed]

3. Bañon, S.; Fernandez, J.A.; Franco, J.A.; Torrecillas, A.; Alarcón, J.J.; Sánchez-Blanco, M.J. Effects of water stress and night temperature preconditioning on water relations and morphological and anatomical changes of Lotus creticus plants. *Sci. Hortic.* **2004**, *101*, 333–342. [CrossRef]

4. Wahid, A.; Gelani, S.; Ashraf, M.; Foolad, M.R. Heat tolerance in plants: An overview. *Environ. Exp. Bot.* **2007**, *61*, 199–223. [CrossRef]

5. Li, S.; Li, F.; Wang, J.; Zhang, W.E.N.; Meng, Q.; Chen, T.H.H.; Lycopersicon, T. Glycinebetaine enhances the tolerance of tomato plants to high temperature during germination of seeds and growth of seedlings. *Plant Cell Environ.* **2011**, *34*, 1931–1943. [CrossRef] [PubMed]

6. Vollenweider, P.; Gu, M.S. Diagnosis of abiotic and biotic stress factors using the visible symptoms in foliage. *Environ. Pollut.* **2005**, *137*. [CrossRef] [PubMed]

7. Bi, A.; Fan, J.; Hu, Z.; Wang, G.; Amombo, E.; Fu, J.; Hu, T. Differential Acclimation of Enzymatic Antioxidant Metabolism and Photosystem II Photochemistry in Tall Fescue under Drought and Heat and the Combined Stresses. *Front. Plant Sci.* **2016**, *7*. [CrossRef]

8. Hu, T.; Liu, S.; Amombo, E.; Fu, J. Stress memory induced rearrangements of HSP transcription, photosystem II photochemistry and metabolism of tall fescue (*Festuca arundinacea* Schreb) in response to high-temperature stress. *Front. Plant Sci.* **2015**, *6*, 1–13. [CrossRef]

9. Xu, W.; Cai, S.; Zhang, Y.; Wang, Y.; Ahammed, G.J.; Xia, X.; Campus, Z. Melatonin enhances thermotolerance by promoting cellular protein protection in tomato plants. *Front. Plant Sci.* **2016**. [CrossRef]

10. Potters, G.; Pasternak, T.P.; Guisez, Y.; Palme, K.J.; Jansen, M.A.K. Stress-induced morphogenic responses: Growing out of trouble? *Trends Plant Sci.* **2007**, *12*. [CrossRef]

11. Sade, B.; Soylu, S.; Yetim, E. Drought and oxidative stress. *Afr. J. Biotechnol.* **2011**, *10*, 11102–11109. [CrossRef]

12. Barnabás, B.; Jäger, K.; Fehér, A. The effect of drought and heat stress on reproductive. *Plant Cell Environ.* **2008**, *31*, 11–38. [CrossRef] [PubMed]

13. Mo, Y.; Liang, G.; Shi, W.; Xie, J. Metabolic responses of alfalfa (*Medicago Sativa* L.) leaves to low and high temperature induced stresses. *Afr. J. Biotechnol.* **2011**, *10*, 1117–1124. [CrossRef]

14. Song, Y.; Lv, J.; Ma, Z.; Dong, W. The mechanism of alfalfa (*Medicago sativa* L.) response to abiotic stress. *Plant Growth Regul.* **2019**. [CrossRef]

15. Aranjuelo, I.; Irigoyen, J.J.; Sánchez-Díaz, M. Effect of increased temperature and drought associated to climate change on productivity of nodulated alfalfa. Quality in lucerne and medics for animal production. In Proceedings of the XIV Eucarpia Medicago spp. Group Meeting, Zaragoza and Lleida, Lleida, Spain, 12–15 January 2001.

16. Aranjuelo, I.; Irigoyen, J.J.; Sanchez-Diaz, M. Effect of elevated temperature and water availability on CO_2 exchange and nitrogen fixation of nodulated alfalfa plants. *Environ. Exp. Bot.* **2007**, *59*, 99–108. [CrossRef]

17. Erice, G.; Irigoyen, J.J.; Sánchez-Díaz, M.; Avice, J.C.; Ourry, A. Effect of drought, elevated CO_2 and temperature on accumulation of N and vegetative storage proteins (VSP) in taproot of nodulated alfalfa before and after cutting. *Plant Sci.* **2007**, *172*, 903–912. [CrossRef]

18. Li, W.; Wei, Z.; Qiao, Z.; Wu, Z.; Cheng, L.; Wang, Y. Proteomics analysis of alfalfa response to heat stress. *PLoS ONE* **2013**, *8*, e82725. [CrossRef]

19. Zhang, Q.; Liu, X.; Zhang, Z.; Liu, N.; Li, D.; Hu, L. Melatonin Improved Waterlogging Tolerance in Alfalfa (*Medicago sativa*) by Reprogramming Polyamine and Ethylene Metabolism. *Front. Plant Sci.* **2019**, *10*, 1–14. [CrossRef]

20. Huang, X.; Shi, H.; Hu, Z.; Liu, A.; Amombo, E.; Chen, L.; Fu, J. ABA Is Involved in Regulation of Cold Stress Response in Bermudagrass. *Front. Plant Sci.* **2017**, *8*, 1–10. [CrossRef]

21. Hu, L.; Bi, A.; Hu, Z.; Amombo, E.; Li, H.; Fu, J. Antioxidant Metabolism, Photosystem II, and Fatty Acid Composition of Two Tall Fescue Genotypes with Different Heat Tolerance Under High Temperature Stress. *Front. Plant Sci.* **2018**, *9*, 1–13. [CrossRef]

22. Bradford, M. A rapid and sensitive method for the quantification of microgram quantities of protein utilizing the principle of protein-dye binding. *Anal. Biochem.* **1976**, *254*, 248–254. [CrossRef]

23. Tian, Z.; Yang, Y.; Wang, F. A comprehensive evaluation of heat tolerance in nine cultivars of marigold. *Hortic. Environ. Biotechnol.* **2015**, *56*, 749–755. [CrossRef]

24. Bita, C.E.; Gerats, T. Plant tolerance to high temperature in a changing environment: Scientific fundamentals and production of heat stress-tolerant crops. *Front. Plant Sci.* **2013**, *4*, 1–18. [CrossRef] [PubMed]

25. Siddiqui, M.H.; Al-Khaishany, M.Y.; Al-Qutami, M.A.; Al-Whaibi, M.H.; Grover, A.; Ali, H.M.; Al-Wahibi, M.S. Morphological and physiological characterization of different genotypes of faba bean under heat stress. *Saudi J. Biol. Sci.* **2015**, *22*, 656–663. [CrossRef] [PubMed]

26. Ashraf, M.; Hafeez, M. Thermotolerance of pearl millet and maize at early growth stages: Growth and nutrient relations. *Biol. Plant.* **2004**, *48*, 81–86. [CrossRef]

27. Wahid, A.; Close, T.J. Expression of dehydrins under heat stress and their relationship with water relations of sugarcane leaves. *Biol. Plant.* **2007**, *51*, 104–109. [CrossRef]

28. Hameed, A.; Goher, M.; Iqbal, N. Heat Stress-Induced Cell Death, Changes in Antioxidants, Lipid Peroxidation, and Protease Activity in Wheat Leaves. *J. Plant Growth Regul.* **2012**, *31*, 283–291. [CrossRef]

29. Fahad, S.; Hussain, S.; Saud, S.; Hassan, S.; Tanveer, M. A combined application of biochar and phosphorus alleviates heat-induced adversities on physiological, agronomical and quality attributes of rice. *Plant Physiol. Biochem.* **2016**, *103*, 191–198. [CrossRef]

30. Sita, K.; Sehgal, A.; Kumar, J.; Kumar, S.; Singh, S.; Siddique, K.H.M.; Nayyar, H. Identification of High-Temperature Tolerant Lentil (*Lens culinaris* Medik.) Genotypes through Leaf and Pollen Traits. *Front. Plant Sci.* **2017**, *8*, 1–27. [CrossRef]

31. Dias, A.S.; Barreiro, M.G.; Campos, P.S.; Ramalho, J.C.; Lidon, F.C. Wheat cellular membrane thermotolerance under heat stress. *J. Agron. Crop Sci.* **2010**, *196*, 100–108. [CrossRef]

32. Kumar, S.; Thakur, P.; Kaushal, N.; Malik, J.A.; Gaur, P.; Nayyar, H. Effect of varying high temperatures during reproductive growth on reproductive function, oxidative stress and seed yield in chickpea genotypes differing in heat sensitivity. *Arch. Agron. Soil Sci.* **2013**, *59*, 823–843. [CrossRef]

33. Awasthi, R.; Bhandari, K.; Nayyar, H. Temperature stress and redox homeostasis in agricultural crops. *Front. Environ. Sci.* **2015**, *3*. [CrossRef]

34. Chakraborty, U.; Pradhan, B. Drought stress-induced oxidative stress and antioxidative responses in four wheat (*Triticum aestivum* L.) varieties. *Arch. Agron. Soil Sci.* **2012**, *58*, 617–630. [CrossRef]

35. Zhang, H.; Liu, X.-L.; Zhang, R.-X.; Yuan, H.-Y.; Wang, M.-M.; Yang, H.-Y.; Liang, Z.-W. Root Damage under Alkaline Stress Is Associated with Reactive Oxygen Species Accumulation in Rice (*Oryza sativa* L.). *Front. Plant Sci.* **2017**, *8*, 1–12. [CrossRef] [PubMed]

36. Bajji, M.; Kinet, J.M.; Lutts, S. The use of the electrolyte leakage method for assessing cell membrane stability as a water stress tolerance test in durum wheat. *Plant Growth Regul.* **2002**, *36*, 61–70. [CrossRef]

37. Chen, J.; Wang, P.; Mi, H.L.; Chen, G.Y.; Xu, D.Q. Reversible association of ribulose-1, 5-bisphosphate carboxylase/oxygenase activase with the thylakoid membrane depends upon the ATP level and pH in rice without heat stress. *J. Exp. Bot.* **2010**, *61*, 2939–2950. [CrossRef] [PubMed]

38. Awasthi, R.; Kaushal, N.; Vadez, V.; Turner, N.C.; Berger, J.; Siddique, K.H.M.; Nayyar, H. Individual and combined effects of transient drought and heat stress on carbon assimilation and seed filling in chickpea. *Funct. Plant Biol.* **2014**, *41*, 1148–1167. [CrossRef]

39. Dutta, S.; Mohanty, S.; Tripathy, B.C. Role of Temperature Stress on Chloroplast Biogenesis and Protein Import in Pea. *Plant Physiol.* **2009**, *150*, 1050–1061. [CrossRef]

40. Janska, A.; Marsik, P.; Zelenkova, S.; Ovesna, J. Cold stress and acclimation: What is important for metabolic adjustment? *Plant Biol.* **2010**, *12*, 395–405. [CrossRef]

41. Gupta, N.K.; Agarwal, S.; Agarwal, V.P.; Nathawat, N.S.; Gupta, S.; Singh, G. Effect of short-term heat stress on growth, physiology and antioxidative defence system in wheat seedlings. *Acta Physiol. Plant.* **2013**, *35*, 1837–1842. [CrossRef]

42. Han, Y.; Fan, S.; Zhang, Q.; Wang, Y. Effect of heat stress on the MDA, proline and soluble sugar content in leaf lettuce seedlings. *Agric. Sci.* **2013**, *4*, 112–115. [CrossRef]

43. Harsh, A.; Sharma, Y.K.; Joshi, U.; Rampuria, S.; Singh, G.; Kumar, S.; Sharma, R. Effect of short-term heat stress on total sugars, proline and some antioxidant enzymes in moth bean (*Vigna aconitifolia*). *Ann. Agric. Sci.* **2016**, *61*, 57–64. [CrossRef]

44. Lang-Mladek, C.; Popova, O.; Kiok, K.; Berlinger, M.; Rakic, B.; Aufsatz, W.; Luschnig, C. Transgenerational inheritance and resetting of stress-induced loss of epigenetic gene silencing in arabidopsis. *Mol. Plant* **2010**, *3*, 594–602. [CrossRef] [PubMed]

45. Khan, S.U.; Din, J.U.; Qayyum, A.; Jan, N.E.; Jenks, M.A. Heat tolerance indicators in Pakistani wheat (*Triticum aestivum* L.) genotypes. *Acta Bot. Croat.* **2015**, *74*, 109–121. [CrossRef]

46. Kumar, S.; Beena, A.S.; Awana, M.; Singh, A. Physiological, Biochemical, Epigenetic and Molecular Analyses of Wheat (*Triticum aestivum*) Genotypes with Contrasting Salt Tolerance. *Front. Plant Sci.* **2017**, *8*, 1–20. [CrossRef] [PubMed]

47. Porcar-Castell, A.; Tyystjärvi, E.; Atherton, J.; Van Der Tol, C.; Flexas, J.; Pfündel, E.E.; Berry, J.A. Linking chlorophyll a fluorescence to photosynthesis for remote sensing applications: Mechanisms and challenges. *J. Exp. Bot.* **2014**, *65*, 4065–4095. [CrossRef] [PubMed]

48. Krause, G.H.; Winter, K.; Krause, B.; Virgo, A. Light-stimulated heat tolerance in leaves of two neotropical tree species, Ficus insipida and Calophyllum longifolium. *Funct. Plant Biol.* **2015**, *42*, 42–51. [CrossRef]

49. Tan, W.; wei Meng, Q.; Brestic, M.; Olsovska, K.; Yang, X. Photosynthesis is improved by exogenous calcium in heat-stressed tobacco plants. *J. Plant Physiol.* **2011**, *168*, 2063–2071. [CrossRef] [PubMed]

50. Maxwell, K.; Johnson, G.N. Chlorophyll fluorescence—A practical guide. *J. Exp. Bot.* **2000**, *51*, 659–668. [CrossRef] [PubMed]

51. Zushi, K.; Kajiwara, S.; Matsuzoe, N. Chlorophyll a fluorescence OJIP transient as a tool to characterize and evaluate response to heat and chilling stress in tomato leaf and fruit. *Sci. Hortic.* **2012**, *148*, 39–46. [CrossRef]

52. Zhao, B.; Wang, J.; Gong, H.; Wen, X.; Ren, H.; Lu, C. Effects of heat stress on PSII photochemistry in a cyanobacterium *Spirulina platensis*. *Plant Sci.* **2008**, *175*, 556–564. [CrossRef]

53. Stefanov, D.; Petkova, V.; Denev, I.D. Screening for heat tolerance in common bean (*Phaseolus vulgaris* L.) lines and cultivars using JIP-test. *Sci. Hortic.* **2011**, *128*, 1–6. [CrossRef]

54. Apostolova, E.L.; Dobrikova, A.G. Effect of high temperature and UV-A radiation on photosystem II. In *Handbook of Plant and Crop Stress*; Pessarakli, M., Ed.; CRC Press: Boca Raton, FL, USA, 2011; pp. 577–591.

55. Wen, X.; Qiu, N.; Lu, Q.; Lu, C. Enhanced thermotolerance of photosystem II in salt-adapted plants of the halophyte Artemisia anethifolia. *Planta* **2005**, *220*, 486–497. [CrossRef] [PubMed]

56. Yusuf, M.A.; Kumar, D.; Rajwanshi, R.; Strasser, R.J.; Tsimilli-Michael, M.; Govindjee; Sarin, N.B. Overexpression of γ-tocopherol methyl transferase gene in transgenic Brassica juncea plants alleviates abiotic stress: Physiological and chlorophyll a fluorescence measurements. *Biochim. Biophys. Acta Bioenerg.* **2010**, *1797*, 1428–1438. [CrossRef]

57. Kalaji, H.M.; Jajoo, A.; Oukarroum, A.; Brestic, M.; Zivcak, M.; Samborska, I.A.; Ladle, R.J. Chlorophyll a fluorescence as a tool to monitor physiological status of plants under abiotic stress conditions. *Acta Physiol. Plant.* **2016**, *38*. [CrossRef]

Article

Transcriptomic Analysis Reveals the Temporal and Spatial Changes in Physiological Process and Gene Expression in Common Buckwheat (*Fagopyrum esculentum* Moench) Grown under Drought Stress

Zehao Hou [1,†], Junliang Yin [1,†], Yifei Lu [1], Jinghan Song [1], Shuping Wang [1], Shudong Wei [2], Zhixiong Liu [3], Yingxin Zhang [1] and Zhengwu Fang [1,*]

1   College of Agriculture, Yangtze University, Jingzhou 434000, Hubei, China;
    201771374@yangtzeu.edu.cn (Z.H.); yinjunliang@yangtzeu.edu.cn (J.Y.); 201872410@yangtzeu.edu.cn (Y.L.);
    201875095@yangtzeu.edu.cn (J.S.); wangshuping@yangtzeu.edu.cn (S.W.); 518009@yangtzeu.edu.cn (Y.Z.)
2   College of Life Science, Yangtze University, Jingzhou 434000, Hubei, China; 500896@yangtzeu.edu.cn
3   College of Horticulture and Gardening, Yangtze University, Jingzhou 434000, Hubei, China;
    zxliu77@yahoo.com
*   Correspondence: fangzhengwu88@yangtzeu.edu.cn; Tel.: +86-137-9732-9339
†   These authors contributed equally to this work.

Received: 17 August 2019; Accepted: 19 September 2019; Published: 20 September 2019

Abstract: Common buckwheat is a traditional alternative crop that originated from the northwest of China and is widely cultivated worldwide. However, common buckwheat is highly sensitive to drought stress, especially at the seedling stage, and the molecular mechanisms underlying the response to drought stress still remain elusive. In this study, we analyzed the stress phenotypes of buckwheat seedlings under drought condition. The results showed the wrinkled cotyledon due to the decrease of relative water content (RWC) in response to the increased activity of antioxidant enzymes. Transcriptomic analysis was further performed to analyze the regulation patterns of stress-responding genes in common buckwheat cotyledons and roots under drought stress conditions. Characterizations of the differentially expressed genes (DEGs) revealed differential regulation of genes involved in the photosynthesis and oxidoreductase activity in cotyledon, and that they were highly related to the post-transcriptional modification and metabolic process in root. There were 180 drought-inducible transcription factors identified in both cotyledons and roots of the common buckwheat. Our analysis not only identified the drought responsive DEGs and indicated their possible roles in stress adaption, but also primarily studied the molecular mechanisms regulating the drought stress response in common buckwheat.

Keywords: common buckwheat; cotyledon; root; drought stress; transcriptome analysis

---

1. Introduction

Among the forms of environmental stress, drought stress has been considered as one of the major constraints in plant growth, survival, and production [1,2]. A lack of water not only disturbs photosynthesis, limits metabolic reactions, and inhibits CO_2 exchange, but also results in stress-related damage to chloroplasts [3–5]. In order to adapt to the extreme environments, plants have evolved several mechanisms (e.g., drought escape, avoidance, and tolerance) to ensure high survival rates under drought stress [5,6]. Specifically, plants recruit a variety of responding mechanisms to deal with drought stress [7–9], such as stomatal closure, leaf rolling, and alteration in biosynthetic and antioxidant pathways, which are highly regulated by complex transcriptional networks [10].

Drought stress affects several physiological and biochemical pathways in plants [11]. Previous research has shown that the water deficits not only affect the chlorophyll biosynthesis, but also the level of malondialdehyde (MDA) and the relative water contents (RWC) of the plant, brining detrimental effects to the lipid peroxidation, and membrane constitution [12,13]. Also, the abiotic stresses can further induce the oxidative stress through generating reactive oxygen species (ROS), a prevalently recognized destroyer in cellular metabolism [14–16]. ROS generate the oxidation of photosynthetic pigments, initiate lipid peroxidation, and degrade proteins in plants, and thereby cause damage to cell structures and metabolism, particularly those associated with photosynthesis [17,18]. To counteract the effects of oxidative stress, plants have developed an efficient detoxification defense system consisting of non-enzymatic scavengers and enzymatic components to scavenge free ROS [19,20]. In terms of the enzymatic scavenging, a series of antioxidative enzymes, including peroxidase (POD), catalase (CAT), superoxide dismutase (SOD), and ascorbate peroxidase (APX), have been reported to play a vital role in reducing the damage effects (i.e., water deficiency) caused by drought stress [21]. There is evidence that keeping a high antioxidative enzyme activity level to reduce the damaging effects caused by water deficit stress may be associated with the osmotic stress tolerance of plants [22], which is also found to be positively related to plant drought tolerance [23].

Presently, there are several drought-inducible genes, including stress responses and resistance, which have been identified through transcriptome analyses [1]. These genes can be divided into two groups according their functions. The first group is composed of function proteins that include the late embryogenesis abundant (LEA) proteins, ROS detoxification enzymes, molecular chaperones, heat shock proteins (HSP), and lipid-transfer proteins [24,25]. The second group is involved in regulatory proteins or transcriptional factors (TFs), which correlate with the signal transduction and stress-responsive gene expressions, for example, the phospholipases and dehydration-responsive elements [26,27]. In order to elucidate the biological functions of these genes, several transgenic plants overexpressing various drought-resistant genes have been generated, which have both shown enhanced drought tolerance and growth retardation [28–31], demonstrating that plants may adapt to the drought environment at the expense of normal growth [1].

Common buckwheat (*Fagopyrum esculentum* Moench) is an important dual-purpose alternativecrops originated from Yunnan Province of China [32] and is widely cultivated around the world, especially in China, Japan, and Russia [33]. Because of its abundant nutrients in seeds, common buckwheat is considered as one of the sources of flour, groats, and whole grain foods. However, common buckwheat is highly sensitive to drought, especially at the seedling stage [34,35], and short-term drought occurs frequently in China, posing a threat to domestic food safety [36]. Thus, it would be important and necessary to study the physiological and molecular bases of osmotic stress tolerance in common buckwheat. We have previously identified *FeDREB1L* (GenBank: JN600617.1), a CBF/DREB homologous gene, from common buckwheat, and overexpression of the *FeDREB1L* gene was found to significantly increase the water deficit resistance of transgenic *Arabidopsis* [31]. In order to further understand the drought-resistant mechanism and identify novel water-deficit-related genes in common buckwheat, a transcriptomics analysis was carried out to investigate the variations in common buckwheat growth under short-term drought treatment, and the phenotypes and biochemical traits of seedlings were also analyzed. Our results may provide more information with regard to the transcriptional control of common buckwheat under the abiotic stresses, and help to identify the novel genes that are potentially valuable for future common buckwheat breeding.

2. Materials and Methods

2.1. Plant Material and Drought Treatments

Common buckwheats (cv. Xi'nong 9976) were germinated in Petri dishes in an incubator (plant growth incubator JY412L, Shanghai, China) in darkness (25 °C) and relative humidity of approximately 60%. After germination for 36 h, when the root length of the seedlings grew to approximately 2 cm, the seedlings were transplanted for hydroponics in an incubator with 12 h photoperiods (25 °C/20 °C, day/night temperature) and relative humidity of approximately 60%. The 7-day old buckwheat seedlings were treated with 15% polyethylene glycol 6000 (PEG 6000) solution for 1 d, 3 d, and 5 d. After treatment, the cotyledons and roots were collected and quickly frozen in liquid nitrogen and stored at −80 °C until used. The seedlings before drought treatment were served as the control. Each treatment was carried out in three biological replicates.

2.2. Physiological Measurement

Relative water content (RWC) was determined according to the formula described by Pan et al. [36]. The chlorophyll content and chlorophyll a/b ratio was calculated using to the method described by Harper et al. [37]. The changes of malondialdehyde (MDA) concentration were determined using the thiobarbituric acid (TBA) reaction [38], and the activities of POD and CAT were detected according to the description of Harper et al. [37]. The Rubisco activities were assayed with Rubisco assay kits (Beijing Solarbio Science and Technology Co., Ltd., Beijing, China) according to the manufacturer's instructions.

2.3. RNA Isolation and Transcriptome Sequencing

Total RNA was isolated from the non-treated control and drought-stressed cotyledon and root samples using EASYspin Plus Plant RNA Kit (Aidlab, Wuhan, China). The RNA quality was checked by Agilent bioanalyzer 2100 (Agilent Technologies, Santa Clara, CA, USA), and Nanodrop 2000 r spectrophotometer (Nanodrop Technologies, Wilmington, DE, USA) was used for RNA quantification.

Transcriptome sequencing was performed at Beijing Allwegene Technology Co. Ltd. (Beijing, China), following manufacturer protocols. Briefly, mRNA was enriched from total RNA using Oligo (dT) magnetic beads, and the mRNA was fragmented into small pieces using a fragmentation buffer. Then, these fragments were used as reverse transcription to synthesized the first- and second-strand cDNA, and the second-stand cDNA were purified with AMPure XP Beads Kit, repaired, poly (A) added, and ligated to paired-end adapters. Finally, the cDNA libraries were sequenced on Illumina HiSeqTM 2500 platform. Each sample had three biological replicates. The raw reads were submitted to the National Center for Biotechnology Information (NCBI) Sequence Read Archive with a Bioproject ID: PRJNA555746.

The raw reads in FASTQ format were processed using in-house Perl scripts, and the high-quality clean data were obtained by removing the low-quality data, which included the reads that contained the adapter, and more than 10% of N nucleotides, and the low-quality reads that contained more than 50% of low quality bases (Q-value ≤ 20). In addition, we calculated the Q30, GC content, and sequence duplication levels for the clean data. Cleaned and qualified reads were aligned against the *F. esculentum* reference genome [39] using Tophat2 software [40]. Then, these sequences were subjected to functional annotation and coding sequence (CDS) prediction [41], and the resulting sequences were called genes. Finally, fragments per kilobase of transcript permillionmapped reads (FPKM) method was used to calculate the gene expression unit.

2.4. Identification and Functional Annotation of Different Expressed Genes (DEGs)

The differential gene expression analysis was carried out using DESeq software, and DEGs were determined by combining a q value cutoff of 0.05 and adjusting to |log2 (fold change)| ≥ 1. For DEG functional annotation, Gene Ontology (GO) enrichment analysis was carried out by GOseq software, and Kyoto Encyclopedia of Genes and Genomes (KEGG) was used to perform pathway enrichment analysis of DEGs. In addition, the gene expression profiles at the pathway were display by MapMan software (version 3.6.0) [33].

2.5. Quantitative Real-Time PCR (qRT-PCR) Analysis

Total RNA from cotyledons and roots of both samples were extracted using EASYspin Plus Plant RNA Kit (Aidlab, Wuhan, China) following the manufacturer's protocols, and the first-strand cDNA for qPCR analysis was synthesized from 500 ng of total RNA using PrimeScript RT Reagent Kit with gDNA Eraser (Takara, Dalian, China) following the manufacturer's instructions, and cDNA was diluted 10-fold and used as the template for qRT-PCR. The primers were designed using Primer Premier 5.0 and beta-actin was used as a reference gene, with the primer information listed in Table S1. qRT-PCR was performed on a CFX 96 real-time PCR system (BioRad, Hercules, CA, USA) using TB Green (TaKaRa), according to the manufacturer's protocols. PCR amplification was conducted in a volume of 20 µL, containing ~100 ng of cDNA template, 0.6 µL of each primer (10 µmol), and 10 µL PCR-mix (2×). The conditions for all reactions were as follows: 30 s at 95 °C, followed by 40 cycles of 10 s at 95 °C, and 30 s at 55 °C, and the melting curve was generated to confirm the PCR specificity. The non-treated control treatment was chosen as the control to standardizing all samples, using the $2^{-\Delta\Delta Ct}$ method to calculate the relative expression levels [42].

2.6. Statistical Analysis

A one-way ANOVA was carried out by SPSS Statistics 19.0 software (IBM Corp, Armonk, NY, USA), and means were compared using the Duncan test to determine significant differences ($p < 0.05$). The results were presented as mean ± SD (standard deviation).

3. Results

3.1. Changes in Phenotype of Common Buckwheat Seedlings at Drought Stress

To investigate the dynamic phenotypic changes of common buckwheat seedlings in response to drought stress treatments, plant height, root length, and relative water content (RWC) were measured under 15% PEG 6000 solution treatments across four time-points (0, 1, 3, and 5 days). There was no significant change in plant height, root length, and RWC of control samples after 1, 3, and 5 days (Table S2). Under drought stress, the buckwheat seedlings showed stress phenotypes of wrinkled cotyledon (Figure 1a), but the plant height and root length did not significantly change during the treatment (Figure 1b,c). The RCW is generally used as an important indicator of plant water status under osmotic conditions, and in this study, the RCW values were clearly decreased in the 3 and 5 day-treated (DPT3d and DPT5d) seedlings, but were not significantly different among the control (CK) and the 1 day-treated (DPT1d) seedlings (Figure 1d).

3.2. Changes in Physiology of Common Buckwheat Seedlings under Drought Conditions

To investigate the physiological changes under different levels of water deficit conditions, the MDA content and the activities of POD and CAT of cotyledons were measured after drought treatment. Compared with the drought treatment, these physiological traits were not significantly changed during the whole treatment under control conditions, in which the content of chlorophyll a and chlorophyll b were increased in the 1, 3, and 5 day control plants; however, the chlorophyll a/b ratios were not significantly changed during the whole treatment under the control condition (Table S3). Under water deficit condition, the MDA content was greatly increased from 0 to 1 days, and then slightly increased until 5 days (Figure 2a). Meanwhile, the activities of POD and CAT were significantly increased under PEG treatment (Figure 2b,c). The chlorophyll a content was elevated in the 3 and 5 dy-treated plants (Figure 2d), while the content of chlorophyll b content was increased in the 1, 3, and 5 day-treated plants (Figure 2e). In addition, there was no difference between the control and 1 and 3 day-treated seedlings in chlorophyll a/b ratios, but marked decreases were observed in the 5 day-treated plants (Figure 2f). These results indicated that there were significant changes in the physiology of the common buckwheat seedlings in response to osmotic stress.

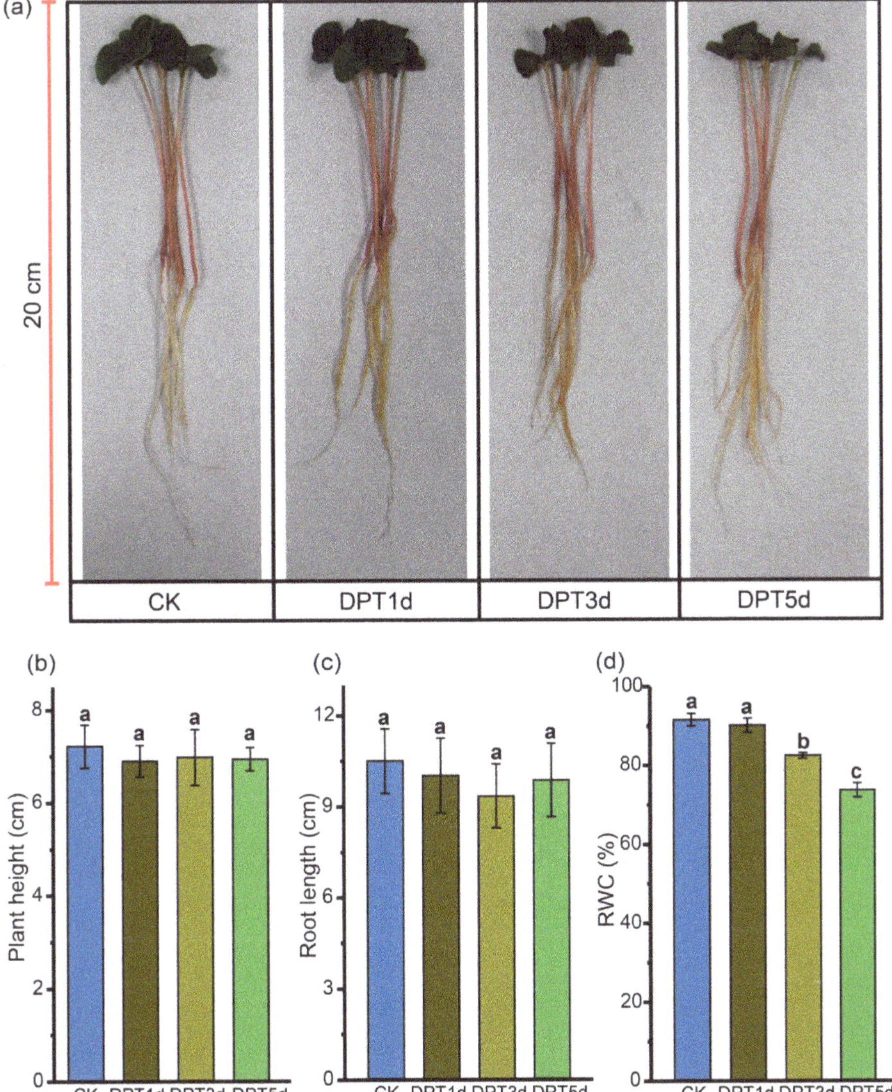

Figure 1. Changes in phenotype of common buckwheat seedlings under drought stress. (**a**) A photograph of common buckwheat seedlings after PEG treatments; CK, non-stressed control; DPT1d, DPT3d, and DPT5d, drought treatment with PEG solution for 1, 3, and 5 days, respectively; bar = 20 cm. The change of (**b**) plant height, (**c**) root length, and (**d**) relative water content of cotyledon during drought stress treatment. Bars represent means of three replicates ± SD (standard deviation). Different letters indicate means that are significantly different at the $p < 0.05$ level among different drought conditions.

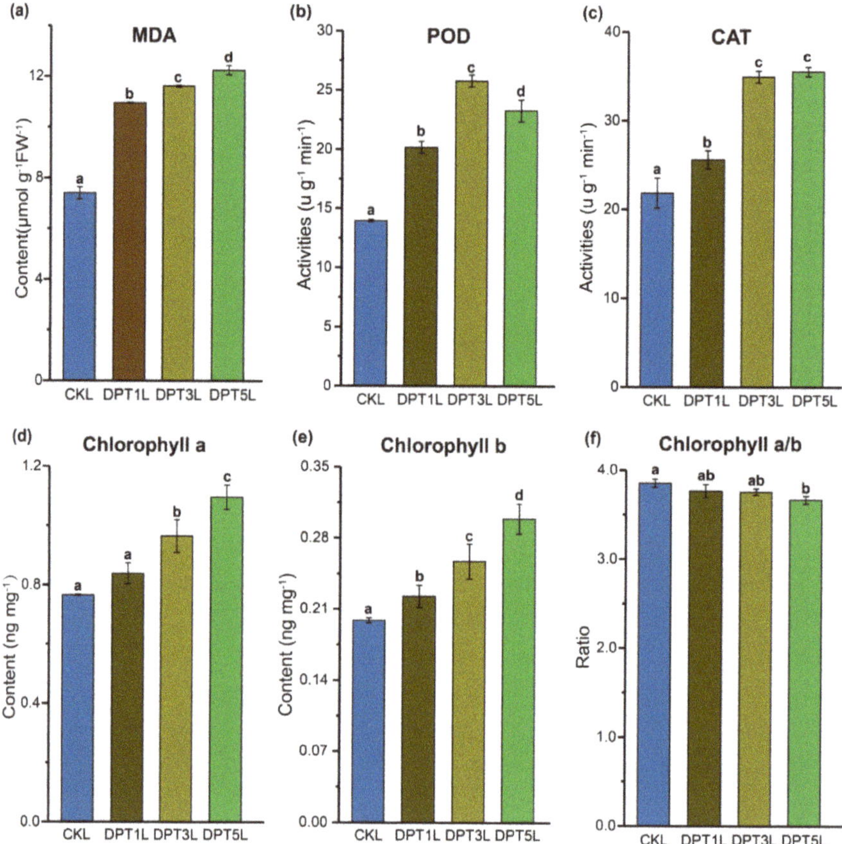

Figure 2. Changes in physiology in cotyledons of common buckwheat seedlings under different drought stress conditions. (**a**) Changes in the malondialdehyde (MDA) content of cotyledons, (**b**) changes in the peroxidase (POD) activities of cotyledons, (**c**) changes in the catalase (CAT) activities of cotyledons, (**d**) chlorophyll a content in the cotyledons, (**e**) chlorophyll b content in the cotyledons, and (**f**) ratios of chlorophyll a/b.

3.3. Overview of the Common Buckwheat Transcriptome and Identification of DEGs

To reveal the expression changes in common buckwheat cotyledons and roots at 0, 1, 3, and 5 days after PEG treatments, 24 common buckwheat samples (including 12 cotyledon samples and 12 root samples) were used for RNA-Seq analysis to further investigate the changes at the transcriptional level. A full-scale sequencing analysis from 24 cDNA samples is shown in Table S4, 88.76 gigabytes (Gb) of clean reads were abstained, and the percentages of the Q30 base of these 24 common buckwheat samples were greater than or equal to 95.70%. Furthermore, there was a highly mapped efficiency between the samples and reference genome (79.46%–87.91%), which met the requirements for information analysis. There were 877,111 CDS (coding sequence)-encoded proteins, and the length of CDS is shown in Figure 3a. Of these CDSs, only a minority (3729 CDSs, 0.43%) were more than 1500 nt, and 90.57% of CDSs appeared with a length ranging from 0 to 500 nt. In addition, the principal component analysis (PCA) was performed and the results demonstrated that the control treatment was clearly separated from the drought stress treatment in the cotyledons or roots (Figure 3b), suggesting that the gene expression pattern of common buckwheat was greatly changed under drought condition.

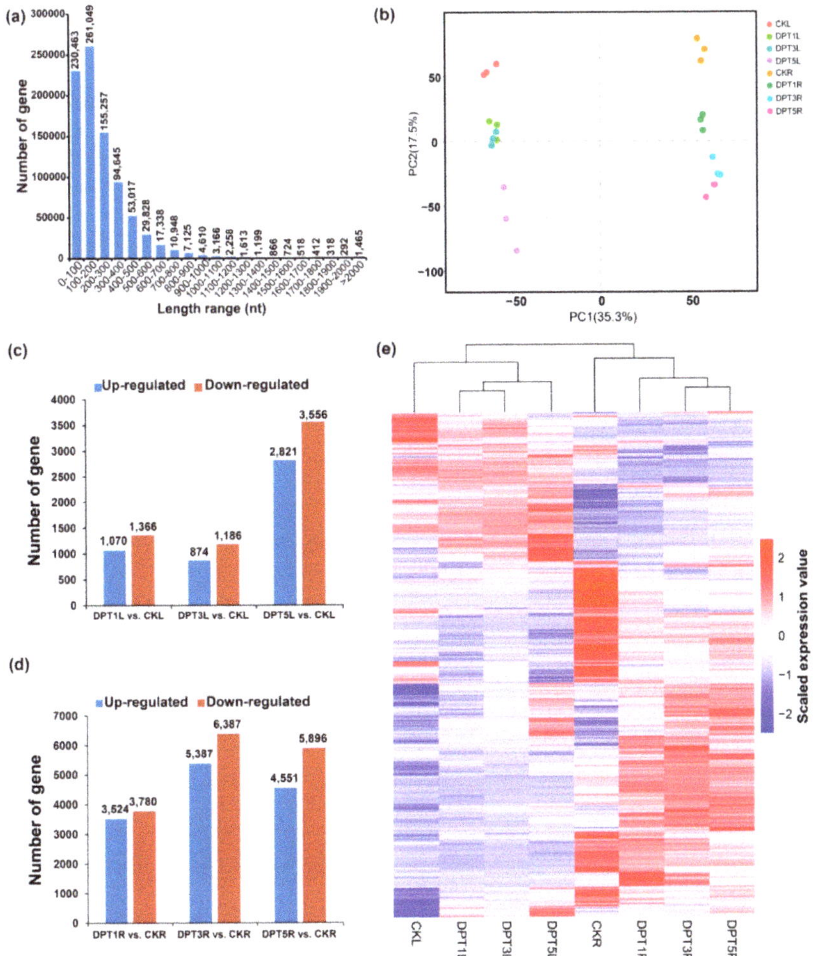

Figure 3. Overview of the transcriptomic results and changes in gene expression profiles in cotyledons and roots after drought stress treatment. (**a**) Predicted length distribution map of coding sequence (CDS)-encoded protein nucleotide (nt). (**b**) Principal component (PC) analysis of gene expression at different drought stress conditions. (**c**) Numbers of differently expressed genes (DEGs) in cotyledons of common buckwheat seedlings at different drought stress conditions in pairwise comparisons. (**d**) DEGs in roots of common buckwheat seedlings at different drought stress conditions in pairwise comparisons. (**e**) Heat-map graphics exhibiting the gene expression levels of total DEGs. CKL and CKR are the cotyledon samples and root samples of the non-stressed control, respectively. DPT1L, DPT3L, and DPT5L are the cotyledon samples of drought treatment for 1, 3, and 5 days, respectively. DPT1R, DPT3R, and DPT5R are the root samples of drought treatment for 1, 3, and 5 days, respectively. DPT1L vs. CKL, DPT3L vs. CKL, and DPT5L vs. CKL are the cotyledon samples of drought treatment for 1, 3, and 5 days compared to the non-stressed control, respectively. DPT1R vs. CKR, DPT3R vs. CKR, and DPT5R vs. CKR are the root samples of drought treatment for 1 day compared to the non-stressed control, respectively. Up-regulated means that genes were up-regulated in drought stress conditions compared to the non-stressed control and down-regulated means that genes were down-regulated in the drought stress condition compared to the non-stressed control.

The gene expression levels were calculated as FPKM values via HTSeq software analysis, and the differential gene expression analysis was carried out using DESeq software. There were 2436, 2060, and 6377 DEGs identified in the cotyledons after drought treated for 1, 3, and 5 days, respectively (Figure 3c). In the roots, compared with the control treatment, 7304, 11,774, and 10,447 DEGs were identified after drought treatment for 1, 3, and 5 days, respectively (Figure 3d). Upon drought stress exposure, more DEGs were identified in the roots than in cotyledons, suggesting that there were different drought stress response mechanisms between roots and leaves in common buckwheat. Furthermore, in order to provide a comprehensive understanding of the change in gene expression of common buckwheat under drought conditions, a heat map was developed, as shown in Figure 3e, to exhibit the overall changes of the gene expression under water deficit conditions.

3.4. Comprehensive Sets of DEGs in the Cotyledons of Drought-Treated Common Buckwheat Seedlings

Venn diagrams showed that the number of genes commonly up-regulated under the DPT1L and DPT5L were greater than the number of genes commonly up-regulated under the DPT3L and DPT5L, and the number of commonly down-regulated genes showed the same trend (Figure 4a,b). Gene Ontology (GO) was used to find the functional significance of the identified DEGs (Figure 4c), and the GO terms related to signaling and DNA modification were detected in the set of genes up-regulated under the drought conditions, while the GO terms related to light harvesting and light reaction were detected in the sets of genes down-regulated under the water deficit conditions. Furthermore, the DEGs related to the light reactions of photosynthesis and the Calvin cycle were visualized through MapMan analysis (Figure 4d), and the expression of most of these DEGs was decreased in the DPT3L and DPT5L, compared with control treatment. RubisCO, as the major photosynthetic enzyme in plants, plays a crucial role in photosynthesis of green plants. In this study, the activities of RubisCO were significantly decreased at 3 days and greatly declined at 5 days under drought treatment (Figure 4e). These results indicate that photosynthesis in the cotyledons of the common buckwheat seedlings decreased under drought stress conditions. The representative genes related to ABA (abscisic acid) metabolism are listed in Table S5 according to their functional description, and most of these (including 6 *NCED* (*9-cis-epoxycarotenoid dioxygenase*), 3 B3 domain-containing protein, and 3 protein phosphatase 2C) were significantly up-regulated in DPT5L, which indicated that common buckwheat seedlings may use the ABA regulatory systems to affect leaf wilting and defense against the water-deficit stress.

To confirm and investigate the transcriptomic data, qRT-PCR was performed to check the expression levels of several genes, and the expression of *LCHb*, a gene encode chlorophyll a/b binding protein, was markedly decreased under DPT3L and DPT5L conditions (Figure 4f). The expression of *NCED*, a gene involved in ABA biosynthesis, was observably increased under drought stress conditions (Figure 4i). Furthermore, the *DREB1L* gene, which encodes a stress tolerance-related protein, was markedly enhanced in its expression level under drought stress treatment, compared with the non-stressed control plants (Figure 4m). The correlation coefficient (R^2) between RNA-Seq data and qPCR results for the 24 total plots was 0.8743 (Figure S1). These analyses of gene expression confirmed that the transcriptomic datasets were efficacious (Figure 4f–n).

3.5. Comprehensive Sets of DEGs in the Roots of Drought Treated Common Buckwheat Seedlings

Three comparison groups were constructed to further understanding the universal response in root of common buckwheat to drought stress. As shown in Figure 5a, 1731 genes were both up-regulated in DPT1R, DPT3R, and DPT5R, compared with CKR. Furthermore, 2857 genes were collectively down-regulated in DPT1R, DPT3R, and DPT5R (Figure 5b). All non-overlapped DEGs in the three comparison groups were subjected to GO enrichment analysis, and 430, 606, and 621 GO accessions classified into three categories comprising "molecular function", "biological process", and "cellular component" were identified in DPT1R vs. CKR, DPT3R vs. CKR, and DPT5R vs. CKR, respectively (Table S6). The drought-induced DEGs were mainly involved in the nucleotide binding,

ATP binding, macromolecule modification, protein phosphorylation, protein modification, protein metabolic process, and cellular protein metabolic process (Figure 5c). According to Mapman software analysis, there were 92 protein modification and phosphorylation-related DEGs that were up-regulated after drought treatment (Table S7), suggesting that they were candidate genes for protein modification in roots of drought-treated seedlings.

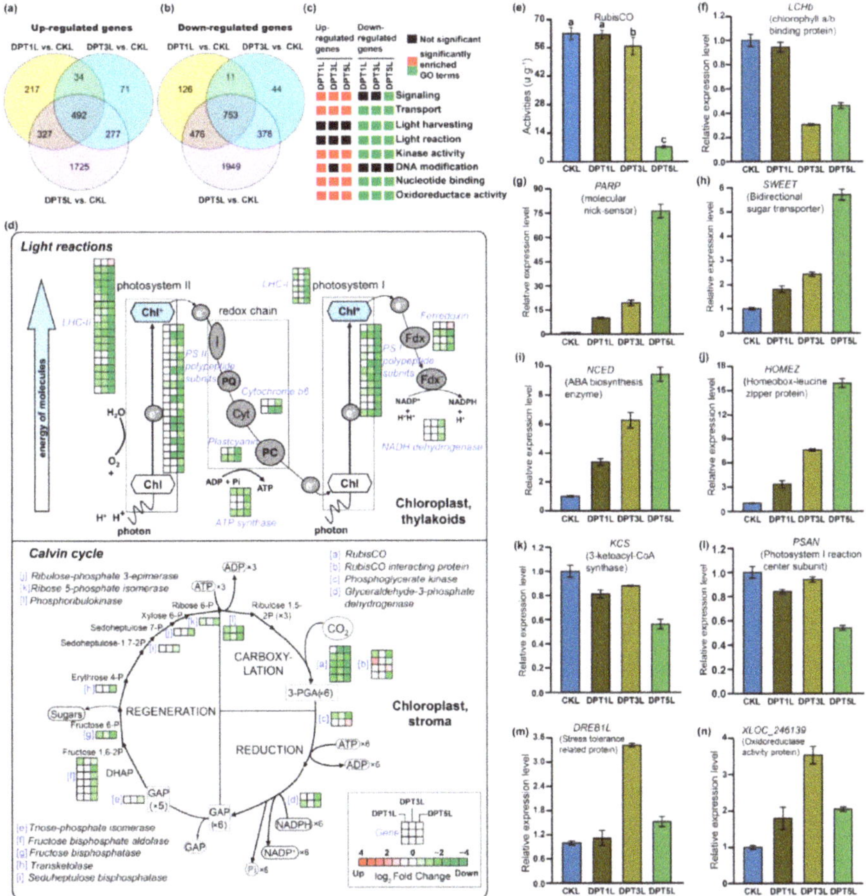

Figure 4. Comprehensive expression patterns of DEGs in the cotyledons of common buckwheat seedlings under drought stress conditions. (**a**) Venn diagrams of the numbers of up-regulated [|log2 (Fold Change)| > 1 and q-value < 0.005] genes acquired through the transcriptome analysis. (**b**) Venn diagrams of the numbers of down-regulated genes acquired through the transcriptome analysis. (**c**) Over-represented Gene Ontology (GO) terms estimated using GOseq software. (**d**) Changes in the expression of photosynthesis-related genes. Pathway diagram of light and dark reactions of photosynthesis with superimposed color-coded squares showing DEGs, drawn using MapMan. (**e**) Changes in the RubisCO activities of cotyledons. (**f–n**) Expression profiles of the selected DEGs, *LCHb* (**f**), *PARP* (**g**), *SWEET* (**h**), *NCED* (**i**), *HOMEZ* (**j**), *KCS* (**k**), *PSAN* (**l**), *DREB1L* (**m**), and *XLOC_246139* (**n**) determined using qRT-PCR analyses.

The expression patterns of several genes related to stress tolerance in the roots were analyzed via qRT-PCR (Figure 5d–k). The expression of *AAO*, a gene encoding L-ascorbate oxidase that plays a crucial role in plant cell growth, was markedly decreased under drought stress conditions (Figure 5h), and the expression of some transcription factors, such as *HOMEZ* and *DREB1L*, were significantly induced by water deficit (Figure 5e,j). In addition, the R^2 between the two experiments was 0.9385 (Figure S2). These results confirmed the effectiveness of the transcriptomic datasets and indicated that drought stress strongly affected the expression level of the genes that are related to stress tolerance in the roots.

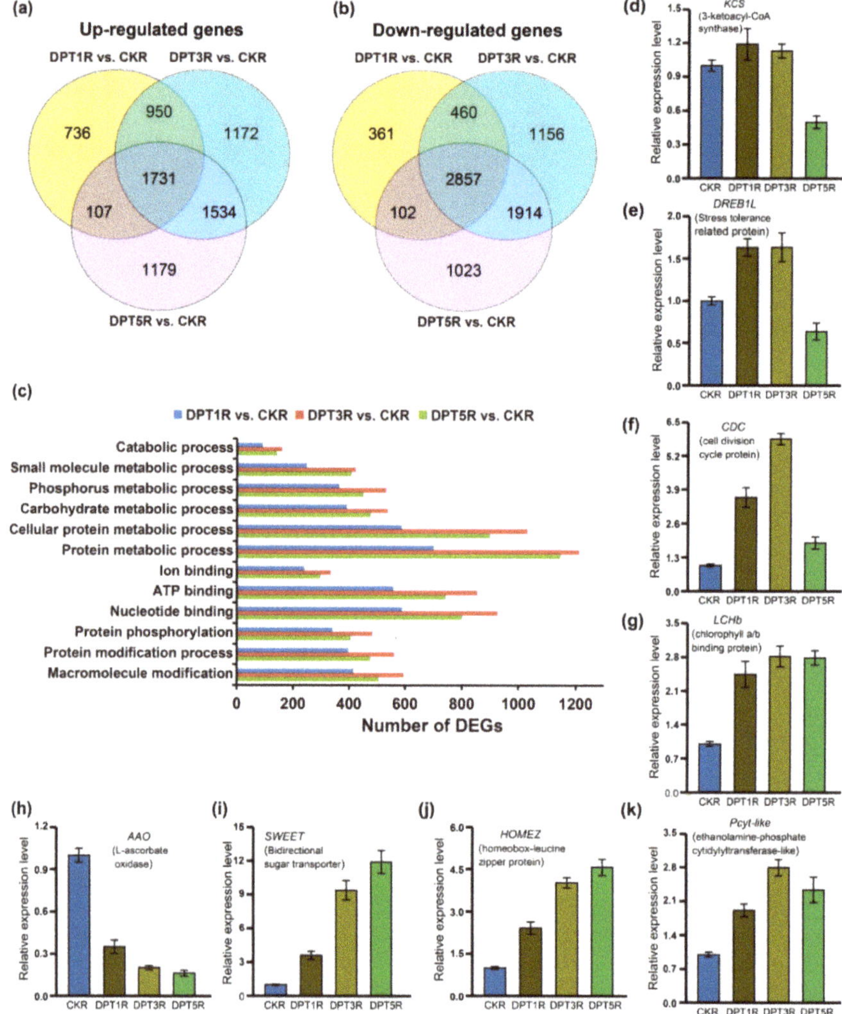

Figure 5. Comprehensive expression patterns of DEGs in the roots of common buckwheat seedlings under drought stress conditions. (**a**) Venn diagrams of the numbers of up-regulated [|log2 (Fold Change)| > 1 and qvalue < 0.005] genes acquired through the transcriptome analysis. (**b**) Venn diagrams of the numbers of down-regulated genes acquired through the transcriptome analysis. (**c**) Classification of DEGs based on metabolism, binding and modification categories. (**d–k**) Expression profiles of the selected DEGs, *KCS* (**d**), *DREB1L* (**e**), *CDC* (**f**), *LCHb* (**g**), *AAO* (**h**), *SWEET* (**i**), *HOMEZ* (**j**), *Pcyt-like* (**k**), determined by qRT-PCR.

3.6. Change in the Expression of Transcription Factors (TFs) Associated with Drought-Stress Response in Common Buckwheat Seedlings

Transcription factors (TFs) are important for regulating plant response to abiotic and biotic stresses. In the roots of the drought treated seedlings, large numbers of TFs were identified as DEGs, compared to the cotyledons (Figure 6a), and there were 180 TFs that were commonly identified in response to drought stress in both cotyledons and roots. Among them, the most differentially expressed TF families were the C2C2 family, follow by MYB, bZIP, HB and AP2/ERF (Figure 6b, Table S8). According to the GO enrichment analysis, 30.0%, 12.2%, and 9.4% of TFs were classified into "biological regulation", "intracellular", and "nucleic acid binding", respectively (Table S8). In addition, in order to reflect the major trends and patterns, 180 TFs were assigned to six clusters on the basis of their expression patterns. Those in cluster 1 and 2 were up-regulated by drought conditions, but the expression levels of the cluster 1 genes were high at cotyledons, while the cluster 2 genes were highly expressed at roots. Meanwhile, the cluster 3 and 4 genes showed the lowest expression level at cotyledons and roots. The cluster 5 genes were down-regulated by the drought condition. In contrast, there were 25 TFs in cluster 6, and these genes were up-regulated by water deficit, and the genes' expression was most high at cotyledons and roots. These results indicate that the expression of TFs was greatly affected by water deficit in the cotyledons and roots, and had different patterns between cotyledon and root tissues.

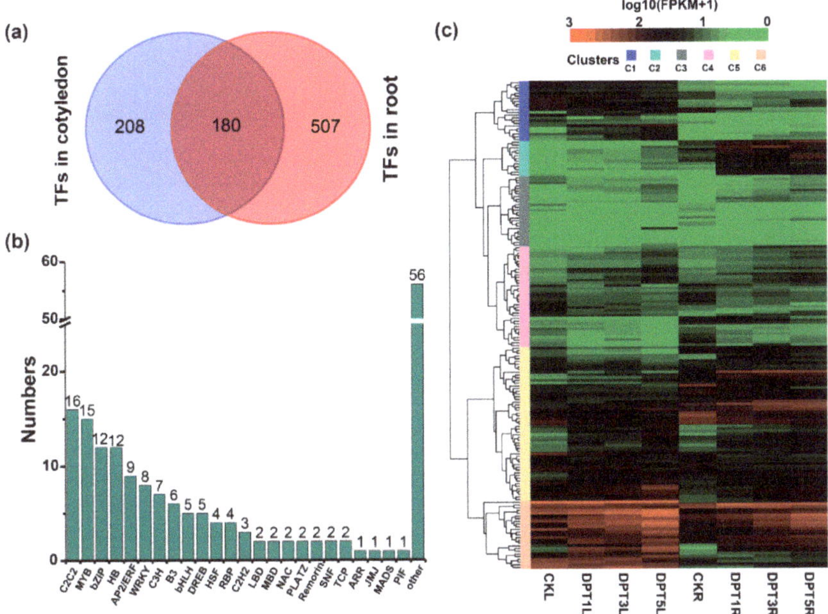

Figure 6. The differentially expressed TFs in cotyledons and roots responsive to drought stress. (**a**) Venn diagrams of TFs between cotyledons and roots. (**b**) Classification of TFs that were commonly identified in both cotyledon and root transcriptome libraries. (**c**) Expression pattern of TFs that were commonly identified in both cotyledons and roots in response to drought stress.

4. Discussion

Drought stress is one of the most detrimental environmental factors disturbing crop growth and production, and therefore understanding the drought-tolerance mechanism is pivotal for crop breeding [4,43]. Currently, RNA sequencing has been widely used to identify the drought-responsive pathway and genes that are activated during the seedling stage when exposed to abiotic stresses [5].

In this study, the phenotypic and physiological alterations of common buckwheat seedlings during drought stress were analyzed and characterized with the transcriptome analysis. Our results indicate that the common buckwheat seedlings relied on complex biological process to tackle the drought stress.

4.1. Morphological and Physiological Characteristics Related to Drought Stress in Common Buckwheat

Under drought stress, plant seedlings exhibit certain physiological and morphological variations [2,8], shown by leaf rolling and wilting [44,45], as well as the decreased RCW and wrinkled leaves (Figure 1a,d). Previous studies have demonstrated that drought stress also inhibits the photosynthesis of plants by affecting chlorophyll biosynthesis and facilitating stomatal closure [46,47], leading to the accumulation of MDA and ROS, which is harmful to the chloroplast photosystem II (PSII) [17,48]. As a result, plants have evolved antioxidant enzyme systems, such as superoxide dismutase (SOD), guaiacol peroxidase (GPX), CAT, and POD, to counteract the damage caused by drought stress [49]. MDA content has been considered important to reflect the drought tolerance ability of plants [12]. An active antioxidant capability in scavenging the cytotoxic ROS is preferred by plants in drought stress [50]. In common buckwheat seedlings, the ratio of chlorophyll a/b and RubisCO activities were significantly decreased compared to the control treatment in 5 days of treatment (Figure 2f, Figure 4e), and the expression levels of DEGs involved in the photosynthesis were correspondingly decreased DPT3L and DPT5L (Figure 4d), which may be due to the decrease in photosynthesis of the common buckwheat seedlings. By contrast, the CAT and POD activities were enhanced (Figure 2b,c). These observations suggest that the drought stress induced evident perturbations in the photosynthesis and ROS scavenging enzyme activities.

4.2. Multiple Biological Processes Are Involved in Drought Stress Responses in Common Buckwheat

Previous studies reported that multiple biological processes, such as oxidoreductase activities, and carbohydrate and protein metabolic processes, could be influenced when the ROS accumulation increased [51,52]. Our results indicated that the oxidoreductase activity, kinase activity, and DNA modification were upregulated by the drought treatment in common buckwheat cotyledons, whereas the genes were classified into "light harvesting" and "light reaction" (Figure 4c), and the expression of *LCHb* (encode chlorophyll a/b binding protein) and *PASN* (encode photosystem I reaction center subunit N protein) were markedly downregulated under drought conditions (Figure 4f,l). Repression of photosynthesis under drought stress also occurred in order to help the plants survive the water deficiency [53]. Cellular water deficit in plants caused by drought stress results in weakened carbon fixation, which may be physiologically ascribed to the stomatal closure and the biochemical inhibition of photosynthetic activities, further impacting the carbohydrate metabolism [54]. In addition, phosphorylation, as one of the reversible post-translational protein modification mechanisms, plays an important role in signaling the plant adaptation to osmotic stress [55,56], and a fine-tuned control of protein activity and function [56,57] has also been found to be altered in the common buckwheat root under drought stress. There were 92 DEGs involved in the protein modification and phosphorylation that were up-regulated (Figure 5c, Table S7), which may provide insights in the future study of the protein phosphorylation and modification events in plant drought stress.

4.3. Genes and Functional Proteins Responsive to Drought Stress

Previous studies have descried cellular changes that occur upon exposure to drought stress in plants, and the gene responses to drought stress have been studied spaciously in various species [1,9,11,53,58]. Moreover, the plant hormone ABA, as a signal-sensing molecule, can control the expression levels of stress-responsive genes, leading to cellular and physiological changes in response to water deficiency [59–61]. Moreover, previous studies have reported two regulatory systems: ABA-dependent and ABA-independent pathways, that play a major role for plant adaptation to drought stress [62]. Genes upstream and within the ABA pathway can be increased under drought conditions, and the *NCED* gene encoded 9-*cis* epoxycarotenoid dioxygenase has been shown to be induced under dehydration stress [61]. Furthermore, the changes of several metabolite levels under the water deficit condition were associated with the changes of biosynthetic gene expression, many of which were regulated by the changes of ABA accumulation levels [1,63]. In this study, a large number of DEGs were involved in ABA signaling and regulation (Table S5), and the detection of the up-regulation of the *NCED* genes in this study set identified the candidate genes for further studies of drought resistance in common buckwheat seedlings.

Recently, many drought-inducible genes involved in stress tolerance and stress responses have already been identified in several plant species [1], revealing that the transcription factors (TFs) play a central role in the biotic/abiotic stress responses [64–66]. In our previous study, we isolated and identified a *FeDREB1L* gene encoding a DREB-like transcription factor, which was simultaneously involved in the cold stress, drought stress, and ABA-mediated regulations [31]. The increased expression level of *FeDREB1L* during the earlier stage of drought stress displayed in this study (Figure 4m, Figure 5e) demonstrated that *FeDREB1L* could be a positive factor underpinning the drought stress resistance. Other TFs, including AP2/ERF, MYB, and bZIP families, were also identified to be differentially expressed in this study, for example, 180 DEGs that encoded TFs were identified in response to drought stress in both cotyledons and roots of the common buckwheat seedling, as well as the members of C2H2, MYB, bZIP, and WRKY families (Figure 6a, Table S8). Further studies are thereby required to elucidate the functions and gene-regulatory mechanisms of these TFs in response to plant drought stress.

5. Conclusions

To summarize, a comprehensive transcriptome profile of common buckwheat seedlings under drought stress was obtained using RNA-Seq technology. Phenotypic and physiological changes were determined, and the differentially expressed genes were analyzed to understand the regulatory mechanism of common buckwheat seedlings in response drought stress. The photosynthesis of the common buckwheat seedlings decreased, and the activities of antioxidant enzymes such as CAT and POD were increased under drought conditions. DEGs derived from important regulatory metabolisms were characterized. The results reflected in this study may provide useful information to better understand the molecular mechanism underlying the drought resistance in common buckwheat.

Supplementary Materials: The following are available online at http://www.mdpi.com/2073-4395/9/10/569/s1, Table S1: List of RT-qPCR primers; Table S2: Phenotypic changes of plant height, root length, and RWC under control conditions across four time points (0, 1, 3, and 5 days); Table S3: Physiological investigation of MDA content, POD and CAT activity, and chlorophyll content under control conditions across four time points (0, 1, 3, and 5 days); Table S4: Summary of the sequencing data of common buckwheat transcriptome; Table S5: Genes related to ABA metabolism in response to drought stress in common buckwheat cotyledons; Table S6: GO enrichment of differentially expressed genes in root transcriptome of common buckwheat seedlings under drought stress; Table S7: Genes related to protein modification and phosphorylation in response to drought stress in common buckwheat roots; Table S8: Classification of TFs that were commonly identified in both cotyledon and root transcriptome libraries; Figure S1: Confirmation of transcriptome data in cotyledons by qPCR analysis; Figure S2: Confirmation of transcriptome data in roots by qPCR analysis.

Author Contributions: Z.F., Z.H., and J.Y. designed the study and wrote the manuscript. Z.F., Y.L., J.S., S.W. (Shuping Wang), Z.L., S.W. (Shudong Wei), Y.Z., Z.H., and J.Y. participated in experiments. Z.H. submitted the raw data to Sequence Read Archive (SRA). Z.F., J.S., Z.L., Z.H., and J.Y. discussed the results and revised the manuscript. All authors have read and approved the final manuscript.

Funding: This research was funded by the National Natural Science Foundation of China (grant No. 31671755 and No.31571736) and the Supported Project of Outstanding Doctoral and Master's Degree Dissertation Cultivation Program of Yangtze University (YS2018032).

Acknowledgments: Authors acknowledge Xiaoyu Xu for critical reading of the manuscript.

Conflicts of Interest: The authors declare no conflict of interest.

References

1. Todaka, D.; Zhao, Y.; Yoshida, T.; Kudo, M.; Kidokoro, S.; Mizoi, J.; Kodaira, K.S.; Takebayashi, Y.; Kojima, M.; Sakakibara, H.; et al. Temporal and spatial changes in gene expression, metabolite accumulation and phytohormone content in rice seedlings grown under drought stress conditions. *Plant J.* **2017**, *90*, 61–78. [CrossRef] [PubMed]

2. Pan, J.; Li, Z.; Wang, Q.; Garrell, A.K.; Liu, M.; Guan, Y.; Zhou, W.; Liu, W. Comparative proteomic investigation of drought responses in foxtail millet. *BMC Plant Biol.* **2018**, *18*, 315. [CrossRef] [PubMed]

3. Xu, K.; Chen, S.; Li, T.; Ma, X.; Liang, X.; Ding, X.; Liu, H.; Luo, L. *OsGRAS23*, a rice GRAS transcription factor gene, is involved in drought stress response through regulating expression of stress-responsive genes. *BMC Plant Biol.* **2015**, *15*, 141. [CrossRef] [PubMed]

4. Chen, G.; Wang, Y.; Wang, X.; Yang, Q.; Quan, X.; Zeng, J.; Dai, F.; Zeng, F.; Wu, F.; Zhang, G.; et al. Leaf epidermis transcriptome reveals drought-Induced hormonal signaling for stomatal regulation in wild barley. *Plant Growth Regul.* **2018**, *87*, 39–54. [CrossRef]

5. Zhang, X.; Lei, L.; Lai, J.; Zhao, H.; Song, W. Effects of drought stress and water recovery on physiological responses and gene expression in maize seedlings. *BMC Plant Biol.* **2018**, *18*, 68. [CrossRef] [PubMed]

6. Turner, N.C.; Stern, W.R.; Evans, P. Water relations and osmotic adjustment of leaves and roots of lupins in response to water deficits. *Crop Sci.* **1987**, *27*, 977. [CrossRef]

7. Beck, E.H.; Fettig, S.; Knake, C.; Hartig, K.; Bhattarai, T. Specific and unspecific responses of plants to cold and drought. *J. Biosci.* **2007**, *32*, 501–510. [CrossRef]

8. Xiang, D.B.; Peng, L.X.; Zhao, J.L.; Zou, L.; Zhao, G.; Song, C. Effect of drought stress on yield, chlorophyll contents and photosynthesis in tartary buckwheat (*Fagopyrum tataricum*). *J. Food Agric. Environ.* **2013**, *11*, 1358–1363. [CrossRef]

9. Eom, S.H.; Baek, S.A.; Kim, J.K.; Hyun, T.K. Transcriptome analysis in Chinese Cabbage (*Brassica rapa* ssp. *pekinensis*) provides the role of glucosinolate metabolism in response to drought stress. *Molecules* **2018**, *23*, 1186. [CrossRef]

10. Joshi, R.; Wani, S.H.; Singh, B.; Bohra, A.; Dar, Z.A.; Lone, A.A.; Pareek, A.; Singla-Pareek, S.L. Transcription factors and plants response to drought stress: Current understanding and future directions. *Front. Plant Sci.* **2016**, *7*, 1029. [CrossRef]

11. Kumar, J.; Gunapati, S.; Kianian, S.F.; Singh, S.P. Comparative analysis of transcriptome in two wheat genotypes with contrasting levels of drought tolerance. *Protoplasma* **2018**, *255*, 1487–1504. [CrossRef] [PubMed]

12. Yin, J.; Jia, J.; Lian, Z.; Hu, Y.; Guo, J.; Huo, H.; Zhu, Y.; Gong, H. Silicon enhances the salt tolerance of cucumber through increasing polyamine accumulation and decreasing oxidative damage. *Ecotoxicol. Environ. Saf.* **2019**, *169*, 8–17. [CrossRef] [PubMed]

13. Luo, W.; Song, F.; Xie, Y. Trade-off between tolerance to drought and tolerance to flooding in three wetland plants. *Weltlands* **2008**, *28*, 866–873. [CrossRef]

14. Zhang, J.; Kirkham, M.B. Antioxidant responses to drought in sunflower and sorghum seedlings. *New Phytol.* **1996**, *132*, 361–373. [CrossRef] [PubMed]

15. Pagter, M.; Bragato, C.; Brix, H. Tolerance and physiological responses of *Phragmites australis* to water deficit. *Aquat. Bot.* **2005**, *81*, 285–299. [CrossRef]

16. Moller, I.M.; Jensen, P.E.; Hansson, A. Oxidative modifications to cellular components in plants. *Annu. Rev. Plant Biol.* **2007**, *58*, 459–481. [CrossRef]

17. Smirnoff, N. The role of active oxygen in the response of plants to water deficit and desiccation. *New Phytol.* **1993**, *125*, 27–58. [CrossRef]

18. Chakhchar, A.; Lamaoui, M.; Aissam, S.; Ferradous, A.; Wahbi, S.; El Mousadik, A.; Ibnsouda-Koraichi, S.; Filali-Maltouf, A.; El Modafar, C. Differential physiological and antioxidative responses to drought stress and recovery among four contrasting *Argania spinosa* ecotypes. *J. Plant Interact.* **2016**, *11*, 30–40. [CrossRef]

19. Prochazkova, D.; Sairam, R.K.; Srivastava, G.C.; Singh, D.V. Oxidative stress and antioxidant activity as the basis of senescence in maize leaves. *Plant Sci.* **2001**, *161*, 765–771. [CrossRef]

20. Kholová, J.; Hash, C.T.; Kočová, M.; Vadez, V. Does a terminal drought tolerance QTL contribute to differences in ROS scavenging enzymes and photosynthetic pigments in pearl millet exposed to drought. *Environ. Exp. Bot.* **2011**, *71*, 99–106. [CrossRef]

21. Apel, K.; Hirt, H. Reactive oxygen species: Metabolism, oxidative stress, and signal transduction. *Annu. Rev. Plant Biol.* **2004**, *55*, 373–399. [CrossRef] [PubMed]

22. Sharma, P.; Dubey, R.S. Drought induces oxidative stress and enhances the activities of antioxidant enzymes in growing rice seedlings. *Plant Growth Regul.* **2005**, *46*, 209–221. [CrossRef]

23. Ajithkumar, I.P.; Panneerselvam, R. ROS scavenging system, osmotic maintenance, pigment and growth status of Panicum sumatrense roth. Under drought stress. *Cell Biochem. Biophys.* **2014**, *68*, 587–595. [CrossRef] [PubMed]

24. Shinozaki, K.; Yamaguchi-Shinozaki, K. Gene Expression and signal transduction in water-stress response. *Plant Physiol.* **1997**, *115*, 327–334. [CrossRef] [PubMed]

25. Shinozaki, K.; Yamaguchi-Shinozaki, K. Gene networks involved in drought stress response and tolerance. *J. Exp. Bot.* **2007**, *58*, 221–227. [CrossRef] [PubMed]

26. Liu, X.; Zhang, R.; Ou, H.; Gui, Y.; Wei, J.; Zhou, H.; Tan, H.; Li, Y. Comprehensive transcriptome analysis reveals genes in response to water deficit in the leaves of *Saccharum narenga* (Nees ex Steud.) hack. *BMC Plant Biol.* **2018**, *18*, 250. [CrossRef]

27. Ji, T.; Li, S.; Li, L.; Huang, M.; Wang, X.; Wei, M.; Shi, Q.; Li, Y.; Gong, B.; Yang, F. Cucumber *Phospholipase D* alpha gene overexpression in tobacco enhanced drought stress tolerance by regulating stomatal closure and lipid peroxidation. *BMC Plant Biol.* **2018**, *18*, 355. [CrossRef]

28. Zhu, Y.; Yin, J.; Liang, Y.; Liu, J.; Jia, J.; Huo, H.; Wu, Z.; Yang, R.; Gong, H. Transcriptomic dynamics provide an insight into the mechanism for silicon-mediated alleviation of salt stress in cucumber plants. *Ecotoxicol. Environ. Saf.* **2019**, *174*, 245–254. [CrossRef]

29. Ito, Y.; Katsura, K.; Maruyama, K.; Taji, T.; Kobayashi, M.; Seki, M.; Shinozaki, K.; Yamaguchi-Shinozaki, K. Functional analysis of rice DREB1/CBF-type transcription factors involved in cold-responsive gene expression in transgenic rice. *Plant Cell Physiol.* **2006**, *47*, 141–153. [CrossRef]

30. Sakuma, Y.; Maruyama, K.; Osakabe, Y.; Qin, F.; Seki, M.; Shinozaki, K.; Yamaguchi-Shinozaki, K. Functional analysis of an *Arabidopsis* transcription factor, *DREB2A*, involved in drought-responsive gene expression. *Plant Cell* **2006**, *18*, 1292–1309. [CrossRef]

31. Fang, Z.; Zhang, X.; Gao, J.; Wang, P.; Xu, X.; Liu, Z.; Shen, S.; Feng, B. A buckwheat (*Fagopyrum esculentum*) DRE-Binding transcription factor gene, *FeDREB1*, enhances freezing and drought tolerance of transgenic *Arabidopsis*. *Plant Mol. Biol. Rep.* **2015**, *33*, 1510–1525. [CrossRef]

32. Germ, M.; Gaberščik, A. Chapter twenty one—The effect of environmental factors on buckwheat. *Mol. Breed. Nutr. Asp. Buckwheat* **2016**, 273–281. [CrossRef]

33. Fang, Z.; Hou, Z.; Wang, S.; Liu, Z.; Wei, S.; Zhang, Y.; Song, J.; Yin, J. Transcriptome analysis reveals the accumulation mechanism of anthocyanins in buckwheat (*Fagopyrum esculentum* Moench) cotyledons and flowers. *Int. J. Mol. Sci.* **2019**, *20*, 1493. [CrossRef] [PubMed]

34. Delpérée, C.; Kinet, J.M.; Lutts, S. Low irradiance modifies the effect of water stress on survival and growth-related parameters during the early developmental stages of buckwheat (*Fagopyrum esculentum*). *Physiol. Plant.* **2003**, *119*, 211–220. [CrossRef]

35. Cawoy, V.; Lutts, S.; Kinet, J.M. Osmotic stress at seedling stage impairs reproductive development in buckwheat (*Fagopyrum esculentum*). *Physiol. Plant.* **2006**, *128*, 689–700. [CrossRef]

36. Pan, L.; Meng, C.; Wang, J.; Ma, X.; Fan, X.; Yang, Z.; Zhou, M.; Zhang, X. Integrated omics data of two annual ryegrass (*Lolium multiflorum* L.) genotypes reveals core metabolic processes under drought stress. *BMC Plant Biol.* **2018**, *18*, 26. [CrossRef]

37. Harper, A.L.; von Gesjen, S.E.; Linford, A.S.; Peterson, M.P.; Faircloth, R.S.; Thissen, M.M.; Brusslan, J.A. Chlorophyllide a oxygenase mrna and protein levels correlate with the chlorophyll a/b ratio in *Arabidopsis Thaliana*. *Photosynth. Res.* **2004**, *79*, 149–159. [CrossRef]

38. Spanic, V.; Viljevac Vuletic, M.; Abicic, I.; Marcek, T. Early response of wheat antioxidant system with special reference to *Fusarium* head blight stress. *Plant Physiol. Biochem.* **2017**, *115*, 34–43. [CrossRef]

39. Yasui, Y.; Hirakawa, H.; Ueno, M.; Matsui, K.; Katsube-Tanaka, T.; Yang, S.J.; Aii, J.; Sato, S.; Mori, M. Assembly of the draft genome of buckwheat and its applications in identifying agronomically useful genes. *DNA Res.* **2016**, *23*, 215–224. [CrossRef]

40. Kim, D.; Pertea, G.; Trapnell, C.; Pimentel, H.; Kelley, R.; Salzberg, S.L. TopHat2: Accurate alignment of transcriptomes in the presence of insertions, deletions and gene fusions. *Genome Biol.* **2013**, *14*, R36. [CrossRef]

41. Lu, Q.-H.; Wang, Y.-Q.; Song, J.-N.; Yang, H.-B. Transcriptomic identification of salt-related genes and de novo assembly in common buckwheat (*F. esculentum*). *Plant Physiol. Biochem.* **2018**, *127*, 299–309. [CrossRef]

42. Yin, J.; Liu, M.; Ma, D.; Wu, J.; Li, S.; Zhu, Y.; Han, B. Identification of circular RNAs and their targets during tomato fruit ripening. *Postharvest Biol. Technol.* **2018**, *136*, 90–98. [CrossRef]

43. Ye, G.; Ma, Y.; Feng, Z.; Zhang, X. Transcriptomic analysis of drought stress responses of sea buckthorn (*Hippophae rhamnoides* subsp. sinensis) by RNA-Seq. *PLoS ONE* **2018**, *13*, e0202213. [CrossRef]

44. Lafitte, H.R.; Price, A.H.; Courtois, B. Yield response to water deficit in an upland rice mapping population: Associations among traits and genetic markers. *Theor. Appl. Genet.* **2004**, *109*, 1237–1246. [CrossRef]

45. Kadioglu, A.; Terzi, R.; Saruhan, N.; Saglam, A. Current advances in the investigation of leaf rolling caused by biotic and abiotic stress factors. *Plant Sci.* **2012**, *182*, 42–48. [CrossRef]

46. Flexas, J.; Bota, J.; Loreto, F.; Cornic, G.; Sharkey, T.D. Diffusive and metabolic limitations to photosynthesis under drought and salinity in C3 plants. *Plant Biol.* **2004**, *6*, 269–279. [CrossRef]

47. Iturbe-Ormaetxe, I.; Escuredo, P.R.; Arrese-Igor, C.; Becana, M. Oxidative damage in pea plants exposed to water deficit or paraquat. *Plant Physiol.* **1998**, *116*, 173–181. [CrossRef]

48. Demmig-Adams, B.; Adams, W.W. Photoprotection and other responses of plants to high light stress. *Annu. Rev. Plant Physiol. Plant Mol. Biol.* **1992**, *43*, 599–626. [CrossRef]

49. Sharma, P.; Jha, A.B.; Dubey, R.S.; Pessarakli, M. Reactive oxygen species, oxidative damage, and antioxidative defense mechanism in plants under stressful conditions. *J. Bot.* **2012**, *2012*, 1–26. [CrossRef]

50. Zaefyzadeh, M.; Quliyev, R.A.; Babayeva, S.M.; Abbasov, M.A. The effect of the interaction between genotypes and drought stress on the superoxide dismutase and chlorophyll content in durum wheat landraces. *Turk. J. Biol.* **2009**, *33*, 1–7. [CrossRef]

51. Hayano-Kanashiro, C.; Calderon-Vazquez, C.; Ibarra-Laclette, E.; Herrera-Estrella, L.; Simpson, J. Analysis of gene expression and physiological responses in three Mexican maize landraces under drought stress and recovery irrigation. *PLoS ONE* **2009**, *4*, e7531. [CrossRef]

52. Min, H.; Chen, C.; Wei, S.; Shang, X.; Sun, M.; Xia, R.; Liu, X.; Hao, D.; Chen, H.; Xie, Q. Identification of drought tolerant mechanisms in maize seedlings based on transcriptome analysis of recombination inbred lines. *Front. Plant Sci.* **2016**, *7*, 1080. [CrossRef]

53. Zhu, Y.; Gong, H.; Yin, J. Role of Silicon in mediating salt tolerance in plants: A Review. *Plants* **2019**, *8*, 147. [CrossRef]

54. Xue, G.P.; McIntyre, C.L.; Glassop, D.; Shorter, R. Use of expression analysis to dissect alterations in carbohydrate metabolism in wheat leaves during drought stress. *Plant Mol. Biol.* **2008**, *67*, 197–214. [CrossRef]

55. Hoyos, M.E.; Zhang, S. Calcium-Independent activation of salicylic acid-induced protein kinase and a 40-kilodalton protein kinase by hyperosmotic stress. *Plant Physiol.* **2000**, *122*, 1355–1363. [CrossRef]

56. Ren, J.; Mao, J.; Zuo, C.; Calderon-Urrea, A.; Dawuda, M.M.; Zhao, X.; Li, X.; Chen, B. Significant and unique changes in phosphorylation levels of four phosphoproteins in two apple rootstock genotypes under drought stress. *Mol. Genet. Genom.* **2017**, *292*, 1307–1322. [CrossRef]

57. Schulze, W.X. Proteomics approaches to understand protein phosphorylation in pathway modulation. *Curr. Opin. Plant Biol.* **2010**, *13*, 279–286. [CrossRef]

58. Zhu, Y.; Jia, J.; Yang, L.; Xia, Y.; Zhang, H.; Jia, J.; Zhou, R.; Nie, P.; Yin, J.; Ma, D.; et al. Identification of cucumber circular RNAs responsive to salt stress. *BMC Plant Biol.* **2019**, *19*, 164. [CrossRef]

59. Bartels, D.; Sunkar, R. Drought and Salt Tolerance in Plants. *Crit. Rev. Plant. Sci.* **2005**, *24*, 23–58. [CrossRef]

60. Schachtman, D.P.; Goodger, J.Q. Chemical root to shoot signaling under drought. *Trends Plant Sci.* **2008**, *13*, 281–287. [CrossRef]

61. Li, J.; Li, Y.; Yin, Z.; Jiang, J.; Zhang, M.; Guo, X.; Ye, Z.; Zhao, Y.; Xiong, H.; Zhang, Z.; et al. *OsASR5* enhances drought tolerance through a stomatal closure pathway associated with ABA and H_2O_2 signalling in rice. *Plant Biotechnol. J.* **2017**, *15*, 183–196. [CrossRef]

62. Yamaguchi-Shinozaki, K.; Shinozaki, K. Transcriptional regulatory networks in cellular responses and tolerance to dehydration and cold stresses. *Annu. Rev. Plant Biol.* **2006**, *57*, 781–803. [CrossRef]

63. Urano, K.; Maruyama, K.; Ogata, Y.; Morishita, Y.; Takeda, M.; Sakurai, N.; Suzuki, H.; Saito, K.; Shibata, D.; Kobayashi, M.; et al. Characterization of the ABA-regulated global responses to dehydration in *Arabidopsis* by metabolomics. *Plant J.* **2009**, *57*, 1065–1078. [CrossRef]

64. Shinozaki, K.; Yamaguchi-Shinozaki, K.; Seki, M. Regulatory network of gene expression in the drought and cold stress responses. *Curr. Opin. Plant Biol.* **2003**, *6*, 410–417. [CrossRef]
65. Kong, X.M.; Zhou, Q.; Luo, F.; Wei, B.D.; Wang, Y.J.; Sun, H.J.; Zhao, Y.B.; Ji, S.J. Transcriptome analysis of harvested bell peppers (*Capsicum annuum* L.) in response to cold stress. *Plant Physiol. Biochem.* **2019**, *139*, 314–324. [CrossRef]
66. Seo, P.J.; Xiang, F.; Qiao, M.; Park, J.Y.; Lee, Y.N.; Kim, S.G.; Lee, Y.H.; Park, W.J.; Park, C.M. The MYB96 transcription factor mediates abscisic acid signaling during drought stress response in *Arabidopsis*. *Plant Physiol.* **2009**, *151*, 275–289. [CrossRef]

Article

Changes in Root Anatomy of Peanut (*Arachis hypogaea* L.) under Different Durations of Early Season Drought

Nuengsap Thangthong [1,2], Sanun Jogloy [1,2,*], Tasanai Punjansing [3], Craig K. Kvien [4], Thawan Kesmala [1,2] and Nimitr Vorasoot [1,2]

1 Department of Agronomy, Faculty of Agriculture, Khon Kaen University, Khon Kaen 40002, Thailand; nuengsap.th@gmail.com (N.T.); thkesmala@gmail.com (T.K.); nvorasoot@gmail.com (N.V.)
2 Peanut and Jerusalem Artichoke Improvement for Functional Food Research Group, Khon Kaen University, Khon Kaen 40002, Thailand
3 Department of Biology, Faculty of Science, Udonthani Rajaphat University, Udonthani 41000, Thailand; tasanaipun@gmail.com
4 Crop & Soil Sciences, National Environmentally Sound Production Agriculture Laboratory (NESPAL), The University of Georgia, Tifton, GA 31793, USA; ckvien@uga.edu
* Correspondence: sjogloy@gmail.com; Tel.: +66-43-364-637

Received: 12 March 2019; Accepted: 23 April 2019; Published: 27 April 2019

Abstract: Changes in the anatomical structure of peanut roots due to early season drought will likely affect the water acquiring capacity of the root system. Yet, as important as these changes are likely to be in conferring drought resistance, they have not been thoroughly investigated. The objective of this study was to investigate the effects of different durations of drought on the root anatomy of peanut in response to early season drought. Plants of peanut genotype ICGV 98305 were grown in rhizoboxes with an internal dimension of 50 cm in width, 10 cm in thickness and 120 cm in height. Fourteen days after emergence, water was withheld for periods of 0, 7, 14 or 21 days. After these drought periods, the first and second order roots from 0–20 cm below soil surface were sampled for anatomical observation. The mean xylem vessel diameter of first- order lateral roots was higher than that of second- order lateral roots. Under early season drought stress root anatomy changes were more pronounced in the longer drought period treatments. Twenty-one days after imposing water stress, the drought treatment and irrigated treatment were clearly different in diameter, number and area of xylem vessels of first- and second-order lateral roots. Plants under drought conditions had a smaller diameter and area of xylem vessels than did the plants under irrigated control. The ability of plants to change root anatomy likely improves water uptake and transport and this may be an important mechanism for drought tolerance. The information will be useful for the selection of drought durations for evaluation of root anatomy related to drought resistance and the selection of key traits for drought resistance.

Keywords: xylem vessel; water stress; root anatomy

---

1. Introduction

In many areas of the tropics, peanut production is mostly in rain-fed and semi-arid areas with low and unpredictable rainfall and rain distribution. In these areas, drought stress can occur at any growth stage, resulting in yield loss of 22–53% [1]. Drought stress also increases *Aspergillus flavus* infection and aflatoxin contamination by 2–17% [2]. However, drought stress at a pre-flowering growth stage sometimes actually increases yield [3]. Irrigation, planting date selection and drought resistant varieties can improve yield and reduce aflatoxin contamination during periods of drought. However, management of irrigation requires an available water source and investment in additional equipment.

Planting date selection, while less expensive than irrigation is not as effective because rainfall and rain distribution are often unpredictable. The use of drought resistant varieties is a promising and sustainable choice in need of further development. When selecting for drought resistance in peanut, yield and biomass during drought are often used as selection criteria. Yet this selection method is complicated by high genotype by environment interaction. Many physiological and morphological traits have been suggested as surrogate traits for drought resistance to increase selection efficiency, yet measurement for these traits are often quite variable.

Root traits are known to improve drought resistance [4] and are therefore important for plant breeding programs. Improving the water acquiring capacity of crops to extract water from the soil profile during drought is one example. Root traits such as large root systems (root dry weight), root length density and the percentage of root length density that respond to drought have been investigated in peanut [1,3,5–7].

Anatomical parameters, such as xylem vessel number and diameter, have been positively correlated with dry matter production under stress in chili (*Capsicum annum* L.) [8]. Drought resistant varieties of several plant species have been reported to have a higher number of vessel cells and a larger xylem cross-section than susceptible varieties of chili (*Capsicum annum* L.) [8], tomato (*Lycopersicon esculentum*) [9] and grape (*Vitis vinifera* L.) [10]. As in the above studies of other plants, it is likely that studies on the fine root structure of peanut, especially under drought stress, will lead to a better understanding of why some peanut genotypes yield better during a drought than others.

Cell-wall ingrowths or phi-thickening have been reported in loquat (*Eriobotrya japonica* Lindl.) root [11], apple (*Pyrus malus*) [12], geranium (*Pelargonium hortorum*) roots [12] Sibipiruna (*Caesalpinia peltophoroides*) [13] with solute movement (salt stress) [12], water logging [13], and drought stress [11]. Although the effect of early season drought on ingrowths and phi-thickenings has not been investigated in peanut and further investigations are necessary to understand phi-thickenings. The response of phi-thickening might be related to the transport processes in the peanut root.

Root anatomy is interesting, and it might play an important role in plant response to drought. The types of lateral roots during root growth were recognized in peanut [14]. The different types and different structures may be related to different functions. The structure of the first order lateral roots helps determine the efficacy of the axial water transport system, yet the structure within the second order lateral roots helps determine the efficacy of the water uptake process. Unfortunately, this useful information has not been thoroughly investigated in peanut. The objective of this study was to investigate the effects of different durations of drought imposition on the root anatomy of peanut in response to early season drought. The information will be useful for selection of drought durations for evaluation of root anatomy related to drought resistance and selection of key traits for drought resistance.

2. Materials and Methods

2.1. Experimental Design and Plant Material

The experiment was conducted under a rainout shelter at the Field Crop Research Station of Khon Kaen University, Khon Kaen, Thailand (16°28' N, 102°48' E, 200 m. above sea level). The peanut genotype—ICGV 98305—was subjected to four water treatments (0, 7, 14 or 21 days without irrigation), each beginning 14 days after plant emergence (DAE). The experiment used a completely randomized design with three replications and was conducted for two seasons during July–September 2013 and March–May 2014.

ICGV 98305 is a drought resistant line from ICRISAT known for high root length density in the deep sub-soil during periods of drought [1,15].

2.2. Preparation and Irrigation of Rhizobox Experiment

The plants were grown in rhizoboxes with internal dimensions of 50 cm in width, 10 cm in thickness and 120 cm in height (Figure 1a). The rhizoboxes were filled with dry soil to obtain bulk density of 1.57 Mg m^{-3} and height of 115 cm, and water was added to achieve field capacity. Peanut seeds were planted in the center of rhizobox, 5 cm below the soil surface. At 3 days after emergence (DAE), the seedlings were thinned to obtain 1 plant per rhizobox. The front side of the rhizoboxes was transparent and covered with black sheet, and all sides of rhizoboxes were then covered with aluminum foil to reduce light absorption and temperature increase (Figure 1b).

The root needle-board method [16] was used for the observation of root growth and distribution with a minor modification for size and spacing of needles. The root system of the plant in the box was held in place by needles attached to back board of rhizoboxes and projecting out to the transparent front. The needle spacing was 5 × 5 cm^2. The needle columns started 2.5 cm from left and right margins and the needle rows were started at 12.5 cm from the top of rhizobox and continued at 5 cm intervals to the bottom of the box (Figure 1c).

Soil moisture contents for field capacity and permanent wilting point were determined to be 11.13% and 3.40%, respectively. Water was supplied to the rhizoboxes through horizontal tubes which were installed at 5, 15, 35, 55, 75, 95 and 115 cm below the soil surface. For each rhizobox, water was first supplied at field capacity and all three drought treatments (7, 14 and 21 days without added water) began 14 DAE. The fourth treatment was kept at field capacity for the entire experimental period. The field capacity was maintained uniformly throughout the soil profile by using the six watering tubes. Drainage holes, 1.5 cm in diameter, were placed at the bottom of the rhizobox. Drained water was replenished at the same amount. Crop evapotranspiration was calculated as the sum of water lost through plant transpiration and soil evaporation, as described by Reference [17];

$$ETcrop = ETo \times Kc \tag{1}$$

where ETcrop is crop water requirement (mm/day), ETo is evapotranspiration of a reference under specified conditions calculated using the pan evaporation method, and Kc is the peanut water requirement coefficient.

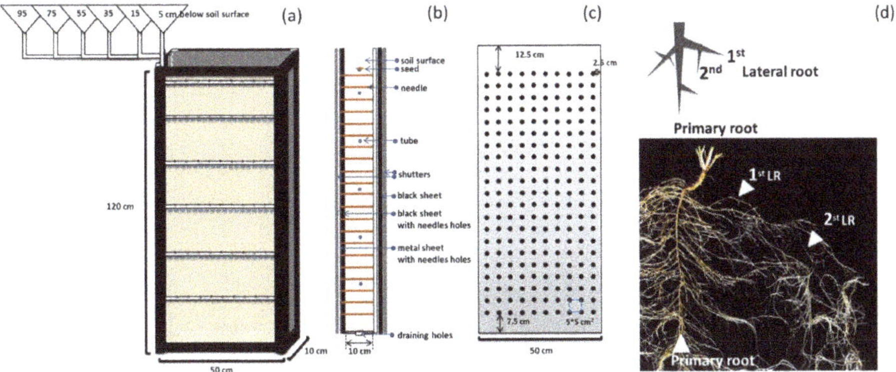

Figure 1. Diagrammatic representation and dimension of rhizobox with six tubes of irrigation (**a**), cross section showing the different elements of the system (**b**), spacing of needle at backside of rhizoboxes (**c**) and taproot system of a rhizobox-grown peanut (**d**).

2.3. Crop Management

Phosphorus as triple superphosphate (Ca(H$_2$PO$_4$)$_2$H$_2$O) (Chia tai company limited, Phranakhonsiayutthaya, Thailand) at the rate of 122.3 kg ha^{-1} and potassium as potassium chloride (KCl;

60% K_2O) (Chia tai company limited, Phranakhonsiayutthaya, Thailand) at the rate of 62.5 kg ha^{-1} were applied to the soil before planting. A water-diluted commercial peat-based inoculum of *Bradyrhizobium* (mixture of strains THA 201 and THA 205; Department of Agriculture, Ministry of Agriculture and Cooperatives, Bangkok, Thailand) was applied 5 cm below the soil surface through the irrigation tubes. Seeds were treated with captan (3a,4,7,7a-tetrahydro-2-[(trichloromethyl) thio]-1H-isoindole-1, 3(2H)-dione, Erawan Agricultural Chemical Co., Ltd., Bangkok, Thailand.) at the rate of 5 g kg^{-1} seeds before planting. Carbosulfan [2-3-dihydro-2,2-dimethylbenzofuran-7-yl (dibutylaminothio) methylcar-bamate 20% (*w/v*) water soluble concentrate] (FMC AG Ltd., Bangkok, Thailand) at 2.5 L ha^{-1} was applied weekly to control thrips, and methomyl [S-methyl-N-((methylcarbamoyl)oxy) thioacetimidateand methomyl [(E,Z)-methyl N-{[(methylamino) carbonyl]oxy}ethanimidothioate] 40% soluble powder (Du Pont Co., Ltd., Bangkok, Thailand) at 1.0 kg ha^{-1} was used to control mites. Weeds were controlled by hand weeding.

2.4. Data Collection

Rainfall, relative humidity, pan evaporation, maximum and minimum temperature and solar radiation were recorded daily from planting to 35 DAE at a weather station located 50 m from the experiment. Soil physical and chemical properties were analyzed before planting. Soil samples for analysis were taken from the mixed pile of soil used for this experiment. The soil's physical properties in the experiment were analyzed for percentage sand, silt and clay. The soil chemical properties were analyzed for pH, organic matter, total N, available P, exchangeable K and exchangeable Ca.

2.5. Soil Moisture Content

Soil moisture content was determined gravimetrically using a micro auger method at 10, 25, 65, and 85 cm soil depths at 14, 21, 28 and 35 DAE. Soil moisture content for each rhizobox was calculated as;

$$\text{Soil moisture content (\%)} = ((\text{wet weight} - \text{dry weight})/\text{dry weight}) \times 100 \qquad (2)$$

2.6. Observation of Root Anatomy

Roots were collected at 7, 14 and 21 days after water withholding began. At the sampling date, the shoot in each box was cut at the soil surface and the roots were carefully washed with a fine spray of tap water to remove soil. Rhizobox needles helped roots maintained the approximate position they were in the soil profile.

Root samples for anatomical observation were taken from 0–20 cm below soil surface. The first-, and second-order lateral roots (Figure 1d) were taken at approximately 5 cm from the root tips from each treatment. The root sampling strategy (5 cm from the tip, and 20 cm deep) was as suggested from a previous rhizotron study [18] in which peanut root growth rates of drought and well-watered treatments were 12.6 and 21.9 cm per week, respectively Therefore, we took the root samples for anatomical study at 5 cm from the root tips, as roots at this position would be expected to be significantly affected by drought. The samples were fixed in a formaldehyde (Sigma-Aldrich; Bangkok, Thailand, 36.5–38% in H_2O)-glacial acetic acid (Fisher Chemical)-40% ethanol-solution (FAA$_{40}$). Dehydration of the samples was accomplished by adding a series of alcohol concentrations at 10% intervals from 40% to 70%. Free-hand cross sections were stained with Safranin O (Dye content $\geq$ 85%; Sigma-Aldrich). Anatomical characteristics of the root samples were observed using a Nikon eclipse 50i optical microscope with ocular and stage micrometers. The microscope's digital camera (Nikon DS-Fi1, Shingawa-ku, Tokyo, Japan) was used for photographs. All transverse sections of roots were measured and recorded for diameter and area of the xylem vessels of first-order and second-order lateral roots. Xylem vessel elements consisted of protoxylem and metaxylem. Although the identification of these xylem tissues was difficult, we were able to classify them into two groups by diameter. Smaller xylem vessels were equal to or smaller than the overall mean diameter of xylem vessels and bigger xylem vessels were larger than the mean diameter of xylem vessels. The cell-wall ingrowths were compared in both the drought and well-watered treatments using the cortical layers of both first-order and second-order lateral roots.

2.7. Data Analysis

The statistical analysis was performed using the statistix-8 program as a completely randomized design. An analysis of variance and least significance difference (LSD) tests were used to compare differences at $p \leq 0.05$.

3. Results

3.1. Meteorological Data and Soil Data

The meteorological details for the two years were collected (data not shown) and are described in Field Crops Research (2016) [19]. Daily air temperatures ranged from 22.7 to 36.8 °C in 2013 and 20.2 to 40.5 °C in 2014. Relative humidity (RH) values ranged between 63–88% in 2013 and 47–87% in 2014. The means of evaporation (E0) were 4.5 mm in 2013 and 5.7 mm in 2014. While rain did not directly fall on the experimental plants, as it was conducted in a rainout shelter; it did affect relative humidity and evapotranspiration.

Differences between years were observed for maximum temperature (T-max) and minimum temperature (T-min) as the trial in 2013 was conducted during the cooler rainy season (May–July) than the 2014 trial conducted from March–May.

3.2. Soil Moisture Content and Relative Water Content

Soil moisture content and relative water content are described in Reference [19]. Soil moisture content measured at field capacity was 11.13% and permanent wilting point was 3.40%. Soil moisture content for non-stress conditions was similar to those at field capacity. However, soil moisture content at field capacity (FC) in the lower soil layers was slightly higher than 11.13% at the initiation of drought stress. Drought and well-irrigated treatments were clearly different at all sampling dates, especially at top soil layers of 10 and 25 cm. The differences between drought and well-irrigated treatments were small in lower soil layers and the treatments became similar at 65 and 85 cm except at 28 and 35 DAE in 2014.

3.3. Observation of Root Anatomy

Peanut has a dicotyledonous root system with a single taproot and branched first-, second-, and higher order lateral roots (Figure 1d). In this study, the anatomy of first- and second- order lateral roots was observed.

3.3.1. First order Lateral Root

Combined analysis of variance for total vessel numbers, bigger vessel numbers, smaller vessel number, total vessel diameter (μm), bigger vessel diameter (μm), smaller vessel diameter (μm), total vessel area (μm^2), bigger vessel area (μm^2), smaller vessel area (μm^2) of the first order lateral root in 2013 and 2014 are shown in Table 1. Significant differences in total vessel numbers, bigger vessel numbers, total vessel area and bigger vessel area were observed in different durations and seasons. The interactions between duration and treatment (D × T) were also significant for total vessel numbers and smaller vessel area traits.

Central cylinders of first order lateral roots had an almost triarch arrangement of the vascular bundles (Figures 2 and 3). Within these bundles, the xylem vessels showed a wide range in size. For ease of discussion, we classified the vessels into two groups (large and small) based on their diameter. Large vessels had a diameter greater than the mean (16.06 μm) of all vessels, and small vessels had a diameter less than the mean.

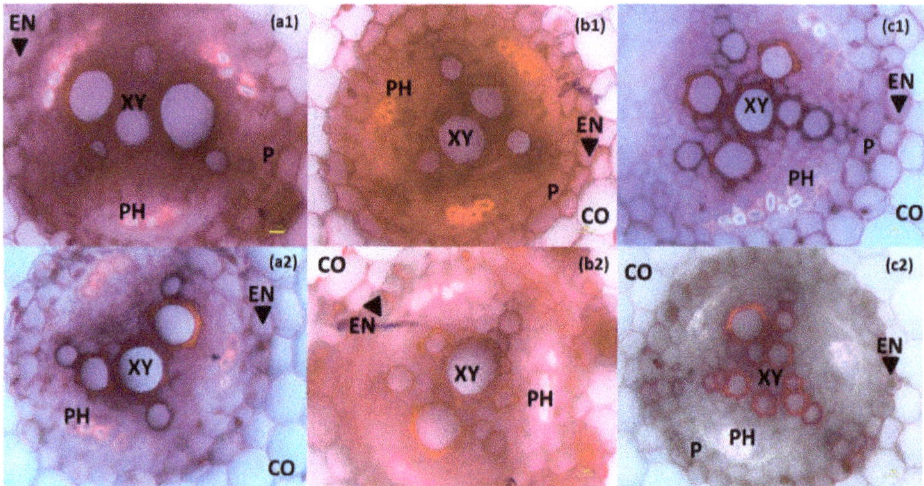

Figure 2. Freehand cross sections of first order lateral roots of peanut under well-irrigated conditions (**a1**, **b1** and **c1**) and drought stress conditions (**a2**, **b2** and **c2**) at 21, 28 and 35 DAE, respectively. CO, cortex; EN, endodermis; P, pericycle; PH, phloem; XY, xylem; Scale bar = 10 μm; 40×.

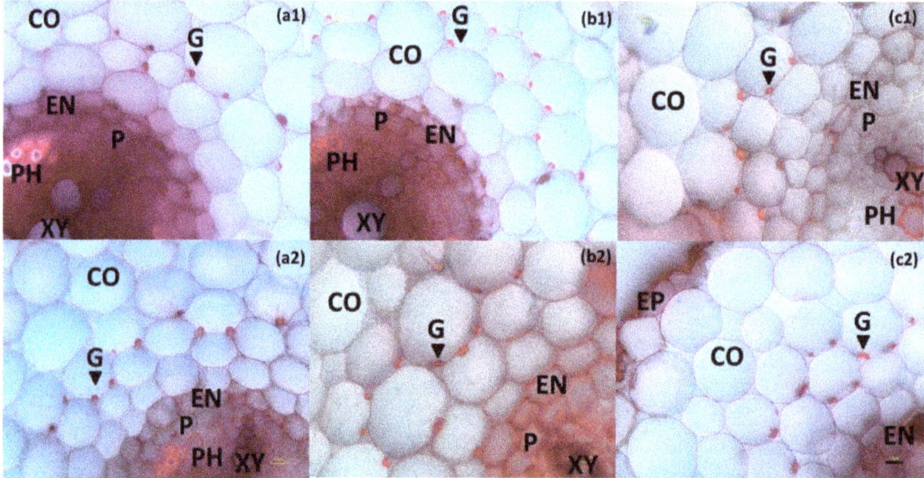

Figure 3. Freehand cross sections of first-order lateral roots under well-irrigated conditions (**a1**, **b1** and **c1**) and drought stress conditions (**a2**, **b2** and **c2**) at 21, 28 and 35 days after plant emergence (DAE). CO, cortex; EN, endodermis; G, phi-thickening or cell wall ingrowth; P, pericycle; PH, phloem; XY, xylem; Scale bar = 10 μm; 40×.

Total xylem numbers per cross-section of first order lateral roots (Figure 4) in the first and second seasons were not significantly different between drought and well-irrigated treatments at 21, 28 and 35 DAE with one exception at 35 DAE in 2014. At 35 DAE in 2014, the drought treatments had higher vessel numbers, in the small diameter vessels, than did well-irrigated treatments. At 35 DAE in 2013, stress and non-stress treatments were not significantly different for the total number of vessels, yet, like in 2014, stress tended to reduce the number of bigger vessels and increase the number of smaller vessel.

Table 1. Mean square from the combined analysis of variance for total vessel numbers, bigger vessel numbers, smaller vessel number, total vessel diameter (μm), bigger vessel diameter (μm), smaller vessel diameter (μm), total vessel area (μm^2), bigger vessel area (μm^2), smaller vessel area (μm^2) of the first order lateral root in 2013 and 2014.

Source	DF	Total Vessel Numbers	Bigger Vessel Numbers	Smaller Vessel Numbers	Total Vessel Diameter (μm)	Bigger Vessel Diameter (μm)	Smaller Vessel Diameter (μm)	Total Vessel Area (μm^2)	Bigger Vessel Area (μm^2)	Smaller Vessel Area (μm^2)
Duration (D)	2	32.028 **	6.19 *	6.91 ns	14.87 ns	8.81 ns	2.04 ns	6,586,518 **	5185627 **	39262 ns
Season (S)	1	42.25 **	11.11 **	1.01 ns	0.79 ns	5.52 ns	0.29 ns	2,160,885 ns	1,053,634 ns	214,114 *
Treatment (T)	1	0.03 ns	4.00 ns	0.01 ns	3.16 ns	30.24 ns	0.07 ns	2,012,582 ns	2,117,714 ns	110,969 ns
D × S	2	3.25 ns	4.30 *	33.47 ns	6.97 ns	43.05 ns	1.32 ns	1,576,123 ns	1,723,383 ns	33,482 ns
D × T	2	8.36 *	0.75 ns	2.40 **	42.04 *	1.39 ns	5.29 ns	1,801,293 ns	3,447,407 *	358,589 **
S × T	1	0.30 ns	1.78 ns	7.65 ns	0.65 ns	2.24 ns	0.97 ns	56,394 ns	11,259 ns	28,413 ns
D × S × T	2	1.36 ns	1.03 ns	5.21 ns	6.77 ns	6.52 ns	2.90 ns	66,096 ns	106,452 ns	91,367 ns
Pooled error	24	1.80	1.13	5.21	10.98	13.75	2.10	566,481	792,286	40,728
Total	35									

ns, *, ** = non-significant and significant at $p < 0.05$ and $p < 0.01$ probability levels, respectively, durations (7, 14 and 21 days without added water), treatments (well-watered and water stress) and seasons (2013 and 2014).

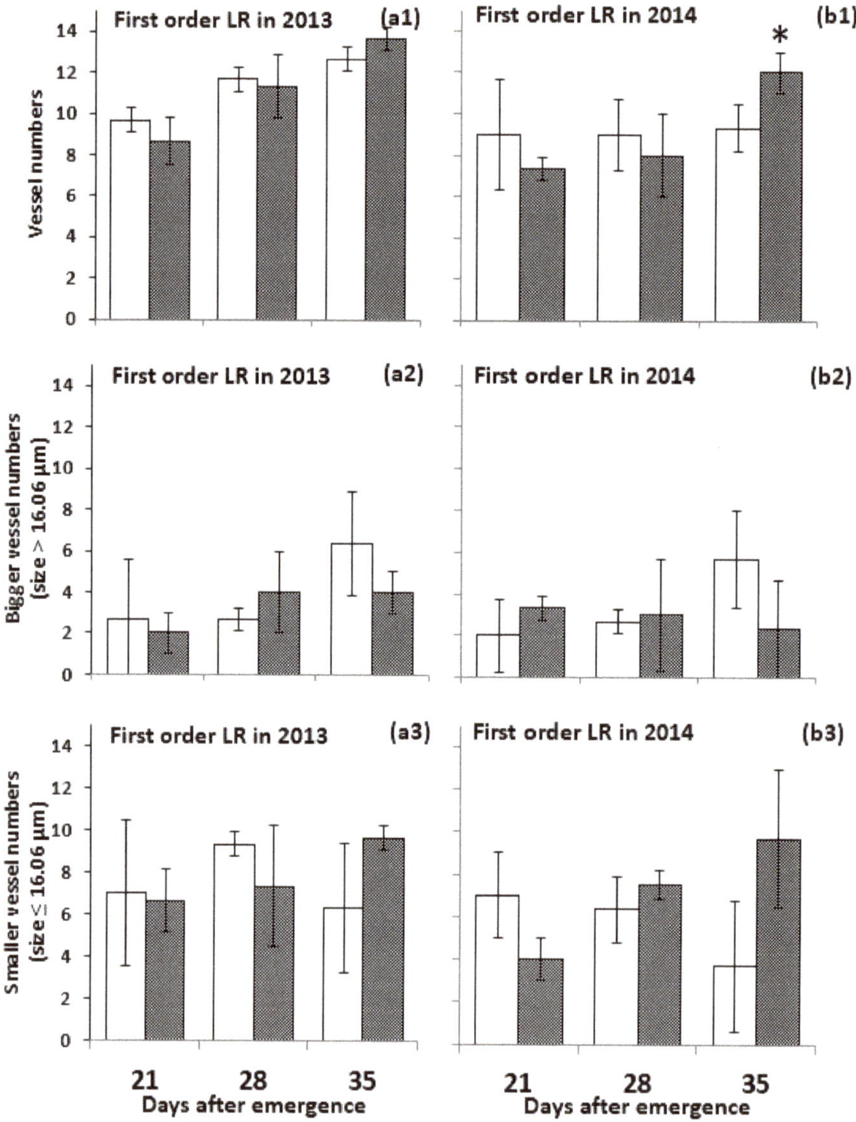

Figure 4. Vessel numbers of first order lateral roots (**a1, b1**), bigger vessel number (**a2, b2**) and smaller vessel number (**a3, b3**) of peanut at 21, 28 and 35 DAE in 2013 (**a**) and in 2014 (**b**); Significant at * $p \leq 0.05$, non-stress treatments (□) and stress treatments (■).

Vessel diameters in first order lateral roots (Figure 5) under non-stress and drought stress treatments varied between 4.03 to 41.09 μm (data not shown, unpublished data). Yet, the total vessel area in smaller vessels increased in both 2013 and 2014 and the total vessel area in the large vessels decreased in 2013 and slightly reduced in 2014 when the stress treatments, were compared to the well-watered control (Figure 6) in both 2013 and 2014. Stress and non- stress treatments were not significantly different for vessel diameter at all durations of drought stress. However, the average vessel diameter of long duration stress at 35 DAE and 21 days after irrigation withholding in each season tended to reduce. Figure 5 showed that the diameter of bigger xylem vessels in each season and the diameter of smaller xylem vessels were not significantly different except for the diameter of smaller xylem vessels at 35 DAE in 2014. The diameters of smaller xylem vessels were smaller in size under long duration stress at 35 DAE and 21 days after irrigation withholding compared to under well-watered treatment in 2014.

A significant reduction was observed in the diameter of the smaller xylem vessels and the diameter of the bigger xylem vessels tended to reduce, ultimately reducing total xylem area per root cross section.

The area of total xylem vessel elements in roots grown under stress conditions was significantly lower than those grown under non-stress conditions and these differences in area increased as the length of stress increased. Non-stress and stress treatments were significantly different for the area of total xylem vessels and the area of bigger vessels at 35 DAE. Stress treatment reduced the area of total vessels in 2013 and to a smaller exert the area tended to reduce in 2014.

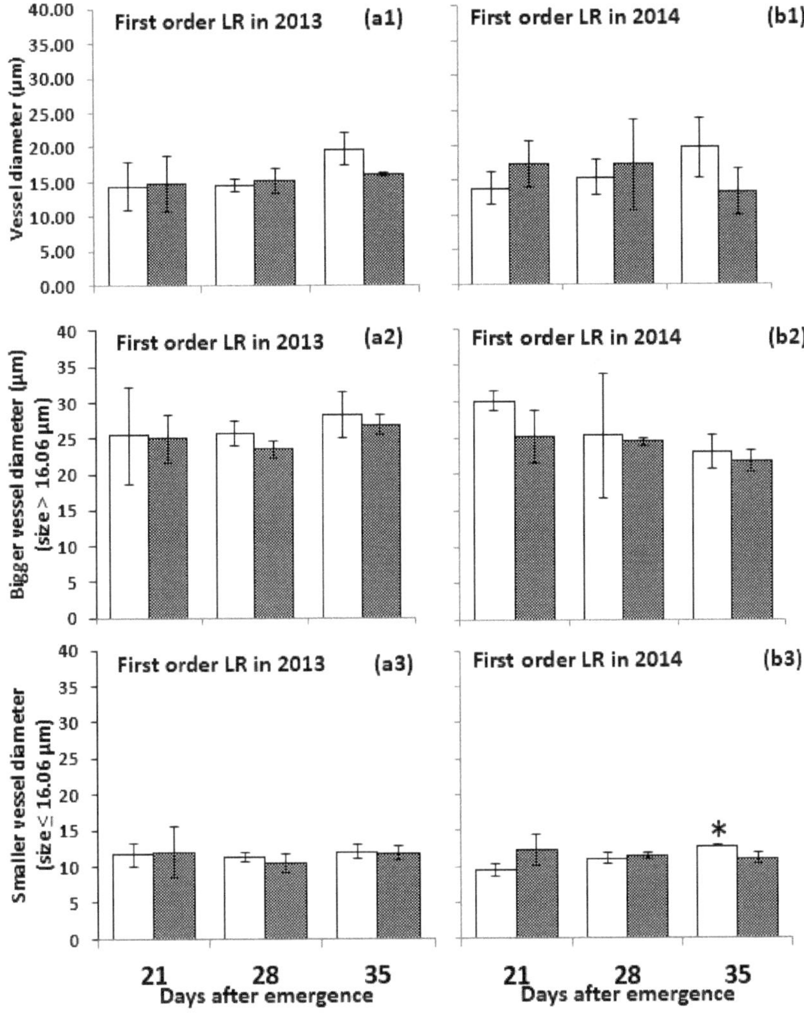

Figure 5. Average vessel diameter of first order lateral roots (**a1**, **b1**), bigger vessel diameter (**a2**, **b2**) and smaller vessel diameter (**a3**, **b3**) of peanut at 21, 28 and 35 DAE under well-irrigate and drought stress in 2013 (**a**) and in 2014 (**b**); Significant at * $p \leq 0.05$, non-stress treatments (□) and stress treatments (■).

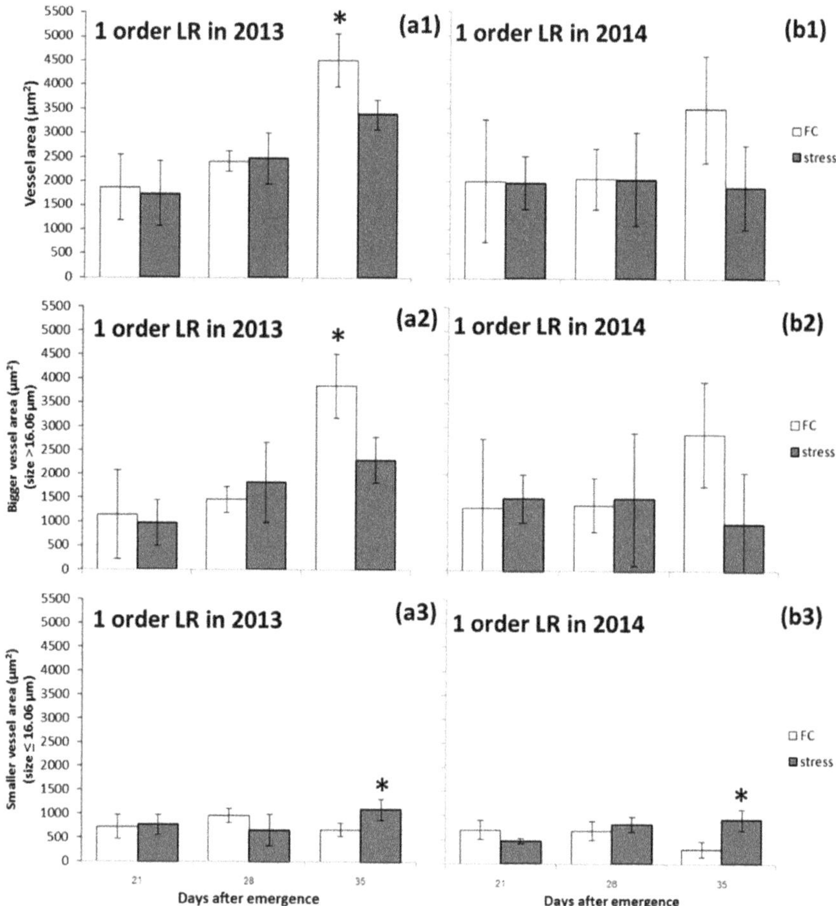

Figure 6. Vessel area (**a1**, **b1**), bigger vessel area (**a2**, **b2**) and smaller vessel area (**a3**, **b3**) of first order lateral roots of peanut at 21, 28 and 35 DAE in 2013 (**a**) and in 2014 (**b**); Significant at * $p \leq 0.05$, non-stress treatments (□) and stress treatments (■).

The cell-wall ingrowths in the first order lateral roots were detected in the cortical cells under both well-watered and drought stress treatments (Figure 3). The cell-wall ingrowths were localized at the opposite side of the intercellular spaces adjacent to the endodermis except in under drought at 28 DAE (Figure 3b2). The cell-wall ingrowths were found in two positions which were on the opposite side of the intercellular spaces and cell-cell conjunction. The 1–2 layers of this cell were found and indicated as the peri-endodermal layer.

3.3.2. Second Order Lateral Root

Combined analysis of variance for total vessel numbers, bigger vessel numbers, smaller vessel number, total vessel diameter (μm), bigger vessel diameter (μm), smaller vessel diameter (μm), total vessel area (μm^2), bigger vessel area (μm^2), smaller vessel area (μm^2) of the second order lateral root in 2013 and 2014 are shown in Table 2. Differences in duration (D) and treatment (T) were significant ($p \leq 0.01$ and $p \leq 0.05$) for most traits. Season (S) was significant for total vessel numbers and bigger vessel numbers. The interactions between duration × treatment (D × T) and duration × season (D × S) were also significant for some traits.

The structure of second order lateral roots differed from that of the first order lateral roots. First order lateral roots are thicker, and the stele and vascular bundle tissues are more extensive than in the second order lateral roots. Second order lateral roots had an almost diarch and triarch organization of vascular bundles (Figures 7 and 8). Average value of vessel diameter was 14.21 µm (data not shown, unpublished data).

Drought and well-irrigated treatments at all durations were not significantly different for number of total xylem per cross-section of second order lateral roots (Figure 9) in 2013 and 2014. Drought and well-watered treatments were also not significantly different in the number of bigger vessels but the number of bigger vessels tended to reduce at 35 DAE, whereas the number of smaller vessels increased at 35 DAE (21 days after water withholding began).

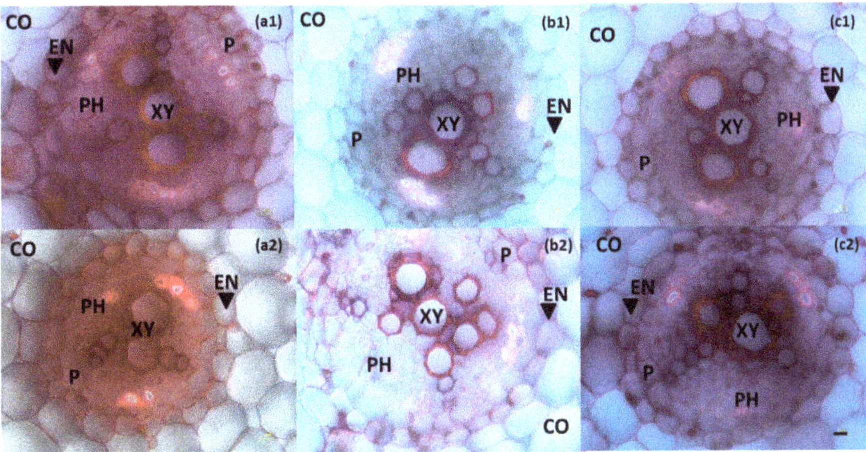

Figure 7. Freehand cross sections of second order lateral roots of peanut under well-irrigated conditions (**a1**, **b1** and **c1**) and drought stress conditions (**a2**, **b2** and **c2**) at 21, 28 and 35 DAE. CO, cortex; EN, endodermis; P, pericycle; PH, phloem; XY, xylem; Scale bar = 10 µm; 40×.

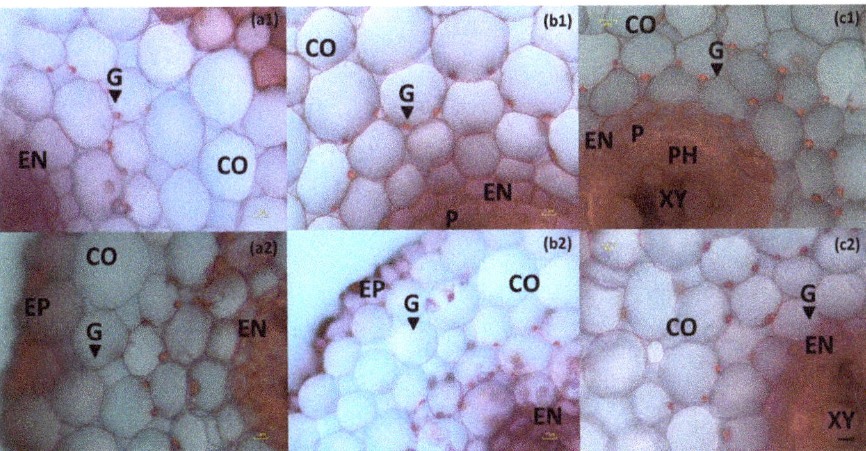

Figure 8. Freehand cross sections of second-order lateral roots under well-irrigated conditions (**a1**, **b1** and **c1**) and drought stress conditions (**a2**, **b2** and **c2**) at 21, 28 and 35 DAE. CO, cortex; EN, endodermis; G, phi-thickening or cell wall ingrowth; P, pericycle; PH, phloem; XY, xylem; Scale bar = 10 µm; 40×.

Table 2. Mean square from the combined analysis of variance for total vessel numbers, bigger vessel numbers, smaller vessel number, total vessel diameter (μm), bigger vessel diameter (μm), smaller vessel diameter (μm), total vessel area (μm^2), bigger vessel area (μm^2), smaller vessel area (μm^2) of the second order lateral root in 2013 and 2014.

Source	DF	Total Vessel Numbers	Bigger Vessel Numbers	Smaller Vessel Numbers	Total Vessel Diameter (μm)	Bigger Vessel Diameter (μm)	Smaller Vessel Diameter (μm)	Total Vessel Area (μm^2)	Bigger Vessel Area (μm^2)	Smaller Vessel Area (μm^2)
Duration (D)	2	11.44 **	6.19 *	3.03 ns	26.52 **	4.63 ns	9.02 **	2,437,286 **	1,689,456 **	59,535 *
Season (S)	1	18.78 **	1.11 **	1.00 ns	1.41 ns	13.96 ns	0.23 ns	1,155,729 *	860,956 ns	24,033 ns
Treatment (T)	1	7.11 *	4.00 ns	31.78 **	34.54 **	35.64 *	0.19 ns	1,747,821 **	2,743,513 **	95,334 *
D $\times$ S	2	1.44 ns	4.36 *	2.08 ns	12.83 **	11.36 ns	2.66 *	1,174,800 **	1,167,283 **	1126 ns
D $\times$ T	2	3.11 ns	0.75 ns	5.86 *	3.55 *	31.43 *	0.00 ns	413,242 ns	669,743 ns	22,934 ns
S $\times$ T	1	1.00 ns	1.78 ns	0.11 ns	3.08 ns	0.23 ns	0.04 ns	158,148 ns	138,356 ns	1298 ns
D $\times$ S $\times$ T	2	1.00 ns	1.02 ns	0.36 ns	3.95 ns	3.39 ns	1.85 ns	285,065 ns	199,055 ns	15,256 ns
Pooled error	24	1.17	1.14	1.33	1.84	7.91	0.08	185,616	202,569	13,384
Total	35									

ns, *, ** = non-significant and significant at $p < 0.05$ and $p < 0.01$ probability levels, respectively, durations (7, 14 and 21 days without added water), treatments (well-watered and water stress) and seasons (2013 and 2014).

Means for the vessel diameter of second-order lateral roots (Figure 10) of all treatments varied between 4.29 to 38.48 µm (data not shown). Stress treatment significantly reduced the vessel diameter of second-order lateral roots at 35 DAE with drought imposition for 21 days in 2013 and slightly reduced the vessel diameter of second-order lateral roots at 35 DAE with drought imposition for 21 days in 2014. Stress treatment significantly reduced the diameter of bigger xylem vessels in 2014 at 35 DAE with drought imposition for 21 days and stress treatment also reduced the diameter of bigger xylem vessels in 2013, although the reduction was not significant. Stress treatment did not significantly affect the diameter of smaller xylem vessel diameter in 2013 and 2014.

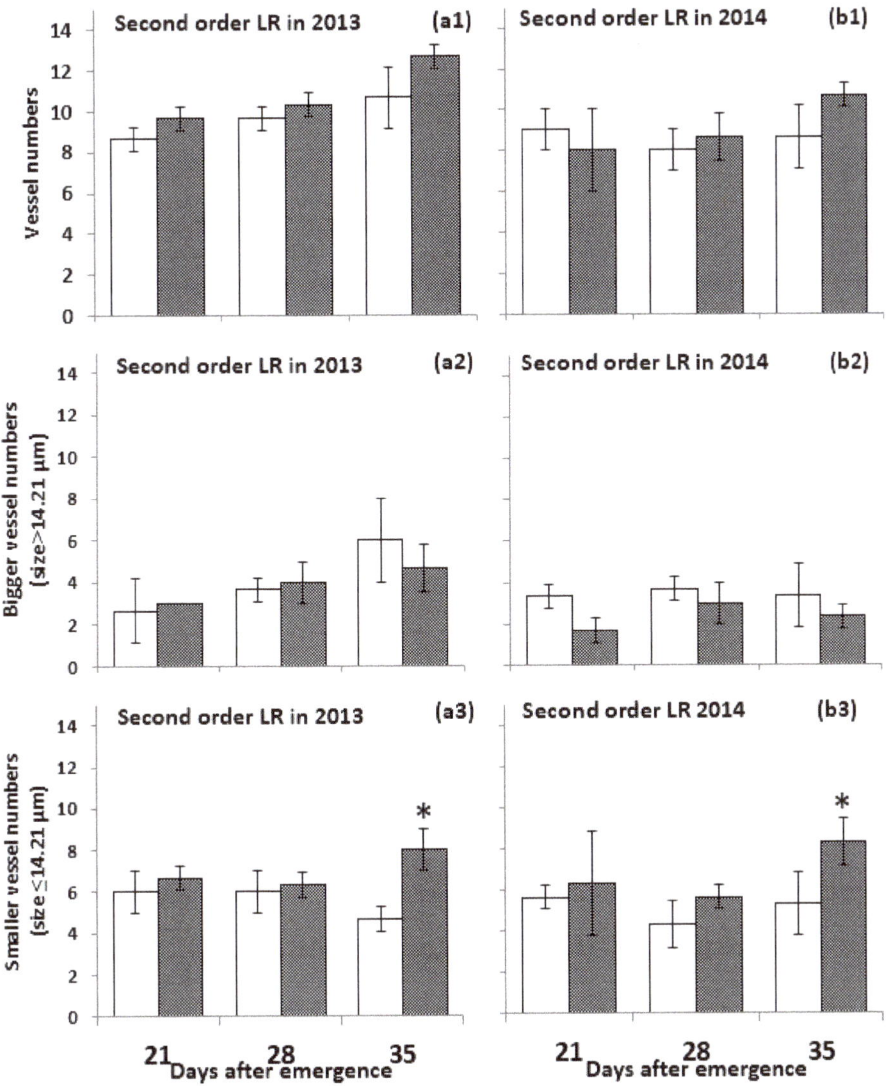

Figure 9. Vessel numbers of second order lateral roots (**a1**, **b1**), bigger vessel number (**a2**, **b2**) and smaller vessel number (**a3**, **b3**) of peanut at 21, 28 and 35 DAE in 2013 (**a**) and in 2014 (**b**); Significant at * $p \leq 0.05$, non-stress treatments (□) and stress treatments (■).

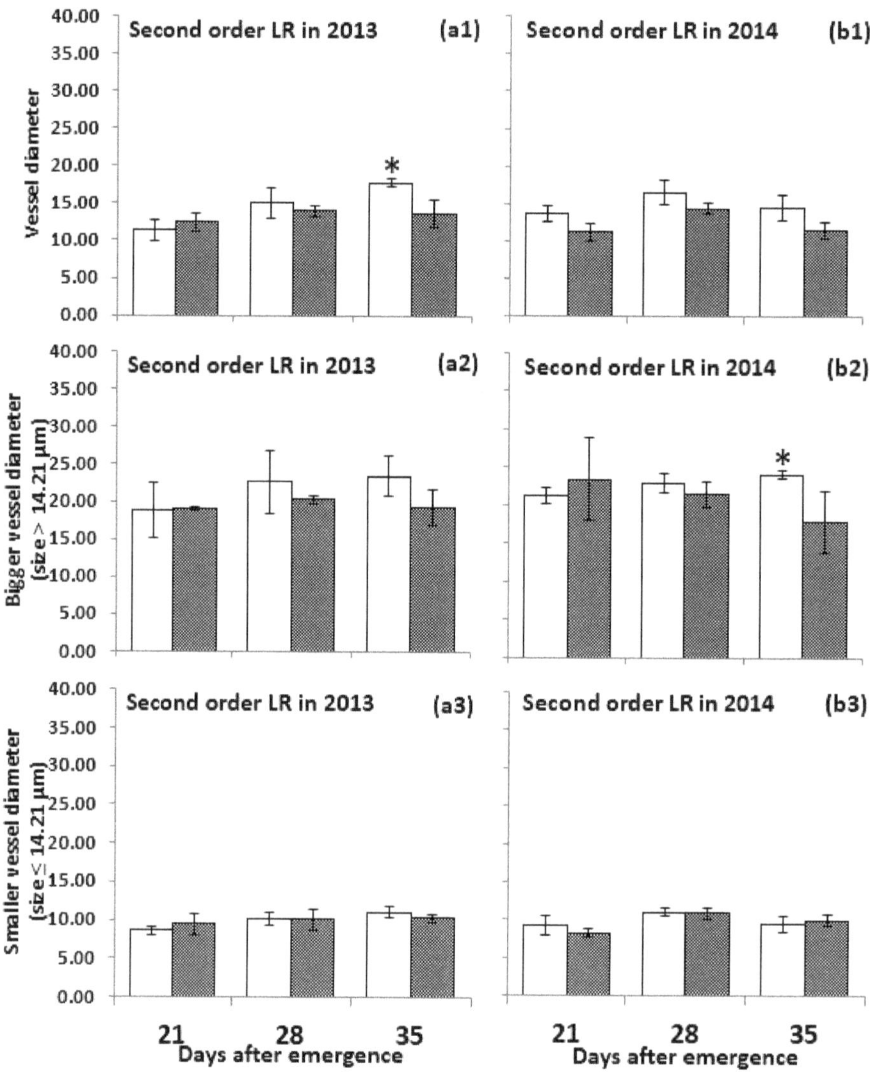

Figure 10. Vessel diameter of second order lateral roots (**a1**, **b1**), bigger vessel diameter (**a2**, **b2**) and smaller vessel diameter (**a3**, **b3**) of peanut at 21, 28 and 35 DAE in 2013 (**a**) and in 2014 (**b**); Significant at * $p \leq 0.05$, non-stress treatments (□) and stress treatments (■).

Because stress treatment reduced the diameters of the average xylem vessels and bigger xylem vessels, the area of vessels per cross section of each season and the area of bigger vessels in 2014 was reduced at 35 DAE, although the reduction was not significant and the area of bigger vessels area was significantly reduced at 35 DAE in 2013 (Figure 11). The area of smaller xylem vessels per cross section under stress treatment was increased.

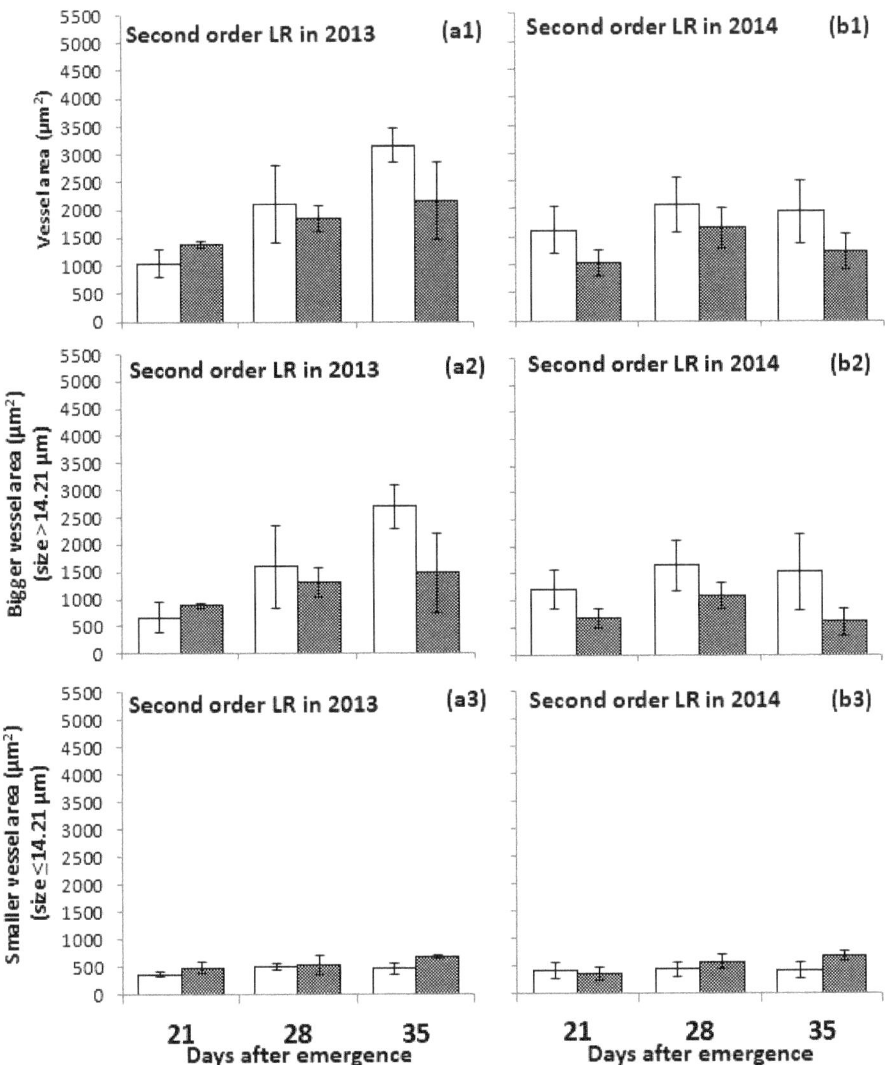

Figure 11. Vessel area (**a1**, **b1**), bigger vessel area (**a2**, **b2**) and smaller vessel area (**a3**, **b3**) of second order lateral roots of peanut at 21, 28 and 35 DAE in 2013 (**a**) and in 2014 (**b**); Significant at * $p \leq 0.05$, non-stress treatments (□) and stress treatments (■).

Cell-wall ingrowths appeared in the cortical cells of the second order lateral root under both conditions (Figure 8). The 1–2 layers of cell-wall ingrowths were found in the peri-endodermal layer.

4. Discussion

Weather conditions may be a key factor affecting the root anatomy of peanut. The experiment was conducted for two years. In the rainy season, air temperature and humidity were low, but in the summer to the early rainy season, air temperature and humidity were rather high. Soil moisture in the drought and well-watered treatments were clearly different in the upper soil layers. Soil moisture content for drought stress treatment at 28 and 35 DAE at the 10 cm of soil layer was less than 3.4%

(the permanent wilting point). However, soil moisture content for drought stress treatment at 65 cm and 85 cm of soil levels was higher than the permanent wilting point. The rate of water loss in 2013 was slower than in 2014, and soil moisture content at 21 days after irrigation withholding in 2013 was similar to those at 14 days after irrigation withholding in 2014.

The responses of plants to water stress depend on many things including timing and the intensity and duration of the drought. Root anatomy and root growth, like other plant parts, are sensitive to drought [20]. In this study, the long duration of the early season drought changed the root anatomical traits of peanut. Long periods of stress caused a significant increase in the number of xylem vessels in first and second order lateral roots but a significant decrease in the vessel diameter and the area of these first and second order lateral roots.

In both seasons, the mean xylem vessel diameters of first order lateral roots was higher than that of the second order lateral roots. The reduction in vessel diameter of first order lateral roots was higher than that of the second order lateral roots and these results may explain the differential root functions. The reduction in vessel diameter of first order lateral roots will better support the transport system's hydraulic conductivity according to Poiseuille's law [21]. In hot pepper, drought stress significantly reduced the diameters of xylem vessels in all of cultivars [8]. Vessel diameter is closely and positively correlated with volume of water flow and therefore it is correlated with the 'safety' of the conductive system [22,23]. The large vessel size under water deficit resulted in xylem cavitation [24]. The narrower diameter of metaxylem vessels maintain the water column, lowers the risk of cavitation, increases water flow resistance and saves water columns in narrower capillaries from damage [25]. Formation of narrower vessels occurring in drought-tolerant dicotyledons (including short-lived perennials and annuals with secondary structure) will likewise be advantageous when the plants are grown under drought [26,27].

Morphometric measurements on xylem vessels showed that the vessels of water-stressed plants had lower sectional areas. These results suggested that the reduction in vessel sectional area due to a diminished growth in response to water stress was the main factor affecting conductivity. Under a water deficit environment, roots develop to help extract soil moisture which being held at greater surface tension [28]. Deep root growth and large xylem diameter in deep roots may also increase the ability of roots to mine more water in deep soil when water in deep soil is abundant [29]. However, small and fine roots with greater specific root length enable plants to efficiently increase water uptake and maintain plant productivity under drought by increasing surface area and root length in contact with soil water, especially at deeper soil with available water [19,29].

The ability of plant to take up water is highly influenced by the number and size of the water conductive elements [25]. The change in number and size of the vessel xylem could help maintain water uptake under water stress [8].

In Ferna'ndez-Garcı'a, Lo'pez-Berenguer, and Olmos book chapter on the role of phi Cells under abiotic stress the authors noted that phi thickening is not the exception in the root anatomy [30]. They noted that the literature has described 16 different families, covering more than 100 species, which present the phi thickening in the roots. The phi thickening is classified into three types based on their root cell location: Type I, the most frequently found phi cell layer, is located in contact with the endodermis. Type II phi cell layer is located in contact with the epidermis and Type III phi cell layers are located in the inner cortical cells but not in contact with either the epidermis or the endodermis. In this study, cell-wall ingrowths were detected in the cortical cells of all first and second order lateral roots under well-watered and drought stress treatments. The 1–2 layers of these cells were localized at the opposite side of the intercellular spaces adjacent to the endodermis. The cell-wall ingrowths layers were indicated as the peri-endodermal layer and also called phi-thickening [31]. In previous studies, phi-thickening was induced under salt stress [30,31] and drought stress [11]. Phi-thickening of loquat roots grown under drought stress developed dramatically compared to normal conditions and the formation of phi-thickening was thought to be a defense mechanism against water stress. As the functions of these cells are difficult to determine precisely, phi thickening would play a role in controlling the water and solute rate of transportation through cell walls [32]. In peanut, cell wall ingrowth development in cortical cells might be a drought resistance mechanism for peanut roots as well. In this study, the 1–2 layers of cell-wall ingrowths were detected in both well-watered and drought stress treatments which were not significantly different for number of cell-wall ingrowths layers. However, the cells could be seen at higher magnification and using an electron microscope.

5. Conclusions

Under early season drought stress, root anatomy changes were more pronounced in the longer drought period treatments. At 21 days after imposing water stress, the drought treatment and irrigated treatment were clearly different in diameter, number and area of xylem vessels of first- and second-order lateral roots. Plants under drought conditions had smaller diameter and area of xylem vessels than did the plants under irrigated control. The ability of plant to change root anatomy likely improves water uptake and transport, and this may be an important mechanism for drought avoidance.

Author Contributions: Conceptualization, N.T., S.J., T.P. and N.V.; methodology, N.T., S.J. and N.V.; validation, N.T., S.J. and N.V.; formal analysis, N.T.; investigation, N.T.; resources, S.J.; data curation, N.T.; writing—original draft preparation, N.T.; writing—review and editing, T.K., C.K.K.; supervision, S.J.; funding acquisition, S.J.

Funding: This research was funded by the Royal Golden Jubilee Ph.D. Program (6.A.KK/ 53/ E.1), Peanut and Jerusalem artichoke Improvement Project for the Functional Food Research Group, and the Thailand Research Fund for providing financial support through the Senior Research Scholar Project of Sanun Jogloy (Project no. RTA6180002).

Acknowledgments: This study was funded by the Royal Golden Jubilee Ph.D. Program (6.A.KK/ 53/ E.1). Assistance was also received from Peanut and Jerusalem artichoke Improvement Project for the Functional Food Research Group, Plant Breeding Research Center for Sustainable Agriculture and the Thailand Research Fund for providing financial support through the Senior Research Scholar Project of Sanun Jogloy (Project no. RTA6180002). Thailand Research Fund (TRF) (IRG 578003), Khon Kaen University (KKU) and Faculty of Agriculture, KKU are acknowledged for providing financial support for training on manuscript preparation. The manuscript was critical reviewed by Ian Charles Dodd.

Conflicts of Interest: The authors declare no conflict of interest.

References

1. Songsri, P.; Jogloy, S.; Vorasoot, N.; Akkasaeng, C.; Patanothai, A.; Holbrook, C.C. Root distribution of drought-resistant peanut genotypes in response to drought. *J. Agron. Crop Sci.* **2008**, *194*, 92–103. [CrossRef]

2. Girdthai, T.; Jogloy, S.; Vorasoot, N.; Akkasaeng, C.; Wongkaew, S.; Holbrook, C.C.; Patanothai, A. Associations between physiological traits for drought tolerance and aflatoxin contamination in peanut genotypes under terminal drought. *Plant Breed.* **2010**, *129*, 693–699. [CrossRef]

3. Jongrungklang, N.; Toomsan, B.; Vorasoot, N.; Jogloy, S.; Boote, K.; Hoogenboom, G.; Patanothai, A. Rooting traits of peanut genotype with different yield response to pre-flowering drought stress. *Field Crops Res.* **2011**, *120*, 262–270. [CrossRef]

4. Russell, R.S. *Plant Root System: Their Function and Interaction with the Soil*; McGRAW-HILL Book Company (UK) Limited: Oxford, UK, 1982.

5. Jongrungklang, N.; Toomsan, B.; Vorasoot, N.; Jogloy, S.; Boote, K.; Hoogenboom, G.; Patanothai, A. Classification of root distribution patterns and their contributions to yield in peanut genotypes under mid-season drought stress. *Field Crops Res.* **2012**, *127*, 181–190. [CrossRef]

6. Koolachart, R.; Jogloy, S.; Vorasoot, N.; Wongkaew, S.; Holbrook, C.; Jongrungklang, N.; Kesmala, T.; Patanothai, A. Rooting traits of peanut genotypes with different yield responses to terminal drought. *Field Crops Res.* **2013**, *149*, 366–378. [CrossRef]

7. Rucker, K.S.; Kvien, C.K.; Holbrook, C.C.; Hook, J.E. Identification of peanut genotypes with improved drought avoidance traits. *Peanut Sci.* **1995**, *22*, 14–18. [CrossRef]

8. Kulkarni, M.; Phalke, S. Evaluating variability of root size system and its constitutive traits in hot pepper (*Capsicum annuum* L.) under water stress. *Scr. Hortic.* **2009**, *120*, 159–166. [CrossRef]

9. Kulkarni, M.; Deshpande, U. Comparative studies in stem anatomy and morphology in relation to drought tolerance in tomato (*Lycopersicon esculentum*). *Am. J. Plant Physiol.* **2006**, *1*, 82–88. [CrossRef]

10. Kulkarni, M.; Borse, T.; Chaphalkar, S. Anatomical variability in grape (*Vitis venifera*) genotypes in relation to water use efficiency (WUE). *Am. J. Plant Physiol.* **2007**, *2*, 36–43. [CrossRef]

11. Pan, C.X.; Nakao, Y.; Nii, N. Anatomical development of Phi thickening and the Casparian strip in loquat roots. *J. Jpn. Soc. Hortic. Sci.* **2006**, *75*, 445–449. [CrossRef]

12. Peterson, C.A.; Emanuel, M.E.; Weerdenburg, C.A. The permeability of phi thickenings in apple (*Pyrus malus*) and geranium (*Pelargonium hortorum*) roots to an apoplastic fluorescent dye tracer. *Can. J. Bot.* **1981**, *59*, 1107–1110. [CrossRef]

13. Henrique, P.D.; Alves, J.D.; Goulart, P.D.P.; Deuner, S.; Silveira, N.M.; Zanandrea, I.; de Castro, E.M. Physiological and anatomical characteristics of sibipiruna plants under hypoxia. *Ciencia Rural* **2010**, *40*, 70–76. [CrossRef]

14. Tajima, R.; Abe, J.; Lee, O.N.; Morita, S.; Lux, A. Developmental changes in peanut root structure during root growth and root-structure modification by nodulation. *Ann. Bot.* **2008**, *101*, 491–499. [CrossRef] [PubMed]

15. Jongrungklang, N.; Toomsan, B.; Vorasoot, N.; Jogloy, S.; Boote, K.; Hoogenboom, G.; Patanothai, A. Drought tolerance mechanisms for yield responses to pre-flowering drought stress of peanut genotypes with different drought tolerant levels. *Field Crops Res.* **2013**, *144*, 34–42. [CrossRef]

16. Kano-Nakata, M.; Inukai, Y.; Wade, L.J.; Siopongco, J.D.; Yamauchi, A. Root development, water uptake, and shoot dry matter production under water deficit conditions in two CSSLs of rice: Functional roles of root plasticity. *Plant Prod. Sci.* **2011**, *14*, 307–317. [CrossRef]

17. Doorenbos, J.; Pruitt, W.O. Calculation of crop water requirement. In *Crop Water Requirements*; FAO of The United Nation: Rome, Italy, 1992; pp. 1–65.

18. Meisner, C.A.; Karnok, K.J. Peanut root response to drought stress. *Agron. J.* **1992**, *84*, 159–165. [CrossRef]

19. Thangthong, N.; Jogloy, S.; Pensuk, V.; Kesmala, T.; Vorasoot, N. Distribution patterns of peanut roots under different durations of early season drought stress. *Field Crops Res.* **2016**, *198*, 40–49. [CrossRef]

20. Boyer, J.S. Leaf enlargement and metabolic rates in corn, soybean, and sunflower at various leaf water potentials. *Plant Physiol.* **1970**, *46*, 233–235. [CrossRef] [PubMed]

21. Steudle, E.; Carol, A.P. How does water get through roots. *J. Exp. Bot.* **1998**, *49*, 775–788. [CrossRef]

22. Carlquist, S. Further concepts in ecological wood anatomy, with comments on recent work in wood anatomy and evolution. *Aliso* **1980**, *9*, 459–553. [CrossRef]

23. Salleo, S.; Lo Gullo, M.A. Xylem cavitation in nodes and internodes of whole *Chorisia insignis* H.B. et K. plants subjected to water stress: Relations between xylem conduit size and cavitation. *Ann. Bot.* **1986**, *58*, 431–441. [CrossRef]

24. Willson, J.C.; Jackson, R.B. Xylem cavitation caused by drought and freezing stress in four co-occurring *Juniperus* species. *Physiol. Plant.* **2006**, *127*, 374–382. [CrossRef]

25. Vasellati, V.; Oesterheld, M.; Medan, D.; Loreti, J. Effects of flooding and drought on anatomy of *Paspalum dialatatum. Ann. Bot.* **2001**, *88*, 355–360. [CrossRef]

26. Carlquist, S. Wood anatomy of Gentianaceae, tribe Helieae, in relation to ecology, habit, systematics, and sample diameter. *Bull. Torrey Bot. Club.* **1985**, *112*, 59–69. [CrossRef]

27. Arnold, D.H.; Mauseth, J.D. Effects of environmental factors on development of wood. *Am. J. Bot.* **1999**, *86*, 367–371. [CrossRef] [PubMed]

28. Comas, L.H.; Mueller, K.E.; Taylor, L.L.; Midford, P.E.; Callahan, H.S.; Beerling, D.J. Evolutionary patterns and biogeochemical significance of angiosperm root traits. *Int. J. Plant Sci.* **2012**, *173*, 584–595. [CrossRef]

29. Comas, L.; Becker, S.; Cruz, V.; Byrne, P.; Dierig, D. Root traits contributing to plant productivity under drought. *Front. Plant Sci.* **2013**, *4*, 442. [CrossRef] [PubMed]

30. Fernandez-Garcia, N.; Lopez-Perez, L.; Hernandez, M.; Olmos, E. Role of phi cells and the endodermis under salt stress in *Brassica oleracea. New Phytol.* **2009**, *181*, 347–360. [CrossRef] [PubMed]

31. López-Pérez, L.; Fernández-García, N.; Olmos, E.; Carvajal, M. The Phi thickening in roots of broccoli plants: An acclimation mechanism to salinity. *Int. J. Plant Sci.* **2007**, *168*, 1141–1149. [CrossRef]

32. Mackenzie, K. The development of the endodermis and phi layer of apple roots. *Protoplasma* **1976**, *100*, 21–32. [CrossRef]

 *agronomy*

Article

A *LEA* Gene from a Vietnamese Maize Landrace Can Enhance the Drought Tolerance of Transgenic Maize and Tobacco

Bui Manh Minh [1], Nguyen Thuy Linh [1], Ha Hong Hanh [1], Le Thi Thu Hien [1,2],
Nguyen Xuan Thang [3], Nong Van Hai [1,2] and Huynh Thi Thu Hue [1,2,*]

[1] Institute of Genome Research, Vietnam Academy of Science and Technology, 18 Hoang Quoc Viet Street,
     Cau Giay District, Hanoi 100000, Vietnam; minhmb@igr.ac.vn (B.M.M.); linhnt@igr.ac.vn (N.T.L.);
     hahonghanh@igr.ac.vn (H.H.H.); hienlethu@igr.ac.vn (L.T.T.H.); vhnong@igr.ac.vn (N.V.H.)
[2] Graduate University of Science and Technology, Vietnam Academy of Science and Technology,
     18 Hoang Quoc Viet Street, Cau Giay District, Hanoi 100000, Vietnam
[3] Maize Research Institute of Vietnam, Vietnam Academy of Agricultural Sciences,
     229 Nguyen Thai Hoc Street, Dan Phuong District, Hanoi 100000, Vietnam; nxthangnmri@gmail.com
*   Correspondence: hthue@igr.ac.vn; Tel.: +84-24-32191173

Received: 20 December 2018; Accepted: 29 January 2019; Published: 31 January 2019

Abstract: Maize (*Zea mays*) is a major cereal crop worldwide, and there is increasing demand for maize cultivars with enhanced tolerance to desiccation. Late embryogenesis abundant (LEA) proteins group 5C is involved in plants' responses to various osmotic stresses such as drought and salt. A putative group 5C LEA gene from *Z. mays* cv. Tevang 1 was isolated, named *ZmLEA14tv*, and cloned into a T-DNA for expression in plants. The deduced amino acid of ZmLEA14tv showed a conserved Pfam LEA_2 domain and a high proportion of hydrophobic residues, characteristic of group 5C LEA proteins. Transgenic tobacco and maize plants expressing *ZmLEA14tv* were generated. During drought simulation conditions, the *ZmLEA14tv*-expressing plants of tobacco showed improved recovery ability, while those of maize enhanced the seed germination in comparison with the non-transgenic control plants. In addition, the survival rate of *ZmLEA14tv* transgenic maize seedlings was twice as high as the control. These results indicated that *ZmLEA14tv* might be involved in the drought tolerance of plants and could be a candidate gene for developing enhanced drought-tolerant crops.

Keywords: drought tolerance; *LEA*; Tevang 1 maize; tobacco

1. Introduction

Late embryogenesis abundant (LEA) proteins are mostly hydrophilic proteins, which can reduce the damage caused by severe environmental conditions. LEA proteins were reported to contribute to various developmental processes and to accumulate in response to drought, low temperature, salt stress, or treatment with the phytohormone ABA [1–4]. The first LEA was reported in cotton seeds [5,6]. LEA proteins accumulated during the late stages of embryogenesis and associated with the desiccation of seeds' embryos [7,8]. The members of the LEA protein family are also expressed during water deficit in bacteria (*Escherichia coli*) and yeast (*Saccharomyces cerevisiae*), suggesting a ubiquitous protective role of these proteins against osmotic stresses [9,10].

Following the Battaglia's classification, LEA proteins are categorized into seven different groups [11]. The LEA proteins of groups 1, 2, 3, 4, 6, and 7 are hydrophilic or typical LEA proteins, which have a low proportion of cysteine and tryptophan residues, and a high proportion of glycine, glutamic acid, lysine, and threonine residues. In contrast, the group 5 LEA protein has high content of hydrophobic residues. Based on amino acid sequences and conserved motifs, the group 5 LEA protein was classified into three subgroups, namely 5A, 5B, and 5C [11]. Subgroup 5C LEA proteins were characterized by a low instability index, low proportion of polar (hydrophilic) and small residues, a higher proportion of non-polar residues, and heat-unstable conformation [11–13]. Moreover, the 5C LEA proteins are folded intrinsically and have more β-sheets than α-helices, which is also different from group 5A and 5B [8,13]. These differences in residue proportion and physical characteristics of group 5C from other LEA protein groups may refer to alternative functions involving stress tolerance.

Recently, due to the development of new sequencing technologies, the whole genome sequences of valuable plants such as rice (*Oryza sativa* L.), maize (*Zea mays* L.), and cotton were published and made available to researchers [14–16]. Based on the conserved domains of LEA proteins, LEA protein families could be identified and characterized through whole-genome prediction approaches. In rice, 34 rice candidate *LEA* (*OsLEA*) genes were identified through a HMMER search (http://hmmer.janelia. org/) [17]. By using a similar method, 242, 136, and 142 candidate DNA regions that encode for LEA proteins were identified in three upland cotton namely *Gossypium hirsutum*, *G. arboreum*, and *G. raimondii*, respectively [18]. The LEA protein profile of maize was also reported with 32 LEA genes distributed non-randomly across chromosomes [19]. The accumulation of LEA profiles in various plants provided fundamental knowledge for functional analysis and *LEA* gene engineering in the future.

A small number of group 5C LEA proteins have been characterized, but their physical characteristics and biological functions are largely unknown. Some members of group 5C were identified in other plants such as cotton LEA14A, soybean D95-4, tomato ER5, hot pepper CaLEA6, *Arabidopsis* AtLEA14A, sweet potato IbLEA14A, rice OSLEA5, foxtail millet SiLEA14A, and wild peanut LEA [8,13,20–26]. The expression of LEA 5C proteins is upregulated by ABA and multiple abiotic stresses including salt and drought [13,23]. Functional studies of group 5C proteins showed that an overexpression of CaLEA6 protein, which originates from hot pepper (*Capsicum annuum*), could improve drought and salt tolerance significantly in tobacco [23]. In addition, the overexpression of other *LEA14A* genes such as *IbLEA14A* and *SiLEA14A* remarkably raised the level of lignification, free proline, and soluble sugar in transgenic sweet potato (*Ipomoea batatas*) calli, *Arabidopsis*, and foxtail millet (*Setaria italica*) [13,23,24]. Recombined OsLEA5 in *E. coli* could protect lactate dehydrogenase from misfolding under different abiotic stresses, resulting in stress tolerance [25].

Maize (*Zea mays* L.) is an important monocot crop worldwide; its production was more than 1.06 billion tons in 2016 [27]. Drought is the major factor that accounts for significant losses in maize productivity. A water reduction of 40% could decrease maize production by 39.3% [28]. Recently, a predicted profile of LEA family in maize has been reported through both bioinformatic and practical approaches [19]. A putative maize *LEA* gene located on chromosome 8 that contained a Pfam LEA_2 domain was described; however, any function of this gene in protecting plants against osmotic stress remains unknown [19]. In the current study, a putative *LEA* gene was isolated from the Tevang 1 landrace, designated as *ZmLEA14tv* and cloned into a T-DNA for expression in plants. The expression patterns of transgenic *ZmLEA14tv* in both tobacco and maize models were investigated to determine the importance of this gene in enhancing drought tolerance in selected plants.

2. Materials and Methods

2.1. Plant Materials and Drought-Stimulating Growth Conditions

Maize seeds (*Z. mays* cv. Tevang 1) were provided by the Maize Research Institute in Vietnam. Tevang 1 cultivar is a well-known landrace maize from the rocky mountain region in Northern Vietnam. The cultivar is well adapted to low annual rainfall and water-deficit cultivation. Maize seeds were germinated and grown under greenhouse conditions at 22/26 °C (night/day) and a photoperiod of 14/10 h (day/night) for two weeks. The genetic background line for transformation was K7, a selected maize that has a higher rate of regeneration and successful transformation through *A. tumefaciens*-mediated methods. The plant material was evaluated for several morphological and physiological traits. For analysis of drought tolerance at the germination stage, 30 seeds of both WT and homozygous *ZmLEA14tv* maize (T2 generation) were germinated on filter paper in a Petri dish wetted with water (as control) or 10% PEG, 20% PEG (*w/v*) solution for one day at 30 °C. Shoot and root lengths were measured after eight days of the treatment, followed by taking photographs. For the drought tolerance assay, the five-leaf-stage maize seedlings were assigned to a withholding water period for 14 days followed by a three-day re-watering. At least five seedlings were grown in each plot. The survival rate, fresh stem weight, and fresh root weight of drought-treated and control seedlings were measured. The experiment was replicated three times.

Seeds of WT tobacco (*Nicotiana tabacum*) cultivar K326 and transgenic tobacco were sterilized with 70% ethanol and 5% bleach, which follows a method described previously [29]. The tobacco explant was germinated in Murashige and Skoog (MS) medium containing 200 mg/μL kanamycin under light/dark cycle conditions of 16/8 h at 25 °C. The 20-day-old T3 tobacco plants were put under drought conditions for 15 days and then re-watered to observe the morphological modification.

2.2. Isolation of the Putative LEA Coding Region in Tevang 1 Landrace Maize

Based on the putative *ZmLEA14A* sequence published in GenBank (accession number EU976614.1) and the flanking sequence of this gene located on maize chromosome 8 in the maizeGBD database (http://www.maizegdb.org/), PCR-specific primers for amplification of this region were designed. The DNA regions containing the putative *ZmLEA14A* open reading frame (ORF) and 5′ UTR of this gene were amplified from genomic DNA of Tevang 1 cultivar using the *ZmLEA14tv_F* forward primer sequence (5′TCACTTCTCTTCCAGCGAGTAC3′) and *ZmLEA14tv_R* reverse primer sequence (5′ TCTCGTACTACTCAAGCAGCAC3′). The PCR product was denoted as *ZmLEA14tv* and purified by Thermo Scientific PCR product purification kit (Cat. number #K0702, Waltham, MA, USA) then cloned into a pJET1.2 cloning vector following the manual of the producer. The cloning vector pJET1.2, which contains the PCR fragment, was transformed into *E. coli* DH10β competent cell by heat shock at 42 °C for 1 min. The colonies harboring the pJET1.2–ZmLEA14tv plasmid were checked by colony PCR with isolation primers and the restriction enzyme *Bgl*II.

2.3. ZmLEA14tv Expression Vector Construction and Transgenic Plant Generation

The coding region of *ZmLEA14tv* was amplified with the *ZmLEA14tv_CloneF* forward primer sequence (5′ ATTACCATGGCGCAGTTGGTG3′) and *ZmLEA14tv_CloneR* reverse primer sequence (5′ ATTAGCGGCCGCGAAGATGCTGG3′) that generates the recognition sites of *Nco*I and *Not*I at the 5′ and 3′ ends of the PCR products, respectively. The thermocycler program (Eppendorf, Hamburg, Germany) starts at 95 °C for 3 min, followed by 30 cycles of amplification (95 °C for 1 min; 56 °C for 30 s; 72 °C for 1 min); the final extension step was 72 °C for 10 min. The 463-bp PCR product was then treated with *Nco*I and *Not*I, purified, and ligated into the pRTRA7/3 vector to generate a 35S promoter-ZmLEA14tv-35S terminator construct. The cassette was cut and combined into the T-DNA region of the pCAMBIA1300 (8958 bp) binary vector. The recombined binary vector pCAMBIA1300/ZmLEA14tv was transformed into *A. tumefaciens* strain EHA 105 after validation by sequencing. The transgenic tobacco and maize plants were regenerated by modified *Agrobacterium*-mediated transformation methods described in the studies of Topping (1998) and Frame et al. (2011), respectively [30,31].

2.4. Sequence Alignment and Gene Evolution Analysis

The sequence of PCR products was verified by 3500 Series Genetic Analyzers (Applied Biosystems, Foster City, CA, USA) followed the Sanger method. A deduced amino acid sequence of ZmLEA14tv protein was generated using the ExPaSy web tool (https://web.expasy.org). The isoelectric point molecular mass, the proportion of amino acid, and grand average of hydropathy (GRAVY) index of the putative ZmLEA14tv peptide were estimated using the ProtParam web tool (https://web.expasy.org/protparam/). Motif analysis was performed using the Pfam program (http://www.ebi.ac.uk/Tools/InterProScan/). The completed amino acid and deduced amino acid sequences of subgroup 5C

LEA proteins were used to construct a phylogenetic tree. Sequence alignment was carried out with ClustalW and adjusted manually. The phylogenetic tree was constructed with the neighbor-joining bootstrap method using the MEGA v6.0 program [32].

3. Results

3.1. Isolation and Vector Construction of ZmLEA14tv Gene

A putative LEA coding sequence from a native maize landrace Tevang 1 (*Z. mays* cv. Tevang 1) was isolated and named *ZmLEA14tv*. The isolated *ZmLEA14tv* had a size of 693 base pairs (bp) with an open reading frame of 459 bp in length encoding for a deduced 152 amino acid (aa) protein. Sequence analysis exhibited the highest similarity at 99% with a *Z. mays LEA14A* coding sequence (accession number NM_001159174), followed by *LEA14A* of *Shorgum bicolor* (XM_002454858.2) and *LEA14A*-like gene of *S. italica* (93% and 87%, respectively). The comparison with the reference sequence number NM_001159174 in GenBank showed two variations, G381A and C456T; however, the deduced amino acid sequence was not changed. The putative protein was predicted at 15.96 kDa in molecular weight with a pI of 6.08. The protein was rich in Val (12.6%), Leu (11.3%), and Gly (9.2%), but contained low quantities of Trp (0.7%), Asn (1.3%), Cys (0.7%), and Gln (1.3%). The GRAVY and instability index of the putative *ZmLEA14tv* were 0.047 and 16.01, respectively, suggesting stability and hydrophobic nature. A conserved "LEA_2" motif (PF03186), which was classified into subgroup 5C according to Battaglia's classification of LEA proteins, was found on the ZmLEA14tv through an InterProScan search [11]. Further analysis showed that ZmLEA14tv contained a low percentage of polar amino acids (23.02%) and a high level of hydrophobic residues (47.02%). The ZmLEA14tv protein displayed diverse homology with other group 5C LEA proteins and broadly matches similar segments in related LEA proteins, indicating a close evolutionary relationship among these proteins (Figure 1A,B). The phylogenetic analysis also showed that the ZmLEA14tv protein has the closest relationship with maize LEA14A protein (accession number NM_001159174), followed by an LEA-like protein from rice, namely Os01g0225600 (accession number NM_001048996), supported by high bootstrap values (99% and 95%, respectively).

A

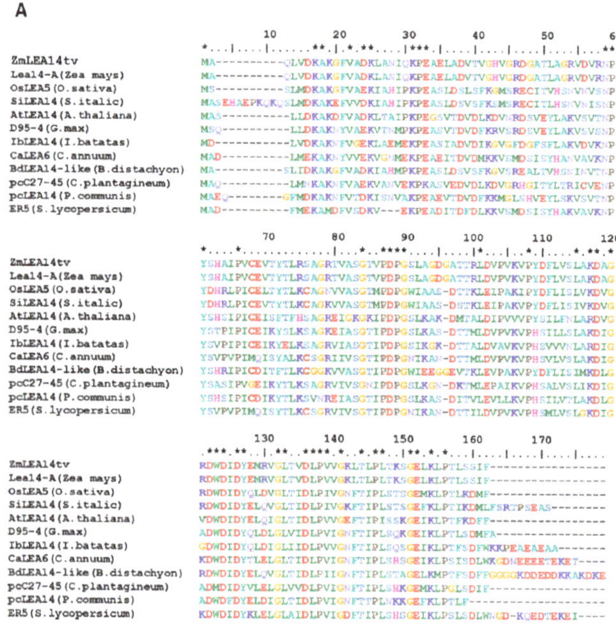

Figure 1. *Cont.*

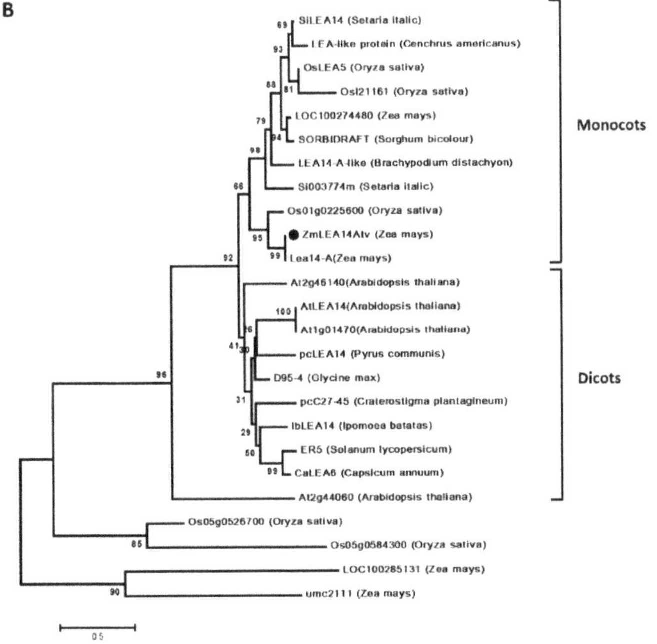

Figure 1. Sequence alignment and phylogenetic relationship for the putative ZmLEA14tv protein and its homologs. (**A**) Multiple sequence alignment of ZmLEA14tv with its homologs (LEA group 5 protein) from various plant species. The conserved positions were marked as stars. (**B**) Neighbor-Joining phylogenetic trees of ZmLEA14tv and its homologs. The clades of monocots and dicots are marked. ZmLEA14tv in *Z. mays* cv. Tevang 1 branch is highlighted by a solid dark circle. The GenBank accession numbers are as follows: SiLEA14 (*S. italic*, KJ767551), AtLEA14 (*Arabidopsis thaliana*, NM100029), Lea14-A (*Z. mays*, NM001159174), D95-4 (*Glycine max*, U08108), IbLEA14 (*Ipomoea batatas*, GU369820), ER5 (*Solanum lycopersicum*, U77719), Lemmi9 (*S. lycopersicum*, Z46654), CaLEA6 (*Capsicum annuum*, AF168168), OsLEA5 (*Oryza sativa*, JF776156), pcC27-45 (*Craterostigma plantagineum*, M62990), pcLEA14 (*Pyrus communis*, AF386513), At1g01470 (*A. thaliana*, BT015111), D95-4 (*G. max*, U08108), At2g46140 (*A. thaliana*, NM130176), Os01g0225600 (*O. sativa*, NM001048996), LEA14-A-like (*Brachypodium distachyon*, XM003567779), BdLEA14-like (*B. distachyon*, XM003567779), LOC100274480 (*Z. mays*, NM001148839), SORBIDRAFT (*Sorghum bicolour*, XM002441543), LEA-like protein (*Cenchrus americanus*, AY823547), OsI21161 (*O. sativa*, CM000130), Os05g0526700 (*O. sativa*, NM001062639), Os05g0584300 (*O. sativa*, NM001062985), At2g44060 (*A. thaliana*, BT024723), LOC100285131 (*Z. mays*, EU970969) and umc2111 (*Z. mays*, NM001155750).

3.2. ZmLEA14tv Gene Expression in Drought Resistance for Transgenic Tobacco

To evaluate the function of the *ZmLEA14tv* transgenic structure in plant osmotic tolerance, the transgenic tobacco plants that expressed ZmLEA14tv under the control of the CaMV 35S promoter were selected for further analysis (Figure 3A). Thirty transgenic plants were obtained, and three homozygous T3 transgenic lines (LEAtv-L1, LEAtv-L3, LEAtv-L7) with high expression levels of ZmLEA14tv (Figure 2B) were chosen for further investigation. To investigate the drought tolerance of the transgenic tobacco, the seedlings were treated with a shortage of water for 15 days. Subsequently, the plants were re-watered and grown for three days. Under normal and drought conditions, there were no significant differences in morphological features such as height, weight, and leaf surface between the transgenic and non-transgenic wild-type (WT) plants. Leaves of the transgenic and control plants became curled and wilted after 15 days of drought. However, 100% of the transgenic tobacco was restored after a three-day re-watering, unlike the control plant (Figure 2C). The fastest recovery was observed in LEAtv-L1, which also expressed the highest level of the *ZmLEA14tv* transgene. This result indicated a correlation between the *ZmLEA14tv* expression and the recovery ability of the plant after the drought conditions.

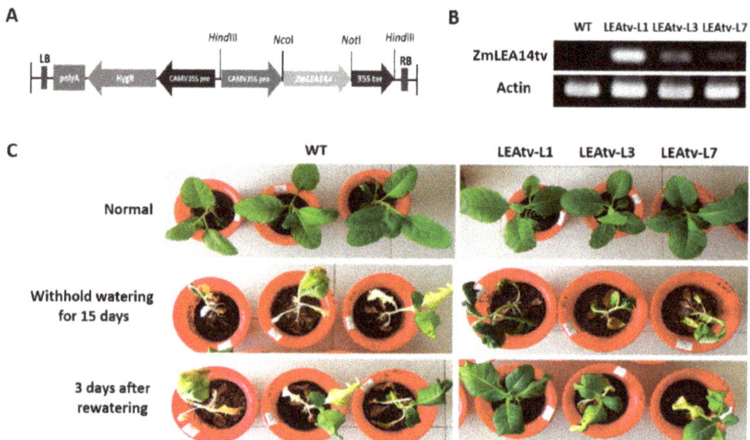

Figure 2. The expression of the *ZmLEA14tv* in tobacco (*Nicotiana tabacum*). (**A**) Schematic description of T-DNA involving in the pCAMBIA1300/ZmLEA14tv plasmid for the expression of *ZmLEA14tv* in plants. LB: left T-DNA border; RB: right T-DNA border; HygR: Hygromycin resistant gene; CaMV35S pro: Cauliflower mosaic virus 35S promoter; 35S ter: 35S terminator; *ZmLEA14A*: coding region of *ZmLEA14tv* gene; *Nco*I, *Not*I, and *Hin*dIII: restriction site of *Nco*I, *Not*I và *Hin*dIII, respectively. (**B**) RT-PCR analysis of ZmLEA14tv in transgenic tobacco lines (LEAtv-L1, LEAtv-L3, LEAtv-L7). WT: the wild type was used as a control. (**C**) The phenotype of transgenic and WT tobacco explants under normal and drought stress conditions.

3.3. Transgenic Maize with ZmLEA14tv Gene in Drought Tolerance

The ZmLEA14tv integration was confirmed by genomic PCR using pairs of primer sets specific to the Hpt and 35S promoter regions, respectively. RT-PCR and qRT-PCR of ZmLEA14tv were performed to validate the expression of the transgenic structure of T2 transgenic maize lines. Three T2 lines, namely L1453, L1482, and L1510, showed the highest expression level (2.7-, 9.2-, and 5.8-fold higher than the control, respectively) and the T2 seeds of these re-watered lines were used for further analysis (Figure 3A,B).

The drought tolerance of transgenic maize seeds during germination was examined. When germinated in water for eight days, the transgenic lines showed better growth than the WT line used as the regeneration material. Under normal conditions (H_2O), no significant differences in shoot height were observed between the transgenic lines and the WT ones. However, the root length of the L1510 and L1482 lines, which have higher ZmLEA14tv transcripts accumulation, was significantly higher than that of the wild type. In comparison with the control group in water, the germination of both the WT and transgenic lines was severely suppressed under 10% and 20% PEG stress (Figure 3C). None of the experimental seeds could germinate in 20% PEG. However, better germination and subsequent development were observed with the seeds of L1510 and L1482, which had higher expression of ZmLEA14tv transcripts in 10% PEG than the wild type (Figure 3D).

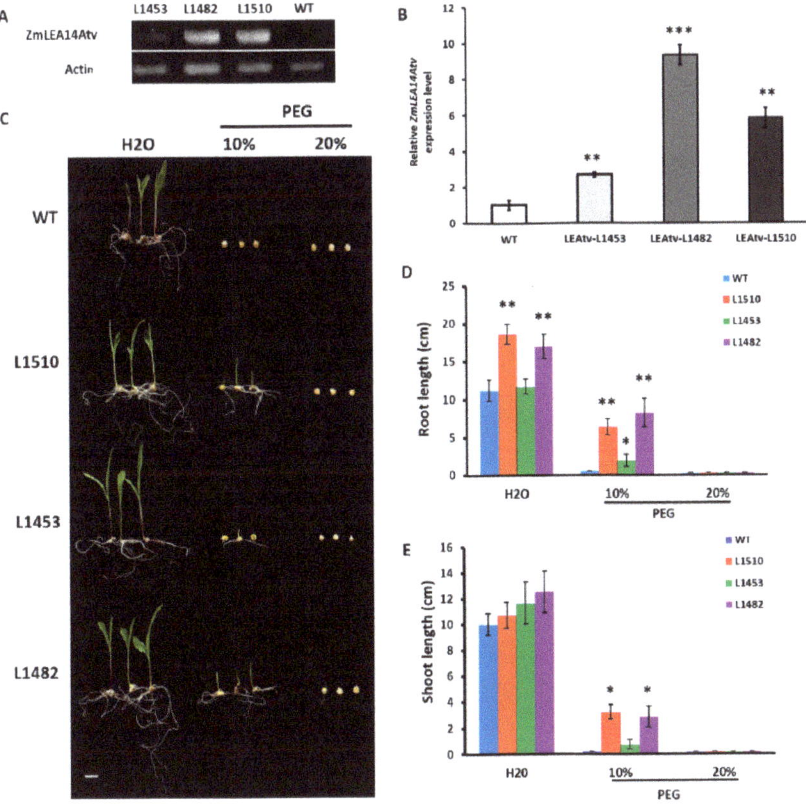

Figure 3. The expression of *ZmLEA14tv* in maize (*Z. mays*). (**A,B**) *ZmLEA14tv* expression in three lines of T2 transgenic maize (L1453, L1482, and L1510) determined by RT-PCR (**A**) and qRT-PCR (**B**). Data in (**B**) represent means and standard errors for three biological replicates. (**C**) The phenotype of transgenic and WT maize under various abiotic stress treatment during the germination stage. The T2 of transgenic seeds were soaked in water (as control) or in 10% PEG, 20% PEG solution for drought simulation for one day at 30 °C and then placed on filter paper in plastic boxes wetted with the same solutions mentioned above for eight days. Each experiment was replicated three times. (**D,E**) The root and shoot length of transgenic and wild-type maize germinated under control (H_2O) and drought-simulating conditions (10% PEG and 20% PEG). Statistical significance was determined by Student's *t*-test. * $p < 0.05$; ** $p < 0.01$.

Furthermore, the drought tolerance of transgenic maize seedlings in soil was examined (Figure 4B–E). No significant differences in survival rate and fresh weight were observed between the transgenic and WT plants under well-wateredconditions (H_2O). However, after drought stress for 14 days and re-watering, only 40% of WT seedlings could be restored, while this ratio in transgenic lines (LEAtv-L1 and LEAtv-L2) was almost doubled (87% and 80%, respectively) (Figure 4C). In addition, the fresh stem weight and fresh root weight of transgenic lines were significantly higher than the K7 wild type, suggesting a better growth rate of these lines in drought condition (Figure 4D,E). Taken together, these results indicate that the *ZmLEA14tv* gene showed improved drought resistance in transgenic maize.

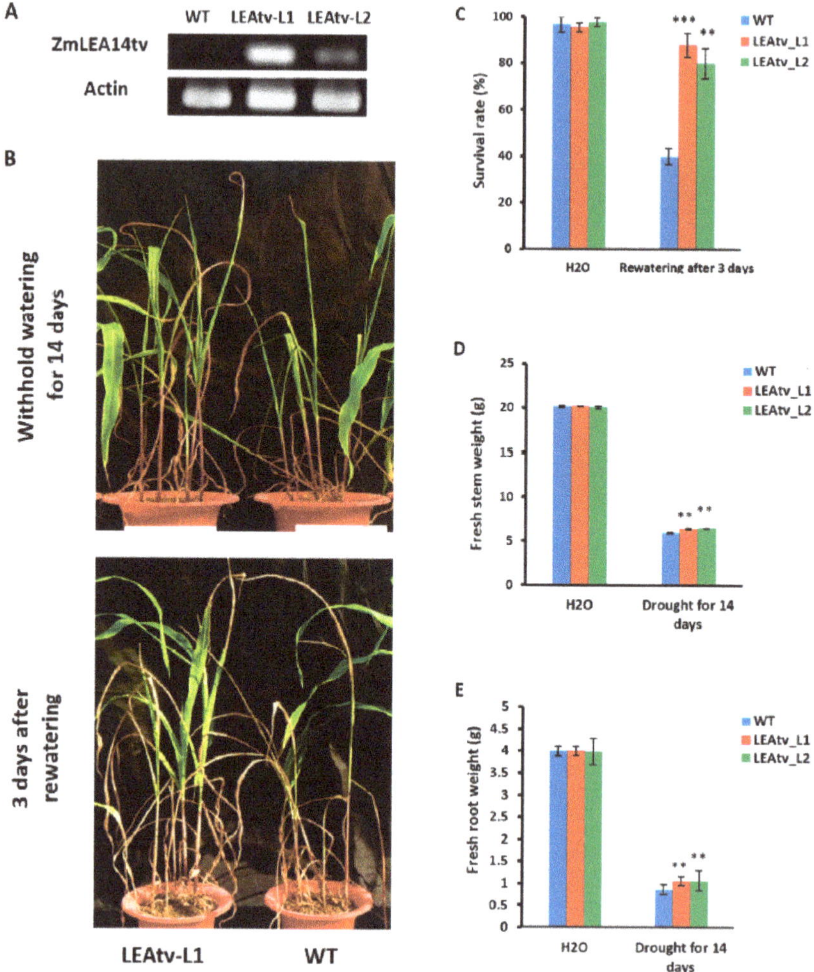

Figure 4. Drought tolerance of maize seedlings overexpressing *ZmLEA14tv*. (**A**) The RT-PCR analysis of *ZmLEA14tv* expression in transgenic maize lines (LEAtv-L1, LEAtv-L2). (**B**) The phenotype of transgenic and WT maize seedlings under drought stress treatment and re-watering conditions. The five-leaf-stage maize seedlings had watering withheld for 14 days, followed by a three-day re-watering. At least five seedlings were grown in each plot and the experiment was replicated three times. WT: wild type. (**C–E**) The survival rate, fresh stem weight, and fresh root weight of WT and transgenic maize seedling after the drought and re-watering treatments. Each experiment was replicated three times. Statistical significance was determined by Student's *t*-test. * $p < 0.05$, ** $p < 0.01$, *** $p < 0.001$.

4. Discussion

The identification and characterization of ZmLEA14tv, a putative atypical LEA group 5C member of the Tevang 1 maize cultivar, were reported in the present study. The deduced amino acid sequence of ZmLEA14tv possessed characteristics of a 5C LEA protein that contains a "LEA_2" domain (Pfam cluster PF03168). The LEA proteins are normally known as hydrophilins with a hydrophilicity index of more than 1 and a glycine (Gly) content more than 6% [11]. Typical LEA proteins can retain water and protect other soluble protein from the aggregation due to their highly hydrophilic properties [33]. Group 5C LEA proteins had a higher proportion of hydrophobic residues than typical LEA proteins; however, they were also involved in various kind of stress tolerance. The estimated GRAVY index of ZmLEA14tv was 0.047, much lower than typical LEA proteins. The ZmLEA14tv protein sequence deduced from the isolated DNA showed a "LEA_2" domain that was characteristic of 5C LEA proteins and a high level of homology with other 5C members.

Functional analysis of 5C LEA proteins showed that the molecular mechanisms of the protective ability against desiccation stress were diverse. The overexpression of AdLEA, a 5C LEA protein from wild peanut, could help maintain the photosynthetic efficiency, reduce the ROS level, and induce the expression of some drought-responsive genes in transgenic tobacco [26]. Meanwhile, overexpressing *SiLEA14* from foxtail millet enhanced a higher level of proline and sugar accumulation in transgenic *Arabidopsis* [13]. The present study showed that the expression of *ZmLEA14tv* in tobacco significantly improved the recoverability of transgenic plants suffering from a short desiccation (Figure 2). This result suggested that *ZmLEA14tv* could function properly in tobacco and enhance its drought tolerance.

One of the methods to generate drought-tolerant maize is enhancing the expression of superior drought-tolerant genes in commercial lines. The overexpression of the *OsSta2* gene, encoding for a AP2/ERF protein, under a maize ubiquitin promoter improved the salt tolerance and grain yield of transgenic rice [34]. Furthermore, enhanced expression of rice dehydrin, namely *OsDhn1*, could increase the tolerance to oxidative stress under salt and drought conditions [35]. Group 5C LEA is well known as an atypical group of proteins that are involved in various abiotic stress response in plants [8,13,24]. In this study, a putative LEA gene namely *ZmLEA14tv* was isolated from the genomic DNA of *Z. mays* cv. Tevang 1, which is well adapted to drought stress in the northern mountains of Vietnam. Our data revealed the potential application of *ZmLEA14tv* in genetic engineering for improving crop performance in the context of climate change. Interestingly, transgenic maize seeds expressing *ZmLEA14tv* showed an improved germination ability in drought-simulated condition in comparison with the WT and did not cause any delay in shoot and root development in transgenic plants.

5. Conclusions

In summary, the coding region of *ZmLEA14tv* was isolated from *Z. mays* cv. Tevang 1 and then cloned into an overexpression cassette. This gene encoded a hydrophobic deduced protein that has a high similarity in structure and a close relationship with other 5C LEA proteins. In addition, the expression of *ZmLEA14tv* in model dicot plants such as tobacco significantly improved the recovery ability, while the enhanced *ZmLEA14tv* transgenic maize showed better germination and growth in drought simulation conditions. These results suggested that *ZmLEA14tv* could act as a potential candidate for genetic engineering to improve drought and other osmotic stress tolerance.

Author Contributions: N.V.H. and H.T.T.H. conceived and designed the experiments; B.M.M., N.T.L., H.H.H., and N.X.T. performed the experiments; B.M.M., L.T.T.H., and H.T.T.H. analyzed the data; B.M.M., N.V.H., and H.T.T.H. prepared the manuscript.

Funding: This research was funded by the Vietnam Ministry of Agriculture and Rural Development (MARD) grant number 16HĐ/KHCN-VP for the period 2014–2018.

Acknowledgments: The authors thank Doan Thi Bich Thao and Nguyen Thu Hoai for participating in some greenhouse experiments.

References

1.   Chandler, P.M.; Robertson, M. Gene expression regulated by abscisic acid and its relation to stress tolerance. *Annu. Rev. Plant Physiol. Plant Mol. Biol.* **1994**, *45*, 113–141. [CrossRef]

2.   Hong-Bo, S.; Zong-Suo, L.; Ming-An, S. LEA proteins in higher plants: Structure, function, gene expression and regulation. *Colloids Surf. B Biointerfaces* **2005**, *45*, 131–135. [CrossRef] [PubMed]

3.   Kimura, M.; Yamamoto Yoshiharu, Y.; Seki, M.; Sakurai, T.; Sato, M.; Abe, T.; Yoshida, S.; Manabe, K.; Shinozaki, K.; Matsui, M. Identification of arabidopsis genes regulated by high light–stress using cDNA microarray. *Photochem. Photobiol.* **2007**, *77*, 226–233.

4.   Tunnacliffe, A.; Wise, M.J. The continuing conundrum of the LEA proteins. *Naturwissenschaften* **2007**, *94*, 791–812. [CrossRef]

5.   Baker, J.; Van Dennsteele, C.; Dure, L. Sequence and characterization of 6 LEA proteins and their genes from cotton. *Plant Mol. Biol.* **1988**, *11*, 277–291. [CrossRef] [PubMed]

6.   Dure, L.; Galau, G.A. Developmental Biochemistry of Cottonseed Embryogenesis and Germination. *Plant Physiol.* **1981**, *68*, 187–194. [CrossRef] [PubMed]

7.   Oliveira, E.; Amara, I.; Bellido, D.; Odena, M.A.; Dominguez, E.; Pages, M.; Goday, A. LC-MSMS identification of *Arabidopsis thaliana* heat-stable seed proteins: Enriching for LEA-type proteins by acid treatment. *J. Mass Spectrom.* **2007**, *42*, 1485–1495. [CrossRef] [PubMed]

8.   Hundertmark, M.; Hincha, D.K. LEA (Late Embryogenesis Abundant) proteins and their encoding genes in *Arabidopsis thaliana*. *BMC Genom.* **2008**, *9*, 118–139. [CrossRef] [PubMed]

9.   Garay-Arroyo, A.; Colmenero-Flores, J.M.; Garciarrubio, A.; Covarrubias, A.A. Highly hydrophilic proteins in prokaryotes and eukaryotes are common during conditions of water deficit. *J. Biol. Chem.* **2000**, *275*, 5668–5674. [CrossRef] [PubMed]

10.   Yale, J.; Bohnert, H.J. Transcript expression in *Saccharomyces cerevisiae* at high salinity. *J. Biol. Chem.* **2001**, *276*, 15996–16007. [CrossRef]

11.   Battaglia, M.; Olvera-Carrillo, Y.; Garciarrubio, A.; Campos, F.; Covarrubias, A.A. The enigmatic LEA proteins and other hydrophilins. *Plant Physiol.* **2008**, *148*, 6–24. [CrossRef] [PubMed]

12.   Wolkers, W.F.; McCready, S.; Brandt, W.F.; Lindsey, G.G.; Hoekstra, F.A. Isolation and characterization of a D-7 LEA protein from pollen that stabilizes glasses in vitro. *Biochim. Biophys. Acta Protein Struct. Mol. Enzymol.* **2001**, *1544*, 196–206. [CrossRef]

13.   Wang, M.; Li, P.; Li, C.; Pan, Y.; Jiang, X.; Zhu, D.; Zhao, Q.; Yu, J. SiLEA14, a novel atypical LEA protein, confers abiotic stress resistance in foxtail millet. *BMC Plant Biol.* **2014**, *14*, 290–305. [CrossRef] [PubMed]

14.   Li, X.; Wu, L.; Wang, J.; Sun, J.; Xia, X.; Geng, X.; Wang, X.; Xu, Z.; Xu, Q. Genome sequencing of rice subspecies and genetic analysis of recombinant lines reveals regional yield- and quality-associated loci. *BMC Biol.* **2018**, *16*, 102. [CrossRef] [PubMed]

15.   Hirsch, C.N.; Hirsch, C.D.; Brohammer, A.B.; Bowman, M.J.; Soifer, I.; Barad, O.; Shem-Tov, D.; Baruch, K.; Lu, F.; Hernandez, A.G.; et al. Draft assembly of elite inbred line ph207 provides insights into genomic and transcriptome diversity in maize. *Plant Cell* **2016**, *28*, 2700–2714. [CrossRef] [PubMed]

16.   Li, F.; Fan, G.; Lu, C.; Xiao, G.; Zou, C.; Kohel, R.J.; Ma, Z.; Shang, H.; Ma, X.; Wu, J.; et al. Genome sequence of cultivated Upland cotton (*Gossypium hirsutum* TM-1) provides insights into genome evolution. *Nat. Biotechnol.* **2015**, *33*, 524–530. [CrossRef] [PubMed]

17.   Wang, X.S.; Zhu, H.B.; Jin, G.L.; Liu, H.L.; Wu, W.R.; Zhu, J. Genome-scale identification and analysis of LEA genes in rice (*Oryza sativa* L.). *Plant Sci.* **2007**, *172*, 414–420. [CrossRef]

18.   Magwanga, R.O.; Lu, P.; Kirungu, J.N.; Lu, H.; Wang, X.; Cai, X.; Zhou, Z.; Zhang, Z.; Salih, H.; Wang, K.; et al. Characterization of the late embryogenesis abundant (LEA) proteins family and their role in drought stress tolerance in upland cotton. *BMC Genet.* **2018**, *19*, 6–37. [CrossRef] [PubMed]

19.   Li, X.; Cao, J. *Late Embryogenesis Abundant (LEA)* gene family in maize: Identification, evolution, and expression profiles. *Plant Mol. Biol. Rep.* **2016**, *34*, 15–28. [CrossRef]

20.   Galau, G.A.; Wang, H.Y.; Hughes, D.W. Cotton *Lea5* and *Lea14* encode atypical late embryogenesis-abundant proteins. *Plant Physiol.* **1993**, *101*, 695–696. [CrossRef]

21.   Maitra, N.; Cushman, J.C. Isolation and characterization of a drought-induced soybean cDNA encoding a D95 family late-embryogenesis-abundant protein. *Plant Physiol.* **1994**, *106*, 805–806. [CrossRef] [PubMed]

22.   Zegzouti, H.; Jones, B.; Marty, C.; Lelievre, J.M.; Latche, A.; Pech, J.C.; Bouzayen, M. ER5, a tomato cDNA encoding an ethylene-responsive LEA-like protein: Characterization and expression in response to drought, ABA and wounding. *Plant Mol. Biol.* **1997**, *35*, 847–854. [CrossRef] [PubMed]

23. Kim, H.S.; Lee, J.H.; Kim, J.J.; Kim, C.H.; Jun, S.S.; Hong, Y.N. Molecular and functional characterization of *CaLEA6*, the gene for a hydrophobic LEA protein from *Capsicum annuum*. *Gene* **2005**, *344*, 115–123. [CrossRef] [PubMed]

24. Park, S.C.; Kim, Y.H.; Jeong, J.C.; Kim, C.Y.; Lee, H.S.; Bang, J.W.; Kwak, S.S. Sweetpotato late embryogenesis abundant 14 (*IbLEA14*) gene influences lignification and increases osmotic- and salt stress-tolerance of transgenic calli. *Planta* **2011**, *233*, 621–634. [CrossRef] [PubMed]

25. He, S.; Tan, L.; Hu, Z.; Chen, G.; Wang, G.; Hu, T. Molecular characterization and functional analysis by heterologous expression in *E. coli* under diverse abiotic stresses for *OsLEA5*, the atypical hydrophobic LEA protein from *Oryza sativa* L. *Mol. Genet. Genom.* **2012**, *287*, 39–54. [CrossRef]

26. Sharma, A.; Kumar, D.; Kumar, S.; Rampuria, S.; Reddy, A.R.; Kirti, P.B. Ectopic expression of an atypical hydrophobic group 5 LEA protein from wild peanut, *Arachis diogoi* confers abiotic stress tolerance in tobacco. *PLoS ONE* **2016**, *11*, e0150609. [CrossRef]

27. FAOSTAT. 2017. Available online: http://www.fao.org/faostat/en/#data/QC (accessed on 6 December 2018).

28. Daryanto, S.; Wang, L.; Jacinthe, P.A. Global synthesis of drought effects on maize and wheat production. *PLoS ONE* **2016**, *11*, e0156362. [CrossRef] [PubMed]

29. Liu, Y.; Wang, L.; Xing, X.; Sun, L.; Pan, J.; Kong, X.; Zhang, M.; Li, D. ZmLEA3, a multifunctional group 3 LEA protein from maize (*Zea mays L.*), is involved in biotic and abiotic stresses. *Plant Cell Physiol.* **2013**, *54*, 944–959. [CrossRef]

30. Topping, J.F. Tobacco transformation. In *Plant Virology Protocols: From Virus Isolation to Transgenic Resistance*; Foster, G.D., Taylor, S.C., Eds.; Humana Press: Totowa, NJ, USA, 1998; pp. 365–372.

31. Frame, B.; Main, M.; Schick, R.; Wang, K. Genetic transformation using maize immature zygotic embryos. In *Plant Embryo Culture: Methods and Protocols*; Thorpe, T.A., Yeung, E.C., Eds.; Humana Press: Totowa, NJ, USA, 2011; pp. 327–341.

32. Tamura, K.; Filipski, A.; Peterson, D.; Stecher, G.; Kumar, S. MEGA6: Molecular Evolutionary Genetics Analysis Version 6.0. *Mol. Biol. Evol.* **2013**, *30*, 2725–2729. [CrossRef]

33. Goyal, K.; Walton, L.J.; Tunnacliffe, A. LEA proteins prevent protein aggregation due to water stress. *Biochem. J.* **2005**, *388*, 151–157. [CrossRef]

34. Kumar, M.; Choi, J.; An, G.; Kim, S.R. Ectopic expression of *OsSta2* enhances salt stress tolerance in rice. *Front. Plant Sci.* **2017**, *8*, 316. [CrossRef] [PubMed]

35. Kumar, M.; Lee, S.C.; Kim, J.Y.; Kim, S.J.; Aye, S.S.; Kim, S.R. Over-expression of dehydrin gene, *OsDhn1*, improves drought and salt stress tolerance through scavenging of reactive oxygen species in rice (*Oryza sativa* L.). *J. Plant Biol.* **2014**, *57*, 383–393. [CrossRef]

Article

Silicon and the Association with an Arbuscular-Mycorrhizal Fungus (*Rhizophagus clarus*) Mitigate the Adverse Effects of Drought Stress on Strawberry

Narges Moradtalab [1,*], **Roghieh Hajiboland** [1], **Nasser Aliasgharzad** [2], **Tobias E. Hartmann** [3] **and Günter Neumann** [3]

[1] Department of Plant Science, University of Tabriz, Tabriz 51666-16471, Iran; ehsan@tabrizu.ac.ir
[2] Department of Soil Science, University of Tabriz, Tabriz 51666-16471, Iran; n-aliasghar@tabrizu.ac.ir
[3] Institute of Crop Science, University of Hohenheim, 70593 Stuttgart, Germany; tobias.hartmann@uni-hohenheim.de (T.E.H.); guenter.neumann@uni-hohenheim.de (G.N.)
* Correspondence: moradtalabnarges@gmail.com; Tel.: +49-711-459-23715; Fax: +49-711-459-23295

Received: 26 November 2018; Accepted: 18 January 2019; Published: 21 January 2019

Abstract: Silicon (Si) is a beneficial element that alleviates the effects of stress factors including drought (D). Strawberry is a Si-accumulator species sensitive to D; however, the function of Si in this species is obscure. This study was conducted to examine the effect of Si and inoculation with an arbuscular mycorrhizal fungus (AMF) on physiological and biochemical responses of strawberry plants under D. Plants were grown for six weeks in perlite and irrigated with a nutrient solution. The effect of Si (3 mmol L^{-1}), AMF (*Rhizophagus clarus*) and D (mild and severe D) was studied on growth, water relations, mycorrhization, antioxidative defense, osmolytes concentration, and micronutrients status. Si and AMF significantly enhanced plant biomass production by increasing photosynthesis rate, water content and use efficiency, antioxidant enzyme defense, and the nutritional status of particularly Zn. In contrast to the roots, osmotic adjustment did not contribute to the increase of leaf water content suggesting a different strategy of both Si and AMF for improving water status in the leaves and roots. Our results demonstrated a synergistic effect of AMF and Si on improving the growth of strawberry not only under D but also under control conditions.

Keywords: silicon; strawberry; total antioxidants; drought; stress responses; arbuscular mycorrhizal fungus (AMF); *Rhizophagus clarus*

1. Introduction

Although silicon (Si) is not considered an essential element for higher plants, numerous studies have demonstrated that Si is a beneficial element that alleviates abiotic and biotic stresses in plants [1–3]. Si is a quasi-essential element for the growth of rice, wheat, sorghum, potato, cucumber, zucchini, and soybean, under various biotic and abiotic stress conditions [4]. According to the Si tissue concentration, plants are classified into Si-accumulators and non-accumulators. The differences in Si accumulation among species can be attributed to the differential ability of roots to take up Si [2].

Drought (D) adversely influences several features of plant growth and development, and a prolonged D severely diminishes plant productivity [5]. Water loss through transpiration is reduced by stomatal closure as an immediate response of plants upon being exposed to D; however, it reduces also nutrient uptake and limits plant ability for dry matter production. In addition, reduced intercellular CO_2 concentration leads to an excess excitation energy that causes enhanced leakage of electrons to molecular oxygen and increases the production of reactive oxygen species (ROS) [6,7]. These cytotoxic ROS destroy normal metabolism through oxidative damage to lipids, proteins, and nucleic acids [8].

Plants have developed complex physiological and biochemical adjustments to tolerate D, including the activation of antioxidative enzymes, maintenance of cell turgor, and water status through the accumulation of organic osmolytes such as soluble carbohydrates and free amino acids, particularly proline [9,10].

Si supplementation of plants alleviates D stress. Several mechanisms including the activation of photosynthetic enzymes [11], the activation of enzymatic antioxidant defense systems, increased water use efficiency [12,13], nutrient uptake [14], root growth and hydraulic conductance [15], and the accumulation of organic osmolytes [16] are involved in Si-mediated growth improvement under D [11,17].

The association of roots with arbuscular mycorrhizal fungi (AMF) is the most abundant symbiosis in the plant kingdom [18]. The colonization of roots by AMF enhances the plant growth by increasing nutrient uptake and plant tolerance to stress [19,20]. Several studies evaluated the effects of AMF-inoculation in horticultural plants such as citrus, apple, and strawberry [21–23]. AMF symbiosis increased the rate of photosynthesis, stomatal conductance, and leaf water potential in colonized plants under D [24]. Moreover, AMF had a significant direct contribution to the uptake of phosphorus (P), zinc (Zn), and copper (Cu) under water stress [25].

Strawberry (*Fragaria x ananasa* Duch.) plants are extremely sensitive to drought because of a shallow root system, large leaf area, and high-water content of fruits. When the strawberry plants are not sufficiently irrigated, both yield and fruit size are reduced [22]. As a Si-accumulating species [26,27], strawberry has both functional influx (Lsi1) and efflux (Lsi2) transporters for Si uptake, and under a constant soluble Si application can absorb 3% Si per dry weight [26]. However, to the best of our knowledge, there is no study on the effect of Si on strawberry under abiotic stresses including D. Another obscure aspect in this regard is Si effect on the association of roots with AMF in this species. Therefore, given the potential of both Si and AMF for mitigation of drought stress effects, the objectives of the present study are (1) to elucidate the influence of Si on photosynthesis, water status, and activity of antioxidative defense system in strawberry plants under D conditions and (2) to investigate the Si effect on the response of mycorrhizal plants when exposed to D stress. We hypothesized the existence of a synergistic effect of Si and AMF on the protection against D in strawberry plants.

2. Materials and Methods

2.1. Preparation of Plant and Fungus Materials

The first-generation strawberry (*Fragaria × ananassa* var. Paros) plantlets of genetically different individuals originating from a strawberry field were prepared as donor mother plants. Second-generation strawberry plantlets from 10 cm stolons of these genetically different mother plants were propagated in a growth chamber. Four independent biological replicates were used per treatment. The offset plants were grown in a standard peat–perlite (1:1) mixture for one week to allow root development.

Inoculum of *Rhizophagus clarus* (Walker & Schüßler; isolated in symbiosis with *Poa annua* L. in a grassland in Cuba) (MUCL 46238–GINCO–BEL; Synonymy: *Glomus clarum* Nicolson & Schenck; [28]) was provided by the Department of Soil Science, University of Tabriz, Iran. Originally, fungi were obtained from Pal Axel Lab, Lund University, Sweden. *R. clarus* was propagated with *Trifolium repens* L. plants in 3.5 L pots containing sterile sandy loam soil. Rorison's nutrient solution, prepared with deionized water [29] with 50% strength of phosphorus, was added to the pots twice a week to bring the soil moisture to water holding capacity (WHC). The pots were incubated in a greenhouse with 28/20 °C day/night and 16/8 h light/dark periods. After four months, the tops of the plants were excised and the pot materials containing soil and mycorrhizal roots were thoroughly mixed and used as fungal inoculum.

2.2. Plant Treatments

The experiment was conducted using a completely randomized design with three factors including irrigation regimes (three levels), Si treatments (two levels), and AMF inoculation (two levels). Each treatment combination was represented by four independent pots as four replicates.

One-week-old strawberry seedlings were transferred to 3 L pots (one plant per pot) filled with washed perlite and containing 60 g autoclaved and non-autoclaved AMF inoculum in $-$AMF and $+$AMF treatments, respectively. The pots were irrigated daily with water or Hoagland nutrient solution at WHC of the perlite after weighing. The total volume of nutrient solution applied to the plants was 200 mL pot^{-1} week^{-1}. To avoid the accumulation of salts in the substrate, electric conductivity in the perlite was measured in samples taken weekly from the bottom of the pots. Si as sodium silicate (Na_2SiO_3, Sigma–Aldrich, Munich, Germany) prepared as the solution (0.6 mM, pH = 6.1) was added to the pots weekly by irrigation leading to a concentration of 3 mmol L^{-1} perlite ($\sim$84 mg L^{-1} perlite) at the end of the experiment after 6 weeks. One week after starting the Si application, the different irrigation regimes (IR) included well-watered (WW, 90% WHC), mild drought (MD, 75% WHC), and severe drought (SD, 35% WHC) and were assigned randomly to the pots, and watering was omitted from D treatments until they reached the respective WHC. This was achieved 4 and 6 days after starting a different IR for the MD and SD treatments, respectively. Well-watered and D plants received the same amount of nutrient solution, and the respective WHC was achieved by adjusting the volume of water used for irrigation.

In order to determine the possible effect of Na as the accompanying ion in the Si salt applied to the plants, an experiment was conducted parallel to the main experiment with an additional control (without the addition of salt or Si) and 6 mmol L^{-1} NaCl containing an equivalent Na with 3 mmol L^{-1} Na$_2$SiO$_3$. The dry weight (g plant^{-1}) of plants under control (0.48 $\pm$ 0.05) and 6 mmol L^{-1} salt (0.51 $\pm$ 0.04) was not significantly different (Tukey test, $p < 0.001$).

Plants were grown under controlled environmental conditions with a temperature regime of 25 °C/18 °C day/night, 14/10 h light/dark periods, a relative humidity of 30%, and at a photon flux density of about 400 μmol m^{-2} s^{-1}.

2.3. Plant Harvest

Six-week-old plants (five weeks after starting Si treatments and four weeks after reaching the respective WHC) were harvested. Shoots and roots were separated, washed with distilled water, and blotted dry on filter paper. After determination of the fresh weight (FW), the dry weight (DW) was determined after drying at 60 °C for 48 h. Subsamples were taken for biochemical analyses before drying. Before harvest, the gas exchange parameters were determined in attached leaves.

For evaluation of the AMF colonization, the fine roots (1 g FW) were cleared in 10% (v/v) KOH and stained with 0.05% (v/v) trypan blue in lacto–glycerin. The colonization rate of the roots (%) was estimated by counting the proportion of root length containing fungal structures (arbuscules, vesicles and hyphae) using the gridline intersect method [30,31]. In brief, stained root segments were spread out evenly in a 10 cm diameter Petri dish. A grid of lines was marked on the bottom of the dish to form 0.5 cm^2. Vertical and horizontal gridlines were observed with a binocular device, and the presence or absence of fungal structures was recorded at each point where the roots intersected a line. Three sets of observations were made recording all the root-gridline intersects. Each of the three replicate records was made on a fresh rearrangement of the same root segments [30,31].

2.4. Leaf Osmotic Potential and Relative Water Content

The leaf osmotic potential (ψ_s) was determined in the second leaves harvested 1 h after the light was turned on in the growth chamber. The leaves were homogenized in a prechilled mortar and pestle and centrifuged at 4000 g for 20 min at 4 °C. The osmotic pressure of the samples was measured by an osmometer (Micro–Osmometer, Herman Roebling Messtechnik, Germany), and the milliosmol data were recalculated to MPa. For the determination of the relative water content (RWC%), the leaf disks (5 mm diameter) were prepared, and after the determination of the fresh weight (FW), they were submerged for 20 h in distilled water; thereafter, they were blotted dry gently on a paper towel, and the turgid weight (TW) was determined. The dry weight (DW) of the samples was determined after drying in an oven at 70 °C for 24 h, and the RWC% was calculated according to the formula (FW − DW)/(TW − DW) × 100.

2.5. Measurements of Photosynthetic Gas Exchange

Before the harvest gas exchange parameters were determined with the attached leaves. The net CO_2 fixation rate (μmol m^{-2} s^{-1}), transpiration rate (mmol m^{-2} s^{-1}), and stomatal conductance (mol m^{-2} s^{-1}) were determined with a calibrated portable gas exchange system (LCA–4, ADC Bioscientific Ltd., Hoddesdon, UK). Water use efficiency (WUE) was calculated as the ratio of photosynthesis/transpiration (μmol mmol^{-1}).

2.6. Biochemical Determinations

For the determination of carbohydrates, leaf and root samples (100 mg) were homogenized in a 100 mM potassium phosphate buffer (pH 7.5) at 4 °C. After centrifugation at 12,000 g for 15 min, the supernatant was used for the determination of total soluble sugars. An aliquot of the supernatant was mixed with an anthrone–sulfuric acid reagent and incubated for 10 min at 100 °C. After cooling, the absorbance was determined at 625 nm. The standard curve was created using glucose (Sigma–Aldrich, Munich, Germany) [32]. The total soluble protein was determined by the Bradford (1976) method using a commercial reagent (Roti®Quant, Roth GmbH, Karlsruhe, Germany) and bovine serum albumin (BSA) as standard. Total free α-amino acids were assayed using a ninhydrin colorimetric method. Glycine (Sigma–Aldrich, Munich, Germany) was used to produce a standard curve [33]. For the determination of proline, samples were homogenized with 3% (*v*/*v*) sulfosalicylic acid and the homogenate was centrifuged at 3000 g for 20 min. The supernatant was treated with acetic acid and acid ninhydrin and boiled for 1 h, and then the absorbance was determined at 520 nm. Proline (Sigma–Aldrich, Munich, Germany) was used to produce a standard curve [34].

2.7. Determination of Enzyme Activities and Concentration Of Oxidants

Fresh leaf samples (100 mg) were ground in liquid nitrogen using a mortar and pestle. Each enzyme assay was tested for linearity between the volume of crude extract and the measured activity. All measurements were undertaken through spectrophotometry (Specord 200, Analytical Jena AG, Jena, Germany) according to optimized protocols described elsewhere [35]. The activity of ascorbate peroxidase (APX, EC 1.11.1.11) was measured by determining the ascorbic acid oxidation; one unit of APX oxidizes ascorbic acid at a rate of 1 μmol min^{-1} at 25 °C. The catalase (CAT, EC 1.11.1.6) activity was assayed by monitoring the decrease in absorbance of H_2O_2 at 240 nm; unit activity was taken as the amount of enzyme which decomposes 1 μmol of H_2O_2 in one min. Peroxidase (POD, EC 1.11.1.7) activity was assayed using the guaiacol test. The enzyme unit was calculated as the enzyme protein required for the formation of 1 μmol tetra–guaiacol for 1 min. The total superoxide dismutase (SOD, EC 1.15.1.1) activity was determined using the mono–formazan formation test. One unit of SOD was defined as the amount of enzyme required to induce a 50% inhibition of nitro blue tetrazolium (NBT) reduction as measured at 560 nm compared with control samples without enzyme aliquot. The concentration of H_2O_2 was determined using KI at 508 nm. Lipid peroxidation was estimated from the amount of malondialdehyde (MDA) formed in a reaction mixture containing thio-barbituric acid (Sigma–Aldrich, Munich, Germany) at 532 nm. The MDA levels were calculated from a 1,1,3,3–tetraethoxypropane (Sigma–Aldrich, Munich, Germany) standard curve [35].

2.8. Mineral Nutrient Analysis

For the determination of the plant nutritional status, 250 mg of dried leaf material was ashed in a muffle furnace at 500 °C for 5 h. After cooling, the samples were extracted twice with 2.5 mL of 3.4 M HNO_3 until dryness to precipitate SiO_2. The ash was dissolved in 2.5 mL of 4 M HCl, subsequently diluted ten times with hot deionized water, and boiled for 2 min. After the addition of a 0.1 mL cesium chloride/lanthanum chloride buffer to the 4.9 mL ash solution, Fe, Mn, and Zn concentrations were measured by atomic absorption spectrometry (AAS, UNICAM 939, Offenbach/Main, Germany) [36].

2.9. Silicon Determination

Dry leaf material (0.2 g) was microwave digested with 3 mL concentrated HNO_3 + 2 mL H_2O_2 for 1 h. Samples were diluted with circa 15 mL deionized H_2O and transferred into 25 mL plastic flasks; 1 mL concentrated Hydrofluoric acid was added and left overnight. After the addition of 2.5 mL 2% (w/v) H_3BO_3, the flask volume was adjusted to 25 mL with deionized H_2O, and Si was determined by ICP–OES (Vista–PRO, Varian Inc., Palo Alto, USA) [36].

2.10. Statistical Analyses

A primary statistical analysis was carried out using the Sigma Plot 11.0 software Systat Software Inc. San Jose, USA. Experimental data were checked for normality using the Shapiro–Wilk test. Where necessary, data were transformed through standard methods to meet the requirements of statistical analysis. In a second analytical step, a so-called insert-and-absorb algorithm was used to truthfully present all relevant significant differences for the main factors and interactions between the main factors. The algorithm was implemented using the SAS 9.4 macro% (Multi factors)based on the work of Piepho, 2012 [37]. The %MULT macro uses output generated from the MIXED, GLIMMIX, or GENMOD procedures. It allows up to three by-variables for factorial experiments but can process the least squares means for one effect only. If Least Squares Means (LSMEANS) are needed for several effects, the linear model procedure must be run several times, each time using only one LSMEANS statement with only one effect. It means each level of one main factor (e.g. IR) was compared separately for each level of the remaining two factors (e.g. AMF and Si) as pairwise comparisons. In our three-factorial analyses (IR, Si and AMF factors), the main effects of the experiment (IR, AMF, Si, IR×AMF, IR×Si, Si×AMF, IR×Si×AMF) were compared using a proc mixed model (MIXED procedures) in the SAS environment at a significance level of $\alpha = 0.05$. LSMEANS of the main and interaction effects were determined.

3. Results

3.1. Effect of Si and Inoculation with AMF on Plants Biomass And Root Colonization

Mycorrhization or Si as single treatments did not significantly affect shoot biomass under well-watered (WW) conditions while the combination of both treatments resulted in a higher shoot biomass suggesting a synergistic effect between AMF and Si. In the plants exposed to mild drought (MD) and severe drought (SD) stresses, in contrast, Si and AMF as single treatments increased the shoot biomass; however, the effect of AMF was not significant in SD plants (Figure 1A). Root biomass was increased by the AMF treatment under WW conditions. Under MD and SD, in comparison, the effect of both Si and AMF as single treatments was significant on root biomass; the effect of AMF was much higher than Si particularly under MD (Figure 1B).

The relative water content decreased with the severity of D. Under WW conditions, there was no significant effect of Si or AMF as single treatments on RWC while the combination of both treatments resulted in higher RWC. In MD and SD plants, in contrast, the effect of single treatments was mainly significant (Figure 1C). The osmotic potential of the leaves and roots was affected by an inverse trend of RWC (Figure 1D). There was a significant interaction among the three main factors including IR, Si, and AMF on the shoot and root biomass, RWC, and osmotic potential where all decreased with the severity of D (IR factor) but were modified by Si and AMF applications (Figure 1).

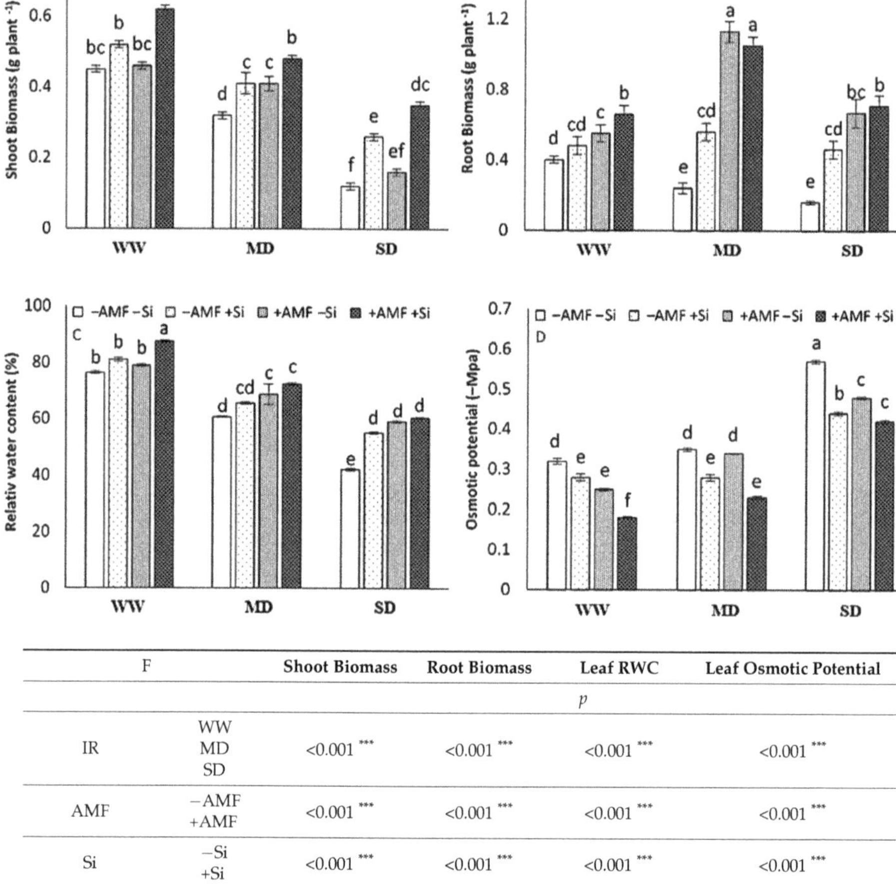

F		Shoot Biomass	Root Biomass	Leaf RWC	Leaf Osmotic Potential
				p	
IR	WW MD SD	<0.001 ***	<0.001 ***	<0.001 ***	<0.001 ***
AMF	−AMF +AMF	<0.001 ***	<0.001 ***	<0.001 ***	<0.001 ***
Si	−Si +Si	<0.001 ***	<0.001 ***	<0.001 ***	<0.001 ***
IR×AMF		0.47 ns	<0.001 ***	<0.001 ***	<0.001 ***
IR×Si		<0.001 ***	0.62 ns	0.21 ns	0.002 **
AMF×Si		0.01 *	<0.001 ***	0.02 *	0.04 *
IR×AMF×Si		0.01 *	0.01 *	<0.001 ***	<0.001 ***

Figure 1. The biomass of shoot (**A**) and root (**B**), the leaf relative water content (RWC) (**C**), and the osmotic potential (**D**) in strawberry plants at harvest after six experimental weeks under three irrigation regimes (IR): well-watered (WW), mild drought (MD), and severe drought (SD) without (−AMF) or with inoculation with arbuscular mycorrhizal fungus (+AMF) *Rhizophagus clarus* (Walker & Schüßler), in the absence (−Si) or presence of silicon (+Si, 3 mmol L^{-1} Na_2SiO_3). The bars show the treatment means (4 replicates) ±SE of the mean. The interactions among the main factors are in the table (**F**); *** $p < 0.001$, ** $p < 0.01$, * $p < 0.05$, and ns is not significant (Tukey test, alpha = 0.05).

There was a low colonization percentage detectable even in −AMF plants, which might be caused by carryover of some fungal populations from the field-grown mother plants or from the peat culture substrate used for the preculture (Table 1). Interestingly, D decreased the hyphal and arbuscular colonization rates (%) in the −AMF plants while not influencing them in the +AMF ones (Table 1). The pairwise comparison indicated that the hyphal colonization percentage of +AMF plants was increased by Si under all IR treatments. The same was true for the frequency of arbuscules that was significant even for the −AMF plants under MD and SD conditions. The frequency of vesicles increased in the −AMF plants under SD conditions. In the +AMF plants, a significant effect was observed under both MD and SD conditions. Si did not affect this parameter. Interestingly, inoculation with AMF decreased the frequency of vesicles in the WW while it increased in the MD and SD plants (Table 1).

Table 1. The root colonization rate (%) in strawberry plants at harvest after six experimental weeks under three irrigation regimes (IR): well-watered (WW), mild drought (MD), and severe drought (SD) without (−AMF) or with inoculation with arbuscular mycorrhizal fungus (+AMF) *Rhizophagus clarus* (Walker & Schüßler), in the absence (−Si) or presence of silicon (+Si, 3 mmol L^{-1} Na_2SiO_3). The numbers show the treatment means (4 replicates) ±SE of the mean. Means with the same letters are not significantly different. The interactions among the main factors include *** $p < 0.001$, ** $p < 0.01$, * $p < 0.05$, and ns as not significant (Tukey test, alpha = 0.05).

			Hyphae	Arbuscules	Vesicles
WW	−AMF	−Si	1.1 ± 0.1 [c]	2.0 ± 0.1 [cd]	0.4 ± 0.0 [c]
		+Si	1.1 ± 0.1 [c]	2.4 ± 0.1 [c]	0.4 ± 0.0 [c]
	+AMF	−Si	1.6 ± 0.1 [b]	4.9 ± 0.2 [b]	0.2 ± 0.1 [d]
		+Si	2.6 ± 0.1 [a]	7.3 ± 0.3 [a]	0.2 ± 0.1 [d]
MD	−AMF	−Si	0.1 ± 0.0 [d]	1.1 ± 0.1 [e]	0.4 ± 0.0 [c]
		+Si	0.1 ± 0.0 [d]	1.7 ± 0.1 [d]	0.4 ± 0.0 [c]
	+AMF	−Si	2.4 ± 0.1 [b]	6.3 ± 0.3 [b]	1.0 ± 0.1 [b]
		+Si	2.6 ± 0.1 [a]	8.1 ± 0.4 [a]	1.0 ± 0.1 [b]
SD	−AMF	−Si	0.1 ± 0.0 [d]	0.0 ± 0.0 [f]	0.9 ± 0.0 [b]
		+Si	0.1 ± 0.0 [d]	1.2 ± 0.1 [e]	0.9 ± 0.0 [b]
	+AMF	−Si	2.4 ± 0.1 [b]	2.8 ± 0.1 [c]	1.9 ± 0.1 [a]
		+Si	2.7 ± 0.1 [a]	3.8 ± 0.2 [b]	2.1 ± 0.1 [a]
				p	
IR	WW MD SD		0.72 [ns]	0.02 *	0.01 *
AMF	−AMF +AMF		<0.001 ***	<0.001 ***	0.04 *
Si	−Si +Si		0.20 [ns]	0.05 *	0.87 [ns]
IR×AMF			0.001 **	<0.001 ***	0.05 *
IR×Si			0.12 [ns]	<0.001 ***	0.06 [ns]
AMF×Si			0.06 [ns]	0.01 *	0.12 [ns]
IR×AMF×Si			0.14 [ns]	<0.001 ***	0.11 [ns]

3.2. Effect of Si and Inoculation with AMF on the Leaf Gas Exchange Parameters

The single application of Si or AMF did not influence the rate of photosynthesis under WW conditions. A significant effect of Si as the single treatment, however, was observed under the MD and SD conditions, and a significant AMF effect was observed under MD conditions. The combined application of Si and AMF, in contrast, increased the rate of photosynthesis under WW, MD, and SD conditions, and the highest photosynthesis rate was obtained under the combination of both treatments with a significant difference with each single treatment (Figure 2A). In the absence of AMF and Si treatments, SD decreased the transpiration rate. This parameter increased by AMF only under SD conditions and by Si under MD and SD conditions. Under WW conditions, in contrast, the rate of transpiration was only increased by the combined application of Si and AMF (Figure 2B). The stomatal conductance showed a similar pattern to the rate of photosynthesis (Figure 2C). The sater use efficiency (WUE) decreased by D irrespective of the AMF or Si treatments. Significant effects of the single treatments were observed in the SD plants for Si and in both MD and SD plants for AMF, and the highest value of WUE was obtained in the combination of both treatments (Figure 2D). There was a significant interaction among the three main factors on photosynthetic activity, transpiration rate, and stomatal conductance. There was not any three-way interaction evident for water use efficiency. Significant differences were observed in IR, AMF, Si, and IR×Si (Figure 2).

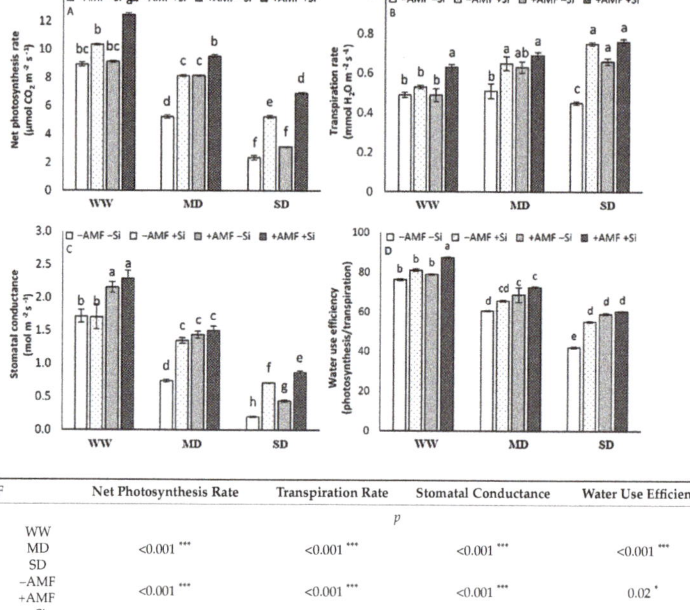

F		Net Photosynthesis Rate	Transpiration Rate	Stomatal Conductance	Water Use Efficiency
			p		
IR	WW MD SD	<0.001 ***	<0.001 ***	<0.001 ***	<0.001 ***
AMF	−AMF +AMF	<0.001 ***	<0.001 ***	<0.001 ***	0.02 *
Si	−Si +Si	<0.001 ***	<0.001 ***	<0.001 ***	<0.001 ***
IR×AMF		<0.001 ***	0.005 **	0.03 *	0.09 ns
IR×Si		<0.001 ***	0.1 ns	0.003 **	0.008 **
AMF×Si		<0.001 ***	0.3 ns	0.09 ns	0.9 ns
IR×AMF×Si		<0.001 ***	0.004 **	0.01 *	0.2 ns

Figure 2. The net photosynthesis rate (**A**), transpiration rate (**B**), stomatal conductance (**C**), and water use efficiency (**D**) of strawberry plants at harvest after six experimental weeks under three irrigation regimes (IR): well-watered (WW), mild drought (MD), and severe drought (SD) without (−AMF) or with inoculation with arbuscular mycorrhizal fungus (+AMF) *Rhizophagus clarus* (Walker & Schüßler), in the absence (−Si) or presence of silicon (+Si, 3 mmol L^{-1} Na$_2$SiO$_3$). The bars show the treatment means (4 replicates) ±SE of the mean. The interactions among the main factors are in the table (**F**); *** $p < 0.001$, ** $p < 0.01$, * $p < 0.05$, and ns is not significant (Tukey test, alpha = 0.05).

3.3. Effect of Si and Inoculation with AMF on the Concentrations of Osmolytes

Under WW conditions, there was no effect of either AMF or Si on the proline concentrations (Table 2). Under MD and SD, in comparison, both Si and AMF treatments decreased leaf proline concentrations; a synergistic effect, however, was observed only under SD conditions (Table 2). The opposite trend of the proline concentration was observed in the root under SD, which was increased by Si and AMF applications where the combined application was not significantly different from the single application. There was a significant interaction among the three main factors including IR, Si, and AMF on the leaf proline concentration (Table 2). D conditions decreased the concentration of proteins while increased the concentration of free amino acids (AA) in leaf and root tissues. The application of Si in the $-$AMF plants increased leaf protein concentrations under D (but not under WW) conditions while decreasing that of free AA. In the roots, in contrast, both protein and free AA concentrations increased by Si in the $-$AMF plants under SD conditions. Similar to Si as a single treatment, AMF application as a single treatment decreased the concentration of free AA in the leaves while increased that in the roots under SD conditions (Table 2). The total free AA concentration of the leaf was significantly affected by all two-way and three-way interactions, while there was not IR$\times$AMF interaction regarding leaf protein concentration. For the roots, there was only an interaction of AMF and Si factors on protein concentrations and of IR and Si on free AA concentration (Table 2).

The concentration of soluble sugars increased under D conditions in both leaves and roots irrespective the AMF and Si treatments. Upon the application of Si and AMF, the soluble sugars concentration decreased in the leaves under MD and SD conditions while increased in the roots of SD plants. The lowest and the highest concentrations of soluble sugars was observed in the leaves and roots in the +AMF+Si plants, respectively. Under WW conditions, the effects of Si and AMF as single treatments were not statistically significant in the leaves and of Si in the roots. There was a three-way interaction among the main factors on shoot sugar concentrations (Table 2).

3.4. Effect of Si and Inoculation with AMF on the Function of Enzymatic Antioxidant Defense

The activity of CAT, SOD, and POD in the leaves and the activity of CAT and SOD in the roots were increased under D conditions irrespective the Si and AMF treatments (Table 3). The highest activity of antioxidative enzymes was observed in the combination of Si and AMF treatments (+AMF+Si). A significant effect of Si and AMF as single treatments was found in SD plants for all analyzed antioxidative enzymes while this effect in the leaves was not significant for POD in MD and for SOD and POD in WW plants. Among all analyzed leaf antioxidative enzymes, only SOD was significantly affected by a three-way interaction. In the roots, the effect of AMF on the CAT and SOD activity was higher than Si as single treatments. There was a significant interaction of the three main factors in CAT but not SOD activities of the root. The activity of root SOD was affected only by IR$\times$AMF (Table 3).

In the absence of Si and AMF, MDA concentration as an indicator of damage to the membrane increased with increasing severity of D. Both Si and AMF treatments decreased the concentration of the leaf MDA that was observed only in D plants. AMF was more effective than Si as single treatment on the reduction of MDA concentrations; the lowest value was observed in +AMF+Si plants. A significant three-way interaction affected the leaf MDA (Table 3).

D treatment led to the accumulation of H_2O_2 in the roots that increased with increasing severity of stress. Si treatment decreased H_2O_2 concentration that was significant only in the D treatments. AMF inoculation caused a significant reduction of the H_2O_2 concentration only under D treatment that was significant in SD plants. The H_2O_2 concentration of the root was decreased by a significant interaction among IR, Si, and AMF factors (Table 3).

3.5. Effect of Si and Inoculation with AMF on the Leaf Concentrations of Nutrients and Si

The Si concentration significantly decreased in SD plants and increased by Si application in the presence or absence of AMF under WW and D conditions (Table 4). The effect of AMF on Si concentration was significant only in +Si plants under WW and in $-$Si ones under SD conditions (Table 4). The interaction effects between two (IR$\times$AMF, IR$\times$Si, and AMF$\times$Si) and among three main factors (IR$\times$AMF$\times$Si) on Si concentration were significant (Table 4).

A significant effect of D on the leaf concentrations of Mn, Fe, and Cu was observed only in the SD treatment while leaf Zn concentration decreased under both MD and SD conditions (Table 4). Si and AMF treatments alone or in combination did not influence the concentrations of Mn, Fe, and Cu. However, Si and AMF significantly increased the leaf Zn concentration under MD conditions (Table 4). Furthermore, significant two-way (IR$\times$Si) and three-way (IR$\times$AMF$\times$Si) interactions were observed for the leaf Zn concentration (Table 4).

Table 2. The concentrations of proline (μg g^{-1} FW), total free amino acids (AA, μg g^{-1} FW), total soluble proteins (mg g^{-1} FW), and soluble sugars (mg g^{-1} FW) in the leaf and roots of strawberry plants at harvest after six experimental weeks under three irrigation regimes (IR): well-watered (WW), mild drought (MD), and severe drought (SD) without ($-$AMF) or with inoculation with arbuscular mycorrhizal fungus (+AMF) *Rhizophagus clarus* (Walker & Schüßler), in the absence ($-$Si) or presence of silicon (+Si, 3 mmol L^{-1} Na$_2$SiO$_3$). The numbers show the treatment means (4 replicates) $\pm$SE of the mean. Means with the same letters are not significantly different. Interactions among the main factors are indicated as *** $p < 0.001$, ** $p < 0.01$, and * $p < 0.05$ and ns as not significant (Tukey test, alpha = 0.05).

			Leaf				Root			
			Proline	Free AA	Protein	Soluble Sugars	Proline	Free AA	Protein	Soluble Sugars
WW	$-$AMF	$-$Si	1.4 $\pm$ 0.0 [e]	2.6 $\pm$ 0.1 [g]	9.0 $\pm$ 0.3 [bc]	5.1 $\pm$ 0.0 [d]	2.9 $\pm$ 0.4 [ab]	1.9 $\pm$ 0.3 [df]	2.9 $\pm$ 0.3 [ab]	1.5 $\pm$ 0.0 [f]
		+Si	1.3 $\pm$ 0.0 [e]	2.5 $\pm$ 0.0 [g]	10.3 $\pm$ 0.3 [b]	4.5 $\pm$ 0.0 [d]	3.0 $\pm$ 0.3 [ab]	1.4 $\pm$ 0.4 [f]	3.0 $\pm$ 0.3 [ab]	2.8 $\pm$ 0.1 [ef]
	+AMF	$-$Si	1.4 $\pm$ 0.0 [e]	2.5 $\pm$ 0.0 [g]	9.1 $\pm$ 0.2 [bc]	4.3 $\pm$ 0.1 [d]	3.4 $\pm$ 0.4 [ab]	1.4 $\pm$ 0.4 [f]	3.4 $\pm$ 0.3 [ab]	2.3 $\pm$ 0.2 [e]
		+Si	1.3 $\pm$ 0.0 [e]	2.9 $\pm$ 0.0 [g]	12.5 $\pm$ 0.3 [a]	2.0 $\pm$ 0.0 [e]	3.7 $\pm$ 0.2 [a]	1.4 $\pm$ 0.3 [f]	3.7 $\pm$ 0.1 [a]	2.6 $\pm$ 0.1 [e]
MD	$-$AMF	$-$Si	39.5 $\pm$ 0.0 [b]	8.2 $\pm$ 0.1 [c]	6.4 $\pm$ 0.1 [d]	9.8 $\pm$ 0.0 [b]	1.7 $\pm$ 0.1 [bc]	4.8 $\pm$ 0.5 [d]	1.7 $\pm$ 0.0 [bc]	3.7 $\pm$ 0.3 [ed]
		+Si	7.4 $\pm$ 0.2 [d]	5.1 $\pm$ 0.1 [e]	8.2 $\pm$ 0.5 [c]	6.5 $\pm$ 0.0 [c]	2.4 $\pm$ 0.3 [b]	5.8 $\pm$ 0.3 [cd]	2.4 $\pm$ 0.0 [b]	5.9 $\pm$ 0.1 [d]
	+AMF	$-$Si	7.7 $\pm$ 0.1 [d]	4.8 $\pm$ 0.1 [e]	8.2 $\pm$ 0.4 [c]	6.3 $\pm$ 0.0 [c]	2.3 $\pm$ 0.1 [b]	6.2 $\pm$ 0.3 [cd]	2.3 $\pm$ 0.2 [b]	4.5 $\pm$ 0.1 [d]
		+Si	5.1 $\pm$ 0.0 [de]	3.3 $\pm$ 0.1 [f]	9.5 $\pm$ 0.1 [b]	6.0 $\pm$ 0.0 [cd]	2.0 $\pm$ 0.2 [b]	8.9 $\pm$ 0.8 [c]	2.0 $\pm$ 0.2 [b]	5.7 $\pm$ 0.4 [d]
SD	$-$AMF	$-$Si	62.1 $\pm$ 2.9 [a]	12.3 $\pm$ 0.1 [a]	2.4 $\pm$ 0.2 [f]	14.8 $\pm$ 0.1 [a]	1.3 $\pm$ 0.2 [c]	11.0 $\pm$ 0.8 [c]	1.3 $\pm$ 0.3 [c]	8.2 $\pm$ 0.4 [c]
		+Si	16.2 $\pm$ 0.1 [c]	9.1 $\pm$ 0.0 [b]	5.3 $\pm$ 0.1 [e]	9.2 $\pm$ 0.0 [b]	2.8 $\pm$ 0.1 [ab]	20.6 $\pm$ 1.1 [a]	2.8 $\pm$ 0.2 [ab]	12.7 $\pm$ 0.5 [b]
	+AMF	$-$Si	18.1 $\pm$ 0.0 [c]	6.3 $\pm$ 0.0 [d]	3.2 $\pm$ 0.1 [f]	10.1 $\pm$ 0.0 [b]	2.6 $\pm$ 0.2 [ab]	15.7 $\pm$ 0.6 [b]	2.6 $\pm$ 0.4 [ab]	12.0 $\pm$ 0.4 [b]
		+Si	9.2 $\pm$ 0.0 [d]	6.1 $\pm$ 0.0 [d]	7.0 $\pm$ 0.2 [d]	7.2 $\pm$ 0.1 [c]	2.9 $\pm$ 0.3 [ab]	23.3 $\pm$ 1.6 [a]	2.9 $\pm$ 0.2 [ab]	15.1 $\pm$ 0.6 [a]
						p				
IR	WW MD SD		<0.001 ***	<0.001 ***	<0.001 ***	<0.001 ***	<0.001 ***	<0.001 ***	<0.001 ***	<0.001 ***
AMF	$-$AMF +AMF		<0.001 ***	<0.001 ***	<0.001 ***	<0.001 ***	0.004 **	0.151 [ns]	0.001 **	<0.001 ***
Si	$-$Si +Si		<0.001 ***	<0.001 ***	<0.001 ***	<0.001 ***	0.006 **	0.061 [ns]	0.002 **	<0.001 ***
IR$\times$AMF			<0.001 ***	<0.001 ***	0.5 [ns]	<0.001 ***	0.2 [ns]	0.1 [ns]	0.2 [ns]	0.01 *
IR$\times$Si			<0.001 ***	<0.001 ***	<0.001 ***	<0.001 ***	0.1 [ns]	0.02 *	0.09 [ns]	0.6 [ns]
AMF$\times$Si			0.05 [ns]	0.01 *	0.01 *	<0.001 ***	0.05 [ns]	0.4 [ns]	0.03 *	<0.001 ***
IR$\times$AMF$\times$Si			<0.001 ***	<0.001 ***	0.01 *	<0.001 ***	0.1 [ns]	0.3 [ns]	0.08 [ns]	0.04 *

Table 3. The activity of catalase (CAT, µmol mg^{-1} protein min^{-1}), superoxide dismutase (SOD, Unit mg^{-1} protein min^{-1}), and peroxidase (POD, (µmol tetra guaicol mg^{-1} protein min^{-1}) and the concentration of malondialdehyde (MDA, nmol g^{-1} FW) in the leaf and the activity of CAT and SOD and the concentration of hydrogen peroxide (H$_2$O$_2$, µmol g^{-1} FW) in the roots of strawberry plants at harvest after six experimental weeks under three irrigation regimes (IR): well-watered (WW), mild drought (MD), and severe drought (SD) without (−AMF) or with inoculation with arbuscular mycorrhizal fungus (+AMF) *Rhizophagus clarus* (Walker & Schüßler), in the absence (−Si) or presence of silicon (+Si, 3 mmol L^{-1} Na$_2$SiO$_3$). The numbers show the treatment means (4 replicates) ±SE of the mean. Means with the same letters are not significantly different. Interactions among the main factors are indicated as *** $p < 0.001$, ** $p < 0.01$, and * $p < 0.05$ and ns as not significant (Tukey test, alpha = 0.05).

			Leaf				Root		
			CAT	SOD	POD	MDA	CAT	SOD	H$_2$O$_2$
WW	−AMF	−Si	36.7 ± 1.7 [g]	4.5 ± 0.5 [d]	1.7 ± 0.3 [f]	0.2 ± 0.0 [i]	2.8 ± 0.1 [e]	3.7 ± 0.2 [e]	0.5 ± 0.0 [e]
		+Si	50.0 ± 4.3 [f]	8.1 ± 0.4 [d]	3.8 ± 0.2 [f]	0.2 ± 0.0 [i]	3.8 ± 0.3 [e]	4.5 ± 0.3 [e]	0.3 ± 0.0 [e]
	+AMF	−Si	54.1 ± 1.6 [f]	5.2 ± 0.5 [d]	3.8 ± 0.4 [f]	0.1 ± 0.0 [i]	4.0 ± 0.1 [e]	4.7 ± 0.4 [e]	0.5 ± 0.0 [e]
		+Si	61.2 ± 0.9 [df]	9.2 ± 0.7 [d]	4.5 ± 0.6 [df]	0.1 ± 0.0 [i]	4.5 ± 0.1 [ed]	5.1 ± 0.4 [ed]	0.3 ± 0.0 [e]
MD	−AMF	−Si	66.1 ± 1.2 [e]	18.0 ± 0.8 [c]	7.1 ± 0.3 [d]	25.2 ± 0.1 [b]	7.2 ± 0.2 [d]	8.2 ± 0.4 [de]	2.5 ± 0.2 [c]
		+Si	84.1 ± 1.9 [d]	34.2 ± 2.8 [b]	9.8 ± 0.9 [d]	10.2 ± 0.0 [e]	8.4 ± 0.2 [cd]	9.4 ± 0.5 [dc]	1.5 ± 0.1 [d]
	+AMF	−Si	86.0 ± 1.0 [d]	35.4 ± 3.1 [b]	9.9 ± 1.0 [d]	7.2 ± 0.0 [g]	14.1 ± 0.7 [c]	13.5 ± 0.4 [c]	2.1 ± 0.0 [c]
		+Si	111.6 ± 4.1 [c]	37.1 ± 3.6 [b]	10.8 ± 0.7 [d]	6.3 ± 0.1 [h]	17.1 ± 1.3 [c]	17.3 ± 0.6 [c]	1.0 ± 0.0 [d]
SD	−AMF	−Si	113.1 ± 1.2 [c]	23.0 ± 1.7 [c]	30.0 ± 3.5 [c]	49.5 ± 0.2 [a]	18.3 ± 1.1 [c]	19.0 ± 1.1 [c]	3.9 ± 0.1 [a]
		+Si	132.9 ± 1.0 [b]	45.1 ± 1.6 [a]	37.8 ± 6.1 [b]	17.5 ± 0.1 [c]	26.3 ± 2.0 [b]	26.3 ± 3.0 [b]	2.7 ± 0.1 [b]
	+AMF	−Si	126.0 ± 2.0 [b]	43.0 ± 1.7 [b]	37.1 ± 4.6 [b]	12.3 ± 0.2 [d]	36.7 ± 2.0 [a]	36.7 ± 4.0 [a]	2.2 ± 0.1 [bc]
		+Si	151.0 ± 1.3 [a]	50.5 ± 2.4 [a]	41.3 ± 1.8 [a]	9.2 ± 0.0 [f]	38.0 ± 2.5 [a]	38.8 ± 4.0 [a]	2.5 ± 0.2 [b]
						p			
IR	WW MD SD		<0.001 ***	<0.001 ***	<0.001 ***	<0.001 ***	<0.001 ***	<0.001 ***	<0.001 ***
AMF	−AMF +AMF		0.006 **	<0.001 ***	0.5 ns	<0.001 ***	0.8 ns	<0.001 ***	<0.001 ***
Si	−Si +Si		0.6 ns	<0.001 ***	0.7 ns	<0.001 ***	0.02 *	<0.001 ***	0.008 **
IR×AMF			<0.001 ***	0.5 ns	0.08 ns	<0.001 ***	<0.001 ***	0.002 **	0.06 ns
IR×Si			<0.001 ***	<0.001 ***	0.2 ns	<0.001 ***	<0.001 ***	0.9 ns	<0.001 ***
AMF×Si			<0.001 ***	<0.001 ***	0.08 ns	<0.001 ***	0.03 *	0.4 ns	0.5 ns
IR×AMF×Si			0.2 ns	<0.001 ***	0.06 ns	0.002 **	<0.001 ***	0.3 ns	0.008 **

Table 4. The concentrations of Si (%), Zn (μg g^{-1} DW), Mn (μg g^{-1} DW), Fe (μg g^{-1} DW), and Cu (μg g^{-1} DW) in the leaf of strawberry plants at harvest after six experimental weeks under three irrigation regimes (IR): well-watered (WW), mild drought (MD), and severe drought (SD) without ($-$AMF) or with inoculation with arbuscular mycorrhizal fungus (+AMF) *Rhizophagus clarus* (Walker & Schüßler), in the absence ($-$Si) or presence of silicon (+Si, 3 mmol L^{-1} Na$_2$SiO$_3$). The numbers show the treatment means (4 replicates) $\pm$SE of the mean. Means with the same letters are not significantly different. Interactions among the main factors are indicated as *** $p < 0.001$, ** $p < 0.01$, and * $p < 0.05$ and ns as not significant (Tukey test, alpha = 0.05).

			Si	Zn	Mn	Fe	Cu
WW	$-$AMF	$-$Si	0.3 ± 0.0 [c]	70.6 ± 4.0 [a]	63.1 ± 9.4 [a]	80.6 ± 14.9 [a]	7.1 ± 0.4 [a]
		+Si	1.4 ± 0.2 [b]	75.3 ± 5.0 [a]	65.1 ± 5.1 [a]	102.5 ± 18.9 [a]	8.0 ± 0.4 [a]
	+AMF	$-$Si	0.4 ± 0.1 [bc]	78.0 ± 2.8 [a]	74.2 ± 6.3 [a]	92.2 ± 15.0 [a]	7.8 ± 0.9 [a]
		+Si	1.9 ± 0.1 [a]	79.4 ± 7.0 [a]	79.4 ± 8.0 [a]	99.4 ± 21.3 [a]	8.7 ± 0.3 [a]
MD	$-$AMF	$-$Si	0.2 ± 0.0 [c]	31.5 ± 4.3 [c]	37.5 ± 5.9 [ab]	32.8 ± 13.0 [ab]	3.9 ± 0.9 [ab]
		+Si	1.1 ± 0.1 [b]	58.1 ± 2.7 [b]	57.9 ± 6.4 [a]	47.6 ± 6.3 [ab]	5.8 ± 0.3 [a]
	+AMF	$-$Si	0.5 ± 0.0 [bc]	47.5 ± 2.5 [bc]	57.5 ± 6.3 [a]	50.0 ± 4.1 [a]	6.0 ± 0.9 [a]
		+Si	0.8 ± 0.1 [b]	61.7 ± 1.7 [b]	64.8 ± 4.9 [a]	69.8 ± 8.2 [a]	6.2 ± 0.2 [a]
SD	$-$AMF	$-$Si	0.1 ± 0.0 [d]	13.4 ± 3.5 [d]	32.2 ± 8.6 [b]	12.2 ± 2.3 [b]	3.5 ± 0.5 [ab]
		+Si	0.5 ± 0.1 [bc]	16.3 ± 4.0 [dc]	26.8 ± 7.3 [b]	19.3 ± 3.5 [b]	3.4 ± 0.6 [b]
	+AMF	$-$Si	0.5 ± 0.0 [bc]	21.0 ± 8.4 [dc]	38.8 ± 6.6 [b]	23.8 ± 5.5 [b]	3.4 ± 0.6 [b]
		+Si	1.0 ± 0.1 [b]	15.5 ± 8.3 [dc]	35.0 ± 2.9 [b]	22.5 ± 6.3 [b]	4.8 ± 0.8 [ab]
					p		
IR	WW MD SD		<0.001 ***	<0.001 ***	<0.001 ***	<0.001 ***	<0.001 ***
AMF	$-$AMF +AMF		<0.001 ***	0.04 *	<0.007 **	<0.01 **	0.02 *
Si	$-$Si +Si		<0.001 ***	0.001 ***	0.3 ns	0.03 **	<0.02 *
IR$\times$AMF			0.03 *	0.5 ns	0.8 ns	0.4 ns	0.7 ns
IR$\times$Si			<0.001 ***	0.02 *	0.2 ns	0.3 ns	0.9 ns
AMF$\times$Si			0.03 *	0.5 ns	0.7 ns	0.4 ns	0.9 ns
IR$\times$AMF$\times$Si			<0.001 ***	0.002 **	0.6 ns	0.6 ns	0.2 ns

4. Discussion

Our results showed that the application of Si and AMF in strawberry might alleviate the adverse effects of D stress in a synergistic manner. Different mechanisms could be involved in this synergistic effect, including Si-mediated improvement of the carbon supply for fungi and likely an increase in the formation of arbuscules. Our results also provide evidence for the effect of Si and AMF on the improvement of strawberry growth under optimum growth conditions through an elevated photosynthesis and water use efficiency.

4.1. Effect of Si and AMF on Growth and Photosynthesis of Plants under Water Stress

Biomass production, water content, and photosynthetic activity of leaves decreased under D conditions in the strawberry plants of this work. Both the Si and AMF treatments alleviated the effects of D and increased leaf water content and photosynthesis rate, leading to a higher biomass production. The observations of gas exchange parameters indicated a D-induced decrease in CO$_2$ assimilation caused by the closure of stomata. The application of Si and AMF increased net photosynthesis rate through an elevation of stomatal conductance. Our results on the effect of Si are in agreement with those of Ma, 2004 [38] for cucumber, Chen et al. 2011 [14] for rice, and Pilon et al. 2013 [39] for potato. Further research has shown that AMF significantly increased leaf area, carboxylation efficiency, chlorophyll content, net photosynthetic rate, and the photochemical efficiency of PS II under water stress [40,41]. Although an improved stomatal conductance upon Si and AMF treatments resulted also in a higher transpiration rate, a greater stimulation of photosynthetic capacity than water loss led to higher water use efficiency in +AMF and +Si plants.

Despite lower photosynthesis activity, soluble sugars accumulated in the leaves of D plants following an impaired growth. It has been stated that water stress triggers sugar accumulation and leads to an adjustment of the rate of photosynthesis [42]. This accumulation of soluble sugars under water stress, in turn, causes an impaired plant metabolism by changing either the composition or the translocation of sugars in the leaves [43]. In the leaves, the concentration of soluble sugars decreased by AMF and Si treatments most likely because of the growth resumption and consumption of carbohydrates for biomass production. Thus, Si and AMF may modulate the accumulation of soluble sugars in water-stressed leaves in a negative feedback mechanism of biochemical limitations.

The same effect of D on the soluble sugars concentration was observed in the roots. However, in contrast to the leaves, the soluble sugars concentration in the roots increased by AMF and Si treatments. This increase may be resulted from an improved net CO_2 assimilation and/or allocation of photosynthates to the roots and may, in turn, contribute to the stimulation of root growth under these conditions. Considering the osmotic effect of soluble carbohydrates, elevated soluble sugars pool may also improve root water uptake capacity from a dry substrate (see below).

4.2. Effect of Si and AMF on the Water Status and Concentration of Organic Osmolytes

The accumulation of organic osmolytes leading to an osmotic gradient with the environment, as a common response in plants under water stress [44], was observed in the strawberry plants in this work for proline, free AA, and soluble sugars, concomitant with the reduction of osmotic potential. The alleviating effect of AMF and Si, however, was not mediated by an osmo-adjustment, and the concentration of organic osmolytes rather decreased in the leaves of +AMF and +Si plants. These results suggest that the Si-mediated increase in leaf water uptake was not due to an increase in the osmotic driving force in strawberry plants under water stress. An increase in the leaf RWC was achieved apparently by an increased capacity for water uptake that in turn hindered triggering the stomatal closure and allowed the maintenance of a high photosynthetic capacity for supporting growth and dry matter production. Increasing levels of organic compounds under osmotic stress are usually thought to adversely affect growth because of the cost associated with their synthesis [45]. Thus, the method of stress alleviation of AMF and Si for an increase in water uptake capacity may be less expensive than the strategy of osmo-adjustment. This result is in contrast with our previous observation on tobacco plants showing a Si-mediated improvement of plant water status through the leaf accumulation of organic osmolytes including soluble sugars, free amino acids, and proline [13].

In contrast to the leaves, the root concentration of organic osmolytes increased by AMF and Si treatments, suggesting a different strategy for the adjustment of the water economy triggered by AMF and Si in the roots than in the leaves of strawberry. In tomatos, water stress did not change the root osmotic potential in Si-treated plants [46], and in cucumbers, the role of the osmotic driving force in the Si-mediated enhancement of water uptake was genotype-dependent [47]. Collectively, these results suggest different strategies for the improvement of water content and uptake capacity under osmotic stress in Si-treated plants depending on plant organ, species, and genotypes. There are reports on the increased root hydraulic conductance by Si, and the increase was attributed to the Si-mediated upregulation of transcription of some aquaporin genes [48].

Under D conditions, proline accumulated in the leaves while the application of AMF and Si reduced leaf proline concentrations. The accumulation of proline in the leaves under water stress is a well-documented phenomenon, but the role of proline in osmotolerance remains controversial. In some studies, the accumulation of proline has been correlated with stress tolerance [49], but other researchers suggest that proline accumulation is a symptom of stress impairment rather than stress tolerance [50]. Our results support the view that proline accumulation under stress is a symptom of stress and, thus, the Si-mediated reduction of proline concentrations is a sign of stress alleviation.

Similarly, the AMF-mediated reduction of the proline concentration suggests that the AMF colonization of plants, to an extent, mitigated the effects of drought stress and reduced proline concentrations in leaves. These results are in agreement with a previous report [51].

An inhibited formation of proteins from amino acids, which could be judged by the accumulation of free AA concomitant with a reduced protein concentration, was observed under water stress of leaves. Both AMF and Si treatments caused the reduction of the free AA pool associated with an increase in soluble proteins. The accumulation of proteins helps the plant to maintain the water-status of leaves, reduce negative effects from active and reactive oxygen species [52] under severe and long-term drought, and maintain the water-status of leaves [10].

4.3. Effect of Si and AMF on the Antioxidative Defense System

Water stress caused the activation of antioxidative defense enzymes in the leaves and roots. However, this activation was not obviously sufficient for the protection of the plants against ROS that was reflected well in the increasing MDA concentrations in the D plants. The application of AMF and Si to the D plants similarly increased the activity of antioxidative defense enzymes (particularly of SOD). However, compared to the stress-induced activation of enzymes, it led to a decline of stress metabolites (MDA, H_2O_2). It may be suggested that AMF and Si contributed to the alleviation of oxidative damage not only by an elevated capacity of defense system but also through less production of the stress metabolites. It has been frequently shown that plants with higher root colonization with AMF exhibit greater enzymatic and non-enzymatic antioxidative defense systems activity [21,35] than non-inoculated plants. A clear biochemical link between Si and antioxidative capacity in stressed plants, however, has not yet been found. It has been argued that the biochemical enhancement of antioxidant defense mechanisms is a beneficial, physical result of Si-deposition in the cell membrane [4]. Several investigators argue that the Si-induced increases in the activity of antioxidant enzymes and the levels of non-enzymatic antioxidative substances in plants exposed to abiotic stress lead to an implication of Si in the plant metabolism [46,47,53]. According to Ma and Yamaji, 2006 [54], the Si-mediated increase in antioxidant defense abilities is a beneficial result of Si rather than a direct effect.

4.4. Effect of Si and AMF on Plants Nutrients Uptake

Water stress reduced the nutritional status of plants, causing deficiencies in Zn, Mn, Cu, and Fe, particularly in more severely stressed plants, but was already partially detectable under MD conditions. The application of AMF and Si led to an improved micronutrient status, equaling or even exceeding the critical deficiency thresholds of Fe, Zn, and Mn. Maksimovi´c et al., 2012 [55] and Pavlovic et al., 2013 [56] found that Si application increased the uptake of Zn and Fe at low concentrations on the rhizoplane. In this work, the effect of Si on nutrient acquisition under D stress was more pronouncedly observed for Zn than other micronutrients. This effect is likely mediated by stimulation of root growth [57] that increases the spatial availability of Zn for plants [58] or by an enhanced concentrations of low molecular weight organic compounds by Si (e.g., citrate) that might contribute to metal uptake and transport from root to shoot, thereby diminishing deficiency symptoms [59]. The higher Zn uptake after the application of Si under D conditions is also likely to result from the effect of the Si on Zn transporters. It has been observed that Si increases the expression levels of the Fe transporters (IRT1 and IRT2) [56] belonging to the ZIP (Zrt/IRT-like protein) family that include also Zn transporters. A limited Zn/Mn availability in the D plants of this work disbalanced Zn/Mn-dependent ROS detoxification systems produced excessive ROS accumulation and caused oxidative damage. The excessive production of ROS can promote oxidative degradation of indoleacetic acid, as was demonstrated in Zn-deficient maize plants under cold stress, which is restored by the Si application [60]. Auxin deficiency is an important factor for growth limitation in Zn-deficient plants [60]. Regarding the role of AMF, plants with a higher root colonization by AMF are more efficient in the uptake and translocation of macro- and micronutrients to the shoot than non-inoculated plants [61,62].

4.5. A Synergistic Effect of Si and AMF

The synergistic effects of Si and AMF as a combined treatment (+AMF+Si) on the low-Si medium used as growth substrate in this work may partly be related to the contribution of AMF to Si uptake observed in this work and in other works [63–66], the Si-induced stimulation in root growth that in turn promotes AMF colonization in the combination treatment, and the effect of Si on an increase in the root soluble sugars pool, which is important for supporting AMF entry, and further establishment in the roots are other probable mechanisms. The mycorrhizal association is completely dependent on the organic carbon supply from their photosynthetic partner since 4 to 20% of the C fixed through photosynthesis is transferred to the AM fungi [67]. Similarly, the Si-induced increase in the percentage of arbuscules formation observed in this work may result from the improved root growth, the enhancement of nutrients uptake and transfer within the plant, and the induced photosynthesis rate that provides more carbon sources for the fungi partner. A significant increase in the percentage of arbuscule formation in response to Si added to a sand substrate has been reported for Banana [65]. In contrast to our results, in a report on the effect of Si on mycorrhizal chickpea [66], an increase was observed in the salinity tolerance by both Si and AMF, but a synergistic effect was not detected.

Another possible explanation for the synergistic effect of AMF and Si is a Si-induced alteration of the AMF-hosts metabolism. In another report, the authors reported an enhanced metabolism of phenolic compounds (flavonoid-type phenolics) influenced by Si [68]. Phenolic compounds such as flavonoids may play a role in facilitating the interactions between fungus and host [69] and have some positive effects on fungal growth parameters, e.g., hyphal growth and branching, germination of spores [70], and formation of secondary spores. Moreover, they play a role during the fungal invasion and arbuscule formation inside the root [71]. The recent identification of strigolactones as host-recognition signals for AM fungi, however, raises the question about the role of flavonoids as general signaling molecules in AMF-plant interactions [72].

4.6. Effect of Si and AMF on Plants Growth in the Absence of Stress

In the well-watered (WW) strawberry plants grown as unstressed controls, Si treatment caused a significant increase in the shoot growth, where the highest biomass production of the shoot was observed in the +AMF+Si treatment. This Si effect under WW conditions disagrees with some of the previous reports [4] describing the beneficial effects of Si on plant growth only under stress conditions. The Si application has been frequently related to the stimulation of enzymatic defense strategies involved in the detoxification of ROS [12]. However, the lower growth of –Si plants under WW conditions in our experiment was not associated with significant changes in the physiological stress indicators, such as MDA and proline. Furthermore, the positive effects of Si on plant growth under WW conditions could also not be attributed to the increased concentrations of the micronutrients. Even in −Si control plants, the nutritional status exceeded the critical levels reported for the respective micronutrient deficiencies. The unexpected positive effects of Si supplementation on the growth of WW plants may be attributed to a significant improvement of the leaf photosynthesis and water content. Considering a higher leaf area in the +Si plants, it is expected that the photosynthesis of the whole strawberry plants is considerably higher than the –Si ones. Improved Si supply may increase the physical stability of the leaves, leading to a more horizontal orientation of the leaves and thereby improving photosynthetic efficiency as previously reported for cucumber [73]. A recent unified model, so-called apoplastic obstruction hypothesis (74), argued for a fundamental role of Si as an extracellular prophylactic agent as opposed to an active cellular agent. In this model, Si, rather than being involved directly in the regulation of gene expression and metabolism, regulates plant metabolism through a cascading effect [74]. Here in our work, the highest growth improvement was observed in the WW plants under the combination of Si with AMF treatments because a Si-induced shoot growth was associated with an AMF-mediated increase in the root growth. The soil-free culture systems that are based on perlite or vermiculite and are being widely used in horticultural practices and are characterized by low plant availability of Si [75]. Thus, the significant effect of Si supplementation in plants cultivated on these potting substrates, in contrast to the soil-grown plants, could be related to supply of plants with Si and meeting their requirement at least in the accumulator species.

5. Conclusions

The findings of the present study suggest that the major factors determining the sensitivity of strawberry plants to D stress are a reduction of micronutrients uptake, particularly Zn, a reduced photosynthesis rate and protein level, a ROS overproduction, and the consequent membrane damage. In this context, the protective effects of Si and AMF treatments seems to be related to an improved micronutrients status, an increased expression of the enzymatic antioxidative defense system, and an elevated water uptake capacity and use efficiency. Our results indicate that Si and AMF alleviated water stress in a synergistic manner. The AMF colonization and formation of fungal structures were increased by Si, and, in turn, Si uptake was increased upon mycorrhization. Other probable interactions at the metabolic levels need to be elucidated. A conceptual model of these proposed roles of Si and AMF, mediating D tolerance in strawberry plants is presented in Figure 3. Our results provide a theoretical basis for the application of Si fertilizers and AMF in water-conserving irrigation systems for strawberry cultivation under field conditions and for greenhouse production, particularly in the soil-free culture systems.

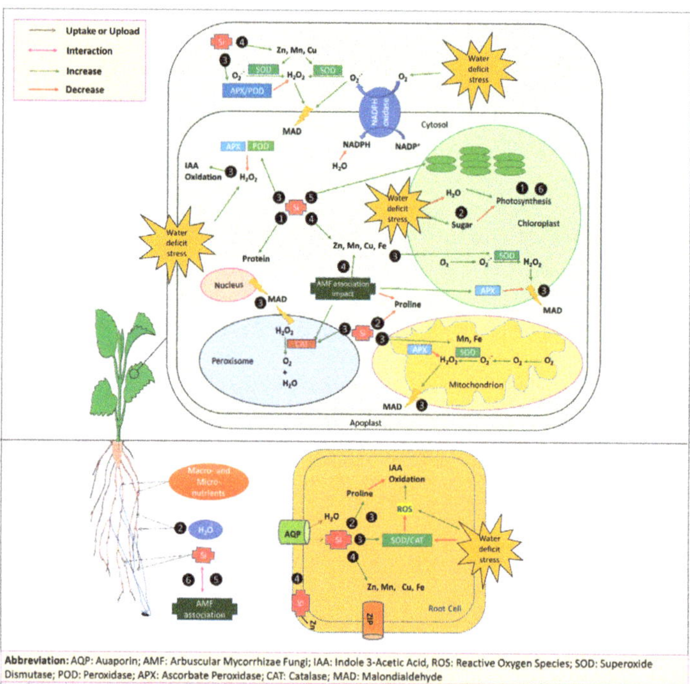

Figure 3. A conceptual model representing the effect of Si and AMF in drought-stressed strawberry plants reverting plant performance to well-watered conditions. Si and AMF 1) enhanced growth and photosynthesis of plants, 2) regulated the water status and concentration of organic osmolytes, 3) promoted the antioxidative defense system, 4) increased plants nutrients uptake, 5) had synergistic effects, and 6) enhanced plant growth even in the absence of stress. Abbreviations: AQP: Aquaporin, AMF: Arbuscular Mycorrhizae Fungi, IAA: Indole 3-Acetic Acid, ROS: Reactive Oxygen Species, SOD: Superoxide Dismutase, POD: Peroxidase, APX: Ascorbate Peroxidase, CAT: Catalase, MAD: Malondialdehyde, NADPH: nicotinamide adenine dinucleotide phosphate hydrogen, NADP$^+$: Nicotinamide adenine dinucleotide phosphate.

Author Contributions: R.H. and N.M. conceived and designed the experiments. N.M. conducted the experiments, performed the analyses, and collected the data. N.A., R.H., and G.N. provided the facilities and advised on the preparation of materials. N.M. wrote the manuscript. N.M. and T.E.H. did the statistics evaluations. G.N. and R.H. read and edited the manuscript. All authors approved the final manuscript.

Funding: This research was funded by the University of Tabriz, Iran [Ph.D project].

Acknowledgments: The University of Tabriz, Iran, is greatly appreciated for its financial support. Thanks to Zarrin Eshaghi (Payame Noor University, Mashhad) for giving support to the analytical facilities and to Filippo Capezzone (Hohenheim University, Biostatistics department) for the support with SAS and statistical analyses. Very special thanks to Hans Lambers (University of Western Australia) for reviewing the manuscript.

Conflicts of Interest: The authors declare no competing financial interests.

References

1. Etesami, H.; Jeong, B.R. Silicon (Si) Review and Future Prospects on the Action Mechanisms in Alleviating Biotic and Abiotic Stresses in Plants. *Ecotoxicol. Environ. Saf.* **2018**, *147*, 881–896. [CrossRef] [PubMed]

2. Broadley, M.; Brown, P.; Cakmak, I.; Ma, J.F.; Rengel, Z.; Zhao, F. Chapter 8—Beneficial Elements Marschner, Petra BT. In *Marschner's Mineral. Nutrition of Higher Plants*, 3rd ed.; Academic Press: San Diego, CA, USA, 2012; pp. 249–269. ISBN 9780123849052.

3. Liang, Y.; Sun, W.; Zhu, Y.G.; Christie, P. Mechanisms of Silicon–mediated Alleviation of Abiotic Stresses in Higher Plants: A Review. *Environ. Pollut.* **2007**, *147*, 4228. [CrossRef]

4. Savvas, D.; Ntatsi, G. Biostimulant Activity of Silicon in Horticulture. *Sci. Hort.* **2015**, *196*, 66–81. [CrossRef]

5. Basu, S.; Ramegoda, V.; Kumar, A.; Pereira, A. Plant Adaptation to Drought Stress. *F1000Res* **2016**, *5*, F1000 Faculty Rev-1554. [CrossRef] [PubMed]

6. Hajiboland, R. Chapter 1—Reactive Oxygen Species and Photosynthesis. In *Oxidative Damage to Plants, Antioxidant Networks and Signaling*; Ahmad, P., Ed.; Elsevier: San Diego, CA, USA, 2014; pp. 1–63, ISBN 978-0-12-799963-0.

7. Singh, R.; Parihar, P.; Singh, S.; Kumar, R. Redox Biology Reactive Oxygen Species Signaling and Stomatal Movement: Current Updates and Future Perspectives. *Redox Biol.* **2017**, *11*, 213–218. [CrossRef] [PubMed]

8. Noctor, G.; Lelarge–Trouverie, C.; Mhamdi, A. The Metabolomics of Oxidative Stress. *Phytochemistry* **2014**, *112*, 33–53. [CrossRef] [PubMed]

9. Singh, M.; Kumar, J.; Singh, S.; Singh, V.P.; Prasad, S.M. Roles of Osmoprotectants in Improving Salinity and Drought Tolerance in Plants: A Review. *Rev. Environ. Sci. Biol.* **2015**, *14*, 407–426. [CrossRef]

10. Anjum, S.A.; Xie, X.; Wang, L.; Saleem, M.F.; Man, C.; Lei, W. Morphological, Physiological and Biochemical Responses of Plants to Drought Stress. *Afr. J. Agric Res.* **2011**, *6*, 2026–2032. [CrossRef]

11. Yin, L.; Wang, S.; Liu, P.; Wang, W.; Cao, D.; Deng, X.; Zhang, S. Silicon-mediated Changes in Polyamine and 1-aminocyclopropane-1-carboxylic Acid are Involved in Silicon-induced Drought Resistance in *Sorghum bicolor* L. *Plant. Physiol. Biochem.* **2014**, *80*, 268–277. [CrossRef] [PubMed]

12. Shen, X.; Zhou, Y.; Duan, L.; Li, Z.; Eneji, E.; Li, J. Silicon Effects on Photosynthesis and Antioxidant Parameters of Soybean Seedlings under Drought and Ultraviolet-B Radiation. *J. Plant Physiol.* **2010**, *167*, 1248–1252. [CrossRef]

13. Hajiboland, R.; Cheraghvareh, L.; Poschenrieder, C. Improvement of Drought Tolerance in Tobacco (*Nicotiana rustica* L.) Plants by Silicon. *J. Plant Nutr.* **2017**, *40*, 1661–1676. [CrossRef]

14. Chen, W.; Yao, X.; Cai, K.; Chen, J. Silicon Alleviates Drought Stress of Rice Plants by Improving Plant Water Status, Photosynthesis and Mineral Nutrient Absorption. *Biol. Trace Elem. Res.* **2011**, *142*, 67–76. [CrossRef]

15. Lux, A.; Luxová, M.; Hattori, T.; Inanaga, S.; Sugimoto, Y. Silicification in Sorghum (*Sorghum bicolor*) Cultivars with Different Drought Tolerance. *J. Plant Physiol.* **2002**, *115*, 87–92. [CrossRef]

16. Ming, D.F.; Pei, Z.F.; Naeem, M.S.; Gong, H.J.; Zhou, W.J. Silicon alleviates PEG-induced Water-deficit Stress in Upland Rice Seedlings by Enhancing Osmotic Adjustment. *J. Agron. Crop. Sci.* **2012**, *198*, 14–26. [CrossRef]

17. Rizwan, M.; Ali, S.; Rizwan, M.; Ali, S.; Ibrahim, M.; Farid, M. Mechanisms of Silicon–mediated Alleviation of Drought and Salt Stress in Plants: A Review. *Environ. Sci. Pollut. Res. Int.* **2015**, *22*, 15416–15431. [CrossRef]

18. Willis, A.; Rodrigues, B.F.; Harris, P.J.C. The Ecology of Arbuscular Mycorrhizal Fungi. *CRC Crit. Rev. Plant Sci.* **2013**, *32*, 1–20. [CrossRef]

19. Abdel, A.; Abdel, H.; Hashem, A.; Rasool, S.; Fathi, E.; Allah, A. Arbuscular Mycorrhizal Symbiosis and Abiotic Stress in Plants: A Review. *J. Plant Biol.* **2016**, *59*, 407. [CrossRef]

20. Hajiboland, R.; Bahrami-Rad, S.; Bastani, S. Phenolics Metabolism in Boron Deficient Tea (*Camellia sinensis* (L.) O. Kuntze) Plants. *Acta Biol. Hung.* **2013**, *64*, 196–206. [CrossRef]

21. Wu, Q.S.; Srivastava, A.K.; Zou, Y.N. AMF-induced Tolerance to Drought Stress in Citrus: A Review. *Sci. Hort.* **2013**, *164*, 77–87. [CrossRef]

22. Krishna, H.; Das, B.; Attri, B.L.; Grover, M.; Ahmed, N. Suppression of Botryosphaeria Canker of Apple by Arbuscular Mycorrhizal Fungi. *Crop. Prot.* **2010**, *29*, 1049–1054. [CrossRef]

23. Boyer, L.R.; Brain, P.; Xu, X.M.; Jeffries, P. Inoculation of Drought-stressed Strawberry with a Mixed Inoculum of Two Arbuscular Mycorrhizal Fungi: Effects on Population Dynamics of Fungal Species in Roots and Consequential Plant Tolerance to Water Deficiency. *Mycorrhiza* **2015**, *25*, 215–227. [CrossRef] [PubMed]

24. Augé, R.M.; Toler, H.D.; Saxton, A.M. Mycorrhizal Stimulation of Leaf Gas Exchange in Relation to Root Colonization, Shoot Size, Leaf Phosphorus and Nitrogen: A Quantitative Analysis of the Literature Using Meta-Regression. *Front. Plant Sci.* **2016**, *7*, 1084. [CrossRef] [PubMed]

25. Smith, F.A.; Smith, S.E. What is the Significance of the Arbuscular Mycorrhizal Colonization of Many Economically Important Crop Plants? *Plant Soil* **2011**, *348*, 63–79. [CrossRef]

26. Ouellette, S.; Goyette, M.H.; Labbé, C.; Laur, J.; Gaudreau, L.; Gosselin, A.; Dorais, M.; Deshmukh, R.K.; Bélanger, R.R. Silicon Transporters and Effects of Silicon Amendments in Strawberry under High Tunnel and Field Conditions. *Front. Plant Sci.* **2017**, *8*, 1–11. [CrossRef] [PubMed]

27. Wang, S.Y.; Galletta, G.J. Foliar Application of Potassium Silicate Induces Metabolic Changes in Strawberry Plants. *J. Plant Nutr.* **1998**, *21*, 157–167. [CrossRef]

28. Nicolson, T.H.; Schenck, N.C. Endogonaceous Mycorrhizal. Endophytes in Florida. *Mycologia* **1979**, *71*, 178–198. [CrossRef]

29. Merryweather, J.W.; Fitter, A.H. A Modified Method for Elucidating the Structure of the Fungal Partner in a Vesicular Arbuscular Mycorrhiza. *Mycol. Res.* **1991**, *95*, 1435–1437. [CrossRef]

30. Giovanetti, M.; Mosse, B. An Evaluation of Techniques for Measuring Vesicular Arbuscular Mycorrhizal Infection in Roots. *New Phytol.* **1980**, *84*, 489–500. [CrossRef]

31. McGonigle, T.P.; Miller, M.H.; Evans, D.G.; Fairchild, G.L.; Swan, J.A. A New Method which Gives an Objective Measure of Colonization of Roots by Vesicular Arbuscular Mycorrhizal Fungi. *New Phytol.* **1990**, *115*, 495–501. [CrossRef]

32. Yemm, E.W.; Willis, A.J. The Estimation of Carbohydrates in Plant Extracts by Anthrone. *Biochem. J.* **1954**, *57*, 508–514. [CrossRef]

33. Yemm, E.W.; Cocking, E.C. The Determination of Amino Acids with Ninhydrin. *Analyst* **1955**, *80*, 209–213. [CrossRef]

34. Bates, L.S.; Waldren, R.P.; Teare, I.D. Rapid Determination of Free Proline for Water–stress Studies. *Plant Soil* **1973**, *39*, 205–207. [CrossRef]

35. Hajiboland, R.; Aliasgharzadeh, N.; Laiegh, S.F.; Poschenrieder, C. Colonization with Arbuscular Mycorrhizal Fungi Improves Salinity Tolerance of Tomato (*Solanum lycopersicum* L.) Plants. *Plant Soil* **2010**, *1*, 313–327. [CrossRef]

36. *VDLUFA Method Book VII Environmental Analysis*; VDLUVA-Verlag: Darmstadt, Germany, 2011; p. 690, ISBN 978-3-941273-10-8.

37. Piepho, H.P. A SAS Macro for Generating Letter Displays of Pairwise Mean Comparisons. *Com. Biom. Crop. Sci.* **2012**, *7*, 4–13.

38. Ma, J.F. Role of Silicon in Enhancing the Resistance of Plants to Biotic and Abiotic Stresses. *J. Soil Sci. Plant Nutr.* **2004**, *50*, 11–18. [CrossRef]

39. Pilon, C.; Soratto, R.P.; Moreno, L.A. Effects of Soil and Foliar Spplication of Soluble Silicon on Mineral Nutrition, Gas Exchange, and Growth of Potato Plants. *J. Crop. Sci.* **2013**, *53*, 1605–1614. [CrossRef]

40. Wu, Q.S.; Xia, R.X. Arbuscular Mycorrhizal Fungi Influence Growth, Osmotic Adjustment and Photosynthesis of Citrus under Well–watered and Water Stress Conditions. *J. Plant Physiol.* **2006**, *163*, 417–425. [CrossRef]

41. Zhu, X.Q.; Wang, C.Y.; Chen, H.; Tang, M. Effects of Arbuscular Mycorrhizal Fungi on Photosynthesis, Carbon Content, and Calorific Value of Black Locust Seedlings. *Photosynthetica* **2014**, *52*, 247–252. [CrossRef]

42. McCormick, A.J.; Cramer, M.D.; Watt, D.A. Regulation of Photosynthesis by Sugars in Sugarcane Leaves. *J. Plant Physiol.* **2008**, *165*, 1817.e29. [CrossRef]

43. Silva, E.N.; Ribeir, R.V.; Ferreira-Silva, S.L.; Vieira, S.; Ponte, L.F.; Silveira, J.G. Coordinate Changes in Photosynthesis, Sugar Accumulation and Antioxidative Enzymes Improve the Performance of *Jatropha curcas* Plants under Drought Stress. *Biomass Bioenergy* **2012**, *45*, 270–279. [CrossRef]

44. Ashraf, M.; Akram, N.A.; Foolad, M.R. Drought Tolerance: Roles of Organic Osmolyts, Growth Regulators, and Mineral Nutrients. *Adv. Agron.* **2011**, *111*, 249–296. [CrossRef]

45. Munns, R. Comparative Physiology of Salt and Water Stress. *Plant Cell Environ.* **2002**, *25*, 239–250. [CrossRef] [PubMed]

46. Shi, Y.; Zhang, Y.; Han, W.; Feng, R.; Hu, Y.; Guo, J.; Gong, H. Silicon Enhances Water Stress Tolerance by Improving Root Hydraulic Conductance in *Solanum lycopersicum* L. *Front. Plant Sci.* **2016**, *7*, 196. [CrossRef]

47. Zhu, Y.X.; Xu, X.B.; Hu, Y.H.; Han, W.H.; Yin, J.L.; Li, H.L.; Gong, H.J. Silicon Improves Salt Tolerance by Increasing Root Water Uptake in *Cucumis sativus* L. *Plant Cell Rep.* **2015**, *34*, 1629–1646. [CrossRef] [PubMed]

48. Liu, P.; Yin, L.N.; Deng, X.P.; Wang, S.W.; Tanaka, K.; Zhang, S.Q. Aquaporin-mediated Increase in Root Hydraulic Conductance is Involved in Silicon-induced Improved Root Water Uptake under Osmotic Stress in *Sorghum bicolor* L. *J. Exp. Bot.* **2014**, *65*, 4747–4756. [CrossRef] [PubMed]

49. Zou, Y.N.; Wu, Q.S.; Huang, Y.M.; Ni, Q.D.; He, X.H. Mycorrhizal Mediated Lower Proline Accumulation in *Poncirus trifoliata* under Drought Derives from the Integration of Inhibition of Proline Synthesis with Increase of Proline Degradation. *PLoS ONE* **2013**, *8*, e80568. [CrossRef] [PubMed]

50. Crusciol, C.C.; Pulz, A.L.; Lemos, L.B.; Soratto, R.P.; Lima, G.P.P. Effects of Silicon and Drought Stress on Tuber Yield and Leaf Biochemical Characteristics in Potato. *Crop. Sci.* **2009**, *49*, 949–954. [CrossRef]

51. Porcel, R.; Aroca, R.; Ruiz-Lozano, J.M. Salinity Stress Alleviation Using Arbuscular Mycorrhizal Fungi. *Agron. Sustain. Dev.* **2012**, *32*, 181–200. [CrossRef]

52. Martinelli, T.; Whittaker, A.; Bochicchio, A.; Vazzana, C.; Suzuki, A.; Masclaux-Daubresse, C. Amino acid Pattern and Glutamate Metabolism during Dehydration Stress in the "Resurrection" Plant *Sporobolus stapfianus*: A Comparison Between Desiccation-sensitive and Desiccation-tolerant Leaves. *J. Exp. Bot.* **2007**, *58*, 3037–3046. [CrossRef] [PubMed]

53. Zhu, Y.; Gong, H. Beneficial Effects of Silicon on Salt and Drought Tolerance in Plants. *Agron. Sustain. Dev* **2014**, *34*, 455–472. [CrossRef]

54. Ma, J.F.; Yamaji, N. Silicon Uptake and Accumulation in Higher Plants. *Trends Plant Sci.* **2006**, *11*, 392–397. [CrossRef] [PubMed]

55. Maksimović, J.D.; Mojović, M.; Maksimović, V.; Römheld, V.; Nikolic, M. Silicon Ameliorates Manganese Toxicity in Cucumber by Decreasing Hydroxylradical Accumulation in the Leaf Apoplast. *J. Exp. Bot.* **2012**, *63*, 2411–2420. [CrossRef] [PubMed]

56. Pavlovic, J.; Samardzic, J.; Maksimović, V.; Timotijevic, G.; Stevic, N.; Laursen, K.H.; Hansen, T.H.; Husted, S.; Schjoerring, J.K.; Liang, Y.; et al. Silicon Alleviates Iron Deficiency in Cucumber by Promoting Mobilization of Iron in the Root Apoplast. *New Phytol.* **2013**, *198*, 1096–1107. [CrossRef] [PubMed]

57. Hattori, T.; Inanaga, S.; Tanimoto, E.; Lux, A.; Luxová, M.; Sugimoto, Y. Silicon-induced Changes in Viscoelastic Properties of Sorghum Root Cell Walls. *Plant Cell Physiol.* **2003**, *44*, 743–749. [CrossRef] [PubMed]

58. Rengel, Z. Availability of Mn, Zn and Fe in the Rhizosphere. *J. Soil. Sci. Plant Nutr.* **2015**, *15*, 397–409. [CrossRef]

59. Hernandez-apaolaza, L. Can Silicon Partially Alleviate Micronutrient Deficiency in Plants? A Review. *Planta* **2014**, *240*, 447–458. [CrossRef] [PubMed]

60. Moradtalab, N.; Weinmann, M.; Walker, F.; Höglinger, B.; Ludewig, U.; Neumann, G. Silicon Improves Chilling Tolerance during Early Growth of Maize by Effects on Micronutrient Homeostasis and Hormonal Balances. *Front. Plant Sci.* **2018**, *9*, 420. [CrossRef]

61. Cakmak, I.; Marschner, H.; Bangerth, F. Effect of Zinc Nutritional Status on Growth, Protein Metabolism and Levels of Indole-3-acetic Acid and Other Phytohormones in Bean (*Phaseolus vulgaris* L.). *J. Exp. Bot.* **1989**, *40*, 405–412. [CrossRef]

62. Rouphael, Y.; Franken, P.; Schneider, C.; Schwarz, D.; Giovannetti, M.; Agnolucci, M.; De Pascale, S.; Bonini, P.; Colla, G. Arbuscular Mycorrhizal Fungi Act as Biostimulants in Horticultural Crops. *Sci. Hortic.* **2015**, *196*, 91–108. [CrossRef]

63. Singh, L.P.; Gill, S.S.; Tuteja, N. Unraveling the Role of Fungal Symbionts in Plant Abiotic Stress Tolerance. *Plant Signal. Behav.* **2011**, *6*, 175–19164. [CrossRef]

64. Clark, R.B.; Zeto, S.K. Mineral Acquisition by Arbuscular Mycorrhizal Plants. *J. Plant Nutr.* **2000**, *23*, 867–902. [CrossRef]

65. Anda, O.C.C.; Opfergelt, S.; Declerck, S. Silicon Acquisition by Bananas (c.V. Grande Naine) is Increased in Presence of the Arbuscular Mycorrhizal Fungus *Rhizophagus irregularis* MUCL 41833. *Plant Soil* **2016**, *409*, 77–85. [CrossRef]

66. Garg, N.; Bhandari, P. Silicon Nutrition and Mycorrhizal Inoculations Improve Growth, Nutrient Status, K+/Na+ Ratio and Yield of *Cicer arietinum* L. Genotypes under Salinity Stress. *Plant Growth Regul.* **2016**, *78*, 371–387. [CrossRef]

67. Smith, S.E.; Read, D.J. *Mycorrhizal Symbiosis*, 3rd ed.; Academic Press: London, UK, 2008; p. 800, ISBN 9780123705266.

68. Rodrigues, F.A.; McNally, D.J.; Datnoff, L.E.; Jones, J.B.; Labbé, C.; Benhamou, N.; Menzies, J.G.; Bélanger, R.R. Silicon Enhances the Accumulation of Diterpenoid Phytoalexins in Rice: A Potential Mechanism for Blast Resistance. *Phytopathology* **2004**, *94*, 177–183. [CrossRef]

69. Mandal, S.M.; Chakraborty, D.; Dey, S. Phenolic Acids Act as Signalling Molecules in Plant-microbe Symbioses. *Plant Signal. Behav.* **2010**, *5*, 359–368. [CrossRef] [PubMed]

70. Steinkellner, S.; Lendzemo, V.; Langer, I.; Schweiger, P.; Khaosaad, T.; Toussaint, J.P.; Vierheilig, H. Flavonoids and Strigolactones in Root Exudates as Signals in Symbiotic and Pathogenic Plant–Fungus Interactions. *Molecules* **2007**, *12*, 1290–1306. [CrossRef]

71. Hassan, S.; Mathesius, U. The Tole of Flavonoids in Rootrhizosphere Signalling: Opportunities and Challenges for Improving Plant-microbe Interactions. *J. Exp. Bot.* **2012**, *63*, 3429–3444. [CrossRef]

72. Abdel-Lateif, K.; Bogusz, D.; Hocher, V. The Role of Flavonoids in the Establishment of Plant Roots Endosymbioses with Arbuscular Mycorrhiza Fungi, Rhizobia and Frankia Bacteria. *Plant Signal. Behav.* **2012**, *7*, 636–641. [CrossRef]

73. Botta, A.; Rodrigues, F.A.; Sierras, N.; Marin, C.; Cerda, J.M.; Brossa, R. Evaluation of Armurox® (cComplex of Peptides with Soluble Silicon) on Mechanical and Biotic Stresses in Gramineae. In Proceedings of the 6th International Conference on Silicon in Agriculture, Stockholm, Sweden, 26–30 August 2014.

74. Coskun, D.; Deshmukh, R.; Sonah, H.; Menzies, J.G.; Reynolds, O.; Ma, J.F.; Kronzucker, H.J.; Bélanger, R.R. The Controversies of Silicon's Role in Plant Biology. *New Phytol.* **2019**, *221*, 67–85. [CrossRef]

75. Reddy, S. Time to Say Sí to Silicon—And Bring Back the Missing Element in Soilless Growing. Available online: http://www.sungro.com/time-say-si-silicon-bring-back-missing-element-soilless-growing (accessed on 12 May 2014).

 *agronomy*

Review

What Has Been Thought and Taught on the Lunar Influence on Plants in Agriculture? Perspective from Physics and Biology

Olga Mayoral [1,2,*], Jordi Solbes [1], José Cantó [1] and Tatiana Pina [1,*]

[1] Department of Science Education, Universitat de València (UV), Avda. Tarongers, 4, 46022 Valencia, Spain; jordi.solbes@uv.es (J.S.); jose.canto@uv.es (J.C.)

[2] Botanical Garden UV, Universitat de València, Calle Quart, 80, 46008 Valencia, Spain

[*] Correspondence: olga.mayoral@uv.es (O.M.); tatiana.pina@uv.es (T.P.); Tel.: +34-961-625-489 (O.M.); +34-961-625-924 (T.P.)

Received: 30 April 2020; Accepted: 17 June 2020; Published: 2 July 2020

Abstract: This paper reviews the beliefs which drive some agricultural sectors to consider the lunar influence as either a stress or a beneficial factor when it comes to organizing their tasks. To address the link between lunar phases and agriculture from a scientific perspective, we conducted a review of textbooks and monographs used to teach agronomy, botany, horticulture and plant physiology; we also consider the physics that address the effects of the Moon on our planet. Finally, we review the scientific literature on plant development, specifically searching for any direct or indirect reference to the influence of the Moon on plant physiology. We found that there is no reliable, science-based evidence for any relationship between lunar phases and plant physiology in any plant–science related textbooks or peer-reviewed journal articles justifying agricultural practices conditioned by the Moon. Nor does evidence from the field of physics support a causal relationship between lunar forces and plant responses. Therefore, popular agricultural practices that are tied to lunar phases have no scientific backing. We strongly encourage teachers involved in plant sciences education to objectively address pseudo-scientific ideas and promote critical thinking.

Keywords: plant growth; agriculture; traditions; pseudo-science; lunar phases; physics; biology; education

---

1. Introduction

This paper addresses the existing dichotomy between what science shows regarding agriculture protocols and past and current agricultural practices in much of Europe and Latin America. More specifically, it focuses on some pseudo-scientific questions and beliefs that impregnate a large part of agricultural traditions and agronomic practices according to which certain lunar phases encourage plant growth while others compromise their development. These beliefs share our lives with scientific and technological advances not reached ever before.

After introducing the main features of the Moon and its phases, as well as the factors that determine plant growth, this study continues with a brief historical overview about what has been thought from the agricultural sector concerning the lunar influence on plants and crops. In this overview, we have included references to both earlier ages and the most recent trends within agriculture, such as biodynamic agriculture, which bases part of its operating on the close relationship between the Moon and plant growth.

Then, we analysed monographs on botany, plant biology and physiology—considered as texts of consolidated science—, searching for any mention about the Moon being a factor influencing plant growth. At the same time, we reviewed physics handbooks, focusing on which aspects or natural processes the Moon has influence on, looking for any mention of some type of effect on living beings and, specifically, on plants.

The paper concludes with a reflection on the implications of the different existing visions lasting over time within both the field of agriculture and citizenship in general, being part of the global desirable scientific literacy.

1.1. The Moon

This section covers the basic key aspects about the Moon required to understand most of the arguments detailed in the subsequent analysis of both scientific literature and explanations provided by some agricultural sectors.

1.1.1. The Gravitational Pull

The Moon is the only natural satellite of our planet describing an elliptical orbit around it with a semi-major axis of 384,000 km, an eccentricity of 0.0549 and an angle of 5°9′ relative to the ecliptic plane. The Moon takes 29.5 days to orbit around the Earth and return to its analogous position with respect to the Sun and the Earth (lunar month or synodic month) [1]. However, it takes 24.8 h for a specific location on the Earth to rotate from one exact point beneath the Moon and back (lunar day) [2,3]. The combined action of these two cycles (lunar month and day) has different effects on the Earth such as changes in tides and in the intensity of illuminance.

So, what is the explanation for the Moon's influence on tides? Tides are due to the difference in gravitational pull (or gravity acceleration) between the part of the oceans which are nearest (A) and farthest (B) to the Moon and the relative acceleration in relation to the Earth's centre of mass (CM) in such points (Figure 1).

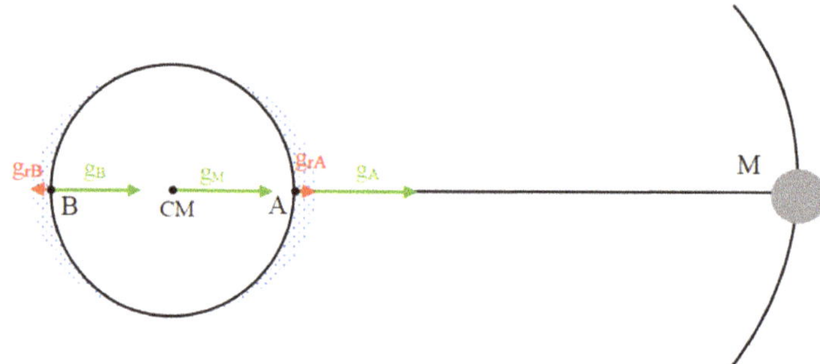

Figure 1. Representation of how tides are produced. In the drawing, g_M represents the acceleration in the Earth's centre of mass (CM) caused by lunar attraction; g_A and g_B are, respectively, the accelerations of points A and B located at both ends of the Earth's surface over the Earth–Moon line, and g_{rA} and g_{rB} are the acceleration in relation to the Earth's CM. Modified from Martínez et al. [1].

From the gravitational point of view, the effect of gravity on the Earth's CM produced by the Moon (g_M) can be calculated by means of the expression $g_M = Gm/r^2$, being G the universal gravitational constant, r the distance Earth–Moon (E–M) and m the mass of the Moon. That is to say, the value of the Moon's gravity on the Earth's surface is approximately 2,951,800 times lower than the Earth's gravity ($g_M = 3.32 \times 10^{-5}$ ms^{-2}). Therefore, the gravitational pull is negligible. Accordingly, the Sun's gravity (g_S) on the Earth is 177 times greater than the Moon's ($g_S = g/1627 = 177\ g_M = 6 \times 10^{-3}$ ms^{-2}).

We can also calculate the Moon's gravity in point A ($g_A \approx g_M(1 + 2\ R/r)$ being M the mass of the Earth and R its radius) and in point B ($g_B \approx g_M(1 - 2\ R/r)$ and their relative accelerations in A (g_{rA}) and B (g_{rB}) regarding Earth's CM ($g_{rA} = g_A - g_M \approx 2\ GmR/r^3 = 2\ RgM/r$ and $g_{rB} = g_B - g_M \approx -2\ GmR/r^3 = -2\ RgM/r$, respectively) (Figure 1). From these calculi, we observe that the relative acceleration in relation to CM depends on the distance between A and B and on the cubed distance between the Earth and the Moon, instead of squared distance, rendering identical values in both A and B points (as r = 60 R, its value is 10^{-6} ms^{-2}, 30 times lower than g_M) but in the opposite direction: the relative acceleration in point A (gr_A) is directed towards the Moon and in point B (gr_B) towards the opposite direction [1,4]. Therefore, there will be high tide in A and B, and low tide in those points located at 90° (Figure 2).

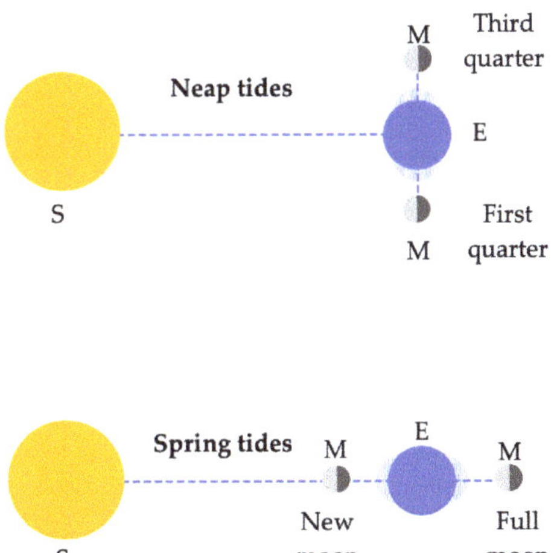

Figure 2. Diagram to illustrate the Sun (S)–Earth (E)–Moon (M) configuration regarding neap and spring tides. Source: designed by the authors.

For this reason, although the Sun's gravity on the Earth is greater than the one from the Moon, as tides are influenced by the inverse of the distance to cube ($1/r^3$), its effect is lower than the one from the Moon, as its distance is much greater. And as tides depend on the size of the object, in the Mediterranean Sea, for instance, these are negligible due to the fact that it is a semi-enclosed, shallow and small sea (with an average depth of R = 1500 m). In contrast, as all the oceans are communicated, we can consider their size being the size of the Earth and, therefore, tides are apparent. In this sense, the tidal effect of the Moon over a 2 m height living being located on the Earth is about 1000 times lower (tidal acceleration = 2 gMh/r = 3×10^{-13} ms^{-2}) than the effect produced by a mass of 1 kg at 1 m height above it (2.67×10^{-10} ms^{-2}) [5].

Thus, because of the daily rotation of the Earth, the tides rise and fall twice each lunar day in most coastal areas and estuaries at intervals of approximately 12.4 h (tidal cycles) reflecting the lunar 24.8 h day. The amplitude of successive tides is also modulated every 14.77 days, or semi-lunar cycle. So, the highest tides, or spring tides, take place when the Sun and the Moon are aligned with the Earth (i.e., full and new moon), and the lowest, or neap tides, occur when the Sun–Earth axis and the Moon–Earth axis are at right angles (90°) to each other (i.e., first and third quarter) (Figure 2) [2,6]. These tidal forces due to the Moon and the Sun are also observed in the atmosphere and the Earth's crust [7].

1.1.2. Illuminance

The moonlight we see from the Earth is the sunlight reflected on the greyish-white surface of the Moon. Since the Moon orbits the Earth and the Earth orbits the Sun, the fraction of the Moon we see changes along the lunar month giving rise to the lunar phases, being new moon, first quarter, full moon and third quarter the main ones. The illuminance (defined as the amount of luminous flux striking a surface per unit area) varies depending on the lunar phase [2,8]. In the case of the Moon, as endpoint cases we can find 0.001 lx for a new moon and 0.25 lx for a full moon to 0.01 lx for a crescent or waning moon (Table 1).

Table 1. Illuminance according to the Moon phase.

Illuminance (lx)	Description
0.001	Clear night sky, new moon
0.01	Clear night sky, crescent or waning moon
0.25	Full moon on a cloudless night
600	Sunrise or sunset on a cloudless day
32,000	Sunlight on an average day (minimum)
100,000	Sunlight on an average day (maximum)

Source: adapted from RCA Corporation [9] and Schlyter [10].

As it can be seen in Table 1, the Moon's maximum illuminance is 128,000 times lower than the minimum of sunlight on an average day or 400,000 times lower than the maximum of sunlight on an average day.

1.2. Factors Influencing Plant Growth and Development

The revision carried out considers plant growth and development from a holistic point of view. This implies all those changes in structure and function of plants and their parts, the course of genesis, assimilation, growth and development, as well as environmental, physiological and chemical modifications, maturation and decline [11–15].

Plant growth and development is regulated by both endogenous and external factors [12,14]. Regarding endogenous factors, phytohormones are in charge of the coordination of metabolic and developmental processes at the molecular and cellular level. Phytohormones can be divided into two groups depending on their functions: (i) those involved in growth-promoting activities; (ii) those in charge of responding to wounds or to biotic and abiotic stresses [16]. Synthesis or changes in the concentration of these phytohormones transduce the perception of environmental stimulus (i.e., radiation, photoperiod, temperature, gravity or stresses as cold, heat, drought or flooding). However, which plant hormones will be triggered will depend on the plant developmental state, the type of external stimulus, the part of the plant exposed, when this stimulus arrives, etc. [14]. Phytohormones, together with external factors, can activate growth and differentiation processes and allow the synchronization of plant development and seasonal changes. Furthermore, they also regulate plant growth (intensity and direction), the metabolic activity and the storage and transport of nutrients. All these endogenous factors are determined by endogenous genetic components (genome structure and gene expression, i.e., plant genotype) [17].

The growth and development of plants can also be affected by external factors such as quality, intensity, direction and duration of radiation, temperature, position with relation to Earth's gravitational field and stresses conferred by wind, water currents or snow cover, apart from other chemical influences. These external factors can initiate, complete and regulate the timing of developmental processes (inductive mode of action) but can also act quantitatively (by altering the speed and extent of growth) and formatively (by affecting morphogenesis and tropisms) [14]. These external factors are the ones which might be affected by a potential effect of the Moon—specifically, the gravitational and illuminance effects—.

Other authors propose to split the factors that determine quality and quantity into biotic and abiotic factors [18,19]. Within the biotic factors, we find arthropods, nematodes, bacteria, fungus and viruses as well as their relationships with other plants and organisms which can be competitive, mutual or parasitic types, among others [20]. In addition, within the abiotic factors, we find soil composition, salinity, pH, temperatures, pollution, humidity (water), wind and ultraviolet radiation, among others. The interaction of biotic and abiotic factors will determine plant growth, development, and productivity. Understanding their interactions is essential in agriculture when searching for the ideal growth conditions for each particular plant. In this sense, stress physiology research is very valuable, as it focuses on whether the full genetic potential of plants will be fulfilled and if plants will attain maximal growth and reproductive potential depending on different factors [21]. In particular, the study of abiotic stress originated from excess or deficit in the physical, chemical and energetic conditions to which plants are exposed provides farmers with guidelines for optimizing their harvests.

The references to the potential influence of the Moon on plants will have to be searched considering this influence as an abiotic factor. The excess or deficit of this factor should be studied taking into account that the Moon is always present, so the search should focus on those moon-derived sub-factors that can undergo substantial changes. The indirect possible effect of the Moon on the biotic components interacting with plants is a matter which falls outside the scope of the revision carried out in this paper.

2. What Has Been Thought on the Influence of the Moon on Plants

This section focuses on all those aspects related to agriculture that, according to some traditions, are determined by the Moon. To do so, we have developed a brief overview of what has been thought and written throughout history about the influence of the Moon on living beings and, in particular, plants. This analysis addresses manuals which have been used and are still used in certain agricultural sectors and information present on websites related to agriculture, gardening, agricultural machinery and so on. A special section is dedicated to biodynamic agriculture which links plant growth and lunar phases.

2.1. Brief Historical Overview

The Moon and the Sun hold a significant place in many mythologies and popular legends throughout the world. In particular, beliefs regarding the relationships between lunar phases and human and other organisms' behaviour are as ancient as human cultural heritage but have hardly ever found any solid scientific support [22].

Assertions concerning the existence of repetitive cycles in the Moon (phases), the Sun (day/night, solstices, and equinoxes/seasons), and Sirius (its heliacal rising) were extremely useful to develop lunar, lunar–solar and solar calendars, and to predict eclipses—just as it happened in the Egyptian, Babylonian, Greek or Chinese world. Such knowledge was continued in different cultures, mainly the Arab or the Mayan, the Aztec, and the Inca in America [23,24]. It is known that the Mayan carried out thorough observations of natural events, finding certain cyclical repetitions which allowed making predictions and organising when to sow or harvest [25].

Botanists and herbalists from the seventeenth century, such as Nicholas Culpeper (1616–1654), believed that plants and ailments were determined by constellations. The Sun ruled our heart, blood circulation and spine, while the Moon had influence on growth, fertility, breasts, stomach, uterus and menstrual flow. In fact, all the body fluids, as the tides, were controlled by the lunar phases. This was the prevalent belief at that time, since astrology was broadly accepted as the key to understanding the universe [26].

Surprisingly, these beliefs are still active, as shown by Phillips [26] in his *Encyclopaedia of Plants in Myth, Legend, Magic and Lore* which includes more than 200 entries linking different plant species or genera with stars or natural elements. This author states, as we have been able to check along with personal interviews made in the agricultural and rural world of the Iberian Peninsula, that garlic (*Allium sativum* L.) has been strongly associated with the Moon, and it was thought to grow stronger as the Moon waned. However, Navazio [27], in his manual dedicated to the organic seed grower, makes no mention of the Moon as an element to consider. Potato growing (*Solanum tuberosum* L.) is also supposed to be influenced by the Moon. According to different beliefs this underground crop should be planted during the black moon, that is to say, when it is waning [26]. But, once again, Navazio [27] does not mention the need to consider this aspect in the organic cultivation. Other examples would be the white clover (*Trifolium repens* L.), which has to be seeded by the darkness of the Moon or "no-Moon"—that is during the 24 h between the waning moon and the crescent moon—if you want it to grow, since if it is seeded under the moonlight, it will not sink into the ground [26]. Or corn (*Zea mays* L.), the seeds of which must be planted by moonlight in order to obtain a good performance [26].

Anglés Farrerons [28], in his work *Influence of the Moon on Agriculture and Other Topics of Main Interest for the Farmer and People from the City*, collects all the existing beliefs among elder farmers regarding the Moon. He dedicates specific chapters to the vine and the wine, the fruit growing, the cereals, the olive tree, several horticultural crops such as chard (*Beta vulgaris* var. *cicla* (L.) K.Koch), artichoke (*Cynara cardunculus* var. *scolymus* (L.) Benth.), garlic, celery (*Apium graveolens* L.), onion (*Allium cepa* L.), etc. Anglés Farrerons [28] also focuses on tree felling, forage harvesting, influence of animal manure or weather forecast. According to these traditions, he states when to sow, prune, harvest, etc., depending on the Moon phase and the crop, being true nonsense in some cases, just as the author suggests in his introduction.

Another work that requires special attention is that of Restrepo [29], a Brazilian agronomist who reflects the beliefs from Latin America and the Caribbean Area. He provides an interesting revision of the calendars of the ancient people and cultures as well as an extensive description of when to carry out all the agricultural practices (e.g., sow, layer, graft, prune, transplant) based on whether they are annual or perennial plants, vegetables, cereals and grains, tubers, bulbs and rhizomes. He also includes a description of how lunar phases and Moon illuminance affect the movement of the sap in plants (Figure 3).

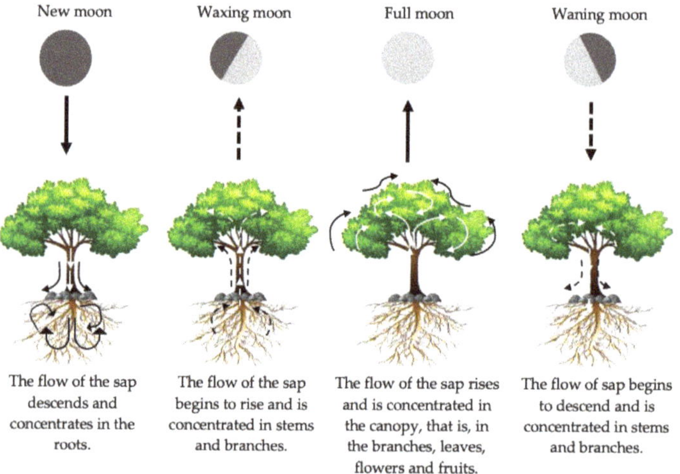

Figure 3. Explanation of how lunar phases affect sap dynamics in plants according to Restrepo [29]. Redrawn and translated from Restrepo [29].

Which is the cause–effect explanation proposed to link lunar phases and sap movement? Restrepo [29] links it with the tides:

"Therefore, in certain positions of the Moon, the water from the oceans rises to reach a maximum height, and then goes down to a minimum level, maintaining this oscillation regularly and successively. It has also been checked that this phenomenon makes itself felt in plant sap". (Translated from Restrepo [29]).

It has already been shown that the effect of the tide of the Moon on a 2 m height living being is absolutely negligible (3×10^{-13} ms^{-2}), compared to the Earth's gravity (9.8 ms^{-2}) [5]. But considering the tides, there are two high tides and two low tides each day, so if the tide caused any effect on a plant, there should be two sap rises and falls per day and none with the lunar phases. If the latter wanted to be introduced, we have already seen that at both new moon and full moon, the tides are a bit stronger due to the fact that the Sun and the Moon are aligned (as the Sun is much farther away, its effect on the tide is much lower), but the effects of the tide are symmetrical making the water rise (Figures 1 and 2). Yet, to make matters more contradictory, Restrepo [29] assigns them different effects: the full moon takes up to the leaves the waters of the plant, and the new moon takes them to the roots. On the other hand, the illuminance is the only thing that surely varies with the Moon phases (Table 1), but it does not generate any force that can cause the movement of the sap.

Finally, it has to be pointed out that there are beliefs and practices contained in many different manuals which we did not pretend to either analyse or introduce in detail in this paper. Only as an example, we outlined best sellers, such as *The Secret Life of Plants* [30], in which certain physical, emotional and spiritual relationships between plants and our species are explained in such an appealing manner that have clearly helped to strengthen several pseudo-scientific beliefs among society and many farmers. This best seller has been explicitly refuted by texts, such as *The Not-So-Secret Life of Plants*, in which the historical and experimental myths about emotional communication between animals and plants are put to rest by researchers such as Galston and Slayman [31] or Horowitz et al. [32].

2.2. Agricultural Astronomy Manuals and Websites

Some authors summarise the situation such as follows [26]:

"There has been a certain amount of interest in planting according to the phase of the Moon. The basic premise being that 'above ground crops' should be planted in the light of the Moon, i.e., on the days between the new moon and the full moon. 'Below ground crops' must be planted in the dark of the Moon, that is between the full moon and the next new moon. Refinements on this require that leaf crops are planted at the new moon and fruit crops or flowers planted at the full moon".

Apart from the agricultural traditions that could explain the use of the Moon as a calendar to organise the crops, the time of seeding, harvesting, reaping, etc., and the advice given by some authors on an individual basis, some companies have taken a step forward by developing a range of documents and even manuals appearing in the guise of scientific advice, which raise popular wisdom to the category of regulated recommendations [33–38]. Such manuals are used as reference books by many farmers at the small and large scales, and they offer a scheduled sequence of agricultural activities and, in many cases, advice regarding health care on the basis of the phases of the Moon and its ascending or descending position in the sky. But in this case, its influence is not obvious since neither gravitation nor illuminance vary.

We can also find these recommendations in the area of gardening with titles such as *Gardening by the Moon Calendar* [39]. This book provides guidance based on the following statements:

"The best rate of germination is achieved just before a full moon, when moonlight and the Moon's gravitational pull are both at their maximum, grafting should be done on a waxing moon, because sap rises in plants during this period and this will help a graft to establish, pruning should be done on a waning moon, because the sap is now falling, and this will help cut surfaces to heal quickly and crops for storage should be harvested while the Moon is waning".

Apart from the Moon, it includes elements about astrology to guide the practice: "the planting of fruit trees and bushes should be done when the Moon is passing through a fire constellation". Besides, this author links the effectiveness of the response to the planetary influence on cultivating without chemicals, pointing out that chemical products desensitize agricultural land.

Many entries to websites link the influence of the Moon to the biodynamic agriculture and the zodiac, which has nothing to do with it since the zodiac are the constellations where the ecliptic passes by (the apparent path of the Sun among the fixed stars).

Some companies within the forestry, agriculture and gardening area also dedicate sections to providing advice on tasks related to the handling of vegetables, fruit trees, etc., on their websites (Table 2).

Table 2. Summary of the tasks recommended by a multinational company supplying forestry, agriculture, and gardening machinery according to the lunar phases.

New Moon	Waxing Moon	Full Moon	Waning Moon
Crop covering with soil	Prune diseased or fruit trees	Prune	Sow root vegetables
Fertilise	Cultivate sandy soils	Plant perennial species	Remove withered leaves
Remove weeds	Sow flowers, leafy vegetables	Transplant	Water flowering plants
Remove withered leaves	Grafting	Vegetative propagation	Fertilise
Sow grass	Avoid watering flowering plants	-	Plant longleaf trees

Source: modified and translated from [40].

Moreover, this link between lunar phases and different aspects of cultivation rank highly in major search engines on the internet and in database image repositories, which raises many of the same ideas as Restrepo [29]—that lunar gravity changes according to the phases is the only way to explain popular beliefs concerning the influence of the Moon on plant growth. However, illuminance is the only thing that varies according to the phases, as it can be seen in Table 1. The Moon's gravitational pull does not generate any force able to cause sap movement.

2.3. Biodynamic Agriculture

Part of the traditions regarding the Moon have been incorporated into biodynamic agriculture, an agricultural management system which is mainly based on the fact that the astronomical bodies influence crop production. As in other forms of organic farming, the use of industrial fertilizers, pesticides and herbicides is avoided. However, the difference lies in the use of plant and mineral preparations as additives to compost and soil sprays—"biodynamic preparations"—and in following a planting schedule for cultivation, sowing and harvesting based on cosmic forces and rhythms and, particularly, on Moon rhythm (Table 3) [41,42].

Table 3. Practices and products used in organic and biodynamic agriculture.

Practice or Product	Organic	Biodynamic
Crop rotation	x	x
Polyculture or intercropping	x	x
Cover cropping	x	x
Low- or no-till	x	x
Green manures	x	x
Biological, cultural, mechanical and physical means of pest control	x	x
Biodynamic preparations that involve alchemy and homeopathy		x
Lunar and astrological calendars for planting, managing and harvesting		x
Stones used for channelling cosmic energy and radiant fields		x
Burning of pests and weeds (pest ashing)		x
Sensitive testing (including biocrystallization or morphochromatography, among others)		x

Source: modified from Chalker-Scott [41].

This variant of organic agriculture, initiated in a series of lectures given by Rudolf Steiner [43] in 1924, is considered by some authors as an alternative approach to modern agriculture (see review in Brock et al. [44]), while others consider it as not being a science-based practice (see review in Chalker-Scott [41]). The latter brands it as a scam of great implantation in countries as advanced as Germany where it has its origin and from where Brussels is pressured to accept its principles. The restoration of soil quality, of the "harmony" of ecosystems, and of biodiversity can be pointed out as the main objective of biodynamic farmers. The US website for biodynamic certification marks an update of the Moon phases with a quotation from the Natural History of Pliny the Elder (23–79 CE) —the first-century Roman naturalist who wrote extensively about tides—, explaining that the Moon "replenishes the Earth; when she approaches it, she fills all bodies, while, when she recedes, she empties them" [45].

Kirchmann [46] suggests that biodynamic agriculture has a mystical origin (called spiritualistic research by Steiner, and based on mediation and clairvoyance) that drove Steiner's research to reject scientific inquiry because as he explained in the sixth lecture of the course, "We do not need any confirmation by circumstances or by external methods. Spiritualism is an extension of scientific thought broadening the prevailing one-sided scientific view, being true and correct". A good example of this can be found in what Rudolf Steiner [47] wrote in 2004 insisting on the special care that should be taken when teaching questions about "moon forces", since conventional science considers them pure superstition or mystical fantasy, being a truth that is still difficult to talk about openly. However, Kirchmann [46] maintains that "Steiner's predictions that can be scientifically tested have been found to be incorrect".

3. What Has Been Taught on the Influence of the Moon on Plants? Analyses of Handbooks and Scientific Literature

This section deals with all those aspects that are known and taught in the agronomic and biological background in relation to plant development. It also analyses the physics books that explain the influence of the Moon on our planet with the intention of clarifying whether they mention a possible relationship with plant development. Consequently, specific sections are devoted to those factors that could depend on or have an influence from the Moon, concretely in relation to the effect of gravity and the light reflected by it, both from the point of view of biology and physics. Likewise, the basic books on botany and plant physiology have been revised, paying special attention to those factors that are determining or causing stress to the development of plants.

The revision of the handbooks has been complemented with the information gathered from the analysis of different scientific articles published in data repositories, such as Web of Knowledge, Scopus or Google Scholar, using the keywords "Moon and plants" or "lunar and plants".

3.1. What Handbooks Say from the Perspective of Physics and Biology

According to traditional beliefs, the influence of the Moon on plant growth is attributed, among other factors, to the attractive forces that the satellite exerts on the Earth and more specifically on its waters. The gravitational theory of the Moon could be attributed for the first time to Kepler (1571–1630), who claimed that the ocean tides were produced by a hidden force from the Moon. Kepler believed it was due to the affinity that the Moon had for water which was one of the four basic elements [48] in [8]. Gravity was also recognised as an agent of lunar influence with the publication of "Principia" by Newton (1643–1727).

The analyses of various physics textbooks (Table 4) commonly used in science and engineering courses reveals that the term Moon appears in most of them linked to different concepts such as the distance from the Earth (as it was calculated in ancient times or as it is calculated today, with laser telemetry), the Moon's gravity, tides, etc. With regard to the origin of tides, there are many possibilities: (i) it is not approached [49]; (ii) it is introduced in a qualitative way [50,51]; (iii) the exact dependence of R/r^3 is provided where r is the distance Earth–Moon, and R is the size of the object on which the tides act—in the case of the oceans, the Earth's radius [5]; (iv) locally, high tides are shown as an effect of the resonance [52] or tidal applications to produce energy are explained [53]. In order to find correct demonstrations of the tides, we have to deal with books on Astronomy (e.g., [4]) which, due to their extraordinary specificity, are beyond our general review.

Another factor that should be considered when approaching sap movement in plants is capillary action or capillarity, described as the spontaneous ability of a liquid to flow against gravity in a narrow space such as a thin tube or pipe (in plants, vascular tissues as xylem and phloem). This rising of liquid is the outcome of two opposing forces: cohesion (the attractive forces among similar molecules or atoms) and adhesion (the attractive forces among dissimilar molecules or atoms). In our case, the contact area between the particles of the liquid and the particles forming the tube. Capillarity is high when adhesion is greater than cohesion and vice versa. There is another important factor in capillarity, which is the contact area, dependent on the diameter of the tube (i.e., vascular tissue). Capillarity interacts with other forces, as gravity, which should be included when considering possible gravitational effects of the Moon on plants. In this sense, Jurin's law is usually introduced, giving information on the height (h) reached when balancing the weight of the column of a liquid and the force h = 2 γ cos θ/ρgr, where γ is the surface tension (Nm^{-1}), θ is the contact angle, ρ is the density of the liquid (kgm^{-3}), g is the gravity acceleration (ms^{-2}) and r is the radius of the pipe (m). Therefore, the Moon's gravity would have to be subtracted from that of the Earth g = g_E − g_M, and since it is 288,000 times smaller, its effect on capillarity is negligible.

Table 4. Revision of some of the reference handbooks on physics in relation to possible mentions of the Moon affecting plants.

Book	Issues Regarding the Moon			
	Gravity	Tides	Capillarity	Luminosity
Feynman [50]	Law of Gravitation	Qualitative explanation of tides	No	Illuminance I = S/r^2
Gettys et al. [52]	Point where g_E = g_M	Tides in the Bay of Fundy (e.g., resonance)	No	No
Giancoli [53]	Law of Gravitation	Tidal energy	Jurin's Law and negative pressure	No
Hewitt [5]	Moon radius, distance Earth–Moon. Law of Gravitation	Compares Sun and Moon tides by distance. It makes approximations to introduce R/r^3 Distinguishes between spring and neap tides. Applied to people. Tides in the ionosphere	Qualitative capillarity from surface tension	No
Holton and Brush [51]	Law of Gravitation	Qualitative explanation of spring and neap tides	No	No
Tipler [49]	Calculation of the g_E on the Moon	No	Jurin's Law	No

Source: authors' review.

Physics books, even those studying applications of physics in biology [53], do not deal with the Moon's influence on plant growth. This may be due to the fact that the Moon's gravity is, as we have seen in the Introduction (Section 1), negligible compared to that of the Earth. Regarding illuminance, since it is a topic addressed in specialized books on optics [54], it is not usually included in physics books (only one of them does, as shown in Table 4), even less lunar illuminance.

The analysis of reference handbooks and monographs dealing with plant growth and development in the background of biology, environmental sciences, forestry, and agronomy is a key issue to understanding the extent to which this is a question that is limited to agricultural practice and/or the scientific and training field. Table 5 shows a summary of six widespread and commonly used books on botany and plant physiology, making a synoptic review of the endogenous and exogenous factors that determine and modulate plant development. In particular, the focus has been placed on those Moon-dependent factors that could be beneficial or stressful for plants, specifically in relation to Moon gravity or to the light reflected by the Moon.

As mentioned in the Introduction (Section 1) and reflected in Table 5, plant growth and development are regulated by endogenous and exogenous factors. The possible effects of the Moon should be considered as abiotic external factors, either if the effect is considered to be due to the light reflected or to gravitation. Regarding light, we searched for possible quotes of the Moon when addressing light effects on seeds, plant development, phototropism, photoperiodism, phototaxis, photonasties and quantity and quality of light, etc. Focusing on gravitational influence, the search was made on different aspects of gravitropism.

Table 5. Revision of some of the reference handbooks on botany and plant physiology in relation to possible references to the Moon's influence on plant growth.

Handbook	Phase/Process	Endogenous Factors	Exogenous or Environmental Factors		Moon Mention
			Biotic	Abiotic	
Arteca [12]	Vital cycle	Endogenous growth substances			No
	Seeds			Water, temp., aeration and light	No
	Flowering			Photoperiod and vernalization	No
	Abscission			Temperature, oxygen and nutrients	No
Evans [13]	Growth			Environment and ecosystem	No
Fosket [55]	Development	Genetics		Light versus darkness	No
	Embryogenesis, germination and development				No
	Apical meristems and development				No
	Plant development		Plant-microbe and symbiotic interactions		No
	Vital cycle			Phototropism and gravitropism	No
Raven et al. [56]	Biological rhythms	Circadian and biological clocks		Track of daylength by length of darkness	No
	Flowering			Daylength as determinant of flowering time	No
	Photoautotrophy			Sunlight	No
Strasburguer [18]	Growth and differentiation	Phytohormones		Temperature, light, gravity, hydromorphosis	No
	Biological rhythms	Circadian rhythms		Photoperiod	No
	Movement			Phototaxis, phototropisms, photonasties	No
	Daylength perception	Circadian rhythms, phytochromes		Light quality, phase setting, interaction of light	No
Thomas and Vince-Prue [56]	Flower timing			Light quantity, moonlight	Yes
	Flower development			Photoperiodism	No
	Bud dormancy				No
	Storage and propagation				No
	Germination				No
	Stem elongation				No
	Leaf growth				No

Source: authors' review.

Considering endogenous factors, we searched for possible interactions of Moon radiation and photoperiod, as well as gravity, on the transduction of the perception of those environmental stimuli as well as the possible determination by endogenous genetic components. An important internal process in plants, animals, fungi and cyanobacteria is that related to circadian rhythms that refer to any biological process that displays oscillation, driven by circadian clocks, synchronized with solar time. Plant circadian rhythms are related to seasons and determine, for example, when to flower to maximize the success of pollinator attraction. Circadian rhythms also determine leaf movement, growth, germination, gas exchange or photosynthetic activity, among others. All monographs reviewed mentioning circadian rhythms refer exclusively to synchronization with the light cycle of the surrounding environments of plants, considering the Sun as light source that can determine or influence these cycles.

This search in what is considered consolidated science and is incorporated to handbooks has revealed practically no mention of the Moon (Table 5). We have only found an anecdotic reference, in relation to the possible influence of moonlight on flowering in Thomas and Vince-Prue [57]. These authors explain the work of Salisbury [58], who had indicated that the effective red-light threshold for flowering is higher than the amount of red light produced by the Moon. In addition, it is important to consider that the shade provided by the leaves of the plant itself can reduce the radiation received to 5–10% of the direct moonlight [59]. Thomas and Vince-Prue [57] state that it seems unlikely that full moon light can influence flowering, even in the most sensitive plants, highlighting the scarcity of research on this issue. In this book the authors mention the work of Kadman-Zahavi and Peiper [60], who carried out research with *Pharbitis nil* (L.) Roth —a very sensitive short-day species—which they exposed to moonlight or shielded for different periods. They concluded that, although it is possible that moonlight is perceived, it had no effect on the experience developed with a short-day species that is particularly sensitive to radiation. The difficulty of isolating the "Moon" factor was highlighted, pointing out the possible influence of shade treatment on plants in other environmental factors that could in turn have an effect on flowering [60]. On the other hand, they indicated that the full moon was only present on very few days of the lunar cycle, so its effect should be negligible under natural conditions.

3.2. What Research Papers Say from the Perspective of Physics and Biology

We consider a reference and starting point for the review of scientific articles, the brief paper published in *Nature* by Cyril Beeson in 1946, entitled "The Moon and Plant Growth" [61]. In this paper, the author writes "Beliefs that phases of the Moon have a differential effect on the rate of development of plants are both ancient and world-wide" and concludes that the research carried out to that date had not been able to demonstrate a correlation between the Moon and vital processes of terrestrial plants pointing out that, if any research does, the relationship was so unclear that it has no implications for agriculture.

In the 1950s, Frank A. Brown [22,62,63] undertook different investigations in which he studied the possible lunar rhythmicity in organisms. Most of this research was carried out on marine organisms closely linked to the tides—such as algae, crustaceans, molluscs—he also studied the physiological aspects of terrestrial plants. Brown et al. [22] studied the persistent rhythms of O_2-consumption in potatoes, carrots (*Daucus carota* L.) and brown seaweed (*Fucus*) and searched for a possible influence of barometric pressure rhythms of primary lunar frequency, noting that they are of much lower amplitude than the solar ones. The study was inconclusive in relation to what external rhythmic forces are involved in the rhythms of O_2-consumption, as many of them exhibit some degree of correlation with barometric pressure. In barometric pressure $p = \rho gh$, as its expression depends on g, we would have the same case as with capillarity: the effect of g_M should be subtracted from g_E and, as we have seen, g_M is approximately 300,000 times lower than g_M, so the effect of the Moon on barometric pressure is negligible. The authors discuss the possibility that some of the responses attributed to external factors are due to endogenous rhythmic components. This connection between internal and external factors is supported by Wolfgang Schad [64], who states that "all chrono-biological rhythms are always exo-endogenous, sharing their autonomous inner clock to some degree with the periodicity of the environment, both sides being connected by the long process of evolution", remaining unanswered, the question of how the balance between endogenous and exogenous factors oscillates.

Some authors mention the influence of the lunar phases in a tangential way, without getting to clarify anything. One example explores the resistance of circadian clocks to transient fluctuations in night light levels in nature (i.e., change in cloud cover or stellar/lunar illumination) [65]. Van Norman et al. [66], when differentiating the circadian and infradian rhythms, indicate that the former are the best characterised with a period of around 24 h, while the infradians have periods of more than 24 h and can be due to the tides, lunar, seasonal, annual or longer. In other publications, the authors actively search, without finding them, for relationships between the Moon and some organisms. A paradigmatic case is the study conducted by Bitzand Sargent [67], who unsuccessfully tries to relate the growth rate of the fungus *Neurospora crassa* Shear & Dodge to the influence of a supposed lunar magnetic field (which, as we explain in detail in this article, is even more negligible than the gravitational field). Recently Mironov et al. [68] mentioned a circalunar growth rhythm in a research carried out with genus *Sphagnum*. They found an acceleration in the growth of the mosses studied near the new moon, and a slowdown in growth near the full moon.

Regarding biodynamic practices in agriculture, Hartmut Spiess carried out chronobiological investigations of crops grown under biodynamic management, developing experiments to test the effects of lunar rhythms on the growth of winter rye (*Secale cereale* L.) and little radish (*Raphanus sativus* L., cv. Parat) [69,70]. Spiess [69,70] tried to clarify some of the varying results that a number of studies conducted in the 1930s and 1940s had left unclear. This author also focused on studies made by M. and M.K. Thun [71] establishing a relationship between the position of the Moon relative to the zodiac (sidereal rhythms), planting dates and crop growth, which served as a basis for the publication of calendars. Spiess' [69,70] results pointed out that the effects of lunar rhythms were weak, and especially the effects of the sidereal rhythms described by Thun and Thun were not apparent. In contrast to these papers, Kollerstrom and Staudenmaier [72], pointed out that, although Spiess' [69,70] experiments were well designed, there was a lack of care in the data analysis. According to these authors, the results published to date of its publication suggested that lunar factors may have a practical significance for agriculture.

Without a doubt, one of the botanists who dedicated the most effort and publications to the search for relationships between the Moon and plants was Peter Barlow. Barlow [73–85] devoted part of his research to decoding the influence of the Moon on biological phenomena. Specifically those aspects that take place in plants [73], such as the movements of leaves [74–76], stem elongation [77], fluctuations in tree stem diameters [78], the growth of roots [79–81], biophoton emissions from seedlings [82–84], and chlorophyll fluorescence [85]. According to Barlow et al. [76], and other works of the same author, at least in the cases analysed, the rhythm of leaf movements seem to have been developed or entrained in synchrony with the exogenous lunisolar rhythm experienced either on the Earth or in Space. Barlow [76] believed that plant movements were related with water movements within the plant: as ocean tides are produced by lunisolar gravitational force, water movement in the pulvinus could be responsible for leaf movement, explanation that we have previously discussed.

From all external factors, the perception of light plays a significant role as it can modify biosynthesis by photostimulation and act as a trigger initiating the different stages of development (Table 6). Reversive responses of plant to changes in light conditions can allow them to adjust their leaf or flower position (photonastic and heliotropic movements, respectively) to modulate the incoming radiation. Germination is also severely affected in some plants by light exposition. In fact, some seeds only germinate when they are exposed to a particular red to far-red ratios (660/730 nm), and in a particular moment [14].

Table 6. Radiation effects on developmental processes in plants.

Process	Mode of Action [1]	Spectral Range [2]	Fluctuation [3]
Seed germination and bud break	I	R/FR, B	P
Stem elongation	Q, F	R/FR	P
Stem orientation	Q, F	B	
Leaf orientation	Q	R/FR	C
Flowering process	I	R/FR	C
Development and filling of storage organs	I	R/FR	P
Dormancy	I	R/FR	P
Enzyme synthesis	I	R/FR	
Enzyme activation	I	R/FR	
Membrane potentials	I	R/FR	

[1] I = Inductive; Q = Quantitative; F = Formative. [2] B = Blue light; R/FR = Red-to-Far-Red ratio. [3] P = Photoperiodism; C = Circadian rhythm. Source: modified from Larcher [14], Kronenberg et al. [86] and Salisbury [87].

Despite light being crucial for plant life, just a few studies have explored the effect of moonlight on plant physiology and their results are not conclusive. Kolisko [88] observed that the period and percentage of germination and subsequent plant growth was influenced by the phase of the Moon at sowing time. And according to Bünning and Moser [59], light intensities as low as 0.1 lx, which correspond approximately to moonlight intensities (see Table 1), may influence photoperiodism in plants and animals whose threshold values of photoperiodic time-measurement is on the order of 0.1 lx. They suggest that light intensity may reach 0.7 lx or even 1 lx when the altitude of the Moon is at 60° or higher altitudes in tropical and subtropical regions (respectively), clearly influencing photoperiodic reactions. However, they observed that in short-day plants such as *Perilla ocymoides* L. and *Chenopodium amaranticolor* H.J.Coste & Reyn., light intensities similar to those of the full moon favoured rather than inhibited flowering [59]. They justified the circadian leaf movements observed in *Glycine*, *Arachis* and *Trifolium* plants as an adaptive mechanism to reduce the intensity of full moon received in the upper surface of the leaf avoiding plant misinterpretations of confounding full moonlight as it would be long day [59]. However, Kadman-Zahavi and Peiper [60] rejected this hypothesis concluding "that in the natural environment moonlight may have at most only a slight delaying effect on the time of flower induction in short-day plants" (p. 621). Furthermore, Raven and Cockell [89] suggested that photosynthesis on Earth can occur in the photosynthetically active radiation (PAR) range of $(10^{-8}$–$8 \times 10^{-3})$ mol of photons m^{-2} s^{-1}, and PAR values of moonlight at full moon goes from $(0.5$–$5) \times 10^{-9}$ mol of photons m^{-2} s^{-1}, suggesting that moonlight is not a significant source of energy for photosynthesis on Earth.

Recently, Breitler et al. [90] described that the photoreceptors present in *Coffea arabica* L. plants are able to perceive full moonlight and this full moonlight PAR is inadequate for photosynthetically supported growth. Plants perceive it as blue light with a very low R/FR ratio, yet this weak light has a great impact on numerous genes. In particular, it affects up to 50 genes related to photosynthesis, chlorophyll biosynthesis and chloroplast machinery at the end of the night. Moreover, full moonlight promotes the modification of the transcription of major rhythmic redox genes, many heat shock proteins and carotenoids genes suggesting that the moonlight seems to be perceived as a stress factor by the plant.

In other cases, full moonlight is correlated with a successful pollination of *Ephedra* species. Rydin and Bolinder [91] observed a correlation between pollination and the phases of the Moon on the gymnosperm *Ephedra foeminea* Forssk., specifically with the full moon of July. During that period, non-mature cones secreted enough pollination drops to apparently attract pollinators that can use the full moon to navigate and also be attracted to the glittering drops in the full moonlight. According to the authors, when insects are not used as pollinators, as it happens in other species of *Ephedra*, the adaptive value of correlating pollinating with the full moon is lost.

In the literature review carried out, some works were found that deal with two different topics that could have relationship with the Moon: polarization and magnetism. According to Semmens [92–94] during certain periods, moonlight is partially polarised, "the maximum effect being with the oblique reflexion of half-moon, or somewhat later for the waxing and earlier for the waning moon" and that polarised light can favour the diastase, which catalyses the hydrolysis, first of starch into dextrin and immediately afterwards into sugar or glucose, to favour germination, as he observed in crushed mustard seeds in the presence of this polarised light. Macht [95] studied the effect of (not lunar) polarized light on seeds of *Lupinus albus* L. and his results were consistent with previous findings of the action of diastase on starch. However, as far as we know, apart from those works no other research papers have been focused on the role of lunar polarized light. Despite, a full body of evidence supports that polarized moonlight has a biological significance in the vision and orientation of nocturnal animals [96,97]. Although we are at the very beginning of understanding the extent to which and why nocturnal animals use the lunar polarization, we do know that the land area over which it is viewable in pristine form is relentlessly shrinking due to human activity. In this sense, Kyba et al. [98] showed that urban skyglow has a great degree of linear polarization and confirmed that its presence diminishes the natural lunar polarization signal. They also observed that the misalignment between the polarization angles of the skyglow and scattered moonlight could explain the reduction of the degree of linear polarization as the Moon rises. Regarding nocturnal animal navigation systems based on perceiving polarized scattered moonlight, these authors highlighted the necessity of considering polarization pollution models in highly light-polluted areas. In any case, there is almost no doubt that the level of polarization of moonlight would be extremely small: so minimal, that its effect would be completely negligible in plants [98].

On the other hand, some studies suggest an influence of the lunar magnetic field. There is evidence that some animals, fungi, some protists and some bacteria seem to be able to react to the variation of the Earth's magnetic field [99–101]. The question that arises is whether plants are also able to respond to these fields and whether the Moon is capable of producing some magnetic field that plants can respond. There is abundant literature discussing magnetoreception in plants [102–106], but no conclusive results have been reported with direct application to agriculture.

Our planet has a magnetic field, called geomagnetic field, with an intensity of approximately $(25–65) \times 10^{-6}$ T, ridiculously small compared to a commercial magnet (about 0.01 T) or a 0.2 T neodymium magnet. Although there are studies that argue that billions of years ago the Moon generated a magnetic field probably even stronger than the current magnetic field of the Earth, the lunar dynamo ended around one billion years ago [107,108]. The intensity of the present-day magnetic field on the lunar surface is $<0.2 \times 10^{-9}$ T, indicating that the Moon currently does not have a global magnetic field [109]. A magnetic field of this numerical value is approximately 225,000 times less than the Earth's, and if divided by the distance Earth–Moon (3.84×10^{10} m), we can easily conclude that the possible effect of a hypothetical lunar magnetic field on the Earth would be much more negligible than that of the gravitational field.

Other theories claim that it is not the lunar magnetic field that affects, but the disturbance in the Earth's electromagnetic field caused by the lunar gravitational changes that take place during the full moon [4]; or also that Moon effects to the Earth's magnetosphere [110]. In both cases, the assumed effects would be (as we have seen in the calculations for the gravity case) completely insignificant.

A general analysis of the above-mentioned literature highlights the heterogeneity in the information sources regarding year of publication and discipline of the journal. On the one hand, there are very recent papers [68,90] but also literature from more than half a century ago [61,92–94]. On the other hand, there are peer-reviewed papers indexed in the Q1 of JCR in specific publications on Plant Science discipline, as *Annals of Botany* [75,79,80], *BMC Plant Biology* [90], *Frontiers in Plant Science* [104], *Journal of Plant Research* [103], *New Phytologist* [81], *Planta* [76], *Physiologia Plantarum* [68], *Plant Cell* [65,66] or *Plant Physiology* [67], with a long and consolidated trajectory in the field and with a pool of reviewers with solid expertise. Other articles are published in the Q2–Q3 of *JCR* in the same category as *Plant Biology* [77] and *Protoplasma* [78,83], or in other categories as Horticulture or Agronomy (e.g., *Biological Agriculture and Horticulture* [69,70,72]). Other papers included belong to other disciplines: *Astrobiology* [89], *Biology Letters* [91], *Icarus* [109], *Philosophical Transactions of the Royal Society B: Biological Sciences* [96], *Nature* [61,92–94], *Naturwissenschaften* [84], indexed in Q1–Q2 JCR lists. Nevertheless, there are also some papers not included in *JCR* lists but in other repositories as *Communicative and Integrative Biology* [73], *Earth, Moon and Planets* [64], *Pathophysiology* [110] and *Star and Furrow* [71].

This analysis also raises the question of the extent to which the authors have a good basis in the physics behind all these phenomena, given that to date Moon has not been proved to affect plant biology regarding consolidated physics.

4. Next Steps from the Perspective of Science Teaching

We are concerned about the insidious spread of pseudo-scientific ideas, not only in the field of plant science (which determines many of the behaviours, habits and techniques of many farmers in rural areas) but into the broader population through both formal and informal education. As science educators, we are especially concerned about the widespread belief in pseudo-science throughout the general populace and especially in science teachers [111–114]. Solbes et al. [114] showed that 64.9% of a sample of 131 future science teachers agree or partially agree with the expression "The phase of the Moon can affect, to some extent, several factors such as health, the birth of children or certain agricultural tasks".

Given this worrying scenario, teachers must promote critical thinking as an essential part of citizenship development. Critical thinking implies being informed about issues or problems, not limiting oneself to the dominant discourses in the media, understanding alternative, well-argued positions and being able to analyse the evidence supporting each of them, studying the problem in its complexity, so that scientific, technical, social, economic, environmental, cultural and ethical dimensions are involved, etc. [115–117]. We believe that it is crucial for teachers to be aware of these beliefs in order to address them from a scientific perspective, as has been demanded for some time [118,119].

One way to approach pseudoscience is to involve students in the proper process of reasoning and knowledge building in science, and research-based teaching is postulated as a suitable teaching methodology to address the problem [120]. In the same way that Lie and Boker [121] analysed the perceptions of complementary therapies of medical students who claimed to have pseudo-scientific beliefs related to health, it would be of great interest to address these issues with agronomic students. In this line, a teaching–research sequence has been developed [122] with future science teachers, in which the strategy followed was to plant seeds of different plant species in each of the phases of the Moon and to measure their growth once a lunar cycle was completed. The participants specified the research question and the initial hypothesis. In addition, they established the experimental design as a whole, fixed the dependent, independent, and constant variables, the materials, the sequence of the sessions, etc. [123,124]. As a result of the proposal, still under analysis, it is expected that students will develop a critical attitude towards pseudoscience and improve their training in research methodologies. Didactic proposals of this type could help, not only in teacher training, but also in any scientific study, promoting critical thinking.

With this work, we wanted to draw attention to one of the many facets of current pseudo-scientific ideas, especially in agriculture. However, we want to emphasize that dismantling these ideas, which, as we have seen, lack in any scientific basis, should not be incompatible with knowing and preserving agricultural traditions that are an important part of an ethnographic and anthropological heritage, as some institutions such as the Food and Agriculture Organization of the United Nations (FAO) claim [125–128]. Many of these traditions (e.g., organic farming, traditional and seasonal crops) allow a harmonious and sustainable coexistence with their natural environments, compatible with the conservation of biodiversity, varieties of certain species.

Furthermore, this paper encourages new research on this long-lasting topic of the possible influence of the Moon on plants in order to clarify many aspects that still remain unanswered or which have not been approached. Considering that modern ecophysiology requires a good understanding of both the molecular aspects of plant processes and the environment, future studies will necessarily have to move to a higher level: scaling from physiology to the Globe [129], considering relationships between plant ecophysiological processes and those occurring at ecosystems but also including social aspects—as traditions, farmers' behaviours and protocols, etc.—that can determine the environment where plants grow.

This review opens the door to possible research that would help to complete the picture of the extent to which certain pseudo-scientific ideas have permeated different sectors of the population. In this sense, it would be interesting to carry out an in-depth analysis of what farmers, as well as students in careers related to the agriculture sector and plant biology, think about the relationship of the Moon with the growth and development of plants.

5. Conclusions

Science has widely established different evidences: (i) the Moon's gravity on the Earth cannot have any effect on the life cycle of plants due to the fact that it is 3.3×10^{-5} ms^{-2}, almost 300,000 times lower that the Earth's gravity; (ii) since all the oceans are communicated and we can consider their size being the size of the Earth, the Moon's influence on the tides is 10^{-6} ms^{-2}, but for a 2 m height plant such value is 3×10^{-13} ms^{-2} and, therefore, completely imperceptible; (iii) the Moon's illuminance cannot have any effect on plant life since it is, at best, 128,000 times lower than the minimum of sunlight on an average day; (iv) the rest of possible effects of the Moon on the Earth (e.g., magnetic field, polarization of light) are non-existent.

The logical consequence of such evidence is that none of these effects appear in physics and biology reference handbooks. However, many of these beliefs are deeply ingrained in both agricultural traditions and collective imagery. This shows that more research should be undertaken on the possible effects observed on plants and assigned to the Moon by the popular belief, addressing their causes, if any. It would also be interesting to address these issues in both compulsory education and formal higher agricultural education in order to address pseudo-scientific ideas and promote critical thinking.

Author Contributions: Conceptualization, O.M. and T.P.; methodology, O.M., J.S., J.C. and T.P.; formal analysis, O.M. and T.P.; investigation, O.M., J.S., J.C. and T.P.; data curation, O.M. and T.P.; writing—original draft preparation, O.M., J.S., J.C. and T.P.; writing—review and editing, O.M., J.S., J.C. and T.P.; supervision, O.M., T.P.; project administration, O.M., J.S. and T.P.; funding acquisition, O.M. and J.S. All authors have read and agreed to the published version of the manuscript.

Funding: This work is framed within the project "Proposal for the Improvement of Science Teacher Training Based on Inquiry and Modelling in Context" (EDU2015-69701-P), funded by the Ministry of Economy, Industry, and Competitiveness and the European Regional Development.

Acknowledgments: A significant part of the research and bibliographic review was carried out by O.M. and T.P. in the different libraries of Harvard University—especially in the excellent library of the Arnold Arboretum of Harvard University. In this sense, we would like to express our gratitude to the Real Colegio Complutense (RCC) —Harvard, who granted two of the authors (O.M. and T.P.) for a stay at Harvard University during 2019.

Conflicts of Interest: The authors declare no conflict of interest.

References

1. Martínez, V.J.; Miralles, J.A.; Marco, E.; Galadí-Enríquez, D. *Astronomía Fundamental*; Universitat de València: Valencia, Spain, 2005.
2. Morgan, E. The moon and life on earth. *Earth Moon Planets* **2001**, *85–86*, 279–290.
3. Rackham, T. *Moon in Focus*; Pergamon: Oxford, UK, 1968.
4. Bakulin, P.I.; Kononovich, E.V.; Moroz, V.I. *Curso de Astronomía General*; Mir: Moscow, Russia, 1987.
5. Hewitt, P.G. *Conceptual Physics*, 9th ed.; Pearson Education: San Francisco, CA, USA, 2002.
6. Neumann, D. Timing in Tidal, Semilunar, and Lunar Rhythms. In *Annual, Lunar and Tidal Clocks: Patterns and Mechanisms of Nature's Enigmatic Rhythms*; Numata, H., Helm, B., Eds.; Springer: Tokyo, Japan, 2014; pp. 3–24.
7. Adushkin, V.V.; Riabova, S.A.; Spivak, A.A. Lunar–solar tide effects in the Earth's crust and atmosphere. *Izv. Phys. Solid Earth* **2017**, *53*, 565–580. [CrossRef]
8. Myers, D.E. Gravitational effects of the period of high tides and the new moon on lunacy. *Int. J. Emerg. Med.* **1995**, *13*, 529–532. [CrossRef]
9. RCA Corporation. *Electro-Optics Handbook*; RCA/Commercial Engineering: Harrison, NJ, USA, 1974.
10. Schlyter, P. (1997–2017) Radiometry and photometry in astronomy. Available online: http://www.stjarnhimlen.se/comp/radfaq.html#13 (accessed on 29 April 2020).
11. Hunt, R. *Basic Growth Analysis: Plant Growth Analysis for Beginners*; Unwin Hyman: London, UK, 2012.
12. Arteca, R.N. *Plant Growth Substances: Principles and Applications*; Springer Science & Business Media: Dordrecht, The Netherlands, 2013.
13. Evans, G.C. *The Quantitative Analysis of Plant Growth*; Blackwell Scientific Publications: Oxford, UK, 1972; Volume 1.
14. Larcher, W. *Physiological Plant Ecology: Ecophysiology and Stress Physiology of Functional Groups*; Springer-Verlag Berlin Heidelberg: Berlin, Germany, 2003.
15. Leopold, A.C. *Plant Growth and Development*; MCGraw-Hill Education: New York, NY, USA, 1964.
16. Srivastava, L.M. *Plant Growth and Development: Hormones and Environment*; Elsevier: San Diego, CA, USA, 2002.
17. Taiz, L.; Zeiger, E.; Møller, I.M.; Murphy, A. *Plant Physiology and Development*, 6th ed.; Sinauer Associates, Inc.: Sunderland, MA, USA, 2015.
18. Bulgari, R.; Franzoni, G.; Ferrante, A. Biostimulants Application in Horticultural Crops under Abiotic Stress Conditions. *Agronomy* **2019**, *9*, 306. [CrossRef]
19. Drobek, M.; Frąc, M.; Cybulska, J. Plant Biostimulants: Importance of the Quality and Yield of Horticultural Crops and the Improvement of Plant Tolerance to Abiotic Stress—A Review. *Agronomy* **2019**, *9*, 335. [CrossRef]
20. Strasburger, E.; Noll, F.; Schenck, H.; Schimper, A. *Tratado de Botánica*; Omega: Barcelona, Spain, 2004.
21. Bhatla, S.C.; Lal, M.A. *Plant Physiology, Development and Metabolism*; Springer: Singapore, 2018.

22. Brown, F.A., Jr.; Freeland, R.O.; Ralph, C.L. Persistent Rhythms of O_2-Consumption in Potatoes, Carrots and the Seaweed, Fucus. *Plant Physiol.* **1955**, *30*, 280. [CrossRef] [PubMed]

23. Ferris, T. *Coming of Age in the Milky Way*; Anchor Books: Morrow, NY, USA, 1988.

24. Solbes, J.; Palomar, R. ¿Por qué resulta tan difícil la comprensión de la astronomía a los estudiantes? *Didáctica Cienc. Exp. Soc.* **2011**, *25*, 187–211.

25. Böckler, C.G. *Donde Enmudecen las Conciencias: Crepúsculo y Aurora en Guatemala*; CIESAS: Mexico City, Mexico, 1986.

26. Phillips, S. *An Encyclopaedia of Plants in Myth, Legend, Magic and Lore*; The Crowood Press Ltd.: Marlborough, UK, 2012.

27. Navazio, J. *The Organic Seed Grower: A farmer's Guide to Vegetable Seed Production*; Chelsea Green Publishing Co.: London, UK, 2012.

28. Anglés Farrerons, J.M. *Influencia de la Luna en la Agricultura y Otros Temas de Especial Interés Para el Campesino y Gentes de la Ciudad*; Dilagro-Ediciones: Lleida, Spain, 1984.

29. Restrepo, J. *La Luna: El sol Nocturno en los Trópicos y su Influencia en la Agricultura*; (No. 630.2233 R436.); Servicio de Información Mesoamericano sobre Agricultura Sostenible: Managua, Nicaragua, 2004.

30. Tompkins, P.; Bird, C. *The Secret Life of Plants*; (No. QK50. T65I 1973.); Harper & Row: New York, NY, USA, 1973.

31. Galston, A.W.; Slayman, C.L. The Not-So-Secret Life of Plants: In which the historical and experimental myths about emotional communication between animal and vegetable are put to rest. *Am. Sci.* **1979**, *67*, 337–344.

32. Horowitz, K.A.; Lewis, D.C.; Gasteiger, E.L. "Plant primary perception": Electrophysiological unresponsiveness to brine shrimp killing. *Science* **1975**, *189*, 478–480. [CrossRef]

33. Bussagli, M. *Calendario Lunar de las Siembras y Labores Agrícolas (Pequeñas Joyas)*; Susaeta: Madrid, Spain, 2019.

34. Calendario Zaragozano. *Calendario Zaragozano. El Firmamento para toda España*; Castillo y Ocsiero, M., Ed.; Zaragozano: Madrid, Spain, 2019.

35. Geiger, P. (Ed.) *Farmers' Almanac for the Year 2020*; Almanac Publishing Company: Lewiston, ME, USA, 2019.

36. Gros, M. *Lunario 2020: Calendario Lunar para el Huerto y el Jardín Ecológico y También para Mantener la Salud*; Artús Porta Manresa: Manresa, Spain, 2019.

37. Leendertz, L. *The Almanac: A Seasonal Guide to 2020*; Mitchell Beazley: London, UK, 2019.

38. Trédoulat, T. *Réussir son Potager Avec la Lune*; Rustica éditions: Paris, France, 2020.

39. Littlewood, M. *A Guide to Gardening by the Moon*; Gaby Bartai: Glasgow, UK, 2009.

40. Descubre cómo las fases lunares pueden afectar a tus cultivos. Available online: https://www.todohusqvarna.com/blog/fases-lunares/ (accessed on 19 June 2020).

41. Chalker-Scott, L. The science behind biodynamic preparations: A literature review. *HortTechnology* **2013**, *23*, 814–819. [CrossRef]

42. Thun, M.; Thun, M.K. *Calendario de Agricultura Biodinámica 2020*; Editorial Rudolf Steiner: Madrid, Spain, 2019.

43. Steiner, R. Agriculture (English translation). 1958. Available online: http://wn.rsarchive.org/Biodynamics/GA327/English/BDA1958/Ag1958_index.html (accessed on 27 April 2020).

44. Brock, C.; Geier, U.; Greiner, R.; Olbrich-Majer, M.; Fritz, J. Research in biodynamic food and farming–a review. *Open Agric.* **2019**, *4*, 743–757. [CrossRef]

45. Demeter Association, Inc. Available online: https://www.demeter-usa.org/about-demeter/biodynamic-certification-marks.asp (accessed on 27 April 2020).

46. Kirchmann, H. Biological dynamic farming—An occult form of alternative agriculture? *J. Agric. Env. Ethics* **1994**, *7*, 173–187. [CrossRef]

47. Steiner, R. *A Modern Art of Education*; Anthroposophic Press: Great Barrington, MA, USA, 2004.

48. Beer, A.; Beer, P. *Kepler, Four Hundred Years*; Pergammon Press: Oxford, UK, 1975.

49. Tipler, P.A. *Physics for Scientist and Engineers*, 3rd ed.; Worth Publishers: New York, NY, USA, 1992.

50. Feynman, R.P.; Leighton, R.B.; Sands, M. *The Feynman Lectures on Physics*; Fondo Educativo Interamericano: Bogotá, Colombia, 1971; Volume 1.

51. Holton, G.; Brush, S. *Introduction to Concepts and Theories in Physical Science*; Addison Wesley: Reading, MA, USA, 1976.

52. Gettys, W.E.; Keller, F.J.; Skove, M.J. *Physics Classical and Modern*; McGraw-Hill: New York, NY, USA, 1989.

53. Giancoli, D.C. *Physics*; Prentice-Hall: Englewood Cliffs, NJ, USA, 1985.

54. Smith, F.G.; Thomson, J.H. *Optics*; Wiley: Hoboken, NJ, USA, 1989.
55. Fosket, D.E. *Plant Growth and Development: A Molecular Approach*; Academic Press Inc.: San Diego, CA, USA, 1994.
56. Raven, P.H.; Evert, R.F.; Eichhorn, S.E. *Biology of Plants*; W.H. Freeman and Company: New York, NY, USA, 2005.
57. Thomas, B.; Vince-Prue, D. *Photoperiodism in Plants*; Academic Press, Inc.: San Diego, CA, USA, 1996.
58. Salisbury, F.B. Plant adaptations to the light environment. In *Plant Production in the North*; Kaurin, A., Junttila, O., Nilsen, J., Eds.; Norwegian University Presss: Tromso, Norway, 1985; pp. 43–61.
59. Bünning, E.; Moser, I. Interference of moonlight with the photoperiodic measurement of time by plants, and their adaptive reaction. *Proc. Natl. Acad. Sci. USA* **1969**, *62*, 1018–1022. [CrossRef] [PubMed]
60. Kadman-Zahavi, A.; Peiper, D. Effects of moonlight on flower induction in *Pharbitis nil*, using a single dark period. *Ann. Bot.* **1987**, *60*, 621–623. [CrossRef]
61. Beeson, C.F.C. The moon and plant growth. *Nature* **1946**, *158*, 572. [CrossRef] [PubMed]
62. Brown, F.A., Jr.; Bennett, M.F.; Marguerite Webb, H. Persistent daily and tidal rhythms of O_2-consumption in fiddler crabs. *J. Cell Comp. Physiol.* **1954**, *44*, 477–505. [CrossRef]
63. Brown, F.A., Jr.; Webb, H.M.; Bennett, M.F.; Sandeen, M.I. Temperature-independence of the frequency of the endogenous tidal rhythm of *Uca*. *Physiol. Zool.* **1954**, *27*, 345–349. [CrossRef]
64. Schad, W. Lunar influence on plants. *Earth Moon Planets* **2001**, *85–86*, 405–409.
65. Covington, M.F.; Panda, S.; Liu, X.L.; Strayer, C.A.; Wagner, D.R.; Kay, S.A. ELF3 modulates resetting of the circadian clock in *Arabidopsis*. *Plant Cell* **2001**, *13*, 1305–1316. [CrossRef]
66. Van Norman, J.M.; Breakfield, N.W.; Benfey, P.N. Intercellular communication during plant development. *Plant Cell* **2011**, *23*, 855–864. [CrossRef]
67. Bitz, D.M.; Sargent, M.L. A failure to detect an influence of magnetic fields on the growth rate and circadian rhythm of *Neurospora crassa*. *Plant Physiol.* **1974**, *53*, 154–157. [CrossRef] [PubMed]
68. Mironov, V.L.; Kondratev, A.Y.; Mironova, A.V. Growth of *Sphagnum* is strongly rhythmic: Contribution of the seasonal, circalunar and third components. *Physiol. Plant* **2020**, *168*, 765–776. [CrossRef] [PubMed]
69. Spiess, H. Chronobiological Investigations of Crops Grown under Biodynamic Management. I. Experiments with Seeding Dates to Ascertain the Effects of Lunar Rhythms on the Growth of Winter Rye (Secale cereale cv. Nomaro). *Biol. Agric. Hortic.* **1990**, *7*, 165–178. [CrossRef]
70. Spiess, H. Chronobiological Investigations of Crops Grown under Biodynamic Management. II. Experiments with Seeding Dates to Ascertain the Effects of Lunar Rhythms on the Growth of Little Radish (*Raphanus sativus*, cv. Parat). *Biol. Agric. Hortic.* **1990**, *7*, 179–189. [CrossRef]
71. Thun, M. Nine years observation of cosmic influences on annual plants. *Star Furrow* **1964**, *22*.
72. Kollerstrom, N.; Staudenmaier, G. Evidence for lunar-sidereal rhythms in crop yield: A review. *Biol. Agric. Hortic.* **2001**, *19*, 247–259. [CrossRef]
73. Chaffey, N.; Volkmann, D.; Baluška, F. The botanical multiverse of Peter Barlow. *Comm. Integr. Biol.* **2019**, *12*, 14–30. [CrossRef]
74. Barlow, P.W.; Klingelé, E.; Klein, G.; Mikulecký Sen, M. Leaf movements of bean plants and lunar gravity. *Plant Signal. Behav.* **2008**, *3*, 1083–1090. [CrossRef]
75. Barlow, P.W. Leaf movements and their relationship with the lunisolar gravitational force. *Ann. Bot.* **2015**, *116*, 149–187. [CrossRef]
76. Fisahn, J.; Klingelé, E.; Barlow, P. Lunar gravity affects leaf movement of *Arabidopsis thaliana* in the International Space Station. *Planta* **2015**, *241*, 1509–1518. [CrossRef]
77. Zajączkowska, U.; Barlow, P.W. The effect of lunisolar tidal acceleration on stem elongation growth, nutations and leaf movements in peppermint (*Mentha × piperita* L.). *Plant Biol.* **2017**, *19*, 630–642. [CrossRef] [PubMed]
78. Barlow, P.W.; Mikulecký, M., Sr.; Střeštík, J. Tree-stem diameter fluctuates with the lunar tides and perhaps with geomagnetic activity. *Protoplasma* **2010**, *247*, 25–43. [CrossRef] [PubMed]
79. Barlow, P.W.; Fisahn, J. Lunisolar tidal force and the growth of plant roots, and some other of its effects on plant movements. *Ann. Bot.* **2012**, *110*, 301–318. [CrossRef] [PubMed]
80. Barlow, P.W.; Fisahn, J.; Yazdanbakhsh, N.; Moraes, T.A.; Khabarova, O.V.; Gallep, C.M. *Arabidopsis thaliana* root elongation growth is sensitive to lunisolar tidal acceleration and may also be weakly correlated with geomagnetic variations. *Ann. Bot.* **2013**, *111*, 859–872. [CrossRef] [PubMed]

81. Fisahn, J.; Yazdanbakhsh, N.; Klingele, E.; Barlow, P. *Arabidopsis thaliana* root growth kinetics and lunisolar tidal acceleration. *New Phytol.* **2012**, *195*, 346–355. [CrossRef] [PubMed]

82. Gallep, C.M.; Moraes, T.A.; Červinková, K.; Cifra, M.; Katsumata, M.; Barlow, P.W. Lunisolar tidal synchronism with biophoton emission during intercontinental wheat-seedling germination tests. *Plant Signal. Behav.* **2014**, *9*, e28671. [CrossRef]

83. Gallep, C.M.; Barlow, P.W.; Burgos, R.C.; van Wijk, E.P.R. Simultaneous and intercontinental tests show synchronism between the local gravimetric tide and the ultra-weak photon emission in seedlings of different plant species. *Protoplasma* **2017**, *254*, 315–325. [CrossRef]

84. Moraes, T.A.; Barlow, P.W.; Klingelé, E.; Gallep, C.M. Spontaneous ultra-weak light emissions from wheat seedlings are rhythmic and synchronized with the time profile of the local gravimetric tide. *Naturwissenschaften* **2012**, *99*, 465–472. [CrossRef]

85. Fisahn, J.; Klingelé, E.; Barlow, P. Lunisolar tidal force and its relationship to chlorophyll fluorescence in *Arabidopsis thaliana*. *Plant Signal. Behav.* **2015**, *10*, e1057367. [CrossRef]

86. Kronenberg, G.H.M.; Kendrick, R.E. The physiology of action. In *Photomorphogenesis in Plants*; Kendrick, R.E., Kronenberg, G.H.M., Eds.; Nijhoff Publ.: Dordrecht, The Netherlands, 1986; pp. 99–114.

87. Salisbury, F.B. *The Flowering Process*; Pergamon Press, Inc.: New York, NY, USA, 1963.

88. Kolisko, L. *The Moon and the Growth of Plants*; Bray-on-Thames, Anthroposophical Agricultural Foundation: London, UK, 1936.

89. Raven, J.A.; Cockell, C.S. Influence on photosynthesis of starlight, moonlight, planetlight, and light pollution (reflections on photosynthetically active radiation in the universe). *Astrobiology* **2006**, *6*, 668–675. [CrossRef]

90. Breitler, J.C.; Djerrab, D.; Leran, S.; Toniutti, L.; Guittin, C.; Severac, D.; Pratlong, M.; Dereeper, A.; Etienne, H.; Bertrand, B. Full moonlight-induced circadian clock entrainment in *Coffea arabica*. *BMC Plant Biol.* **2020**, *20*, 1–11. [CrossRef] [PubMed]

91. Rydin, C.; Bolinder, K. Moonlight pollination in the gymnosperm *Ephedra* (Gnetales). *Biol. Lett.* **2015**, *11*, 20140993. [CrossRef] [PubMed]

92. Semmens, E.S. Effect of Moonlight on the Germination of Seeds. *Nature* **1923**, *111*, 49–50. [CrossRef]

93. Semmens, E.S. Hydrolysis in Green Plants by Moonlight. *Nature* **1932**, *130*, 243. [CrossRef]

94. Semmens, E.S. Chemical Effects of Moonlight. *Nature* **1947**, *159*, 613. [CrossRef]

95. Macht, D.I. Concerning the influence of polarized light on the growth of seedlings. *J. Gen. Physiol.* **1926**, *10*, 41–52. [CrossRef]

96. Cronin, T.W.; Marshall, J. Patterns and properties of polarized light in air and water. *Philos. Trans. R. Soc. Lond. B Biol. Sci.* **2011**, *366*, 619–626. [CrossRef]

97. Nowinszky, L.; Szabó, S.; Tóth, G.; Ekk, I.; Kiss, M. The effect of the moon phases and of the intensity of polarized moonlight on the light-trap catches. *Zeitschrift Angewandte Entomologie* **1979**, *88*, 337–353. [CrossRef]

98. Kyba, C.C.; Ruhtz, T.; Fischer, J.; Hölker, F. Lunar skylight polarization signal polluted by urban lighting. *J. Geophys. Res.* **2011**, *116*, D24106. [CrossRef]

99. Begall, S.; Malkemper, E.P.; Cerveny, J.; Nemec, P.; Burda, H. Magnetic alignment in mammals and other animals. *Mammal. Biol.* **2013**, *78*, 10–20. [CrossRef]

100. Ritz, T.; Wiltschko, R.; Hore, P.J.; Rodgers, C.T.; Stapput, K.; Thalau, P.; Timmel, C.R.; Wiltschko, W. Magnetic compass of birds is based on a molecule with optimal directional sensitivity. *Biophys. J.* **2009**, *96*, 3451–3457. [CrossRef] [PubMed]

101. Yan, L.; Zhang, S.; Chen, P.; Liu, H.; Yin, H.; Li, H. Magnetotactic bacteria, magnetosomes and their application. *Microbiol. Res.* **2012**, *167*, 507–519. [CrossRef] [PubMed]

102. Belyavskaya, N.A. Biological effects due to weak magnetic field on plants. *Adv. Space Res.* **2004**, *34*, 1566–1574. [CrossRef]

103. Galland, P.; Pazur, A. Magnetoreception in plants. *J. Plant Res.* **2005**, *118*, 371–389. [CrossRef]

104. Maffei, M.E. Magnetic field effects on plant growth, development, and evolution. *Front. Plant Sci.* **2014**, *5*, 445. [CrossRef] [PubMed]

105. Nyakane, N.E.; Markus, E.D.; Sedibe, M.M. The effects of magnetic fields on plants growth: A comprehensive review. *Int. J. Food Eng.* **2019**, *5*, 79–87. [CrossRef]

106. Vian, A.; Davies, E.; Gendraud, M.; Bonnet, P. Plant responses to high frequency electromagnetic fields. *Biomed. Res. Int.* **2016**, *2016*, 1830262. [CrossRef]

107. Mighani, S.; Wang, H.; Shuster, D.L.; Borlina, C.S.; Nichols, C.I.; Weiss, B.P. The end of the lunar dynamo. *Sci. Adv.* **2020**, *6*, eaax0883. [CrossRef]

108. Tikoo, S.M.; Weiss, B.P.; Shuster, D.L.; Suavet, C.; Wang, H.; Grove, T.L. A two-billion-year history for the lunar dynamo. *Sci. Adv.* **2017**, *3*, e1700207. [CrossRef]

109. Mitchell, D.L.; Halekas, J.S.; Lin, R.P.; Frey, S.; Hood, L.L.; Acuña, M.H.; Binder, A. Global mapping of lunar crustal magnetic fields by Lunar Prospector. *Icarus* **2008**, *194*, 401–409. [CrossRef]

110. Bevington, M. Lunar biological effects and the magnetosphere. *Pathophysiology* **2015**, *22*, 211–222. [CrossRef] [PubMed]

111. Eve, R.A.; Dunn, D. Psychic powers, astrology and creationism in the classroom? Evidence of pseudoscientific beliefs among high school biology & life science teachers. *Am. Biol. Teach.* **1990**, *52*, 10–21. [CrossRef]

112. Happs, J.C. Challenging pseudoscientific and paranormal beliefs held by some pre-service primary teachers. *Res. Sci. Educ.* **1991**, *21*, 171–177. [CrossRef]

113. Kaplan, A.O. Research on the pseudo-scientific beliefs of preservice science teachers: A sample from astronomy-astrology. *J. Balt. Sci. Educ.* **2014**, *13*, 381–393.

114. Solbes, J.; Palomar, R.; Dominguez-Sales, M.C. To what extent do pseudosciences affect teachers? A look at the mindset of science teachers in training. *Mètode Sci. Stud. J. Ann. Rev.* **2018**, *8*, 188–195. [CrossRef]

115. Halpern, D. Teaching critical thinking for transfer across domains. *Am. Psychol.* **1998**, *53*, 449–455. [CrossRef]

116. Torres, N.; Solbes, J. Contribuciones de una intervención didáctica usando cuestiones sociocientíficas para desarrollar el pensamiento crítico. *Enseñanza Cienc.* **2016**, *34*, 43–65. [CrossRef]

117. Yager, R.E. Science and critical thinking. In *Teaching Critical Thinking: Reports from Across the Curriculum*; Clarke, J.H., Biddle, A.W., Eds.; Prentice Hall: Englewood Cliffs, NJ, USA, 1993.

118. Bates, J.; Culpepper, W. Using Pseudoscience to Teach Science: Encouraging Skepticism of Paranormal Powers in the Classroom. *J. Coll. Sci. Teach.* **1991**, *21*, 106–111. [CrossRef] [PubMed]

119. Wilson, J.A. Reducing pseudoscientific and paranormal beliefs in university students through a course in science and critical thinking. *Sci. Educ.* **2018**, *27*, 183–210. [CrossRef]

120. Chinn, C.A.; Malhotra, B.A. Epistemologically authentic inquiry in schools: A theoretical framework for evaluating inquiry tasks. *Sci. Educ.* **2002**, *86*, 175–218. [CrossRef]

121. Lie, D.; Boker, J. Development and validation of the CAM Health Belief Questionnaire (CHBQ) and CAM use and attitudes amongst medical students. *BMC Med Educ.* **2004**, *4*, 2. [CrossRef] [PubMed]

122. Pedaste, M.; Mäeots, M.; Siiman, L.A.; De Jong, T.; Van Riesen, S.A.; Kamp, E.T.; Constantinos, C.; Zacharias, M.; Tsourlidaki, E. Phases of inquiry-based learning: Definitions and the inquiry cycle. *Educ. Res. Rev.* **2015**, *14*, 47–61. [CrossRef]

123. Pina, T.; Mayoral, O.; Solbes, J. ¿Influye la Luna en el crecimiento de las plantas? Indagación para favorecer el pensamiento crítico. In *Propuestas de Educación Científica Basadas en la Indagación y Modelización en Contexto*; Solbes, J., Jiménez, M.R., Pina, T., Eds.; Tirant lo Blanch: Valencia, Spain, 2019; pp. 121–143.

124. Pina, T.; Mayoral, O.; Solbes, J. Do lunar phases influence the growth of plants? Scientific inquiry to encourage critical thinking in the classroom. In Proceedings of the ESERA conference, Bologna, Italy, 26–30 August 2019.

125. Dixon, J.A.; Gibbon, D.P.; Gulliver, A. *Farming Systems and Poverty: Improving Farmers' Livelihoods in a Changing World*; FAO: Rome, Italy; World Bank: Washington, DC, USA, 2001.

126. Kuhnlein, H.V.; Erasmus, B.; Spigelski, D. *Indigenous Peoples' Food Systems: The Many Dimensions of Culture, Diversity and Environment for Nutrition and Health*; FAO: Rome, Italy, 2009.

127. Kuhnlein, H.V. Biodiversity and sustainability of indigenous peoples' foods and diets. Sustainable diets and biodiversity. In *Sustainable Diets and Biodiversity: Directions and Solutions for Policy, Research and Action. Proceedings of the International Scientific Symposium, Biodiversity and Sustainable Diets United Against Hunger, Rome, Italy, 3–5 November 2010*; Burlingame, B., Dernini, S., Eds.; FAO Headquarters: Rome, Italy, 2012.

128. Sinclair, F.; Wezel, A.; Mbow, C.; Chomba, S.; Robiglio, V.; Harrison, R. *The Contribution of Agroecological Approaches to Realizing Climate-Resilient Agriculture*; GCA: Rotterdam, The Netherlands, 2019.

129. Lambers, H.; Chapin, F.S., III; Pons, T.L. *Plant Physiological Ecology*; Springer Science + Business Media: Berlin, Germany; LLC: New York, NY, USA, 2008.

 *agronomy*

Review

Adaptation of Plants to Salt Stress: Characterization of Na^+ and K^+ Transporters and Role of CBL Gene Family in Regulating Salt Stress Response

Toi Ketehouli, Kue Foka Idrice Carther, Muhammad Noman, Fa-Wei Wang, Xiao-Wei Li * and Hai-Yan Li *

College of Life Sciences, Engineering Research Center of the Chinese Ministry of Education for Bioreactor and Pharmaceutical Development, Jilin Agricultural University, Changchun 130118, China; stanislasketehouli@yahoo.com (T.K.); kuefokaidricecarther@yahoo.com (K.F.I.C.); mohmmdnoman@gmail.com (M.N.); fw-1980@163.com (F.-W.W.)
* Correspondence: xiaoweili1206@163.com (X.-W.L.); hyli99@163.com (H.-Y.L.); Tel.: +86-0431-84533428 (X.-W.L.); +86-0431-84532885 (H.-Y.L.)

Received: 14 September 2019; Accepted: 21 October 2019; Published: 28 October 2019

Abstract: Salinity is one of the most serious factors limiting the productivity of agricultural crops, with adverse effects on germination, plant vigor, and crop yield. This salinity may be natural or induced by agricultural activities such as irrigation or the use of certain types of fertilizer. The most detrimental effect of salinity stress is the accumulation of Na^+ and Cl^- ions in tissues of plants exposed to soils with high NaCl concentrations. The entry of both Na^+ and Cl^- into the cells causes severe ion imbalance, and excess uptake might cause significant physiological disorder(s). High Na^+ concentration inhibits the uptake of K^+, which is an element for plant growth and development that results in lower productivity and may even lead to death. The genetic analyses revealed K^+ and Na^+ transport systems such as SOS1, which belong to the CBL gene family and play a key role in the transport of Na^+ from the roots to the aerial parts in the *Arabidopsis* plant. In this review, we mainly discuss the roles of alkaline cations K^+ and Na^+, Ion homeostasis-transport determinants, and their regulation. Moreover, we tried to give a synthetic overview of soil salinity, its effects on plants, and tolerance mechanisms to withstand stress.

Keywords: salinity; sodium; potassium; ion homeostasis-transport determinants; CBL gene family

---

1. Introduction

The adverse effects of salinity on plant growth are generally associated with the low osmotic potential of the soil solution and the high level of toxicity of sodium (and chlorine for some species) that causes multiple disturbances to metabolism, growth, and plant development at the molecular, biochemical, and physiological levels [1,2]. In vitro experiments have shown that the enzymes extracted from the halophyte plants *Triplex spongeosa* or *Suaeda maritima (L.)* are sensitive to NaCl to the same degree as those extracted from the glycophyte plants [3,4]. These experiments suggest that tolerance to salinity is not limited to a metabolic response in tolerant plants. Generally, sodium begins to have an inhibitory effect on enzymatic activity from a concentration of 100 mmol/L. Thus, the ability of plants to reduce sodium levels in the cytoplasm appears to be one of the decisive factors in salinity tolerance [5,6]. However, although chloride ions are micro-elements necessary as co-factors, for enzymatic activity, photosynthesis, and the regulation of cell turgor, pH, and electrical membrane potential, they remain no less toxic than Na^+ ions if their concentration reaches the critical threshold tolerated by plants [7]. Ionic cellular homeostasis is an essential and vital phenomenon for all organisms. Most cells maintain a high level of potassium and a low level of sodium in the cytoplasm through the coordination and regulation of different transporters and channels. There are two main strategies that plants use to cope with salinity—The compartmentalization of toxic ions within the vacuole and their exclusion outside the cell [5,6]. On the other hand, plants modify the composition of their sap; they can accumulate Na^+ and Cl^- ions to adjust the water potential of tissues necessary to maintain growth [6]. This accumulation should be consistent with a metabolic tolerance of the resulting concentration or with compartmentalization between the various components of the cell or plant. It requires relatively little energy expenditure. If this accumulation does not take place, the plant synthesizes organic solutes to adjust its water potential. It will require a large amount of biomass to ensure the energy expenditure necessary for such a synthesis. Therefore, one adaptation strategy consists of synthesizing osmoprotective agents, mainly amino compounds and sugars, and accumulating them in the cytoplasm and organelles [8,9]. These osmolytes, usually of a hydrophilic nature, are slightly charged but polar and highly soluble molecules [10], suggesting that they can adhere to the surface of proteins and membranes to protect them from dehydration. Another function attributed to these osmolytes is protection against the action of oxygen radicals following salt stress [11]. Under high sodium concentration levels, whether the latter is compartmentalized within the vacuole or excluded from the cell, the osmotic potential of the cytoplasm must be balanced with that of the vacuole and the external environment in order to maintain the cell turgor and the water absorption necessary for cell growth. This requires an increase in osmolyte levels in the cytoplasm, either by the synthesis of solutes (compatible with cellular metabolism) or by their uptake of the soil solution [12,13]. Among these synthesized compounds are some polyols, sugars, amino acids, and betaines, which, energetically, are very expensive to be produced by the cell [14]. The main role of these solutes is to maintain a low water potential inside the cells to generate a suction force for water absorption. Furthermore, the involvement of solutes such as glycine betaine, sorbitol, mannitol, trehalose, and proline in improving tolerance to abiotic stress has been demonstrated by genetic engineering and plant transgenesis [6,14,15]. On the other hand, salt stress induces the production of active forms of oxygen following the alteration of metabolism in the mitochondria and chloroplasts. These active forms of oxygen cause oxidative stress whose adverse effects are reflected in various cellular components such as membrane lipids, proteins, and nucleic acids [16]. As a result, the reduction of these oxidative damages through the deployment of a range of antioxidants could contribute to improving plant tolerance to stress [17]. Early events in plant stress adaptation begin with mechanisms of perception and signaling via signal and messenger transduction to activate various physiological and metabolic responses, including the expression of stress response genes. The main pathways activated during the salt stress signaling include calcium, abscisic acid (ABA), mitogen-activated protein kinases (MAPKinases), salt overly sensitive proteins (SOS), and ethylene [12]. In this chapter, we mainly discuss roles of alkaline cations K^+ and Na^+, ion homeostasis-transport determinants, and their regulation. Furthermore, we tried to give a hypothetical overview of soil salinity, its effects on plants, and tolerance mechanisms that allow the plants to withstand stress. A fundamental biological understanding and knowledge of the effects of salt stress on plants is needed to provide additional information for the study of the plant response to salinity and try to find other way for improving the impact of salinity in plants and accordingly enhance crop yields to cope with the starvation that persists in some parts of the world

2. Roles of Alkaline Cations K^+ and Na^+ in Plants

Potassium (K) is the third of the three primary nutrients required by plants, along with nitrogen (N) and phosphorus (P). Potassium, with about 100 to 200 mM concentration in the cytosol, is the major inorganic cation of the cytoplasm in plant and animal cells. The reasons for its preferential accumulation compared to Na^+ is probably due to the fact that Na^+ is more "chaotropic" (because of its smaller size and stronger electric field on its surface) [18].

Na^+ is not an essential nutrient for higher plants. For a high concentration of Na^+ in the soil, this cation becomes even toxic to the plant. At lower concentrations, the plant can use it beneficially as a vacuolar osmoticum.

2.1. Physiological Roles

As most inorganic cations are abundant in the cytoplasm, the potassium is involved in critical cell functions. In addition to its role in the neutralization of the net electric charge of biomolecules, the potassium participates, for example, in membrane transport processes, enzyme activation, and osmotic potential. In plants, in conjunction with osmotic potential [19], K^+ is involved in the control of the turgor pressure [20] and related functions, cell elongation and cell movement. Finally, K^+ plays a direct or indirect role, in the regulation of enzyme activities, the protein synthesis, photosynthesis and homeostasis of the cytoplasmic pH.

These different roles at the cellular level involve potassium in essential functions at the level of the whole plant, for example gas exchange control via regulation of the opening and closing of the stoma, the xylem sap ascension by root thrust, installation of potential osmotic gradient carrying phloem sap flow from original organs to hole organs or even port of herbaceous species.

2.2. Effect of K^+ Deficiency on Plants Physiology

In K^+ deficiency, the sap flow is disturbed, with spontaneous reduction of the phloem sap velocity of circulation. The photoassimilates then accumulate inside of the leaves. Symptoms of chlorosis and necrosis from the photooxidation of the photosynthetic system are frequently observed. It is well settled that K^+ deficiency induces the acidification of the extracellular medium. Minjian et al. [21], showed that root K^+ absorption depends on the activity of the proton pumps (H^+-ATPases) and the occurrence of K^+ transporters on the cellular membrane. The level of H^+ expulsion can be used as a criterion of tolerance to K^+ deficiency. Chen and Gabelman [22] observed in tomato strains that K^+ uptake efficiency is associated with a high net K^+ influx coupled with low pH around root surfaces. The proton-electrochemical gradient may contribute to energizing K^+ uptake, and indeed it is used by some KT/ KUP/HAK transporters, which co-transport K^+ and H^+ [23].

2.3. Toxicity of Na^+ in the Cytoplasm

In plants, the concentration of Na^+ in the cytosol is maintained at a lower value than that of K^+ in animals. In animal cells, the concentration of Na^+ is closely regulated to 10^{-2} mol L^{-1} value [24]. In plant cells, the concentration of Na^+ does not seem to be subjected to narrow homeostasis. When the plant grows in salinity conditions, the accumulation of Na^+ in the cytoplasm beyond a certain threshold becomes toxic, but this threshold is not clearly determined.

The toxicity of Na^+ in the cytosol would result from its "chaotropic" character by comparison with K^+ [18]. The toxicity of Na^+ would also probably mean its ability to compete with K^+ during the process of fixing important proteins. More than 50 enzymes require K^+ to be active, and Na^+ would not provide the same function [25]. Therefore, a high concentration of Na^+ in the cytoplasm inhibits the activity of many enzymes and proteins, leading to cell dysfunctions. In addition, protein synthesis requires a high concentration of K^+ for tRNA attachment to ribosomes [26], so the translation would also be affected.

2.4. Na^+ Acts as Osmoticum

If the plant cell cannot substitute Na^+ to K^+ in its cytosol, it can do it so in the vacuoles and use Na^+ as osmoticum. Different studies have actually shown that moderate amounts of Na^+ can improve the growth of many plant species [27]. For example, a beneficial "nutritious" effect of Na^+ has been described in tomato [28,29].

It is likely that the beneficial effect of Na^+ can especially be observed in conditions of K^+ deficiency. In these circumstances, a controlled build-up of Na^+ probably helps to ensure the regulation of cell turgor pressure [30,31]. Similarly, a moderate absorption of Na^+ can be beneficial if it helps the plant, for example, to quickly adjust their osmotic potential from the beginning of salt stress.

Despite these physiological observations, the genetic determinants of improving the growth of plants by sodium and genes may be involved in these processes, however, they are still poorly characterized. Research on rice [32] concerning the function in planta of a transporter HKT family provided genetic proof on the fact that an accumulation of Na^+ in K^+ deficiency conditions can promote the growth of the plant.

3. Interaction between K^+ and Na^+ Transport and Adaptation to Salt Stress

The adaptation of the plant to the presence of salt in the soil and salt stress involves various processes, occurring at different levels, from the cell to the whole organism, such as a modification of the metabolic activity leading to the accumulation of organic osmolytes [33], or morphological and developmental changes of the leaves [34]. Within this very complex network of responses, the control of membrane transport activities occurring through a variety of mechanisms, a selective accumulation of K^+ and an exclusion of Na^+ [25,35], appear as a central process. Thus, in a large number of models, from isolated cell culture to the whole plant, adaptation to salt stress appears to be correlated with the ability to selectively remove K^+, to control the Na^+ entrance, and maintain the K^+/Na^+ ratio of the internal contents of these two cations at a high level. In this context, the molecular and functional characterization of membrane transport systems of K^+ and Na^+ is, therefore, a priority objective. It is probable that the capacity of the channels and transporters to discriminate K^+ from Na^+ is essentially based on the difference of intensity of the electric field at the surface of these two ions, which results from their difference in size and hydration energies. The crystallographic resolution of the bacterial potassium channel structure [36,37] provides an example for understanding how carbonyl groups of the polypeptide chain can be spatially distributed along the permeation pathway to substitute, without energy barrier to the hydration shell of the ion.

4. Physiology of K^+ and Na^+ Transport in Plants

4.1. Structure–Function Relationship of the Root

The movement of the mineral elements by the roots and their transfer to the aerial parts involves at least two membrane steps—Ions *sensu stricto* absorption from the soil solution by the epidermal cells, cortical, and contingently endodermic, and the secretion inside the vessels at the level of xylem parenchyma cells. The ions radial movement from the orbital cells of the root to the stela can, in theory, take three paths [38]—The apoplastic pathway (through cell wall), the symplastic pathway (through cytoplasm), or a mixed path passing the ions alternately from apoplastic compartment to the symplastic compartment (Figure 1). Above the cell differentiation zone, the apoplastic path is interrupted by the endoderm of the root. The walls of these cells are impregnated with lignin and suberin. This deposition of hydrophobic compounds forms the framework of Caspary and constitutes a barrier that blocks water and solutes movement. The very close association of the endodermal cell membrane with the Caspary framework forces the ions and water to undergo membrane control to pass the endodermal barrier and migrate into the stele. However, at several levels in the root, the ions can take a direct apoplastic path from the external environment to the xylem: at the apex, where the endodermis is not yet suberized, at the level of endoderm discontinuity, induced by the appearance of the secondary roots [39], and in some species, at the level of some non-suberized endodermal cells, called passage cells which are thought to serve as cellular gatekeepers, controlling access to the root interior [40].

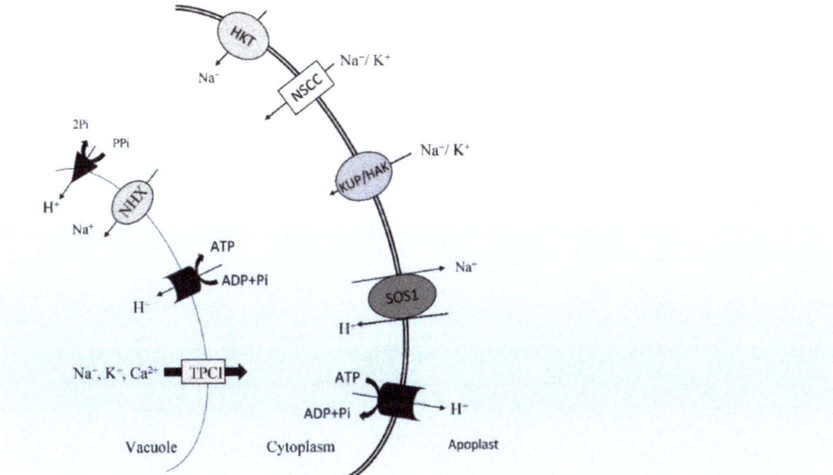

Figure 1. Sodium transport at the cellular level. Schematic representation of transport systems involved in Na$^+$ transport at the plants through the plasma membrane or the tonoplast. Primary transport systems consisting of proton pump ATPases on the plasmalemma and the tonoplast and a pyrophosphatase on the tonoplast create a pH gradient and a potential difference electric on both sides of the membranes (cytosolic side more alkaline and charged more negatively). Proton concentration gradients allow Na$^+$ excretion of cytoplasm towards the outside environment or the vacuole via the operation of antiports Na$^+$/H$^+$ (appealed SOS1 (Salt Overly Sensitive protein 1) on the plasmalemma or NHX1 (K$^+$, Na$^+$/H$^+$ antiporter), on the vacuole). Potential gradients electric created by the pumps cause the entry of Na$^+$ in the cytoplasm of the cell since the external environment or the vacuole via non-selective cationic channels (NSCC) (CNGC (Cyclic Nucleotide Gated Channels) on the plasma membrane? TPC1 (Two-Pore Channel 1) on the tonoplast) or possibly carriers of the HKT (High-Affinity K$^+$ Transporters type in some species. At high external concentration, Na$^+$ could also enter the cell by borrowing K$^+$ carriers KUP/HAK (K$^+$ uptake/High-Affinity K$^{+.}$) type.

In the mature root areas of the majority of plants, a second concentric barrier to that formed by the endodermis is formed at the root periphery on the exoderm, subepidermal cell layer. The suberization of the exoderm would occur later during root development than that of the endoderm and would be accelerated in case of drought [40] or salt stress.

4.2. Structure–Function Relationship of Root and Salt Stress

The current data about root structure and function, as discussed above, indicate that sodium ions can take a direct apoplastic path from the outer medium to the xylem at several levels of the root because endodermal suberization is not yet in place in the young roots area, and leaks remain in secondary roots appearance, which induces a brief discontinuity of the endoderm [41]. The relative contributions of the apoplastic and symplastic pathways of Na$^+$ transport is therefore largely conditioned by root anatomy and are likely to alter according to plant species and soil salinity. The apoplastic pathway (also called apoplastic leak) could be predominant in Na$^+$ transport under salt stress conditions.

Studies carried out on rice have shown that there is a strong correlation between sodium transport and the apoplastic tracer. In two different lines of rice, one more tolerant to salt than the other, a significant difference between the proportions of sodium amount and accumulated PTS in their aerial parts was observed [42,43]. This phenomenon results from the fact that the Na$^+$ entrance into the rice is essentially by free migration in the apoplast up to the stele in spots where the endoplasmic barrier is not functional. This apoplastic leak could occur at the lateral root connection points, at root's apex before complete differentiation of rhizodermis and endodermis, and even in mature areas with differentiated endoderm because of the inherent permeability of the parietal broad outline [44].

It has been shown in rice that the control of apoplastic leakage of Na^+ into the roots is a critical determinant of salinity tolerance. The addition in the culture medium of silicon in sodium silicate partially blocked the apoplastic pathway and considerably improved the growth and photosynthesis of rice plants under salt stress, especially in the GR4 variety [44,45]. This improvement is correlated with the reduction of the Na^+ concentration in plant aerial parts. Furthermore, the authors found that the addition of sodium silicate in the culture medium reduced the accumulation of Na^+ in the aerial parts of sensitive and tolerant varieties at the same level [44,45].

The apoplastic pathway importance in the overall balance of Na^+ inflow varies with species. Garcia et al. [46] estimated that the contribution of the apoplastic pathway is 10 times greater in rice than in wheat. Moreover, it is important to emphasize that halophytes have root anatomy that can limit the entry of Na^+ via the apoplastic pathway. Indeed, the Caspary band in halophytes is 2–3 times thicker than in glycophytes, and the inner layer of cortical cells in halophytes can differentiate to form the second endoderm [2]. In cotton, considered as salinity-tolerant plant among cultivated species, salinity also accelerates the formation of the Caspary band and induces the formation of an additional exodermal layer [47].

All these findings show that there is a correlation between plant tolerance to salinity and the ability to control the apoplastic influx of Na^+ into the roots. It is, therefore, possible to postulate that reducing apoplastic leakage in sensitive species such as rice is a strategy for increasing plant tolerance to salinity. In this perspective, it is important to write down that complete blockage of apoplastic leakage is not likely to significantly affect water inflow and nutrient ion uptake because this leakage contributes little (less than 6%) in rice) to the incoming flows in the roots [46,48]. Some authors have estimated that the apoplastic flow contributes to the xylem flow feeding in a proportion that cannot exceed 1 to 5% [49]. This means that, concerning K^+, the symplasmic transport ensures the essential translocation of this ion from soil solution to the xylem vessels of the stele.

5. Potassium Availability in the Soil and Its Absorption by Plants

K^+ is an important cofactor in many biosynthetic processes, and in the vacuole, it plays key roles in cell volume regulation [50].

The concentration of K+ in the soil solution is generally between a few tens of μmol. L^{-1} and a few mmol. L^{-1} (i.e., approximately 10 to 10^3 times lower than that of the cell). The roots are thus confronted with a wide concentration range and the plants possess transport systems allowing them to grow over concentration ranges of K^+, ranging from 10^{-6} to 10^{-1} mol. L^{-1} [51].

An enhancement of the absorption capacity of K^+ by the root is observed when the availability of this ion in the soil is limited [52]. In wheat, K^+ deprivation increases the high-affinity transport efficiency, without altering the characteristics of low-affinity transport. This type of response has also been observed in barley and ryegrass [53]. This reaction is not general, but there are many proteins involved in high-affinity potassium transport. However, in Arabidopsis, two proteins have been identified as the most important transporters in this process. Interestingly, one of these transporters, AtHAK5, is a carrier protein and is thought to mediate active transport of potassium into plant roots, whereas the other protein, AKT1, is a channel protein and likely mediates a passive transport mechanism with an increased affinity for K^+ under conditions of potassium limitation [54,55].

Several different natural phenomena could be involved in root absorption capacity enhancement observed in response to K^+ deficiency in the soil. An initial model to account for this response proposes an allosteric regulation of the absorption capacity in terms of the cytosolic concentration of K^+, resulting in an inhibition by "feedback" of the transporters when the availability of this ion in the area is high, leading to an increase in its concentration in the cytoplasm [56]. Under this model, the K^+ availability diminution in the area leads to a decrease of K^+ concentration in the cytoplasm, which would lift the allosteric inhibition of transport, thus causing absorption capacity augmentation. Another hypothesis, non-exclusive of the previous one, is based on the observation of modifications of the membrane polypeptide equipment when the plants are cultivated in a weakly concentrated potassium area, confirming the installation of new transport systems in barley [57], especially high-affinity transporters in barley [58], wheat [59], and *Arabidopsis thaliana*, [55,60]. In *Arabidopsis*, studies using the patch-clamp technical revealed that K^+ deficiency increases the activity of IRK-type channels (inward rectifying K^+ channel). This augmentation may reflect a corresponding gene(s) expression enhancement or the existence of a post-translational regulation mechanism (e.g., by dephosphorylation). However, the physiological meaning of the channels activity stimulation—And thus of passive transport systems in response to K^+ concentration diminution in the area—Is unclear, even though it is possible that channels may participate in the absorption function from a relatively low external K^+ concentration. Membrane potentials have indeed been found to be negative enough to be able to involve channels in the influx of potassium from an external solution of which K^+ concentration is less than 10 μM [61].

6. Long-range Transport in Xylem and Phloem

6.1. Transport into the Xylem

The Na^+ content of the roots appears to be relatively constant during salt stress. This steady-state probably results in part from root cells' ability to discharge Na^+ in the external area. It also results from Na^+ translocation in the stele and xylem vessels to the aerial parts. The sodium levels of the xylem and phloem may alter during the flow of plant sap. An increase of Na^+ concentration in xylem sap has been described in an "includer" type plant (definition below) *Plantago maritima* [48].

In opposition to this, a decrease of Na^+ concentration in xylem sap has been reported in "excluder" plants type—The sodium contained in the xylem is reabsorbed by roots during the ascent of the sap, and re-excreted toward the outside environment [48]. The amount of sodium that reaches the leaves via xylem sap can be controlled during transport in xylem vessels.

Unfortunately, there is a lack of knowledge about the mechanisms of Na^+ transport in the xylem. However, in *Arabidopsis* under moderate salt stress conditions (40 mM NaCl), *Sos1* mutants (having lost an H^+/Na^+ antiport system) accumulate fewer Na^+ in the aerial parts than wild-type plants [34,62]. This suggests that SOS1 plays a role in the transport of Na^+ from the roots to the aerial parts. However, the use of a reporter gene reveals that in the roots, SOS1 is expressed preferentially in the parenchymal cells around the xylem vessels [62]. Together, these data suggest that SOS1 has been involved in Na^+ secretion in xylem sap from stele parenchymal cells under moderate salt stress conditions.

In some plants, there is a reduction of Na^+ accumulation in the aerial parts. This reduction could be explained by sodium removal from the xylem before it reaches the foliar system. The existence of this strategy in plants has been clearly demonstrated by the research work of Adem et al. [63]. The authors have shown that in barley, the Na^+ concentration of the xylem sap varies together with the stem height (10 mM at the base of the stem and only 2 mM at the 8th leaf). This difference of concentration is important particularly for maintaining the photosynthetic activity of young leaves, which in return allows the formation and growth of new leaves. Molecular mechanisms of Na^+ removal from xylem sap ("desalting" of xylem sap) are beginning to be documented. In particular, the genetic analyses revealed that two transporters of the HKT family, *AtHKT1* in *Arabidopsis* and *OsHKT8* in rice, are involved in this desalting process.

The majority of plants maintain a high K^+/Na^+ ratio in their aerial parts, so it appears that the selectivity to the benefit of K^+ is ensured during the secretion. The ions are excreted in the xylem bundles via xylem transfer cells that can promote, or delay, the efflux of Na^+ in this vessel. The control of the Na^+ concentration in the xylem can also be carried out all along the stem by reabsorption of the sodium in exchange of potassium in the raw sap by the parenchyma cells [3]. H^+-ATPases of the plasma membrane would ensure the energization of the various transports resulting in the exchange of Na^+ against K^+. The H^+ gradient created by these pumps would allow the secretion of K^+ via an antiport H^+/K^+, and a uniporter of Na^+ would ensure sodium reabsorption.

Concerning potassium, the ions absorbed at the level of the plasma membrane of the root superficial cells (epidermal and cortical) are transported towards the tissues of the stele by diffusion from one cell to another through plasmodesmata (symplastic pathway). After migration beyond the endodermal barrier, the ions leave the symplasm crossing a second plasma membrane at the level of the last living cells that border the vessels (xylem parenchyma). Once in the apoplast stellar, the ions are driven by the centripetal flow of water to the vessels, and the convection flow of the raw sap (water and mineral units) carried by transpiration and/or root thrust then exports them to the aerial parts [64].

The inner position of xylem parenchyma cells in the root makes the electrophysiology analyses using microelectrodes difficult. As a result, the mechanisms of secretion of ions in the xylem have been less studied than the mechanisms of absorption. It has been acknowledged that the stela's tissues are not able to accumulate ions and that these ions, inflated at the entrance of the symplasm, passively diffuse to the vessels. This passive diffusion was thought to be the consequence of an oxygen deficiency in the central tissues of the root that results in cell depolarization [65]. The stele cells in hypoxic conditions were then unable to retain the ions. However, Zhu et al. have shown that aeration of root pivotal tissues allows cells sufficient oxygenation. CCCP instantly blocks efflux in the xylem of the $^{36}Cl^-$ accumulated in advance but not efflux to the area through the epidermis [66]. These results show that the CCCP affects the existing system at the level of the stele and not the one located in the cells of the epidermis. Since the 1970s, it has been clearly established that the ions efflux in the stellar apoplast depends on specific transporters located on the plasma membrane of xylem parenchyma cells. Several experimental data indicate that absorption and secretion are controlled separately.

In general, the secretion of nutrient ions in the stellar apoplast could in many cases be a passive phenomenon, catalyzed by channels. For example, in *Arabidopsis*, the SKOR potassium channel of the Shaker family plays an important role in K^+ secretion in xylem sap [67]. The knowledge at the molecular level on the mechanisms of secretion of nutrient ions in the xylem sap is, however, still rather small.

6.2. Transport into the Phloem

The growth and development of the plant require distribution of photosynthesis products. These molecules synthesized in the so-called "source" organs (mature leaves) must then be relocated to the growing organs and non-photosynthetic plant tissues (organs called "wells," young leaves, flowers, seeds, fruits, roots). This relocation requires selective long-distance transport, which is provided by the phloem system.

Data obtained from barley show that the sodium contents of xylem and phloem sap are altered throughout transport in the vessels of the aerial parts [68]. The sodium contained in xylem would be absorbed and stored into leaf cells during its movement, and there would also be a translocation of a part of the sodium from xylem to phloem in the leaf, so that the sodium concentration in phloem sap has increased, as it moves from the top of the leaf to its base. Foliar anatomy, particularly in the area of young veins, suggests that such a transfer could occur either directly from apoplast to symplasm of phloem cells, or by symplasmic transport from parenchymal cells [68]. This recirculation of ions from xylem to phloem thus makes it possible to significantly reduce the salt content of the leaves. This has also been observed in some species such as Lupin [69], pepper [70], corn [71], and barley [13].

Perez-Alfocea et al. [72] have found that Na^+ translocation in the phloem of *Lycopersicon pennellii*, a wild type tomato that is tolerant to salinity, is more important than that observed in domesticated tomatoes. This suggests that the translocation of Na^+ into the phloem would be an adaptation strategy in plants. However, the Na^+ translocation direction and the conditions under which it occurs are probably critical. Indeed, it seems crucial that translocation by phloem does not transport Na^+ to the young tissues—Otherwise, it would completely inhibit their growth. In other words, translocation by the phloem should essentially redirect Na^+ to the roots. In the pepper plant, it has been shown that the translocation of Na^+ from the aerial parts to the roots only occurs when Na^+ is removed from the nutrient solution, i.e. when there is a favorable gradient between phloem and roots [70]. In *Arabidopsis*, it has been shown that the sodium transporter *AtHKT1*, expressed in phloem tissues, assure Na^+ recirculation from the leaves to the roots through phloem by removing Na^+ from the rising stream of raw sap at the aerial parts. This system thus plays the role of controlling the Na^+ accumulation in the leaves and plant resistance to salt stress [73].

With regard to potassium, the phloem loading and its discharge contribute to the establishment of the osmotic potential gradients (and therefore hydric) created between the source organs (high concentration of sugars and ions in the phloem sap) and the well organs (lower concentrations). The osmotic gradient is initiated at the level of the source organs by the creation of an electrochemical potential due to the H^+-ATPases activity of the fellow cells that are in direct electrical contact with the cells of the screened canals (making the phloem vessels) via plasmodesmata. This energization of the membrane allows the influx of sugars (essentially sucrose) and potassium into the cells. In summary, the available data indicate that control of K^+ transport in phloem tissues of source and well organs contribute to three main functions: (i) the phloem cells membrane potential regulation, tending to bring its value closer to that of equilibrium potential of K^+ (E_K), (ii) the installation of the osmotic gradient responsible for the sap flow between the source and well organs, and (iii) well organs (including seeds and fruits) potassium supply.

The electrophysiological characterization of the potassium conductance of phloem cells is still poorly advanced because of the difficulty to obtain protoplasts. This difficulty is less with corn roots, whose stele is easy to separate from the cortex. Phloem cells can then be obtained by dissection in which potassium conductance has been identified. They are close to the IRKs in their selectivity and responses to inhibitors, but they show a small correlation. It means that they allow an entrance or output of potassium according to the membrane potential value. In *Arabidopsis*, the AKT2 gene from the Shaker potassium channel family could code this type of conductance [74,75].

In *Arabidopsis thaliana*, the use of plants expressing the GFP gene under the AtSUC2 gene promoter (active specifically in phloem fellow cells) made it possible to isolate protoplasts from these cells and to identify two potassium conductance—An outgoing conductance of the ORK type and an incoming conductance of the IRK type. However, conductance that may be specific to cells located either in the source regions or in the well regions has not yet been demonstrated.

7. Adaptive Strategies of Plants to Na^+: Exclusion and Inclusion

The ability of a plant to compartmentalize Na^+ at the cellular level induces a difference of Na^+ management in the whole plant. We can distinguish two ways of plants responses to salt (exclusion and inclusion). "These strategies characterize behavior patterns that are not mutually exclusive" (Levigneron et al. reviewed in [76]). Excluder type plants are generally salinity-sensitive and are unable to control the level of cytoplasmic Na^+. This ion is transported in the xylem, conveyed to the leaves by transpiration stream, and then partly "re-circulated" by the phloem to be brought back to the roots. These sensitive species, therefore, contain little Na^+ in the leaves and an excess in the roots. Includer plants, which are resistant to NaCl, accumulate Na^+ in the leaves where it is sequestered (in the vacuole, foliar epidermis, and old limbs). However, excluder plants also accumulate Na^+ in the vacuole of root and stem cells. Of course, these two types of behavior are extreme, and some species can incorporate behaviors characteristic of both types of strategy.

7.1. K^+ and Na^+ Transport Systems in Plants

The kinetic characteristics of K^+ transport systems were studied (since 1950) using the ^{42}K and ^{86}Rb tracers, in particular by Epstein et al. The incorporation rate analysis of the tracer into the excised barley roots, in terms of the external concentration, reveals complex kinetics that presents two phases [77]. This kinetics, which can be analyzed according to the Michaelis-Menten formalism, suggests the existence of two absorption mechanisms. The first mechanism corresponds to a high-affinity saturable system (Km $\approx$ 20 μM), which allows the influx of K^+ from low concentrations in the area (less than

1 mM). The second mechanism corresponds to a low-affinity system (Km ≈10 mM) responsible for ions absorption from high concentrations. The second absorption mechanism differs from the first by the fact that it is not selective for K^+ (vis-a-vis of Na^+), and its ability to transport K^+ depends on the nature of the accompanying anion [78]. Electrophysiological data obtained on roots suggest that H^+-K^+ symports are responsible for high-affinity K^+ transport [79]. Low-affinity absorption is passive and involves channels.

For sodium, it is established that its initial entrance from the external environment into the cytoplasm of the roots cortical cells is passive [48], either via non-selective voltage-dependent cation channels (NSCCs) [2] or probably via some family members of sodium transporter [80,81] (Figure 2).

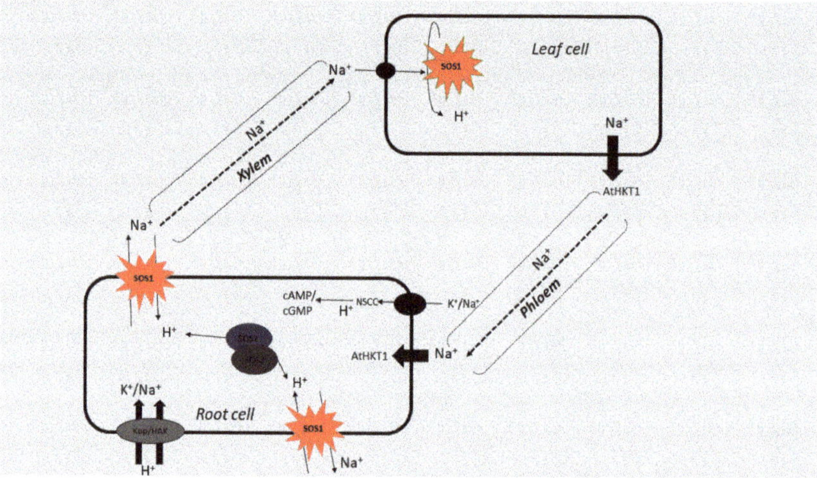

Figure 2. Na^+ transport at the level of the whole plant. Sodium ions can enter the cells of the root through non-selective channels (NSCCs) not formally identified at the molecular level, some of which appear to be inactivated by cyclic nucleotides (cAMP and cGMP; Maathuis and Sanders 2001), transporters HKT and high concentration of Na, KUP/HAK carriers. Na excretion cell roots to the soil solution or to the vessels of the xylem involve the antiport H^+/Na^+ SOS1, whose activity is regulated by the SOS3 CBL protein associated with the SOS2 kinase of the HKT conveyors allow desalinization of xylem sap and phloem loading in Na^+ at the leaf level.

Several families of channels and transporters involved in K^+ and Na^+ transport have been identified at the molecular level in plants.

7.2. Channels

7.2.1. Shaker Channels

These channels exist in plants, fungi, bacteria, and animals. The first members of this family were identified in animals.

These channels are formed with four subunits, which are organized around a central pore. The hydrophobic region of each subunit includes six transmembrane segments (TMS). A membrane loop (called P, for pore) between the fifth and sixth TMS participates in wall constitution of the central pore. Subunits can gather into homotetramers or heterotetramers. These channels are all voltage-regulated and active on the plasma membrane. They are very selective of K^+ beside Na^+. In higher plants, several Shaker channels have been cloned and characterized. There are nine members in *Arabidopsis*, with different functional properties, expression patterns, and localization [82]. The first two Shaker channels identified in plants are AKT1 and KAT1, cloned in 1992 in *Arabidopsis* [83].

In a very interesting way, the characterization of these functional systems has shown that they act as inward rectifying channels [55,84], despite strong homology with voltage-dependent, highly selective K^+ channels, which act as outgoing channels. This observation has generated a lot of interest and triggered numerous studies on the structure–function relationship of these channels, with the main objectives of understanding the mechanisms of opening-closing of the pore and the regulation by the voltage.

There are three main functional types of Shaker channels—incoming rectification channels (KAT, AKT1, and ATKC1 families), outgoing rectification channels (SKOR family), and low rectification channels (AKT2 family). The fourth TMS, carrying positively charged residues (R or K), is the cause of the channel sensitivity to the voltage. The pore loop (P) studies by controlled mutagenesis have identified a motif (TxGYG) involved in ionic selectivity.

The role in the plant of several *Arabidopsis* Shakers was analyzed by reverse genetics. In a general way, these channels allow the massive exchanges of K^+ (influx or efflux), between the symplast and the apoplast (K^+ entrance of the cell for the incoming channels, exit for the outgoing channels), entry, and exit by low rectification channels. They play a role in the removal of K^+ from the soil solution (AKT1 and AtKC1 channels in Arabidopsis), long-distance K^+ transport in the xylem and phloem (SKOR and AKT2 channels), or in the transport of K^+ in the guard cells at the origin of stomatal movements (GORK, KAT1, and KAT2 channels) [85,86]. The shaker channels, outstanding the potassium conductance of the plasmalemma, participate in parallel to the regulation of cellular potassium concentration, to the control of the membrane potential, and to the osmotic potential regulation.

7.2.2. KCO Channels

KCO (or TPK) is the second family of specific K^+ channels identified in plants. These channels are probably composed of either two subunits (family KCO-2P) or four subunits (family KCO-1P), which are organized around a central pore associating four domains (P for pore). In *Arabidopsis*, the KCO-2P family (two P domains per subunit) has five members, and the KCO-1P family (one pore domain per subunit) has only one member [85]. The first member of the KCO-2P family, KCO1, was discovered in silico via the use of the highly conserved GYGD motif in Shaker channels [87]. It has been expressed in insect cells, where it acts as a selective channel of K^+. At the subcellular level, *AtKCO1* has been localized at the level of the tonoplast [61,88], suggesting that it plays a different role from that of the Shaker channels in transporting K^+ through intracellular membranes. Electrophysiological analysis of vacuolar currents on invalidated mutant kco1 suggests that KCO1 contributes to SV type currents, which are outgoing and slow vacuolar currents [61].

7.2.3. Non-Selective Cationic Channels (NSCCs)

These channels, less selective of K^+ than Shaker, have been characterized in different cell types [89]. NSCCs include CNGCs and GLRs, which are still poorly characterized. Obviously, all transporters have not significant role in potassium or sodium uptake, thus recent studies on GLRs showed that their expression throughout the plant, open up the possibility that GLR receptors could have a pervasive role in plants as non-specific amino acid sensors in diverse biological processes [90]. There has been no progress in elucidating their role in potassium and sodium uptake for the last two decades.

An indication of the CNGCs involvement in Na^+ influx is that the addition of similar cyclic nucleotides in the environment inhibits Na^+ influx and non-selective cation channel activity [91]. In animals, CNGCs are non-selective cationic channels involved in signal transduction in response to different stimuli. They are permeable to Ca^{2+}, Na^+, and K^+ [92]. Activation of these channels leads to membrane depolarization and cytoplasmic calcium concentration enhancement, thereby activating the signaling pathways dependent on this ion. These channels have a similar structure to the Shaker-type voltage-dependent potassium channels, a hydrophobic domain formed by 6 TMSs (named S1 to S6), and a P domain forming the pore between the fifth and the sixth TMS. In their hydrophilic N- and

C-terminal ends, they have respectively a calmodulin binding domain (CaMBD) and a cyclic nucleotide binding domain (CNBD).

In plants, a family of ion channels homologous to CNGCs animal channels was identified in the late 1990s. The first cDNA encoding a channel belonging to this family was cloned in barley by screening an expression library by searching for calmodulin-interacting proteins and was named *HvCBT1* [93]. The second member of the family was isolated from tobacco by the same approach [94]. This cDNA, named *NtCBP4*, has 61.2% identity with *HvCBT1*. In *Arabidopsis*, 20 family members, named CNGC-1 to 20, have been identified in silico by sequence analogy [95].

CNGCs are a class of nonselective cation channels that are permeable to monovalent and divalent cations such as Na^+, K^+, and Ca^{2+} [89,96,97]. Although their down-regulation can prevent Na^+ uptake, it can potentially be concomitantly harmful to the plants, as the uptake of other elements will be compromised. However, in rice root, the downregulation of the rice (*Oryza sativa*) *OsCNGC1* contributed to the superior tolerance of the cultivar FL478 to salt stress [25], as it could avert toxic Na^+ influx, in contrast to the sensitive cultivar, in which the gene was up-regulated by salinity stress. Also, *Arabidopsis thaliana* null mutants, *Atcngc10*, were found to have enhanced growth under salt stress compared to wild-type plants [98]. Furthermore, Atcngc3 T-DNA insertion mutants showed an increase in tolerance to high levels of NaCl and KCl [99]. With regard to the correlation between CNGC down-regulation and stress tolerance, Mekawy et al. (2015) evaluated the relative tolerance of two rice cultivars, Egyptian Yasmine and Sakha 102. They observed that the greater tolerance of Egyptian Yasmine was partially attributable to the down-regulation of OsCNGC1, with the concomitant up-regulation of plasma membrane protein 3 (PMP3), a plasma membrane protein involved in the inhibition of excess Na^+ uptake at the level of the root [100].

Also, some observations show that, in *Arabidopsis*, the *AtCNGC1 and AtCNGC2* channels introduced into yeast expression plasmids appear to complement a defective yeast mutant for K^+ transport [95]. In tobacco, over-expression of *NtCBP4* confers transgenic plants nickel tolerance and tin hypersensitivity that decrease Ni^{2+} accumulation and increase Pb^{2+} accumulation [94]. Subsequently, it has been shown that NtCBP4 is expressed on the plasma membrane of tobacco cells [94]. The hypothesis is that *NtCBP4* would be a transport system (perhaps permeable to Ca^{2+}) allowing Pb^{2+} entry into the cell.

The data in planta on the function of a CNGC were obtained indirectly following genetic analysis on an altered *Arabidopsis* mutant in response to a pathogen [101]. This study has made it possible, for the first time, to highlight the involvement of a CNGC ion channel in a signaling pathway. In general, CNGCs are probably involved, like their homologs in animal cell signaling [89,102]. They would be permeable to monovalent and/or Ca^{2+} cations and regulated by cyclic nucleotides and calmodulin. In plant CNGCs, the cyclic nucleotide-binding domain and the calmodulin-binding domain are both located in the C-terminal cytoplasmic region, where they overlap slightly [102].

7.3. Transporters

The KUP/HAK/KT family. A transporter belonging to a new family of K^+ transport systems has been identified in *Escherichia coli* (KUP1) [103] and in yeast *Schwanniomyces occidentalis* (SoHAK1) [104]. The *SoHAK1*expression in a mutated strain of *S. cerevisiae* for K^+ uptake systems restored growth onto a low K^+ concentration environment [104], SoHAK1 seems to be a high-affinity K^+ transporter. The homologs in plants, named KUP, HAK, or KT (for "K^+ uptake," "High-Affinity K^+ transporter," and K^+ Transporter, respectively), form a large family containing at least 17 members in rice [105]. The structure of these transporters is poorly known. The hydrophobicity profiles suggest that they have 12 TMSs and a long cytoplasmic loop between the second and third segments.

In plants, the first gene of the HAK/KT/KUP family, named *HvHAK*, was cloned in barley by qRT-PCR, with corresponding primers to conserved regions of *E. coli* KUP1 transporters and SoHAK1 [58]. In *Arabidopsis*, the first members identified in the HAK/KT/KUP family were cloned by complementation of a yeast mutant [106] or by the search for homologous sequences to KUP1 and HvHAK in the data banks [60]. Overexpression of AtKUP1 and AtKUP2 cDNAs induces an $^{86}Rb^+$

influx in yeast or in *Arabidopsis* growth cells [60,106]. For AtKUP1, the absorption kinetics in terms of concentration shows a Michaellian style in the low concentration range (less than 100 μmol. L^{-1}), raising the kinetics associated with the mechanism I in roots [60,106]. This similarity suggested that KUP-type systems are responsible for active K$^+$ transport with high affinity in plant cells. However, the analysis of absorption kinetics by the AtKUP1 system as a function of K$^+$ concentration also reveals low-affinity transport activity [60]. In other words, the AtKUP1 system alone can generate biphasic absorption kinetics, which evokes the kinetics observed in the roots (mechanism I plus mechanism II). The duality of transport kinetics by AtKUP1 could reflect two different modes of operation for this system. No current was detected by heterologous expression of AtKUP1 in the Xenopus oocyte, and the transport mechanisms unable to be determined [60]. However, the K$^+$ influx generated by *HvHAK1* and *AtKUP1* proteins in yeast is inhibited by the presence of Na$^+$ in the environment [58,106]. The localization of *AtKUP1* gene expression analyzed by northern blot, has led to variable results in which, the mRNA is undetectable in the roots but present in the aerial parts [60], mainly expressed in roots [58,106] or undetectable throughout the plant [107]. These variations could be associated with differences in plant growth conditions. This would mean that the accumulation of *AtKUP1 mRNA* is highly dependent on environmental conditions.

By a classic genetic approach based on the search for altered mutants in absorbent hairs growth, the authors of reference [108] have isolated another family member, named *TRH1* or *AtKUP4*. The trh1 mutant shows a decrease in ^{86}Rb$^+$ uptake. The phenotype of absorbent hair growth of the mutant plants is not restored when they are grown in an environment containing 50 mM of K$^+$. The high-affinity K$^+$ transporter function of *TRH1* has been demonstrated by the complementation of yeast mutant *trk1*. TRH1 is expressed in the roots and in the aerial parts. It could be involved in the absorbent hair formation by allowing the influx of K$^+$ necessary for the growth and the elongation of these cells.

In general, all these HAK/KT/KUP transporters are not sufficiently characterized at the functional level, because of difficulties in expressing them in a heterologous system (a few rare members, however, express themselves in the yeast *S. cerevisiae* and/or in *E. coli* bacteria). In plants, they are present in many cell types and seem to be found on both the plasma membrane and the vacuolar membrane [105].

7.3.1. HKT Transporters

HKT transporters have homologs in fungi (TRK) and bacteria. Their predicted global structure, based on sequence analyses, is similar to that of potassium channels (at 2 TMS) that exist for example in bacteria. The hydrophobic region of the HKT polypeptides comprises four repeats of the (1 TMS/1 P/1 TMS) module. In the functional protein, the four loops are arranged to form a central pore [109].

All HKT transporters characterized so far in plants are permeable to Na$^+$, and some are also permeable to K$^+$. The role of these transporters in planta of K$^+$ transport has not yet been clarified. Several studies have demonstrated the role of these systems in planta in the transport of Na$^+$ and revealed that HKTs are involved in the tolerance of plants to salinity.

The protein sequence of *TaHKT1* has about 20% homology with the TRK systems identified in yeast and its structure would integrate 10 to 12 hydrophobic regions likely to correspond to TMS.

The *TaHKT1* expression in the Xenopus oocyte causes an activated current by the addition of K$^+$ or other cations to the external medium. The intensity of this current increases when the pH of the external medium is lowered. However, the analysis of transgenic plants overexpressing *TaHKT1* did not make it possible to highlight a contribution of this system to the absorption function of K$^+$ by the root [110]. Subsequent analyses revealed a sensitivity of the transport to the presence of Na$^+$ in the area. These data suggested that *TaHKT1* would rather function as a high-affinity Na$^+$: K$^+$ symport for K$^+$ (ca = 10 μM), energized by the electrochemical gradient of Na$^+$ across the membrane [111], which is completely unexpected energy coupling mechanism in plants. Moreover, this type of operation is limited to conditions of low external concentration of Na$^+$.

When the Na$^+$ concentration is higher, the transport of K$^+$ by *TaHKT1* would be blocked and this system would function as a low-affinity Na$^+$ transporter (*Km* close to 5 mM) [111]. The physiological significance of this result remains unclear since in vivo K$^+$-transport analyses in higher plants have never revealed Na$^+$-K$^+$ symport activity (e.g., the addition of Na$^+$ in the medium does not stimulate K$^+$ uptake).

The only member of the HKT family in *Arabidopsis*, orthologue of the wheat *TaHKT1* gene, has been identified and designated as *AtHKT1*. The expression of this gene in yeast strains lacking the Na$^+$ efflux system aggravates their sensitivity to Na$^+$, but it does not suppress K$^+$ transport deficiency in *trk1* and *trk2* mutants that have difficulty to absorb potassium [112]

When expressed in the *Xenopus* oocyte, *AtHKT1* exhibits strictly selective Na$^+$ transport activity, without any permeability to K$^+$. Similarly, *AtHKT1*expression does not complement a type of *E. coli* mutant unable to absorb K$^+$, which helps to show that *AtHKT1* carries only Na$^+$.

AtHKT1 is expressed in the vascular tissues of the root and the aerial parts, at the level of the phloem and the xylem parenchyma [113].

While the *AtHKT1* gene is unique in *Arabidopsis*, it is interesting to note that the HKT family in rice has 7–9 members, depending on the cultivars [114]. The analysis of the polypeptide sequences of the transporters encoded by these genes shows a rather significant difference between the members—apart from two pairs of highly homologous transporters (OsHKT3/OsHKT9 and OsHKT1/OsHKT2, 93 and 91% identity, respectively), the percentage of identity between the different transporters is between 40 and 50%. Nipponbare (japonica), *Ni-OsHKT2*, and *Ni-OsHKT5* probably do not encode functional transporters due to large deletions or the presence of "stop codons" in the reading frame. However, *OsHKT2* is identified in another cultivar (*indica*) and codes for a functional transporter, Po-OsHKT2 [115].

Localization studies by analysis of transformed plants with a promoter (GUS fusion) has shown that these two HKTs are expressed at the vascular tissue level. Specifically, all of the available data (including in situ hybridization analyses) reveal that *OsHKT1* is localized in foliar vascular tissue but also in the root cortex and endoderm [32], whereas *OsHKT8* is mainly localized at the level of the xylem parenchyma, in the roots and in the leaves [116].

The most detailed data at the functional level concerns OsHKT1. This system is one of the closest counterparts in rice of the first HKT characterized, TaHKT1 (wheat), which is a transporter of K$^+$ and Na$^+$ (*OsHKT1* and *TaHKT1* have 67% identity). *OsHKT1* has been characterized by three different teams, leading to conflicting results. Expressed in the Xenopus oocyte, OsHKT1 is described as a cationic transport system, with little discrimination with respect to the different alkaline cations [117], or as a very selective transporter of Na$^+$ [115]. Expressed in yeast, it is described either as a K$^+$ permeable transport system [117] or as a Na$^+$ transport system blocked by K$^+$ [114]. OsHKT1 expression in *S. cerevisiae* yeast mutants deficient for K$^+$ transport did not allow growth on medium poor in K$^+$ (0.1 mM KCl). The growth inhibition test on *S. cerevisiae* G19 yeast strains, highly sensitive to Na$^+$ following the disruption of ENA genes (which code for Na$^+$ excretory ATPases), revealed that the cells expressing OsHKT1 exhibited more sensitivity to Na$^+$ than those expressing TaHKT1 in the presence of 50 and 100 mM NaCl.

7.3.2. CHX Transporters (Monovalent Cation H$^+$ Exchanger)

These transport systems have been identified in plants on the basis of their homology with systems previously characterized in other organisms, such as bacteria, yeasts or algae. Only transporters involved in sodium compartmentalization in the plant vacuole are now relatively well known.

As in unicellular organisms, transports through the tonoplast is activated by an H$^+$-ATPase pump that establishes a proton gradient [118]. The operation of the CHXs is electron-based and thus does not disturb the potential difference across the membrane. These systems are probably involved in both monovalent cation homeostasis and cytoplasmic and/or vacuolar pH regulation [119].

From a biochemical point of view, tonoplast antiport Na$^+$/H$^+$ activity, which may be involved in sodium vacuolar compartmentalization, was initially demonstrated by the Blumwald group in several species [120]. This Na$^+$/H$^+$ antiport activity was associated with a 170 kDa vacuolar protein identified in *Beta vulgaris*, whose accumulation is increased by NaCl treatments [121]. Antibodies planned against this protein inhibited the Na^+/H^+ antiport activity. This protein was, therefore, a good candidate for the antiport activity detected on the tonoplast but its coding gene remains unknown.

From the molecular point of view, an *Arabidopsis* cDNA, named *AtNHX1*, related to the yeast ScNHX1 protein, constituted the first characterized system. Only this tonoplast antiport Na^+/H^+ of Arabidopsis antigen has yet clearly been involved in sodium vacuolar compartmentalization [5,120,122,123]. The expression of this plant cDNA complements defective yeasts in the Na^+/H^+ transporter present in the vacuolar membrane [123]. In *Arabidopsis*, *AtNHX1* overexpression confers to transgenic plant tolerance to external Na^+ concentrations above 200 mM [5]. *AtNHX1* is expressed in all plant tissues and is found on the internal system tonoplast and on the membranes (RER, Glogi). Systematic sequencing of the *Arabidopsis thaliana* genome has identified 35 genes that can encode proteins being similar to antiport Na^+/H^+. Constitutive overexpression of *AtNHX1* improves salinity tolerance also in tomato [124], *Brassica napus* [125], and soybean [126]. Fukuda et al. have identified an *AtNHX1* homologue in rice, *OsNHX1*. OsNHX1 expression is induced into the roots and into the aerial parts during salt stress. The authors found that OsNHX1overexpression enhances the salinity tolerance of transgenic cells and plants [127].

Within the *CHX* family, some members may be good candidates for K^+ transport. This is the case in *Arabidopsis* for AtKEAs that resemble the K^+/H^+ bacterial antigens KefB and KefC. However, no experimental data for these systems are available, except for expression data in *Arabidopsis* tissues. Of the 28 KEA genes in this plant, 18 are specifically expressed during the microgametogenesis phase or in sporophytic tissues, suggesting that CHXs are involved in the regulation of potassium homeostasis in the pollen growth phase and germination [128]. Two CHXs have been characterized in more detail. *AtCHX17* appears to be preferentially expressed in roots under stress conditions, such as high salt concentrations, low external pH, low external K^+ concentration, and/or basic acid treatment [125]. The analysis of the mutant *AND-T atnhk17* suggests that this gene has a function in potassium homeostasis since the mutant plants accumulate less potassium than the wild ones. When expressed in yeast, *AtNHX17* co-localizes with markers of the *Golgi apparatus* and complements the pH sensitivity of a *kha1* mutant yeast strain [129], suggesting a role in potassium homeostasis and pH regulation under stress conditions. Loss of function mutants of this gene showed alteration in the ultrastructure of the chloroplast with a sharp decrease in chlorophyll level in the leaves, and an increase in cytosolic pH in the guard cells. The growth of *atnhx23* mutants was enhanced by the addition of high concentrations of potassium in the environment but altered by the addition of NaCl [130]. All these data suggest that *AtNHX23* is an antiport K^+ (Na^+/H^+) active at the level of the chloroplast envelope and involved in potassium homeostasis and perhaps in regulating the pH of the stroma.

The Na^+/H^+ antiport systems of the plasma membrane are still poorly characterized. The only information relates to the SOS1 protein in *Arabidopsis*, which has a homologous sequence with antiport Na^+/H^+ and would be involved in sodium efflux at the plasmalemma level [62]. Evidence has been provided that SOS1 does have antiport Na^+/H^+ activity [62].

The sodium hypersensitive *Arabidopsis* mutant *sos1* exhibits, when cultivated in presence of moderate NaCl concentrations (40 mM), higher Na^+ content in its roots than those observed in the plant control of wild-type genotype. Moreover, using the reporter gene system, the authors have highlighted the localization of SOS1 in epidermal cells at the root end. These results suggest the involvement of *SOS1* in Na^+ efflux from the roots in the environment. In addition, it is interesting to note that SOS1 overexpression in *Arabidopsis* significantly improves plants tolerance to salinity. AtSOS1 is, therefore, an important determinant of salt sensitivity in plants. *AtSOS1* activity is controlled by *AtSOS2* and *AtSOS3*. AtSOS3 (a Ca^{2+} affine protein belonging to the CBL family) directly interacts with AtSOS2 which a serine/threonine protein kinase is [131]. The interaction of *AtSOS3 and AtSOS2* triggers *AtSOS2* protein kinase activity, which phosphorylates and activates SOS1. Moreover, CBL/CIPK perceive cytosolic Ca^{2+} signals resulting from salt stress and have important roles in regulating salt stress response and ion homeostasis [132].

7.4. Ion Transporters Mediating Role in Salt Tolerance

In Arabidopsis roots, *AtCNGC3* is thought to be involved in Na$^+$ fluxes. It has been reported that a null mutation in *AtCNGC3* would reduce the net Na$^+$ uptake during the early stages of NaCl exposure (40–80 mM). However, longer exposure of wild type (WT) and mutant seedlings to NaCl (80–120 mM), induces the accumulation of similar Na$^+$ concentrations in both plants [99].

These results indicate the involvement of *AtCNGC3* in Na$^+$ uptake during the early stages of salt stress. In salt-tolerant rice varieties, *OsCNGC1* is negatively more regulated than in salt-sensitive varieties subjected to salt stress conditions [133]. *Arabidopsis thaliana AtHKT1;1*, facilitates the influx of Na$^+$ into heterologous expression systems [134]. Apparently, there is a determinant of salt stress tolerance that controls the influx of Na$^+$ into the roots, resulting in lower accumulation of Na$^+$ in *athkt1* mutants than in WT plants [135]. Horie et al. demonstrated that *OsHKT2;1*, regulates the influx of Na$^+$ into root cells [32]. Plants lacking the *OsHKT2;1* gene, when exposed to 0.5 mM Na$^+$ in the absence of K$^+$, exhibit lower Na$^+$ accumulation and reduced growth [32]. *OsHKT2;2/1*, a new isoform of HKT isolated from the rice plant roots that is no more than an intermediate between *OsHKT2;1* and *OsHKT2;2*, was supposed to confer salt tolerance to the Nona Bokra rice cultivar by allowing the absorption of K$^+$ in roots under salt stress [136]. It has now been shown that OsHKT2;2/1 regulates the influx of Na$^+$ into the roots of plants exposed to salt stress [137]. Note that the constitutive overexpression of *AtNHX1* in Arabidopsis improves salt tolerance [138]. Besides, overexpression of NHX1 in various transgenic plants, such as Brassica [139], cotton [17], maize [140], rice [141], tobacco [142], tomato [143], and wheat [144], exposed to NaCl concentrations ranging from 100 mM to 200 mM improve their tolerance to salt stress. The induction of NHX1 and NHX2 in response to salt stress depends on ABA [145,146]. It is widely known that, during salt stress, NHX activity increases, which promotes salt stress tolerance in many plants [147]. AtCHX21, expressed in the endodermal cells of the roots and its mutants, subjected to salt stress, accumulate less Na$^+$ in their xylem and leaves sap, indicating that CHX21 could be involved in the transport of Na$^+$ through the endoderm in the stele [148]. Under moderately saline conditions, SOS1 most likely occurs in the xylem load of Na$^+$, due to the fact that Na$^+$ accumulates to a lesser extent in sos1 mutants [149]. In high salinity conditions, the xylem load of Na$^+$ is probably a passive process because a high concentration of cytosolic Na$^+$ in xylem parenchyma cells and a comparatively depolarized plasma membrane would favor the movement of Na$^+$ in the xylem [150]. Plants can recover xylem Na$^+$ from root cells to avoid high concentrations of Na$^+$ in aerial tissues [151]. This recovery has been observed in the basal regions of the roots and shoots of plants such as maize, beans, and soybeans [2,65]. In *Arabidopsis*, the *HKT1* mutation renders the mutants hypersensitive to salt stress and causes a greater accumulation of Na$^+$ in the leaves [152–154]. Inactivation lines have higher levels of Na$^+$ but low levels of K$^+$ in shoots. These results show that AtHKT1 is involved in the recovery of Na$^+$ from xylem while directly stimulating the load of K$^+$. This is one of the mechanisms to maintain a higher K$^+$/Na$^+$ ratio in shoots during salt stress in plants [155]. Synergistic effects of *SOS1*, *HKT1;5*, and *NHX1* have been proposed to regulate Na$^+$ homeostasis in *Puccinellia tenuiflora*, a halophytic plant [156]. The NaCl stress–induced vacuolar compartmentalization of its xylem load has been attributed to regulation by the differential expression of *NHX1* and *HKT1;5*. The NaCl stress-induced expression of SOS1 and NHX1 in the roots would also have been more effective in excluding Na$^+$ and Cl$^-$ in the intertidal population of *Suaeda salsa* [157]. The genetic or environmental variation of salt tolerance among halophyte populations is related to the differential expression of Na$^+$ efflux channels. Detailed structural analysis of HKT1;5 was performed in *Triticum sp.* [158]. Variations in its amino acid sequences result in a change in Na$^+$ affinity and a subsequent change in salt tolerance in two species of *Triticum*. Comparative analysis of antioxidant mechanisms in *Cynodon dactylon* (salt-tolerant grass) and *Oryza sativa* (salt-sensitive plant) was corroborated by the high expression levels of SOS 1 and NHX1 transporters in Cynodon [159]. Salt tolerance in barley has been attributed to the regulation of Na$^+$ loading in root xylem elements [160]. This is controlled by a cross between reactive oxygen species (ROS), nicotinamide adenine dinucleotide phosphate oxidase (NADPH oxidase), Ca^{2+}, and K$^+$.

8. Calcineurin B–Like Proteins (CBL) and CBL-Interacting Protein Kinases (CIPK) and Salt Tolerance in Plants

Calcium serves as a pivotal messenger in many adaptation and developmental processes. Cellular calcium signals are detected and transmitted by sensor molecules such as calcium-binding proteins. In plants, the calcineurin B-like protein (CBL) family seems to be a unique group of calcium sensors and plays a key role in decoding calcium transients by specifically interacting with and regulating a family of protein kinases (CIPKs) [161]. Several CBL proteins appear to be targeted to the plasma membrane by processes of dual lipid modification by myristoylation and S-acylation. Additionally, CBL/CIPK complexes have been identified in other cellular localizations, suggesting that this network may confer spatial specificity in Ca^{2+} signaling.

Molecular genetics analysis of loss-of-function mutants involves various CBL proteins and CIPKs as important components of abiotic stress responses, hormone reactions, and ion transport processes. The event of CBL and CIPK proteins appears not to be restricted to plants, raising the question about the function of these Ca^{2+} decoding components in non-plant species.

8.1. Organization of the CBL–CIPK Network

CBL proteins have been initially identified from *Arabidopsis thaliana* [162]. Bioinformatics and comparative genomic analysis in plants have provided details about the sequence specificity, conservation, function, and complexity, and ancestry of CBL and CIPK proteins families from lower plants to higher plants. Bioinformatics research reports showed that *Arabidopsis thaliana* has 10 CBLs and 26 CIPKs [163], while in other plants, *Populus trichocarpa* has 10 CBLs and 27 CIPKs [164], *Oryza sativa* has 10 CBLs and 31 CIPKs [165], *Zea mays* has 8 CBLs and 43 CIPKs [165], *Vitis vinifera* has eight CBLs and 21 CIPKs [166], *Sorghum bicolor* has 6 CBLs and 32 CIPKs [166], *Glycine max* has 52 CIPKs [62], and *Brassica rapa L.* (Chinese cabbage) has 17 CBL genes [167].

All CBL proteins share a rather conserved core region consisting of four EF-hand calcium-binding domains that are separated by spacing regions encompassing an absolutely conserved number of amino acids in all CBL Proteins [161].

In contrast to CNB from animals and fungi, CBLs do not interact with a PP2B-type phosphatase that appears to be absent in plants.

Instead, CBL proteins interact with a group of serine-threonine protein kinases that evolutionary belong to the superfamily of CaM-dependent kinases (CaMKs) and form a phylogenetically separate cluster within the group of SNF1 related kinases. Therefore, this group has also been indicated as *Snf1* related kinase group 3 (*SnRK3*; [168]. As in other kinases of the CaMK group, the kinase domain in CIPKs is segregated by a domain called "junction domain" from the less-conserved C-terminal regulatory domain. Within the regulatory region of CIPKs, a conserved NAF domain (designated according to the prominent amino acids N, A and F) mediates binding of CBL proteins and simultaneously functions as an auto-inhibitory domain [169]. Binding of CBLs to the NAF motif removes the auto-inhibitory domain from the kinase domain, thereby conferring auto-phosphorylation and activation of the kinase [170]. Additional phosphorylation of the activation loop within the kinase domain by a yet unknown kinase further contributes to the activation of CIPKs [171].

Kinases related to CIPKs, like the AMP-activated protein kinase (AMPK), are dephosphorylated by type 2C protein phosphatases (PP2C) [172]. Interestingly, CIPKs can associate with PP2Cs like ABI1 and ABI2 via a C-terminal protein-phosphatase interaction (PPI) domain [173]. Currently, it is not known if PP2Cs may dephosphorylate CIPKs or if phosphorylation of PP2Cs by CIPKs occurs in vivo. Alternatively, the generation of CIPK/PP2C complexes could serve the formation of signaling kinase/phosphatase modules allowing for rapid alternating phosphorylation/dephosphorylation of target proteins.

In this regard, crystallization studies of CBL4 in complex with the regulatory domain of CIPK24 suggest that either CBLs or PP2Cs may mutually exclusively interact with the regulatory domain of CIPKs, and that formation of a trimeric complex is unlikely [174]. Therefore, it is tempting to speculate that PP2C interaction with the PPI domain of CIPKs leads to competitive replacing of the CBL protein, which binds to the NAF and partly to the PPI domain. Dissociation of the CBL protein would release the otherwise masked auto-inhibitory domain of the CIPK resulting in inactivation of the kinase. Alternative Ca^{2+}-dependent binding of CBL proteins to the CIPK would favor phosphorylation of a given substrate by out-competing the PP2C from the complex. However, as the target stowage domains are still unknown for CIPKs and PP2Cs, such models can currently not consider the influence of substrate binding.

Interestingly, the PPI domain was shown to be structurally related to the kinase-associated domain1 (KA1) of the kinase KIN2/PAR-1/MARK subfamily [174,175]. Moreover, SnRK1, the SNF1 homologous in plants, also contains such a structural domain [175]. Although the function of this domain is not known, this finding may point to a mechanism of protein regulation that is conserved from animals to plants [174,175].

8.2. Mechanisms of CBL-CIPK Pathway

Structural characteristics of CBL and CIPK proteins provide the basis for their interaction. The crystal structure of the complex of Ca^{2+}-CBL4 with the C-terminal regulatory domain of CIPK24 was first resolved [174]. It reveals how the CBL-CIPK complex decodes intracellular Ca^{2+} signals provoked by extracellular stimulation [176]. The CBL protein harbors four elongation factor hands (EF-hands), and each EF-hand contains a conserved α-helix-loop-α-helix structure responsible for Ca^{2+} binding [163]. The EF-hands are organized in fixed spaces that are 22, 25, and 32 amino acids distant from EF1 to EF4 in turn [177,178]. The loop region is characterized by a consensus sequence of 12 residues DKDGDGKIDFEE [163]. Amino acids in positions 1 (X), 3 (Y), 5 (Z), 7 (−X), 9 (−Y), and 12 (−Z) are responsible for Ca^{2+} coordination [176]. EF1 contains an insertion of two amino acid residues between position X and position Y. Variation of amino acids in these positions causes the change of Ca^{2+}-binding affinity [163]. Amino acid residues of CBL4 at positions X, Y, Z, and −Z bind Ca^{2+} depending on side-chain donor oxygen, while backbone carbonyl oxygen atom and water facilitation are used at positions −Y and −X, respectively [176].

The CIPK protein consists of two domains, one is the conserved N-terminal kinase catalytic domain, which comprises a phosphorylation site-containing activation loop, and the other is the highly variant C-terminal regulatory domain harboring NAF/FISL motif and a phosphatase interaction motif (PPI) [170]. The NAF motif, named by its highly conserved amino acids Asn (N), Ala (A), Phe (F), Ile (I), Ser (S), and Leu (L), is necessary for binding CBL protein. This motif is necessary for sustaining the interaction between CIPK24 and CBL4 and is able to attach the C-terminal regulatory domain of CIPK24 to cover its activation loop for keeping the kinase in an auto-inhibited state (Figure 3) [179]. Attachment of Ca^{2+} by EF-hands leads to the modification of molecular surface properties of CBL4 [176] and supports CBL4 interact with CIPK24 via the NAF motif. The interaction triggers the conformational changes of CIPK24 and exposes its activation loop [180]. Once the activation loop is free, the auto-inhibited CIPK24 is phosphorylated by an unknown upstream kinase and activates CIPK24. Subsequently, the activated CIPK24 phosphorylates the Na^+/H^+ exchanger SOS1 on the PM to exclude the excess Na^+ from the cell (Figure 3a) [180]. Abscisic acid-insensitive 2 (ABI2), a member of protein phosphatase 2C (PP2C), was identified as a CIPK24-interacting phosphatase [179]. The salt-tolerant phenotype of *abi2* indicated that ABI2 is a negative regulator of CIPK24 in the SOS pathway. Up to now, the blocking mechanism of ABI2 in the CBL4-CIPK24 pathway is not yet elucidated. It is assumed that ABI2 might function in the process of dephosphorylating SOS1 (Figure 3b) or CIPK24 (Figure 3c) [179].

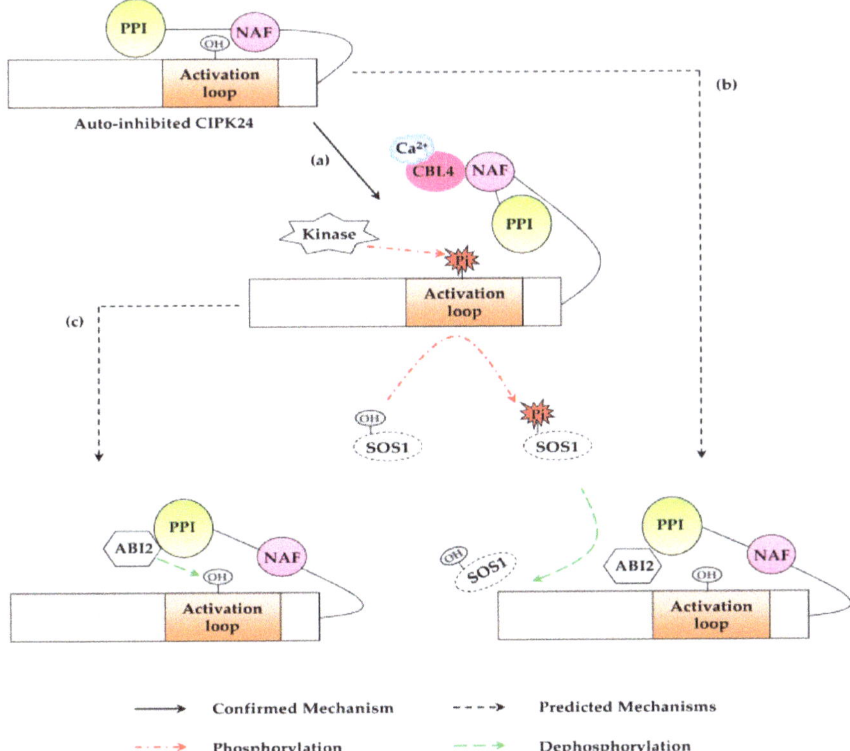

Figure 3. Mechanism of Calcineurin B-like protein 4 (CBL4)-CBL–interacting protein kinase (CIPK24) signaling pathway. (**a**) The Ca^{2+}-binding CBL4 interacts with the NAF motif of CIPK24 and changes the conformation of CIPK24. CIPK24 exposes its activation loop and then is phosphorylated by an unknown upstream protein kinase. Activated CIPK24 phosphorylates and stimulates salt overly sensitive 1 (SOS1), subsequently. (**b**) Abscisic acid-insensitive 2 (ABI2) binds to the phosphatase interaction (PPI) domain of CIPK24 and dephosphorylates SOS1 which was phosphorylated by CIPK24. (**c**) Activated CIPK24 is dephosphorylated by ABI2, and its activity is inhibited. (Adapted from Mao et al. (2016)).

8.3. Physiological Roles of CBLs and CIPKs in Plant Responses to Abiotic Stress Signals

The physiological roles of CBL and CIPK were firstly uncovered in salt overly sensitive (SOS) pathway (Figure 4) [180]. The *Arabidopsis* mutants *sos1*, *sos2*, and *sos3* produced the same salt-sensitive phenotype under high-salt stress [181]. SOS3 and SOS2, also known as CBL4 and CIPK24 respectively, were demonstrated to synergistically up-regulate the activity of plasma membrane (PM)-located Na^+/H^+ exchanger SOS1 in *Arabidopsis*, leading to the Na^+ efflux from cells in the high-salt environment [180].

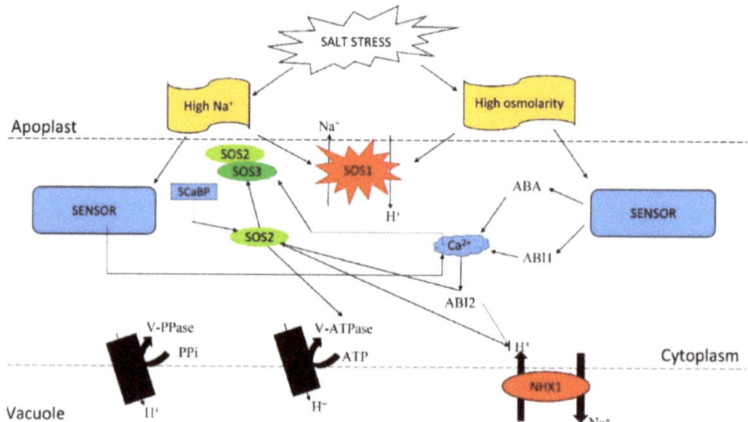

Figure 4. Signaling pathways responsible for Na$^+$ extrusion in Arabidopsis under salt stress. Excess Na$^+$ and high osmolarity are separately sensed by unknown sensors at the plasma membrane level, which then induce an increase in cytosolic Ca^{2+}. This increase is sensed by SOS3, which activates SOS2. The activated SOS3-SOS2 protein complex phosphorylates SOS1, the plasma membrane Na$^+$/H+ antiporter, resulting in the efflux of Na$^+$ ions. SOS2 can regulate NHX1 antiport activity and V-H+-ATPase activity independently of SOS3, possibly by SOS3-like Ca^{2+}-binding proteins (SCaBP) that target it to the tonoplast. Salt stress can also induce the accumulation of ABA, which, by means of ABI1 and ABI2, can negatively regulate SOS2 or SOS1 and NHX1.

It has been found that CBL-CIPK pathways work as regulators in nutrients transport systems, regulating sodium (Na$^+$) [180], potassium(K$^+$) [182], magnesium (Mg^{2+}) [183], nitrate (NO$_3^-$) [184], and proton (H$^+$) homeostasis [185]. Recently, in some reviews, particular attention to the possible involvement of the CBLs and CIPKs in different ions sensitivity has been drawn [186,187].

As calcium sensor relieves in plants, calcineurin B–like (CBL) proteins provide an important contribution to decoding Ca^{2+} signatures elicited by a variety of abiotic stresses. Currently, it is well known that CBLs perceive and transmit the Ca^{2+} signals mainly to a group of serine/threonine protein kinases called CBL-interacting protein kinases (CIPKs).

In the year 2016, Cho et al. reported that the CBL10 member of this family has a novel interaction partner besides the CIPK proteins. Yeast two-hybrid screening with CBL10 as bait identified an *Arabidopsis* cDNA clone encoding a TOC34 protein, which is a member of the translocon of the outer membrane of chloroplasts (TOC) complex and possesses the GTPase activity. Bimolecular fluorescence complementation (BiFC) analysis verified that the CBL10–TOC34 interaction takes place at the outer membrane of chloroplasts in vivo and thus decreases its GTPase activity in *Arabidopsis* [188].

These findings indicate that a member of the CBL family, CBL10, can modulate not only the CIPK members but also TOC34, allowing the CBL family to relay the Ca^{2+} signals in more diverse ways than currently known.

In tomato, the calcium sensor Cbl10 and its interacting protein kinase *Cipk6* define a signaling pathway in plant immunity [189]. Ca^{2+} signaling is an early and necessary event in plant immunity. The tomato (*Solanum lycopersicum*) kinase *Pto* triggers localized programmed cell death (PCD) upon recognition of *Pseudomonas syringae* effectors AvrPto *or* AvrPtoB. In a virus-induced gene silencing screen in *Nicotiana benthamiana*, Fernando and al. identified two components of a Ca^{2+}-signaling system, Cbl10 (for calcineurin B-like protein) and *Cipk6* (for calcineurin B-like interacting protein kinase), as their silencing inhibited Pto/AvrPto-elicited PCD. *N. benthamiana* Cbl10 and Cipk6 are also required for PCD triggered by other plant resistance genes and virus, oomycete, and nematode effectors and for host susceptibility to two *P. syringae* pathogens.

Tomato *Cipk6* interacts with Cbl10 and its in vitro kinase activity is enhanced in the presence of *Cbl10* and Ca^{2+}, suggesting that tomato Cbl10 and Cipk6 constitute a Ca^{2+}-regulated signaling module. Overexpression of tomato *Cipk6* in *N. benthamiana* leaves causes accumulation of reactive oxygen species (ROS), which requires the respiratory burst homolog *RbohB*. Tomato Cbl10 and Cipk6 interact with *RbohB* at the plasma membrane. Finally, Cbl10 and Cipk6 contribute to ROS generated during effector-triggered immunity in the interaction of *P. syringae* pv tomato DC3000 and *N. benthamiana*. The role of the Cbl/Cipk signaling module in PCD has been identified, establishing a mechanistic link between Ca^{2+} and ROS signaling in plant immunity [189].

Xu et al. showed that the protein kinase CIPK23, encoded by the Arabidopsis Low-K^+-sensitive 1 (*LKS1*) gene, regulates K^+ uptake under low K^+ conditions. Lesion of *LKS1* has reduced K^+ uptake and caused leaf chlorosis and growth inhibition, whereas overexpression of *LKS1* significantly enhanced K^+ uptake and tolerance to low K^+. They demonstrated that *CIPK23* directly phosphorylates the K^+ transporter *AKT1* and further found that CIPK23 is activated by the binding of two calcineurin B-like proteins, CBL1 and CBL9 [55]. Further research on protein kinase *CIPK23* in *Arabidopsis* has revealed that CIPK23 is expressed in a variety of cell types and tissues and regulates distinct physiological processes including the opening/closing of stomata in the leaves, and the potassium uptake in the roots [190]. In addition, the authors showed that CIPK23 kinase interacts and functions with both CBL1 and CBL9 calcium sensors, providing a molecular link between intracellular calcium fluctuations and the regulation of transpiration and nutrient uptake. CBL1 and CBL9 can both recruit CIPK23 on the plasma membrane, suggesting that CIPK23-CBL complexes associated with the plasma membrane modulate the membrane on which the target proteins are located, including the AKT1 potassium channel [190,191] by proteins phosphorylation. Cheong et al provided more information on the mechanistic aspects of calcium signaling by plants. According to their finds, the combination of CIPK23 with a specific set of other components in the guard cells results in the regulation of the stomatal response to ABA, while CIPK23 and another set of components in the root tissues participate in the regulation of potassium absorption. Since CIPK23 is also present in other tissues, such as vascular tissues of roots, stems, and leaves, the authors hypothesized that CIPK23 could also be associated with other components of these tissues, for example during long-distance transport and distribution of K^+ throughout the plant. They showed that the other components that interact with CIPK23 include the CBL1 and CBL9 calcium sensors that functionally overlap in regulating stomatal movement and K^+ uptake. It is possible that other CBLs may also interact with CIPK23 in regulating K^+ nutrition. Such selective and overlapping interactions can encode unique responses that are different from any CBL–CIPK interaction. Among the CBLs that regulate a specific CIPK in the same process, some may play a more dominant role than others. For example, the functions of CIPK23 in stomatal response and K^+ absorption appear to be primarily regulated by CBL1 and CBL9, each functioning in other processes by regulating other CIPKs [190,192].

Hashimoto et al. have identified a novel general regulatory mechanism of CBL-CIPK complexes in that CBL phosphorylation at their flexible C-terminus probably induces conformational changes that enhance specificity and activity of CBL-CIPK complexes toward their target proteins. The phosphorylation status of CBLs does not appear to influence the stability, localization, or CIPK interaction of these calcium sensor proteins in general. However, proper phosphorylation of CBL1 is absolutely required for the in vivo activation of the AKT1, K^+ channel by CBL1-CIPK23 and CBL9-CIPK23 complexes in oocytes [190,193]. Moreover, the authors have shown that, by combining CBL1, CIPK23, and AKT1, the reconstituted CBL-dependent enhancement of phosphorylation of target proteins by CIPKs in vitro. In addition, they reported that phosphorylation of CBL1 by CIPK23 is also required for the CBL1-dependent enhancement of CIPK23 activity toward its substrate.

Recent studies have uncovered the crucial functions of CBL-CIPK complexes in an increasing number of biological processes like salt tolerance, potassium transport, nitrate sensing, and stomatal regulation [194]. CBL proteins determine the cellular localization of their interacting protein kinases in vivo and are essential for the activity of the resulting CBL-CIPK complexes toward their target proteins [55,184]. Despite the established importance of CBL-CIPK complexes in regulating the activity of ion channels and transporters like SOS1, AKT1, AKT2, and NRT1 [195], only very few target phosphorylation sites of CIPKs have been clearly identified. The occurrence of phosphorylation of CBLs by CIPKs appears not to be restricted to the model organism *Arabidopsis*.

In 2017, it was reported that *BdCIPK31*, a CIPK gene from *Brachypodium distachyon*, functions positively to drought and salt stress through the ABA signaling pathway [196]. Overexpressing *BdCIPK31* functions in stomatal closure, ion homeostasis, ROS scavenging, osmolyte biosynthesis, and transcriptional regulation of stress-related genes. In fact, it appears that transgenic tobacco plants overexpressing *BdCIPK31* presented improved drought and salt tolerance and displayed hypersensitive response to exogenous ABA [196]. Further investigations revealed that *BdCIPK31* functioned positively in ABA-mediated stomatal closure, and transgenic tobacco exhibited reduced water loss under dehydration conditions compared with the controls. *BdCIPK31* also affected Na^+/K^+ homeostasis and root K^+ loss, which contributed to maintaining intracellular ion homeostasis under salt conditions. Moreover, the reactive oxygen species scavenging system and osmolyte accumulation were enhanced by *BdCIPK31* overexpression, which was conducive for alleviating oxidative and osmotic damages. Additionally, overexpression of *BdCIPK31* could elevate several stress-associated gene expressions under stress conditions [196].

In 2013, *TaCIPK14* and *TaCIPK29* were found to confer single or multiple stress tolerance in transgenic tobacco [197]. Transgenic tobaccos overexpressing TaCIPK14 exhibited higher contents of chlorophyll and sugar, higher catalase activity, while decreased amounts of H_2O_2 and malondialdehyde (MDA), and lesser ion leakage under cold and salt stresses. In addition, overexpression also enhanced the seed germination rate, root elongation and decreased Na^+ content in the transgenic lines under salt stress. Higher expression of stress-related genes was observed in lines overexpressing *TaCIPK14* compared to controls under stress conditions [197].

Under conditions of high salinity, *TaCIPK25* expression was markedly down-regulated in wheat roots [198]. Overexpression of *TaCIPK25* resulted in hypersensitivity to Na^+ and superfluous accumulation of Na^+ in transgenic wheat lines. The *TaCIPK25* expression did not decline in transgenic wheat and remained at an even higher level than that in wild-type wheat controls under high-salinity treatment. Furthermore, the transmembrane Na^+/H^+ exchange was impaired in the root cells of transgenic wheat. These results suggested that *TaCIPK25* negatively regulated salt response in wheat [198].

9. Conclusions and Perspectives

The data available on the CHX family in *Arabidopsis* and other plants clearly highlight a novel and original mechanism involved in plants' tolerance to the salinity. This mechanism, which was previously not demonstrated in plants, allows detoxification of Na^+ in leaves by recirculation of this ion to the roots via the phloem. Plants face a dilemma regarding the transport of sodium. Sodium absorption is useful for lowering osmotic potential, being able to absorb water and maintaining turgor, but excess sodium is toxic. Many studies have focused on the toxic role of Na^+ in the plant during salt stress and the elucidation of the mechanisms of tolerance to this stress.

The role of Na^+ at lower concentrations is not well known. The current consensus is that the energization of the cell membrane is based solely on a proton gradient. However, the available data for some *CHXs* encourage us to continue to imagine that Na^+ (at non-toxic concentrations) can lead to symport systems and energize active K^+ uptake. Several indices seem to support this hypothesis, for example, AtNHX23 an antiport K^+ (Na^+/H^+) active at the level of the chloroplast envelope and involved in potassium homeostasis and perhaps in regulating the pH of the stroma. However, the Na^+/H^+ antiport systems of the plasma membrane are still poorly characterized. The available information is only related to the SOS1 protein in *Arabidopsis*, which has a homologous sequence with antiport Na^+/H^+ and would be involved in sodium efflux at the plasmalemma level. *SOS1* overexpression in *Arabidopsis* significantly improves plants' tolerance to salinity. AtSOS1 is, therefore, an important determinant of salt sensitivity in plants. AtSOS1 activity is controlled by AtSOS2 and AtSOS3. AtSOS3 (a Ca^{2+} affine protein belonging to the CBL family) directly interacts with AtSOS2, which is a serine/threonine protein kinase.

Studies on CBLs and CIPKs over the past decade have greatly advanced our knowledge of the function of single proteins in distinct physiological processes. Major advances in understanding this signaling system were through the identification of an increasing number of targets regulated by the CBL-CIPK complexes. The progress of the research on the CBL and CIPK families in different plant species other than *Arabidopsis thaliana* is still at an infant stage; in most cases, it is limited to interaction studies and expression analysis of these families.

The CBL-CIPK signaling model emphasizes the importance of future research that focuses on the molecular mechanisms underlying the regulation of transporters that allow us to better understand plant's response to abiotic stress such as salt stress and also establish a proficient method of identifying molecular targets for genetically engineered resistant crops with enhanced tolerance to various environmental stresses. Therefore, the most important challenge for future research is not only functional characterization but also the elucidating of the details of synergistic functions in this interaction network and revealing the molecular mechanisms of the complexes regulating target proteins.

Author Contributions: Conceptualization, H.-Y.L. and T.K.; T.K. wrote this manuscript. X.-W.L., F.-W.W., K.F.I.C., and M.N. participated in the writing and modification of this manuscript. Validation, X.-W.L. All authors read and approved the final manuscript.

Funding: This research was supported by the Special Program for Research of Transgenic Plants (2016ZX08010-002), the National Natural Science Foundation of China (31601323), the National Key Research and Development Program (2016YFD0101005), and the Natural Science Foundation of Science Technology Department of Jilin Province (20170101015JC, 20180101028JC, 20190201259JC).

Acknowledgments: The authors are grateful to Prof. Li Haiyan for the critical discussion of this article.

Conflicts of Interest: The authors declare no conflict of interest.

References

1.	Munns, R. Comparative physiology of salt and water stress. *Plant Cell Env.* **2002**, *25*, 239–250. [CrossRef]
2.	Tester, M.; Davenport, R. Na^+ tolerance and Na^+ transport in higher plants. *Ann. Bot.* **2003**, *91*, 503–527. [CrossRef] [PubMed]
3.	Flowers, T.J.; Colmer, T.D. Plant salt tolerance: Adaptations in halophytes. *Ann. Bot.* **2015**, *115*, 327–331. [CrossRef] [PubMed]
4.	Ruan, C.J.; da Silva, J.A.T.; Mopper, S.; Qin, P.; Lutts, S. Halophyte Improvement for a Salinized World. *Crit. Rev. Plant Sci.* **2010**, *29*, 329–359. [CrossRef]
5.	Apse, M.P.; Blumwald, E. Na^+ transport in plants. *Febs Lett.* **2007**, *581*, 2247–2254. [CrossRef]
6.	Munns, R. Genes and salt tolerance: Bringing them together. *New Phytol.* **2005**, *167*, 645–663. [CrossRef]
7.	Teakle, N.L.; Tyerman, S.D. Mechanisms of Cl-transport contributing to salt tolerance. *Plant Cell Environ.* **2010**, *33*, 566–589. [CrossRef]
8.	Ashraf, M.; Foolad, M. Roles of glycine betaine and proline in improving plant abiotic stress resistance. *Environ. Exp. Bot.* **2007**, *59*, 206–216. [CrossRef]
9.	Ksouri, R.; Falleh, H.; Megdiche, W.; Trabelsi, N.; Mhamdi, B.; Chaieb, K.; Bakrouf, A.; Magné, C.; Abdelly, C. Antioxidant and antimicrobial activities of the edible medicinal halophyte Tamarix gallica L. and related polyphenolic constituents. *Food Chem. Toxicol.* **2009**, *47*, 2083–2091. [CrossRef]
10.	Sairam, R.; Tyagi, A. Physiology and molecular biology of salinity stress tolerance in plants. *Curr. Sci.* **2004**, *86*, 407–421.
11.	Adler, G.; Blumwald, E.; Bar-Zvi, D. The sugar beet gene encoding the sodium/proton exchanger 1 (BvNHX1) is regulated by a MYB transcription factor. *Planta* **2010**, *232*, 187–195. [CrossRef] [PubMed]
12.	Chinnusamy, V.; Jagendorf, A.; Zhu, J.-K. Understanding and improving salt tolerance in plants. *Crop Sci.* **2005**, *45*, 437–448. [CrossRef]

13. Shabala, S.; Shabala, S.; Cuin, T.A.; Pang, J.; Percey, W.; Chen, Z.; Conn, S.; Eing, C.; Wegner, L.H. Xylem ionic relations and salinity tolerance in barley. *Plant J.* **2010**, *61*, 839–853. [CrossRef] [PubMed]

14. Majumder, A.; Parida, A.; Sankar, K.; Qureshi, Q. Utilization of food plant species and abundance of hanuman langurs (*Semnopithecus entellus*) in Pench Tiger Reserve, Madhya Pradesh, India. *Taprobanica J. Asian Biodivers.* **2011**, *2*, 105–108. [CrossRef]

15. Rontein, D.; Dieuaide-Noubhani, M.; Dufourc, E.J.; Raymond, P.; Rolin, D. The metabolic architecture of plant cells stability of central metabolism and flexibility of anabolic pathways during the growth cycle of tomato cells. *J. Biol. Chem.* **2002**, *277*, 43948–43960. [CrossRef] [PubMed]

16. Cuin, T.A.; Tian, Y.; Betts, S.A.; Chalmandrier, R.; Shabala, S. Ionic relations and osmotic adjustment in durum and bread wheat under saline conditions. *Funct. Plant Biol.* **2009**, *36*, 1110–1119. [CrossRef]

17. He, C.; Yan, J.; Shen, G.; Fu, L.; Holaday, A.S.; Auld, D.; Blumwald, E.; Zhang, H. Expression of an Arabidopsis vacuolar sodium/proton antiporter gene in cotton improves photosynthetic performance under salt conditions and increases fiber yield in the field. *Plant Cell Physiol.* **2005**, *46*, 1848–1854. [CrossRef]

18. Clarkson, D.T.; Hanson, J.B. The mineral nutrition of higher plants. *Ann. Rev. Plant Physiol.* **1980**, *31*, 239–298. [CrossRef]

19. White, P.J. Ion Uptake Mechanisms of Individual Cells and Roots: Short-Distance Transport. In *Marschner's Mineral Nutrition of Higher Plants*; Elsevier: Amsterdam, The Netherlands, 2012; pp. 7–47.

20. Gierth, M.; Mäser, P. Potassium transporters in plants–involvement in K^+ acquisition, redistribution and homeostasis. *FEBS Lett.* **2007**, *581*, 2348–2356. [CrossRef]

21. Cao, M.; Yu, H.; Yan, H.; Jiang, C. Difference in tolerance to potassium deficiency between two maize inbred lines. *Plant Prod. Sci.* **2007**, *10*, 42–46.

22. Chen, J.; Gabelman, W.H. Morphological and physiological characteristics of tomato roots associated with potassium-acquisition efficiency. *Sci. Hortic.* **2000**, *83*, 213–225. [CrossRef]

23. Rodríguez-Navarro, A. Potassium transport in fungi and plants. *Biochim. Et Biophys. ActaRev. Biomembr.* **2000**, *1469*, 1–30. [CrossRef]

24. Thomson, S.J.; Hansen, A.; Sanguinetti, M.C. Identification of the intracellular Na^+ sensor in Slo2. 1 potassium channels. *J. Biol. Chem.* **2015**, *290*, 14528–14535. [CrossRef] [PubMed]

25. Assaha, D.V.; Ueda, A.; Saneoka, H.; Al-Yahyai, R.; Yaish, M.W. The role of Na^+ and K^+ transporters in salt stress adaptation in glycophytes. *Front. Physiol.* **2017**, *8*, 509. [CrossRef]

26. Blaha, G.; Stelzl, U.; Spahn, C.M.T.; Agrawal, R.K.; Frank, J.; Nierhaus, K.H. Preparation of functional ribosomal complexes and effect of buffer conditions on tRNA positions observed by cryoelectron microscopy. *Methods Enzymol.* **2000**, *397*, 292–306.

27. Maathuis, F.J. Sodium in plants: Perception, signalling, and regulation of sodium fluxes. *J. Exp. Bot.* **2013**, *65*, 849–858. [CrossRef]

28. Figdore, S.S.; Gabelman, W.; Gerloff, G. The Accumulation and Distribution of Sodium in Tomato Strains Differing in Potassium Efficiency When Grown under Low-K Stress. In *Genetic Aspects of Plant Mineral Nutrition*; Springer: Berlin, Germany, 1987; pp. 353–360.

29. Tahal, R.; Mills, D.; Heimer, Y.; Tal, M. The relation between low K^+/Na^+ ratio and salt-tolerance in the wild tomato species Lycopersicon pennellii. *J. Plant Physiol.* **2000**, *157*, 59–64. [CrossRef]

30. Marschner, H.; Kuiper, P.; Kylin, A. Genotypic differences in the response of sugar beet plants to replacement of potassium by sodium. *Physiol. Plant.* **1981**, *51*, 239–244. [CrossRef]

31. Subbarao, G.; Wheeler, R.M.; Stutte, G.W.; Levine, L.H. How far can sodium substitute for potassium in red beet? *J. Plant Nutr.* **1999**, *22*, 1745–1761. [CrossRef]

32. Horie, T.; Costa, A.; Kim, T.H.; Han, M.J.; Horie, R.; Leung, H.-Y.; Miyao, A.; Hirochika, H.; An, G.; Schroeder, J.I. Rice OsHKT2; 1 transporter mediates large Na^+ influx component into K^+-starved roots for growth. *EMBO J.* **2007**, *26*, 3003–3014. [CrossRef]

33. Ghars, M.A.; Parre, E.; Debez, A.; Bordenave, M.; Richard, L.; Leport, L.; Bouchereau, A.; Savouré, A.; Abdelly, C. Comparative salt tolerance analysis between Arabidopsis thaliana and Thellungiella halophila, with special emphasis on K^+/Na^+ selectivity and proline accumulation. *J. Plant Physiol.* **2008**, *165*, 588–599. [CrossRef] [PubMed]

34. Munns, R.; Tester, M. Mechanisms of salinity tolerance. Annu. *Rev. Plant Biol.* **2008**, *59*, 651–681. [CrossRef] [PubMed]

35. Brini, F.; Hanin, M.; Mezghani, I.; Berkowitz, G.A.; Masmoudi, K. Overexpression of wheat $Na^+/H+$ antiporter TNHX1 and H^+-pyrophosphatase TVP1 improve salt-and drought-stress tolerance in Arabidopsis thaliana plants. *J. Exp. Bot.* **2007**, *58*, 301–308. [CrossRef] [PubMed]

36. Demidchik, V.; Straltsova, D.; Medvedev, S.S.; Pozhvanov, G.A.; Sokolik, A.; Yurin, V. Stress-induced electrolyte leakage: The role of K^+-permeable channels and involvement in programmed cell death and metabolic adjustment. *J. Exp. Bot.* **2014**, *65*, 1259–1270. [CrossRef]

37. Qie, L.; Lewis, S.L.; Sullivan, M.J.P.; Lopez-Gonzalez, G.; Pickavance, G.C.; Sunderland, T.; Ashton, P.; Hubau, W.; Abu Salim, K.; Aiba, S.-I.; et al. Long-term carbon sink in Borneo's forests halted by drought and vulnerable to edge effects. *Nat. Commun.* **2017**, *8*, 1966. [CrossRef]

38. Clarkson, D.T. Roots and the delivery of solutes to the xylem. Philosophical Transactions of the Royal Society of London. *Ser. B Biol. Sci.* **1993**, *341*, 5–17.

39. Hossain, M.R.; Pritchard, J.; Ford-Lloyd, B.V. Qualitative and quantitative variation in the mechanisms of salinity tolerance determined by multivariate assessment of diverse rice (*Oryza sativa* L.) genotypes. *Plant Genet. Resour.* **2016**, *14*, 91–100. [CrossRef]

40. Robbins, N.E.; Trontin, C.; Duan, L.; Dinneny, J.R. Beyond the barrier: Communication in the root through the endodermis. *Plant Physiol.* **2014**, *166*, 551–559. [CrossRef]

41. Flowers, T.J.; Munns, R.; Colmer, T.D. Sodium chloride toxicity and the cellular basis of salt tolerance in halophytes. *Ann. Bot.* **2014**, *115*, 419–431. [CrossRef]

42. Anil, V.S.; Krishnamurthy, H.; Mathew, M. Limiting cytosolic Na^+ confers salt tolerance to rice cells in culture: A two-photon microscopy study of SBFI-loaded cells. *Physiol. Plant.* **2007**, *129*, 607–621. [CrossRef]

43. Flam-Shepherd, R.; Huynh, W.Q.; Coskun, D.; Hamam, A.M.; Britto, D.T.; Kronzucker, H.J. Membrane fluxes, bypass flows, and sodium stress in rice: The influence of silicon. *J. Exp. Bot.* **2018**, *69*, 1679–1692. [CrossRef] [PubMed]

44. Yeo, A.; Flowers, S.; Rao, G.; Welfare, K.; Senanayake, N.; Flowers, T. Silicon reduces sodium uptake in rice (*Oryza sativa* L.) in saline conditions and this is accounted for by a reduction in the transpirational bypass flow. *Plant Cell Environ.* **1999**, *22*, 559–565. [CrossRef]

45. Farooq, M.A.; Saqib, Z.A.; Akhtar, J. Silicon-mediated oxidative stress tolerance and genetic variability in rice (*Oryza sativa* L.) grown under combined stress of salinity and boron toxicity. *Turk. J. Agric. For.* **2015**, *39*, 718–729. [CrossRef]

46. Garcia, A.; Rizzo, C.; Ud-Din, J.; Bartos, S.; Senadhira, D.; Flowers, T.; Yeo, A. Sodium and potassium transport to the xylem are inherited independently in rice, and the mechanism of sodium: Potassium selectivity differs between rice and wheat. *Plant Cell Environ.* **1997**, *20*, 1167–1174. [CrossRef]

47. Davis, L.; Sumner, M.; Stasolla, C.; Renault, S. Salinity-induced changes in the root development of a northern woody species, Cornus sericea. *Botany* **2014**, *92*, 597–606. [CrossRef]

48. Keisham, M.; Mukherjee, S.; Bhatla, S. Mechanisms of sodium transport in plants—progresses and challenges. *Int. J. Mol. Sci.* **2018**, *19*, 647. [CrossRef]

49. Yeo, A.; Yeo, M.; Flowers, T. The contribution of an apoplastic pathway to sodium uptake by rice roots in saline conditions. *J. Exp. Bot.* **1987**, *38*, 1141–1153. [CrossRef]

50. Barragán, V.; Leidi, E.O.; Andrés, Z.; Rubio, L.; De Luca, A.; Fernández, J.A.; Cubero, B.; Pardo, J.M. Ion exchangers NHX_1 and NHX_2 mediate active potassium uptake into vacuoles to regulate cell turgor and stomatal function in Arabidopsis. *Plant Cell* **2012**, *24*, 1127–1142.

51. Wang, Y.; Wu, W.-H. Potassium transport and signaling in higher plants. *Annu. Rev. Plant Biol.* **2013**, *64*, 451–476. [CrossRef]

52. Glass, A.D.; Fernando, M. Homeostatic processes for the maintenance of the K^+ content of plant cells: A model. *Isr. J. Bot.* **1992**, *41*, 145–166.

53. Glass, A.D.; Dunlop, J. The influence of potassium content on the kinetics of potassium influx into excised ryegrass and barley roots. *Planta* **1978**, *141*, 117–119. [CrossRef] [PubMed]

54. Pyo, Y.J.; Gierth, M.; Schroeder, J.I.; Cho, M.H. High.-affinity K^+ transport in Arabidopsis: AtHAK5 and AKT1 are vital for seedling establishment and postgermination growth under low-potassium conditions. *Plant Physiol.* **2010**, *153*, 863–875. [CrossRef] [PubMed]

55. Xu, J.; Li, H.-D.; Chen, L.-Q.; Wang, Y.; Liu, L.-L.; He, L.; Wu, W.-H. A protein kinase, interacting with two calcineurin B-like proteins, regulates K^+ transporter AKT1 in Arabidopsis. *Cell* **2006**, *125*, 1347–1360. [CrossRef] [PubMed]

56. Coskun, D.; Britto, D.T.; Kronzucker, H.J. Regulation and mechanism of potassium release from barley roots: An in planta42K$^+$ analysis. *New Phytol.* **2010**, *188*, 1028–1038. [CrossRef]

57. Zeng, J.; Quan, X.; He, X.; Cai, S.; Ye, Z.; Chen, G.; Zhang, G. Root and leaf metabolite profiles analysis reveals the adaptive strategies to low potassium stress in barley. *BMC Plant Biol.* **2018**, *18*, 187. [CrossRef] [PubMed]

58. Santa-María, G.E.; Rubio, F.; Dubcovsky, J.; Rodríguez-Navarro, A. The HAK1 gene of barley is a member of a large gene family and encodes a high-affinity potassium transporter. *Plant Cell* **1997**, *9*, 2281–2289.

59. Wang, T.-B.; Gassmann, W.; Rubio, F.; Schroeder, J.I.; Glass, A.D. Rapid up-regulation of HKT1, a high-affinity potassium transporter gene, in roots of barley and wheat following withdrawal of potassium. *Plant Physiol.* **1998**, *118*, 651–659. [CrossRef]

60. Kim, E.J.; Kwak, J.M.; Uozumi, N.; Schroeder, J.I. AtKUP1: An Arabidopsis gene encoding high-affinity potassium transport activity. *Plant Cell* **1998**, *10*, 51–62. [CrossRef]

61. Sharma, T.; Dreyer, I.; Riedelsberger, J. The role of K$^+$ channels in uptake and redistribution of potassium in the model plant Arabidopsis thaliana. *Front. Plant Sci.* **2013**, *4*, 224. [CrossRef]

62. Zhu, K.; Chen, F.; Liu, J.; Chen, X.; Hewezi, T.; Cheng, Z.-M.M. Evolution of an intron-poor cluster of the CIPK gene family and expression in response to drought stress in soybean. *Sci. Rep.* **2016**, *6*, 28225. [CrossRef]

63. Adem, G.D.; Roy, S.J.; Zhou, M.; Bowman, J.P.; Shabala, S. Evaluating contribution of ionic, osmotic and oxidative stress components towards salinity tolerance in barley. *Bmc Plant Biol.* **2014**, *14*, 113. [CrossRef] [PubMed]

64. Nardini, A.; Salleo, S.; Jansen, S. More than just a vulnerable pipeline: Xylem physiology in the light of ion-mediated regulation of plant water transport. *J. Exp. Bot.* **2011**, *62*, 4701–4718. [CrossRef] [PubMed]

65. De Boer, A.; Volkov, V. Logistics of water and salt transport through the plant: Structure and functioning of the xylem. *Plant Cell Environ.* **2003**, *26*, 87–101. [CrossRef]

66. Zhu, J.; Liang, J.; Xu, Z.; Fan, X.; Zhou, Q.; Shen, Q.; Xu, G. Root aeration improves growth and nitrogen accumulation in rice seedlings under low nitrogen. *Aob Plants* **2015**, *7*, plv131. [CrossRef]

67. Kim, H.Y.; Choi, E.-H.; Min, M.K.; Hwang, H.; Moon, S.-J.; Yoon, I.; Byun, M.-O.; Kim, B.-G. Differential gene expression of two outward-rectifying shaker-like potassium channels OsSKOR and OsGORK in rice. *J. Plant Biol.* **2015**, *58*, 230–235. [CrossRef]

68. Shabala, S. Learning from halophytes: Physiological basis and strategies to improve abiotic stress tolerance in crops. *Ann. Bot.* **2013**, *112*, 1209–1221. [CrossRef]

69. Munns, R.; Lorraine Tonnet, M.; Shennan, C.; Anne Gardner, P. Effect of high external NaCl concentration on ion transport within the shoot of *Lupinus albus*. II. Ions in phloem sap. *PlantCell Environ.* **1988**, *11*, 291–300. [CrossRef]

70. Blom-Zandstra, M.; Vogelzang, S.A.; Veen, B.W. Sodium fluxes in sweet pepper exposed to varying sodium concentrations. *J. Exp. Bot.* **1998**, *49*, 1863–1868. [CrossRef]

71. Lohaus, G.; Hussmann, M.; Pennewiss, K.; Schneider, H.; Zhu, J.J.; Sattelmacher, B. Solute balance of a maize (*Zea mays* L.) source leaf as affected by salt treatment with special emphasis on phloem retranslocation and ion leaching. *J. Exp. Bot.* **2000**, *51*, 1721–1732. [CrossRef]

72. Alfocea, F.P.; Balibrea, M.E.; Alarcón, J.J.; Bolarín, M.C. Composition of xylem and phloem exudates in relation to the salt-tolerance of domestic and wild tomato species. *J. Plant Physiol.* **2000**, *156*, 367–374. [CrossRef]

73. An, D.; Chen, J.-G.; Gao, Y.-Q.; Li, X.; Chao, Z.-F.; Chen, Z.-R.; Li, Q.-Q.; Han, M.-L.; Wang, Y.-L.; Wang, Y.-F. AtHKT1 drives adaptation of Arabidopsis thaliana to salinity by reducing floral sodium content. *PLoS Genet.* **2017**, *13*, e1007086. [CrossRef] [PubMed]

74. Cuin, T.; Dreyer, I.; Michard, E. The role of potassium channels in Arabidopsis thaliana long distance electrical signalling: AKT2 modulates tissue excitability while GORK shapes action potentials. *Int. J. Mol. Sci.* **2018**, *19*, 926. [CrossRef] [PubMed]

75. Gajdanowicz, P.; Michard, E.; Sandmann, M.; Rocha, M.; Corrêa, L.G.G.; Ramírez-Aguilar, S.J.; Gomez-Porras, J.L.; González, W.; Thibaud, J.-B.; Van Dongen, J.T.; et al. Potassium (K$^+$) gradients serve as a mobile energy source in plant vascular tissues. *Proc. Natl. Acad. Sci. USA* **2011**, *108*, 864–869. [CrossRef] [PubMed]

76. Renault, S.; Lait, C.; Zwiazek, J.J.; MacKinnon, M. Effect of high salinity tailings waters produced from gypsum treatment of oil sands tailings on plants of the boreal forest. *Environ. Pollut.* **1998**, *102*, 177–184. [CrossRef]

77. Epstein, E.; Rains, D.; Elzam, O. Resolution of dual mechanisms of potassium absorption by barley roots. *Proc. Natl. Acad. Sci. USA* **1963**, *49*, 684. [CrossRef]

78. Hamouda, S.B.; Touati, K.; Amor, M.B. Donnan dialysis as membrane process for nitrate removal from drinking water: Membrane structure effect. *Arab. J. Chem.* **2017**, *10*, S287–S292. [CrossRef]

79. Nieves-Cordones, M.; Alemán, F.; Martínez, V.; Rubio, F. K^+ uptake in plant roots. The systems involved, their regulation and parallels in other organisms. *J. Plant Physiol.* **2014**, *171*, 688–695. [CrossRef]

80. Ali, A.; Maggio, A.; Bressan, R.A.; Yun, D.-J. Role and functional differences of HKT1-type transporters in plants under salt stress. *Int. J. Mol. Sci.* **2019**, *20*, 1059. [CrossRef]

81. Haro, R.; Bañuelos, M.A.; Senn, M.E.; Barrero-Gil, J.; Rodríguez-Navarro, A. HKT1 mediates sodium uniport in roots. Pitfalls in the expression of HKT1 in yeast. *Plant Physiol.* **2005**, *139*, 1495–1506. [CrossRef]

82. Pilot, G.; Pratelli, R.; Gaymard, F.; Meyer, Y.; Sentenac, H. Five-group distribution of the Shaker-like K^+ channel family in higher plants. *J. Mol. Evol.* **2003**, *56*, 418–434. [CrossRef]

83. Anderson, J.A.; Huprikar, S.S.; Kochian, L.V.; Lucas, W.J.; Gaber, R.F. Functional expression of a probable Arabidopsis thaliana potassium channel in Saccharomyces cerevisiae. *Proc. Natl. Acad. Sci. USA* **1992**, *89*, 3736–3740. [CrossRef] [PubMed]

84. Ahmad, I.; Mian, A.; Maathuis, F.J. Overexpression of the rice AKT1 potassium channel affects potassium nutrition and rice drought tolerance. *J. Exp. Bot.* **2016**, *67*, 2689–2698. [CrossRef] [PubMed]

85. Lebaudy, A.; Véry, A.-A.; Sentenac, H. K^+ channel activity in plants: Genes, regulations and functions. *Febs Lett.* **2007**, *581*, 2357–2366. [CrossRef] [PubMed]

86. Saponaro, A.; Porro, A.; Chaves-Sanjuan, A.; Nardini, M.; Rauh, O.; Thiel, G.; Moroni, A. Fusicoccin activates KAT1 channels by stabilizing their interaction with 14-3-3 proteins. *Plant Cell* **2017**, *29*, 2570–2580. [CrossRef]

87. Czempinski, K.; Zimmermann, S.; Ehrhardt, T.; Müller-Röber, B. New structure and function in plant K^+ channels: KCO1, an outward rectifier with a steep Ca^{2+} dependency. *Embo. J.* **1997**, *16*, 2565–2575. [CrossRef]

88. Czempinski, K.; Gaedeke, N.; Zimmermann, S.; Müller-Röber, B. Molecular mechanisms and regulation of plant ion channels. *J. Exp. Bot.* **1999**, *50*, 955–966. [CrossRef]

89. Demidchik, V.; Maathuis, F.J. Physiological roles of nonselective cation channels in plants: From salt stress to signalling and development. *New Phytol.* **2007**, *175*, 387–404. [CrossRef]

90. Forde, B.G.; Roberts, M.R. Glutamate receptor-like channels in plants: A role as amino acid sensors in plant defence? *F1000prime Rep.* **2014**, *6*, 37. [CrossRef]

91. Zhao, N.; Zhu, H.; Zhang, H.; Sun, J.; Zhou, J.; Deng, C.; Zhang, Y.; Zhao, R.; Zhou, X.; Lu, C. Hydrogen sulfide mediates K^+ and Na^+ homeostasis in the roots of salt-resistant and salt-sensitive poplar species subjected to NaCl stress. *Front. Plant Sci.* **2018**, *9*, 1366. [CrossRef]

92. Gamel, K.; Torre, V. The Interaction of Na^+ and K^+ in the Pore of CyclicNucleotide-Gated Channels. *Biophys. J.* **2000**, *79*, 2475–2493. [CrossRef]

93. Schuurink, R.C.; Shartzer, S.F.; Fath, A.; Jones, R.L. Characterization of a calmodulin-binding transporter from the plasma membrane of barley aleurone. *Proc. Natl. Acad. Sci. USA* **1998**, *95*, 1944–1949. [CrossRef] [PubMed]

94. Arazi, T.; Sunkar, R.; Kaplan, B.; Fromm, H. A tobacco plasma membrane calmodulin-binding transporter confers Ni^{2+} tolerance and Pb^{2+} hypersensitivity in transgenic plants. *Plant J.* **1999**, *20*, 171–182. [CrossRef] [PubMed]

95. Köhler, C.; Merkle, T.; Neuhaus, G. Characterisation of a novel gene family of putative cyclic nucleotide-and calmodulin-regulated ion channels in Arabidopsis thaliana. *Plant J.* **1999**, *18*, 97–104. [CrossRef] [PubMed]

96. Hanin, M.; Ebel, C.; Ngom, M.; Laplaze, L.; Masmoudi, K. New insights on plant salt tolerance mechanisms and their potential use for breeding. *Front. Plant Sci.* **2016**, *7*, 1787. [CrossRef] [PubMed]

97. Mian, A.A.; Senadheera, P.; Maathuis, F.J. Improving crop salt tolerance: Anion and cation transporters as genetic engineering targets. *Plant Stress* **2011**, *5*, 64–72.

98. Jin, Y.; Jing, W.; Zhang, Q.; Zhang, W. Cyclic nucleotide gated channel 10 negatively regulates salt tolerance by mediating Na^+ transport in Arabidopsis. *J. Plant Res.* **2015**, *128*, 211–220. [CrossRef] [PubMed]

99. Gobert, A.; Park, G.; Amtmann, A.; Sanders, D.; Maathuis, F.J. Arabidopsis thaliana cyclic nucleotide gated channel 3 forms a non-selective ion transporter involved in germination and cation transport. *J. Exp. Bot.* **2006**, *57*, 791–800. [CrossRef]

100. Mekawy, A.M.M.; Assaha, D.V.; Yahagi, H.; Tada, Y.; Ueda, A.; Saneoka, H. Growth, physiological adaptation, and gene expression analysis of two Egyptian rice cultivars under salt stress. *Plant Physiol. Biochem.* **2015**, *87*, 17–25. [CrossRef]

101. Moon, J.Y.; Belloeil, C.; Ianna, M.L.; Shin, R. Arabidopsis CNGC Family Members Contribute to Heavy Metal Ion Uptake in Plants. *Int. J. Mol. Sci.* **2019**, *20*, 413. [CrossRef]

102. Very, A.-A.; Sentenac, H. Molecular mechanisms and regulation of K$^+$ transport in higher plants. *Ann. Rev. Plant Biol.* **2003**, *54*, 575–603. [CrossRef]

103. Schleyer, M.; Bakker, E.P. Nucleotide sequence and 3′-end deletion studies indicate that the K(+)-uptake protein kup from Escherichia coli is composed of a hydrophobic core linked to a large and partially essential hydrophilic C terminus. *J. Bacteriol.* **1993**, *175*, 6925–6931. [CrossRef]

104. Banuelos, M.; Klein, R.; Alexander-Bowman, S.; Rodríguez-Navarro, A. A potassium transporter of the yeast Schwanniomyces occidentalis homologous to the Kup system of Escherichia coli has a high concentrative capacity. *EMBO J.* **1995**, *14*, 3021–3027. [CrossRef]

105. Banuelos, M.A.; Garciadeblas, B.; Cubero, B.; Rodriguez-Navarro, A. Inventory and functional characterization of the HAK potassium transporters of rice. *Plant Physiol.* **2002**, *130*, 784–795. [CrossRef]

106. Fu, H.-H.; Luan, S. AtKUP1: A dual-affinity K$^+$ transporter from Arabidopsis. *Plant Cell* **1998**, *10*, 63–73.

107. Quintero, F.J.; Blatt, M.R. A new family of K$^+$ transporters from Arabidopsis that are conserved across phyla. *Febs Lett.* **1997**, *415*, 206–211. [CrossRef]

108. Rigas, S.; Debrosses, G.; Haralampidis, K.; Vicente-Agullo, F.; Feldmann, K.A.; Grabov, A.; Dolan, L.; Hatzopoulos, P. TRH1 encodes a potassium transporter required for tip growth in Arabidopsis root hairs. *Plant Cell* **2001**, *13*, 139–151. [CrossRef]

109. Durell, S.R.; Guy, H.R. Structural models of the KtrB, TrkH, and Trk1, 2 symporters based on the structure of the KcsA K$^+$ channel. *Biophys. J.* **1999**, *77*, 789–807. [CrossRef]

110. Laurie, S.; Feeney, K.A.; Maathuis, F.J.M.; Heard, P.J.; Brown, S.J.; Leigh, R.A. A role for HKT1 in sodium uptake by wheat roots. *Plant J.* **2002**, *32*, 139–149. [CrossRef]

111. Rodríguez-Navarro, A.; Rubio, F. High.-affinity potassium and sodium transport systems in plants. *J. Exp. Bot.* **2006**, *57*, 1149–1160. [CrossRef]

112. Locascio, A.; Andrés-Colás, N.; Mulet, J.M.; Yenush, L. Saccharomyces cerevisiae as a Tool to Investigate Plant. Potassium and Sodium Transporters. *Int. J. Mol. Sci.* **2019**, *20*, 2133. [CrossRef]

113. Tada, Y. The HKT Transporter Gene from Arabidopsis, AtHKT1; 1, Is Dominantly Expressed in Shoot Vascular Tissue and Root Tips and Is Mild Salt Stress-Responsive. *Plants* **2019**, *8*, 204. [CrossRef]

114. Garciadeblás, B.; Senn, M.E.; Bañuelos, M.A.; Rodríguez-Navarro, A. Sodium transport and HKT transporters: The rice model. *Plant J.* **2003**, *34*, 788–801. [CrossRef]

115. Horie, T.; Yoshida, K.; Nakayama, H.; Yamada, K.; Oiki, S.; Shinmyo, A. Two types of HKT transporters with different properties of Na$^+$ and K$^+$ transport in Oryza sativa. *Plant J.* **2001**, *27*, 129–138. [CrossRef]

116. Ren, Z.-H.; Gao, J.-P.; Li, L.-G.; Cai, X.-L.; Huang, W.; Chao, D.-Y.; Zhu, M.-Z.; Wang, Z.-Y.; Luan, S.; Lin, H.-X. A rice quantitative trait locus for salt tolerance encodes a sodium transporter. *Nat. Genet.* **2005**, *37*, 1141. [CrossRef]

117. Golldack, D.; Su, H.; Quigley, F.; Kamasani, U.R.; Muñoz-Garay, C.; Balderas, E.; Popova, O.V.; Bennett, J.; Bohnert, H.J.; Pantoja, O. Characterization of a HKT-type transporter in rice as a general alkali cation transporter. *Plant J.* **2002**, *31*, 529–542. [CrossRef]

118. Haruta, M.; Gray, W.M.; Sussman, M.R. Regulation of the plasma membrane proton pump (H$^+$-ATPase) by phosphorylation. *Curr. Opin. Plant Biol.* **2015**, *28*, 68–75. [CrossRef]

119. Chanroj, S.; Padmanaban, S.; Czerny, D.D.; Jauh, G.-Y.; Sze, H. K$^+$ transporter AtCHX17 with its hydrophilic C tail localizes to membranes of the secretory/endocytic system: Role in reproduction and seed set. *Mol. Plant* **2013**, *6*, 1226–1246. [CrossRef]

120. Blumwald, E.; Poole, R.J. Na$^+$/H$^+$ antiport in isolated tonoplast vesicles from storage tissue of Beta vulgaris. *Plant Physiol.* **1985**, *78*, 163–167. [CrossRef]

121. Barkla, B.J.; Blumwald, E. Identification of a 170-kDa protein associated with the vacuolar Na$^+$/H$^+$ antiport of Beta vulgaris. *Proc. Natl. Acad. Sci. USA* **1991**, *88*, 11177–11181. [CrossRef]

122. Garbarino, J.; DuPont, F.M. NaCl induces a Na$^+$/H$^+$ antiport in tonoplast vesicles from barley roots. *Plant Physiol.* **1988**, *86*, 231–236. [CrossRef]

123. Gaxiola, R.A.; Rao, R.; Sherman, A.; Grisafi, P.; Alper, S.L.; Fink, G.R. The Arabidopsis thaliana proton transporters, AtNhx1 and Avp1, can function in cation detoxification in yeast. *Proc. Natl. Acad. Sci. USA* **1999**, *96*, 1480–1485. [CrossRef] [PubMed]

124. Zhang, P.; Senge, M.; Dai, Y. Effects of salinity stress at different growth stages on tomato growth, yield, and water-use efficiency. *Commun. Soil Sci. Plant Anal.* **2017**, *48*, 624–634. [CrossRef]

125. Chakraborty, K.; Bose, J.; Shabala, L.; Shabala, S. Difference in root K^+ retention ability and reduced sensitivity of K^+-permeable channels to reactive oxygen species confer differential salt tolerance in three Brassica species. *J. Exp. Bot.* **2016**, *67*, 4611–4625. [CrossRef] [PubMed]

126. Nguyen, N.T.; Vu, H.T.; Nguyen, T.T.; Nguyen, L.-A.T.; Nguyen, M.-C.D.; Hoang, K.L.; Nguyen, K.T.; Quach, T.N. Co-expression of Arabidopsis AtAVP1 and AtNHX1 to Improve Salt Tolerance in Soybean. *Crop Sci.* **2019**, *59*, 1133–1143. [CrossRef]

127. Fukuda, A.; Nakamura, A.; Tagiri, A.; Tanaka, H.; Miyao, A.; Hirochika, H.; Tanaka, Y. Function, intracellular localization and the importance in salt tolerance of a vacuolar Na^+/H^+ antiporter from rice. *Plant Cell Physiol.* **2004**, *45*, 146–159. [CrossRef] [PubMed]

128. Padmanaban, S.; Czerny, D.D.; Levin, K.A.; Leydon, A.R.; Su, R.T.; Maugel, T.K.; Zou, Y.; Chanroj, S.; Cheung, A.Y.; Johnson, M.A. Transporters involved in pH and K^+ homeostasis affect pollen wall formation, male fertility, and embryo development. *J. Exp. Bot.* **2017**, *68*, 3165–3178. [CrossRef]

129. Maresova, L.; Sychrova, H. Arabidopsis thaliana CHX17 gene complements the kha1 deletion phenotypes in Saccharomyces cerevisiae. *Yeast* **2006**, *23*, 1167–1171. [CrossRef]

130. Song, C.-P.; Guo, Y.; Qiu, Q.; Lambert, G.; Galbraith, D.W.; Jagendorf, A.; Zhu, J.-K. A probable Na^+ $(K^+)/H^+$ exchanger on the chloroplast envelope functions in pH homeostasis and chloroplast development in Arabidopsis thaliana. *Proc. Natl. Acad. Sci. USA* **2004**, *101*, 10211–10216. [CrossRef]

131. Halfter, U.; Ishitani, M.; Zhu, J.-K. The Arabidopsis SOS_2 protein kinase physically interacts with and is activated by the calcium-binding protein SOS_3. *Proc. Natl. Acad. Sci. USA* **2000**, *97*, 3735–3740. [CrossRef]

132. Ren, X.L.; Qi, G.N.; Feng, H.Q.; Zhao, S.; Zhao, S.S.; Wang, Y.; Wu, W.H. Calcineurin B-like protein CBL 10 directly interacts with AKT 1 and modulates K^+ homeostasis in Arabidopsis. *Plant J.* **2013**, *74*, 258–266. [CrossRef]

133. Senadheera, P.; Singh, R.; Maathuis, F.J. Differentially expressed membrane transporters in rice roots may contribute to cultivar dependent salt tolerance. *J. Exp. Bot.* **2009**, *60*, 2553–2563. [CrossRef] [PubMed]

134. Uozumi, N.; Kim, E.J.; Rubio, F.; Yamaguchi, T.; Muto, S.; Tsuboi, A.; Bakker, E.P.; Nakamura, T.; Schroeder, J.I. The Arabidopsis HKT1 gene homolog mediates inward Na^+ currents in Xenopus laevis oocytes and Na^+ uptake in Saccharomyces cerevisiae. *Plant Physiol.* **2000**, *122*, 1249–1260. [CrossRef] [PubMed]

135. Rus, A.; Yokoi, S.; Sharkhuu, A.; Reddy, M.; Lee, B.-h.; Matsumoto, T.K.; Koiwa, H.; Zhu, J.-K.; Bressan, R.A.; Hasegawa, P.M. AtHKT1 is a salt tolerance determinant that controls Na^+ entry into plant roots. *Proc. Natl. Acad. Sci. USA* **2001**, *98*, 14150–14155. [CrossRef] [PubMed]

136. Oomen, R.J.; Benito, B.; Sentenac, H.; Rodríguez-Navarro, A.; Talón, M.; Véry, A.A.; Domingo, C. HKT2; 2/1, a K^+-permeable transporter identified in a salt-tolerant rice cultivar through surveys of natural genetic polymorphism. *Plant J.* **2012**, *71*, 750–762. [CrossRef] [PubMed]

137. Suzuki, K.; Costa, A.; Nakayama, H.; Katsuhara, M.; Shinmyo, A.; Horie, T. OsHKT2; 2/1-mediated Na^+ influx over K+ uptake in roots potentially increases toxic Na^+ accumulation in a salt-tolerant landrace of rice Nona Bokra upon salinity stress. *J. Plant Res.* **2016**, *129*, 67–77. [CrossRef] [PubMed]

138. Rodríguez-Rosales, M.P.; Gálvez, F.J.; Huertas, R.; Aranda, M.N.; Baghour, M.; Cagnac, O.; Venema, K. Plant NHX cation/proton antiporters. *Plant Signal. Behav.* **2009**, *4*, 265–276.

139. Zhang, H.-X.; Hodson, J.N.; Williams, J.P.; Blumwald, E. Engineering salt-tolerant Brassica plants: Characterization of yield and seed oil quality in transgenic plants with increased vacuolar sodium accumulation. *Proc. Natl. Acad. Sci. USA* **2001**, *98*, 12832–12836. [CrossRef]

140. Yin, X.-Y.; Yang, A.-F.; Zhang, K.-W.; Zhang, J.-R. Production and analysis of transgenic maize with improved salt tolerance by the introduction of AtNHX1 gene. *Acta Bot. Sin.-Engl. Ed.* **2004**, *46*, 854–861.

141. Ohta, M.; Hayashi, Y.; Nakashima, A.; Hamada, A.; Tanaka, A.; Nakamura, T.; Hayakawa, T. Introduction of a Na^+/H^+ antiporter gene from Atriplex gmelini confers salt tolerance to rice. *Febs Lett.* **2002**, *532*, 279–282. [CrossRef]

142. LÜ, S.Y.; JING, Y.X.; SHEN, S.H.; ZHAO, H.Y.; MA, L.Q.; ZHOU, X.J.; REN, Q.; LI, Y.F. Antiporter gene from Hordum brevisubulatum (Trin.) link and its overexpression in transgenic tobaccos. *J. Integr. Plant Biol.* **2005**, *47*, 343–349.

143. Zhang, H.-X.; Blumwald, E. Transgenic salt-tolerant tomato plants accumulate salt in foliage but not in fruit. *Nat. Biotechnol.* **2001**, *19*, 765. [CrossRef] [PubMed]

144. Xue, Z.-Y.; Zhi, D.-Y.; Xue, G.-P.; Zhang, H.; Zhao, Y.-X.; Xia, G.-M. Enhanced salt tolerance of transgenic wheat (*Tritivum aestivum* L.) expressing a vacuolar Na^+/H^+ antiporter gene with improved grain yields in saline soils in the field and a reduced level of leaf Na^+. *Plant Sci.* **2004**, *167*, 849–859. [CrossRef]

145. Shi, H.; Zhu, J.-K. SOS_4, a pyridoxal kinase gene, is required for root hair development in Arabidopsis. *Plant Physiol.* **2002**, *129*, 585–593. [CrossRef] [PubMed]

146. Yokoi, S.; Bressan, R.A.; Hasegawa, P.M. Salt stress tolerance of plants. *Jircas Work. Rep.* **2002**, *23*, 25–33.

147. Silva, P.; Gerós, H. Regulation by salt of vacuolar H^+-ATPase and H^+-pyrophosphatase activities and Na^+/H^+ exchange. *Plant Signal. Behav.* **2009**, *4*, 718–726. [CrossRef] [PubMed]

148. Hall, D.; Evans, A.; Newbury, H.; Pritchard, J. Functional analysis of CHX21: A putative sodium transporter in Arabidopsis. *J. Exp. Bot.* **2006**, *57*, 1201–1210. [CrossRef]

149. Shi, H.; Ishitani, M.; Kim, C.; Zhu, J.-K. The Arabidopsis thaliana salt tolerance gene SOS_1 encodes a putative Na^+/H^+ antiporter. *Proc. Natl. Acad. Sci. USA* **2000**, *97*, 6896–6901. [CrossRef]

150. Wegner, L.H.; De Boer, A.H. Properties of two outward-rectifying channels in root xylem parenchyma cells suggest a role in K^+ homeostasis and long-distance signaling. *Plant Physiol.* **1997**, *115*, 1707–1719. [CrossRef]

151. Lacan, D.; Durand, M. Na^+-K^+ exchange at the xylem/symplast boundary (its significance in the salt sensitivity of soybean). *Plant Physiol.* **1996**, *110*, 705–711. [CrossRef]

152. Berthomieu, P.; Conéjéro, G.; Nublat, A.; Brackenbury, W.J.; Lambert, C.; Savio, C.; Uozumi, N.; Oiki, S.; Yamada, K.; Cellier, F. Functional analysis of AtHKT1 in Arabidopsis shows that Na^+ recirculation by the phloem is crucial for salt tolerance. *EMBO J.* **2003**, *22*, 2004–2014. [CrossRef]

153. Davenport, R.J.; Muñoz-mayor, A.; Jha, D.; Essah, P.A.; Rus, A.; Tester, M. The Na^+ transporter AtHKT1; 1 controls retrieval of Na^+ from the xylem in Arabidopsis. *Plant Cell Environ.* **2007**, *30*, 497–507. [CrossRef]

154. Mäser, P.; Eckelman, B.; Vaidyanathan, R.; Horie, T.; Fairbairn, D.J.; Kubo, M.; Yamagami, M.; Yamaguchi, K.; Nishimura, M.; Uozumi, N. Altered shoot/root Na^+ distribution and bifurcating salt sensitivity in Arabidopsis by genetic disruption of the Na^+ transporter AtHKT1. *Febs Lett.* **2002**, *531*, 157–161.

155. Horie, T.; Hauser, F.; Schroeder, J.I. HKT transporter-mediated salinity resistance mechanisms in Arabidopsis and monocot crop plants. *Trends Plant Sci.* **2009**, *14*, 660–668. [CrossRef] [PubMed]

156. Zhang, W.-D.; Wang, P.; Bao, Z.; Ma, Q.; Duan, L.-J.; Bao, A.-K.; Zhang, J.-L.; Wang, S.-M. SOS_1, HKT1; 5, and NHX1 synergistically modulate Na^+ homeostasis in the halophytic grass Puccinellia tenuiflora. *Front. Plant Sci.* **2017**, *8*, 576.

157. Liu, Q.; Liu, R.; Ma, Y.; Song, J. Physiological and molecular evidence for Na^+ and Cl^- exclusion in the roots of two Suaeda salsa populations. *Aquat. Bot.* **2018**, *146*, 1–7. [CrossRef]

158. Xu, B.; Waters, S.; Byrt, C.S.; Plett, D.; Tyerman, S.D.; Tester, M.; Munns, R.; Hrmova, M.; Gilliham, M. Structural variations in wheat HKT1; 5 underpin differences in Na^+ transport capacity. *Cell. Mol. Life Sci.* **2018**, *75*, 1133–1144. [CrossRef] [PubMed]

159. Roy, S.; Chakraborty, U. Role of sodium ion transporters and osmotic adjustments in stress alleviation of Cynodon dactylon under NaCl treatment: A parallel investigation with rice. *Protoplasma* **2018**, *255*, 175–191. [CrossRef]

160. Zhu, M.; Zhou, M.; Shabala, L.; Shabala, S. Physiological and molecular mechanisms mediating xylem Na^+ loading in barley in the context of salinity stress tolerance. *Plant Cell Environ.* **2017**, *40*, 1009–1020. [CrossRef]

161. Batistic, O.; Kudla, J. Integration and channeling of calcium signaling through the CBL calcium sensor/CIPK protein kinase network. *Planta* **2004**, *219*, 915–924. [CrossRef]

162. Kudla, J.; Xu, Q.; Harter, K.; Gruissem, W.; Luan, S. Genes for calcineurin B-like proteins in Arabidopsis are differentially regulated by stress signals. *Proc. Natl. Acad. Sci. USA* **1999**, *96*, 4718–4723. [CrossRef]

163. Kolukisaoglu, Ü.; Weinl, S.; Blazevic, D.; Batistic, O.; Kudla, J. Calcium sensors and their interacting protein kinases: Genomics of the Arabidopsis and rice CBL-CIPK signaling networks. *Plant Physiol.* **2004**, *134*, 43–58. [CrossRef] [PubMed]

164. Zhang, H.; Yin, W.; Xia, X. Calcineurin B-Like family in Populus: Comparative genome analysis and expression pattern under cold, drought and salt stress treatment. *Plant Growth Regul.* **2008**, *56*, 129–140. [CrossRef]

165. Chen, X.-F.; Gu, Z.-M.; Feng, L.; Zhang, H.-S. Molecular analysis of rice CIPKs involved in both biotic and abiotic stress responses. *Rice Sci.* **2011**, *18*, 1–9. [CrossRef]

166. Weinl, S.; Kudla, J. The CBL–CIPK Ca^{2+}-decoding signaling network: Function and perspectives. *New Phytol.* **2009**, *184*, 517–528. [CrossRef] [PubMed]

167. Jung, H.-J.; Kayum, M.A.; Thamilarasan, S.K.; Nath, U.K.; Park, J.-I.; Chung, M.-Y.; Hur, Y.; Nou, I.-S. Molecular characterisation and expression profiling of calcineurin B-like (CBL) genes in Chinese cabbage under abiotic stresses. *Funct. Plant Biol.* **2017**, *44*, 739–750. [CrossRef]

168. Crozet, P.; Margalha, L.; Confraria, A.; Rodrigues, A.; Martinho, C.; Adamo, M.; Elias, C.A.; Baena-González, E. Mechanisms of regulation of SNF1/AMPK/SnRK1 protein kinases. *Front. Plant Sci.* **2014**, *5*, 190. [CrossRef]

169. Albrecht, V.; Ritz, O.; Linder, S.; Harter, K.; Kudla, J. The NAF domain defines a novel protein–protein interaction module conserved in Ca^{2+}-regulated kinases. *Embo J.* **2001**, *20*, 1051–1063. [CrossRef]

170. Guo, Y.; Halfter, U.; Ishitani, M.; Zhu, J.-K. Molecular characterization of functional domains in the protein kinase SOS_2 that is required for plant salt tolerance. *Plant Cell* **2001**, *13*, 1383–1400. [CrossRef]

171. Gong, D.; Guo, Y.; Jagendorf, A.T.; Zhu, J.-K. Biochemical characterization of the Arabidopsis protein kinase SOS_2 that functions in salt tolerance. *Plant Physiol.* **2002**, *130*, 256–264. [CrossRef]

172. Bhattacharyya, M.; Stratton, M.M.; Going, C.C.; McSpadden, E.D.; Huang, Y.; Susa, A.C.; Elleman, A.; Cao, Y.M.; Pappireddi, N.; Burkhardt, P. Molecular mechanism of activation-triggered subunit exchange in Ca^{2+}/calmodulin-dependent protein kinase II. *Elife* **2016**, *5*, e13405. [CrossRef]

173. Ohta, M.; Guo, Y.; Halfter, U.; Zhu, J.-K. A novel domain in the protein kinase SOS_2 mediates interaction with the protein phosphatase 2C ABI2. *Proc. Natl. Acad. Sci. USA* **2003**, *100*, 11771–11776. [CrossRef] [PubMed]

174. Sánchez-Barrena, M.J.; Fujii, H.; Angulo, I.; Martínez-Ripoll, M.; Zhu, J.-K.; Albert, A. The structure of the C-terminal domain of the protein kinase $AtSOS_2$ bound to the calcium sensor $AtSOS_3$. *Mol. Cell* **2007**, *26*, 427–435.

175. Akaboshi, M.; Hashimoto, H.; Ishida, H.; Saijo, S.; Koizumi, N.; Sato, M.; Shimizu, T. The crystal structure of plant-specific calcium-binding protein AtCBL2 in complex with the regulatory domain of AtCIPK14. *J. Mol. Biol.* **2008**, *377*, 246–257. [CrossRef] [PubMed]

176. Sánchez-Barrena, M.; Martínez-Ripoll, M.; Albert, A. Structural biology of a major signaling network that regulates plant abiotic stress: The CBL-CIPK mediated pathway. *Int. J. Mol. Sci.* **2013**, *14*, 5734–5749. [CrossRef] [PubMed]

177. Nagae, M.; Nozawa, A.; Koizumi, N.; Sano, H.; Hashimoto, H.; Sato, M.; Shimizu, T. The crystal structure of the novel calcium-binding protein AtCBL2 from Arabidopsis thaliana. *J. Biol. Chem.* **2003**, *278*, 42240–42246. [CrossRef]

178. Sánchez-Barrena, M.J.; Martínez-Ripoll, M.; Zhu, J.-K.; Albert, A. The structure of the Arabidopsis thaliana SOS_3: Molecular mechanism of sensing calcium for salt stress response. *J. Mol. Biol.* **2005**, *345*, 1253–1264. [CrossRef]

179. Mao, J.; Manik, S.; Shi, S.; Chao, J.; Jin, Y.; Wang, Q.; Liu, H. Mechanisms and physiological roles of the CBL-CIPK networking system in Arabidopsis thaliana. *Genes* **2016**, *7*, 62. [CrossRef]

180. Qiu, Q.-S.; Guo, Y.; Dietrich, M.A.; Schumaker, K.S.; Zhu, J.-K. Regulation of SOS_1, a plasma membrane Na^+/H^+ exchanger in Arabidopsis thaliana, by SOS_2 and SOS_3. *Proc. Natl. Acad. Sci. USA* **2002**, *99*, 8436–8441. [CrossRef]

181. Zhu, J.-K.; Liu, J.; Xiong, L. Genetic analysis of salt tolerance in Arabidopsis: Evidence for a critical role of potassium nutrition. *Plant Cell* **1998**, *10*, 1181–1191. [CrossRef]

182. Li, M.O.; Sanjabi, S.; Flavell, R.A. Transforming growth factor-β controls development, homeostasis, and tolerance of T cells by regulatory T cell-dependent and -independent mechanisms. *Immunity* **2006**, *25*, 455–471. [CrossRef]

183. Tang, R.-J.; Zhao, F.-G.; Garcia, V.J.; Kleist, T.J.; Yang, L.; Zhang, H.-X.; Luan, S. Tonoplast CBL–CIPK calcium signaling network regulates magnesium homeostasis in Arabidopsis. *Proc. Natl. Acad. Sci. USA* **2015**, *112*, 3134–3139. [CrossRef] [PubMed]

184. Ho, C.-H.; Lin, S.-H.; Hu, H.-C.; Tsay, Y.-F. CHL1 functions as a nitrate sensor in plants. *Cell* **2009**, *138*, 1184–1194. [CrossRef] [PubMed]

185. Fuglsang, A.T.; Guo, Y.; Cuin, T.A.; Qiu, Q.; Song, C.; Kristiansen, K.A.; Bych, K.; Schulz, A.; Shabala, S.; Schumaker, K.S. Arabidopsis protein kinase PKS5 inhibits the plasma membrane H+-ATPase by preventing interaction with 14-3-3 protein. *Plant Cell* **2007**, *19*, 1617–1634. [CrossRef] [PubMed]

186. Manik, S.; Shi, S.; Mao, J.; Dong, L.; Su, Y.; Wang, Q.; Liu, H. The calcium sensor CBL-CIPK is involved in plant's response to abiotic stresses. *Int. J. Genom.* **2015**, *2015*, 1–10. [CrossRef] [PubMed]

187. Thoday-Kennedy, E.L.; Jacobs, A.K.; Roy, S.J. The role of the CBL–CIPK calcium signalling network in regulating ion transport in response to abiotic stress. *Plant Growth Regul.* **2015**, *76*, 3–12. [CrossRef]

188. Cho, J.H.; Lee, J.H.; Park, Y.K.; Choi, M.N.; Kim, K.-N. Calcineurin B-like protein CBL10 directly interacts with TOC34 (Translocon of the Outer membrane of the Chloroplasts) and decreases its GTPase activity in Arabidopsis. *Front. Plant Sci.* **2016**, *7*, 1911. [CrossRef]

189. de la Torre, F.; Gutiérrez-Beltrán, E.; Pareja-Jaime, Y.; Chakravarthy, S.; Martin, G.B.; del Pozo, O. The tomato calcium sensor Cbl10 and its interacting protein kinase Cipk6 define a signaling pathway in plant immunity. *Plant Cell* **2013**, *25*, 2748–2764. [CrossRef]

190. Cheong, Y.H.; Pandey, G.K.; Grant, J.J.; Batistic, O.; Li, L.; Kim, B.G.; Lee, S.C.; Kudla, J.; Luan, S. Two calcineurin B-like calcium sensors, interacting with protein kinase CIPK23, regulate leaf transpiration and root potassium uptake in Arabidopsis. *Plant J.* **2007**, *52*, 223–239. [CrossRef]

191. Li, L.; Kim, B.G.; Cheong, Y.H.; Pandey, G.K.; Luan, S. A Ca^{2+} signaling pathway regulates a K(+) channel for low-K response in Arabidopsis. *Proc. Natl. Acad. Sci. USA* **2006**, *103*, 12625–12630. [CrossRef]

192. D'Angelo, C.; Weinl, S.; Batistic, O.; Pandey, G.K.; Cheong, Y.H.; Schültke, S.; Albrecht, V.; Ehlert, B.; Schulz, B.; Harter, K.; et al. Alternative complex formation of the Ca^{2+}-regulated protein kinase CIPK1 controls abscisic acid-dependent and independent stress responses in Arabidopsis. *Plant J.* **2006**, *48*, 857–872.

193. Hashimoto, K.; Eckert, C.; Anschutz, U.; Scholz, M.; Held, K.; Waadt, R.; Reyer, A.; Hippler, M.; Becker, D.; Kudla, J. Phosphorylation of calcineurin B-like (CBL) calcium sensor proteins by their CBL-interacting protein kinases (CIPKs) is required for full activity of CBL-CIPK complexes toward their target proteins. *J. Biol. Chem.* **2012**, *287*, 7956–7968. [CrossRef] [PubMed]

194. Kudla, J.; Batistič, O.; Hashimoto, K. Calcium signals: The lead currency of plant information processing. *Plant Cell* **2010**, *22*, 541–563. [CrossRef] [PubMed]

195. Quintero, F.J.; Martinez-Atienza, J.; Villalta, I.; Jiang, X.; Kim, W.-Y.; Ali, Z.; Fujii, H.; Mendoza, I.; Yun, D.-J.; Zhu, J.-K. Activation of the plasma membrane Na/H antiporter Salt-Overly-Sensitive 1 (SOS_1) by phosphorylation of an auto-inhibitory C-terminal domain. *Proc. Natl. Acad. Sci. USA* **2011**, *108*, 2611–2616. [CrossRef] [PubMed]

196. Luo, Q.; Wei, Q.; Wang, R.; Zhang, Y.; Zhang, F.; He, Y.; Zhou, S.; Feng, J.; Yang, G.; He, G. BdCIPK31, a calcineurin B-like protein-interacting protein kinase, regulates plant response to drought and salt stress. *Front. Plant Sci.* **2017**, *8*, 1184. [CrossRef]

197. Deng, X.; Zhou, S.; Hu, W.; Feng, J.; Zhang, F.; Chen, L.; Huang, C.; Luo, Q.; He, Y.; Yang, G. Ectopic expression of wheat TaCIPK14, encoding a calcineurin B-like protein-interacting protein kinase, confers salinity and cold tolerance in tobacco. *Physiol. Plant.* **2013**, *149*, 367–377.

198. Jin, X.; Sun, T.; Wang, X.; Su, P.; Ma, J.; He, G.; Yang, G. Wheat CBL-interacting protein kinase 25 negatively regulates salt tolerance in transgenic wheat. *Sci. Rep.* **2016**, *6*, 28884. [CrossRef]

Review

Physiological Responses of Selected Vegetable Crop Species to Water Stress

Eszter Nemeskéri * and Lajos Helyes

Institute of Horticulture, Szent István University, H-2100 Gödöllő, Hungary
* Correspondence: Nemeskeri.Eszter@mkk.szie.hu; Tel.: +36-28-522071; Fax: +36-28-410804

Received: 21 July 2019; Accepted: 11 August 2019; Published: 13 August 2019

Abstract: The frequency of drought periods influences the yield potential of crops under field conditions. The change in morphology and anatomy of plants has been tested during drought stress under controlled conditions but the change in physiological processes has not been adequately studied in separate studies but needs to be reviewed collectively. This review presents the responses of green peas, snap beans, tomatoes and sweet corn to water stress based on their stomatal behaviour, canopy temperature, chlorophyll fluorescence and the chlorophyll content of leaves. These stress markers can be used for screening the drought tolerance of genotypes, the irrigation schedules or prediction of yield.

Keywords: vegetable crops; stomatal conductance; canopy temperature; chlorophyll fluorescence; SPAD; water stress

1. Introduction

As a result of climate change, the increasing atmospheric CO_2 enhances the photosynthesis capacity and improves water use efficiency therefore the amount of yield will increase in most of vegetable crops, however its advantage cannot be shown under limited water and nitrogen deficiency. The high temperature during reproductive growth is harmful for many important vegetable crops, such as tomatoes, peppers, beans and sweet corn, and yield reduction will probably occur [1]. The frequency of drought periods decreases vegetable yield and quality, however soluble solid content of produce may be increased by water deficiency in some crops [2,3]. Nevertheless, the occurrence of excess precipitation causes waterlogging in soils, the symptoms of which are similar to water deficit. Soil waterlogging impedes the oxygen supply and respiration of roots, water uptake and hydraulic conductance which results in stomatal closure [4,5]. Under these conditions the stomatal closure results in a reduction of net photosynthesis which is due to the decrease in stomatal conductance, chlorophyll fluorescence and chlorophyll content of leaves [6]. Excess water causes a decline in grain filling and grain weight of corn leading to decreased yield [7]. However, water stress commonly refers to water deficits not excess water.

The selection of the vegetable crops grown under field conditions for the investigation was based on their production in the world and Europe. During the last twenty years from 1997 to 2017 the growing area of tomatoes increased intensively, that of green peas increased moderately while the growing area of snap beans and sweet corn increased slowly in the world. During this time in Europe the growing area of tomatoes and snap beans decreased from 650.4 to 496.2 thousand ha while that of green peas increased slightly from 208 to 212 thousand ha and sweet corn's increased intensively (from 50.5 to 110.1 thousand ha) (FAOSTAT 2017 [8]. In Hungary, the production of green peas and sweet corn is performed in large field growing areas (19.5 thousand hectares and 34.5 thousand hectares, respectively) while snap beans are grown in smaller ones (1.6 thousand hectares) (FAOSTAT 2017) [8]. The other aspect of the selection was the sensitivity of plant species to water stress.

Corn, soybeans, beans and peas are considered to be moderately water stress sensitive while tomatoes belong to the extremely drought sensitive group [9,10]. The responses of plant species significantly depend on the intensity and duration of stress and their stages of development. The spring-sown green pea utilizes the precipitation well (if there is any) and requires a low temperature during vegetative growth but during the flowering and seed development periods it is sensitive to water deficiency. The crops require a warm temperature, even though they have different ripening times, snap beans have short (60 days), sweet corn has medium (75–90 days) and tomatoes have long ripening times (110–130 days), their generative stages of development coincided with dry June and July, thus they require irrigation. Irrigation scheduling and the amount of irrigation water are determined by the water stress tolerance and water use of the plant varieties. The evaluation of drought tolerance in field conditions is difficult because low soil moisture and high air temperature stress generally occur together, and it is difficult to evaluate the responses separately. Drought under field conditions promotes the evapotranspiration and affects the photosynthesis, which leads to reduced yield [11]. Use of remote-sensing methods makes the measurement of physiological responses of varieties to various strong water stresses easy. These non-destructive methods help the breeder to select drought tolerant genotypes and the growers to measure the water deficit of plants and decide the time of irrigation.

The selection for water stress tolerance in traditional breeding is based on the suitability of performance under a series of environmental conditions using extensive statistical methods. This progress could be improved by the introduction of traits which contribute to the prediction of yield in the drought-prone environments. In this study, the effect of water stress on the plants and those physiological traits which influenced the yield are mainly demonstrated. Information was gathered on the physiological responses of selected vegetable species to drought to analyse their use in breeding for high and stable yield.

2. Water Stress during Growth of Vegetable Crops

Sensitivity of plants to water stress such as snap beans and green peas differs with the stages of development. During the early stages of vegetative growth most crops are less sensitive to water scarcity [2,3,12,13], but during the generative stage the water deficiency results in changes of many physiological traits [2,14–16], causing the disturbance of fertility and reduction of yield. During flowering of legume plants water stress increases the ratio of flower drop [17], decreases the pod numbers and seed abortion in the pods [18,19] and increases the ratio of curved pods [20]. Under water deficiency, bean plants produce shorter shoots and smaller leaves and decrease the length of pods [21]. Semi-leafless pea varieties have reduced leaf area that is presumed to have a low water use and they have higher water use efficiency (WUE) than traditional varieties with normal leaves [22]. In sweet corn, ear differentiation begins at the six- or eight-leaf stage growth when the water deficiency decreases the length of ears and the numbers of ear rows [23], but during tasselling the water deficiency causes significant yield reduction [24,25]. Tomatoes are most sensitive to water deficiency at fruit setting and intensive fruit development periods [3], when the increasing water stress resulted in a 25 to 50% decrease in the yield [10,26–29]. During early flowering of tomatoes, water scarcity causes flower shedding and lack of fertilization [30], and during fruit setting, plants with small sized fruits are produced [10,31].

The effect of water stress on morphology and anatomy of plants has been studied by several researchers under controlled conditions [6,32–35], however, the changes in physiological responses have been less investigated under field conditions. The physiological characteristics that have an important role in the defence against drought can be measured by remote sensing techniques using non-destructive methods in open field conditions. The leaf photosynthetic activity of plants can be monitored with measurement of chlorophyll content using a portable chlorophyll meter and chlorophyll fluorescence while the measurement of stomatal conductance indicates the severity of water stress [2,3,14]. Spectral vegetation indices such as the green normalized vegetation index

(GNDVI) and the normalized differential vegetation index (NDVI) were used for monitoring the growth of the plant to detect the water stress and for yield prediction [36–39]. Crop water stress index is determined by an infrared thermometry technique to indicate the change in canopy temperature of plants under water stress conditions. More physiological indices such as leaf water potential, relative water content, turgor potential, osmotic adjustment, difference between canopy and air temperature can also be used as a screening tools for testing the water stress tolerance of genotypes [40]. Studies have focused on the identification of drought tolerance-related traits using Quantitative Trait Locus (QTL)s and Marker Assisted Selection (MAS) techniques [41–44], however, the identification of the most relevant loci controlling drought tolerance and drought-related traits could be achieved by the integration of molecular genetics with physiology [45].

3. Drought Tolerance

Adaptive mechanisms promoting the survival of plants have been grouped into three categories; drought escape, drought avoidance and drought tolerance [46]. Drought escape is the ability of plants to accomplish their life cycle before the development of soil and plant water deficit. The varieties with early flowering and short maturity are able to escape drought [47], however, they are not drought tolerant in every case. The varieties with moderate drought sensitivity developed different defence strategies to avoid short- and long-term water stress which prevents the water loss in their cells and tissues. The essential defence mechanism against drought operating in the plants is the maintenance of the water status and the reduction of tissue water loss (Figure 1).

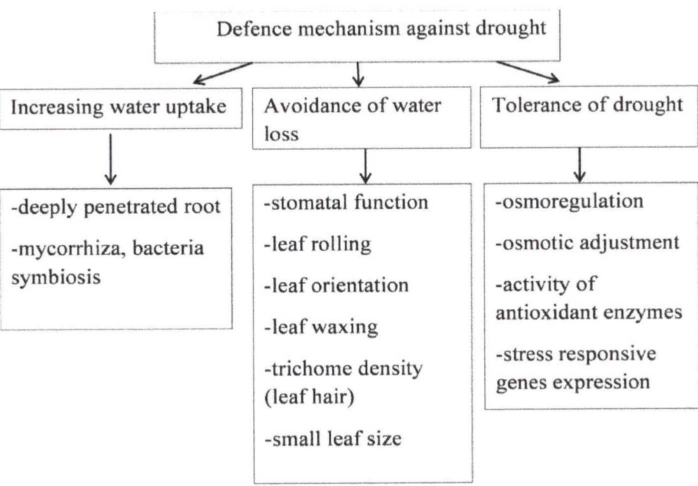

Figure 1. Defence mechanism against drought (Leonardis et al. [46]).

A well-developed deeply penetrating root system provides the water uptake and maintenance of water circulation inside the plant despite the low soil-moisture content. Nevertheless, in dry soil the lives of microorganisms are retarded when the activity of mycorrhiza living symbiotically with root nodules of legumes is low, which results in a decrease in the nitrogen uptake [48,49], therefore the growth of the plant is retarded. Long-term drought of soil accelerates the senescence of root nodules and production of reactive oxygen species (ROS) [50,51], therefore the nodule weight, root and shoot weight are decreased [52]. Water stress results in a change in the proportion of root weight as the ratio of root to shoot increases [53]. Under permanent low moisture content of soil, a 27–42% decrease in leaf weight and 12–27% decrease in specific leaf area of snap bean varieties were found [54]. Tomatoes are able to survive prolonged periods of low soil water content by the development of a deep root system [28,55]. In dry years, tomatoes inoculated with mycorrhiza easily endured the water scarcity, for example larger weight fruits and higher yield were produced by deficit irrigation than under non-irrigated conditions [56].

4. Reduction of Water Loss

Drought avoidance is the ability of plants to maintain high tissue water potential despite the deficiency of soil moisture. The mechanisms developed for the reduction of water loss are related to the duration of water stress.

During short-term water deficiency the leaf movement, deep penetrating roots with strong suction force and partial or total stomatal closure provide a decrease in the water loss. Leaf movement of plants not only protects from the photodamage caused by high irradiation but reduces the effective leaf area for transpiration [57]. Paraheliotropic movement of leaves occurs mainly in beans while leaf rolling is typical for maize. Fernandez and Castrillo [58] found that the extent of leaf rolling is linearly correlated with the water potential. During leaf rolling of maize the transpiration, stomatal conductance, intracellular CO_2 concentration and net photosynthetic rate decreased [59]. Pastenes et al. [60] found that the degree of paraheliotropic leaf movement was larger in the water stressed plants because of lower water potential, however, it also occurred in the water supplied crops. Deep, thick and dense roots intensively promote the use of available water and the optimal development of aboveground parts. During short-term water stress (<7 nap), abscisic acid (ABA) is produced in the roots then it is transported into the leaves where ABA induces the stomatal closure and thus decreases the water loss [61,62]. Partial or total stomatal closure restricts the transpiration therefore the water and nutrient uptake is decreased, which results in a decrease in photosynthesis and growth of plants [63]. Stomatal responses of legume species are different; under water deficiency, beans have a rapid and complete stomatal closure generating the stomatal conductivity and photosynthesis decreases significantly, whereas in cowpeas (*Vigna unguiculata*), the stomata remain partially open and have a lower decrease in their net photosynthetic rate under the same conditions [64]. Under moderate water deficit conditions, the growth of snap beans was already retarded, and their leaf area decreased while the leaf area index (LAI) of sweet corn did not change [3,15]. Nevertheless, water scarcity did not influence the leaf area of tomatoes [65] but heat and water stress up to 6 days already significantly decreased the weight of shoots and roots of tomato seedlings under a controlled environment [66].

During long-term water deficiency, plants try to prevent the dehydration of cells of vegetative and generative organs with some morphological and physiological changes. Trichome density (leaf hairs) on the leaf protects the tissues from sunlight injury, decreases the water loss by evaporation and enhances the transpiration resistance [67]. Under water stress conditions, a lower number of trichomes was found only on the basal zone of leaves on both surfaces in comparison with irrigated plants [68]. However, according to Nobel [69], the length of trichomes can be more important than their frequency. The epicuticular wax layer of the leaf controls the water flow across the cuticle and protects from high radiation and prevents damage caused by UV light. Water stress induced the increase in the wax layer on the leaf surface of peas and the wax-rich varieties had significantly lower canopy temperature [70].

Drought tolerance is the ability of the plants to endure the long-term moisture deficit and survive the water loss. When the morphological changes seem to be insufficient to avoid the water deficiency, biochemical processes of plants are activated to maintain the osmotic adjustment and the structure of cell membranes in order to avoid cell dehydration. Decreasing the water potential of leaves induces the accumulation of different osmotic compounds such as sugars, amino acids and quaternary ammonium compounds. The osmotic pressure of cells is increased by the accumulation of osmotic compounds because water movement into the cells and tissues provides the maintenance of turgor [71]. It was found that peas and castor beans exposed to water deficit accumulate a significant amount of soluble sugars and proline [72,73], and the raffinose and sucrose level of leaves are significantly increased by water stress during flowering of snap beans [74]. Action of enzymatic and non-enzymatic antioxidants is intensified to alleviate the oxidative damages in the tissues. Concerted operation of numerous water soluble antioxidant compounds (ACW) contributes to the adaptation of plants to environmental stresses. The level of ACW in the leaves is influenced by stomatal closure because it is related to ascorbic acid redox potential of guard cells [75]. In snap bean genotypes that have a high ACW level in leaves during the flowering and pod development periods, this provides a defence against water deficiency [74].

4.1. Regulation of Water Circulation in Plants under Drought

Many physiological processes are activated to mitigate the water loss of plants (Table 1). Transpiration is restrained as a result of stomatal closure and by decreasing leaf area. Stomata play an important role in the regulation of transpiration and CO_2 uptake. Use of light energy gathered by photosynthesis determines the growth and biomass production of plants. In these processes, the stomatal characteristics such as stomatal size, number and ratio of stomata on abaxial and adaxial surfaces significantly affect the C assimilation and water use efficiency (WUE) [76,77]. The higher stomatal density on the abaxial surface of the leaf is related to a higher water use efficiency [78], while those existing on upper epidermis (adaxial surface) of the leaf influenced the water use of plants [15]. Nevertheless, the number of stomata on both epidermis of leaves changes significantly depending on the variety and water supply.

Table 1. Physiological traits relevant for response to drought.

Physiological Traits	Effect Relevant for Yield	Alteration under Stress	References
Size and density of stomata	relation to leaf water potential and water consumption	increase/decrease depending on species	Hardy et al. [79], Nemeskéri et al. [14,15]
Leaf temperature	relation to transpiration	increase	Helyes et al. [31], DeJonge et al. [80]
Stomatal conductance	correlation with water consumption, decrease in individual yield	decrease in diffusion of CO_2, stomatal resistance increases	Jones [81], Nemeskéri et al. [14,15]
Photosynthetic capacity	modulation of activity enzymes of Calvin cycle	reduction under stress	Lawlor and Cornic [82]
Change in chlorophyll fluorescence	alteration of quantum yield of PSII photosystem	decrease in Fv/Fm under severe drought	Flagella et al. [83], Pol et al. [84], Yordanov et al. [85]
Chlorophyll content of leaf	decrease in photosynthesis	decrease under stress, relative chlorophyll content (SPAD value) can increase	Nankishore and Farrell [32], Nemeskéri et al. [16]
Reduced growth rate		leaf area reduces, biomass decreases	Ghanbari et al. [86], Guida et al. [87]

4.1.1. Stomatal Characteristics

More stomata (134–195 stomata mm^{-2}) were observed on the abaxial surface of tomato leaves but it was significantly less (40–62 mm^{-2}) on the adaxial surface of leaves [76]. A significant difference can be demonstrated in stomatal density of leaves between snap beans, green peas and sweet corn grown under non-irrigated and deficit irrigated (50% water deficiency) conditions (Table 2). On the basis of 3 year experiments, on the lower epidermis of leaves the stomata density was significantly higher for snap beans under moderate and severe water stress and it was similarly high for sweet corn only in severe water deficiency, but no difference could be shown for green peas in comparison with the optimal water supplied plants [14,15,88]. On the upper epidermis of leaves more and larger sized stomata can be found for snap beans exposed to drought while there were fewer similar sized stomata for the green peas compared to the irrigated plants (Table 2). However, under water scarcity, significant differences in stomata number and size can be detected between the varieties. Under non-irrigated conditions, the size of stomata on the upper (adaxial) surface of leaves of green-podded bean varieties was smaller by 5–12%, but more of them were found than on optimal water supplied plants. Nevertheless, yellow-podded snap bean varieties had 13–18% larger sized stomata on the adaxial surface of leaves of plants exposed to water deficiency in comparison with the irrigated plants [15]. A larger stomatal density was observed for late ripening green pea varieties [14] and late ripening sweet corn hybrids under water scarcity [88] than for the short duration ones. Nevertheless, the distribution and size of stomata can be different on both areas and surfaces of the same leaf. Various number and sized stomata were detected on different areas of leaves of tomatoes; on the abaxial surface of leaves and their apical and middle areas, larger sized (32–34 μm) and more stomata were found than that on the adaxial surface. The stomata on the apical areas of leaves responded sensitively to water deficiency in that they showed fewer and larger sized stomata on the adaxial surface of leaves than for well-watered plants [68]. A significant correlation between the stomatal density and stomatal conductance ($r^2 = 0.958$) was established in tomatoes. According to this correlation, 130 stomata mm^{-2} was associated with high stomatal conductance (0.1 mol H$_2$O m^{-2} s^{-1}) [76]. Others [89] found that the relationship between stomatal density and WUE was positive and the size of stomata correlated negatively with the WUE for grass peas.

Table 2. Size and density of stomata measured during generative stages of vegetable crops under different water supplies Source: modified from Nemeskéri et al. [14,15,88].

Species	Water Supply	Lower Epidermis		Upper Epidermis	
		Stomata mm^{-2} *	Size of Stomata μ	Stomata mm^{-2}	Size of Stomata μ
Snap bean	I0	387.79	23.72	104.81	30.51
	DI	374.17	-	93.41	-
	WI	331.22	24.90	78.61	29.64
	average	364.39	24.31	92.28	30.08
Green pea	I0	214.29	25.82	165.70	25.79
	DI	214.65	25.48	170.86	24.68
	WI	214.74	24.35	194.72	25.21
	average	214.56	25.22	177.10	25.23
Sweet corn	I0	145.61	-	95.23	-
	DI	140.79	-	94.98	-
	WI	136.13	-	93.73	-
	average	140.84	50.04	94.65	53.22

* Based on average of years [14,15,88], μ = micron, I0 = non-irrigation, DI = deficit irrigation, WI = optimal water supply.

4.1.2. Canopy Temperature-Transpiration

Under high photosynthetically active radiation (PAR) water deficit combined with high temperature results in an increase in leaf temperature and air temperature oscillation (±3–4 °C) due to the opening and closing of stomata [53]. Stomata closure triggers the decrease in the transpiration which contributes to the increase in canopy temperature of plants. One of the tasks of transpiration is to keep the temperature of plants at a favourable level for life processes. Decreasing transpiration causes the temperature of plants to increase. If soil water content is sufficient for the plant stand, the difference between canopy temperature and air temperature is zero or negative, but if the plants suffer from water stress this value is positive. An increase of 1 °C in canopy temperature related to a 10% decrease in the transpiration [31]. Size and stomatal density of genotypes are different thereby the transpiration varies in intensity which correlates with the difference of the canopy temperature of plants. Changes in canopy temperature have often been used to signal water stress [90] to evaluate the drought tolerance of bean genotypes and the difference in canopy temperature and air temperature was used for the real time irrigation [91–94]. During the daytime the canopy temperature rises along with the daily air temperature and radiation as the available soil water changes. The lowest value of crop water stress index (CWSI) of maize was measured at 10:00 and 11:00 and it was the largest between 12:00 and 13:00 [95]. Under water deficiency, the canopy temperature of both snap beans and tomatoes was higher than the air temperature from 09:00 to 15:00 however, that of tomatoes was higher than the air temperature only at 12:00 and 13:00 [96]. Under water stress conditions, between 09:00 and 15:00, the canopy temperature of snap beans was higher by 3.8 °C than the air temperature while it was lower by 1.6 °C in well-watered plants [93]. When the amount of available moisture in the soil for the plants decreases, then the transpiration is limited depending on the air temperature, which results in an increasing canopy temperature. Under moderate water deficiency, at 25–50% available soil water the canopy temperature of snap beans almost coincided with the air temperature (Figure 2a) that denoted the need for irrigation [93]. Nevertheless, the available soil water below 25% was not able to satisfy the water demands of plants. In this case the cooling of the canopy was not shown by transpiration and the temperature on the foliage surface was higher than the air temperature by 2.5 °C on average, indicating the plants suffered from water stress (Figure 2b) [93].

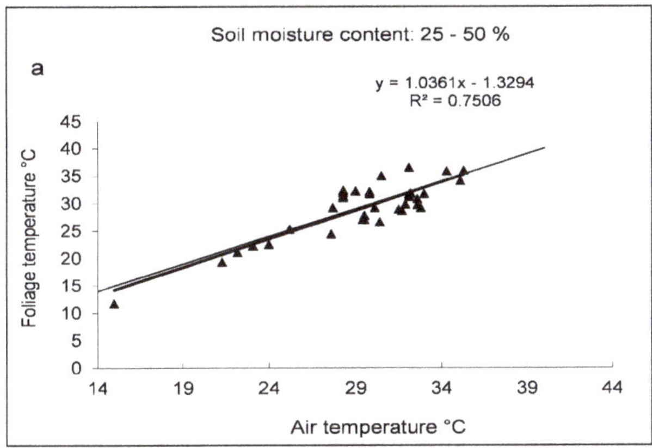

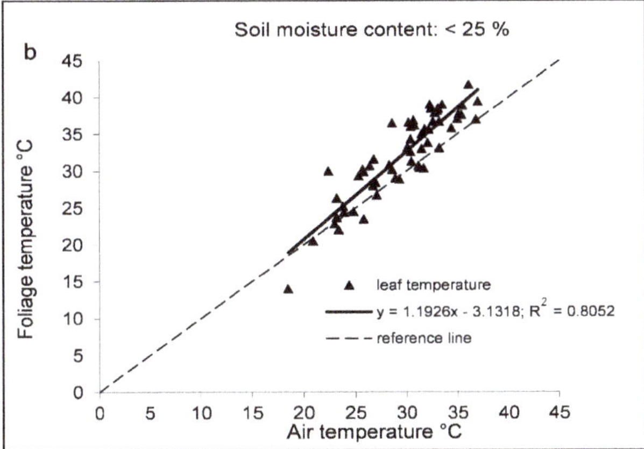

Figure 2. Relationship between air and canopy temperature for snap beans under water deficit (**a**) and severe water stress (**b**). The thick line shows the increase in leaf temperature compared to air temperature (broken line) Source: Helyes et al. [93].

Tomatoes seemed to better use deep soil moisture with deep, strong suction force roots than the shallow rooted snap beans. Under water stress conditions, the canopy temperature of tomatoes was only higher than the air temperature by 1.8 °C, while it was significantly lower (0.6 °C) under optimal water supply conditions [92]. Air temperature had a small impact on the canopy temperature of tomatoes grown under regular irrigation and cut-off stand (i.e., irrigation was stopped 30 days before harvest) ($r^2 = 0.60$; $r^2 = 0.55$), however, the canopy temperature of water stressed plants increased with rising air temperature ($r^2 = 0.59$) (Figure 3) [31]. A close correlation between canopy temperature and leaf water potential of maize was established [80] and the lowest CWSI values were measured between 10:00 and 11:00 and the highest ones between 12:00 and 13:00 [95].

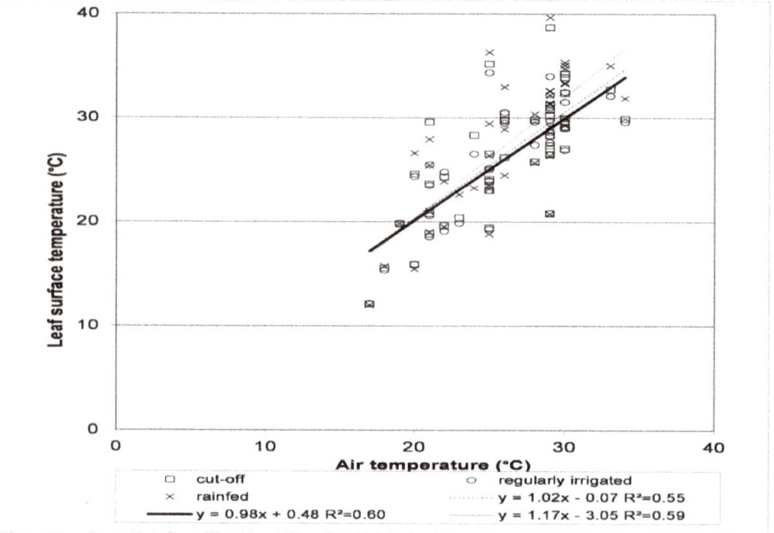

Figure 3. Relationship between the air and canopy temperature for the Kecskeméti jubileum tomato variety under rain-fed (thin line), cut-off (broken line) and regularly irrigated (thick line) conditions Source: Helyes et al. [31].

4.1.3. Stomatal Conductance

Stomatal conductance indicates the speed of water vapour evaporation that depends on more plant-specific characteristics such as stomatal density, leaf age and size, guard cell and cell turgor [97]. Stomatal conductance is related to the photosynthetic assimilation rates ensuring an appropriate balance between CO_2 uptake for photosynthesis and water loss through transpiration [98]. Variability in photosynthesis capacity can be explained by the CO_2 diffusion through stomata and leaf mesophyll which was influenced by the mesophyll thickness and porosity and size of stomata. In drought-acclimated tomato plants the decrease in mesophyll CO_2 conductance was due to an increased cell wall thickness [76]. Water stress significantly decreased the transpiration rate (37%) and stomatal conductance (26%) of maize [99]. Nevertheless, the extent of decrease in stomatal conductance depends on the growing period when the water deficiency occurred; at 7 days after anthesis of maize cultivars stomatal conductance decreased by 35% on average but at 21 days after anthesis this decrease was significantly larger (74%) under water deficiency than in well-watered cultivars [100]. In the case of tomatoes grown under non-irrigated conditions, stomatal conductance decreases from 14 to 73% depending on the weather and variety in comparison with the well-watered plants [3,32,87,101] (Table 3).

Table 3. Physiological traits related to water use and photosynthesis for vegetable crops under optimal water supply (OW) and water stress (WS) conditions.

Traits	Crops	Units	OW	WS	Difference %	References
Stomatal resistance	green pea	s cm^{-1}	2.87	3.22	12.2	Nemeskéri et al. [14]
	snap bean	s cm^{-1}	1.33	2.54	90.9	Nemeskéri et al. [15]
	sweet corn	s cm^{-1}	2.13	2.85	33.8	Nemeskéri et al. [88]
Stomatal conductance	green pea	mmol m^{-2} s^{-1}	0.57	0.32	−43.9	Gurumurthy et al. [35]
	tomato	mmol m^{-2} s^{-1}	1200	125	−89.6	Nankishore and Farrell [32]
		μmol m^{-2} s^{-1}	457.26	394.95	−13.6	Nemeskéri et al. [3]
		mol m^{-2} s^{-1}	20.2–37.9	6.3–10.2	−68.8 −73.1	Helyes et al. [101]
Chlorophyll fluorescence	snap bean	Fv/Fm	0.80	0.78	−2.5	Tari et al. [102]
	maize	Fv/Fm	0.810	0.695	−14.2	Yan et al. [103]
	tomato	Fv/Fm	0.785	0.745	−5.1	Nankishore and Farrell [32]
		Fv/Fm	0.748	0.696	−7.0	Nemeskéri et al. [3]
		Fq'/Fm'	0.4	0.25	−37.5	Zhou et al. [66]
Chlorophyll content	green pea	SPAD *	48.16	49.02	1.8	Nemeskéri et al. [14]
	snap bean	SPAD	34.57	38.94	12.6	Nemeskéri et al. [16]
	sweet corn	SPAD	47.48	44.67	−5.9	Nemeskéri et al. [2]
	tomato	SPAD	50.97	52.63	3.3	Nemeskéri et al. [3]
Vegetation index	green pea	NDVI	0.679	0.693	2.3	Nemeskéri et al. [14]
	snap bean	NDVI	0.778	0.681	−12.5	Nemeskéri et al. [16]
	sweet corn	NDVI	0.743	0.711	−4.3	Nemeskéri et al. [2]

* SPAD = relative chlorophyll content of leaves; NDVI = normalized differential vegetation index.

Under water scarcity, stomatal conductance for both water and CO_2 flow decreased by closing the stomata [104], thus it can be said that stomatal resistance increased. The extent of stomatal resistance mainly gives information about the speed of water vapour. Under severe water deficit conditions, stomatal resistance increased by 91% for snap beans, 34% for sweet corn and 12% for green peas in comparison with the well-watered plants (Table 3). The studies shown in Table 3 proved that snap beans responded more intensively to severe water stress than sweet corn and green peas. Flowering and pod development periods of legume crops are the most sensitive to water stress when stomatal resistance changes depending on the varieties and the degree of water stress. Under moderate water deficiency, the late ripening green pea varieties had high stomatal resistance (>3.0 s cm^{-1}), while that of green-podded snap bean varieties was relatively low (0.8–1.2 s cm^{-1}) and yellow-podded snap beans showed different values depending on the varieties (1.0–1.43 s cm^{-1}) [14,15]. During tasselling, the late ripening sweet corn hybrids responded with higher stomatal resistance (3 s cm^{-1}) to medium water deficiency than during the silking period [88].

4.2. Photosynthesis in Drought

The aspects of photosynthesis of selected vegetable crops which can be measured by remote sensing methods and used for the evaluation for drought tolerance of genotypes have to be taken into consideration. In the photosynthesis process the light capture and conversion of light energy to chemical energy is made by photosynthetic pigments in the photochemistry photosystems (PSI, PSII) of leaves. The light energy in the leaf that is not used for photosynthesis is either emitted as fluorescence or released as heat [105]. The efficiency of photosynthesis can be measured by the efficiency of PSII photochemistry or by the amount of photosynthetic pigments [106].

4.2.1. Chlorophyll Fluorescence

Intense dry conditions of soil cause stomatal closure, reduced CO_2 mesophyll conductance [107] and decreasing activity of PSII [108], which contributes to the decrease in photosynthesis. Photosystem II (PSII) is highly sensitive to light and drought [60] and the maximum quantum yield of PSII photochemistry (Fv/Fm) indicates an undisturbed or deficient operation of photosynthesis. Chlorophyll *a* fluorescence is considered to be suitable for the measurement of activity of photosynthesis because environmental stresses significantly affect the emission of chlorophyll fluorescence [109]. For example, UV-B radiation decreased the chlorophyll fluorescence of green peas [110] and ozone stress decreases the Fv/Fm and chlorophyll *a* concentration of leaves [111]. In higher plants, Fv/Fm fluorescence ranged from 0.78 to 0.84 [112], however this change depended on the variety and intensity of water stress.

In snap beans, the Fv/Fm ratio was relatively high (0.82–0.83) under optimal water supply conditions and it only decreased to 0.80 in the drought sensitive genotype under water stress conditions [102], which proved that chlorophyll fluorescence was not highly sensitive to water deficit.

In dry years, tomatoes grown under non-irrigated conditions had low photosynthetic activity (Fv/Fm = 0.662) and under moderate and optimal water supply conditions the Fv/Fm value ranged between 0.753 and 0.758 [3]. Likewise, the above-mentioned results from Nankishore and Farrell [32] showed a small decrease in Fv/Fm (5.1%) in tomatoes under drought (Table 3).

The maximum efficiency of PSII (Fv/Fm) of well-watered maize plants stayed constant while that of drought stressed plants stayed at control level during the first 2 days then decreased sharply as the soil became drier [103].

Use of Fv/Fm to evaluate the drought tolerance of crops is contradictory. Under controlled conditions, Fv/Fm for pot-grown grapevines decreased when water potential dropped but it seemed to be a good indicator to distinguish the moderate and severe drought stress in the field [113]. Drought stress affected the Fv/Fm parameter of the asparagus bean (*Vigna unguiculata* L.) [114]. Contrary to these results, no change was detected in the Fv/Fm for strawberries [115] and soybeans [116] grown under drought. Others [117,118] stated that PSII activity expressed by the Fv/Fm of drought tolerant tomato genotypes was less decreased under intensive water stress than sensitive ones. Likewise, Li et al. [119] found that Fv/Fm in drought tolerant barley varieties was higher than those of the drought sensitive group under drought stress. Under 4 day waterlogging conditions, the chlorophyll fluorescence (Fv/Fm) of flooding stress tolerant wax maize hybrids did not change significantly, while the photosynthesis efficiency of sensitive hybrids was relatively low and the Fv/Fm value decreased by 5.2% in comparison with the control [6]. The measurement of chlorophyll fluorescence as a rapid non-destructive method can be easily carried out in the field, thus it can be recommended for screening for drought tolerance [120].

4.2.2. Photosynthetic Pigments

Environmental stresses change not only the activity of the photochemistry apparatus but the chlorophyll concentration in the leaf due to metabolic disturbance [121], whereupon the light absorption decreases. Water also absorbs the radiation in the infrared wavelength of the spectrum and as the water content of leaf decreases, the light absorption decreases and reflectance increases due to the radiative attributes of water [122,123]. Therefore, the water content of leaves and the amount of photosynthetic pigments in leaves both influence the light absorption by leaves. The light absorption of the leaf can be indirectly measured by portable chlorophyll meter. In this way the calculated SPAD values correlated with the chlorophyll content of leaves [124] expresses the efficiency of photosynthesis by the intercepted photosynthetic active radiation. The high SPAD value indicates the low water and chlorophyll concentration simultaneously in the leaf, resulting in a decrease in light absorption and increase in reflectance that is larger in extent in snap beans and smaller in green peas and tomatoes (Table 3). Iturbe-Ormaetxe et al. [125] stated that the decrease in chlorophyll *a* concentration of leaves was larger (−30%) than that of chlorophyll *b* (−20.8%) for green peas exposed to water stress than in well-watered plants.

5. Relationship between Drought Stress Markers and Yield

During reproductive periods of plants that are most sensitive to water deficiency, the changes in physiological responses can be used to screen the water stress tolerance of genotypes. During this time the water supply determines the yield production. Stomatal resistance and the relative chlorophyll content of leaves (SPAD) of the individual plants indicate the disturbance of water circulation and photosynthesis. Spectral vegetation indices indicate the absorption of solar energy of the canopy in the visible light spectrum. Health status and water deficit of vegetation can be monitored by different vegetation indices and it can also determine the need for irrigation [126–129]. The normalized differential vegetation index (NDVI) expresses the ratio of spectral reflectance on the canopy in the infrared and red region and it is used to monitor the effect of water stress on plant growth and forecast biomass [130,131].

The question is how closely the physiological traits are related to water circulation and photosynthesis and can be used to predict the expected yield. Nevertheless, the physiological traits measured during the generative stages of plant species are different (Table 4). On the basis of long-term experiments, stomatal resistance measured during flowering of snap beans and tomatoes correlated with the pod yield of individual plants and weight of tomato fruits under severe drought. A close correlation between the relative chlorophyll content of leaves (SPAD) and weight of tomato fruits and final yield was detected under both mild and severe water deficiency which can be used for selection of genotypes with water stress tolerance. During tasselling of sweet corn, the expected yield of plants can be less predicted by stomatal resistance (47%) and to a higher extent by spectral traits (58–68%) under moderate water deficiency. During flowering of green peas, stomatal resistance and chlorophyll content of leaves showed a close correlation with the expected yield only under severe drought (Table 4).

Table 4. Correlation coefficients between physiological traits measured during flowering and yield under drought.

| Crops | Water Supply | NI | | DI | |
	Traits	Yield g plant^{-1}	Yield t ha^{-1}	Yield g plant^{-1}	Yield t ha^{-1}
Green peas	SR	0.3885	0.7648	0.3541	0.4371
	SPAD	0.4685	0.7027	0.6378	0.5301
	NDVI	0.5550	0.7192	0.6200	0.2891
Snap beans	SR	0.6075	0.4687	0.5249	0.7163
	SPAD	0.4326	0.4671	0.6567	0.4385
	NDVI	0.4251	0.7300	0.3356	0.7665
Sweet corns *	SR	0.6184	0.5756	0.6866	0.6214
	SPAD	0.5346	0.4614	0.8221	0.6250
	NDVI	0.6804	0.4619	0.7648	0.4907
Tomato	gs	0.6851 [y]	0.7153	0.3026 [y]	0.3018
	SPAD	0.8655 [y]	0.8405	0.9256 [y]	0.8482
	Fv/Fm	0.4505 [y]	0.3669	0.1103 [y]	0.0961

* during tasselling [y] = fruit weight (g), gs =stomatal conductance, SR = stomatal resistance, NI = non-irrigation, DI = deficit irrigation Source: [2,3,14–16].

Other researchers used the normalized differential vegetation index (NDVI) for yield prediction; it was successful for castor beans [132], soybeans [133] and beans [134]. According to Spitkó et al. [38], a medium correlation ($r = 0.5$–0.6) was detected between NDVI and final yield at 15 days after flowering of maize but not during the flowering period. Different stress indices such as stress degree days (SDD) or crop water stress index (CWSI) can be used to evaluate the water stress tolerance of genotypes [25] for scheduling of irrigation [93] and maybe for prediction of yield. In the case of sweet corn, significant negative correlation was detected between the CWSI and chlorophyll content of leaves ($r = 0.802$) but for the CWSI, a significant positive ($r = 0.478$) correlation was observed with the yield [25].

Helyes et al. [31] found a close correlation between the stress degree days (SDD) and yield of tomatoes. If the canopy temperature exceeded the air temperature (at noon), transpiration was reduced, which indicated water stress and resulted in yield reduction and quantity. Figure 4 shows the interrelation between the canopy and air temperature difference values and the yield. In our experiments the correlation was significant at $p = 0.01$ level with $r^2 = 0.57$ correlation coefficient. High yield per hectare can be achieved if the difference between the cumulative canopy and the air temperature is negative during the growing season.

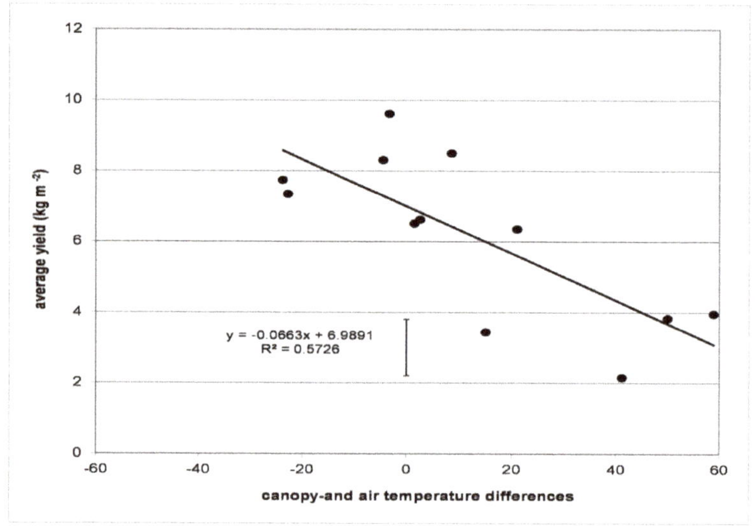

Figure 4. Correlation between canopy and air temperature differences and yield Source: [31].

6. Use of Physiological Characteristics

The use of physiological traits in a breeding program, either directly by selection or stress markers, depends on their genetic correlation with the yield, heritability and genotype × environment interaction [11,135]. Under water stress, high heritability of stomatal resistance, photosynthesis rate and transpiration rate ($h^2 = 0.91$–0.99) was found for *Vigna mungo* that gives a possibility for successful selection of genotypes [35]. Under severe drought, stomatal conductance and relative chlorophyll content of leaves (SPAD) measured during flowering correlated with the expected yield therefore they are suitable for the selection of individual genotypes for green peas and tomatoes, while the use of these traits for the selection of sweet corn can be efficient only under moderate water stress (Table 4). In the case of snap beans, because the water deficiency has a significant effect on leaf area, the normalized differential vegetation index (NDVI) measured during flowering can predict the expected yield more efficiently than the SPAD value of the leaves of individual plants.

Application of remote sensing techniques makes monitoring the water status of plants and real time irrigation easy [39,136]. The trend in the canopy temperature and the difference between the leaf temperature and air temperature (SDD) can be considered to be the water stress markers of plants [92]. The relationships between the canopy temperature, air temperature and transpiration involving the atmospheric and soil conditions and plant characteristics [40], was used to develop the crop water stress index (CWSI), indicating the need for irrigation. During drought, the decrease in NDVI occurred to a larger extent in snap beans, while it was less in sweet corn and hardly changed in green peas in comparison with optimal water supply conditions (Table 3). This explained why the NDVI was used as spectral indicator for irrigation scheduling mainly in snap beans [136].

In summary, some of the physiological traits influencing the decrease of water loss and biomass production of plants can be used to evaluate the water status of vegetable crops and the water stress tolerance of genotypes. During the generative period, under water deficit conditions, the changes in the stomatal conductance and chlorophyll content of leaves for individual plants is suitable for the estimation the productivity of genotypes. Nevertheless, leaf area of crops should be taken into consideration as they determine the transpiration and their chlorophyll density influences the intensity of photosynthesis and finally the yield. Water stress indices and spectral vegetation indices seemed to be more appropriate in the detection of perceived water deficiency than for the prediction of final yield.

Author Contributions: E.N. planned and wrote the first draft of the review. L.H. contributed to the writing and reviewed the final draft.

Funding: The publication was supported by the Ministry of Human Capacities grant Higher Education Institutional Excellence Program in framework of the water related research of Szent István University and grant number TUDFO/51757/2019-ITM FEKUTSTRAT and EFOP-3.6.3-VEKOP-16-2017-00008. The project is co-financed by the European Union and the European Social Fund.

Conflicts of Interest: The authors declare no conflict of interest.

References

1. Ray, P. Hi-Tech Horticulture and Climate Change. In *Climate Dinamics in Horticultural Science, Principles and Applications*; Choudhary, M.L., Patel, V.B., Siddiqui, M.W., Mahdi, S.S., Eds.; Apple Academic Press: Oakville, ON, Canada; CRC Press Taylor & Francis Group: Boca Raton, FL, USA, 2015; Volume 1, pp. 1–22.

2. Nemeskéri, E.; Molnár, K.; Rácz, C.; Dobos, A.C.; Helyes, L. Effect of Water Supply on Spectral Traits and Their Relationship with the Productivity of Sweet Corns. *Agronomy* **2019**, *9*, 63. [CrossRef]

3. Nemeskéri, E.; Neményi, A.; Bőcs, A.; Pék, Z.; Helyes, L. Physiological Factors and their Relationship with the Productivity of Processing Tomato under Different Water Supplies. *Water* **2019**, *11*, 586. [CrossRef]

4. Aroca, R.; Porcel, R.; Ruiz-Lozano, J.M. Regulation of root water uptake under abiotic stress conditions. *J. Exp. Bot.* **2012**, *63*, 43–57. [CrossRef] [PubMed]

5. Limami, A.; Diab, H.; Lothier, J. Nitrogen metabolism in plants under low oxygen stress. *Planta* **2014**, *239*, 531–541. [CrossRef] [PubMed]

6. Zhu, M.; Li, F.H.; Shi, Z.S. Morphological and photosynthetic response of waxy corn inbred line to waterlogging. *Photosynthetica* **2016**, *54*, 636–640. [CrossRef]

7. Tian, L.; Bi, W.; Liu, X.; Sun, L.; Li, J. Effects of waterlogging stress on the physiological response and grain-filling characteristics of spring maize (*Zea mays* L.) under field conditions. *Acta Physiol. Plant.* **2019**, *41*, 63. [CrossRef]

8. Food and Agriculture Organization. FAOSTAT Crops Production. 2017. Available online: http://www.fao.org/faostat/en/#data/QC (accessed on 9 July 2018).

9. Heszky, L. Szárazság és a növény kapcsolata. *Agrofórum* **2007**, *18*, 37–41. (In Hungarian)

10. Patanè, C.; Tringali, S.; Sortino, O. Effects of deficit irrigation on biomass, yield, water productivity and fruit quality of processing tomato under semi-arid Mediterranean climate conditions. *Sci. Hortic.* **2011**, *129*, 590–596. [CrossRef]

11. Mir, R.R.; Zaman-Allah, M.; Sreenivasulu, N.; Trethowan, R.; Varshney, R.K. Integrated genomics, physiology and breeding approaches for improving drought tolerance in crops. *Theor. Appl. Genet.* **2012**, *125*, 625–645. [CrossRef] [PubMed]

12. Kang, S.; Shi, W.; Zhang, J. An improved water-use efficiency for maize grown under regulated deficit irrigation. *Field Crop. Res.* **2000**, *67*, 207–214. [CrossRef]

13. Kirda, C. *Deficit Irrigation Scheduling Based on Plant Growth Stages Showing Water Stress Tolerance*; Deficit Irrigation Practice Water Report 22; FAO: Rome, Italy, 2002; pp. 3–10.

14. Nemeskéri, E.; Molnár, K.; Vígh, R.; Nagy, J.; Dobos, A. Relationships between stomatal behaviour, spectral traits and water use and productivity of green peas (Pisum sativum L.) in dry seasons. *Acta Physiol. Plant.* **2015**, *37*, 1–16. [CrossRef]

15. Nemeskéri, E.; Molnár, K.; Pék, Z.; Helyes, L. Effect of water supply on the water use-related physiological traits and yield of snap beans in dry seasons. *Irrig. Sci.* **2018**, *36*, 143–158. [CrossRef]

16. Nemeskéri, E.; Molnár, K.; Helyes, L. Relationships of spectral traits with yield and nutritional quality of snap beans (Phaseolus vulgaris L.) in dry seasons. *Arch. Agron. Soil Sci.* **2018**, *64*, 1222–1239. [CrossRef]

17. Fang, X.; Turner, N.C.; Yan, G.; Li, F.; Siddique, K.H.M. Flower numbers, pod production, pollen viability, and pistil function are reduced and flower and pod abortion increased in chickpea (Cicer arietinum L.) under terminal drought. *J. Exp. Bot.* **2010**, *61*, 335–345. [CrossRef] [PubMed]

18. Behboudian, M.H.; Ma, Q.; Turner, N.C.; Palta, J.A. Reactions of chickpea to water stress: Yield and seed composition. *J. Sci. Food Agric.* **2001**, *81*, 1288–1291. [CrossRef]

19. Boutraa, T.; Sanders, F.E. Influence of Water Stress on Grain Yield and Vegetative Growth of Two Cultivars of Bean (Phaseolus vulgaris L.). *J. Agron. Crop. Sci.* **2001**, *187*, 251–257. [CrossRef]

20. Beshir, H.; Bueckert, R.; Tar'An, B. Effect of temporary drought at different growth stages on snap bean pod quality and yield. *Afr. Crop. Sci. J.* **2016**, *24*, 317–330. [CrossRef]

21. Durigon, A.; Evers, J.; Metselaar, K.; Lier, Q.D.J.V. Water Stress Permanently Alters Shoot Architecture in Common Bean Plants. *Agronomy* **2019**, *9*, 160. [CrossRef]

22. Baigorri, H.; Antolín, M.C.; Sánchez-Díaz, M. Reproductive response of two morphologically different pea cultivars to drought. *Eur. J. Agron.* **1999**, *10*, 119–128. [CrossRef]

23. Moser, S.B.; Feil, B.; Jampatong, S.; Stamp, P. Effects of pre-anthesis drought, nitrogen fertilizer rate, and variety on grain yield, yield components, and harvest index of tropical maize. *Agric. Water Manag.* **2006**, *81*, 41–58. [CrossRef]

24. Öktem, A. Effect of water shortage on yield, and protein and mineral compositions of drip-irrigated sweet corn in sustainable agricultural systems. *Agric. Water Manag.* **2008**, *95*, 1003–1010. [CrossRef]

25. Uçak, A.B.; Öktemb, A.; Sezerc, C.; Cengizc, R.; İnald, B. Determination of arid and temperature resistant sweet corn (Zea mays saccharata Sturt) lines. *Int. J. Environ. Agric. Res.* **2016**, *2*, 79–88.

26. Helyes, L.; Varga, G. Irrigation demand of tomato according to the results of three decades. *Acta Hortic.* **1994**, *376*, 323–328. [CrossRef]

27. Helyes, L.; Varga, G.; Dimény, J.; Pék, Z. The simultaneous effect of variety, irrigation and weather on tomato yield. *Acta Hortic.* **1999**, *487*, 499–506. [CrossRef]

28. Patane, C.; Cosentino, S.; Cosentino, S. Effects of soil water deficit on yield and quality of processing tomato under a Mediterranean climate. *Agric. Water Manag.* **2010**, *97*, 131–138. [CrossRef]

29. Pires, R.C.D.M.; Furlani, P.R.; Ribeiro, R.V.; Junior, D.B.; Sakai, E.; Lourenção, A.L.; Neto, A.T. Irrigation frequency and substrate volume effects in the growth and yield of tomato plants under greenhouse conditions. *Sci. Agric.* **2011**, *68*, 400–405. [CrossRef]

30. Bahadur, A.; Chatterjee, A.; Kumar, R.; Singh, M.; Naik, P.S. Physiological and biochemical basis of drought tolerance in vegetables. *Veg. Sci.* **2011**, *38*, 1–16.

31. Helyes, L.; Bőcs, A.; Pék, Z. Effect of water supply on canopy temperature, stomatal conductance and yield quantity of processing tomato (Lycopersicon esculentum Mill.). *Int. J. Hortic. Sci.* **2010**, *16*, 13–15. [CrossRef]

32. Nankishore, A.; Farrell, A.D. The response of contrasting tomato genotypes to combined heat and drought stress. *J. Plant Physiol.* **2016**, *202*, 75–82. [CrossRef]

33. Agbna, G.H.; Dongli, S.; Zhipeng, L.; Elshaikh, N.A.; Guangcheng, S.; Timm, L.C. Effects of deficit irrigation and biochar addition on the growth, yield, and quality of tomato. *Sci. Hortic.* **2017**, *222*, 90–101. [CrossRef]

34. Rodriguez-Ortega, W.; Martinez, V.; Rivero, R.; Cámara-Zapata, J.-M.; Mestre, T.; Garcia-Sanchez, F. Use of a smart irrigation system to study the effects of irrigation management on the agronomic and physiological responses of tomato plants grown under different temperatures regimes. *Agric. Water Manag.* **2017**, *183*, 158–168. [CrossRef]

35. Gurumurthy, S.; Sarkar, B.; Vanaja, M.; Lakshmi, J.; Yadav, S.K.; Maheswari, M. Morpho-physiological and biochemical changes in black gram (Vigna mungo L. Hepper) genotypes under drought stress at flowering stage. *Acta Physiol. Plant.* **2019**, *41*, 42. [CrossRef]

36. Stone, P.J.; Wilson, D.R.; Jamieson, P.D.; Gillespie, R.N. Water deficit effects on sweet corn. Part II. Canopy development. *Aust. J. Agric. Res.* **2001**, *54*, 115–126. [CrossRef]

37. Genc, L.; Inalpulat, M.; Kizil, U.; Mirik, M.; Smith, S.E.; Mendes, M. Determination of water stress with spectral reflectance on sweet corn (Zea mays L.) using classification tree (CT) analysis. *Zemdirbyste-Agriculture* **2013**, *100*, 81–90. [CrossRef]

38. Spitkó, T.; Nagy, Z.; Zsubori, Z.T.; Szőke, C.; Berzy, T.; Pintér, J.; Márton, L. Connection between normalized difference vegetation index and yield in maize. *Plant Soil Environ.* **2016**, *62*, 293–298. [CrossRef]

39. Zhou, J.; Khot, L.R.; Boydston, R.A.; Miklas, P.N.; Porter, L. Low altitude remote sensing technologies for crop stress monitoring: A case study on spatial and temporal monitoring of irrigated pinto bean. *Precis. Agric.* **2018**, *19*, 555–569. [CrossRef]

40. Chatterjee, A.; Solankey, S. Functional Physiology in Drought Tolerance of Vegetable Crops: An Approach to Mitigate Climate Change Impact. In *Climate Dynamics in Horticultural Science, Principles and Applications*; Choudhary, M.L., Patel, V.B., Siddiqui, M.W., Mahdi, S.S., Eds.; Apple Academic Press: Oakville, ON, Canada; CRC Press Taylor & Francis Group: Boca Raton, FL, USA, 2015; Volume 1, pp. 149–171.

41. Cattivelli, L.; Baldi, P.; Crosatti, C.; Di Fonzo, N.; Faccioli, P.; Grossi, M.; Mastrangelo, A.M.; Pecchioni, N.; Stanca, A.M. Chromosome regions and stress-related sequences involved in resistance to abiotic stress in Triticeae. *Plant Mol. Biol.* **2002**, *48*, 649–665. [CrossRef]

42. Ramanjulu, S.; Bartels, D. Drought- and desiccation-induced modulation of gene expression in plants. *Plant Cell Environ.* **2002**, *25*, 141–151. [CrossRef]

43. Zhang, J.; Zheng, H.G.; Aarti, A.; Pantuwan, G.; Nguyen, T.T.; Tripathy, J.N.; Sarial, A.K.; Robin, S.; Babu, R.C.; Nguyen, B.D.; et al. Locating genomic regions associated with components of drought resistance in rice: Comparative mapping within and across species. *Theor. Appl. Genet.* **2001**, *103*, 19–29. [CrossRef]

44. Lanceras, J.C.; Pantuwan, G.; Jongdee, B.; Toojinda, T. Quantitative Trait Loci Associated with Drought Tolerance at Reproductive Stage in Rice. *Plant Physiol.* **2004**, *135*, 384–399. [CrossRef]

45. Cattivelli, L.; Rizza, F.; Badeck, F.-W.; Mazzucotelli, E.; Mastrangelo, A.M.; Francia, E.; Maré, C.; Tondelli, A.; Stanca, A.M. Drought tolerance improvement in crop plants: An integrated view from breeding to genomics. *Field Crop. Res.* **2008**, *105*, 1–14. [CrossRef]

46. De Leonardis, A.M.; Petrarulo, M.; Vita, P.D.; Mastrangelo, A.M. Genetic and Molecular Aspects of Plant Response to Drought in Annual Crop Species. In *Advances in Selected Plant Physiology Aspects*; Giuseppe, M., Dichio, B., Eds.; InTech Publisher: Rijeka, Croatia, 2012; pp. 45–74.

47. Kumar, R.; Solankey, S.S.; Singh, M. Breeding for drought tolerance in vegetables. *Veg. Sci.* **2012**, *39*, 1–15.

48. Augé, R.M.; Sylvia, D.M.; Park, S.; Buttery, B.R.; Saxton, A.M.; Moore, J.L.; Cho, K. Partitioning mycorrhizal influence on water relations of Phaseolus vulgaris into soil and plant components. *Can. J. Bot.* **2004**, *82*, 503–514. [CrossRef]

49. Mnasri, B.; Aouani, M.E.; Mhamdi, R. Nodulation and growth of common bean (Phaseolus vulgaris) under water deficiency. *Soil Biol. Biochem.* **2007**, *39*, 1744–1750. [CrossRef]

50. Dalton, D.A.; Moran, J.F.; Iturbe-Ormaetxe, I.; Matamoros, M.A.; Rubio, M.C.; Iturbe-Ormaetxe, I. Reactive oxygen species and antioxidants in legume nodules. *Physiol. Plant.* **2000**, *109*, 372–381.

51. Collados, C.; Barea, J.M.; Ruiz-Lozano, J.M.; Azcón, R. Arbuscular mycorrhizal symbiosis can alleviate drought-induced nodule senescence in soybean plants. *New Phytol.* **2001**, *151*, 493–502.

52. Esfahani, M.N.; Mostajeran, A. Rhizobial strain involvement in symbiosis efficiency of chickpea–rhizobia under drought stress: Plant growth, nitrogen fixation and antioxidant enzyme activities. *Acta Physiol. Plant.* **2011**, *33*, 1075–1083. [CrossRef]

53. Nicholas, S. Plant resistance to environmental stress. *Curr. Opin. Biotechnol.* **1998**, *9*, 214–219.

54. Nemeskéri, E. Water deficiency resistance study on soya and bean cultivars. *Acta Agron. Hung.* **2001**, *49*, 83–93. [CrossRef]

55. Marouelli, W.A.; Silva, W.L.C. Water tension thresholds for processing tomatoes under drip irrigation in Central Brazil. *Irrig. Sci.* **2007**, *25*, 411–418. [CrossRef]

56. Nemeskéri, E.; Horváth, K.; Pék, P.; Helyes, L. Effect of mycorrhizal and bacterial products on the traits related to photosynthesis and fruit quality of tomato under water deficiency conditions. *Acta Hortic.* **2019**, 61–66. [CrossRef]

57. Nilsen, E.T.; Forseth, E.N., Jr. The role of leaf movements for optimizing photosynthesis in relation to environmental variation. In *Advance in Photosynthesis and Respiration, The Leaf: A Platform for Performing Photosynthesis*; Adams, W.W., Terashima, I., Eds.; Springer: Basel, Switzerland, 2018; Volume 4, pp. 401–423.

58. Fernandez, D.; Castrillo, M. Maize leaf rolling initiation. *Photosynthetica* **1999**, *37*, 493–497. [CrossRef]

59. Sağlam, A.; Kadioglu, A.; Demiralay, M.; Terzi, R. Leaf Rolling Reduces Photosynthetic Loss in Maize Under Severe Drought. *Acta Bot. Croat.* **2014**, *73*, 315–332. [CrossRef]

60. Pastenes, C.; Pimentel, P.; Lillo, J. Leaf movements and photoinhibition in relation to water stress in field-grown beans. *J. Exp. Bot.* **2005**, *56*, 425–433. [CrossRef]

61. Gomes, A.M.M.; Lagoa, A.M.M.A.; Medina, C.L.; Machado, E.C.; Machado, M.A. Interactions between leaf water potential, stomatal conductance and abscisic acid content of orange trees submitted to drought stress. *Braz. J. Plant Physiol.* **2004**, *16*, 155–161. [CrossRef]

62. Parry, A.D.; Horgan, R. Abscisic acid biosynthesis in roots. *Planta* **1992**, *187*, 185–191. [CrossRef]

63. Reynolds-Henne, C.E.; Langenegger, A.; Mani, J.; Schenk, N.; Zumsteg, A.; Feller, U. Interactions between temperature, drought and stomatal opening in legumes. *Environ. Exp. Bot.* **2010**, *68*, 37–43. [CrossRef]

64. De Carvalho, M.H.C.; Laffray, D.; Louguet, P. Comparison of the physiological responses of Phaseolus vulgaris and Vigna unguiculata cultivars when submitted to drought conditions. *Environ. Exp. Bot.* **1998**, *40*, 197–207. [CrossRef]

65. Garcia, A.L.; Marcelis, L.; Garcia-Sanchez, F.; Nicolas, N.; Martinez, V. Moderate water stress affects tomato leaf water relations in dependence on the nitrogen supply. *Biol. Plant.* **2007**, *51*, 707–712. [CrossRef]

66. Zhou, R.; Kong, L.; Wu, Z.; Rosenqvist, E.; Wang, Y.; Zhao, L.; Zhao, T.; Ottosen, C.O. Physiological response of tomatoes at drought, heat and their combination followed by recovery. *Physiol. Plant.* **2019**, *165*, 144–154. [CrossRef]

67. Du, W.-J.; Yu, D.-Y.; Fu, S.-X. Analysis of QTLs for the trichome density on the upper and downer surface of leaf blade in soybean [Glycine max (L.) Merr.]. *Agric. Sci. China* **2009**, *8*, 529–537. [CrossRef]

68. Nobel, P.S. *Physicochemical and Environmental Plant Physiology*; Academic Press: Cambridge, MA, USA, 1991; p. 635.

69. Sam, O.; Jeréz, E.; Dell'Amico, J.; Ruiz-Sanchez, M.C. Water stress induced changes in anatomy of tomato leaf epidermis. *Biol. Plant.* **2000**, *43*, 275–277. [CrossRef]

70. Sánchez, F.J.; Manzanares, M.; De Andres, E.F.; Tenorio, J.L.; Ayerbe, L. Residual transpiration rate, epicuticular wax load and leaf colour of pea plants in drought conditions. Influence on harvest index and canopy temperature. *Eur. J. Agron.* **2001**, *15*, 57–70. [CrossRef]

71. Gomes, F.P.; Oliva, M.A.; Mielke, M.S.; Almeida, A.-A.F.; Aquino, L.A. Osmotic adjustment, proline accumulation and cell membrane stability in leaves of Cocos nucifera submitted to drought stress. *Sci. Hortic.* **2010**, *126*, 379–384. [CrossRef]

72. Sánchez, F.J.; Manzanares, M.; De Andres, E.F.; Tenorio, J.L.; Ayerbe, L. Turgor maintenance, osmotic adjustment and soluble sugar and proline accumulation in 49 pea cultivars in response to water stress. *Field Crop. Res.* **1998**, *59*, 225–235. [CrossRef]

73. Babita, M.; Maheswari, M.; Rao, L.; Shanker, A.K.; Rao, D.G. Osmotic adjustment, drought tolerance and yield in castor (Ricinus communis L.) hybrids. *Environ. Exp. Bot.* **2010**, *69*, 243–249. [CrossRef]

74. Nemeskéri, E.; Sárdi, E.; Remenyik, J.; Kőszegi, B.; Nagy, P. Study of defensive mechanisms against drought of French bean (Phaseolus vulgaris L.) varieties. *Acta Physiol. Plant.* **2010**, *32*, 1125–1134. [CrossRef]

75. Chen, Z.; Gallie, D.R. The Ascorbic Acid Redox State Controls Guard Cell Signaling and Stomatal Movement. *Plant Cell* **2004**, *16*, 1143–1162. [CrossRef]

76. Galmés, J.; Ochogavía, J.M.; Gago, J.; Roldán, E.J.; Cifre, J.; Conesa, M.A. Leaf responses to drought stress in Mediterranean accessions of Solanum lycopersicum: Anatomical adaptations in relation to gas exchange parameters. *Plant Cell Environ.* **1913**, *36*, 920–935. [CrossRef]

77. Muir, C.D.; Conesa, M.À.; Galmés, J. Independent evolution of ab- and adaxial stomatal density enables adaptation. *bioRxiv* **2015**, 1–25. [CrossRef]

78. Galmés, J.; Conesa, M.A.; Manuel Ochogavía, J.; Alejandro Perdomo, J.; Francis, D.M.; Ribas-Carbo, M.; Save, R.; Flexas, J.; Medrano, H.; Cifre, J. Physiological and morphological adaptations in relation to water use efficiency in Mediterranean accessions of *Solanum lycopersicum*. *Plant Cell Environ.* **2011**, *34*, 245–260. [CrossRef]

79. Hardy, J.P.; Anderson, V.J.; Gardner, J.S. Stomatal characteristics, conductance ratios, and drought-induced leaf modifications of semiarid grassland species. *Am. J. Bot.* **1995**, *82*, 1–7. [CrossRef]

80. Dejonge, K.C.; Taghvaeian, S.; Trout, T.J.; Comas, L.H. Comparison of canopy temperature-based water stress indices for maize. *Agric. Water Manag.* **2015**, *156*, 51–62. [CrossRef]

81. Jones, H.G.; Jones, H. Use of thermography for quantitative studies of spatial and temporal variation of stomatal conductance over leaf surfaces. *Plant Cell Environ.* **1999**, *22*, 1043–1055. [CrossRef]

82. Cornic, G.; Lawlor, D.W. Photosynthetic carbon assimilation and associated metabolism in relation to water deficits in higher plants. *Plant Cell Environ.* **2002**, *25*, 275–294.

83. Flagella, Z.; Campanile, R.G.; Stoppelli, M.C.; De Caro, A.; Di Fonzo, N. Drought tolerance of photosynthetic electron transport under CO_2-enriched and normal air in cereal species. *Physiol. Plant.* **1998**, *104*, 753–759. [CrossRef]

84. Pol, M.; Gołębiowska, D.; Miklewska, J. Influence of enhanced concentration of carbon dioxide and moderate drought on fluorescence induction in white clover (Trifolium repens L.). *Photosynthetica* **1999**, *37*, 537–542. [CrossRef]

85. Yordanov, I.; Velikova, V.; Tsonev, T. Plant Responses to Drought, Acclimation, and Stress Tolerance. *Photosynthetica* **2000**, *38*, 171–186. [CrossRef]

86. Ghanbari, A.A.; Shakiba, M.R.; Toorchi, M.; Choukan, R. Morpho-physiological responses of common bean leaf to water deficit stress. *Eur. J. Exp. Biol.* **2013**, *3*, 487–492.

87. Guida, G.; Sellami, M.H.; Mistretta, C.; Oliva, M.; Buonomo, R.; De Mascellis, R.; Patanè, C.; Rouphael, Y.; Albrizio, R.; Giorio, P. Agronomical, physiological and fruit quality responses of two Italian long-storage tomato landraces under rain-fed and full irrigation conditions. *Agric. Water Manag.* **2017**, *180*, 126–135. [CrossRef]

88. Nemeskéri, E.; Molnár, K.; Dobos, A.C. Csemegekukorica (Zea mays L. convar. saccharata) sztóma működése, és hatása a növekedésre és terméskomponensekre eltérő vízellátás alatt (Stomatal behaviour and its influence on the growing and yield components of sweet corn (Zea mays L. convar. saccharata). *Növénytermelés* **2017**, *66*, 75–95.

89. Yang, H.M.; Zhang, X.Y.; Wang, G.X. Relationships between stomatal character, photosynthetic character and seed chemical composition in grass pea at different water availabilities. *J. Agric. Sci.* **2004**, *142*, 675–681. [CrossRef]

90. Gonzalez-Dugo, M.P.; Moran, M.S.; Mateos, L.; Bryant, R. Canopy temperature variability as an indicator of crop water stress severity. *Irrig. Sci.* **2006**, *24*, 233–240. [CrossRef]

91. Bonanno, A.R.; Mack, H.J. Use of canopy—Air temperature differentials as a method for scheduling irrigation in snap beans. *J. Am. Hortic. Sci.* **1983**, *108*, 826–831.

92. Helyes, L. Relations among the water supply, foliage temperature and the yield of tomato. *Acta Hortic.* **1990**, *277*, 115–122. [CrossRef]

93. Helyes, L.; Dimény, J.; Varga, G. Az öntözés tervezése a lombfelszín-hőmérséklet alapján (Scheduling of irrigation with canopy temperature). *Növénytermelés* **2005**, *54*, 341–350.

94. Dufková, R. Difference in canopy and air temperature as an indicator of grassland water stress. *Soil Water Res.* **2006**, *1*, 127–138.

95. Taghvaeian, S.; Chávez, J.L.; Hansen, N.C. Infrared Thermometry to Estimate Crop Water Stress Index and Water Use of Irrigated Maize in Northeastern Colorado. *Remote Sens.* **2012**, *4*, 3619–3637. [CrossRef]

96. Helyes, L. A zöldségnövények vízellátottságának és öntözési igényének meghatározása a lombhőmérséklettel. Ph.D. Thesis, Agricultural Unversity of Gödöllő, Gödöllő, Hungary, 1991; p. 123. (In Hungarian).

97. Jones, H.G. *Plants and Microclimate*, 2nd ed.; Cambridge University Press: Cambridge, UK, 1992; p. 428.

98. Lawson, T.; Terashima, I.; Fujita, T.; Wang, Y. Coordination between Photosynthesis and Stomatal Behavior. In *The Leaf: A Platform for Performing Photosynthesis*; Adams, W.W., Terashima, I., Eds.; Springer: Basel, Switzerland, 2018; pp. 142–156.

99. Anjum, S.A.; Wang, L.C.; Farooq, M.; Hussain, M.; Xue, L.L.; Zou, C.M. Brassinolide Application Improves the Drought Tolerance in Maize through Modulation of Enzymatic Antioxidants and Leaf Gas Exchange. *J. Agron. Crop. Sci.* **2011**, *197*, 177–185. [CrossRef]

100. Sabagh, A.E.; Barutçular, C.; Islam, M.S. Relationships between stomatal conductance and yield under deficit irrigation in maize (Zea mays L.). *J. Exp. Biol. Agric. Sci.* **2017**, *5*, 15–21. [CrossRef]

101. Helyes, L.; Szuvandzsiev, P.; Neményi, A.; Pék, Z.; Lugasi, A. Different water supply and stomatal conductance correlates with yield quantity and quality parameters. *Acta Hortic.* **2013**, *971*, 119–125. [CrossRef]

102. Tari, I.; Camen, D.; Coradini, G.; Csiszár, J.; Fediuc, E.; Gémes, K.; Lazar, A.; Madosa, E.; Mihacea, S.; Poór, P.; et al. Changes in chlorophyll fluorescence parameters and oxidative stress responses of bush bean genotypes for selecting contrasting acclimation strategies under water stress. *Acta Biol. Hung.* **2008**, *59*, 335–345. [CrossRef]

103. Yan, H.; Wu, L.; Filardo, F.; Yang, X.; Zhao, X.; Fu, D. Chemical and hydraulic signals regulate stomatal behavior and photosynthetic activity in maize during progressive drought. *Acta Physiol. Plant.* **2017**, *39*, 125. [CrossRef]

104. Sing, S.K.; Reddy, K.R. Regulation of photosynthesis, fluorescence, stomatal conductance and water-use efficiency of cowpea [Vigna unguiculata (L.) Walp.] under drought. *J. Photochem. Photobiol. B* **2011**, *105*, 40–50. [CrossRef]

105. Lambrev, P.H.; Miloslavina, Y.; Jahns, P.; Holzwarth, A.R. On the relationship between non-photochemical quenching and photoprotection of Photosystem II. *Biochim. Biophys. Acta* **2012**, *1817*, 760–769. [CrossRef]

106. Bauerle, W.L.; Weston, D.J.; Bowden, J.D.; Dudley, J.B.; Toler, J.E. Leaf absorptance of photosynthetically active radiation in relation to chlorophyll meter estimates among woody plant species. *Sci. Hortic.* **2004**, *101*, 169–178. [CrossRef]

107. Flexas, J.; Bota, J.; Escalona, J.M.; Sampol, B.; Medrano, H. Effects of drought on photosynthesis in grapevines under field conditions: An evaluation of stomatal and mesophyll limitations. *Funct. Plant Biol.* **2002**, *29*, 461–471. [CrossRef]

108. Filek, M.; Łabanowska, M.; Kościelniak, J.; Biesaga-Kościelniak, J.; Kurdziel, M.; Szarejko, I.; Hartikainen, H. Characterization of barley leaf tolerance to drought stress by chlorophyll fluorescence and electron paramagnetic resonance studies. *J. Agron. Crop. Sci.* **2015**, *201*, 228–240. [CrossRef]

109. Guo, Y.; Tan, J. Recent advances in the application of chlorophyll a fluorescence from photosystem II. *Photochem. Photobiol.* **2015**, *91*, 1–14. [CrossRef]

110. Nogués, S.; Allen, D.J.; Morison, J.I.L.; Baker, N.R. Ultraviolet-B radiation effects on water relations, leaf development, and photosynthesis in droughted pea plants. *Plant Physiol.* **1998**, *117*, 173–181. [CrossRef]

111. Ismail, I.; Basahi, J.; Hassan, I. Gas exchange and chlorophyll fluorescence of pea (Pisum sativum L.) plants in response to ambient ozone at a rural site in Egypt. *Sci. Total Environ.* **2014**, *497*, 585–593. [CrossRef]

112. Demmig, B.; Björkman, O. Photon yield of O2 evolution and chlorophyll fluorescence characteristics at 77 K among vascular plants of diverse origins. *Planta* **1987**, *170*, 489–504.

113. Zulini, L.; Rubinigg, M.; Zorer, R.; Bertamini, M. Effects of drought stress on chlorophyll fluorescence and photosynthetic pigments in grapevine leaves (vitis vinifera cv. 'white riesling'). *Acta Hortic.* **2007**, *754*, 289–294. [CrossRef]

114. Wang, B.; Liu, Y.; Wu, X.; Lu, Z.; Li, G. The relationship of chlorophyll fluorescence parameters and drought tolerance in asparagus bean seedlings under drought stress. *Acta Agric. Zhejiang* **2009**, *21*, 246–249.

115. Razavi, F.; Pollet, B.; Steppe, K.; Van Labeke, M.C.; Labeke, M.C. Chlorophyll fluorescence as a tool for evaluation of drought stress in strawberry. *Photosynthetica* **2008**, *46*, 631–633. [CrossRef]

116. Ohashi, Y.; Nakayama, N.; Saneoka, H.; Fujita, K. Effects of drought stress on photosynthetic gas exchange, chlorophyll fluorescence and stem diameter of soybean plants. *Biol. Plant.* **2006**, *50*, 138–141. [CrossRef]

117. Srinivasa Rao, N.K.; Bhatt, R.M.; Mascarenhas, J.B.D.; Naren, A. Influence of moisture stress on leaf water status, osmotic potential, chlorophyll fluorescence and solute accumulation in field grown tomato cultivars. *Veg. Sci.* **1999**, *26*, 129–132.

118. Bahadur, A.; Kumar, R.; Mishra, U.; Rai, A.; Singh, M. Physiological approaches for screening of tomato genotypes for moisture stress tolerance. In Proceedings of the National Conference of Plant Physiology (NCPP-2010) BHU, Varanasi, India, 25–27 November 2010; p. 142.

119. Li, R.-H.; Guo, P.-G.; Michael, B.; Stefania, G.; Salvatore, C. Evaluation of Chlorophyll Content and Fluorescence Parameters as Indicators of Drought Tolerance in Barley. *Agric. Sci. China* **2006**, *5*, 751–757. [CrossRef]

120. Woo, N.S.; Badger, M.R.; Pogson, B.J. A rapid, non-invasive procedure for quantitative assessment of drought survival using chlorophyll fluorescence. *Plant Methods* **2008**, *4*, 27. [CrossRef]

121. Carter, G.A.; Knapp, A.K. Leaf optical properties in higher plants: Linking spectral characteristics to stress and chlorophyll concentration. *Am. J. Bot.* **2001**, *88*, 677–684. [CrossRef]

122. Bowman, W.D. The relationship between leaf water status, gas exchange, and spectral reflectance in cotton leaves. *Remote Sens. Environ.* **1989**, *30*, 249–255. [CrossRef]

123. Huntjr, E.; Rock, B. Detection of changes in leaf water content using Near- and Middle-Infrared reflectances. *Remote Sens. Environ.* **1989**, *30*, 43–54. [CrossRef]

124. Yadava, U.L. A rapid and nondestructive method to determine chlorophyll in intact leaves. *Hortic. Sci.* **1986**, *1*, 1449–1450.

125. Iturbe-Ormaetxe, I. Oxidative Damage in Pea Plants Exposed to Water Deficit or Paraquat. *Plant Physiol.* **1998**, *116*, 173–181. [CrossRef]

126. Anderson, M.C.; Zolin, C.A.; Sentelhas, P.C.; Hain, C.R.; Semmens, K.; Yilmaz, M.T.; Gao, F.; Otkin, J.A.; Tetrault, R. The Evaporative Stress Index as an indicator of agricultural drought in Brazil: An assessment based on crop yield impacts. *Remote Sens. Environ.* **2016**, *174*, 82–99. [CrossRef]

127. Xu, H.; Zhu, S.; Ying, Y.; Jiang, H. Application of multispectral reflectance for early detection of tomato disease. In Proceedings of the Optics for Natural Resources, Agriculture, and Foods, Boston, MA, USA, 1–4 October 2006; Volume 6381. [CrossRef]

128. Zwart, S.J.; Leclert, L.M.C. A remote sensing-based irrigation performance assessment: A case study of the Office du Niger in Mali. *Irrig. Sci.* **2010**, *28*, 371–385. [CrossRef]

129. Chawade, A.; Van Ham, J.; Blomquist, H.; Bagge, O.; Alexandersson, E.; Ortiz, R. High-Throughput Field-Phenotyping Tools for Plant Breeding and Precision Agriculture. *Agronomy* **2019**, *9*, 258. [CrossRef]

130. Romano, G.; Zia, S.; Spreer, W.; Sanchez, C.; Cairns, J.; Araus, J.L.; Müller, J. Use of thermography for high throughtput phenotyping of tropical maize adaptation in water stress. *Comput. Electron. Agric.* **2011**, *79*, 67–74. [CrossRef]

131. Thenkabail, P.S.; Smith, R.B.; De Pauw, E. Hyperspectral Vegetation Indices and Their Relationships with Agricultural Crop Characteristics. *Remote Sens. Environ.* **2000**, *71*, 158–182. [CrossRef]

132. Li, G.; Zhang, H.; Wu, X.; Shi, C.; Huang, X.; Qin, P. Canopy reflectance in two castor bean varieties (Ricinus communis L.) for growth assessment and yield prediction on coastal saline land of Yancheng District, China. *Ind. Crop. Prod.* **2011**, *33*, 395–402. [CrossRef]

133. Ma, B.L.; Dwyer, L.M.; Costa, C.; Cober, E.R.; Morrison, M.J. Early Prediction of Soybean Yield from Canopy Reflectance Measurements. *Agron. J.* **2001**, *93*, 1227–1234. [CrossRef]

134. Gutiérrez-Rodríguez, M.; Escalante-Estrada, J.A.; Gonzalez, M.T.R.; Reynolds, M.P. Canopy reflectance indices and its relationship with yield in common bean plants (Phaseolus vulgaris L.) with phosphorous supply. *Int. J. Agric. Biol.* **2006**, *8*, 203–207.

135. Scully, B.; Wallace, D.; Viands, D. Heritability and Correlation of Biomass, Growth Rates, Harvest Index, and Phenology to the Yield of Common Beans. *J. Am. Soc. Hortic. Sci.* **1991**, *116*, 127–130. [CrossRef]

136. Köksal, E.S. Hyperspectral reflectance data processing through cluster and principal component analysis for estimating irrigation and yield related indicators. *Agric. Water Manag.* **2011**, *98*, 1317–1328. [CrossRef]

Review

Phytohormone-Mediated Stomatal Response, Escape and Quiescence Strategies in Plants under Flooding Stress

Kazi Khayrul Bashar [1,*], Md. Zablul Tareq [2], Md. Ruhul Amin [1], Ummay Honi [1], Md. Tahjib-Ul-Arif [3], Md. Abu Sadat [1] and Quazi Md. Mosaddeque Hossen [1]

[1] Bangladesh Jute Research Institute, Manik Mia Avenue, Dhaka 1207, Bangladesh; shironbge03@gmail.com (M.R.A.); lethe_0000@yahoo.com (U.H.); sadat@snu.ac.kr (M.A.S.); mosaddequebjri@gmail.com (Q.M.M.H.)

[2] Bangladesh Jute Research Institute, Jagir, Manikganj 1800, Bangladesh; zablulbarj@gmail.com

[3] Department of Biochemistry and Molecular Biology, Bangladesh Agricultural University, Mymensingh 2202, Bangladesh; arif1002215@gmail.com

* Correspondence: kazi.khayrulbashar@gmail.com

Received: 8 January 2019; Accepted: 16 January 2019; Published: 22 January 2019

Abstract: Generally, flooding causes waterlogging or submergence stress which is considered as one of the most important abiotic factors that severely hinders plant growth and development. Plants might not complete their life cycle even in short duration of flooding. As biologically intelligent organisms, plants always try to resist or survive under such adverse circumstances by adapting a wide array of mechanisms including hormonal homeostasis. Under this mechanism, plants try to adapt through diverse morphological, physiological and molecular changes, including the closing of stomata, elongating of petioles, hollow stems or internodes, or maintaining minimum physiological activity to store energy to combat post-flooding stress and to continue normal growth and development. Mainly, ethylene, gibberellins (GA) and abscisic acid (ABA) are directly and/or indirectly involved in hormonal homeostasis mechanisms. Responses of specific genes or transcription factors or reactive oxygen species (ROS) maintain the equilibrium between stomatal opening and closing, which is one of the fastest responses in plants when encountering flooding stress conditions. In this review paper, the sequential steps of some of the hormone-dependent survival mechanisms of plants under flooding stress conditions have been critically discussed.

Keywords: flood; plants; hormonal homeostasis; physiological activity

1. Introduction

Several oxygen limiting factors, such as flooding, waterlogging, and partial or full submergence are detrimental for normal growth and development of plants [1]. Sea, river belt and low land areas experience limited or reduced crop production due to the flooding stress. Plants try to adapt to these adverse conditions by applying several strategies, like the storage of energy, elongation of the petiole or internodes, maintenance of water level by regulating stomatal movements, the formation of adventitious roots, development of aerenchyma etc. [2–5]. Crop plants simultaneously activate various biochemical reactions, molecular and signaling pathways, and physiological processes to cope with this oxygen-limiting condition [1,6].

Phytohormone plays a central role in all morphological, anatomical, biochemical, molecular and signaling mechanisms for plant survival under oxygen-limiting stress conditions. Predominantly, ethylene, gibberellins (GA) and abscisic acid (ABA) play the most crucial roles during submergence stress conditions in plants [7]. Ethylene directly and/or indirectly induces GA expression that aids plants in carrying out the escape and/or quiescence strategy. Petiole/internode elongation and storage of carbohydrates results from the escape and quiescence strategy, respectively in plants under submergence stress [8]. GAs are directly involved in escaping submergence stress in both DELLA (N-terminal D-E-L-L-A amino acid sequence) dependent and independent pathways [9]. Furthermore, ethylene and ABA are directly responsible for the stomatal closure under waterlogging or post-waterlogging stress [10]. On the other hand, GAs are responsible for the stomatal opening under the same stress conditions [11]. So, balanced expression of these three hormones is highly indispensable to maintain the ratio of stomatal opening and closing in plants under excess water stress conditions.

Based on the above facts, this review paper proposes flowcharts of sequential steps for stomatal closing and escaping and quiescencing strategies to offer the best possible explanations. This might help plant scientists to consider several factors during the development of waterlogging-tolerant plant varieties.

2. Ethylene, GA and ABA Interactions in Plants under Submergence Stress

Under submergence stress, ethylene, GA and ABA play influential roles in the survivability of the submerged plants, where ABA biosynthesis is reduced and GA signaling is induced for shoot elongation, especially in rice plants (variety: C9285) [12,13].

Plants have evolved two types of strategies, i.e., escape strategy and quiescence strategy, to survive under flooding stress. In the escape strategy, the rice plant elongates its internodes under slow progressive flooding conditions. On the other hand, rice plants reserve energy under deep transient flash flooding conditions to escape the unfavorable conditions, which is termed the quiescence strategy [14,15]. It is quite interesting that two different functions of two distinct gene families under the same subgroup of a transcription factor are involved in submergence tolerance in plants. This is a complex process, but one that is interesting to study, and required to unveil this mechanism by functional genomics. Under the *AP2/ERF* (*Apelata2/Ethylene response factor*) transcription factor (TF) subgroup, the *Snorkel (SK)* gene is responsible for internode elongation, whereas *Sub1A* (*Submergence 1A*) is related to shoot elongation restriction [16].

In low land rice varieties, the *ERF* transcription factor, *Sub1*, is considered as a major player in submergence tolerance [17,18]. As a quiescence strategy in rice genotypes, ethylene directly enhances *Sub1A* expression. *Sub1A* induces the over accumulation of GA signaling repressors, *Slender Rice-1 (SLR1)* and *SLR1 Like-1 (SLRL1)*. *Sub1A,* a group under VII *AP2/ERF* transcription factor, restricts shoot elongation by suppressing the *SLR1* and *SLRL2* for saving energy that is necessary for growth and development under desubmergence conditions. Therefore, ethylene is indirectly responsible for the induction of these GA signaling repressors and the reduction of GA responsive gene expression under submergence stress through the *Sub1A*-dependent pathway. *Sub1A* may also play an important role in limiting ethylene production during submergence stress conditions, resulting in the restriction of ethylene-induced enhancement of GA responsiveness in submergence tolerant varieties [17]. The function of *Sub1A1* is also regulated by MPK3 (Mitogen-activated protein kinase3). MPK3-dependent phosphorylated *Sub1A1* binds to the G-box of the MPK3 promoter to regulate its activity [19].

On the other hand, ethylene can directly enhance GA responsive shoot elongation in rice genotypes as an escape strategy of submergence stress, which is not only a highly energy-consuming process, but also requires the continuous production of energy. Due to the lack of *Sub1A* in submergence-susceptible genotypes, plants' survivability in the desubmergence stage is very low or limited [20]. Thus, *Sub1A* acts as a limiting factor for ethylene-promoted GA responsive shoot elongation in tolerant genotypes during submergence conditions to store or save the energy required for normal physiological and biochemical activities [21].

Moreover, in rice, *Sub1A* actively participates in maintaining chlorophyll contents and carbohydrate reserves in photosynthetic tissues [22]. *Sub1A* increases brassinosteroid (BR) levels in rice plants under submergence stress. BR induces GA catabolic *SLR1* proteins which restrict shoot elongation under oxygen-limiting conditions. In addition, BR enhances the expression of *GA2ox7* (*GA 2 oxidase7*) as an early response (within 1 d after submergence), which is responsible for catabolic GA degradation of endogenous GA_4 [23] (Figure 1). *OsAP2-39*, an *Apelata 2* (*AP2*) transcription factor, directly regulates the ABA biosynthetic gene *OsNCED1* (*9-cis-epoxycarotenoid dioxygenase 1*) and the GA repressing *EUI* (*Elongation of uppermost internode I*) gene. Over-expression of *OsAP2-39* has been shown to enhance drought resistance in rice by producing more ABA and degrading GA [24], which supports the antagonistic crosstalk between ABA and GA as a crucial mechanism to control plant growth and development under abiotic stress conditions [25]. From the above discussion, it can be said that *OsAP2-39* might have great scope to restrict GA signaling in plants under waterlogging or submergence conditions, and stomatal control through ABA signaling pathways under desubmergence conditions.

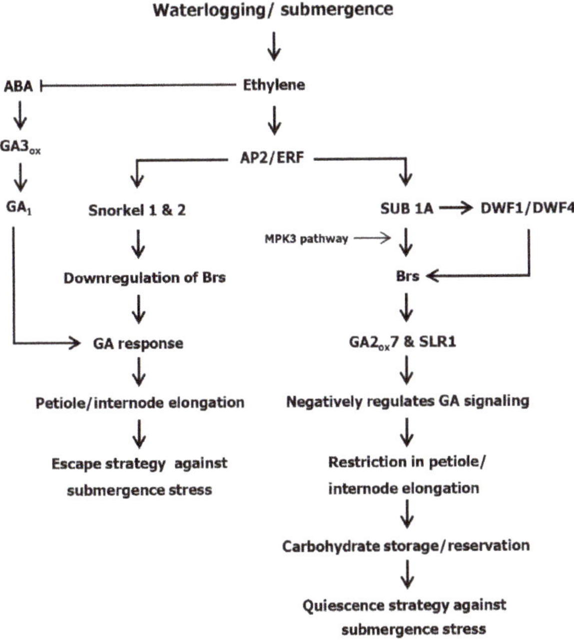

Figure 1. Ethylene-mediated escape and quiescence strategies in plants against submergence stress. Enhanced ethylene expression under submergence stress induces *Snorkel* and *Sub1A* genes for escape and quiescence strategies respectively. Ethylene suppresses ABA expression, which triggers GA_1 expression for escape strategy in plants. ABA: Abscisic acid; AP2/ERF: Apelata2/Ethylene response factor; Brs: Brassinosteroids; DWF1/4: Dwarf 1/4; GA_1: Gibberellins 1; GA2ox7: GA 2 oxidase7; GA3ox: GA 3 oxidase; Sub1A: Submergence1A; SLRL1: SLR1 Like-1.

In deep water rice, ethylene enhances the expression of *Snorkel1* (*SK1*) and *Snorkel2* (*SK2*) which are responsible for significant internode elongation via GA signaling pathways [26]. *Snorkel* genes are only present in lowland deepwater rice accessions for their internode elongation through downregulation of BR biosynthesis as an escape mechanism under submergence stress [1] (Figure 2).

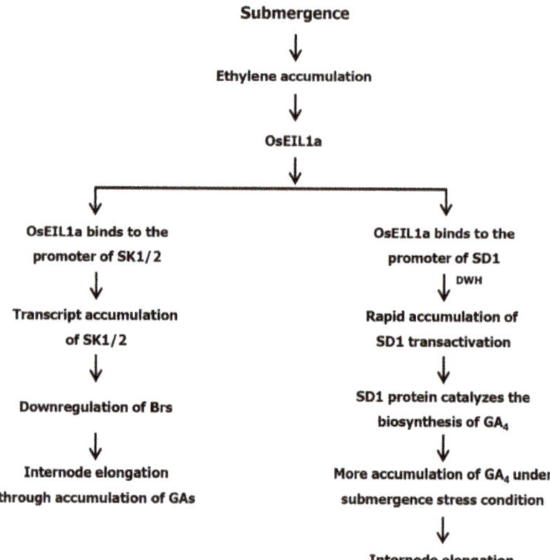

Figure 2. Proposed model for rice internode elongation under submergence stress conditions. Modified figure [13]. *EIL1a* binds with the both promoters of *SK1/2* and *SD1* in rice under submergence stress conditions. *SK1/2* downregulates Brs that induce internode elongation through the accumulation of GAs. On the other hand, *SD1* induces internode elongation through the accumulation of GA_4. Brs: brassinosteroids; EIL1a: Ethylene Insensitive 3-like 1a; GA: Gibberellins; SD1: Semidwarf1; SK1/2: Snorkel 1/2.

In rice, *Snorkel* dependent (variety: C9285) and independent (variety: T65) internode elongation for escaping the submergence stress conditions was discovered by [13]. In this escaping strategy the accumulated ethylene enhances *OsEIL1a* (*Ethylene Insensitive 3-like 1a*) in rice plants. In the *Snorkel* dependent escaping strategy, *OsEIL1a* binds to the promoter of *Snorkel1/2* to accumulate the transcript of *Snorkel1/2* [27]. This leads to the downregulation of BR to induce GA-mediated (mainly GA_1) internode elongation [22,25]. GA response enhances the expression of cyclins transcription factor, which leads to rapid cell division in lotus (*Nelumbo nucifera*) under submergence stress conditions [28]. However, as a *Snorkel* independent escaping strategy, *OsEIL1a* binds to the promoter of *SD1* (*Semidwarf1*) for DWH (deepwater rice-specific haplotype) mediated rapid amplification of *SD1* transactivation. The SD1 protein catalyzes the biosynthesis of bioactive GA species, GA_4 that increases GA_4 level in addition to GA_1 after submergence. GA_4 is more capable of internode elongation than GA_1 [13]. So, *Snorkel* independent *SD1* mediated internode elongation in rice is comparatively faster than that of the *Snorkel* dependent pathway (Figure 2).

Rumex plants showed *Snorkel* independent petiole elongation. In flood-tolerant *Rumex palustris*, ethylene reduces *RpNCED* expression which inhibits ABA biosynthesis. Thus, *R. palustris* elongates its petiole by degrading ABA into phaseic acid and enhancing GA_1-mediated gene expression in an ethylene-mediated pathway under oxygen-limited condition [29]. *R. palustris* maintains gas exchange between the submerged tissues and the atmosphere by elongating shoots under long-term flooding stress condition [29,30]. In this stress condition, the accumulation of ethylene not only breaks down ABA into phaseic acid, but also downregulates ABA expression by inhibiting 9-cis-epoxycarotenoid dioxygenase expression. Elevated content of ethylene independently degrades ABA through ABA 8′ hydroxylase pathway under submergence stress conditions [31]. Inhibition of ABA stimulates GA 3-oxidase to produce bioactive GA (GA_1). The downstream function of GA is to mobilize food materials by the breakdown of starch and cell wall loosening, which ultimately elongates internode or leaf sheath to escape submergence or waterlogging stress conditions [30]. In rice, a similar type of ABA-dependent GA expression was also reported [32]. Both the quiescence strategy and escape strategy are considered as the survival mechanisms of plants under submergence stress, but escape strategy is considered as a yield limiting factor in rice plants under de-submergence stress [33].

3. DELLA-Dependent GA Expression under Submergence Stress

Gibberellin-insensitive dwarf 1 (GID1), a soluble receptor for GA signaling, is involved in GA-mediated signaling pathways in plants under stress conditions, especially during abiotic stress conditions [34,35]. GA binds to GID1 and generates a GID1-GA complex which has the ability to interact with different growth repressors like DELLA or SLENDER1 (SLR1), a DELLA ortholog in rice [33]. Five types of DELLA are present in the model plant *Arabidopsis thaliana*, namely Gibberellin-insensitive (GAI), Repressor of GA1-3 (RGA), RGA-like1 (RGL1), RGL2 and RGL3 [36]. GID1-GA complex facilitates GA-mediated interaction between GID1 and DELLA protein, which is responsible for conformational changes in DELLA proteins. The *Sleepy1* (*SLY1*) gene contain the F-box domain, which is a positive regulator of GA signaling in *Arabidopsis*. *Sleepy1* (*SLY1*) in *Arabidopsis* and GID2 in rice are capable of recognizing this change of DELLA proteins where *SCF* (*SLY1*) (Skp1, Cullin, F-box), E3 ubiquitin ligase ubiquitinates DELLA protein, targeting DELLA for degradation through proteolysis by the 26S proteasome [34,35,37]. Thus GA expression is continued in plants by suppressing the expression of DELLA protein (Figure 3A).

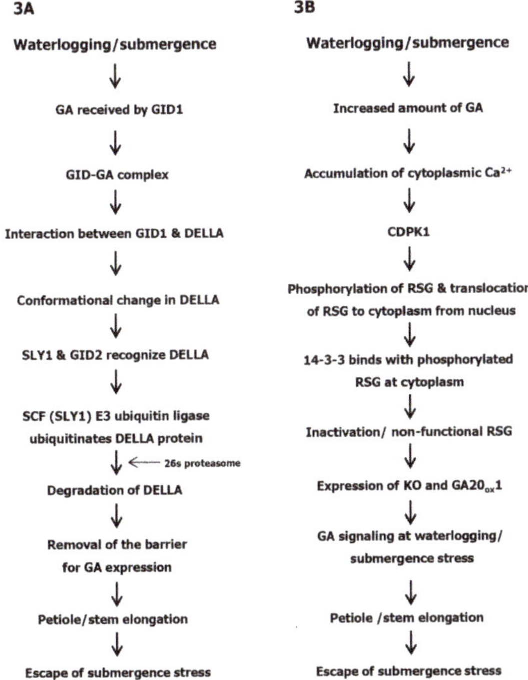

Figure 3. DELLA-dependent and -independent GA signaling in plants under submergence stress conditions. (**3A**) DELLA-dependent escape strategy in plants; GID-GA complex makes conformational change in DELLA. After that SCF ubiquitinates DELLA protein which induces GA expression for petiole/stem elongation. (**3B**) DELLA-independent escape strategy in plants; Ca^{2+} accumulation enhances CDPK1 production for the phosphorylation of RSG to translocate it into the cytoplasm from nucleus. 14-3-3 inactivates RSG as a GA-mediated submergence escaping strategy. CDPK1: Ca2+-dependent protein kinase 1; DELLA: N-terminal D-E-L-L-A amino acid sequence; GA: Gibberellins; GA20ox1: GA 20 oxidase1; GID1: Gibberellin-insensitive dwarf 1; GID2: Gibberellin-insensitive dwarf 2; RSG: Repression of shoot growth; SCF: Skp1, Cullin, F-box; SLY1: Sleepy1.

4. DELLA Independent GA Expression under Submergence Stress

GAs can increase cytoplasmic Ca^{2+} very rapidly during various cellular processes occurring inside the cell. However, the mechanism of increasing GA-mediated cytoplasmic Ca^{2+} in a DELLA-independent manner is still unknown. The GA-mediated increase of Ca^{2+} and degradation of DELLAs are completely independent processes [38]. An elevated level of cytoplasmic Ca^{2+} activates Ca^{2+}-dependent protein kinase (NtCDPK1) via a DELLA-independent GA pathway in tobacco [39]. NtCDPK1 is responsible for the translocation of Repression of shoot growth (RSG) from the nucleus to the cytoplasm [40,41]. RSG represses the expression of two important GA biosynthetic genes, *NtKO* and *NtGA20ox1*, which are considered as GA enhancing genes [42,43]. RSG binds to the promoter region of *NtGA20ox1* through GA-mediated approach while RSG binds to *NtKO* promoter in the independent manner of GA concentrations [43]. NtCDPK1 acts as a RSG kinase and phosphorylates RSG, which promotes 14-3-3 to bind with RSG at cytoplasm [40,44,45]. In rice plants, CDPK is induced under low oxygen stress for survivability under anaerobic conditions [46]. 14-3-3 proteins directly bind to the RSG in the cytoplasm and regulate RSG function negatively, making non-functional RSG [41,47]. So, inactivation of RSG by 14-3-3 proteins indirectly helps to express both *NtKO* and *NtGA20ox1* to continue GA biosynthesis for different cellular processes (Figure 3B).

5. ABA in Plants under Waterlogging Stress

Imbalanced conditions between leaf transpiration and root water uptake creates dehydration (physiological drought) stress in plants, that is noticed in plants under waterlogging stress with partially or fully damaged root tissues. The plant hormone ABA has the ability to modify root hydraulic properties [48]. For example, ABA downregulation and upregulation in tomato plants expresses lower and higher hydraulic conductance, respectively [49,50]. Depending on both the flooding duration and plant species, ABA differentially responds in leaves and in the roots of plants. *Malus sieversii* is considered as less tolerant to hypoxia than *Malus hupehensis*, showing a larger increase in ABA in both the leaf and root tissues [51]. In *Gerbera jamesonii*, ABA levels were increased in both leaf and root, where a transient increase in root follows a sharp decrease in the recovery period [52]. ABA is also increased in roots and leaves of *Triticum aestivum* L., but gradually decreases after reaching certain levels [53].

6. Stomatal Regulation at Waterlogging Stress

Stomata is a specialized epidermal pore-like structure consisting of two guard cells through which plants exchange both CO_2 and O_2 with the environment [54,55]. Stomatal conductance was insignificantly affected by flooding stress despite a significant reduction of photosynthesis in both flood-tolerant and flood-sensitive poplar genotypes [56]. A similar finding was also reported in maize plants where flooding resulted in a significant decrease in photosynthesis and ribulose-1,5-bisphosphate carboxylase activity without a noticeable reduction in the rates of stomatal conductance [57]. In GID1 mutated rice plant, increased chlorophyll content has been found, which is subsequently responsible for the increase of carbohydrate production and decrease of reactive oxygen species (ROS) accumulation under submergence stress [58]. Reduction in leaf transpiration is a common phenomenon of flooding stress, which affects the lowered stomatal aperture [59–61]. It is difficult to believe that flooding paradoxically causes leaf dehydration in plants [62–64]. Stomatal closing is operated by the turgor pressure and volume of guard cells. ABA-dependent signaling effluxes of anions, potassium ions and conversion of malate into starch trigger the reduction of turgor pressure, as well as changing the volume of guard cells close the stomata [65]. Under hypoxic stress, *hypoxia responsive universal stress protein 1* (*HRU1*) activates *AtrbohD* (*Respiratory burst oxidase homolog protein D*), following interaction between *ROP2* (*Rho of plants 2*) and *AtrbohD* in *Arabidopsis* under low oxygen stress conditions. NADPH oxidase *AtrbohD* is responsible for alcoholic fermentation and ABA-dependent stomatal closure in plants under abundant water conditions [66,67]. It has been reported that barley plants swiftly close the stomata after flooding stress imposition [68]. A similar phenomenon was also observed in pea plants under flooding conditions, where a prompt closure of stomata from older leaves was recorded. Wilting of younger leaves was protected by older pea leaves by increasing ABA-dependent stomatal closure [69]. Unwilted younger pea leaves might result from the ABA transportation from older to younger leaves or de novo biosynthesis of ABA in the younger leaves [70]. Though *Populus deltoides* is considered as a waterlogging tolerant plant species, it showed a significant reduction in stomatal conductance at both waterlogging and de-waterlogging stress, with an exception at recovery period after 90 days of waterlogging stress [71].

Hormonal Regulation in Stomatal Closing of Plants under Waterlogging Stress

Several waterlogging related experiments showed a positive correlation between ABA accumulation and increase in ROS, in soybean roots [72], barley roots and leaves [73], maize leaves [74], bread wheat roots [75]. The enzymatic mechanisms responsible for ABA-triggered ROS generation in guard cells at the molecular level are little known at present. ABA initiates H_2O_2 generation by using the plasma membrane NADPH oxidase [76]. H_2O_2 activates plasma membrane Ca^{2+} channels, resulting in an increase in Ca^{2+} level in guard cells. [77]. Inhibition of inward K^+ channels in guard cells is the result of increased Ca^{2+} level in the cytoplasm of guard cells [78,79]. This results in reduced solute accumulation following a reduced amount of water entrance in the guard cells and ultimately leading to stomata closure [80]. In *Arabidopsis*, H_2O_2 can also stimulate NO (nitric oxide) production to induce stomatal closure [81]. Ethylene and ABA activate CuAO (copper amine oxidase) in *Vicia faba* [82,83]. Oxidation of putrescine by CuAO produces H_2O_2, follows stomatal closure in *V. faba* [82] (Figure 4). ABA is also responsible for stomatal closing in plants as a survival mechanism under post-waterlogging stress [84].

Moreover, H_2O_2 is also produced by extracellular calmodulin (ExtCaM) which is activated by heterotrophic G protein [85]. In rice, accumulated ethylene induces G protein for aerenchyma formation under flooding stress [86]. G protein induces H_2O_2 production for epidermal cell death in rice under submergence stress [87]. Inactivation of CTR1 (Constitutive triple response 1) is induced by the binding of ethylene to ETR1 (Ethylene receptor 1), ERS1 (Ethylene response sensor 1) and EIN4 (Ethylene insensitive 4), resulting in the activation of G_{alpha} (G protein alpha subunit). G_{alpha} promotes H_2O_2 production in plants via NADPH oxidases. ETR1 and ERS1 translocate the signals of H_2O_2 to EIN2, EIN3 and ARR2 (2-component response regulator), which are essential for stomatal closure functioning [88]. Activated G proteins may inhibit inward K^+ channels via an elevated level of cytoplasmic Ca^{2+} in the guard cells [78,79]. From the above discussion, a proposed signaling pathway may work for stomatal closure in plants under waterlogging stress (Figure 4).

In *Arabidopsis thaliana*, flooding stress operates H_2O_2-mediated stomatal closure followed by an increase in antioxidant enzyme activities [89]. Improvement of anoxia tolerance was confirmed by applying exogenous ABA in different plants including maize, citrus, lettuce and *Arabidopsis* [90–94]. Waterlogged pea plant restricted its leaf ABA to translocate in shoot, while in non-waterlogged plants, ABA moves readily from shoots to roots [95,96]. This extra ABA in pea leaves was responsible for reducing leaf transpiration by the closing of stomata [96].

Stomata closing may occur in an ABA-independent manner. As for example, ABA was increased in citrus after 3 weeks of flooding, indicating that closing of stomata and increase of ABA are independent in citrus plants upon flooding stress, and not only that ABA was transported to younger leaves from older leaves, rather than, as expected, being transportated from roots to shoots [97].

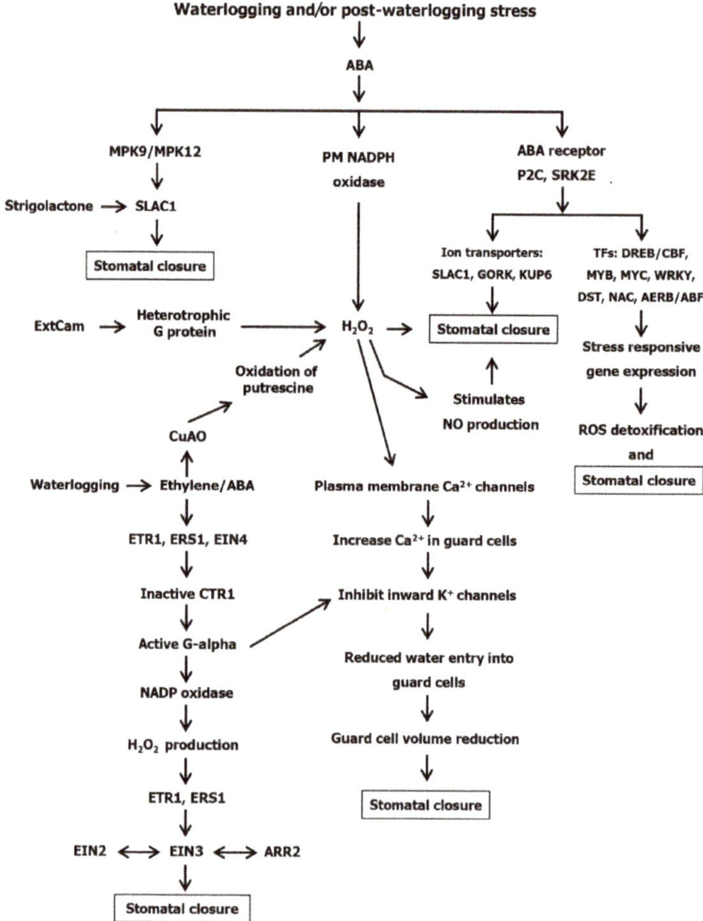

Figure 4. Proposed model for hormone mediated stomatal closure under waterlogging and/or post-waterlogging stress conditions. Modified figure [84]. Ethylene and/or ABA are directly or indirectly enhance H_2O_2 production under waterlogging and/or post-waterlogging stress conditions. H_2O_2 is directly responsible for closing of stomata or indirectly increasing Ca^{2+} in the guard cell for stomatal closure by reducing the volume of the guard cell. ABA: Abscisic acid; AREB/ABF: (ABA)-responsive element binding proteins/ABRE(ABA-responsive element)-binding factors; ARR2: 2-component response regulator; CTR1: Constitutive triple response 1; CuAO: Copper amine oxisase; DREB/CBF: Dehydration-responsive element-binding/C-repeat binding factor; DST: Drought and salt tolerance; EIN: Ethylene Insensitive; ERS1: Ethylene response sensor 1; ETR1: Ethylene receptor 1; ExtCaM: Extracellular Calmodulin; GORK: Guard cell outward-rectifying K+; KUP6: K+ uptake transporter 6; MPK3/9: Mitogen-activated protein kinase3/9; MYB: **My**eloblastosis; MYC2: **My**elocytomatosis; NAC: NAM (No apical meristem), ATAF (Arabidopsis transcription activation factor) and CUC (Cup-shaped cotyledon) transcription factor; P2C: Protein phosphatase 2C; ROS: Reactive oxygen species; SLAC1: Slow anion channel-associated 1; SRK2E: SNF1 (Sucrose nonfermenting 1)-related protein kinase 2; WRKY: W-tryptophan, R-arginine, K-lysine, Y-tyrosine.

7. Conclusions

From this above discussion, it can be concluded that ethylene directly or indirectly regulates the expression of gibberellins (GAs) and abscisic acid (ABA) in plants under flooding stress conditions. However, ABA plays a major role in stomatal closing, whereas escape and quiescence strategies are controlled by the expression of GA. Finally, it can be concluded that plants maintain their internal homeostasis by balancing hormonal cross talk under excess water stress.

Author Contributions: Conceptualization and writing-original draft preparation, K.K.B.; Writing-review and editing, K.K.B.; M.Z.T, M.R.A., U.H. and M.T.-U.-A.; Supervision, M.A.S. and Q.M.M.H.

Funding: This research received no external funding.

Conflicts of Interest: The authors declare no conflict of interest.

References

1. Tamang, B.G.; Fukao, T. Plant adaptation to multiple stresses during submergence and following desubmergence. *Int. J. Mol. Sci.* **2015**, *16*, 30164–30180. [CrossRef] [PubMed]
2. Bailey-Serres, J.; Fukao, T.; Ronald, P.; Ismail, A.; Heuer, S.; Mackill, D. Submergence tolerant rice: SUB1's journey from landrace to modern cultivar. *Rice* **2010**, *3*, 138–147. [CrossRef]
3. Verboven, P.; Pedersen, O.; Quang, T.H.; Nicolai, B.M.; Colmer, T.D. The mechanism of improved aeration due to gas films on leaves of submerged rice. *Plant Cell Environ.* **2014**, *37*, 2433–2452. [CrossRef] [PubMed]
4. Eysholdt-Derzsó, E.; Sauter, M. Hypoxia and the group VII ethylene response transcription factor HRE2 promote adventitious root elongation in *Arabidopsis*. *Plant biol.* **2018**, *1*, 103–108. [CrossRef] [PubMed]
5. Yamauchi, T.; Colmer, T.D.; Pedersen, O.; Nakazono, M. Regulation of root traits for internal aeration and tolerance to soil waterlogging–flooding Stress. *Plant Physiol.* **2018**, *176*, 1118–1130. [CrossRef]
6. Lin, I.; Wu, Y.; Chen, C.; Chen, G.; Hwang, S.; Jauh, G.; Tzen, J.T.C.; Yang, C. AtRBOH I confers submergence tolerance and is involved in auxin-mediated signaling pathways under hypoxic stress. *Plant Growth Regul.* **2017**, *83*, 277–285. [CrossRef]
7. Phukan, U.J.; Mishra, S.; Shukla, R.K. Waterlogging and submergence stress: Affects and acclimation. *Crit. Rev. Biotechnol.* **2015**, *36*, 956–966. [CrossRef] [PubMed]
8. Xiang, J.; Wu, H.; Zhang, Y.; Zhang, Y.; Wang, Y.; Li, Z.; Lin, H.; Chen, H.; Zhang, J.; Zhu, D. Transcriptomic analysis of gibberellin- and paclobutrazol-treated rice seedlings under submergence. *Int. J. Mol. Sci.* **2017**, *18*, 2225. [CrossRef]
9. Colebrook, E.H.; Thomas, S.G.; Phillips, A.L.; Hedden, P. The role of gibberellin signaling in plant responses to abiotic stress. *J. Exp. Biol.* **2014**, *217*, 67–75. [CrossRef] [PubMed]
10. Else, M.A.; Janowiak, F.; Atkinson, C.J.; Jackson, M.B. Root signals and stomatal closure in relation to photosynthesis, chlorophyll a fluorescence and adventitious rooting of flooded tomato plants. *Ann. Bot.* **2009**, *103*, 313–323. [CrossRef]
11. Göring, H.; Koshuchowa, S.; Deckert, C. Influence of gibberellic acid on stomatal movement. *Biochem. Physiol. Pflanzen* **1990**, *186*, 367–374. [CrossRef]
12. Bailey-Serres, J.; Voesenek, L.A.C.J. Life in the balance: A signaling network controlling survival of flooding. *Curr. Opin. Plant Biol.* **2010**, *13*, 489–494. [CrossRef] [PubMed]
13. Kuroha, T.; Nagai, K.; Gamuyao, R.; Wang, D.R.; Furuta, T.; Nakamori, M.; Kitaoka, T.; Adach, K.; Minami, A.; Mori, Y.; et al. Ethylene-gibberellin signaling underlies adaptation of rice to periodic flooding. *Science* **2018**, *361*, 181–186. [CrossRef] [PubMed]
14. Bailey-Serres, J.; Fukao, T.; Gibbs, D.J.; Holdsworth, M.J.; Lee, S.C.; Licausi, F.; Perata, P.; Voesenek, L.A.C.J.; Dongen, J.T.V. Making sense of low oxygen sensing. *Trends Plant Sci.* **2012**, *17*, 129–138. [CrossRef] [PubMed]
15. Fukao, T.; Xiong, L. Genetic mechanisms conferring adaptation to submergence and drought in rice: Simple or complex? *Curr. Opin. Plant Biol.* **2013**, *16*, 196–204. [CrossRef] [PubMed]
16. Jung, K.; Seo, Y.; Walia, H.; Cao, P.; Fukao, T.; Canlas, P.E.; Amonpant, F.; Bailey-Serres, J.; Ronald, P.C. The submergence tolerance regulator Sub1A mediates stress-responsive expression of AP2/ERF transcription factors. *Plant Physiol.* **2010**, *152*, 1674–1692. [CrossRef] [PubMed]

17. Fukao, T.; Xu, K.; Ronald, P.C.; Bailey-Serres, J. A variable cluster of ethylene response factor-like genes regulates metabolic and developmental acclimation responses to submergence in rice. *Plant Cell* **2006**, *18*, 2021–2034. [CrossRef] [PubMed]

18. Xu, K.; Xu, X.; Fukao, T.; Canlas, P.; Maghirang-Rodriguez, R.; Heuer, S.; Ismail, A.M.; Bailey-Serres, J.; Ronald, P.C.; Mackill, D.J. Sub1A is an ethylene-response-factor-like gene that confers submergence tolerance to rice. *Nature* **2006**, *442*, 705–708. [CrossRef] [PubMed]

19. Singh, P.; Sinha, A.K. A positive feedback loop governed by SUB1A1 interaction with Mitogen-activated protein kinase3 imparts submergence tolerance in rice. *Plant Cell.* **2016**, *28*, 1127–1143. [CrossRef] [PubMed]

20. Locke, A.M.; Gregory, A.B., Jr.; Sathnur, S.; Larive, C.K.; Bailey-Serres, J. Rice SUB1A constrains remodeling of the transcriptome and metabolome during submergence to facilitate post-submergence recovery. *Plant Cell Environ.* **2018**, *41*, 721–736. [CrossRef] [PubMed]

21. Fukao, T.; Bailey-Serres, J. Submergence tolerance conferred by Sub1A is mediated by SLR1 and SLRL1 restriction of gibberellin responses in rice. *PNAS* **2008**, *105*, 16814–16819. [CrossRef] [PubMed]

22. Fukao, T.; Yeung, E.; Bailey-Serres, J. The submergence tolerance gene SUB1A delays leaf senescence under prolonged darkness through hormonal regulation in rice. *Plant Physiol.* **2012**, *160*, 1795–1807. [CrossRef] [PubMed]

23. Schmitz, A.J.; Folsom, J.J.; Jikamaru, Y.; Ronald, P.; Walia, H. SUB1A-mediated submergence tolerance response in rice involves differential regulation of the brassinosteroid pathway. *New Phytol.* **2013**, *198*, 1060–1070. [CrossRef] [PubMed]

24. Yaish, M.W.; El-kereamy, A.; Zhu, T.; Beatty, P.H.; Good, A.G.; Bi, Y.; Rothstein, S.J. The APETALA-2-Like transcription factor OsAP2-39 controls key interactions between abscisic acid and gibberellin in rice. *PLoS Genet.* **2010**, *6*, e1001098. [CrossRef] [PubMed]

25. Seo, M.; Hanada, A.; Kuwahara, A.; Endo, A.; Okamoto, M.; Yamauchi, Y.; North, H.; Marion-Poll, A.; Sun, T.; Koshiba, T.; et al. Regulation of hormone metabolism in Arabidopsis seeds: Phytochrome regulation of abscisic acid metabolism and abscisic acid regulation of gibberellin metabolism. *Plant J.* **2006**, *48*, 354–366. [CrossRef] [PubMed]

26. Hattori, Y.; Nagai, K.; Furukawa, S.; Song, X.; Kawano, R.; Sakakibara, H.; Wu, J.; Matsumoto, T.; Yoshimura, A.; Kitano, H.; et al. The ethylene response factors *SNORKEL1* and *SNORKEL2* allow rice to adapt to deep water. *Nature* **2009**, *460*, 1026–1030. [CrossRef] [PubMed]

27. Voesenek, L.A.; Bailey-Serres, J. Flood adaptive traits and processes: An overview. *New Phytol.* **2015**, *206*, 57–73. [CrossRef] [PubMed]

28. Wang, B.; Jin, Q.; Zhang, X.; Mattson, N.S.; Ren, H.; Cao, J.; Wang, Y.; Yao, D.; Xu, Y. Genome-wide transcriptional analysis of submerged lotus reveals cooperative regulation and gene responses. *Sci. Rep.* **2018**, *8*, 918. [CrossRef]

29. Benschop, J.J.; Jackson, M.B.; Guhl, K.; Vreeburg, R.A.M.; Croker, S.J.; Peeters, A.J.; Voesenek, L.A.C.J. Contrasting interactions between ethylene and abscisic acid in *Rumex* species differing in submergence tolerance. *Plant J.* **2005**, *44*, 756–768. [CrossRef]

30. Benschop, J.J.; Bou, J.; Peeters, A.J.; Wagemaker, N.; Gühl, K.; Ward, D.; Hedden, P.; Moritz, T.; Voesenek, L.A.C.J. Long-term submergence-induced elongation in *Rumex palustris* requires abscisic acid-dependent biosynthesis of gibberellin. *Plant Physiol.* **2006**, *141*, 1644–1652. [CrossRef]

31. Saika, H.; Okamoto, M.; Miyoshi, K.; Kushiro, T.; Shinoda, S.; Jikumaru, Y.; Fujimoto, M.; Arikawa, T.; Takahashi, H.; Ando, M.; et al. Ethylene promotes submergence-induced expression of *OsABA8ox1*, a gene that encodes ABA 8-hydroxylase in rice. *Plant Cell Physiol.* **2007**, *48*, 287–298. [CrossRef] [PubMed]

32. Bailey-Serres, J.; Voesenek, L.A.C.J. Flooding stress: Acclimations and genetic diversity. *Ann. Rev. Plant Biol.* **2008**, *59*, 313–339. [CrossRef] [PubMed]

33. Perata, P. The rice SUB1A gene: Making adaptation to submergence and post-submergence possible. *Plant Cell Environ.* **2018**, *41*, 717–720. [CrossRef]

34. Ueguchi-Tanaka, M.; Ashikari, M.; Nakajima, M.; Itoh, H.; Katoh, E.; Kobayashi, M.; Chow, T.; Hsing, Y.C.; Kitano, H.; Yamaguchi, I.; et al. Gibberellin insensitive dwarf1 encodes a soluble receptor for gibberellin. *Nature* **2005**, *437*, 693–698. [CrossRef] [PubMed]

35. Griffiths, J.; Murase, K.; Rieu, I.; Zentella, R.; Zhang, Z.L.; Powers, S.J.; Gong, F.; Phillips, A.L.; Hedden, P.; Sun, T.; et al. Genetic characterization and functional analysis of the GID1 gibberellin receptors in Arabidopsis. *Plant Cell* **2006**, *18*, 3399–3414. [CrossRef]

36. Sun, T.P.; Gubler, F. Molecular mechanism of gibberellin signaling in plants. *Annu. Rev. Plant Biol.* **2004**, *55*, 197–223. [CrossRef]

37. Ariizumi, T.; Lawrence, P.K.; Steber, C.M. The role of two F-box proteins, SLEEPY1 and SNEEZY, in Arabidopsis gibberellin signaling. *Plant Physiol.* **2011**, *155*, 765–775. [CrossRef]

38. Okada, K.; Ito, T.; Fukazawa, J.; Takahashi, Y. Gibberellin induces an increase in cytosolic Ca^{2+} via a DELLA-independent signaling pathway. *Plant Physiol.* **2017**, *175*, 1536–1542. [CrossRef]

39. Ito, T.; Okada, K.; Fukazawa, J.; Takahashi, Y. DELLA-dependent and -independent gibberellin signaling. *Plant Signal Behav.* **2018**, *13*, e1445933. [CrossRef]

40. Ishida, S.; Yuasa, T.; Nakata, M.; Takahashi, Y. A tobacco calcium-dependent protein kinase, CDPK1, regulates the transcription factor REPRESSION OF SHOOT GROWTH in response to gibberellins. *Plant Cell* **2008**, *20*, 3273–3288. [CrossRef]

41. Ishida, S.; Fukazawa, J.; Yuasa, T.; Takahashi, Y. Involvement of 14-3-3 signaling protein binding in the functional regulation of the transcriptional activator REPRESSION OF SHOOT GROWTH by gibberellins. *Plant Cell* **2004**, *16*, 2641–2651. [CrossRef] [PubMed]

42. Fukazawa, J.; Sakai, T.; Ishida, S.; Yamaguchi, I.; Kamiya, Y.; Takahashi, Y. Repression of shoot growth, a bZIP transcriptional activator, regulates cell elongation by controlling the level of gibberellins. *Plant Cell* **2000**, *12*, 901–915. [CrossRef] [PubMed]

43. Fukazawa, J.; Nakata, M.; Ito, T.; Yamaguchi, S.; Takahashi, Y. The transcription factor RSG regulates negative feedback of *NtGA20ox1* encoding GA 20-oxidase. *Plant J.* **2010**, *62*, 1035–1045. [CrossRef] [PubMed]

44. Ito, T.; Nakata, M.; Fukazawa, J.; Ishida, S.; Takahashi, Y. Phosphorylation-independent binding of 14-3-3 to NtCDPK1 by a new mode. *Plant Signal. Behav.* **2014**, *9*, e977721. [CrossRef] [PubMed]

45. Ito, T.; Ishida, S.; Oe, S.; Fukazawa, J.; Takahashi, Y. Autophosphorylation affects substrate-binding affinity of tobacco Ca^{2+}-dependent protein kinase1. *Plant Physiol.* **2017**, *174*, 2457–2468. [CrossRef] [PubMed]

46. Morello, L.; Giani, S.; Breviario, D. The Influence of anaerobiosis on membrane-associated rice (*O. sativa* L.) protein kinase activities. *J. Plant Physiol.* **1994**, *144*, 500–504. [CrossRef]

47. Igarashi, D.; Ishida, S.; Fukazawa, J.; Takahashi, Y. 14-3-3 proteins regulate intracellular localization of the bZIP transcriptional activator RSG. *Plant Cell* **2001**, *13*, 2483–2497. [CrossRef]

48. Olaetxea, M.; Mora, V.; Bacaicoa, E.; Garnica, M.; Fuentes, M.; Casanova, E.; Zamarreño, A.M.; Iriarte, J.C.; Etayo, D.; Ederra, I.; et al. Abscisic acid regulation of root hydraulic conductivity and aquaporin gene expression is crucial to the plant shoot growth enhancement caused by rhizosphere humic acids. *Plant Physiol.* **2015**, *169*, 2587–2596. [CrossRef]

49. Nagel, O.W.; Konings, H.; Lambers, H. Growth-rate, plant development and water relations of the ABA-deficient tomato mutant sitiens. *Physiol. Plant.* **1994**, *92*, 102–108. [CrossRef]

50. Thompson, A.J.; Andrews, J.; Mullholland, B.J.; McKee, J.M.T.; Hilton, H.W.; Horridge, J.S.; Farquhar, G.D.; Smeeton, R.C.; Smillie, I.R.A.; Black, C.R.; et al. Overproduction of abscisic acid in tomato increases transpiration efficiency and root hydraulic conductivity and influences leaf expansion. *Plant Physiol.* **2007**, *143*, 1905–1917. [CrossRef]

51. Bai, T.; Yin, R.; Li, C.; Ma, F.; Yue, Z.; Shu, H. Comparative analysis of endogenous hormones in leaves and roots of two contrasting *Malus* species in response to hypoxia stress. *J. Plant Growth Regul.* **2011**, *30*, 119–127. [CrossRef]

52. Olivella, C.; Biel, C.; Vendrell, M.; Save´, R. Hormonal and physiological responses of *Gerbera jamesonii* to flooding stress. *HortScience* **2000**, *35*, 222–225.

53. Nan, R.; Carman, J.G.; Salisbury, F.B. Water stress, CO_2 and photoperiod influence hormone levels in wheat. *J. Plant Physiol.* **2002**, *159*, 307–312. [CrossRef]

54. Cowan, I.R.; Troughton, J.H. The relative role of stomata in transpiration and assimilation. *Planta* **1971**, *97*, 325–336. [CrossRef]

55. Raschke, K.; Zeevaart, J.A. Abscisic acid content, transpiration, and stomatal conductance as related to leaf age in plants of *Xanthium strumarium* L. *Plant Physiol.* **1976**, *58*, 169–174. [CrossRef]

56. Peng, Y.; Zhou, Z.; Zhang, Z.; Yu, X.; Zhang, X.; Du, K. Molecular and physiological responses in roots of two full-sib poplars uncover mechanisms that contribute to differences in partial submergence tolerance. *Sci. Rep.* **2018**, *8*, 12829. [CrossRef]

57. Yordanova, R.Y.; Popova, L.P. Flooding-induced changes in photosynthesis and oxidative status in maize plants. *Acta Physiol. Plant* **2007**, *29*, 535–541. [CrossRef]

58. Du, H.; Chang, Y.; Huang, F.; Xiong, L. GID1 modulates stomatal response and submergence tolerance involving abscisic acid and gibberellic acid signaling in rice. *J. Integr. Plant Biol.* **2015**, *57*, 954–968. [CrossRef]

59. Blanke, M.M.; Cooke, D.T. Effects of flooding and drought on stomatal activity, transpiration, photosynthesis, water potential and water channel activity in strawberry stolons and leaves. *Plant Growth Regul.* **2004**, *42*, 153–160. [CrossRef]

60. Yetisir, H.; Caliskan, M.E.; Soylu, S.; Sakar, M. Some physiological and growth responses of watermelon [*Citrullus lanatus* (Thumb.) Matsum. And Nakai] grafted onto *Langenaria siceraria* to flooding. *Environ. Exper. Bot.* **2006**, *58*, 1–8. [CrossRef]

61. Atkinson, C.J.; Harrison-Murray, R.S.; Taylor, J.M. Rapid-flood induced stomatal closure accompanies xylem sap transportation of root-derived acetaldehyde and ethanol in Forsythia. *Environ. Exp. Bot.* **2008**, *64*, 196–205. [CrossRef]

62. Ruiz-Sánchez, M.C.; Domingo, R.; Morales, D.; Torrecillas, A. Water relations of Fino lemon plants on two rootstocks under flooded conditions. *Plant Sci.* **1996**, *120*, 119–125. [CrossRef]

63. Domingo, R.; Pérez-Pastor, A.; Ruiz-Sánchez, C. Physiological responses of apricot plants grafted on two different rootstocks to flooding conditions. *J. Plant Physiol.* **2002**, *159*, 725–732. [CrossRef]

64. Nicolás, E.; Torrecillas, A.; Dell'Amico, J.; Alarcón, J.J. The effect of short-term flooding on the sap flow, gas exchange and hydraulic conductivity of young apricot trees. *Trees* **2005**, *19*, 51–57. [CrossRef]

65. Eisenach, C.; Angeli, A.D. Ion transport at the vacuole during stomatal movements. *Plant Physiol.* **2017**, *174*, 520–530. [CrossRef] [PubMed]

66. Gonzali, S.; Loreti, E.; Cardarelli, F.; Novi, G.; Parlanti, S.; Pucciariello, C.; Bassolino, L.; Banti, V.; Licausi, F.; Perata, P. Universal stress protein HRU1 mediates ROS homeostasis under anoxia. *Nat. Plants* **2015**, *1*, 15151. [CrossRef] [PubMed]

67. Sun, L.; Ma, L.; He, S.; Hao, F. AtrbohD functions downstream of ROP2 and positively regulates waterlogging response in *Arabidopsis*. *Plant Signal. Behav.* **2018**, *13*, e1513300. [CrossRef]

68. Yordanova, R.Y.; Uzunova, A.N.; Popova, L.P. Effects of short-term soil flooding on stomata behavior and leaf gas exchange in barley plants. *Biol. Plant.* **2005**, *49*, 317–319. [CrossRef]

69. Zang, J.; Zang, X. Can early wilting of old leaves account for much of the ABA accumulation in flooded pea plants? *J. Exp. Bot.* **1994**, *45*, 1335–1342. [CrossRef]

70. Ashraf, M.A. Waterlogging stress in plants: A review. *Afr. J Agric. Res.* **2012**, *7*, 1976–1981. [CrossRef]

71. Miao, L.F.; Yang, F.; Han, C.Y.; Pu, Y.J.; Ding, Y.; Zhang, L.J. Sex-specific responses to winter flooding, spring waterlogging and post-flooding recovery in *Populus deltoides*. *Sci. Rep.* **2017**, *7*, 2534. [CrossRef] [PubMed]

72. Vantoai, T.T.; Bolles, C.S. Postanoxic injury in soybean (*Glycine max*) seedlings. *Plant Physiol.* **1991**, *97*, 588–592. [CrossRef]

73. Kalashnikov, Y.E.; Balakhnina, T.I.; Zakrzhevsky, D.A. Effect of soil hypoxia on activation of oxygen and the system of protection from oxidative destruction in roots and leaves of *Hordeum vulgare*. *Russ. J. Plant Physiol.* **1994**, *41*, 583–588.

74. Yan, B.; Da, Q.; Liu, X.; Huang, S.; Wang, Z. Flooding induced membrane damage, lipid oxidation and activated oxygen generation in corn leaves. *Plant Soil* **1996**, *179*, 261–268. [CrossRef]

75. Biemelt, S.; Keetman, U.; Mock, H.P.; Grimm, B. Expression and activity of isoenzymes of superoxide dismutase in wheat roots in response to hypoxia and anoxia. *Plant Cell Environ.* **2000**, *23*, 135–144. [CrossRef]

76. Kwak, J.M.; Mori, I.C.; Pei, Z.M.; Leonhardt, N.; Torres, M.A.; Dangl, J.L.; Bloom, R.E.; Bodde, S.; Jones, J.D.G.; Schroeder, J.I. NADPH oxidase AtrbohD and AtrbohF genes function in ROS-dependent ABA signaling in *Arabidopsis*. *EMBO J.* **2003**, *22*, 2623–2633. [CrossRef]

77. Coelho, S.M.; Taylor, A.R.; Ryan, K.P.; Sousa-Pinto, I.; Brown, M.T.; Brownlee, C. Spatiotemporal patterning of reactive oxygen production and Ca^{2+} wave propagation in *Fucus rhizoid* cells. *Plant Cell* **2002**, *14*, 2369–2381. [CrossRef]

78. Fairley-Grenot, K.; Assmann, S.M. Evidence for G-protein regulation of inward K^+ channel current in guard-cells of fava-bean. *Plant Cell* **1991**, *3*, 1037–1044. [CrossRef]

79. Wu, W.H.; Assmann, S.M. A membrane-delimited pathway of G protein regulation of the guard-cell inward K^+ channel. *Proc. Natl. Acad. Sci. USA* **1994**, *91*, 6310–6314. [CrossRef]

80. Fischer, R.A. Stomatal opening: Role of potassium uptake by guard cells. *Science* **1968**, *160*, 784–785. [CrossRef]

81. Bright, J.; Desikan, R.; Hancock, J.T.; Weir, I.S.; Neill, S.J. ABA-induced NO generation and stomatal closure in Arabidopsis are dependent on H_2O_2 synthesis. *Plant J.* **2006**, *45*, 113–122. [CrossRef]

82. An, Z.; Jing, W.; Liu, Y.; Zhang, W. Hydrogen peroxide generated by copper amine oxidase is involved in abscisic acid-induced stomatal closure in *Vicia faba*. *J. Exp. Bot.* **2008**, *59*, 815–825. [CrossRef] [PubMed]

83. Song, X.G.; She, X.P.; Yue, M.; Liu, Y.E.; Wang, Y.X.; Zhu, X.; Huang, A.X. Involvement of Copper Amine Oxidase (CuAO)-Dependent Hydrogen Peroxide Synthesis in Ethylene-Induced Stomatal Closure in *Vicia faba*. *Russ. J. Plant Physiol.* **2014**, *61*, 390–396. [CrossRef]

84. Bashar, K.K. Hormone dependent survival mechanisms of plants during post-waterlogging stress. *Plant Signal Behav.* **2018**, *13*, e1529522. [CrossRef] [PubMed]

85. Chen, Y.; Huang, R.; Xiao, Y.; Lü, P.; Chen, J.; Wang, X. Extracellular calmodulin-induced stomatal closure is mediated by heterotrimeric G protein and H_2O_2. *Plant Physiol.* **2004**, *136*, 4096–4103. [CrossRef] [PubMed]

86. Sasidharan, R.; Voesenek, L.A.C.J. Ethylene-mediated acclimations to flooding stress. *Plant Physiol.* **2015**, *169*, 3–12. [CrossRef]

87. Steffens, B.; Sauter, M. Heterotrimeric G protein signaling is required for epidermal cell death in rice. *Plant Physiol.* **2009**, *151*, 732–740. [CrossRef] [PubMed]

88. Ge, X.M.; Cai, H.L.; Lei, X.; Zhou, X.; Yue, M.; He, J.M. Heterotrimeric G protein mediates ethylene-induced stomatal closure via hydrogen peroxide synthesis in Arabidopsis. *Plant J.* **2015**, *82*, 138–150. [CrossRef]

89. Liu, P.; Sun, F.; Gao, R.; Dong, H. RAP2.6L overexpression delays waterlogging induced premature senescence by increasing stomatal closure more than antioxidant enzyme activity. *Plant Mol. Biol.* **2012**, *79*, 609–622. [CrossRef] [PubMed]

90. Hwang, S.-Y.; VanToai, T.T. Abscisic acid induces anaerobiosis tolerance in corn. *Plant Physiol.* **1991**, *97*, 593–597. [CrossRef] [PubMed]

91. Gómez-Cadenas, A.; Tadeo, F.R.; Talón, M.; Primo-Millo, E. Leaf abscission induced by ethylene in water-stressed intact seedlings of Cleopatra mandarin requires previous abscisic acid accumulation in roots. *Plant Physiol.* **1996**, *112*, 401–408. [CrossRef] [PubMed]

92. Kato-Noguchi, H. Abscisic acid and hypoxic induction of anoxia tolerance in roots of lettuce seedlings. *J. Exp. Bot.* **2000**, *51*, 1939–1944. [CrossRef] [PubMed]

93. Ellis, M.H.; Dennis, E.S.; Peacock, W.J. Arabidopsis roots and shoots have different mechanisms for hypoxic stress tolerance. *Plant Phyisol.* **1999**, *119*, 57–64. [CrossRef]

94. Dat, J.F.; Capelli, N.; Folzer, H.; Bourgeade, P.; Badot, P.M. Sensing and signaling during plant flooding. *Plant Physiol. Biochem.* **2004**, *42*, 273–282. [CrossRef] [PubMed]

95. Hocking, T.J.; Hillman, J.R.; Wilkins, M.B. Movement of abscisic acid in *phaseolus vulgaris* plants. *Nat. New Biol.* **1972**, *235*, 124–125. [CrossRef]

96. Jackson, M.B.; Hall, K.C. Early stomatal closure in waterlogged pea plants is mediated by abscisic acid in the absence of foliar water deficits. *Plant Cell Environ.* **1987**, *10*, 121–130. [CrossRef]

97. Rodriguez-Gamir, J.; Ancillo, G.; Gonzalez-Mas, M.C.; Primo-Millo, E.; Iglesias, D.J.; Forner-Giner, M.A. Root signaling and modulation of stomatal closure in flooded citrus seedlings. *Plant Physiol. and Biochem.* **2011**, *49*, 636–645. [CrossRef]

MDPI

St. Alban-Anlage 66

4052 Basel

Switzerland

Tel. +41 61 683 77 34

Fax +41 61 302 89 18

www.mdpi.com

Agronomy Editorial Office

E-mail: agronomy@mdpi.com

www.mdpi.com/journal/agronomy

9 783039 434589